# Improve Your Grade!

## Access included with any new book.

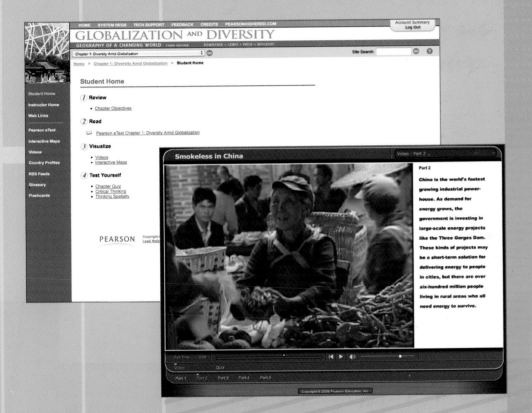

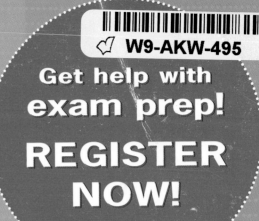
## Get help with exam prep! REGISTER NOW!

### Registration will let you:

– Access your text online 24/7.

– Prepare for exams by taking online quizzes.

– Watch videos and explore interactive maps to broaden your knowledge of world regional geography.

# www.mygeoscienceplace.com

## TO REGISTER

1. Go to www.mygeoscienceplace.com

2. Click "Register."

3. Follow the on-screen instructions to create your login name and password.

Your Access Code is:

**PSWRGD-SHALL-GOOSY-CANNA-ABBOT-ROPES**

Note: If there is no silver foil covering the access code, it may already have been redeemed, and therefore may no longer be valid. In that case, you can purchase access online using a major credit card. To do so, go to www.mygeoscienceplace.com, click on "Buy Access," and follow the on-screen instructions.

## TO LOG IN

1. Go to www.mygeoscienceplace.com

2. Click "Log In."

3. Pick your book cover.

4. Enter your login name and password.

Hint:
Remember to bookmark the site after you log in.

Technical Support:
http://247pearsoned.custhelp.com

# GLOBALIZATION

## AND DIVERSITY

### GEOGRAPHY OF A CHANGING WORLD

THIRD EDITION

**LES ROWNTREE**

University of California, Berkeley

**MARTIN LEWIS**

Stanford University

**MARIE PRICE**

George Washington University

**WILLIAM WYCKOFF**

Montana State University

**Prentice Hall**

Boston  Columbus  Indianapolis
New York  San Francisco  Upper Saddle River
Amsterdam  Cape Town  Dubai  London
Madrid  Milan  Munich  Paris  Montréal  Toronto
Delhi  Mexico City  São Paulo  Sydney
Hong Kong  Seoul  Singapore  Taipei  Tokyo

**Acquisitions Editor:** Christian Botting
**Editor in Chief, Geosciences and Chemistry:** Nicole Folchetti
**Marketing Manager:** Maureen McLaughlin
**Project Editor:** Tim Flem
**Assistant Editor:** Jennifer Aranda
**Editorial Assistant:** Christina Ferraro
**Marketing Assistant:** Nicola Houston
**Managing Editor, Geosciences and Chemistry:** Gina M. Cheselka
**Project Manager, Science:** Maureen Pancza
**Design Director:** Mark Stuart Ong
**Interior and Cover Design:** Jill Little
**Art Project Manager:** Connie Long
**Senior Operations Supervisor:** Nick Sklitsis
**Operations Supervisor:** Maura Zaldivar
**Senior Media Producer:** Angela Bernhardt
**Senior Media Production Supervisor:** Liz Winer
**Associate Media Project Manager:** David Chavez
**Photo Research Manager:** Elaine Soares
**Photo Researcher:** Kristin Piljay
**Composition/Full Service:** GGS Higher Education Resources
**Production Editor, Full Service:** Kelly Keeler
**Copy Editor:** Catharine Wilson
**Illustrations:** Spatial Graphics, Kevin Lear

**Cover Photograph:** National Stadium, Beijing, People's Republic of China. Wang Song, Xinhua Press/Corbis

Credits and acknowledgments borrowed from other sources and reproduced, with permission, in this textbook appear on the appropriate page within the text or on p. 441.

Library of Congress Cataloging-in-Publication Data

Globalization and diversity: geography of a changing world / Les Rowntree ... [et al.]. — 3rd ed.
    p. cm.
  Includes index.
  ISBN 978-0-321-65152-5 (pbk.)
 1. Economic geography. 2. Globalization. 3. Cultural pluralism. I. Rowntree, Lester, 1938-
 HF1025.G59 2011
 330.9—dc22

                                      2009033279

Printed in the United States
10 9 8 7 6 5 4 3 2

**Prentice Hall** is an imprint of

www.pearsonhighered.com

ISBN 0-32-165152-9 / 978-0-32-165152-5 [Student]
ISBN 0-32-169804-5 / 978-0-32-169804-9 [Books á la Carte]

# BRIEF CONTENTS

# About Our Sustainability Initiatives

This book is carefully crafted to minimize environmental impact. The materials used to manufacture this book originated from sources committed to responsible forestry practices. The paper is FSC certified. The binding, cover, and paper come from facilities that minimize waste, energy consumption, and the use of harmful chemicals.

Pearson closes the loop by recycling every out-of-date text returned to our warehouse. We pulp the books, and the pulp is used to produce items such as paper coffee cups and shopping bags. In addition, Pearson aims to become the first climate neutral educational publishing company.

The future holds great promise for reducing our impact on Earth's environment, and Pearson is proud to be leading the way. We strive to publish the best books with the most up-to-date and accurate content, and to do so in ways that minimize our impact on Earth.

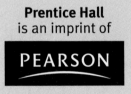

# CONTENTS

# 12 SOUTH ASIA   343

# 13 SOUTHEAST ASIA   373

# 14 AUSTRALIA AND OCEANIA   403

# PREFACE

*Globalization and Diversity: Geography of a Changing World,* Third Edition, is an issues-oriented textbook for college and university world regional geography classes that discusses and analyzes the geographic changes accompanying globalization. With this focus, we join the many who argue that globalization is the most fundamental reorganization of the planet's socioeconomic, cultural, and geopolitical structure since the Industrial Revolution.

As geographers, we think it essential for students to understand two interactive tensions. First, they need to appreciate and critically ponder the consequences of converging environmental, cultural, political, and economic systems resulting from globalization, and, second, they need to deepen their understanding of the creation and persistence of geographic diversity. These opposing and interactive forces form a current running throughout our book and are reflected in the title, *Globalization and Diversity.*

## New to the Third Edition

- **Expanded Treatment of Globalization.** Globalization has become increasingly important during the history of this book and, as a result, globalization is now more central to the study of world regional geography than ever before. Thus we have expanded and revised our introductory discussion in Chapter 1 to present varying perspectives and viewpoints on globalization.

- **New Material on Regions and Globalization, and Transregional Connections.** Each regional chapter also has new information on globalization and its many-faceted implications. Transregional links are further explored in new boxed essays titled "**Exploring Global Connections.**"

- **Increased Emphasis on Global Warming and Climate Change.** There is little question that global warming is one of the major environmental, political, and economic problems facing the world and, because of the importance of this complicated issue, we have expanded our coverage of the projected effects of climate change on different regions of the world, along with the different policies and measures taken to mitigate the human and economic costs of this problem.

- **Updated Maps/Graphics and New Design.** The maps and illustrations for the third edition have been updated with an open, visually appealing approach while retaining the accuracy and pedagogical focus of previous edition maps and illustrations. In addition, the text design has been refreshed to capture students' attention with inviting visual content.

- **Current Data and Information.** All text, tables, and maps have been updated with the most current information.

- **Online Multimedia and Assessment at www.mygeoscience-place.com.** Numerous geography video clips, interactive maps, activities, assessments, flashcards, RSS feeds, a fully-integrated eBook, and additional resources make this website an essential learning and teaching tool.

## Chapter Organization

*Globalization and Diversity,* Third Edition, is structured to explain and describe the major world regions of Asia, Africa, the Americas, Europe, and Oceania. Our twelve regional chapters, however, depart somewhat from traditional world regional textbooks that consist primarily of a country-by-country tour of each region. Instead of filling these regional chapters with descriptions of individual countries, we place most of that important material on the textbook website (www.mygeoscienceplace.com). This leaves us free to develop five important themes as the structure for each regional chapter.

We begin with *Environmental Geography,* discussing the physical geography of each region as well as current environmental issues. Next, we assess the *Population and Settlement* geography, in which demography, land use, and settlement (including cities) are discussed. Third is a section titled *Cultural Coherence and Diversity* that examines the geography of language and religion and also explores current cultural tensions resulting from the interplay of globalization and diversity. Following that is a section on each region's *Geopolitical Framework,* which treats the dynamic political geography of the region, including micro-regionalism, separatism, ethnic conflicts, and global terrorism. We then conclude each regional chapter with *Economic and Social Development,* in which we analyze each region's economic framework as well as its social geography, including gender issues.

This regional treatment follows two substantive introductory chapters that provide the conceptual and theoretical framework of human and physical geography necessary to understand our dynamic world. In the first chapter, students are introduced to the notion of globalization and are asked to ponder the costs and benefits of the globalization process, a critical perspective that is becoming increasingly common and important to understand. Following this, the geographical foundation for each of the five thematic sections is examined. This discussion draws heavily on the major concepts fundamental to an introductory university geography course. The second chapter, *The Changing Global Environment,* briefly presents the themes and concepts of global environmental geography—climate, global warming, biogeography, and agriculture.

## Chapter Features

Within each regional chapter, several unique features complement the thematic pedagogy of our approach:

- **Comparable Maps.** Of the many maps in each regional chapter, 7 are constructed on the same theme and with similar data so that readers can easily draw comparisons between different regions. Thus, in every regional chapter readers will find maps of physical geography and place-names, climate, environmental issues, population density, language geography, religion, and a map showing the geopolitical issues of the region.

- **Other Maps.** In addition, each regional chapter also has 7 or 8 other maps illustrating such major themes as urban growth, ethnic tensions, social development, and linkages to the global economy.

- **Comparable Regional Data Sets.** Again, to facilitate comparison between regions, as well as to provide important insight into the characteristics of each region, each chapter contains two tables. The first provides population data, including population density, level of urbanization, total fertility rates, proportion of the population under 15 and over 65 years of age, and net migration rates for each country within the region. The second table presents economic and social development data for each country including GNI per capita, GDP growth, life expectancy, percentage of the population living on less than $2 per day, infant mortality rates, and the UN gender equity index.

- **Sidebars.** Within each chapter, two sidebars elaborate on important themes: *Setting the Boundaries* sidebars give an overview of the region, discuss its role in the world today, raise questions about its geographic coherence by accentuating its diversity, and look at the vexatious issue of regional boundaries. A second sidebar, *Exploring Global Connections*, contains case studies illuminating the interconnections between regions in today's globalized world, starting in Chapter 2 with the linkages between the European Union's fishing policies off the African coast and the increased smuggling of bushmeat, and concluding in Chapter 14 with the South Pacific island of Nauru's varied strategies to profit from the dynamic world economy.

- **Key Words and Glossary.** Throughout each chapter, key words and concepts are in boldface type, explained in context, then listed again at the end of the chapter as a study guide. The whole array of key words is collected in the book's glossary.

- **Review and Research Questions.** The *Thinking Geographically* questions at the end of each chapter give students an opportunity to apply chapter concepts and material within an active learning research framework.

## Acknowledgments

We have many people to thank for their help in the conceptualization, writing, rewriting, and production of *Globalizaton and Diversity, 3e*. First, we'd like to thank the thousands of students in our world regional geography classes who have inspired us with their energy, engagement, and curiosity; challenged us with their critical insights; and demanded a textbook that better meets their need to understand the people and places of our dynamic world.

Next, we are deeply indebted to many professional geographers and educators for their assistance, advice, inspiration, encouragement, and constructive criticism as we labored through the different stages of this book. Among the many who provided invaluable comments on various drafts and editions of *Globalization and Diversity* are:

Dan Arreola, Arizona State University
Bernard BakamaNume, Texas A&M University
Brad Baltensperger, Michigan Technological University
Max Beavers, Samford University

Laurence Becker, Oregon State University
James Bell, University of Colorado
William H. Berentsen, University of Connecticut
Kevin Blake, Kansas State University
Michelle Calvarese, California State University, Fresno
Craig Campbell, Youngstown State University
Elizabeth Chacko, George Washington University
Philip Chaney, Auburn University
David B. Cole, University of Northern Colorado
Malcolm Comeaux, Arizona State University
Jonathan C. Comer, Oklahoma State University
Catherine Cooper, George Washington University
Jeremy Crampton, George Mason University
Kevin Curtin, University of Texas at Dallas
James Curtis, California State University, Long Beach
Dydia DeLyser, Louisiana State University
Francis H. Dillon, George Mason University
Jason Dittmer, Georgia Southern University
Jerome Dobson, University of Kansas
Caroline Doherty, Northern Arizona University
Vernon Domingo, Bridgewater State College
Roy Doyon, Ball State University
Jane Ehemann, Shippensburg University
Doug Fuller, George Washington University
Gary Gaile, University of Colorado
Sherry Goddicksen, California State University, Fullerton
Reuel Hanks, Oklahoma State University
Steven Hoelscher, University of Texas, Austin
Thomas Howard, Armstrong Atlantic State University
Peter J. Hugil, Texas A&M University
Eva Humbeck, Arizona State University
Ryan S. Kelly, University of Kentucky
Richard H. Kesel, Louisiana State University
Rob Kremer, Front Range Community College
Robert C. Larson, Indiana State University
Mathias Le Bosse, Kutztown University of Pennsylvania
Alan A. Lew, Northern Arizona University
Catherine Lockwood, Chadron State College
John F. Looney, Jr., University of Massachusetts: Boston
Max Lu, Kansas State University
Luke Marzen, Auburn University
Kent Matthewson, Louisiana State University
James Miller, Clemson University
Bob Mings, Arizona State University
Sherry D. Morea-Oakes, University of Colorado, Denver
Anne E. Mosher, Syracuse University
Tim Oakes, University of Colorado
Nancy Obermeyer, Indiana State University
Karl Offen, Oklahoma University
Joseph Palis, University of North Carolina: Chapel Hill
Jean Palmer-Moloney, Hartwick College
Bimal K. Paul, Kansas State University
Michael P. Peterson, University of Nebraska–Omaha
Richard Pillsbury, Georgia State University
Brandon Plewe, Brigham Young University
Patricia Price, Florida International University
Erik Prout, Texas A&M University
David Rain, United States Census Bureau

Rhonda Reagan, Blinn College
Craig S. Revels, Portland State University
Scott M. Robeson, Indiana State University
Paul A. Rollinson, Southwest Missouri State University
Yda Schreuder, University of Delaware
Kay L. Scott, University of Central Florida
J. Duncan Shaeffer, Arizona State University
Dimitrii Sidorov, California State University, Long Beach
Susan C. Slowey, Blinn College
Andrew Sluyter, Louisiana State University
Joseph Spinelli, Bowling Green State University
William Strong, University of Northern Alabama
Philip W. Suckling, University of Northern Iowa
Curtis Thomson, University of Idaho
Benjamin Timms, California Polytechnic State University
Suzanne Traub-Metlay, Front Range Community College
Nina Veregge, University of Colorado
Gerald R. Webster, University of Alabama
Keith Yearman, College of DuPage
Emily Young, University of Arizona
Bin Zhon, Southern Illinois University at Edwardsville
Henry J. Zintambila, Illinois State University

In addition, we wish to thank the many publishing professionals who have been involved with this project. It has been a privilege to work with you. We thank Paul F. Corey, President of Pearson's Science division, for his early—and continued—support for this book project; Geography Editor Christian Botting, for his professional guidance; Project Manager Tim Flem, for his daily miracles and steady hand on the tiller; Media Producer Ziki Dekel, for his work on the book's accompanying website; Assistant Editor Jennifer Aranda and Editorial Assistant Christina Ferraro, for gracefully taking care of the many tasks necessary to this project; Marketing Manager Maureen McLaughlin, for her sales and promotion work; production editors Maureen Pancza and Kelly Keeler, for their work to somehow turn thousands of pages of manuscript into a finished book; copyeditor Sharon O'Donnell, for her eagle eyes and graceful hands; and photo researcher Kristin Piljay, for her creative solutions to indulging four geographers in our quest for outstanding pictures from every part of the world. To all of you, your professionalism is truly inspirational and very, very much appreciated.

Finally, the authors want to thank that special group of friends and family who were there when we needed you most—early in the morning and late at night; in foreign countries and familiar places; when we were on the verge of crying, yet needed to laugh; for your love, patience, companionship, inspiration, solace, understanding, and enthusiasm: Eugene Adogia; Elizabeth Chacko; Meg Conkey; Rob Crandall; Karen Wigen; and Linda, Tom, and Katie Wyckoff. Words cannot thank you enough.

Les Rowntree

Martin Lewis

Marie Price

William Wyckoff

# THE TEACHING AND LEARNING PACKAGE

In addition to the text itself, the authors and publisher have worked with a number of talented people to produce an excellent instructional package. This package includes the traditional supplements as well as new electronic media.

## Instructor

### Instructor Manual Download (0-32-166780-8)
Includes Learning Objectives, detailed Chapter Outlines, Key Terms, For Thought and Discussion topics to generate classroom motivation, and Exercise Activities that link to features in the students' Mapping Workbook and the Interactive Maps. www.pearsonhighered.com/irc

### TestGen®/Test Bank (0-32-166779-4)
TestGen® is a computerized test generator that lets instructors view and edit Test Bank questions, transfer questions to tests, and print the test in a variety of customized formats. This Test Bank includes approximately 1000 multiple choice, true/false, and short answer/essay questions. Questions correlate to the revised U.S. National Geography Standards and Bloom's Taxonomy to help instructors better map the assessments against both broad and specific teaching and learning objectives. The Test Bank is also available in Microsoft Word®, and is importable into Blackboard and WebCT. www.pearsonhighered.com/irc

### Instructor Resource Center on DVD (0-32-166740-9)
Everything instructors need where they want it. The Pearson Prentice Hall Instructor Resource Center helps make instructors more effective by saving them time and effort. All digital resources can be found in one, well-organized, easy to-access place. The IRC on DVD includes:

- All textbook images as JPEGs, PDFs, and PowerPoint™ presentations
- Pre-authored Lecture Outline PowerPoint™ presentations, which outline the concepts of each chapter with embedded art and can be customized to fit instructors' lecture requirements
  — CRS "Clicker" Questions in PowerPoint™ format , which correlate to the U.S. National Geography Standards and Bloom's Taxonomy
- The TestGen software, Test Bank questions, and answers for both MACs and PCs
- Electronic files of the *Instructor Manual* and *Test Bank*
  — Approximately 120 Geography Video Clips

This Instructor Resource content is also available completely online via the Instructor Resources section of www.mygeoscienceplace.com and www.pearsonhighered.com/irc.

### Television for the Environment *Earth Report* Geography Videos on DVD (0-32-166298-9)
This three-DVD set is designed to help students visualize how human decisions and behavior have affected the environment, and how individuals are taking steps toward recovery. With topics ranging from the poor land management promoting the devastation of river systems in Central America to the struggles for electricity in China and Africa, these 13 videos from Television for the Environment's global *Earth Report* series recognize the efforts of individuals around the world to unite and protect the planet.

### Television for the Environment *Life* World Regional Geography Videos on DVD (0-13-159348-X)
From the Television for the Environment's global *Life* series this two-DVD set brings globalization and the developing world to the attention of any world regional geography course. These 10 full length video programs highlight matters such as the growing number of homeless children in Russia, the lives of immigrants living in the United States trying to aid family still living in their native countries, and the European conflict between commercial interests and environmental concerns.

### Television for the Environment *Life* Human Geography Videos on DVD (0-13-241656-5)
This three-DVD set is designed to enhance any human geography course. These DVDs include 14 full-length video programs from Television for the Environment's global *Life* series, covering a wide array of issues affecting people and places in the contemporary world, including the serious health risks of pregnant women in Bangladesh, the social inequalities of the 'untouchables' in the Hindu caste system, and Ghana's struggle to compete in a global market.

### Aspiring Academics: A Resource Book for Graduate Students and Early Career Faculty (0-13-604891-9)
Drawing on several years of research, this set of essays is designed to help graduate students and early career faculty start their careers in geography and related social and environmental sciences. This teaching aid stresses the interdependence of teaching, research, and service—and the importance of achieving a healthy balance in professional and personal life—in faculty work, and does not view it as a collection of unrelated tasks. Each chapter provides accessible, forward-looking advice on topics that often cause the most stress in the first years of a college or university appointment.

### Teaching College Geography: A Practical Guide for Graduate Students and Early Career Faculty (0-13-605447-1)
Provides a starting point for becoming an effective geography teacher from the very first day of class. Divided in two parts, the first set of

chapters addresses "nuts-and-bolts" teaching issues in the context of the new technologies, student demographics, and institutional expectations that are the hallmarks of higher education in the 21st century. The second part explores other important issues: effective teaching in the field; supporting critical thinking with GIS and mapping technologies; engaging learners in large geography classes; and promoting awareness of international perspectives and geographic issues.

### AAG Community Portal for Aspiring Academics & Teaching College Geography

This website is intended to support community-based professional development in geography and related disciplines. Here you will find activities providing extended treatment of the topics covered in both books. The activities can be used in workshops, graduate seminars, brown bags, and mentoring programs offered on campus or within an academic department. You can also use the discussion boards and contributions tool to share advice and materials with others. www.pearsonhighered.com/aag/

### Course Management Systems

Pearson Prentice Hall offers content specific to *Globalization and Diversity* within the BlackBoard and CourseCompass course management system platforms. Each of these platforms lets the instructor easily post his or her syllabus, communicate with students online or off-line, administer quizzes, and record student results and track their progress. www.pearsonhighered.com/elearning/

## Student

### Premium Website

A dedicated Premium Website with eText offers a variety of resources for students and professors, including an eText version of the textbook with linked/integrated multimedia, Interactive Maps with activities, geography videos with associated assessments, Flashcards, RSS Feeds, weblinks, annotated resources for further exploration, and Class Manager & GradeTracker Gradebook functionality for instructors. www.mygeoscienceplace.com

### Mapping Workbook (0-32-166739-5)

This workbook, which can be used in conjunction with either the main text or an atlas, features political and physical base maps of every global region, printed in black and white. These maps, along with the names of the regional key locations and physical features, are the basis for identification exercises. Conceptual exercises are included to further exam students' comprehension of the Key Points presented in the main text chapters. These exercises review both the physical and human environment of the regions.

### Goode's World Atlas 22nd Edition (0-32-165200-2)

Goode's World Atlas has been the world's premiere educational atlas since 1923, and for good reason. It features over 250 pages of maps, from definitive physical and political maps to important thematic maps that illustrate the spatial aspects of many important topics. The 22nd edition includes 160 pages of new, digitally-produced reference maps, as well as new thematic maps on global climate change, sea level rise, $CO_2$ emissions, polar ice fluctuations, deforestation, extreme weather events, infectious diseases, water resources, and energy production.

### Encounter World Regional Geography Workbook & Website (0-32-168175-4)

*Encounter World Regional Geography* provides rich, interactive explorations of world regional geography concepts through Google Earth™ explorations. All chapter explorations are available in print format as well as online quizzes, accommodating different classroom needs. All worksheets are accompanied with corresponding Google Earth™ media files, available for download from www.mygeoscienceplace.com.

### Encounter Earth: Interactive Geoscience Explorations Workbook & Website (0-32-158129-6)

Ideal for professors who want to integrate Google Earth™ in their classrooms, *Encounter Earth* gives students a way to visualize key topics in their introductory geoscience courses. Each exploration consists of a worksheet, available in the workbook and as a PDF file, and a Google Earth™ KMZ file, containing placemarks, overlays, and annotations referred to in the worksheets. The accompanying *Encounter Earth* Premium Website is located at www.mygeoscienceplace.com

### Dire Predictions (0-13-604435-2)

Periodic reports from the Intergovernmental Panel on Climate Change (IPCC) evaluate the risk of climate change brought on by humans. But the sheer volume of scientific data remains inscrutable to the general public, particularly to those who may still question the validity of climate change. In just over 200 pages, this practical text presents and expands upon the essential findings of the 4th Assessment Report in a visually stunning and undeniably powerful way to the lay reader. Scientific findings that provide validity to the implications of climate change are presented in clear-cut graphic elements, striking images, and understandable analogies.

# ABOUT THE AUTHORS

**Les Rowntree** is a Visiting Scholar at the University of California, Berkeley, where he researches and writes about environmental issues. This career change comes after three decades of teaching both Geography and Environmental Studies at San Jose State University in California. As an environmental geographer, Dr. Rowntree's interests focus on international environmental issues, biodiversity conservation, and human-caused global change. He sees world regional geography as a way to engage and inform students by giving them the conceptual tools needed to critically assess global issues. Dr. Rowntree has done research in Iceland, Alaska, Morocco, Mexico, Australia, and Europe, as well as in his native California. Current writing projects include a book on the natural history of California's coast, as well as textbooks in geography and environmental science.

**Martin Lewis** is a Senior Lecturer in History at Stanford University. He has conducted extensive research on environmental geography in the Philippines and on the intellectual history of global geography. His publications include *Wagering the Land: Ritual, Capital, and Environmental Degradation in the Cordillera of Northern Luzon, 1900–1986* (1992), and, with Karen Wigen, *The Myth of Continents: A Critique of Metageography* (1997). Dr. Lewis has traveled extensively in East, South, and Southeast Asia. His current research focuses on the geographical dimensions of globalization. In April 2009 Dr. Lewis was recognized by *Time Magazine,* as a favorite lecturer.

**Marie Price** is a Professor of Geography and International Affairs at George Washington University. A Latin American specialist, Marie has conducted research in Belize, Mexico, Venezuela, Cuba, and Bolivia. She has also traveled widely throughout Latin America and Sub-Saharan Africa. Her studies have explored human migration, natural resource use, environmental conservation, and regional development. She is a non-resident fellow of the Migration Policy Institute, a non-partisan think tank that focuses on immigration. Dr. Price brings to *Globalization and Diversity* a special interest in regions as dynamic spatial constructs that are shaped over time through both global and local forces. Her publications include the co-edited book, *Migrants to the Metropolis: The Rise of Immigrant Gateway Cities* (2008, Syracuse University Press) and numerous academic articles and book chapters.

**William Wyckoff** is a geographer in the Department of Earth Sciences at Montana State University specializing in the cultural and historical geography of North America. He has written and co-edited several books on North American settlement geography, including *The Developer's Frontier: The Making of the Western New York Landscape* (1988), *The Mountainous West: Explorations in Historical Geography* (1995) (with Lary M. Dilsaver), *Creating Colorado: The Making of a Western American Landscape 1860–1940* (1999), and *On the Road Again: Montana's Changing Landscape* (2006). In 2003 he received Montana State's Cox Family Fund for Excellence Faculty Award for Teaching and Scholarship. A World Regional Geography instructor for 26 years, Dr. Wyckoff emphasizes in the classroom the connections between the everyday lives of his students and the larger global geographies that surround them and increasingly shape their future.

# GLOBALIZATION AND DIVERSITY

## GEOGRAPHY OF A CHANGING WORLD

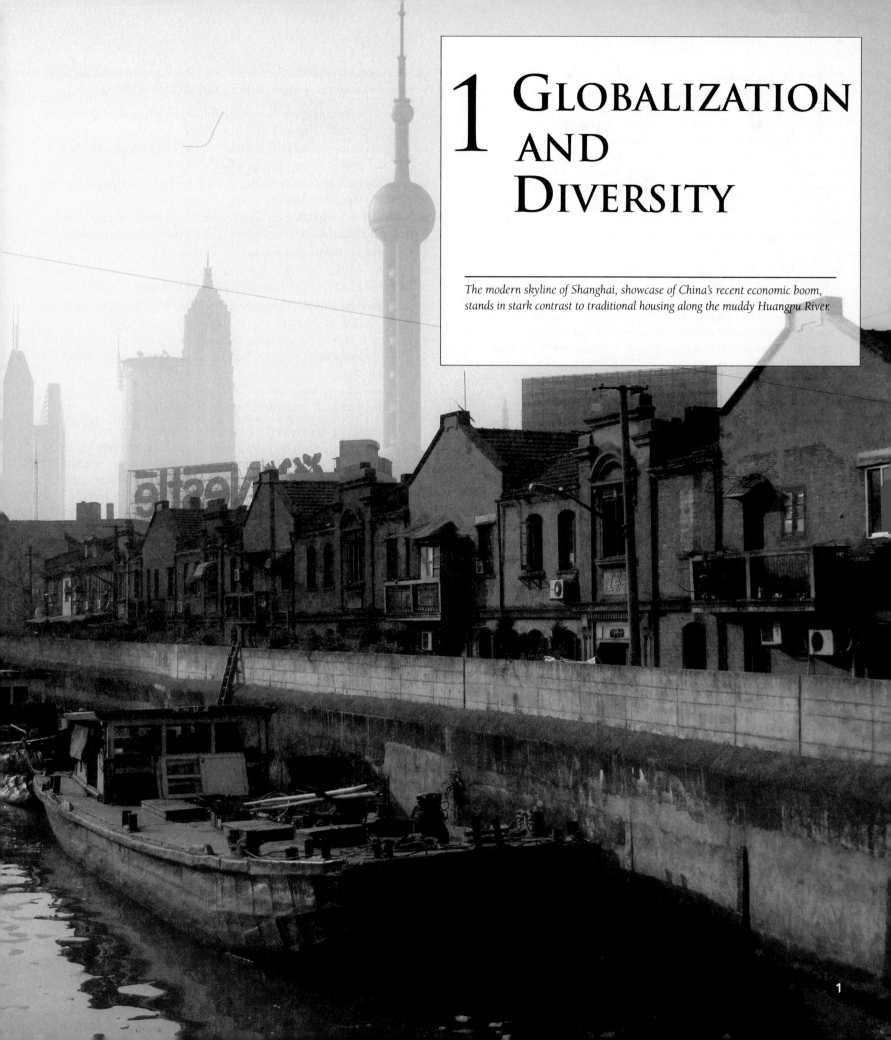

# 1 GLOBALIZATION AND DIVERSITY

*The modern skyline of Shanghai, showcase of China's recent economic boom, stands in stark contrast to traditional housing along the muddy Huangpu River.*

The most important challenges facing the world in the 21st century are associated with **globalization**, the increasing interconnectedness of people and places through converging processes of economic, political, and cultural change. Once-distant regions and cultures are now increasingly linked through commerce, communications, and travel. Although earlier forms of globalization existed, especially during Europe's colonial period, the current degree of planetary integration is stronger than ever. In fact, many observers argue that contemporary globalization is the most fundamental reorganization of the planet's socioeconomic structure since the Industrial Revolution. While few dispute the widespread changes brought about by globalization, not everyone agrees on whether the benefits outweigh the costs.

Although economic activities may be the main force behind globalization, the consequences affect all aspects of land and life. Cultural patterns, political arrangements, and social development are all undergoing profound change. Because natural resources are now global commodities, the planet's physical environment is affected by globalization. Financial decisions made thousands of miles away alter local ecosystems, and the cumulative effects of these far-ranging activities often have negative consequences for the world's climates, oceans, waterways, and forests.

These immense, widespread global changes make understanding our world a challenging yet necessary task. Although globalization cuts across many academic disciplines, world regional geography is a fundamental starting point because of its focus on regions, environment, geopolitics, culture, and economic and social development. This book seeks to impart knowledge of globalization by carefully outlining the basic patterns of world geography and showing how they are being constantly reorganized by global interconnections (Figure 1.1).

## Converging Currents of Globalization

Most scholars agree that the major component of globalization is the economic reorganization of the world. Although different forms of a world economy have existed for centuries, a well-integrated and truly global economy is primarily the product of the past several decades. The attributes of this system, although familiar, bear listing:

- Global communication systems that link all regions on the planet instantaneously (Figure 1.2)
- Transportation systems capable of moving goods quickly by air, sea, and land
- Transnational corporate strategies that have created global corporations more powerful than many sovereign nations
- New and more flexible forms of capital accumulation and international financial institutions that make 24-hour trading possible
- Global agreements that promote free trade
- Market economies that have replaced state-controlled economies and privatized firms and services formerly operated by governments
- An abundance of planetary goods and services that have arisen to fulfill consumer demand, real or imaginary (Figure 1.3)
- Economic disparities between rich and poor regions and countries that drive people to migrate, both legally and illegally, in search of a better life
- An army of international workers, managers, and executives who give this powerful economic force a human dimension

As a result of this global reorganization, economic growth in some areas of the world has been unprecedented during recent decades. International corporations, along with their managers and executives, have amassed vast amounts of wealth, profiting from the new opportunities unleashed by globalization. However, not everyone has profited from economic globalization, nor have all world regions shared equally in the benefits. While globalization is often touted as benefiting everyone through trickle-down economics, there is mounting evidence that this trickling process is not happening in all places or for all peoples.

## Globalization and Cultural Change

Economic changes trigger cultural changes. The spread of a global consumer culture that threatens to reduce local diversity often accompanies globalization, frequently setting up deep and serious social

**FIGURE 1.1 Global Communications** The impacts of globalization, often through global TV, are everywhere, even in remote villages in developing countries. Here, in a small village in southwestern India, a rural family earns a few dollars a week by renting out viewing time on its globally linked television set.

social values also are dispersed globally. Changing expectations about human rights, the role of women in society, and the intervention of nongovernmental organizations are also expressions of globalization that may have far-reaching effects on cultural change.

It would be a mistake, however, to view cultural globalization as a one-way flow that spreads from the United States and Europe into the corners of the world. In actuality, when forms of U.S. popular culture spread abroad, they are typically melded with local cultural traditions in a process known as *hybridization*. The resulting cultural "hybridities," such as world beat music or Asian food, can themselves echo across the planet, adding yet another layer to globalization.

In addition, ideas and forms from the rest of the world are also having a great impact on U.S. culture. The growing internationalization of American food, the multiple languages spoken in the United States, and the spread of Japanese comic book culture among U.S.

**FIGURE 1.2  Global-to-Local Connections**   A controversial aspect of globalization is the outsourcing of jobs from developed to developing countries. One example is the relocation of customer call centers away from Europe and North America to India, where educated English-speaking employees work for relatively low wages. This photo shows a call center in Bangalore, India.

tensions between traditional cultures and new, external globalizing currents. Global TV, movies, Facebook, Twitter, and videos promote Western values and culture that are imitated by millions throughout the world. NBA T-shirts, sneakers, and caps now are found in small villages and large world cities alike.

Fast-food franchises are changing—some would say corrupting—traditional diets, with the explosive growth of McDonald's, Burger King, and Kentucky Fried Chicken outlets in the world's cities. While these changes may seem harmless to North Americans because of their familiarity, they are expressions of the deep cultural changes the world is experiencing through globalization.

Although the media give much attention to the rapid spread of Western consumer culture, nonmaterial culture is also becoming more dispersed and homogenized through globalization. Language is an obvious example. Many a Western tourist in Russia or Thailand has been startled by locals speaking an English made up largely of Hollywood movie phrases. But far more than speech is involved, as

**FIGURE 1.3  Global Shopping Malls**   Once a fixture only of suburban North America, the shopping mall is now found throughout the world. This mall is in downtown Kunming, the capital city of Yunnan province, China.

**FIGURE 1.4 Global Culture in the United States**   The multilingual welcome offered by a public library in Montgomery County, Maryland, not only speaks to the many different languages spoken by people in the suburbs of Washington, DC, but also reminds us that expressions of globalization are found throughout North America.

children are all examples of globalization's effects within the United States (Figure 1.4).

## Globalization and Geopolitics

Globalization has important geopolitical components. To many, an essential dimension of globalization is that it is not restricted by territorial or national boundaries. For example, the creation of the United Nations following World War II was a step toward creating an international governmental structure in which all nations could find representation. The simultaneous emergence of the Soviet Union as a military and political superpower led to a rigid division into Cold War blocs that slowed further geopolitical integration. With the peaceful end of the Cold War in the late 1980s and early 1990s, the former communist countries of eastern Europe and the Soviet Union were opened almost immediately to global trade and cultural exchange. These political developments coincided with the economic and technological changes we now see as the early waves of globalization.

## Environmental Concerns

The expansion of a globalized economy is creating and intensifying environmental problems throughout the world. **Transnational firms**, which do global business through international subsidiaries, disrupt local ecosystems in their incessant search for natural resources and manufacturing sites. Landscapes and resources previously used by only small groups of local peoples are now thought of as global commodities to be exploited and traded on the world marketplace. As a result, native peoples are often deprived of their traditional resource base and displaced into marginal environments. On a larger scale, economic globalization is aggravating worldwide environmental problems such as climate change, air pollution, water pollution, and deforestation. And yet it is only through global cooperation, such as the UN treaties on biodiversity protection or global warming, that these problems can be addressed.

## Social Dimensions

Globalization has a clear demographic dimension. Although international migration is not new, increasing numbers of people from all parts of the world are crossing national boundaries, often permanently (Figure 1.5). Migration from Latin America and Asia has drastically changed the demographic configuration of the United States, and migration from Africa and Asia has transformed western Europe. Countries such as Japan and South Korea that have long been perceived as ethnically homogeneous now have substantial immigrant populations. Even a number of relatively poor countries, such as Nigeria and the Ivory Coast, encounter large numbers of immigrants coming from even poorer countries, such as Burkina Faso. Although international migration is still curtailed by the laws of every country—much more so, in fact, than the movement of goods or capital—it is still

**FIGURE 1.5 International Migration**   Workers from southern India dig a hole to install a street sign in Dubai, United Arab Emirates. This Persian Gulf emirate is experiencing a massive construction boom as it shifts from an oil-based economy to an economy based on real estate, tourism, and international finance. As a result, temporary migrant workers from India and Pakistan constitute much of the labor force.

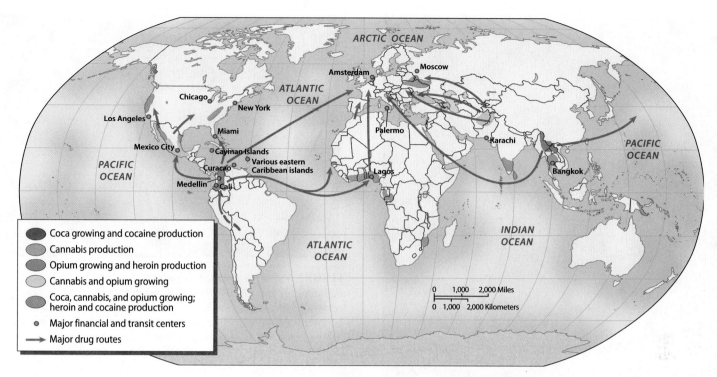

**FIGURE 1.6   The Global Drug Trade**   The cultivation, processing, and transshipment of coca (cocaine), opium (heroin), and cannabis (marijuana) are global issues. The most important cultivation centers are Colombia, Mexico, Afghanistan, and northern Southeast Asia, and the major drug financing centers are located mostly in the Caribbean, the United States, and Europe. In addition, Nigeria and Russia also play significant roles in the global transshipment of illegal drugs.

rapidly mounting, propelled by the uneven economic development associated with globalization.

Finally, there also is a significant criminal element to contemporary globalization, including terrorism (discussed later in this chapter), drugs, pornography, slavery, and prostitution. Illegal narcotics, for example, are definitely a global commodity (Figure 1.6). Some of the most remote parts of the world, such as the mountains of northern Burma, are thoroughly integrated into the circuits of global exchange through the production of opium and, therefore, into the world heroin trade. Even many areas that do not directly produce drugs are involved in their global sale and transshipment. Nigerians often occupy prominent positions in the international drug trade, as do members of the Russian mafia. Many Caribbean countries have seen their economies become reoriented to drug transshipments and the laundering of drug money. Prostitution, pornography, and gambling have also emerged as highly profitable global businesses. Over the past decades, for example, parts of eastern Europe have become major sources of both pornography and prostitution, finding a lucrative but morally questionable niche in the new global economy.

## Advocates and Critics of Globalization

Globalization, especially in its economic form, is one of today's most contentious issues (Figure 1.7). Supporters generally believe that it results in greater economic efficiency that will eventually result in rising prosperity for the entire world. In contrast, critics think that globalization will largely benefit those who are already prosperous, leaving most of the world poorer than before, while the rich exploit the less fortunate.

Economic globalization is generally applauded by corporate leaders and economists, and it has substantial support among the leaders of both major political parties in the United States. Beyond North America, moderate and conservative politicians in most countries generally support free trade and other aspects of economic globalization. Opposition to economic globalization is widespread in the labor and environmental movements, as well as among many student groups worldwide. Hostility toward globalization is sometimes deeply felt, as massive protests at World Bank and World Trade Organization meetings have made obvious.

**The Pro-globalization Stance**   Advocates of globalization argue that globalization is a logical and inevitable expression of contemporary international capitalism that will benefit all nations and all peoples. Economic globalization can work wonders, they contend, by enhancing competition, allowing the flow of capital to poor areas, and encouraging the spread of beneficial new technologies and ideas. As countries reduce their barriers to trade, inefficient local industries will be forced to become more efficient in order to compete with the new flood of imports, thereby enhancing overall national productivity. Those that cannot adjust will most likely go out of business.

Every country and region of the world, moreover, ought to be able to concentrate on those activities for which it is best suited in the global economy. Enhancing such geographic specialization, the pro-globalizers argue, creates a more efficient world economy. Such economic restructuring is made increasingly possible by the free flow of capital to those areas that have the greatest opportunities. By making access to capital more readily available throughout the world, economists contend, globalization should eventually result in a certain global **economic convergence**, meaning that the world's poorer countries will gradually catch up with the more advanced economies.

**FIGURE 1.8  The Global "Electronic Herd"**   One facet of economic globalization is the rapid movement of capital as bond traders, currency speculators, and fund managers direct money into or out of the economies of developing countries. In this picture, traders and clerks work on the Eurodollar Futures floor of the Chicago Mercantile Exchange.

the argument goes, they can gradually improve conditions and move into more sophisticated, better-paying industries. Proponents of economic globalization also commonly argue that multinational firms based in North America or Europe often offer better pay and safer working conditions than do local firms and thus contribute to worker well-being. Extreme pro-globalizers go so far as to assert that poor countries ought to take advantage of their generally lax environmental laws in order to attract highly polluting industries from the wealthy countries because acquiring such industries would enhance their overall economic positions.

The pro-globalizers generally strongly support the large multinational organizations that facilitate the flow of goods and capital across international boundaries. Three such organizations are particularly important: the World Bank, the International Monetary Fund (IMF),

**FIGURE 1.7  Protests Against Globalization**   Meetings of international groups such as the World Trade Organization, World Bank, and International Monetary Fund (IMF) commonly draw large numbers of protesters against globalization. This demonstration took place at a Washington, DC, meeting of the World Bank and IMF.

Thomas Friedman, one of the most influential advocates of economic globalization, argues that the world has not only shrunk but has also become flat in a sense, so that financial capital, goods, and services flow freely from place to place. For example, the need to attract capital from abroad forces countries to adopt new economic policies. Friedman describes the great power of the global "electronic herd" of bond traders, currency speculators, and fund managers who either direct money to or withhold it from developing economies, resulting in economic winners and losers (Figure 1.8).

To the committed pro-globalizer, even the global spread of **sweatshops**—crude factories in which workers sew clothing, assemble sneakers, and perform other labor-intensive tasks for extremely low wages—is to be applauded (Figure 1.9). People go to work in sweatshops because the alternatives in the local economy are even worse. Once countries achieve full employment through sweatshops,

**FIGURE 1.9  Global Sweatshops**   One of the most debated aspects of economic globalization is the crude factories, or sweatshops, in which workers sew clothing, assemble sneakers, stitch together soccer balls, and perform other labor-intensive work for low wages.

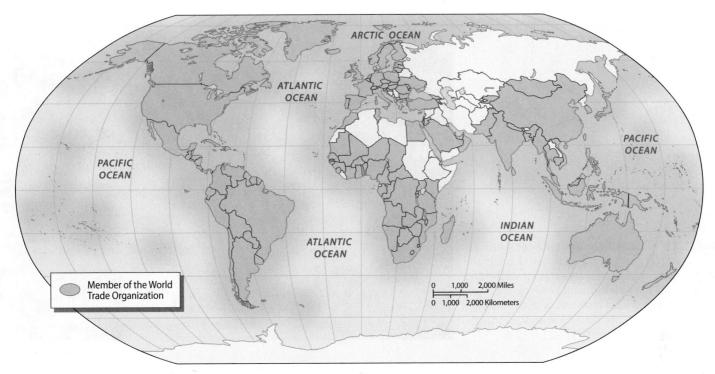

**FIGURE 1.10 World Trade Organization**    One of the most powerful institutions of economic globalization is the World Trade Organization (WTO), which was created in 1995 to oversee trade agreements, encourage open markets, enforce trade rules, and settle disputes. The WTO currently consists of 151 member countries. In addition to these member countries, more than 30 states have "observer status," including Russia, Iran, and Iraq.

and the World Trade Organization (WTO). The primary function of the World Bank is to make loans to poor countries so that they can invest in infrastructure and build more modern economic foundations. The IMF is concerned with making short-term loans to countries that are in financial difficulty—those having trouble, for example, making interest payments on the loans that they had previously taken. The WTO, a much smaller organization than the other two, works to reduce trade barriers between countries to enhance economic globalization. It also tries to mediate between countries and trading blocs that are engaged in trade disputes (Figure 1.10).

To support their claims, pro-globalizers argue that countries that have been highly open to the global economy have generally had much more economic success than those that have isolated themselves by seeking self-sufficiency. The world's most isolated countries, such as Burma (Myanmar) and North Korea, have become economic disasters, with little growth and rampant poverty, whereas those that have opened themselves to global forces, such as Singapore and Thailand, in the same period have seen rapid growth and substantial reductions in poverty.

**Critics of Globalization**    Virtually all of the claims of the pro-globalizers are strongly contradicted by the anti-globalizers. Opponents often begin by arguing that globalization is not a "natural" process. Instead, it is the product of an explicit economic policy promoted by free-trade advocates, capitalist countries (mainly the United States, but also Japan and the countries of Europe), financial interests, international investors, and multinational firms. Opponents also point out that the processes of globalization today are more pervasive than those of the past, even during the period of European colonialism.

Thus, while global economic and political linkages have been around for centuries, their modern expression is unprecedented.

Because the globalization of the world economy appears to be creating greater inequity between rich and poor, the "trickle-down" model of developmental benefits for all people in all regions has yet to be validated. On a global scale, the richest 20 percent of the world's people consume 86 percent of the world's resources, whereas the poorest 80 percent use only 14 percent. The growing inequality of this age of globalization is apparent on both global and national scales. Globally, the wealthiest countries have grown much richer over the past two decades, while the poorest have become more impoverished. Nationally, even in developed countries such as the United States, the wealthiest 10 percent of the population have reaped almost all of the gains that globalization has offered; the poorest 10 percent have either seen their income decline in recent decades as wages have remained static or have become unemployed as jobs have been lost to outsourcing.

Opponents also contend that globalization promotes free-market, export-oriented economies at the expense of localized, sustainable activities. World forests, for example, are increasingly cut for export timber rather than serving local needs. As part of their economic structural adjustment package, the World Bank and the IMF encourage developing countries to expand their resource exports so they will have more hard currency to make payments on their foreign debts. This strategy, however, usually leads to overexploitation of local resources. Opponents also note that the IMF often requires developing countries to adopt programs of fiscal austerity that often entail substantial reductions in public spending for education, health, and food subsidies. By adopting such policies, critics warn, poor countries will end up with even more impoverished populations than before.

**FIGURE 1.11 Global Economic Recession**
Unused shipping containers pile up on parking lots and vacant space near a residential area in northwest Hong Kong because of the downturn in China's export economy during the global recession of 2009. During the height of the economic downturn, about 25% of the world's container ships were idled.

Anti-globalizers also dispute the empirical evidence on national development offered by the pro-globalizers. Highly successful developing countries such as South Korea, Taiwan, and Malaysia, they argue, have indeed been engaged with the world market, but they have generally done so on their own terms rather than those of the IMF and other advocates of full-fledged economic globalization. These countries have actually protected many of their domestic industries from foreign competition and have, at various times, controlled the flow of capital.

Furthermore, anti-globalizers contend that the "free-market" economic model commonly promoted for developing countries is not the one that Western industrial countries used for their own economic development. In Germany, France, and even to some extent the United States, governments historically have played a strong role in directing investment, managing trade, and subsidizing chosen sectors of the economy.

Those who challenge globalization also worry that the entire system—with its instantaneous transfers of vast sums of money over nearly the entire world on a daily basis—is inherently unstable. The noted critic John Gray, for example, thinks that the same "electronic herd" that Thomas Friedman applauds is a dangerous force because it is susceptible to "stampedes." International managers of capital tend to panic when they think their funds are at risk; when they do so, the entire intricately linked global financial system can quickly become destabilized, leading to a crisis of global proportions. The rapid downturn of the global economy in late 2008 serves as an example (Figure 1.11).

Even when the "herd" spots opportunity, trouble may still ensue. As vast sums of money flow into a developing country, they may create a speculatively inflated **bubble economy** that cannot be sustained. Such a bubble economy emerged in Thailand and many other parts of Southeast Asia in the mid-1990s. Analysts have also used the concept of bubble economy to explain the collapse of the economies of Iceland and Ireland in 2009 (Figure 1.12).

**A Middle Position?** A number of experts, not surprisingly, argue that both the anti-globalization and the pro-globalization stances are exaggerated. Those in the middle ground tend to argue that economic globalization is indeed unavoidable; even the anti-globalization movement, they point out, is made possible by the globalizing power of the Internet and is, therefore, itself an expression of globalization. They further contend that while globalization holds both promises and pitfalls, it can be managed, at both the national and international levels, to reduce economic inequalities and protect the natural environment. These experts stress the need for strong yet efficient national governments, supported by international institutions (such as the UN, World Bank, and IMF) and globalized networks of environmental, labor, and human rights groups.

Globalization is one of the most important issues of the day—and one of the most complicated. While this book does not pretend to resolve the controversy, it does encourage readers to reflect on these critical points as they apply to different world regions.

**FIGURE 1.12 Economic Turmoil in Iceland**    Protesters burn an effigy of Iceland's prime minister during a demonstration against the government's handling of the 2009 economic crisis. Given the size of Iceland's economy, the collapse of its bubble economy was the largest suffered by any country in history.

**FIGURE 1.13 Local Cultures** This family in Ghana reminds us that although few places are beyond the reach of globalization, many unique local landscapes, economies, and cultures still exist. For example, most of the belongings of the family in this photograph are local in origin and come from long-standing cultural tradition.

## Diversity in a Globalizing World

As globalization increases, many observers foresee a world far more uniform and homogeneous than today's. The optimists among them imagine a universal global culture uniting all humankind into a single community untroubled by war, ethnic strife, or resource shortage—a global utopia of sorts.

A more common view, however, is that the world is becoming blandly homogeneous as different places, peoples, and environments lose their distinctive character and become indistinguishable from their neighbors. While diversity may be difficult for a society to live with, it also may be dangerous to live without. Nationality, ethnicity, cultural distinctiveness—all are the legitimate legacy of humanity. If this diversity is blurred, denied, or repressed through global homogenization, humanity loses one of its defining traits.

But even if globalization is generating a certain degree of homogenization, the world is still a highly diverse place (Figure 1.13). One

still finds marked differences in culture (including language, religion, architecture, foods, and many other attributes of daily life), economy, and politics—as well as in the natural environment. Such diversity is so vast that it cannot readily be extinguished, even by the most powerful forces of globalization. In fact, globalization often provokes a strong reaction on the part of local people, making them all the more determined to maintain what is distinctive about their way of life. Thus, globalization is understandable only if one also examines the diversity that continues to characterize the world and, perhaps most importantly, the tension between these two forces—the homogenization of globalization and the reaction against it in terms of protecting cultural and political diversity. Unfortunately, this tension often takes unpleasant, violent forms, as illustrated today by radical Islam and tribal warfare in Nigeria and Sudan (Figure 1.14).

The politics of diversity also demands increasing attention as we try to understand worldwide tensions over terrorism, ethnic separateness, regional autonomy, and political independence. Groups of people throughout the world seek self-rule of territory they can call their

**FIGURE 1.14 Tribal Warfare** Rebels from the Sudan Liberation Movement (SLM) declared war against the Sudanese government in the province of Darfur. The two groups that originated the rebellion have now morphed into dozens of warring factions, complicating the search for peace amid a profound humanitarian crisis.

**FIGURE 1.15 World Regions**   These regions are the basis for the 12 regional chapters in this book. Countries or areas within countries that are treated in more than one chapter are designated on the map with a striped pattern. For example, western China is discussed in both Chapter 10, on Central Asia, and Chapter 11, on East Asia. Also, three countries on the South American continent are discussed as part of the Caribbean region because of their close cultural similarities to the island region.

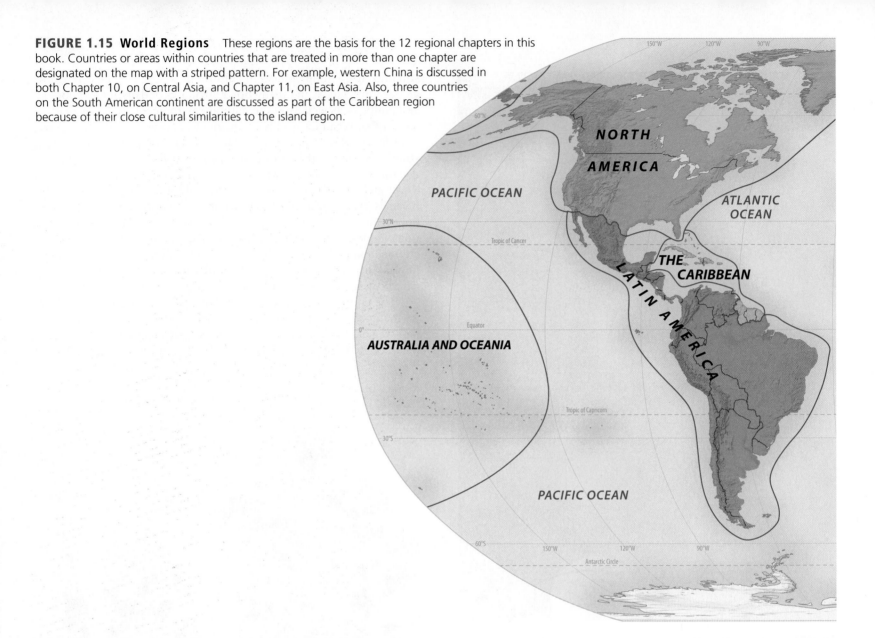

own. Today, most wars are fought *within* countries not *between* them. As a result, our concern with geographic diversity takes many forms and goes far beyond simply celebrating traditional cultures and unique places. People have many ways of making a living throughout the world, and it is important to recognize this fact as the globalized economy becomes increasingly focused on mass-produced retail goods. Furthermore, a stark reality of today's economic landscape is unevenness: While some people and places prosper, others suffer from unrelenting poverty. Unfortunately, this also is a form of diversity amid globalization.

In summary, globalization can be defined as the increasing interconnectedness of people and places through converging processes of economic, political, and cultural change. Globalization is pervasive and unrelenting and is a dominant feature of the contemporary world. Furthermore, its benefits are uneven. Although economic restructuring is a prime cause, globalization is not simply the growth and expansion of international trade. It is also represented by converging and homogenizing forces that many people resist. Thus, an equally important theme is how geographic diversity comes in conflict with globalization—how globalization in different places and at different times is suppressed, renegotiated, hybridized, protected, preserved, or extinguished. Because of the importance of this theme, globalization and diversity

should be examined as inseparable—often in conflict but at other times complementary.

## Themes and Issues in World Regional Geography

Following two introductory chapters, this book adopts a regional perspective, grouping all of Earth's countries into a framework of world regions (Figure 1.15). We begin with a region familiar to most of our readers, North America, and then move to Latin America, the Caribbean, Africa, the Middle East, Europe, Russia, and the different regions of Asia, before concluding with Australia and Oceania. Each of the 12 regional chapters employs the same five-part thematic structure—environmental geography, population and settlement, cultural coherence and diversity, geopolitical framework, and economic and social development. The concepts and data central to each theme are discussed in the following sections.

Chapter 2 provides background on global environmental geography, outlining the global environmental elements fundamental to human settlement—climate, vegetation, and hydrology—and discussing the linkages between environmental issues and globalization.

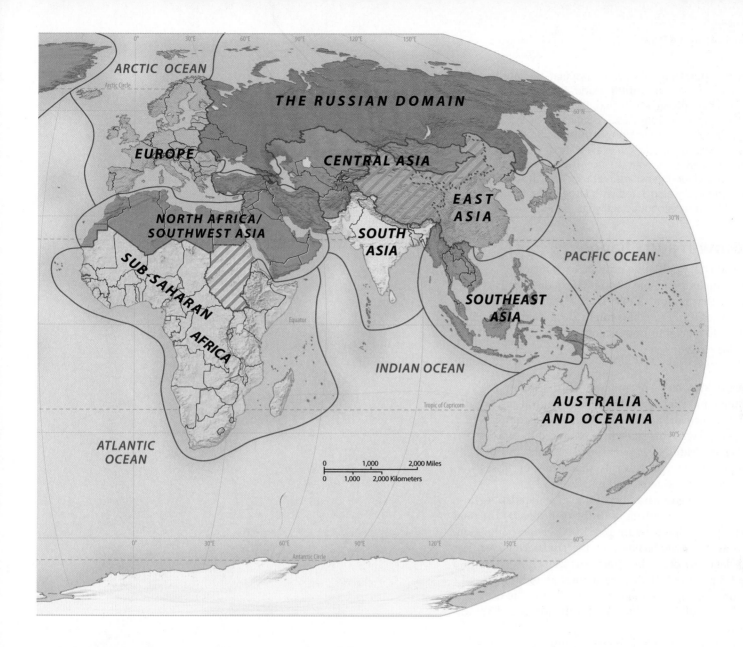

# Population and Settlement: People on the Land

The human population is far larger than it has ever been, and there is considerable debate about whether continued growth will benefit or harm the human condition. This concern is not only about the total population of the world but also about the geographic patterns of human settlement. As the world map shows, while some parts of the world are densely populated, other places remain almost empty (Figure 1.16).

With more than 6.7 billion humans on Earth, we currently add about 80 million people each year, at a rate of 15,000 births per hour. About 90 percent of this population growth takes place in the developing countries of Africa, South and East Asia, and Latin America. Because of rapid growth in developing countries, perplexing questions dominate discussions of many global issues. Can these countries absorb the rapidly increasing population and still achieve the necessary economic and social development to ensure some level of well-being and stability for their populations? What role, if any, should the developed countries of North America, Europe, and East Asia play in helping developing countries with their population problems? While

population is a complex and contentious topic, several points help focus the issues:

- Very different rates of population growth are found in different regions of the world. While some countries are growing rapidly, others have no natural growth at all. Instead, any population growth comes from in-migration. India is an example of the former; Italy of the latter.

- Population planning takes many forms, from the fairly rigid one-child policies of China to the "more children, please" programs of western Europe (Figure 1.17).

- Not all attention should be focused on natural growth because migration is increasingly the root cause of population change in the globalized world. Most international migration is driven by a desire for a better life in the richer regions of the developed world. But there are also millions of migrants who are refugees from civil strife, persecution, and environmental disasters.

- The greatest migration in human history is now going on, as people move from rural to urban environments. By 2009, more than half the world's population lived in towns and cities. As with natural population growth, the developing countries of Africa, Latin America, and Asia are experiencing the most rapid changes in urbanization.

**FIGURE 1.16 World Population**   This world population map shows the differing densities of population in the regions of the world. East Asia stands out as the most populated region, with high densities in Japan, Korea, and eastern China. The second most populated region is South Asia, dominated by India, which is second only to China in population. In North Africa and Southwest Asia, population clusters are often linked to the availability of water for irrigated agriculture, as is apparent with the population cluster along the Nile River. Higher population densities in Europe, North America, and other countries are usually associated with large cities, their extensive suburbs, and nearby economic activities.

## Population Growth and Change

Because of the centrality of population growth, each regional chapter in this book includes a table of population data for the countries within that region (Table 1.1). Although at first glance the statistics in these tables might seem daunting, this information is crucial to understanding the population geography of the regions.

**Natural Population Increase**   A common starting point for measuring demographic change is the **rate of natural increase (RNI)**, which depicts the annual growth rate for a country or region as a percentage. This statistic is produced by subtracting the number of deaths in a given year from the births (Figure 1.18). Important to remember is that gains or losses through migration are not considered in the RNI.

Further, instead of using raw numbers for a large population, demographers divide the gross numbers of births or deaths by the total population, thereby producing a figure per 1,000 of the population. This is referred to as the *crude birthrate* or the *crude death rate*. For example, in 2009 data, the crude birthrate was 21 per 1,000, with a crude death rate of 8 per 1,000. Thus, the natural growth rate was 12 per 1,000. Converting that figure to a percentage allows us to express it as the RNI; thus, the rate of natural increase for the world in 2009 was 1.2 percent per year.

Because birthrates vary greatly between countries (and even between regions of the world), rates of natural increase also vary greatly.

### TABLE 1.1   POPULATION INDICATORS OF THE 10 LARGEST COUNTRIES

| Country | Population (millions) 2009 | Population Density (per square kilometer) | Rate of Natural Increase (RNI) | Total Fertility Rate | Percent Urban | Percent <15 | Percent >65 | Net Migration (Rate per 1,000) 2005–2010* |
|---|---|---|---|---|---|---|---|---|
| China | 1,331.4 | 139 | 0.5 | 1.6 | 46 | 19 | 8 | −0.3 |
| India | 1,171.0 | 356 | 1.6 | 2.7 | 29 | 32 | 5 | −0.2 |
| United States | 306.8 | 32 | 0.6 | 2.1 | 79 | 20 | 13 | 3.3 |
| Indonesia | 243.3 | 128 | 1.5 | 2.5 | 43 | 29 | 6 | −0.6 |
| Brazil | 191.5 | 22 | 1.0 | 2.0 | 84 | 28 | 6 | −0.2 |
| Pakistan | 180.8 | 227 | 2.3 | 4.0 | 35 | 38 | 4 | −1.6 |
| Bangladesh | 162.2 | 1,127 | 1.6 | 2.5 | 25 | 32 | 4 | −0.7 |
| Nigeria | 152.6 | 165 | 2.6 | 5.7 | 47 | 45 | 3 | −0.4 |
| Russia | 141.8 | 8 | −0.3 | 1.5 | 73 | 15 | 14 | 0.4 |
| Japan | 127.6 | 338 | −0.0 | 1.4 | 86 | 13 | 23 | 0.2 |

Source: Population Reference Bureau, *World Population Data Sheet, 2009.*

*Net Migration Rate from the United Nations, Population Division, *World Population Prospects: The 2008 Revision Population Database.*

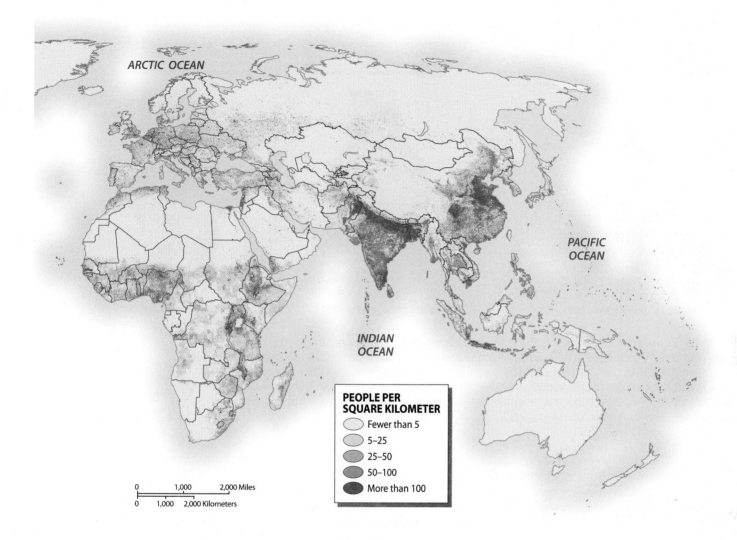

PEOPLE PER
SQUARE KILOMETER

Fewer than 5

5–25

25–50

50–100

More than 100

0        1,000        2,000 Miles

0     1,000    2,000 Kilometers

**FIGURE 1.18 Fertility and Mortality**   Birthrates and death rates vary widely around the world. Fertility rates result from an array of variables, including state family-planning programs and the level of a woman's education. This family is in Bangladesh, a country that is working hard to reduce its birthrate through education programs.

**FIGURE 1.17 Family Planning Policies**   Many countries in the developing world have concluded that unrestrained population growth may keep them from realizing their development goals and have therefore put in place family planning policies. This poster in Vietnam urges smaller families.

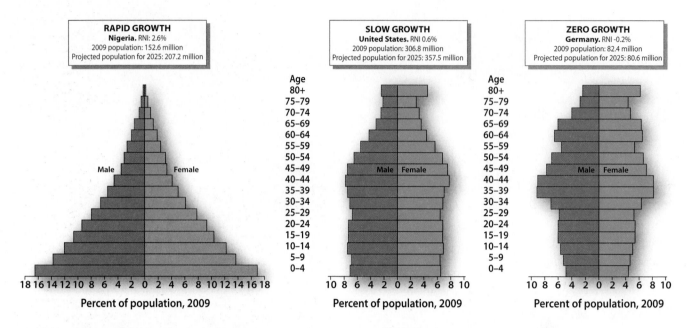

**FIGURE 1.19 Population Pyramids**   The term *population pyramid* comes from the form assumed by a rapidly growing country such as Nigeria, when data for age and sex are plotted graphically as a percentage of the total population. The broad base illustrates the high percentage of young people in the country's population, which indicates that rapid growth will probably continue for at least another generation. This pyramidal shape contrasts with the narrow bases of the slow- and negative-growth countries, the United States and Germany, which have fewer people in the childbearing years.

In Africa, for example, many countries have crude rates of more than 40 births per 1,000 people. Because their death rates are generally less than 20 per 1,000, their RNIs are greater than 3 percent per year; these are among the highest population growth numbers found anywhere in the world.

**Total Fertility Rate**   The crude birthrate gives some insight into current conditions in a country, but demographers place more emphasis on the **total fertility rate (TFR)** to predict future growth. The TFR is a synthetic hypothetical number that measures the fertility of a statistically fictitious yet average group of women moving through their childbearing years. If women marry early and have many children over a long span of years, the TFR will be a relatively high number. Conversely, if data show that women marry late and have few children, the number will be correspondingly small. Important to note is that any number less than 2.1 implies that a population is not growing at all because it takes a minimum of two children to replace their parents. From population data collected in the past decade, the current TFR for the world is 2.6. While that is the average for the whole world, the variability between regions is striking. To illustrate, the current TFR for Africa is 4.8, whereas in slow- and no-growth Europe it is only 1.5.

**Young and Old Populations**   One of the best indicators of the momentum (or lack of) for continued population growth is the youthfulness of a population because these data show the proportion of a population about to enter their prime reproductive years. The common statistic for this measure is the *percentage of a population under age 15.* As a global average, 27 percent of the population is younger than age 15, but in fast-growing Africa, that figure is 41 percent, with several African countries approaching 50 percent (Figure 1.18). This suggests strongly that rapid population growth in Africa will continue for at

least another generation, despite the tragedy of the AIDS epidemic. In contrast, Europe has only 15 percent of its population under 15, and North America 20 percent.

The other end of the age spectrum is also important, and it is measured by the *percentage of a population over age 65.* That figure is useful for inferring the needs of a society in providing social services for its senior citizens and pensioners. Japan and most European countries, for example, have a relatively high proportion of people over age 65. As a result, concerns are raised about whether there will be enough wage earners to support the social security needs of a large elderly population.

The structure of a population, which includes the percentage of young and old, is presented graphically as a **population pyramid.** This graph plots the percentage of all different age groups along a vertical axis that divides the population into male and female (Figure 1.19). The large percentage of young people in a fast-growing population provides a wide base and the small percentage of elderly a narrow tip, thereby giving the graph its pyramidal shape. Older populations, in contrast, with fewer young people and an aging population, give the graph a very different shape, with a narrow base and broader upper segment.

**Life Expectancy**   Another demographic indicator that contains information about health and well-being in a society is *life expectancy,* which is the average length of a life expected at the birth of a typical male or female in a specific country. Because a large number of social factors, such as health services, nutrition, and sanitation, influence life expectancy, these data are often used as an indicator of the level of social development in a country. Because this book uses life expectancy as a social indicator, life expectancy data are found in the economic and social development tables, not with other population statistics (Table 1.2).

Not surprisingly, because social conditions vary widely around the world, so do life expectancy figures. In general, though, life expectancy has been increasing over the decades, implying that the conditions supporting life and longevity are improving. To illustrate, in 1975, the average life expectancy figure for the world was 58 years, whereas today it is 69. In Sub-Saharan Africa, however, life expectancy has changed very little over the last 30 years because of the HIV/AIDS epidemic. As a result, the life expectancy for the region is the same (51) as it was in 1975. In Russia, life expectancy has also fallen in the past two decades because of the deterioration of social services accompanying economic restructuring in the post-Soviet era.

## The Demographic Transition

The historical record shows that population growth rates commonly change over time. More specifically, in Europe, North America, and Japan, growth declined as countries became increasingly industrialized and urbanized. From this historical data, demographers generated the **demographic transition model**, a four-stage conceptualization that tracks changes in birthrates and death rates through time as a population urbanizes (Figure 1.20).

In the demographic transition model, Stage 1 is characterized by both high birthrates and death rates, leading to a very slow rate of natural increase. Historically, this stage is associated with Europe's preindustrial period, a period that also predated common public health measures such as sewage treatment, the scientific understanding of disease transmission, and the most fundamental aspects of modern medicine. Not surprisingly, death rates were high and life expectancy was short. Unfortunately, these conditions are still found in some parts of the world.

In Stage 2, death rates fall dramatically while birthrates remain high, producing a rapid rise in the RNI. Again, in both historical and modern times, this decrease in death rates is usually associated with the development of public health measures and modern medicine. One of the assumptions of the demographic transition model is that these services become increasingly available after some degree of economic development and urbanization takes place.

However, even as death rates fall, it takes time for people to respond with lower birthrates, which begin in Stage 3. This, then, is the transitional stage in which people apparently become aware of the advantages of smaller families in an urban and industrial setting.

In Stage 4, a very low RNI results from a combination of low birthrates and very low death rates. As a result, there is very little natural population increase. Today, the United States, Japan, and most European countries, along with other industrial nations, are clearly in Stage 4 of the demographic transition model.

## Migration Patterns

Never before in human history have so many people been on the move. Today, more than 190 million people live outside the country of their birth and thus are officially designated as migrants by international agencies. Much of this international migration is directly linked to the new globalized economy because half of the migrants live either in the developed world or in developing countries with vibrant industrial, mining, or petroleum extraction economies. In the oil-rich countries of Kuwait and Saudi Arabia, for example, the labor force is composed primarily of foreign migrants. In total numbers, fully one~~...~~ live in seven industrial countries: Japan, ~~Germany, France, Canada,~~ the United States, Italy, and the United ~~Kingdom. Moreover, most of~~ these migrants have moved to cities; in fact, 20 percent of migrants live in just 20 world cities. Further, because industrial countries usually have very low birthrates, immigration accounts for a large proportion of their population growth. For example, about one-third of the annual growth in the United States is due to in-migration.

*Conceptualization*

But not all migrants move for economic reasons. War, persecution, famine, and environmental destruction cause people to flee to safe havens elsewhere. Accurate data on refugees are often difficult to obtain for a number of reasons (such as individuals not legally crossing international boundaries or countries deliberately obscuring the number for political reasons), but UN officials estimate that some 35 million people should be considered refugees. More than half of these are in Africa and western Asia (Figure 1.21).

**Push and Pull Forces**   While its causes may be complicated, the migration process can often be understood through three interactive concepts. First, *push* forces, such as civil strife, environmental degradation, or unemployment, drive people from their homelands. Second, *pull*

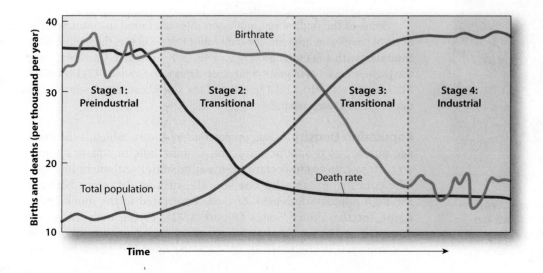

**FIGURE 1.20 Demographic Transition**   As a country goes through industrialization, its population moves through the four stages in this diagram, referred to as the *demographic transition*. In Stage 1, population growth is low because high birthrates are offset by high death rates. Rapid growth takes place in Stage 2, as death rates decline. Stage 3 is characterized by a decline in birthrates. The transition ends with low growth once again in Stage 4, resulting from a relative balance between low birthrates and death rates.

## SETTING THE BOUNDARIES
# The Metageography of World Regions

One of the perplexing issues of world regional geography is the definition and demarcation of regions. For example, historically Europe has been separated from Asia by the Straits of Bosporus, the narrow waterway that divides Turkey into European Turkey to the west and Asiatic Turkey to the east. But because Turkey is unified by language, religion, and culture, any traveler to that country can see that this traditional division is misleading.

The problem of dividing the world into a small number of regional units has been apparent since the earliest days of geography. The early Greeks, for example, conceptualized the world as divided into three continents. As seafaring explorers, they chose to divide the great land masses of Europe, Asia, and Africa based upon seas and waterways. As a result, the Bosporus and the Red Sea became the landmarks separating those continent-based world regions. Later, European explorers agreed with this threefold division even

though a large country—Russia—had to be separated into European Russia west of the Ural Mountains and Asiatic Russia to the east.

Despite this problem, European explorers continued to divide the world into continents: North and South America were referred to as separate continents although they were clearly joined by the narrow isthmus of Panama (or pan-America, as some call it); Australia was added as a continent even though many argued that it was simply a large island. But if Australia was a continent, shouldn't Madagascar also be a continent? Not according to European explorers, who were quite comfortable with considering that large island nation part of the African continent. Finally, Antarctica was given continental status even though it was totally uninhabited.

This seven-part "myth of the continents" seemed primarily useful for allowing Europeans to differentiate themselves from other cultures, specifically Asian and African cultures, for it was easier to draw cultural and

ethnic boundaries if Asians and Africans were placed on distinct continents, separated from Europe by mountains, seas, and waterways. This scheme, however, makes little sense in today's world for there is no such thing as an Asian culture. Nor, for that matter, is there a single African culture.

World War II and the subsequent aid programs focused greater emphasis on cultural and geopolitical regions than on continents. Asia was subdivided into a handful of regions roughly comparable to historically defined civilizations; thus, East Asia, centered on China and Japan, was differentiated from South Asia, anchored by India. Though progress was made by moving away from the myth of continents, there is still a good deal of quibbling about where to draw the boundaries for different world regions. Each regional chapter in this book includes a sidebar titled "Setting the Boundaries" that examines in more detail this complicated issue of regional definition and boundary drawing.

---

forces, such as better economic opportunity or health services, attract migrants to certain locations, within or beyond their national boundaries. Connecting the two are the *informational networks* of families, friends, and sometimes labor contractors who provide information on the mechanics of migration, transportation details, and housing and job opportunities.

**FIGURE 1.21 Refugees**   About 35 million people are refugees from ethnic warfare and civil strife. Most of these refugees are in Africa and western Asia. These women wait in line for water at a refugee camp in southern Sudan.

**Net Migration Rates**   The amount of immigration (in-migration) and emigration (out-migration) is measured by the **net migration rate**, a statistic that depicts whether more people are entering or leaving a country. A positive figure means the population is growing because of in-migration, whereas a negative number means more people are leaving than arriving. As with other demographic indicators, the net migration rate is provided for the number of migrants per 1,000 of a base population. To illustrate, the net migration rate for the United States is 3.3 per 1,000 people, whereas Canada, which receives even more immigrants than the United States, has a net migration rate of 8. In contrast, Mexico, the source of many migrants to North America, has a net migration rate of –4.

Some of the highest net migration rates are found in countries that depend heavily on migrants for their labor force, such as the United Arab Emirates, with a net migration rate of 48, or Kuwait at 8. Countries with the highest negative migration rates are American Samoa, –17, Federated Micronesia, also –15, and a number of the Caribbean island states, with net migration rates around.

**Population Density**   Data on *population density*, which is the average number of people per area unit (square mile or square kilometer), often conveys important information about settlement in a specific country. In Table 1.1 one sees the striking difference between the high population density of India contrasted to the much lower figure for the United States (Figure 1.22). Flying over these two countries and looking down at the settlement patterns would explain this contrast. While much of the United States is covered by farms

**FIGURE 1.22 Contrasting Settlement Density: U.S. and India** In the U.S. the population density is 32 per square kilometer, contrasted with 356 per square kilometer in India. Besides the different in the size of each country's population, another factor in the contrasting densities is the difference in settlement patterns. In the U.S. people commonly settle on dispersed farms on large acreage, as illustrated by the landscape in Iowa, whereas as in India there is dense settlement in both towns and rural landscapes.

covering hundreds of acres, with houses and barns several miles from their neighbors, the landscape of India is made up of small villages, distanced from each other by only a mile or so. This results in a population density three times higher in India than in the United States. Japan has a similar settlement pattern to India, even though it is primarily an urban, industrial country. Bangladesh, one of the most densely settled countries in the world, must squeeze its large and rapidly growing population into a limited amount of dry land built by the delta of two large rivers, the Ganges and the Brahmaputra.

While a population density figure is calculated for the total area of a country, that figure can vary greatly between rural and urban settlement. Many of the world's cities, for example, have densities of more than 30,000 people per square mile (10,344 per square kilometer), with areas of Paris, Mumbai (Bombay), and Hong Kong twice as dense. Most North American cities, in contrast, have densities of fewer than 10,000 people per square mile (3,861 per square kilometer) because of the cultural preference for single-family dwellings on individual lots instead of the high-rise apartment buildings that characterize most of the world's cities.

**An Urban World** Cities are the focal points of the contemporary, globalizing world, the fast-paced centers of deep and widespread economic, political, and cultural change. Because of this vitality and the options cities offer to impoverished and uprooted rural peoples, they are also magnets for migration. The scale and rate of growth of some world cities is staggering: Estimates are that both Mexico City and São Paulo (Brazil), cities of more than 20 million, are adding about 10,000 new people each week (Figure 1.23). Urban planners predict that these two cities may actually double in size within the next 15 years.

Assuming that predictions about the migration rate to cities are correct, the world has approached the point where more than half its people are urban dwellers. Evidence comes from data on the **urbanized population,** which is the percentage of a country's population living in cities. Further, many demographers predict that the world will be 60 percent urbanized by the year 2025.

Tables in this book's regional chapters include data on the urbanization rate for each country. To illustrate, more than 80 percent of the populations of Europe, Japan, Australia, and the United States live in cities. Generally speaking, most countries with such high rates of urbanization are also highly industrialized because manufacturing tends to cluster around urban centers. In contrast, the urbanized rate for developing countries is usually less than 50 percent, with figures closer to 40 percent not uncommon. Urbanization figures also show where there is high potential for urban migration. If the urbanized population is relatively small, as in Zimbabwe (Africa), where only some 34 percent of the population lives in cities, the probability of high rates of urban migration in the next decades is great (Figure 1.24).

## Cultural Coherence and Diversity: The Geography of Tradition and Change

If culture is the weaving that keeps the world's diverse social fabric together, one glance at the daily news suggests that this global tapestry is unraveling because of the frequency of cultural conflict. With the recent rise of global communication systems (satellite TV, films, videos, etc.), stereotypical Western culture is spreading at a rapid pace, and while some cultures accept these new cultural influences willingly, others resist this new form of cultural imperialism through protests, censorship, and even terrorism.

The geography of cultural cohesion and diversity, then, entails an examination of tradition and change, of language and religion, of

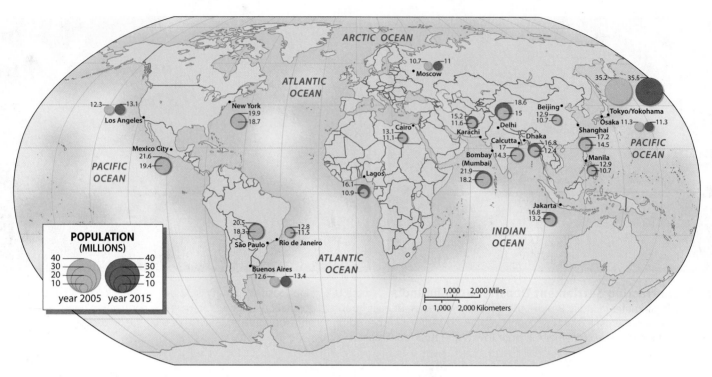

**FIGURE 1.23 Growth of World Cities** This map shows the 20 largest cities in the world, along with predicted growth by the year 2015. Note that the greatest population gains are expected in the large cities of the developing world, such as Lagos, Nigeria; Karachi, Pakistan; and Mumbai (Bombay), India. In contrast, large cities in the developed world are predicted to grow slowly over the next decades. Tokyo, for example, will add fewer than 1 million people.

group belonging and identity, and of the complex and varied currents that underlie 21st-century ethnic factionalism and separatism (Figure 1.25).

## Culture in a Globalizing World

Given the diversity of cultures around the world, coupled with the dynamic changes associated with globalization, traditional definitions of culture must be stretched somewhat to provide a viable conceptual framework. **Culture** is learned, not innate, and is shared behavior held in common by a group of people, empowering them with what could be called, for lack of a better term, a "way of life."

In addition, culture has both abstract and material dimensions: speech, religion, ideology, livelihood, and value systems, and also technology, housing, foods, and music. These varied expressions of culture are relevant to the study of world regional geography because they tell us much about the way people interact with their environment, each other, and the larger world. Finally, especially given the widespread influences of globalization, it is best to think of culture as dynamic rather than static. That is, culture is a process, not a condition, something that is constantly adapting to new circumstances. As a result, there are always tensions between the conservative, traditional elements of a culture and the newer, more progressive forces promoting change (Figure 1.26).

## When Cultures Collide

Cultural change often takes place within the context of international tensions. Sometimes one cultural system will replace another; at other times, resistance will stave off change. More commonly, however, a newer, hybrid form of culture results that is an amalgamation of two cultural traditions. Historically, colonialism was the most important perpetuator of these cultural collisions; today, globalization in its varied forms can be thought of as the major vehicle of cultural tensions and change.

**Cultural imperialism** is the active promotion of one cultural system at the expense of another. Although there are still many expressions of cultural imperialism today, the most severe examples occurred in the colonial period, when European cultures spread worldwide, sometimes overwhelming, eroding, and even replacing indigenous cultures. During this period, Spanish culture spread widely in Latin America; French culture diffused into parts of Africa; and British culture entered India. New languages were mandated, new education systems were implanted, and new administrative institutions took the place of the old. Foreign dress styles, diets, gestures, and organizations were added to existing cultural systems. Many vestiges of colonial culture are still evident today. In India, the makeover was so complete that many are fond of saying, with only slight exaggeration, that "the last true Englishman will be an Indian."

Today's cultural imperialism is seldom linked to an explicit colonizing force but more often comes as a fellow traveler with economic globalization. Though many expressions of cultural imperialism carry a Western (even U.S.) tone—such as McDonald's, MTV, Marlboro cigarettes, and the use of English as the dominant language of the Internet—these facets result more from a search for new consumer markets than from deliberate efforts to spread modern U.S. culture throughout the world.

The reaction against cultural imperialism, **cultural nationalism**, is the process of protecting and defending a cultural system against

**FIGURE 1.25 Ethnic Tensions** Unfortunately, much contemporary cultural change is characterized by violence between different ethnic groups. In this photo, a commuter in the Indian state of Gujarat passes by a car burned during rioting between Muslims and Hindus.

diluting or offensive cultural expressions while at the same time actively promoting national and local cultural values. Often, cultural nationalism takes the form of explicit legislation or official censorship that simply outlaws unwanted cultural traits. Examples of legislated cultural nationalism are common. France has long fought the Anglicization of its language by banning "Franglais," the use of English words such as "Le weekend," in official French. More recently, France has sought to protect its national music and film industries by legislating that radio DJs play a certain percentage of French songs and artists each broadcast day (40 percent in 2009). In addition, many Muslim countries censor Western cultural influences by restricting and censoring international TV, which they consider the source of many undesirable cultural influences. Most Asian countries, as well, are increasingly protective of their cultural values, and many are demanding changes to tone down the sexual content of MTV and other international TV networks.

As noted, the most common product of cultural collision is the blending of forces to form a new, synergistic form of culture, called **cultural syncretism or hybridization** (Figure 1.27). To characterize India's culture as British, for example, would be to grossly oversimplify and exaggerate England's colonial influence. Instead, Indians have adapted many British traits to their own circumstances, infusing them with their own meanings. India's use of English, for example, has produced a unique form of "Indlish" that often befuddles visitors to South Asia. Nor should we forget that India has added many words to the English vocabulary—*khaki, pajamas, veranda,* and *bungalow,* among others. Clearly, both the Anglo and Indian cultures have been changed by the British colonial presence in South Asia.

**FIGURE 1.24 Squatter Settlements** Because of the massive migration of people to world cities, adequate housing for the rapidly growing population becomes a daunting problem. Often, migrant housing needs are filled with illegal squatter settlements, such as this one in New Delhi, India.

(a)

(b)

(c)

(d)

**FIGURE 1.26 Folk, Ethnic, Popular, and Global Culture** A broad spectrum of cultural groups characterizes the world's contemporary cultural geography. Representing global culture (a) is a Korean woman in downtown Washington, DC, with mobile phone and laptop computer. In Amsterdam, the Netherlands, two visitors (b) wear the universal clothes of popular culture, while at the opposite end of the spectrum are the ethnic cultures of (c) Garifuna women in Honduras and (d) men in Peru performing the Chonguinada dance.

## Language and Culture in Global Context

Language and culture are so intertwined that, in the minds of many, language is the characteristic that best differentiates and defines cultural groups (Figure 1.28). Furthermore, because language is the primary means for communication, it obviously folds together many other aspects of cultural identity, such as politics, religion, commerce, and customs. In addition, language is fundamental to cultural cohesiveness. It not only brings people together but also sets them apart; it can be an important component of national or ethnic identity and a means for creating and maintaining boundaries of regional identity.

Because most languages have common historical (and even prehistorical) roots, linguists have grouped the thousands of languages found throughout the world into a handful of language families. This is simply a first-order grouping of languages into large units, based on common ancestral speech. For example, about half of the world's people speak languages of the Indo-European family, which includes not only European languages such as English and Spanish but also Hindi and Bengali, widespread languages of South Asia.

Within language families are smaller units that also give clues to the common history and geography of peoples and cultures.

*Language branches and groups* (also called *subfamilies*) are closely related subsets within a language family, in which there are usually similar sounds, words, and grammar. Well known are the similarities between German and English and between French and Spanish. Because of their similarities, these languages are placed into the same linguistic groupings.

Individual languages often have very distinctive forms associated with specific regions. Such different forms of speech are called *dialects*. Although dialects of the same language have their own unique pronunciation and grammar (think of the distinctive differences, for example, between British, North American, and Australian English), they are—sometimes with considerable effort—mutually intelligible.

When people from different cultural groups cannot communicate directly in their native languages, they often agree on a third language to serve as a common tongue, or **lingua franca**. Swahili has long served that purpose between the many tribal languages of eastern Africa, and French was historically the lingua franca of international politics and diplomacy. Today, English is increasingly the common language of international communications (Figure 1.29).

**FIGURE 1.27  Cultural Hybridity**   The hybridization of culture is clearly visible in this photo of South Asians (India vs. Pakistan) playing cricket, which was, before the colonial period, a uniquely English sport. Note, too, that this traditional sport now carries the logos and advertisements of global culture.

## A Geography of World Religions

Another extremely important defining trait of cultural groups is religion (Figure 1.30). Indeed, in this era of a totalizing global culture, religion is becoming increasingly important in defining cultural identity. Recent ethnic violence and unrest in far-flung places such as the Balkans, Afghanistan, and Indonesia illustrate the point.

**Universalizing religions**, such as Christianity, Islam, and Buddhism, attempt to appeal to all peoples, regardless of location or culture; these religions usually have a proselytizing or missionary program that actively seeks new converts. In contrast are the **ethnic religions**, which remain identified closely with a specific ethnic, tribal, or national group. Judaism and Hinduism, for example, are usually regarded as ethnic religions because they normally do not actively seek new converts. Instead, people are born into ethnic religions.

Christianity, a universalizing religion, is the world's largest religion in both areal extent and number of adherents. Although fragmented into separate branches and churches, Christianity as a whole has 2.1 billion adherents, or about one-third of the world's population. The largest number of Christians can be found in Europe, Africa, Latin America, and North America. Islam, which has spread from its origins on the Arabian Peninsula as far east as Indonesia and the Philippines, has about 1.3 billion members.

While not as severely fragmented as Christianity, Islam should not be thought of as a homogeneous religion because it is also split into separate groups. The major branches are Shiite Islam, which constitutes about 11 percent of the total Islamic population and represents a majority in Iran and southern Iraq, and the dominant Sunni Islam, which is found from the Arab-speaking lands of North Africa to Indonesia. Both of these forms of Islam are experiencing fundamentalist revivals in which proponents are interested in maintaining purity of faith distanced from Western influences.

## EXPLORING GLOBAL CONNECTIONS

Globalization comes in many shapes and forms, as it connects far-flung people and places. While many of these interactions are expected and well known, such as the global reach of multinational corporations, others are more surprising. Who would expect to find Australian fire fighters dowsing California wildfires, Russians investing in Thailand's coastal resorts, Bronx musicians influencing Arab hip-hop music, or African wildlife threatened by Europe's fishing policies?

Indeed, global connections are complex and ubiquitous—so much so that an understanding of the many different shapes, forms, and scales of these interactions is a key component of the study of global geography. To complement that study, each chapter contains an "Exploring Global Connections" sidebar that examines a wide variety of topics. More specifically they are:

- Chapter 2: The world of bushmeat and animal poaching
- Chapter 3: The globalization of firefighting
- Chapter 4: Latin Americans bound for Europe
- Chapter 5: Cuba's medical diplomacy
- Chapter 6: China's investment in Africa
- Chapter 7: Arab hip-hop challenges regional traditions
- Chapter 8: Europe's ties to Russian natural gas
- Chapter 9: Chinese immigrants in Russia's Far East
- Chapter 10: Foreign military bases in Central Asia
- Chapter 11: China and the global scrap paper trade
- Chapter 12: The Tata Group
- Chapter 13: Russian investment in Pattaya, Thailand
- Chapter 14: Nauru and the mixed benefits of Globalization

**FIGURE 1.28 World Languages** Most languages of the world belong to a handful of major language families. About 50 percent of the world's population speaks a language belonging to the Indo-European language family, which includes languages common to Europe, but also major languages in South Asia, such as Hindi. They are in the same family because of their linguistic similarities. The next largest family is the Sino-Tibetan family, which includes languages spoken in China, the world's most populous country. *(Adapted from Rubenstein, 2005*, The Cultural Landscape: An Introduction to Human Geography, *8th ed., Upper Saddle River, NJ: Prentice Hall)*

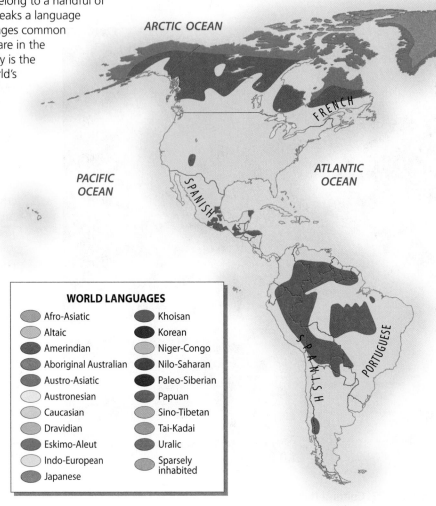

**WORLD LANGUAGES**

- Afro-Asiatic
- Altaic
- Amerindian
- Aboriginal Australian
- Austro-Asiatic
- Austronesian
- Caucasian
- Dravidian
- Eskimo-Aleut
- Indo-European
- Japanese
- Khoisan
- Korean
- Niger-Congo
- Nilo-Saharan
- Paleo-Siberian
- Papuan
- Sino-Tibetan
- Tai-Kadai
- Uralic
- Sparsely inhabited

**FIGURE 1.29 English as Global Language**
A bilingual traffic sign in Dubai, United Arab Emirates, is a reminder that English has become a universal form of communication for transportation, business, and science.

Judaism, the parent religion of Christianity, is also closely related to Islam. Although tensions are often high between Jews and Muslims, as illustrated in the Israeli–Palestinian conflict, these two religions, along with Christianity, actually share historical and theological roots in the Hebrew prophets and leaders. Judaism now numbers about 14 million adherents, having lost perhaps one-third of its total population due to the systematic extermination of Jews by the Nazis during World War II.

Hinduism, which is closely linked to India, has about 900 million adherents. Outsiders often regard Hinduism as polytheistic because Hindus worship many deities. Most Hindus argue, however, that all of their faith's gods are merely representations of different aspects of a single divine, cosmic unity. Historically, Hinduism is linked to the caste system, with its segregation of peoples based on ancestry and occupation. Today, however, because India's democratic government is committed to reducing the social distinctions between castes, the connections between religion and caste are now much less explicit than in the past.

Buddhism, which originated as a reform movement within Hinduism 2,500 years ago, is widespread in Asia, extending from Sri Lanka to Japan and from Mongolia to Vietnam (Figure 1.31). In its spread, Buddhism came to coexist with other faiths in certain areas, making it difficult to accurately estimate the number of its adherents. Estimates of the total Buddhist population range from 350 million to 900 million people.

Finally, in some parts of the world, religious practice has declined significantly, giving way to **secularization**, in which people consider

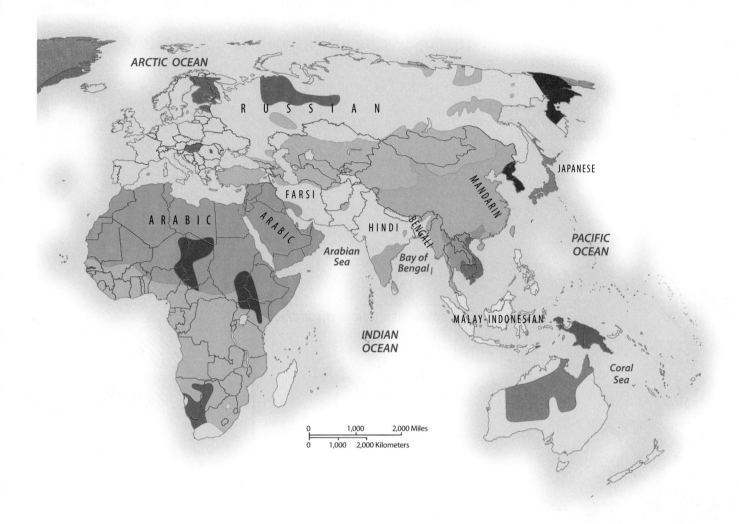

ARCTIC OCEAN

R U S S I A N

JAPANESE

FARSI

MANDARIN

ARABIC

ARABIC

HINDI

BENGALI

PACIFIC OCEAN

*Arabian Sea*

*Bay of Bengal*

MALAY-INDONESIAN

*INDIAN OCEAN*

*Coral Sea*

0   1,000   2,000 Miles

0   1,000   2,000 Kilometers

themselves either nonreligious or outright atheistic. Though secularization is difficult to measure, social scientists estimate that about 1.1 billion people fit into this category worldwide. Perhaps the best example of secularization comes from the former communist lands of Russia and eastern Europe, where, historically, there was overt hostility between government and church. Since the demise of Soviet communism in the 1990s, however, many of these countries have experienced modest religious revivals.

Secularization has grown more pronounced in recent years in western Europe. Although historically and to some extent still culturally a Roman Catholic country, France now reportedly has more people (mostly migrants) attending Muslim mosques on Fridays than it has attending Christian churches on Sundays. Japan and the other countries of East Asia are also noted for their high degree of secularization.

## Geopolitical Framework: Fragmentation and Unity

The term *geopolitics* is used to describe and explain the close link between geography and politics. More specifically, geopolitics focuses on the interactivity between power, territory, and space on all scales, from the local to the global. There is little question that one of the dominant characteristics of the past several decades has been the speed, scope, and character of political change in various regions of the world.

With the end of the Soviet Union in 1991 came opportunities for self-determination and independence in eastern Europe and Central

Asia, resulting in fundamental changes to economic, political, and even cultural alignments (Figure 1.32). Religious freedom helps drive national identities in some new Central Asian republics, but eastern Europe seems primarily concerned with economic and political links to western Europe. Russia wavers perilously between different geopolitical pathways. Phrases and terms common to an earlier generation may disappear; for example, the "Cold War" between the two superpowers, the United States and Soviet Union, has already become a historical artifact.

While wide-ranging international conflicts remain a nagging concern of diplomats and governments, perhaps more common are strife and tension within—rather than between—nation-states. Civil unrest, tribal tensions, terrorism, and religious factionalism have created a new fabric and scale of political tension.

## Nation-States

A traditional reference point for examining political geography is the concept of the nation-state. The hyphen links two concepts: the term *state*, which is a political entity, with territorial boundaries recognized by other countries and internally governed by an organizational structure, and the term *nation*, referring to a large group of people who share numerous cultural elements such as language, religion, tradition, and simple cultural identity. A **nation-state**, then, is ideally a relatively homogeneous cultural group with its own fully independent political territory.

Nations, however, do not always fit neatly into the boundaries of actual states. In fact, perfect nation-state congruence is relatively rare. While some large cultural groups may consider themselves to be

**FIGURE 1.30 Major Religious Traditions** This map shows the major religions throughout the world. For most people, religious tradition is a major component of cultural and ethnic identity. While Christians of different sorts account for about 34 percent of the world's population, this religious tradition is highly fragmented. Within Christianity, there are about twice as many Roman Catholics as Protestants. Islam accounts for about 20 percent of the world's population; Hindus make up about 14 percent.

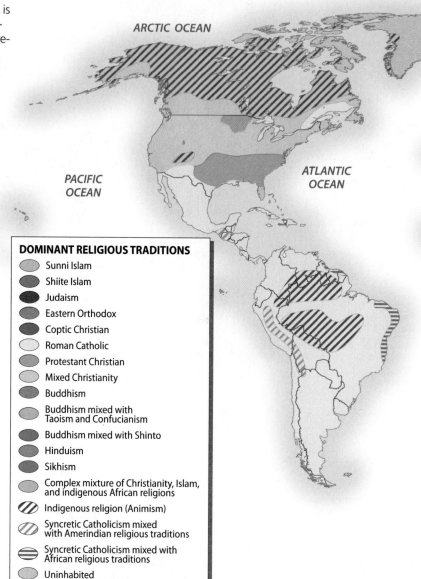

**DOMINANT RELIGIOUS TRADITIONS**

- Sunni Islam
- Shiite Islam
- Judaism
- Eastern Orthodox
- Coptic Christian
- Roman Catholic
- Protestant Christian
- Mixed Christianity
- Buddhism
- Buddhism mixed with Taoism and Confucianism
- Buddhism mixed with Shinto
- Hinduism
- Sikhism
- Complex mixture of Christianity, Islam, and indigenous African religions
- Indigenous religion (Animism)
- Syncretic Catholicism mixed with Amerindian religious traditions
- Syncretic Catholicism mixed with African religious traditions
- Uninhabited

**FIGURE 1.31 Religious Landscapes** The varied expressions of a culture's religion commonly appear in the landscape—the mosques and minarets of Islam, the churches and cathedrals of Christianity, the synagogues of Judaism, the shrines of Hinduism, or the temples and statues of Buddhism. An example is this statue of the sea deity *Kwun Yum* at the Tin Hau Temple in Repulse Bay, Hong Kong.

nations lacking recognized, self-governed territory (for example, the Kurdish people of Southwest Asia), many—if not most—independent states include within their established boundaries cultural and ethnic groups that seek autonomy and self-rule (Figure 1.33). In Spain, for example, both the Catalans and the Basques think of themselves as forming separate, non-Spanish nations and thus seek political autonomy from the central government. On a world scale, out of the more than 200 different countries that now make up the global geopolitical fabric, only several dozen would qualify as true nation-states in the narrow sense of the word.

## Centrifugal and Centripetal Forces

Cultural and political forces acting to weaken or divide an existing state are called **centrifugal forces** because they pull away from the center. Many have already been mentioned: a linguistic minority, ethnic separatism, the desire for territorial autonomy, and disparities in income and well-being. Separatist tendencies in French-speaking Quebec (Canada) and the Basque region in Spain are good examples

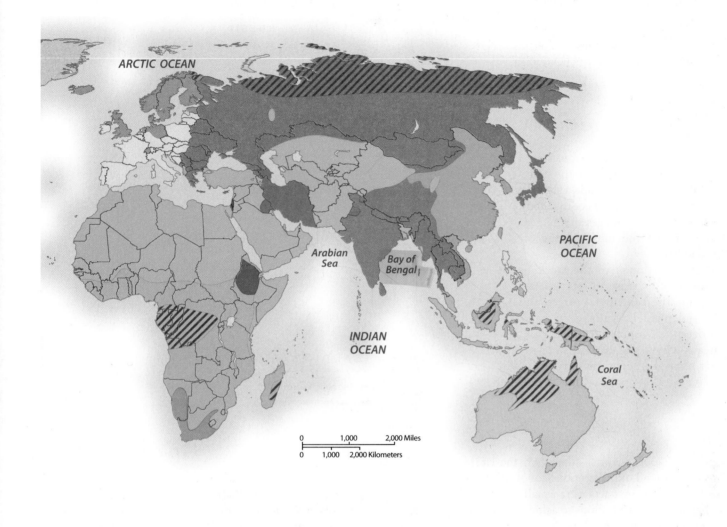

**FIGURE 1.32 The End of the Cold War** With the fall of Soviet communism, many nations rushed to remove the symbols of their former governments to forget the past and make room for a new future. In this photo taken in 1990, a monument to Lenin is toppled in Bucharest, Romania.

**FIGURE 1.33 A Nation Without a State** Not all nations or large cultural groups control their own political territories and thus are without a state. As this map shows, the Kurdish people of Southwest Asia occupy a larger cultural territory that is in four different political states—Turkey, Iraq, Syria, and Iran. As a result of this political fragmentation, the Kurds are considered a minority in each of these four countries.

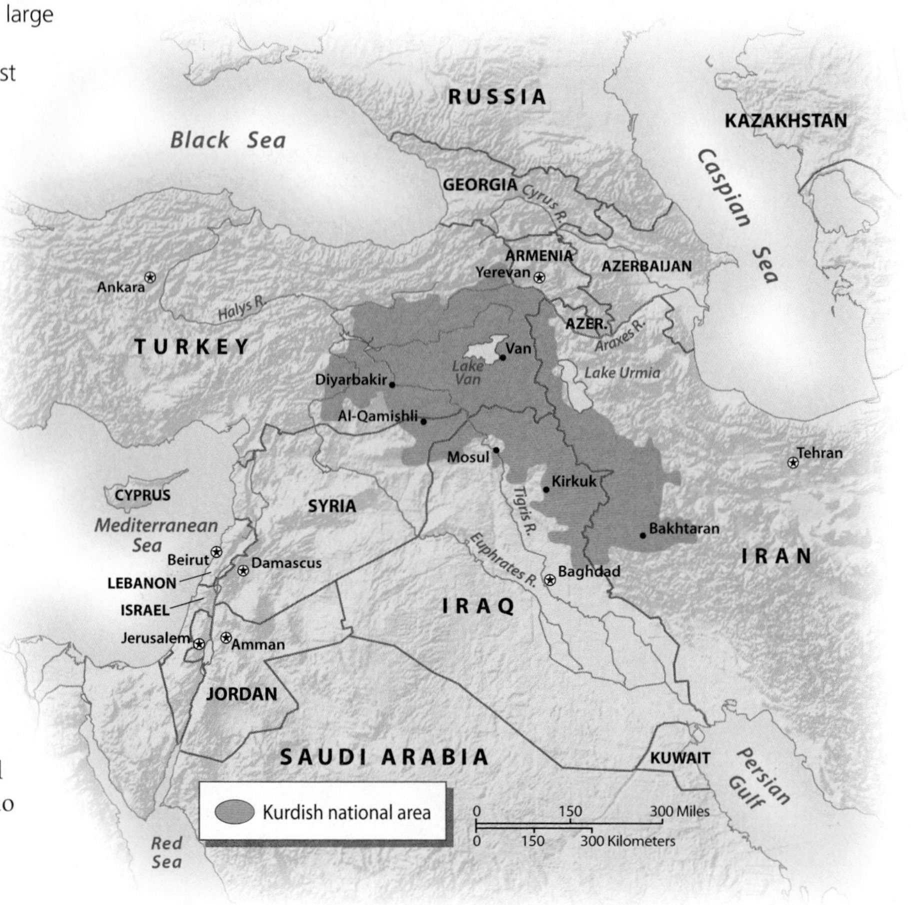

(Figure 1.34). Counteracting these dissipating forces are forces that promote political unity and reinforce the state structure. These **centripetal forces** could include a shared sense of history, a need for military security, a coordinated economic structure, or simply the advantages of a larger political organization to maintain the infrastructure of highways, airports, and schools.

The overriding question in much of the world is whether centrifugal forces will increase in strength and furor so that they overcome the centripetal forces, thus leading to new, smaller independent units. Also important is whether this process takes place through violent struggle or peaceful means. Given the current widespread nature of these internal tensions, separatist struggles will continue to dominate regional geopolitics for decades to come.

## Global Terrorism

In many ways, the September 2001 terrorist attacks on the United States underscore the need to expand our conceptualization of globalization and geopolitics. Previous terrorist acts were usually connected to nationalist or regional geopolitical aspirations to achieve independence or autonomy; the attacks on the World Trade Center and the Pentagon were different. Unlike Irish Republican Army (IRA) terrorist bombings in Great Britain or those of the extremist wing of Basque nationalists in Spain, the September 2001 terrorism went beyond conventional geopolitics, as a small group of religious fanatics attacked the symbols of Western culture, power, and finance. Moreover, these attacks were a stark reminder of the increasingly close and often unpredictable interconnections between political activity, cultural identity, and the economic linkages that bind together our contemporary world (Figure 1.35).

**FIGURE 1.34 Basque Separatism** The demands for independence of the Spanish Basques in northeastern Spain are just one example of the ethnic separatism common in much of the world.

**FIGURE 1.35 Global Terrorism** Innocent people throughout the world have fallen victim to extremists attempting to advance their causes through acts of terrorism. England, for example, has suffered numerous attacks by groups advocating independence for Northern Ireland, as well as more recent attacks by terrorists allied with Islamic extremism.

In geopolitical terms, global terrorism demands a new way of looking at the world to understand the attacks and assess their implications. More to the point, many experts argue that global terrorism is both a product of and a reaction to globalization. Unlike earlier geopolitical conflicts, the geography of global terrorism is not defined by a war between well-established political states. Instead, the Al Qaeda terrorists appear to belong to a web of small, well-organized cells located in many different countries. These cells are linked in a decentralized network that provides guidance, financing, and political autonomy and that has used the tools and means of globalization to its advantage. Members communicate instantaneously via mobile phones and the Internet. Transnational members travel between countries quickly and frequently. The network's activities are financed through a complicated array of holding companies and subsidiaries that traffic in a range of goods, including honey, diamonds, and opium. To recruit members and political support, the network feeds on the unrest and inequities (real and imagined) resulting from economic globalization. The network's terrorist acts then target symbols of those modern global values and activities it opposes. Even though the 9/11 attacks focused terror on the United States, the casualties and resultant damage were international, as citizens from more than 80 countries were killed in the World Trade Center tragedy.

The military and political responses to global terrorism demand an expanded conceptualization of geopolitics. In Afghanistan, the military muscle of a nation-state superpower—the United States—is directed not at another nation-state but rather at a confederation of religious extremists, the Taliban and Al Qaeda. Military strategists refer to this kind of fighting as **asymmetrical warfare**, a term that aptly describes the differences between a superpower's military technology and strategy and the lower-level technology and guerilla tactics used by Al Qaeda and the Taliban. Most superpower military strategists agree that the war on terrorism will be fought with this sort of political and battlefield asymmetry.

## Colonialism and Decolonialization

One of the overarching themes in world regional geography is the waxing and waning of European colonial power over much of the world. **Colonialism** refers to the formal establishment of rule over a foreign population. A colony has no independent standing in the world community but instead is seen only as an appendage of the colonial power. Generally speaking, the main period of colonialization by European states was from 1500 through the mid-1900s, though even today a few colonies remain (Figure 1.36).

**Decolonialization** refers to the process of a colony's gaining (or regaining) control over its territory and establishing a separate, independent government. As was the case with the Revolutionary War in the United States, this process often begins as a violent struggle. As wars of independence became increasingly prevalent in the mid-20th century, some colonial powers recognized the inevitable and

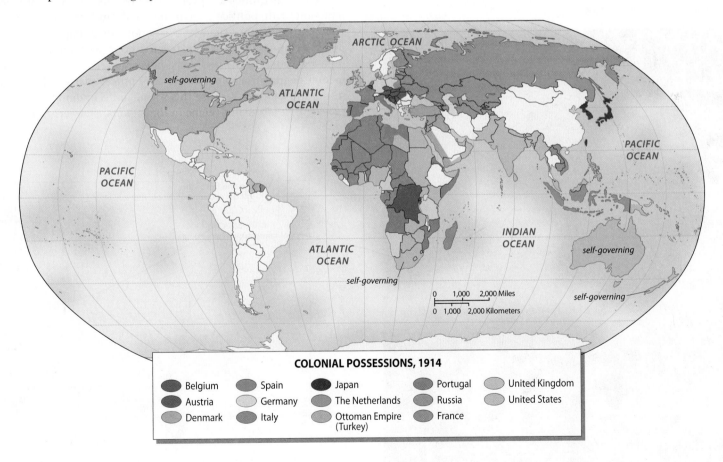

**FIGURE 1.36 The Colonial World, 1914**   This world map shows the extent of colonial power and territory just prior to World War I. At that time, most of Africa was under colonial control, as were Southwest Asia, South Asia, and Southeast Asia. Australia and Canada were very closely aligned with England. Also note that in Asia, Japan controlled colonial territory in Korea and northeastern China, which was known as Manchuria at that time.

began working toward peaceful disengagement from their colonies. In the late 1950s and early 1960s, for example, Britain and France granted independence to most of their former African colonies, often after periods of warfare or civil unrest. This process was nearly completed in 1997, when Hong Kong was peacefully restored to China by the United Kingdom.

But decades and even centuries of colonial rule are not easily erased; thus, the influence of colonialism is still commonly found in the new nations' governments, education, agriculture, and economies. While some countries may enjoy special status with, and receive continued aid from, their former colonial masters, others remain disadvantaged because of a reduced resource base. Some scholars regard the continuing economic ties between certain imperial powers and their former colonies as a form of exploitative neocolonialism. On the other hand, some remaining colonies, such as the Dutch Antilles in the Caribbean, have made it clear that they have no wish for independence, finding both economic and political advantages in continued dependency. Because the consequences of colonialism differ greatly from place to place, the final accounting of its effects is far from complete (Figure 1.37).

**FIGURE 1.37 Colonial Vestiges in Vietnam** The red star flag of communist Vietnam flies in front of the Hotel de Ville in Ho Chi Minh City (formerly Saigon). This juxtaposition of the contemporary government's symbol and an artifact of the French colonial period captures the process of decolonialization and independence.

# Economic and Social Development: The Geography of Wealth and Poverty

The pace of global economic change and development has accelerated dramatically in the past several decades, first ascending rapidly, then, in late 2008, dropping precipitously as the world fell into an economic recession (Figure 1.38). The overarching question, though, is whether the positive changes of economic globalization outweigh the negative. Answers vary and are often elusive and incomplete; a place to begin in world regional geography is to link economic change to social development.

Economic development is commonly accepted as desirable because it generally brings increased prosperity to people, regions, and nations. By conventional thinking, at least, this usually translates into social improvements such as better health care, improved education systems, and more progressive labor practices. However, one of the most troubling expressions of recent economic growth has been the geographic unevenness of prosperity and social improvement. While some regions prosper, others languish and, in fact, fall farther behind more developed countries. As a result, the gap between rich and poor regions has increased over the past several decades, and this economic and social unevenness has, unfortunately, become a signature of globalization. According to the World Bank, about half the people in the world live on less than $2 a day, the commonly accepted definition of poverty (Figure 1.39).

These inequities are problematic because of their inseparable interaction with political, environmental, and social issues. Political instability and civil strife within a nation, for example, are often driven by the economic disparity between a poor periphery and an affluent, industrial core. In the periphery, poverty, social tensions, and environmental degradation often cause civil unrest that ripples through the rest of the country. Because this social and economic unevenness is so prevalent, it is a major theme in the regional chapters of this book.

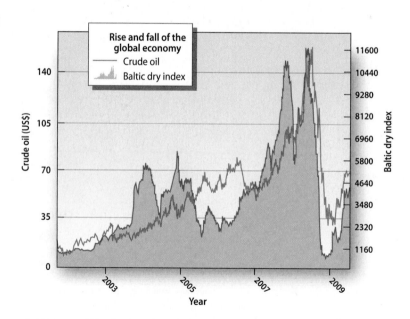

**FIGURE 1.38 The Global Recession** Two indicators of the global economic activity, the price of crude oil and the Baltic Dry Index (BDI), which is linked to world shipping activity, reflect the expansion and, more recently, the contraction of the global economy.

**FIGURE 1.39 Living on Less than $2 a Day** Over half the world's people live on less than $2 a day, which is the UN's definition of poverty. In South Asia, where these workers are making bricks, about three-quarters of the population is considered impoverished.

## More- and Less-Developed Countries

Until the 20th century, economic development was centered in North America, Japan, and Europe, while most of the rest of the world remained gripped in poverty. This uneven distribution of economic power led scholars to devise a **core–periphery model** of the world. According to this scheme, the United States, Canada, western Europe, and Japan constituted the global economic core of the north, whereas most of the areas to the south made up a less-developed global periphery. Although oversimplified, this core–periphery dichotomy does contain some truth. All the G8 countries—the exclusive club of the world's major industrial nations, made up of the United States, Canada, France, England, Germany, Italy, Japan, and Russia—are located in the Northern Hemisphere. In addition, many critics postulate that the developed countries achieved their wealth primarily by exploiting the poorer countries of the southern periphery, historically through colonial relationships and today with economic imperialism.

As a result, much is made today of "north–south tensions," a phrase implying that the rich and powerful countries of the Northern Hemisphere are still at odds with the poor and less powerful countries of the south. Over the past several decades, however, the global economy has grown much more complicated. A few former colonies of the south, most notably Singapore, have become very wealthy. In addition, a few northern countries, namely Russia, have experienced economic decline in recent decades. Further, the developed countries of Australia and New Zealand never fit into the north–south division because of their Southern Hemisphere location. For these reasons, many global experts conclude that the designation *north–south* is outdated and, thus, should be avoided. We agree.

The *third world* is another term often used to refer to the developing world. This phrase suggests a low level of economic development, unstable political organizations, and a rudimentary social infrastructure. Historically, the term came from the Cold War vocabulary used to describe countries that were independent and not allied with either the capitalist and democratic first world or the communist second world superpowers of the Soviet Union and China. Today, however, because the Soviet Union no longer exists and China has changed its economic orientation, the term *third world* has lost its original meaning. In this book, therefore, we also avoid the term and instead use relational terms that capture the complex spectrum of economic and social development—*more-developed country (MDC)* and *less-developed country (LDC)*. The global pattern of more- and less-developed countries can be inferred from a map of gross national income (Figure 1.40), one of several indicators used to assess development and economic wealth.

## Indicators of Economic Development

The terms *development* and *growth* are often used interchangeably when referring to international economic activities. There is, however, value in keeping them separate. *Development* has both qualitative and quantitative dimensions. Common dictionary definitions use phrases such as "expanding or realizing potential" and "bringing gradually to a fuller or better state." When we talk about economic development, then, we usually imply structural changes, such as a shift from agricultural to manufacturing activity that also involves changes in the allocation of labor, capital, and technology. Along with these changes are assumed improvements in standard of living, education, and political organization. The structural changes experienced by Southeast Asian countries such as Thailand and Malaysia in the past several decades illustrate this process.

*Growth*, in contrast, is simply the increase in the size of a system. The agricultural or industrial output of a country may grow, as it has for India in the past decade, and this growth may—or may not—have positive implications for development. Many growing economies, in fact, have actually experienced increased poverty with economic expansion. When something grows, it gets bigger; when it develops, it improves. Critics of the world economy are often heard to say that we need less growth and more development.

In this book, each of the regional chapters includes a table of economic and development indicators (see Table 1.2). A few introductory comments are necessary to explain these data.

**Gross Domestic Product and Income** The traditional measure for the size of a country's economy is the value of all final goods and services produced within its borders, called its **gross domestic product (GDP)**. When combined with net income from abroad, this domestic income constitutes a country's **gross national income (GNI)** (formerly referred to as gross national product [GNP]). Although the term is commonly used, GNI is an incomplete and sometimes misleading economic indicator because it ignores nonmarket economic activity such as bartering or household work and also because it does not take into account ecological degradation or depletion of natural resources. For example, if a country were to clear-cut its forest, an activity that could severely limit future growth if forest resources were in short supply, this cutting would nevertheless increase the GNI for that particular year. Further, diverting educational funds to purchase military weapons might increase a country's GNI in the short run, but the economy would likely suffer in the future because of its less-well-educated population. In other words, GNI is a snapshot of a country's economy for a specific period, not an infallible indicator of continued vitality or social well-being.

Because GNI data vary widely between countries and are commonly expressed in mind-boggling numbers such as trillions and billions of

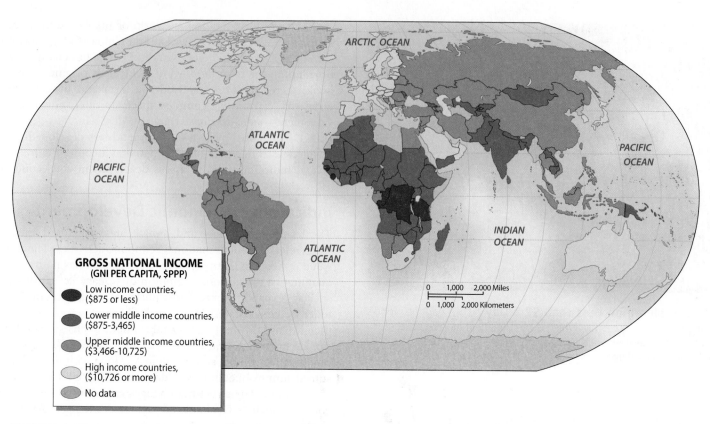

**FIGURE 1.40 More- and Less-Developed Countries** Based upon GNI per capita, PPP adjusted, one sees the global pattern of more- and less-developed countries (MDCs and LDCs). Africa and the different parts of Asia stand out as the regions with the greatest number of developing countries.

### TABLE 1.2 DEVELOPMENT INDICATORS OF THE 10 LARGEST COUNTRIES

| Country | GNI per capita, PPP 2007 | GDP Average Annual % Growth 2000–07 | Human Development Index (2006)# | Percent Population Living Below $2 a Day | Life Expectancy* 2009 | Under Age 5 Mortality Rate 1990 | Under Age 5 Mortality Rate 2007 | Gender Equity 2007 |
|---|---|---|---|---|---|---|---|---|
| China | 5,420 | 10.3 | 0.762 | 36.3 | 73 | 45 | 22 | 100 |
| India | 2,740 | 7.8 | 0.609 | 75.6 | 64 | 117 | 72 | 91 |
| United States | 45,840 | 2.6 | 0.950 | | 78 | 11 | 8 | 100 |
| Indonesia | 3,570 | 5.1 | 0.726 | | 71 | 91 | 31 | 98 |
| Brazil | 9,270 | 3.3 | 0.807 | 12.7 | 73 | 58 | 22 | 103 |
| Pakistan | 2,540 | 5.6 | 0.562 | 60.3 | 66 | 132 | 90 | 78 |
| Bangladesh | 1,330 | 8.3 | 0.524 | 81.3 | 65 | 151 | 61 | 103 |
| Nigeria | 1,760 | 6.6 | 0.499 | 83.9 | 53 | 230 | 189 | 84 |
| Russia | 14,330 | 6.6 | 0.806 | <2 | 68 | 27 | 15 | 99 |
| Japan | 34,750 | 1.7 | 0.956 | | 83 | 6 | 4 | 100 |

Source: World Bank, *World Development Indicators, 2009*.
United Nations, *Human Development Index, 2008*. #
Population Reference Bureau, *World Population Data Sheet, 2009*\*

Gender Equity – Ratio of female to male enrollments in primary and secondary school. Numbers below 100 have more males in primary/secondary school, numbers above 100 have more females in primary/secondary schools.

dollars, a better comparison is to divide GNI by the country's population, thereby generating a **gross national income (GNI) per capita** figure. This way, one can compare large and small economies in terms of how they may (or may not) be benefiting the population. For example, the annual GNI for the United States is over $10 trillion. Dividing that figure by the population of 306.8 million results in a GNI per capita of almost $46,000. Japan has a GNI about half the size of the United States; however, because that country also has a much smaller population of 128 million, its unadjusted GNI per capita is about $39,000; thus, one could conclude that the two economies are almost comparable.

An important qualification to these GNI per capita data is the concept of adjustment through **purchasing power parity (PPP)**, an adjustment that takes into account the strength or weakness of local currencies. When not adjusted by PPP, GNI data are based on the market exchange rate for a country's national currency as compared to the U.S. dollar. As a result, the GNI data might be inflated or undervalued, depending on the strength or weakness of that currency. If the Japanese yen were to fall overnight against the dollar as a result of currency speculation, Japan's GNI would correspondingly drop, despite the fact that Japan had experienced no real decline in economic output. Because of these possible distortions, the PPP adjustment provides a more accurate sense of the local cost of living. To illustrate, when Japan's GNI per capita is adjusted for PPP, which takes out the inflationary factor, the figure is $34,750, which is lower than the unadjusted GNI per capita figure.

### Economic Growth Rates

A country's rate of economic growth is measured by the average annual growth of its GDP over a five-year period, a statistic called *GDP average annual percent growth*. The average growth rate for developing countries such as China, India, and Nigeria is considerably higher than those of the developed countries of the United States and Japan (see Table 1.2). This difference in economic growth rates is expected, given that developing countries are just that—developing; thus one would expect a higher annual growth rate than from a mature, developed economy like that of the United States.

## Indicators of Social Development

Although economic growth is a major component of development, equally important are the conditions and quality of human life. As noted earlier, the standard assumption is that economic development and growth will spill over into the social infrastructure, leading to improvements in public health, gender equity, and education. Unfortunately, even the briefest glance at the world reveals that poverty, disease, illiteracy, and gender inequity are still widespread, despite the booming global economy. Even in China, which has experienced unprecedented economic growth in recent decades, half the population is still impoverished; in Pakistan, two-thirds live below the poverty line of $2 per day, and in some African countries, that figure approaches 90 percent.

However, there are hints of improvement in some developing countries. For example, the percentage of those living in deep poverty, which the UN defines as living on less $1 per day, has fallen from 28 percent in 1990 to just under 20 percent today. Life expectancy, too, has increased in those countries, rising from 60 in 1990 to 65. Female literacy has also increased in developing countries, rising from 62 percent in 1990 to 70 percent in 2007.

For the past three decades, the United Nations has tracked social development in the world's countries through the **Human Development Index (HDI)**, which combines data on life expectancy, literacy, educational attainment, gender equity, and income (Figure 1.41). In a December 2008 analysis, the 179 countries that provided data to the UN are ranked from high to low, with Iceland and Norway tying for the high score (somehow overlooking Iceland's economic meltdown in late 2008) of .968. At the low end of the HDI are a number of African countries, including the Democratic Republic of Congo (.361), Central African Republic (.352), and Sierra Leone (.329).

Although the HDI is criticized for using national data that overlook the diversity of development within a country, overall, the HDI conveys a reasonably accurate sense of a country's human and social development; thus, we include country-by-country data in social development tables throughout this book.

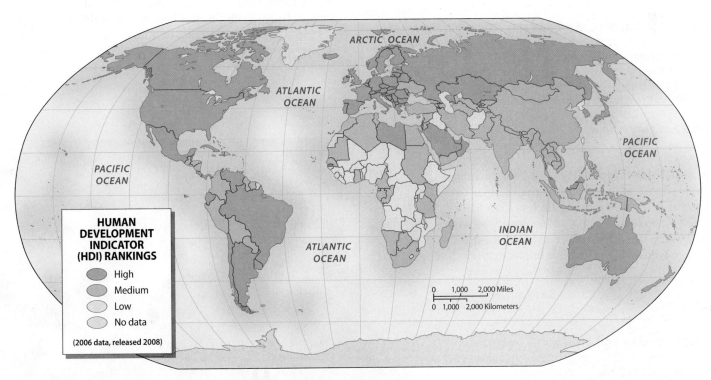

**FIGURE 1.41 Human Development Index**  The United Nations tracks social development through the Human Development Index (HDI), which combines information on literacy, education, gender equity, income, and life expectancy. This map depicts the most recent rankings assigned to three categories. In the numerical tabulation, Norway and Iceland have the highest rankings, while several African countries are lowest in the scale.

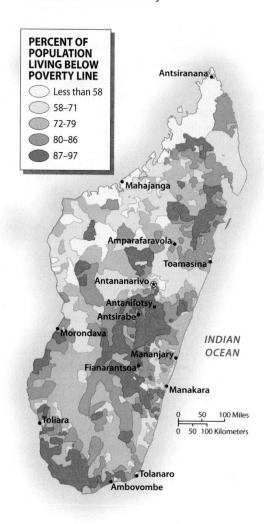

**PERCENT OF POPULATION LIVING BELOW POVERTY LINE**

- Less than 58
- 58–71
- 72-79
- 80–86
- 87–97

Antsiranana

Mahajanga

Amparafaravola

Toamasina

Antananarivo

Antanifotsy

Antsirabe

Morondava

Mananjary

Fianarantsoa

Manakara

INDIAN OCEAN

0    50    100 Miles

0  50  100 Kilometers

Toliara

Tolanaro
Ambovombe

**FIGURE 1.42 The Landscape of Poverty** As is true of most other countries, the distribution of poverty within Madagascar is uneven, with clusters of abject poverty contrasting with regions that are less poor. This map shows that the highest rates of poverty are in the central highlands, where the country's population is concentrated and the density is highest. In contrast, the northern and eastern lowlands are less poor. *(Map adapted from* Where the Poor Are: An Atlas of Poverty. *Center for International Earth Science Information Network, 2006)*

**Poverty and Infant Mortality** The international definition of *poverty* is living on less than $2 per day. *Deep poverty* is defined as existing on less than $1 per day. Granted, the cost of living varies greatly around the world, but the United Nations has found these definitions, when adjusted for local conditions, work well

for measuring poverty and its associated social conditions. While poverty data are usually presented at the national level (see Table 1.2), the UN and other agencies are attempting to compile data at a local scale in order to better understand—and, hopefully, improve—the economic landscape within a country. The patterns of poverty in the African country of Madagascar provide an instructive example (Figure 1.42).

*Under age five mortality* depicts the number of children who die per 1,000 persons within that age bracket, and is another widely used indicator of social conditions. Aside from the tragedy of infant death, child mortality reflects the wider conditions of a society—namely, the availability of food, health services, and public sanitation. If those factors are lacking, children under age five suffer most; thus their death rate is taken as an indication of whether a country has the necessary social infrastructure to sustain life. In the social development tables throughout this book, child mortality data are given for two points in time, 1990 and 2007, to indicate whether the social structure improved over that period. Although most countries have seen improvement over the past decades, disturbing differences in child mortality rates still exist on a global level.

**Gender Equity** The ratio of male to female students enrolled in primary and secondary schools is called *gender equity*. If the gender equity ratio is below 100, then more males are enrolled in schools than females. Conversely, a ratio over 100 means more females are enrolled than males in primary and secondary schools. Used by the UN as a measure of equality in education between the sexes, this statistic is linked closely to literacy rates, which in turn are linked to social development. The assumption is that if females are not enrolled in school, a high rate of female illiteracy results, which has negative consequences for social development. To illustrate, gender equity is high in the United States, Russia, and Brazil, and it is lower in countries such as India, Pakistan, and Nigeria (see Table 1.2).

Although sex discrimination takes other forms, such as the preference for male children in India and China or the barriers erected against women in business and politics, the United Nations is currently

**FIGURE 1.43 Women and Literacy** Gender inequities in education lead to higher rates of illiteracy for women. However, when there is gender equity in education, female literacy has a number of positive outcomes in a society. For example, educated women have a higher participation rate in family planning, which usually results in lower birthrates.

focusing on correcting the gender inequity in education (Figure 1.43). Specifically, the UN has made neutralizing educational disparity (illiteracy is much higher for women than men) one of its Millennium Development Goals. Gender equity now has higher priority because increasing women's literacy has positive outcomes that complement social development. For example, some studies indicate a strong correlation between lower birthrates and female literacy, probably because literate women are more open to family planning measures.

Besides gender equity, there are many other linkages between economic and social development. This introductory chapter has discussed only a few and on a general scale; the regional chapters of this book examine other issues, in more detail.

## SUMMARY

- As the world becomes increasingly interdependent, globalization is driving a fundamental reorganization of economies and cultures through instantaneous global communication, the growth of transnational corporations, and the spread of Western consumer habits.

- One issue—terrorism—tops the list of global geopolitical issues and can be triggered by local groups with limited agendas (such as Basque autonomy), as well as larger, more complex terrorist networks (such as Al Qaeda) with wider goals. Besides terrorism, there are many other kinds of geopolitical tensions, including ethnic strife, border disputes, and movements for regional autonomy and independence.

- In most regions of the developing world, population and settlement issues revolve around four issues: rapid population growth, family planning (or its absence), migration to new centers of economic activity (both within and outside the region), and rapid urbanization.

- A major theme in cultural geography is the tension between the forces of global cultural homogenization and the countercurrents of local cultural and ethnic identity. Throughout the world, small groups are setting themselves apart from larger national cultures with renewed interest in ethnic traits, languages, religion, territory, and shared histories. Not merely a matter of colorful folklore and local customs, this cultural diversity is translated into geopolitics, with outspoken calls for regional autonomy or separatism.

- Cultural tensions have caused the geopolitical issues of many world regions to be dominated by matters of global terrorism, ethnic strife and territorial disputes within nation-states, border tensions between neighbors of different cultural traditions, and new military approaches to deal with the ever-changing nature of national security.

- The theme of economic and social development is dominated by one issue—the increasing disparity between rich and poor, between countries and regions that already have wealth and are getting even richer through globalization and those that remain impoverished. Often, blatant inequities in social development, schools, health care, and working conditions accompany these disparities in wealth.

## KEY TERMS

asymmetrical warfare (p. 27)
bubble economy (p. 8)
centrifugal forces (p. 24)
centripetal forces (p. 26)
colonialism (p. 27)
core–periphery model (p. 29)
cultural imperialism (p. 18)
cultural nationalism (p. 18)
cultural syncretism or
    hybridization (p. 19)

culture (p. 18)
decolonialization (p. 27)
demographic transition model
    (p. 15)
economic convergence (p. 5)
ethnic religion (p. 21)
globalization (p. 2)
gross domestic product (GDP)
    (p. 29)
gross national income (GNI)
    (p. 29)

gross national income
    (GNI) per capita (p. 30)
Human Development Index
    (HDI) (p. 31)
lingua franca (p. 21)
nation-state (p. 23)
net migration rate (p. 16)
population pyramid (p. 14)
purchasing power parity (PPP)
    (p. 31)

rate of natural increase (RNI)
    (p. 12)
secularization (p. 22)
sweatshop (p. 6)
total fertility rate (TFR) (p. 14)
transnational firm (p. 4)
universalizing religion (p. 21)
urbanized population (p. 17)

# THINKING GEOGRAPHICALLY

1. Select an economic, political, or cultural activity in your city and discuss how it has been influenced by globalization.

2. Choose a specific country or region of the world and examine the benefits and liabilities that globalization has posed for that country or region. Remember to look at different facets of globalization, such as the environment and cultural cohesion and conflict, as well as the economic effects on different segments of the population.

3. Drawing on information in current newspapers and magazines and on TV and the Internet, apply the concepts of cultural imperialism, nationalism, and syncretism to a region or place experiencing cultural tensions.

4. Choose a nation-state and elucidate the different centrifugal and centripetal forces within that nation-state. Based on your findings, speculate on how that nation-state might change in the next 10 years.

5. Using the tables of social indicators in the regional chapters of this book, identify traits shared by countries in which there is a high percentage of female illiteracy. What general conclusions do you reach, based on your inquiry?

Log in to **www.mygeoscienceplace.com** for videos, interactive maps, RSS feeds, case studies, and self-study quizzes to enhance your study of Globalization and Diversity.

# 2 THE CHANGING GLOBAL ENVIRONMENT

*Global warming is one of the world's most pressing environmental problems. While there are many implications from human-caused climate change, one is the predicted flooding of low-lying islands and coastal areas from sea level rise as glaciers and ice caps melt from warming temperatures. This photo was taken in southwestern Greenland.*

The human imprint is everywhere on Earth, from the highest mountains to the deepest ocean depths; from dry deserts to lush tropical forests; from frozen Arctic ice caps to the atmospheric layers we breathe. Hundreds of spent oxygen canisters clutter the heights of Mt. Everest, and castoff plastic water bottles litter coastlines and oceans worldwide (Figure 2.1). As deserts bloom in North America with irrigated cotton fields, tropical forests in Brazil are laid waste by logging to create pastures for cattle.

Although many changes to the global environment are intentional and have improved human life, other changes are inadvertent and have proven harmful to both human welfare and the environment. Because of the importance of these environmental issues, a study of the changing global environment is central to the study of world regional geography. These environmental issues are also deeply intertwined with globalization and diversity. The destruction of tropical rain forests, for example, is a response to international demand for wood products, beef, soybeans, and biofuels. Similarly, global warming is closely linked to patterns of world industry, commerce, and consumption. The nearly 7 billion people on Earth interact with and change the natural environment in an almost endless number of ways (see "Exploring Global Connections: The World of Bushmeat and Animal Poaching").

## Global Climates: An Uncertain Forecast

Human settlement and food production are closely linked to patterns of local weather and climate. The life and landscape of the arid parts of Southwest Asia differ considerably from the wet tropical areas of Southeast Asia. People in different parts of the world adapt to weather and climate in widely varying ways, depending on their culture, economy, and technology. Although some desert areas of California, for example, are covered with high-value irrigated agriculture that produces vegetables for the global marketplace year-round, most of the world's arid regions support very little agriculture and barely participate in global commerce.

Further, one of the most pressing problems facing the world today is that human activities are changing Earth's climate through global warming. Just what the future will bring is not entirely clear. However, even if the forecast is uncertain, there is little question that many forms of life—including humans—will face difficulties adjusting to the changes brought about by global warming (Figure 2.2).

### World Climate Regions

Where similar atmospheric conditions prevail, boundaries are drawn cartographically around what is called a **climate region**. Knowing the climate type for a given part of the world not only conveys a sense of

**FIGURE 2.1 The Human Imprint** Trash dumped at sea litters an isolated beach in the Outer Hebrides islands of the Atlantic, attesting to the fact that there's probably not an environment on Earth untouched and unaffected by human activities.

**FIGURE 2.2 Polar Bears Threatened by Global Warming** Ice floes are an important part of the polar bear's habitat. Because Arctic ice is melting more rapidly now due to global warming, the 20,000 bears living in the wild face an uncertain future. Recently, polar bears have been reported drowning trying to swim long distances between ice sheets.

average rainfall and temperatures but also allows inferences about human activities and settlement. If an area is categorized as desert, we can infer that rainfall is so limited that any agricultural activities require irrigation. In contrast, if we see an area characterized by the climatic designation for tropical monsoon, we know there is enough rainfall for agriculture.

A standard scheme of climate types is used throughout this book, and each regional chapter contains a map showing the different climate regions in some detail. Figure 2.3 shows these climatic regions generalized to a world scale. The regional climate maps also contain **climographs,** graphs of average high and low temperatures and precipitation for an entire year. In Figure 2.3 the climate of Cape Town, South Africa, is presented as an example. Besides temperatures, climographs also contain bar graphs depicting average monthly precipitation. Not only is the total amount of rain and snowfall important, but the seasonality of this precipitation provides valuable information for making inferences about agriculture and food production.

## Global Warming

Human activities connected with economic development and industrialization are changing the world's climate in ways that will have significant consequences for all living organisms, be they plants, animals, or human. More specifically, **anthropogenic,** or human-caused, pollution of the lower atmosphere is increasing the natural greenhouse effect so that worldwide **global warming**—an increase in the temperature of Earth's atmosphere—is taking place. This warming, in turn, may change rainfall patterns so that some marginal agricultural areas become increasingly arid; melt polar ice caps, causing a rise in sea levels that will threaten coastal settlement; increase the number of heat waves that strike urban areas, causing death from heat stroke; and possibly lead to a greater intensity in tropical storms. In addition, plants and animals, both terrestrial and oceanic, will also experience changes to their environment. While some will become extinct, others may have to migrate to more familiar conditions.

Moreover, because of climate change from global warming, the world may experience a dramatic change in food production regions. Whereas some prime agricultural areas, such as the lower Midwest of North America, may suffer because of increased dryness, other areas, such as the Russian steppes, may actually become better suited for agriculture. Although it is too early to tell when these changes may occur, there is little question that these climate changes will have a dramatic effect on human settlement and world food supplies.

**Causes of Global Warming** The natural **greenhouse effect** provides us with a warm atmospheric envelope that supports human life (Figure 2.4). This warmth comes from the trapping of incoming and outgoing (or re-radiated) solar radiation by an array of natural greenhouse gases in the atmospheric layer closest to Earth. Although natural greenhouse gases have varied somewhat over long periods of geologic time, they were relatively stable until recently. However, within the past 130 years, coincidental with the Industrial Revolution in Europe and North America, the composition and amount of those greenhouse gases have changed dramatically, primarily from the burning of fossil fuels. As a

## EXPLORING GLOBAL CONNECTIONS
# The World of Bushmeat and Animal Poaching

Today, as a result of globalization, plants and animals in one part of the world are commonly threatened by the desires and decisions of people thousands of miles away. A middle-aged Beijing gentleman's sex life, for example, may determine whether an endangered African rhino lives or dies; similarly, Europe's ravenous appetite for fish endangers Africa's treasured lions, leopards, and great apes.

In Africa, the market for wild animal meat—"bushmeat" in local parlance—is a major conservation issue. Justin Brashares, a University of California–Berkeley conservation biologist, has revealed a close connection between international fishing fleets along Africa's western coast and the increased killing of bushmeat on land. The relationship is simple: When West Africans cannot find affordable fish in local markets, they turn to bushmeat for their protein supply. And the problem has worsened considerably with the recent appearance of European fishing fleets off Africa's coastline. More specifically, the European fish catch off West Africa has increased 20-fold in the past several decades, resulting in fewer fish for local fishermen and thus leading to a higher kill rate of terrestrial bushmeat. Often, threatened and endangered species—great apes, lions, leopards, elephants, and hippos—are poached from national parks and game reserves to fulfill the food needs of West Africa's growing population.

Also taking a heavy toll on the world's wildlife are global poaching networks that constitute a black-market activity comparable to the trade in illegal narcotics and weapons. Much of the activity in the trafficking of poached animal parts is thought to be run by international drug cartels and is estimated to be worth around $20 billion a year.

In Brazil, for example, more than 20 million animals are taken out of the country illegally each year, making that country one of the top sources for smuggled fauna. According to government officials, poaching has brought 208 species to the brink of extinction. A blue macaw, one of Brazil's most threatened species, with only 200 left in South America, will sell for over $50,000 in Europe. Further, a toucan is worth $7,000 in the United States, and a collection of rain forest butterflies sells for $3,000 in China.

As for that gentleman in Beijing, he can find a whole range of traditional sexual stimulants in his local marketplace, most of which come from poached animals—gall bladders from black bears, grizzly bears, or giant pandas; powder ground from rhino horns; and extracts from tiger paws and gonads. Although China is working to stop this illegal trade by subsidizing bear farms where the trade in gall bladders can be controlled, conservation experts say that, while it's a step in the right direction, so far this program has had little effect on the huge black market for poached animal parts.

**FIGURE 2.3 World Climate Regions** A standard scheme, called the *Köppen system*, after the Austrian geographer who devised it in the early 20th century, is used to describe the world's diverse climates. Combinations of letters refer to the general climate types, and precipitation and temperature characteristics are given for each type. Specifically, the *A* climates are tropical, the *B* climates are dry, the *C* climates are generally moderate and are found in the middle latitudes, and the *D* climates are associated with continental and high-latitude locations.

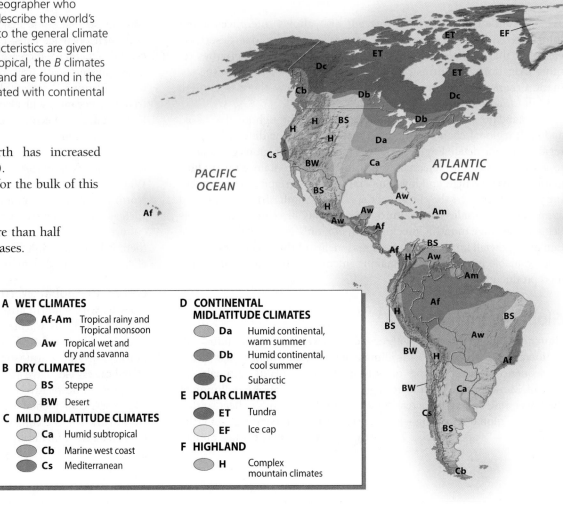

result, the average temperature for Earth has increased markedly over the past decades (Figure 2.5).

Four major greenhouse gases account for the bulk of this change:

1. *Carbon dioxide (CO₂)* accounts for more than half of the human-generated greenhouse gases. The increase in atmospheric $CO_2$ is primarily a result of burning fossil fuels such as coal and petroleum. To illustrate, in 1860 atmospheric $CO_2$ was measured at 280 parts per million (ppm). Today it is more than 380 ppm, and it is projected to exceed 450 ppm by 2050, assuming that fossil fuels remain a major source for the world's energy needs.

2. *Chlorofluorocarbons (CFCs)* make up nearly 25 percent of the human-generated greenhouse gases and come mainly from widespread use of aerosol sprays and refrigeration (including air conditioning). Although CFCs have been banned in North America and most of Europe, they are still increasing in the atmosphere at the rate of 4 percent each year. These gases are highly stable and reside in the atmosphere for a long time, perhaps as long as 100 years. As a result, their role in global warming is highly significant. Recent research has shown that a molecule of CFC absorbs 1,000 times more infrared radiation from Earth than a molecule of $CO_2$.

3. *Methane (CH₄)* has increased 151 percent since 1750 as a result of vegetation burning associated with rain forest clearing, anaerobic activity in flooded rice fields, cattle and sheep effluent, and leakage of pipelines and refineries connected with natural gas production. Currently, $CH_4$ accounts for about 15 percent of anthropogenic greenhouse gases.

**A WET CLIMATES**

**Af-Am** Tropical rainy and Tropical monsoon

**Aw** Tropical wet and dry and savanna

**B DRY CLIMATES**

**BS** Steppe

**BW** Desert

**C MILD MIDLATITUDE CLIMATES**

**Ca** Humid subtropical

**Cb** Marine west coast

**Cs** Mediterranean

**D CONTINENTAL MIDLATITUDE CLIMATES**

**Da** Humid continental, warm summer

**Db** Humid continental, cool summer

**Dc** Subarctic

**E POLAR CLIMATES**

**ET** Tundra

**EF** Ice cap

**F HIGHLAND**

**H** Complex mountain climates

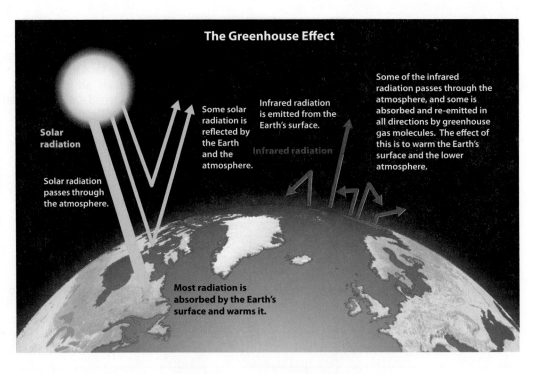

**The Greenhouse Effect**

Solar radiation

Solar radiation passes through the atmosphere.

Some solar radiation is reflected by the Earth and the atmosphere.

Infrared radiation is emitted from the Earth's surface.

Infrared radiation

Some of the infrared radiation passes through the atmosphere, and some is absorbed and re-emitted in all directions by greenhouse gas molecules. The effect of this is to warm the Earth's surface and the lower atmosphere.

Most radiation is absorbed by the Earth's surface and warms it.

**FIGURE 2.4 Solar Energy and the Greenhouse Effect** The greenhouse effect is the trapping of solar radiation in the lower atmosphere, resulting in a warm envelope surrounding Earth.

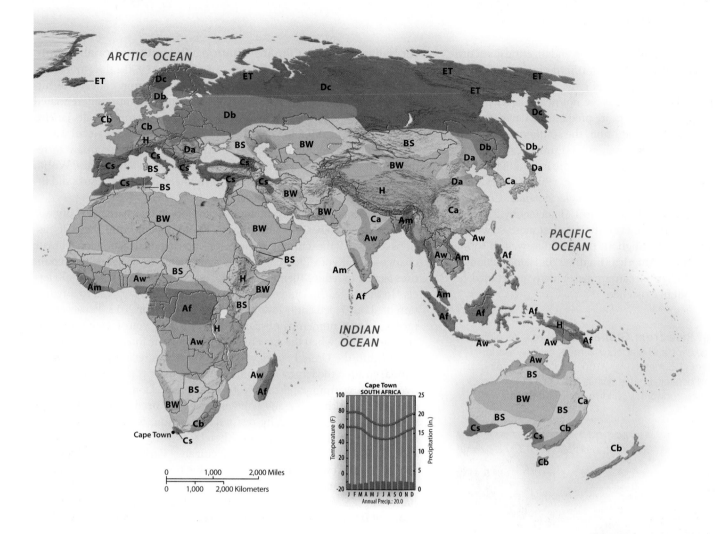

Cape Town
SOUTH AFRICA

Temperature (F) | Precipitation (in.)

J F M A M J J A S O N D
Annual Precip.: 20.0

4. *Nitrous oxide* ($N_2O$) is responsible for just over 5 percent of human-caused greenhouse gases and results mainly from the increased use of industrial fertilizers.

**Effects of Global Warming**   The complexity of the global climate system leaves some uncertainty about exactly how the world's climate may change as a result of human-caused global warming. Increasingly, though, climate scientists using high-powered computer models are reaching consensus on the probable effects of global warming. Unless countries of the world drastically reduce their emission of greenhouse gases in the next few years, computer models predict that average global temperatures will increase 2 to 4°F (1 to 2°C) by 2030. This may not seem dramatic at first glance, but it is about the same magnitude of change as the cooling that caused the ice age glaciers to cover much of Europe and North America 30,000 years ago. More problematic is that this temperature increase is projected to double by 2100.

As mentioned earlier, this climate change could cause a shift in major agricultural areas. For example, the Wheat Belt in the United States might

**FIGURE 2.5   Increase in $CO_2$ and Temperature**   These two graphs show the relationship between the rapid increase of $CO_2$ in the atmosphere and the associated rise in average annual temperature for the world. The graphs go back 1,000 years and show both $CO_2$ and temperature to have been relatively stable until the recent industrial period, when the burning of fossil fuels (coal and oil) began on a large scale.

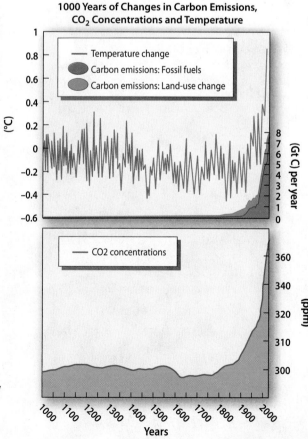

**1000 Years of Changes in Carbon Emissions, $CO_2$ Concentrations and Temperature**

— Temperature change
◼ Carbon emissions: Fossil fuels
◼ Carbon emissions: Land-use change

(°C) | (Gt C) per year

— CO2 concentrations

(ppm)

1000 1100 1200 1300 1400 1500 1600 1700 1800 1900 2000
Years

receive less rainfall and become warmer and drier, endangering grain production as we know it today. While more northern countries, such as Canada and Russia, might experience a longer growing season because of global warming, the soils in these two areas are not nearly as fertile as in the United States. As a result, food experts are predicting a decrease in the world's grain production by 2030. In addition, the southern areas of the United States and the Mediterranean region of Europe can expect a warmer and drier climate that will demand even more irrigation for crops.

Warmer global temperatures will also cause rising sea levels, as oceans warm and polar ice sheets and mountain glaciers melt. Currently, projections are for an increase of at least 3.3 feet (1 meter) by 2100. Although this may not seem like much, even the smallest increase will endanger low-lying island nations throughout the world as well as coastal areas in Europe, Asia, and North America (Figure 2.6). Island nations in the Pacific Ocean and the Indian Ocean are particularly concerned because they may be flooded out of existence.

**FIGURE 2.6  Sea Level Rise from Global Warming**   As global temperatures rise, polar ice caps and glaciers will melt, causing world oceans to rise by over 3 feet (1.5 meters) by 2100, leading to flooding of low-lying coastal areas and oceanic islands. While this child in China may find floodwaters a novel playground, his children may find them a troublesome problem.

**Globalization and Climate Change: The International Debate on Limiting Greenhouse Gases**   By the early 1990s, the United Nations recognized that climate change was "a common concern of humankind" and that all nations have a "common but differentiated responsibility" for combating global warming. At the Rio de Janeiro Earth summit in 1992, the first international agreement on global warming was presented. Signatories were bound by international law to reduce their greenhouse gas emissions by agreed-upon target dates. Within a year, 167 countries had signed. Unfortunately, these signatories did little to actually reduce their emissions.

Consequently, another attempt to develop an international treaty was made at a December 1997 meeting in Kyoto, Japan. From these discussions emerged the current vehicle for international action on climate change—the Kyoto Protocol. In this agreement, 38 industrialized countries agreed to reduce their emissions of greenhouse gases below 1990 levels. To illustrate, the United States, Japan, and Canada agreed to a reduction of about 7 percent below 1990 levels by the year 2012. However, the protocol faced numerous obstacles between the Kyoto meeting in 1997 and early 2005, when the agreement finally became international law.

Within the United States, which, until it was recently overtaken by China, emitted the largest amount of greenhouse gas per year, the debate over global warming and the Kyoto Protocol was contentious, partisan, and divisive. Although many politicians saw the need for greenhouse gas controls, others emphatically resisted any action, fearing that emission controls might constrain business, thereby slowing the economy and increasing the cost of living for U.S. citizens. Early in 2001, President George W. Bush went on record as opposing the Kyoto Protocol. Not only did he voice concern that restrictions on atmospheric emissions would harm the U.S. economy, but President Bush also objected to the Kyoto Protocol on the grounds that large developing countries, namely China

**FIGURE 2.7  Auto Emissions in China**   In 2006, China surpassed the United States as the country that emits the most $CO_2$ into the atmosphere. While the main reason for China's huge emissions is the widespread use of coal to generate power, the rapid increase in private auto ownership is an increasing concern because car traffic has become a major cause of smoggy skies in China's urban areas. This traffic jam in Beijing and the leaden skies are expressions of that problem.

### TABLE 2.1   THE WORLD'S MAJOR CO$_2$ POLLUTERS

| Country | Total CO$_2$ Emissions in MMT[a] (2006) | Percent Change (1996–2006) | Percent of World Total, CO$_2$ Emissions | Per Capita Emissions of CO$_2$, Metric Tons |
|---|---|---|---|---|
| China | 6,017 | 104 | 21 | 4.58 |
| United States | 5,903 | 7 | 20 | 19.8 |
| Russia | 1,704 | 5 | 6 | 12.0 |
| India | 1,293 | 55 | 4.4 | 1.2 |
| Japan | 1,246 | 9 | 4.3 | 9.9 |
| Germany | 857 | −4 | 3 | 10.4 |
| **World Total** | **29,195** | | | 4.58 |

[a] *Million metric tons.*
*The most current and compatible international data from the Energy Information Administration (EIA) are for 2006.*

and India, were not yet bound to specific greenhouse gas reductions (Figure 2.7). As a result of his opposition, the United States was the only industrialized country to not ratify the Kyoto Protocol. More recently, President Obama has pledged that the United States will become a full member of the Kyoto Protocol and reduce its greenhouse emissions accordingly. However, the pathway to this goal remains politically and economically problematic.

There are also unresolved tensions between developed and less-developed countries over the Kyoto Protocol. Historically, emissions in the industrial world (North America, Europe, Japan) created the global warming problem, and these countries still today contribute about half of the total human-caused greenhouse gases (Table 2.1). As a result, many international experts argue that these same developed parts of the world should be required not only to take stringent steps to curb their own emissions but also should subsidize and underwrite emission controls in developing countries. Understandably, the less-developed countries are reluctant to sign an emission control agreement that will constrain their economic future when, as they argue, they have contributed very little to the global warming problem up to now.

To become international law, the Kyoto Protocol had to be ratified by those countries emitting a total of 55 percent of the world's greenhouse gases. With the United States rejecting the agreement, the protocol stalled for several years while Russia, the world's third-largest source of greenhouse gas emissions, pondered the costs and benefits of ratification. Finally, in late 2004, Russia signed the agreement, thus making Kyoto international law. In total, more than 100 countries have ratified the agreement. Although the Kyoto Protocol will not solve the global warming problem, it is an important first step toward international cooperation in mitigating climate change. A second step will be taken in 2012, when the second phase of the Kyoto emission reduction process begins.

## Human Impacts on Plants and Animals: The Globalization of Nature

One aspect of Earth's uniqueness compared to other planets in our solar system is the rich diversity of plants and animals covering its continents. Geographers and biologists think of the cloak of vegetation as the "green glue" that binds together life, land, and atmosphere. Humans are very much a part of this interaction. Not only are we evo-

lutionary products of a specific **bioregion**, or assemblage of local plants and animals (most probably the tropical savanna of Africa), but also our human prehistory included the domestication of certain plants and animals. From this process has come agriculture. Further, humans have changed the natural pattern of plants and animals dramatically by plowing grasslands, burning woodlands, cutting forests, and hunting animals. The pace of such change has accelerated in recent decades, and these actions have led to a crisis in the biological world, as ecosystems are devastated and entire species are exterminated. Many argue that the very vitality of our life-giving biosphere is threatened in many parts of the world.

Many of these problems can be explained by the globalization of nature and of local ecologies. Until the last half century, tropical forests were primarily homes for small populations of indigenous peoples who made modest demands on local environments for their sustenance and subsistence. Today, however, these same tropical forests are capital for multinational corporations that clear-cut forests for international trade in wood products or search out plants and animals to meet the needs of far-removed populations. For example, Japanese lumber companies cut South American rain forests; German pharmaceutical corporations harvest medicinal plants in Africa; poachers kill North American bears and then sell their gallbladders on the Asian black market as elixirs and sexual stimulants. Unfortunately, the list of human impacts on nature is a long one.

## Biomes and Bioregions

**Biome** is the biogeographic term used to describe a grouping of the world's flora and fauna into a large ecological province or region. In this book we use the terms *biome* and *bioregion* interchangeably. Biomes are closely connected with climate regions because the major characteristics of a climate region—temperature, precipitation, and seasonality—are also the major factors influencing the distribution of natural vegetation and animals. A brief overview of the most important bioregions follows (Figure 2.8).

## Tropical Forests and Savannas

Tropical forests are found in the equatorial climate zones of high average annual temperatures, long days of sunlight throughout the year, and heavy amounts of rainfall. This bioregion covers about 7 percent

**FIGURE 2.8 World Bioregions**    Although global vegetation has been greatly modified by clearing land for agriculture and settlements and cutting forests for lumber and paper pulp, there is still a recognizable pattern to the world's bioregions, ranging from tropical forests to arctic tundra. Each bioregion has its own array of ecosystems, containing plants, animals, and insects. These species constitute the biodiversity necessary for robust gene pools. Think of biodiversity as the genetic library that informs life on Earth. *(Adapted from Clawson and Fisher, 2004,* World Regional Geography, *8th ed., Upper Saddle River, NJ: Prentice Hall)*

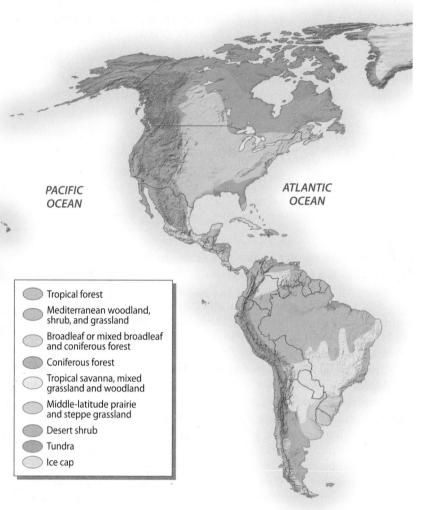

Tropical forest

Mediterranean woodland, shrub, and grassland

Broadleaf or mixed broadleaf and coniferous forest

Coniferous forest

Tropical savanna, mixed grassland and woodland

Middle-latitude prairie and steppe grassland

Desert shrub

Tundra

Ice cap

**FIGURE 2.9 Tropical Rain Forest**    As fragile as they are diverse, tropical rain forest environments feature a complex, multilayered canopy of vegetation. Plants on the forest floor are well adapted to receiving very little direct sunlight.

of the world's land area (roughly the size of the contiguous United States) in Central and South America, Sub-Saharan Africa, Southeast Asia, Australia, and on many tropical Pacific islands. More than half of the known plant and animal species live in the tropical forest bioregion, making it the most diverse of all biomes.

The dense tropical forest vegetation is usually arrayed in three distinct levels that are adapted to decreasing amounts of sunlight from the canopy to the forest floor (Figure 2.9). The tallest trees, around 200 feet (61 meters) high, receive open sunlight; the middle level (around 100 feet, or 31 meters) gets filtered sunlight; the third level is the forest floor, where plants can survive with very little direct sunlight. Even though much organic material accumulates on the forest floor in the form of falling leaves, tropical forest soils tend to be very low in stored nutrients. The nutrients are stored instead in the living plants. As a result, tropical forest soils are not well suited for intensive agriculture.

## Deforestation in the Tropics

For a variety of reasons, tropical forests are being devastated at an unprecedented rate, creating a crisis that tests our political, economic, and ethical systems as the world searches for solutions to this pressing

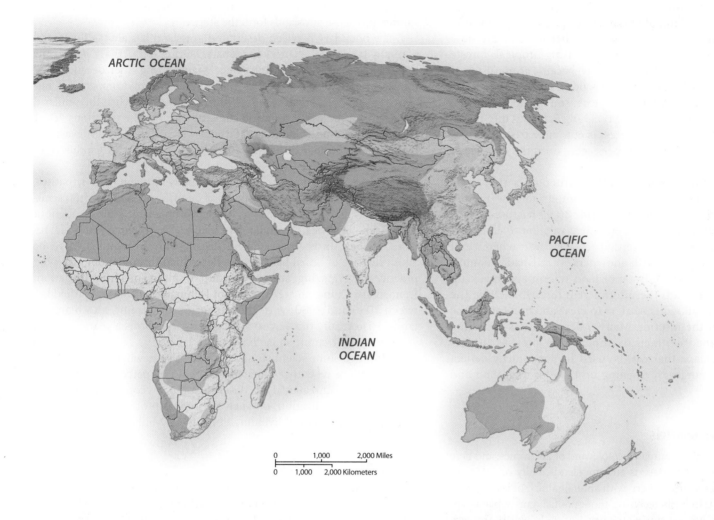

problem. Although deforestation rates differ from region to region, each year an area of tropical forest about the size of Wisconsin or Pennsylvania is denuded. Spatially, almost half of this activity is in the Amazon Basin of South America. However, deforestation appears to be occurring faster in Southeast Asia, where some estimates suggest that logging is occurring three times faster than in the Amazon. If these estimates are accurate, Southeast Asia might be completely stripped of forests within 15 years (Figure 2.10).

An important side effect of tropical deforestation is the release of $CO_2$ into the atmosphere. Current estimates suggest that fully 20 percent of all human-caused greenhouse gas emissions result from cutting and burning tropical forests.

Behind this widespread cutting of tropical forests lies the recent globalization of commerce in international wood products. Reportedly, Japan was the first to globalize its timber business by reaching far beyond its national boundaries to purchase timber in North and South America and Southeast Asia. This was a two-pronged strategy for meeting the increased demand for wood products in Japan while at the same time protecting its own rather limited forests for recreational use. Currently, about one-half of all tropical forest timber is destined for Japan. Unfortunately, much of it is used for throwaway items such as chopsticks and newspapers.

**FIGURE 2.10 Tropical Forest Destruction** There are many ill effects resulting from tropical rain forest destruction including loss of plants and animals, the loss of homeland for native peoples, and the release of large amounts of global warming greenhouse gases. This photo is of a new cacao plantation on Borneo Island, East Malaysia.

Another factor contributing to the rapid destruction of tropical rain forests is the world's seemingly insatiable appetite for beef. Cattle species originally bred to survive the hot weather of India are now raised on grassland pastures created by cutting tropical forests throughout the tropical world. Unfortunately, because tropical forest soils are poor in nutrients, cattle ranching is not a sustainable activity in these new grassland areas. After a few years, soil nutrients become exhausted, making it necessary to move ranching activities into newly cleared forestland. As a result, more forest is cut, more pasture is created, more soil is destroyed, and the process goes on at the expense of forestlands.

A third factor behind tropical forest destruction is that these forest areas are often the last settlement frontiers for the rapidly growing population of the developing world. Brazil has used much of its interior Amazon rain forest for settlement in order to ease population pressure along its densely settled eastern coast, where rural lands are controlled by powerful landowners. Brazil had a choice: Either address the troublesome issue of land reform and break up the coastal estates or open up the interior rain forests. It chose the path of least resistance, allowing settlers to clear and homestead the tropical forestlands of the interior. Many other countries are doing the same, looking at the vast tracts of tropical rain forest as a safety valve of sorts, a landscape that can be used to temporarily deflect the pressures of land hunger by opening up land for migration and settlement.

## Deserts and Grasslands

Large areas of arid and semiarid climate lie poleward of the tropics, and here are found the world's extensive deserts and grasslands. Fully one-third of Earth's land area qualifies as true desert, with annual rainfall of less than 10 inches (25 centimeters). In areas receiving more rainfall, grassy plants appear, often forming a lush cover during the wet season. In North America, the midsection of both Canada and the United States is covered by grassland known as **prairie**, which is characterized by thick, long grasses. In

other parts of the world, such as Central Asia, Russia, and Southwest Asia, shorter, less dense grasslands form the **steppe**.

The boundary between desert and grassland has always fluctuated naturally because of changes in climate. During wet periods, grasslands might expand, only to contract once again during drier decades. The transition zone between the two is a precarious environment for humans, as the United States learned during the 1930s, when the semiarid grasslands of the western prairie lands turned into the notorious "Dust Bowl." At that time, thousands of farmers watched their fields devastated by wind erosion and drought—a disaster that led to an exodus from these once-productive lands.

Farming marginal lands may actually worsen the situation, leading to **desertification**, or a spread of desertlike conditions into semiarid areas (Figure 2.11). This has happened on a large scale throughout the world—in Africa, Australia, and South Asia, to name just a few regions. In fact, in the past several decades, an area estimated to be about the size of Brazil has become desertified through poor cropping practices, overgrazing, and the buildup of salt in soils from irrigation. In northern China, an area the size of Denmark became desertified between 1950 and 1980, with the expansion of farming into marginal lands. Historically this region averaged 3 sandstorms a year; today 25 such storms are common each year.

Although some scientists say the case is somewhat overstated, the United Nations recently estimated that about 60 percent of the world's rangelands are threatened by desertification. According to this estimate, the desertification of such an amount of rangeland would threaten the agricultural livelihood of some 1.2 billion people.

## Temperate Forests

The large tracts of forests found in middle and high latitudes are called *temperate forests*. Their vegetation is different from the low-latitude forests found in the equatorial regions. In temperate forests, two major tree types dominate. One type is softwood coniferous evergreen trees, such as pine, spruce, and fir, which are found in higher elevations and

**FIGURE 2.11 Desertification** Climatic fluctuations and human misuse combine to produce desertification in certain localities, where marginal lands are overcropped or grazed heavily, resulting in expansion of nearby deserts. Globally, many rangelands remain threatened by desertification.

higher latitudes. The second category comprises deciduous trees, which drop their leaves during the winter. Examples are elm, maple, and beech. Because these trees are hardwood, hence harder to mill, they are generally less favored by the timber industry than softwood species.

In North America, conifers dominate the mountainous West, Alaska, and Canada's western mountains, while deciduous trees are found on the Eastern Seaboard of the United States north to New England. There the two tree types intermix before giving way to the softwood forests of Maine and the Maritime provinces of eastern Canada.

In the coniferous forests of western North America, the struggle between timber harvesting and environmental concerns remains controversial. Whereas timber interests argue that increased cutting is the only way to meet high demand for lumber and other wood products, environmental concerns over the protection of habitat for endangered species (the spotted owl, for example) have caused the government to make large tracts of forest off-limits to commercial logging (Figure 2.12).

Further complicating the future of western forests are global market forces. Many Japanese and Chinese timber firms pay premium prices for logs cut from U.S. and Canadian forests, outbidding domestic firms for these scarce resources. Because these trees are often cut from public lands, this facet of globalization raises an interesting question about the appropriate use of public forests that are maintained by tax money.

In Europe, hardwoods were the natural tree cover in western countries such as France and Great Britain before these once-extensive forests were cleared away for agriculture. In the higher latitudes of Norway and Sweden, coniferous species prevail in the remaining forests. These conifers also populate extensive forests across Germany and Eastern Europe, as well as through Russia into Siberia, creating an almost unbroken landscape of dense forests.

The Siberian forest remains a resource that could become a major source of income for financially strapped Russia. Some argue that if the Siberian forests are put on the market for global trade, it will reduce logging pressure on North America's western forests, which, in turn, could make it easier to enact and enforce comprehensive environmental protection in the United States. This example illustrates once again how the multifaceted forces of globalization are intertwined with the fate of local ecosystems.

## Food Resources: Environment, Diversity, and Globalization

If the human population continues to grow at expected rates, food production must double by 2025 just to provide each person in the world with a basic subsistence diet. Every minute of each day, 258 people are born who need food; during that same minute, about 10 acres of existing cropland are lost because of environmental problems such as soil erosion and desertification. Many experts argue that food scarcity will be the defining issue of the next several decades (Figure 2.13).

**FIGURE 2.12 Clear-Cut Forest** Commercial logging in Washington's Olympic Peninsula has dramatically reshaped this landscape. Throughout the Pacific Northwest, environmental lobbies have successfully restricted logging to protect habitat for endangered species and recreation.

**FIGURE 2.13 Industrial Farming in Iowa** The harvesting of corn in Iowa embodies the dilemma of modern agriculture. While high yields result from mechanization and intensive use of chemical fertilizers, questions are raised as to the sustainability of these methods because of their high energy and environmental costs.

## The Green Revolutions

The world's population doubled during the last 40 years of the 20th century and, remarkably, during that same period, global food production also doubled to keep pace with this population explosion. This increase in food production came primarily from the expansion of intensive industrial agriculture into areas that previously produced subsistence crops through extensive and traditional means.

More specifically, since 1950, with the increases in global food production have come changes associated with the **Green Revolution**, which involved new agricultural techniques using genetically altered seeds coupled with high inputs of chemical fertilizers and pesticides. The first stage of the Green Revolution combined three processes: first, the change from traditional mixed crops to monocrops, or single fields, of genetically altered, high-yield rice, wheat, and corn seeds; second, intensive applications of water, fertilizers, and pesticides; and, third, additional increases in the intensity of agriculture by reduction in the fallow, or field-resting time, between seasonal crops.

Since the 1970s, a second stage of the Green Revolution has evolved. This phase emphasizes new strains of fast-growing wheat and rice specifically bred for tropical and subtropical climates (Figure 2.14). When combined with irrigation, fertilizers, and pesticides, these new varieties allow farmers to grow two or even three crops a year on a parcel that previously supported only one. Using these methods, India actually doubled its food production between 1970 and 1992.

Many argue that the practices associated with the Green Revolution carry high environmental and social costs. Because these crops draw heavily on fossil fuels, there has been a 400 percent increase in the agricultural use of fossil fuels during the past several decades. As a result, Green Revolution agriculture now consumes almost 10 percent of the world's annual oil output. Earlier, cheap oil prices facilitated much of this increased agricultural usage of fossil fuel, but today, higher oil prices raise questions about the sustainability of these revolutionary methods and whether higher energy costs will translate into food costs beyond the reach of the inhabitants of developing countries.

The environmental costs of the Green Revolution have also been significant, resulting in damage to habitat and wildlife from diversion of natural rivers and streams for irrigation, pollution of rivers and water sources by pesticides and chemical fertilizers applied in heavy amounts to fields, and increased regional air pollution from factories and chemical plants that produce these agricultural chemicals.

There is also evidence that the Green Revolution can bring high social costs and disruption to some areas. Because the financial costs to farmers participating in the Green Revolution are higher than with traditional farming, successful farmers must have access to capital or bank loans to purchase hybrid seeds, fertilizers, pesticides, and even new machinery. For those with high social standing, a family support system, or good credit, the rewards can be great, but for those without access to loans or family support, the toll can be heavy. Usually traditional farmers cannot compete against those raising Green Revolution crops in the regional marketplace. As a result, they often struggle at a poverty level while Green Revolution farmers prosper.

In some wheat-growing areas of India, the social and economic distance between the well-off Green Revolution farmers and poor traditional farmers has become a major source of economic, social, and political tension. An area where once all shared a common plight has now become a highly stratified society of rich and poor. These social costs can be condoned only because of the pressing need for food in this rapidly growing country.

## Problems and Projections

Even though agriculture has been able to keep up with population growth in the past decades, few experts are confident that this pace will continue in the 21st century. Although the regional chapters of this book include fuller discussion of food and agriculture issues, four general points offer a place to start understanding the world's food security issues:

- While overall food production remains an important global issue, it is in fact local and regional problems that often keep people from obtaining needed food supplies. To many experts, the issue is not so much global food production as the widespread poverty and civil unrest at local levels that keep people from growing, buying, or receiving adequate food.

- Political problems are usually more responsible for food shortages and famines than are natural events such as drought and flooding. Distribution of world food supplies is often highly politicized, starting at the global level and continuing down to the local. Usually food aid goes to political friends and allies, while those allied with enemies go without.

- Globalization is causing dietary preferences to change worldwide, and the implications of these changes could be profound. Currently two-thirds of the world population is primarily vegetarian, eating only small portions of meat because it is usually beyond their food budgets. However, because of recent economic booms in some developing countries, an increasing number of people in developing countries are now eating meat, having moved up the food chain from a subsistence diet based on local resources to a global

**FIGURE 2.14 The Green Revolution in India**    Because of its large and expanding population, India is a region where food supplies may be a problem in the near future. The country doubled its food production between 1970 and 1992, primarily due to the expansion of rice and wheat yields. Nonetheless, the future of food security in this large country is uncertain. In this photo, a woman is planting rice seedlings in the Rishi Valley of Andhra Pradesh state.

**FIGURE 2.15 Cattle Ranching in Brazil**   Growing global demands for meat are reshaping the world's agricultural landscapes. Cattle ranching in western Brazil's rain forest reflects these changing market conditions, and the practice is also introducing new environmental problems.

diet containing a greater amount of meat (Figure 2.15). In many cases this change in diet comes not just from new economic prowess but also from changing cultural tastes and values, as people are exposed to new products through economic and cultural globalization.

There are, however, constraints on the world's ability to supply these new tastes because of the agricultural and environmental support system needed to produce and market meat. According to some food experts, the global food production system could sustain only half the current world population if everyone ate the meat-rich diet of North America, Europe, and Japan; this diet contains three or four times as much meat as the diet of people in developing countries.

■ Most food supply experts agree that the two world regions of greatest concern are Africa and South Asia. Until 1970, Africa was self-sufficient in food, but since that time there have been serious disruptions and even breakdowns in the food supply system. These problems in

Africa stem from rapid population increase and civil disruption from tribal warfare. As a result, one of every four people faces food shortages in Sub-Saharan Africa. (Chapter 6 contains more detail on these issues.) South Asia's future is also problematic. The United Nations predicts that by the year 2010, almost 200 million people in South Asia will suffer from chronic undernourishment. (Further discussion of these problems is found in Chapter 12.)

There is some good news in this otherwise bleak picture. Because population growth rates are generally declining in the industrializing areas of East Asia and food production there is still increasing, the United Nations predicts that the percentage of undernourished people in that region will actually drop by almost 5 percent in the next 10 years. Similar gains are being made in Latin America because of lower population growth and higher agricultural production. World food experts predict that the proportion of chronically hungry in Latin America will fall to about 6 percent in 2010, which is half the 1990 rate.

# SUMMARY

■ Global environmental change is driven by human activities. While some changes are intentional and are largely beneficial for humankind, such as the irrigation of arid lands to increase the world's food supplies, other environmental changes have been unanticipated and have numerous negative consequences. Global warming from fossil fuel consumption is a prime example.

■ Globalization is both a help and a hindrance to world environmental problems. As a positive force, globalization is central to sharing information and increasing public awareness of environmental issues. In addition, many would argue that globalization facilitates a new willingness of countries to work together under the umbrella of international agreements to resolve environmental problems. Such cooperation has led to international treaties on ocean pollution, global warming, and protection of wildlife species.

- Critics say economic globalization is responsible for increased environmental degradation because the responsibility and accountability for environmental problems is masked and disguised by layers of impenetrable global interactions and connections. The destruction of Indonesia's tropical rain forests to provide biofuels so Europe and the United States can reduce their global warming emissions is an example.

## KEY TERMS

anthropogenic (p. 37)
biome (p. 41)
bioregion (p. 41)

climate region (p. 36)
climograph (p. 37)
desertification (p. 44)

global warming (p. 37)
Green Revolution (p. 46)
greenhouse effect (p. 37)

prairie (p. 44)
steppe (p. 44)

## THINKING GEOGRAPHICALLY

1. What are the most threatening natural hazards in your region—earthquakes, tornadoes, hurricanes, floods, drought? Research local agencies or the public library to learn about disaster preparedness plans for your community.

2. Which climate region do you live in? What are the major weather problems faced by people in your area? How do they adjust to these problems?

3. Visit some of the many Internet websites that relate to global warming. Once you have an overview of the different positions, concentrate on one or two of the most contentious issues, such as the debate within the United States between environmentalists and business interests, or the difference in opinion between developed and developing countries. Discuss the different stakeholders and vested interests holding contrasting views.

4. How has the vegetation in your area been changed by human activities in the past 100 years? Have these changes led to the extinction of any plants or animals or placed them on the endangered list? If so, what is being done to protect them or restore their habitat?

5. Conduct a detailed analysis of the food issues in a foreign country of your choice and answer the following questions. In general, has food supply kept up with population growth? Has the country suffered recently from food shortages or famine? If so, were these shortages the result of natural causes, such as drought or floods, or distribution problems resulting from civil disruption? Which segments of the population have food security, and which segments have recurring problems obtaining adequate food? How is this inequity best explained? Finally, have food preferences or diets changed recently? If so, how and why?

Log in to **www.mygeoscienceplace.com** for videos, interactive maps, RSS feeds, case studies, and self-study quizzes to enhance your study of the changing global environment.

# 3 NORTH AMERICA

## GLOBALIZATION AND DIVERSITY

North America plays a pivotal role in economic globalization, both as a leading driver of global change and as a growing participant in an increasingly interconnected international economy. The region's role as a destination for varied global immigrants has also made it one of the world's most diverse cultural settings, home to a relatively new amalgam of people from every corner of the earth.

### ENVIRONMENTAL GEOGRAPHY

Fewer ski days in southern Quebec, widespread drought in the U.S. Southwest, and an increased frequency of major hurricanes in the Gulf of Mexico may be only a few of the many consequences of global climate change within the North American region.

### POPULATION AND SETTLEMENT

Sprawling suburbs characterize the expanding peripheries of hundreds of North American cities, creating a multitude of challenges, including lengthier commutes for urban workers and the loss of prime agricultural lands along the edge of metropolitan areas.

### CULTURAL COHERENCE AND DIVERSITY

Cultural pluralism remains strong in North America. Currently, there are more than 43 million immigrants living in the United States alone. The tremendous growth in Hispanic and Asian immigrants since 1970 has fundamentally reshaped the region's cultural geography.

### GEOPOLITICAL FRAMEWORK

Cultural pluralism continues to shape political geographies in the region. American immigration policy remains hotly contested in the United States, and persisting regional and native rights issues confront Canadians.

### ECONOMIC AND SOCIAL DEVELOPMENT

The geographic impacts of the late 2000s economic crisis in North America were very uneven. Particularly hard hit were portions of the Sun Belt in Arizona, Nevada, and California, where speculative home construction between 2000 and 2007 was replaced by a period of increasing foreclosures, falling prices, and rising unemployment.

*Los Angeles remains one of North America's most spectacular examples of urban sprawl.*

Globalization has fundamentally reshaped North America. A walk down any busy street in Toronto, Tucson, or Toledo reveals how international products, foods, culture, and economic connections shape the everyday scene. From farmers to computer workers, most North Americans have jobs that are either directly or indirectly linked to the global economy. That close relationship was dramatically illustrated in the global economic downturn of 2008–2009, as many North American workers saw their jobs disappear. Similarly, overseas businesses were hurt by declining demand from North American consumers.

Large foreign-born populations in each nation also provide direct links to every part of the world. Tourism brings in millions of additional foreign visitors, and billions of dollars that are spent everywhere from Las Vegas to Disney World. In more subtle ways, North Americans see globalization in their everyday lives. They consume ethnic foods, tune in to international sporting events on television, enjoy the sounds of salsa and Senegalese music, and surf the Internet from one continent to the next.

Globalization is also a two-way street, and North American capital, culture, and power are ubiquitous. By any measure of multinational corporate investment and global trade, the region plays a dominant role that far outweighs its population of 340 million residents. North American automobiles, consumer goods, information technology, and investment capital circle the globe. In addition, North American foods and popular culture are diffusing globally at a rapid pace. North American music, cinema, and fashion have also spread rapidly around the world. Soaring downtown skylines from Mumbai to Beijing increasingly resemble their North American counterparts.

North America includes the United States and Canada, a culturally diverse and resource-rich region that has seen tremendous human modification of its landscape and extraordinary economic development over the past two centuries (Figure 3.1; see also "Setting the Boundaries"). The result is that North America is one of the world's wealthiest regions, where two highly urbanized and mobile countries are associated with the processes of globalization and the highest rates of resource consumption on Earth. Indeed, the region exemplifies a postindustrial economy in which human geographies are shaped by modern technology, by innovative financial and information services, and by a popular culture that dominates both North America and the world beyond.

Politically, North America is home to the United States, the last remaining global superpower. Such status brings the country onto center stage in times of global tensions, whether they are in the Middle East, South Asia, or West Africa. In addition, North America's largest metropolitan area, New York City (22 million people), is home to the United Nations and other global political and financial institutions. North of the United States, Canada is the other political unit within the region. While slightly larger in area than the United States (3.83 million square miles [9.97 million square kilometers] versus 3.68 million square miles [9.36 million square kilometers]), Canada's population is only about 10 percent that of the United States.

Contemporary North America displays both the bounty and the price of the development process. On one hand, the region shares the benefits of modern agriculture, globally competitive industries, excellent transport and communications infrastructure, and two of the most highly urbanized societies in the world. The cost of development, however, has been high: Native populations were all but eliminated by European settlers, forests have been logged, grasslands have been converted into farms, valuable soils have been eroded, numerous species have been threatened with extinction, great rivers have been diverted, and natural resources have often been wasted. Today, although home to only about 5 percent of the world's population, the region consumes 25 percent of the world's commercial energy budget and produces carbon dioxide emissions at a per person rate 15 times that of India.

Nevertheless, economic growth has vastly improved the standard of living for many North Americans, who enjoy high rates of consumption and varied urban amenities that are the envy of the less-developed world. Satellite dishes, sushi, and shopping malls are within easy reach of most North American residents. Amid this material abundance, however, there are continuing differences in income and in the quality of life. Poor rural and inner-city populations still struggle to match the affluence of their wealthier neighbors.

North America's unique cultural character also defines the region. The cultural characteristics that hold this region together include a common process of colonization, a heritage of Anglo dominance, and a shared set of civic beliefs in representative democracy and individual freedom. But the history of the region has also juxtaposed Native Americans, Europeans, Africans, and Asians in fresh ways, and the results are two societies unlike any other (Figure 3.2). Adding to the mix is a popular culture that today exerts a powerful homogenizing influence on North American society.

## Setting the Boundaries

The United States and Canada are commonly referred to as *North America*, but that regional terminology can sometimes be confusing. In some geography textbooks, the realm is called *Anglo America* because of its close connections with Britain and its Anglo-Saxon cultural traditions. But processes of globalization and immigration have added increasing cultural diversity to North America, discouraging the use of the term *Anglo America*. While more culturally neutral, the term *North America* also has its problems. As a physical feature, the North American continent commonly includes Mexico, Central America, and, often, the Caribbean. Culturally, however, the United States–Mexico border seems a better dividing line. However, even that regional division is problematic, given the growing Hispanic presence in the U.S. Southwest, as well as ever-closer economic links across the border. While connections between Mexico and the United States are likely to grow even stronger in the future, our coverage of the "North American" realm concentrates on Canada and the United States, two of the world's largest and wealthiest nation-states.

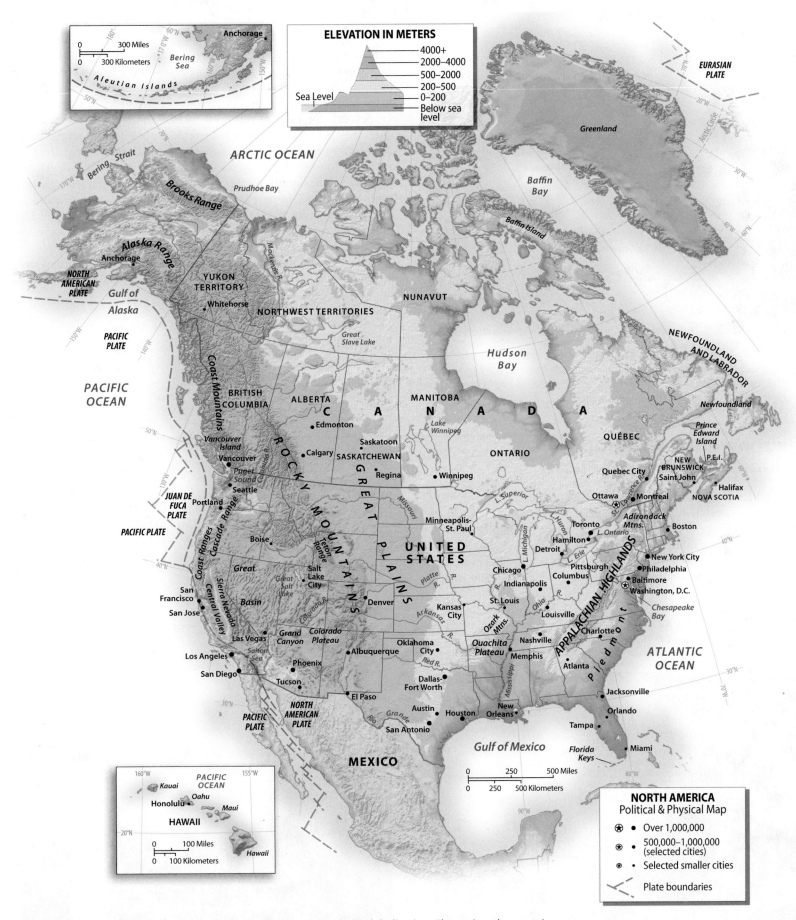

**ELEVATION IN METERS**

4000+
2000–4000
500–2000
200–500
0–200
Sea Level
Below sea level

**NORTH AMERICA**
Political & Physical Map

⊛ ● Over 1,000,000
⊛ ● 500,000–1,000,000 (selected cities)
⊛ • Selected smaller cities
╌╌ Plate boundaries

**FIGURE 3.1 North America** North America plays a key role in globalization. The region also contains one of the world's most highly urbanized and culturally diverse populations. With 340 million people and extensive economic development, North America is also one of the largest consumers of natural resources on the planet.

**FIGURE 3.2 Toronto's Cultural Landscape**    Toronto's varied ethnic population is celebrated during the Caribana Parade through the city. Powerful forces of globalization have reshaped the cultural and economic geographies of dozens of North American cities.

## Environmental Geography: A Threatened Land of Plenty

North America's physical and human geographies are enormously diverse and intricately linked. Hurricane Katrina's tumultuous arrival along the Gulf Coast in August 2005 exemplifies the complexities and significance of those interconnections (Figure 3.3). Three intertwined variables came into play. First, the storm itself was large and powerful, sweeping inland across southern Louisiana and coastal Mississippi

**FIGURE 3.3 Hurricane Katrina**    Approaching from the Gulf of Mexico (a), Hurricane Katrina put much of New Orleans under water (b) and inflicted particular devastation on the city's poor, black population (c).

a)

with winds of more than 120 miles per hour. The U.S. Census Bureau estimates that almost 10 million Americans in the region experienced hurricane-force winds (more than 75 miles per hour) the morning the storm came ashore.

Second, the built environment across the region was extensive, with many population centers vulnerable to the effects of a large hurricane. New Orleans was by far the largest urban area in the path of the storm. Unfortunately, for generations, city planners and developers downplayed the hazards of the city's low-lying, bowl-like setting. Aging levees were breached near Lake Pontchartrain and along several commercial canals, putting 80 percent of the city under water (Figure 3.3). To the east, populated coastal areas of southern Mississippi were also devastated by the storm.

Third, the region's social geography made evacuating urban areas difficult in the face of the storm (Figure 3.3). Poverty rates across the

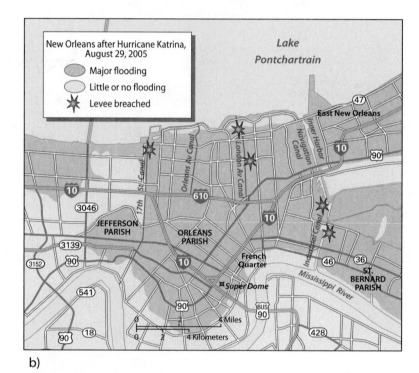

b)

c)

region are almost double the national norm. Some of the worst-hit areas of New Orleans were more than 80 percent black. The city's poor and elderly residents were its least mobile, forced to ride out the storm in their flooded neighborhoods or in the overcrowded Superdome. Added to this were the problems of coordinating a viable emergency response to the disaster. The resulting deaths, damage, and slow economic recovery were a powerful reminder of how the costs and impacts of a "natural" environmental disaster are inevitably wed to a region's cultural, social, and economic characteristics.

## The Costs of Human Modification

Katrina's story is a reminder that North Americans have modified their physical setting in many ways. Processes of globalization and accelerated urban and economic growth have transformed North America's landforms, soils, vegetation, and climate. Indeed, problems such as acid rain, nuclear waste storage, groundwater depletion, and toxic chemical spills are all manifestations of a way of life unimaginable only a century ago (Figure 3.4). Energy consumption in the region

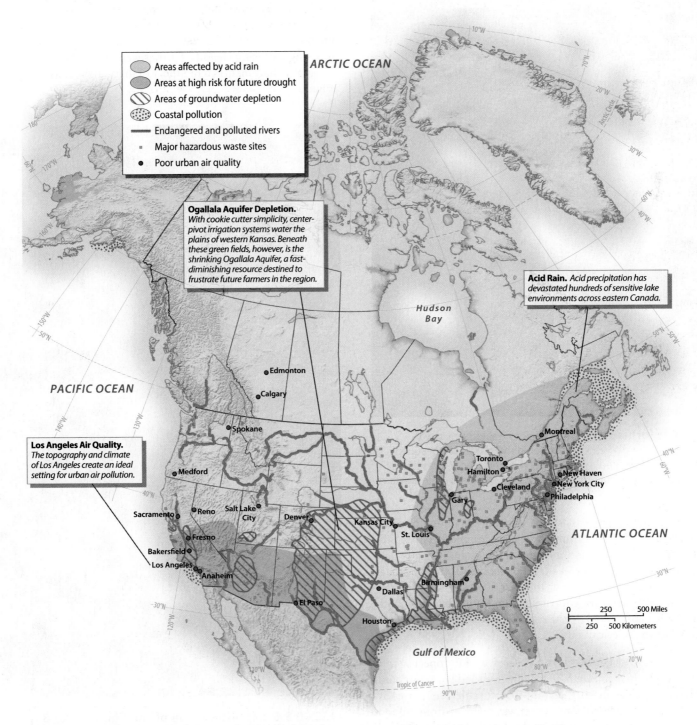

**FIGURE 3.4 Environmental Issues in North America**  Many environmental issues threaten North America. Acid rain damage is widespread in regions downwind from industrial source areas. Elsewhere, air and water pollution present health dangers and economic costs to residents of the region. Since 1970, however, both Americans and Canadians have become increasingly responsive to these environmental challenges.

**FIGURE 3.5  California Aqueduct**   Thirsty Californians have extensively modified the geography of water in their state. Large aqueducts have dramatically reconfigured the distribution of this precious resource, promoting tremendous agricultural and metropolitan expansion in western North America.

also remains extremely high, imposing a growing list of environmental and economic costs both within North America and beyond. Still, many environmental initiatives in the United States and Canada have addressed local and regional problems. For example, the improved water quality of the Great Lakes over the past 30 years is an achievement to which both nations contributed and that benefits both. Tougher air quality standards have also reduced certain types of emissions in many North American cities.

**Transforming Soils and Vegetation**   The arrival of Europeans to the North American continent affected the region's flora and fauna as countless new species were introduced, including wheat, cattle, and horses. As the number of settlers increased, forest cover was removed from millions of acres. Grasslands were plowed under and replaced with grain and forage crops not native to the region. Widespread soil erosion was increased by unsustainable cropping and ranching practices, and many areas of the Great Plains and South suffered lasting damage.

**Managing Water**   North Americans consume huge amounts of water. While conservation efforts and technology have slightly reduced per capita rates of water use over the past 25 years, city dwellers still use an average of more than 175 gallons daily. Indirectly, through other forms of food and industrial consumption, Americans may each consume more than 1,400 gallons of water per day. Many places in North America are threatened by water shortages. Metropolitan areas such as New York City struggle with outdated municipal water supply systems. Beneath the Great Plains, the waters of the Ogallala Aquifer are being depleted. Center-pivot irrigation systems have steadily lowered water tables across much of the region by as much as 100 feet (30 meters) in the past 50 years; and the costs of pumping are rising steadily. Farther west, California's complex system of water management is a reminder of that state's ever-growing demands (Figure 3.5).

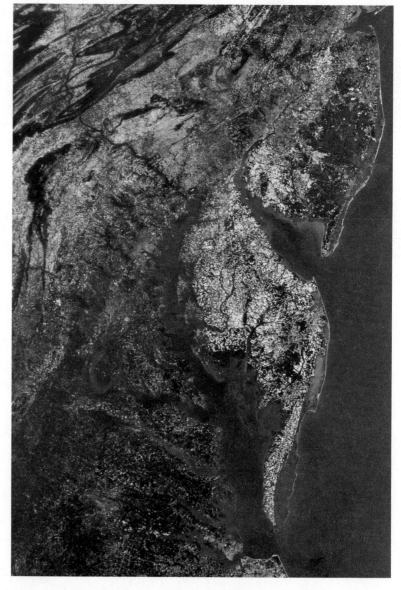

**FIGURE 3.6  Satellite Image of the Chesapeake Bay**
This view of the Middle Atlantic Coast reveals the complex shoreline of the Chesapeake Bay (lower center). The coastal area is characterized by drowned river valleys, barrier islands, and sandy beaches. The Piedmont zone and Appalachian Highlands appear to the northwest.

Water quality is also a major issue (Figure 3.4). North Americans are exposed to water pollution every day, and even environmental laws and guidelines, such as the U.S. Clean Water Act or Canada's Green Plan, cannot eliminate the problem. Mining operations and industrial users such as chemical, paper, and steel plants generate toxic wastewater or metals that enter surface water and groundwater supplies. Metropolitan areas also produce vast amounts of raw sewage that annually costs billions to repurify. America's most toxic places include the petroleum-rich Texas and Louisiana Gulf coasts, the older industrial centers of the Northeast and Midwest, and nuclear fuel and chemical warfare storage areas such as Hanford, Washington.

**Altering the Atmosphere**    North Americans modify the very air they breathe; in doing so, they change local and regional climates as well as the chemical composition of the atmosphere. The development associated with urban settings often produces nighttime temperatures some 9 to 14°F (5 to 8°C) warmer than nearby rural areas. At the local level, industries, utilities, and automobiles contribute carbon monoxide, sulfur, nitrogen oxides, hydrocarbons, and particulates to the urban atmosphere. While some of the region's worst offenders are U.S. cities such as Houston and Los Angeles, Canadian cities such as Toronto, Hamilton, and Edmonton also experience significant problems of air quality.

On a broader scale, North America is plagued by **acid rain**, industrially produced sulfur dioxide and nitrogen oxides in the atmosphere that damage forests, poison lakes, and kill fish. Many acid rain producers are located in the Midwest and southern Ontario, where industrial plants, power-generating facilities, and motor vehicles contribute atmospheric pollution. Prevailing winds transport the pollutants and deposit damaging acid rain and snow across the Ohio Valley, Appalachia, the northeastern United States, and eastern Canada (Figure 3.4).

Air pollution is also going global, freely drifting into Canada and the United States on prevailing winds. One estimate suggests that at least 30 percent of the region's ozone comes from beyond its borders. Both China and Mexico are major contributors of airborne pollutants.

## A Diverse Physical Setting

The North American landscape is dominated by vast interior lowlands bordered by more mountainous topography in the western portion of the region (see Figure 3.1). In the eastern United States, extensive coastal plains stretch from southern New York to Texas and include a sizable portion of the lower Mississippi Valley. The Atlantic coastline is complex, made up of drowned river valleys, bays, swamps, and low barrier islands (Figure 3.6). The nearby Piedmont, which is the transition zone between nearly flat lowlands and steep mountain slopes, consists of rolling hills and low mountains that are much older and less easily eroded than the lowlands. West and north of the Piedmont are the Appalachian Highlands, an internally complex zone of higher and rougher country reaching altitudes from 3,000 to 6,000 feet (915 to 1,829 meters). Far to the southwest, Missouri's Ozark Mountains and the Ouachita Plateau of northern Arkansas resemble portions of the southern Appalachians. Much of the North American interior is a vast lowland extending east–west from the Ohio River valley to the Great Plains, and north–south from west central Canada to the coastal lowlands near the Gulf of Mexico. Glacial forces, particularly north of the Ohio and Missouri rivers, have actively carved and reshaped the landscapes of this lowland zone.

In the West, mountain-building (including large earthquakes and volcanic eruptions), alpine glaciation, and erosion produce a regional topography quite unlike that of eastern North America. The Rocky Mountains reach more than 10,000 feet (3,048 meters) in height and stretch from Alaska's Brooks Range to northern New Mexico's Sangre de Cristo Mountains (Figure 3.7). West of the Rockies, the Colorado Plateau is characterized by highly colorful sedimentary rock eroded into spectacular buttes and mesas. Nevada's sparsely settled basin and range country features north–south-trending mountain ranges alternating with structural basins with no outlet to the sea. North America's western border is marked by the mountainous and rain-drenched coasts of southeast Alaska and British Columbia; the Coast Ranges of Washington, Oregon, and California; the lowlands of the Puget Sound (Washington), Willamette Valley (Oregon), and Central Valley (California); and the complex uplifts of the Cascade Range and Sierra Nevada.

## Patterns of Climate and Vegetation

North America's climates and vegetation are highly diverse, mainly due to the region's size, latitudinal range, and varied terrain (Figure 3.8). Much of North America south of the Great Lakes is characterized by a long growing season, 30 to 60 inches (76.2 to 152.4 centimeters) of precipitation annually, and a deciduous broadleaf forest (later cut down and replaced by crops).

**FIGURE 3.7 Rocky Mountains**    Montana's Glacier National Park reveals the characteristic signatures of alpine glaciation that are found in many portions of the Rocky Mountain region, in both the United States and Canada.

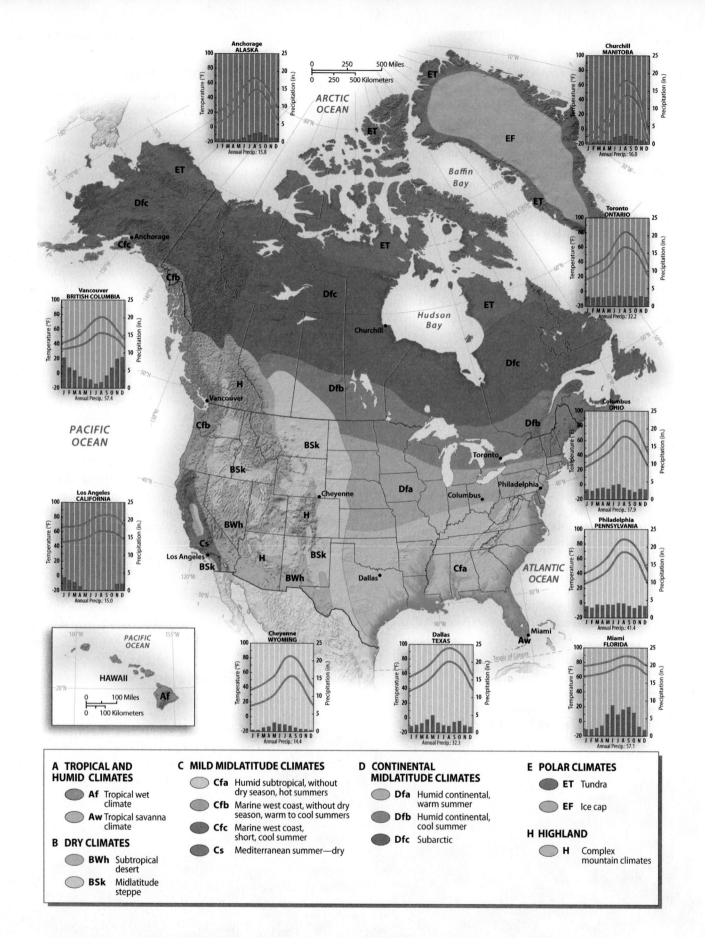

**FIGURE 3.8 Climate Map of North America** North American climates include everything from tropical savanna (Aw) to tundra (ET) environments. Most of the region's best farmland and densest settlements lie in the mild (C) or continental (D) midlatitude climate zones.

**A TROPICAL AND HUMID CLIMATES**

- **Af** Tropical wet climate
- **Aw** Tropical savanna climate

**B DRY CLIMATES**

- **BWh** Subtropical desert
- **BSk** Midlatitude steppe

**C MILD MIDLATITUDE CLIMATES**

- **Cfa** Humid subtropical, without dry season, hot summers
- **Cfb** Marine west coast, without dry season, warm to cool summers
- **Cfc** Marine west coast, short, cool summer
- **Cs** Mediterranean summer—dry

**D CONTINENTAL MIDLATITUDE CLIMATES**

- **Dfa** Humid continental, warm summer
- **Dfb** Humid continental, cool summer
- **Dfc** Subarctic

**E POLAR CLIMATES**

- **ET** Tundra
- **EF** Ice cap

**H HIGHLAND**

- **H** Complex mountain climates

From the Great Lakes north, the coniferous evergreen forest, or **boreal forest**, dominates the continental interior. Near Hudson Bay and across harsher northern tracts, trees give way to **tundra**, a mixture of low shrubs, grasses, and flowering herbs that grow briefly in the short growing seasons of the high latitudes. Drier continental climates found from west Texas to Alberta feature large seasonal ranges in temperature and unpredictable precipitation that averages between 10 and 30 inches (25.4 and 76.2 centimeters) annually. The soils of much of this region are fertile and originally supported **prairie** vegetation, dominated by tall grasslands in the East and by short grasses and scrub vegetation in the West. Western North American climates and vegetation are greatly complicated by the region's many mountain ranges. The Rocky Mountains and the intermontane interior experience the typical seasonal variations of the middle latitudes, but patterns of climate and vegetation are greatly modified by the effects of topography. Farther west, marine West

Coast climates dominate north of San Francisco, while a dry summer Mediterranean climate occurs across central and southern California.

## North America and Global Climate Change

North America's present-day climates and vegetation are merely a snapshot in time, reflecting dynamic natural processes as well as the growing accumulation of human impacts. Accelerating rates of human-influenced climate change appear destined to complicate the situation further. High-latitude and alpine environments are particularly vulnerable. Changes in arctic temperatures, sea ice, and sea levels have increased coastal erosion, affected migrating whale and polar bear populations, and stressed traditional ways of life for Eskimo and Inuit populations. In western North American mountains, expanding

## EXPLORING GLOBAL CONNECTIONS
# The Globalization of Firefighting

In the smoky summer of 2008, you might have heard a curious question on the California fire lines, something like *"owyergoinmateorright?"* ("How are you going, mate? All right?"). Dozens of highly trained firefighters and fire management specialists from Australia were in the California "outback" to fight hundreds of wildfires that had broken out after another dry year in the Golden State (Figure 3.1.1). This new global connection has increasingly made firefighting an international affair. In fact, in 2001, the United States signed a formal agreement with Australia for mutual assistance in fighting fires all the way from western Montana to the interior of New South Wales.

Since 2001, additional sharing of firefighting personnel and fire management strategies has brought together global expertise from the United States, Canada, Australia, and New Zealand. Similar command structures, physical requirements, and training procedures have made the exchanges easier, and the seasonal nature of the trans-hemispheric travel often works well. As a result, North American fire officials can venture south in December, January, and February, when local conditions are quiet. Beginning in 2003, many American fire specialists have found themselves in the Australian bush, establishing backfires in remote areas or directing aircraft to stubborn hot spots. Fires in 2009 devastated the Australian state of Victoria, killing more than 200 residents and destroying or damaging

more than 2,000 homes. The conflagration triggered an international appeal for help, and many North American fire experts from the U.S. Forest Service, the Bureau of Land Management, and the Bureau of Indian Affairs answered the call.

Similarly, Aussie expertise has been available in the Northern Hemisphere's dry summer months. Australian firefighters, aviation specialists, and burn-recovery experts are among the best in the world, and they are in increasing demand in the fire-prone American West. A group of local officials in California and western Nevada consulted with

their Australian counterparts, exploring more effective "stay and defend" approaches for rural residents facing the flames.

Global climate change is likely to further fan the connections between these southern and northern lands. Hotter and drier summers as well as larger, more numerous wildfires are likely for both the American West and much of the Australian interior. This new global link illustrates how sharing knowledge and expertise between nations, regions, and hemispheres can be a vital strategy in coping with a changing geography of natural and human-caused hazards in the decades to come.

**FIGURE 3.1.1 California Fires, 2008**

mountain pine beetle populations are rapidly infesting lodgepole and whitebark pine forests in the northern Rocky Mountains. A 2009 study of 76 stands of old-growth forest throughout western North America confirmed that tree mortality rates have dramatically risen in the past 15 years, and higher temperatures were cited as the primary culprit. Elsewhere, Mexican butterflies are breeding in Texas, armadillos (earlier natives of Texas) are marching into southern Illinois, and exotic fish species are migrating north into the Chesapeake Bay and offshore southern California.

The cumulative long-term consequences of climate change for North Americans are enormous. Basic redistributions of native plants, animals, and crops are already under way. For example, high-latitude distributions of tundra and permafrost environments may shift more than 300 miles (480 km) poleward by 2050, dramatically affecting wildlife populations and human settlements in these regions. For northern U.S. and Canadian farmers, longer growing seasons may open up new possibilities for agriculture. The Great Lakes region may become wetter by the end of the century. But both hotter and drier conditions are likely in the Southwest, portending more western wildfires and growing resource conflicts in that rapidly growing region of the United States (see "Exploring Global Connections: The Globalization of Firefighting").

Ironically, North America's vulnerability to the effects of global climate change is magnified by the economic affluence of the region's population. In the past 50 years, millions of North Americans have been attracted to precisely those scenic, high-amenity areas that are especially vulnerable to environmental disruptions. For example, resorts, hotels, and condominiums dot thousands of miles of North America's spectacular coastline. Events such as tropical hurricanes, coastal storms, heavy winter surf, and beach erosion thus present more hazards to human populations than they did a century or two ago. Future global warming will raise global sea levels further, increasing these hazards even more dramatically in coming centuries, particularly for low-lying zones in the arctic, along the East Coast, and in the Gulf of Mexico.

## Population and Settlement: Reshaping a Continental Landscape

The North American landscape is the product of four centuries of extraordinary human change. During that period, Europeans, Africans, and Asians arrived in the region, disrupted Native American peoples, and created new patterns of human settlement. Today, 340 million people live in the region, and they are some of the world's most affluent and highly mobile populations (Table 3.1).

## Modern Spatial and Demographic Patterns

Large metropolitan areas (including both central cities and suburbs) dominate North America's population geography, producing very uneven patterns of settlement across the region (Figure 3.9). The largest number of cities and the densest collection of rural settlements are found south of central Canada and east of the Great Plains. Within this broad region, Canada's "Main Street" corridor contains most of that nation's urban population, led by the cities of Toronto (5.4 million) and Montreal (3.7 million). **Megalopolis**, the largest settlement cluster in the United States, includes Washington, DC (5.3 million), Baltimore (2.7 million), Philadelphia (6.4 million), New York City (22 million), and Boston (4.5 million). Beyond these two national core areas, other sprawling urban centers cluster around the southern Great Lakes, in various parts of the South, and along the Pacific Coast.

## Occupying the Land

Europeans began occupying North America about 400 years ago. They were not settling an empty land. North America was populated for at least 12,000 years by peoples as culturally diverse as the Europeans who came to conquer them. Native Americans were broadly distributed across the region and adapted to its many natural environments. Cultural geographers estimate Native American populations in 1500 CE at 3.2 million for the continental United States and another 1.2 million for Canada, Alaska, Hawaii, and Greenland. In many areas, European diseases and disruptions reduced these Native American populations by more than 90 percent as contacts increased.

The first stage of a dramatic new settlement geography began with a series of European colonies, mostly in the coastal regions of eastern North America (Figure 3.10). Established between 1600 and 1750, these regionally distinct societies were anchored on the north by the French settlement of the St. Lawrence Valley and extended south along the Atlantic Coast, including separate English colonies. Scattered developments along the Gulf Coast and in the Southwest also appeared before 1750.

The second stage in the Europeanization of the North American landscape took place between 1750 and 1850, and it was highlighted by settlement of much of the better agricultural land in the eastern half of the continent. Following the American Revolution (1776) and a series of Indian conflicts, pioneers surged across the Appalachians. They found much of the Interior Lowlands region almost ideal for agricultural settlement. Much of southern Ontario, or Upper Canada, was also opened to widespread development after 1791.

### TABLE 3.1 POPULATION INDICATORS

| Country | Population (millions) 2009 | Population Density (per square kilometer) | Rate of Natural Increase (RNI) | Total Fertility Rate | Percent Urban | Percent <15 | Percent >65 | Net Migration (Rate per 1000) 2005–10* |
|---|---|---|---|---|---|---|---|---|
| Canada | 33.7 | 3 | 0.4 | 1.6 | 81 | 17 | 14 | 6.3 |
| United States | 306.8 | 32 | 0.6 | 2.1 | 79 | 20 | 13 | 3.3 |

Source: Population Reference Bureau, *World Population Data Sheet, 2009.*

*Net Migration Rate from the United Nations, Population Division, *World Population Prospects: The 2008 Revision Population Database.*

The third stage in North America's settlement expansion picked up speed after 1850 and continued until just after 1910. During this period, most of the region's remaining agricultural lands were settled by a mix of native-born and immigrant farmers. Farmers were challenged and sometimes defeated by drought, mountainous terrain, and short northern growing seasons. In the American West, settlers were attracted by opportunities in California, the Oregon country, Mormon Utah, and the Great Plains. In Canada, thousands occupied southern portions of Manitoba, Saskatchewan, and Alberta. Gold and silver discoveries led to initial development in areas such as Colorado, Montana, and British Columbia's Fraser Valley.

Incredibly, in a mere 160 years, much of the North American landscape was occupied as expanding populations sought new land to settle and as the global economy demanded resources to fuel its

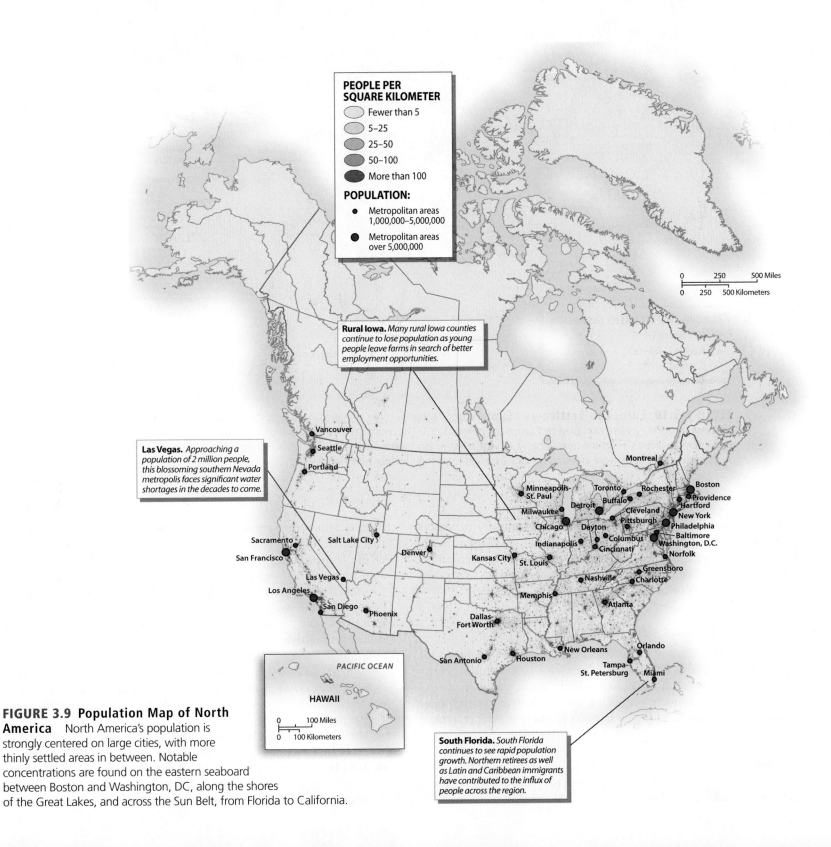

**FIGURE 3.9 Population Map of North America** North America's population is strongly centered on large cities, with more thinly settled areas in between. Notable concentrations are found on the eastern seaboard between Boston and Washington, DC, along the shores of the Great Lakes, and across the Sun Belt, from Florida to California.

**PEOPLE PER SQUARE KILOMETER**
- Fewer than 5
- 5–25
- 25–50
- 50–100
- More than 100

**POPULATION:**
- Metropolitan areas 1,000,000–5,000,000
- Metropolitan areas over 5,000,000

**Rural Iowa.** *Many rural Iowa counties continue to lose population as young people leave farms in search of better employment opportunities.*

**Las Vegas.** *Approaching a population of 2 million people, this blossoming southern Nevada metropolis faces significant water shortages in the decades to come.*

**South Florida.** *South Florida continues to see rapid population growth. Northern retirees as well as Latin and Caribbean immigrants have contributed to the influx of people across the region.*

PACIFIC OCEAN

HAWAII

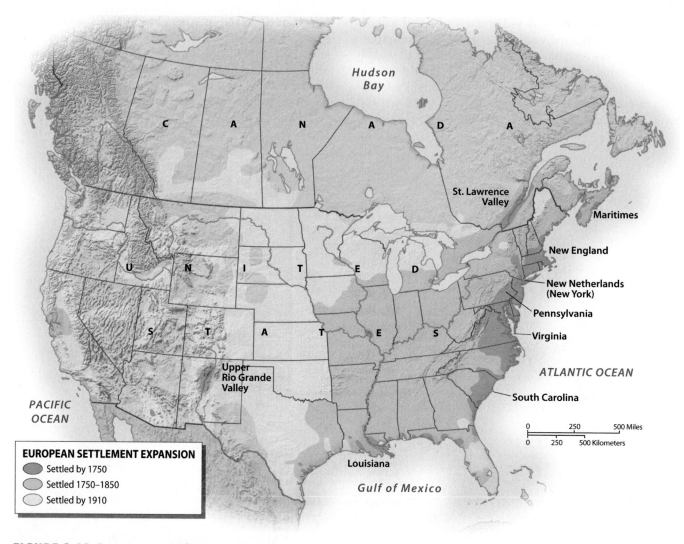

**FIGURE 3.10 European Settlement Expansion** Large portions of North America's East Coast and the St. Lawrence Valley were occupied by Europeans before 1750. The most remarkable surge of settlement, however, occurred during the next century, as vast areas of land were opened to European dominance and Native American populations were exterminated or expelled from their former homelands.

growth. It was one of the largest and most rapid transformations of the landscape in the history of the human population. This European-led advance forever reshaped North America in its own image, and in the process it also changed the larger globe in lasting ways by creating a "New World" destined to reshape the old.

## North Americans on the Move

From the mythic days of Davy Crockett and Daniel Boone to the 20th-century sojourns of John Steinbeck and Jack Kerouac, North Americans have been on the move. Indeed, almost one in every five Americans moves annually, suggesting that residents of the region are quite willing to change addresses in order to improve their income or their quality of life. Several trends dominate the picture.

**Westward-Moving Populations** The most persistent regional migration trend in North America has been the tendency for people to move west. Indeed, the dominant thrust in the past two centuries has

been to follow the setting sun, and many North Americans continue that pattern to the present. By 1990, more than half of the population of the United States lived west of the Mississippi River, a dramatic shift from colonial times.

Much of the extraordinary growth in the Mountain States between 2000 and 2007 was fueled by new job creation in high-technology industries and services, as well as by the region's scenic, recreational, and retirement amenities (Figure 3.11). Larger metropolitan areas such as Salt Lake City, Phoenix, and Denver saw substantial growth. In addition, many smaller nonmetropolitan centers, such as St. George, Utah; Coeur d'Alene, Idaho; and Kalispell, Montana, attracted a flood of new migrants, including many outward-bound Californians.

In the late 2000s, however, the economic slowdown hit the region hard. The contraction of the construction and leisure-time industries in settings such as Las Vegas and Phoenix slowed growth in these metropolitan areas to their lowest rates in decades. California was especially hard hit by the housing crisis, and many communities

**FIGURE 3.11 Intermountain West Growth, 2000–2005** Nevada (20.8 percent growth) and Arizona (15.8 percent growth) were the two fastest-growing states between 2000 and 2005. Large cities, smaller amenity-rich towns, and recreational areas all attracted migrants. Recent economic challenges, however, may slow the pace of future population gains. *(Modified from the U.S. Census,* State and Metropolitan Area Data Book: 2006*)*

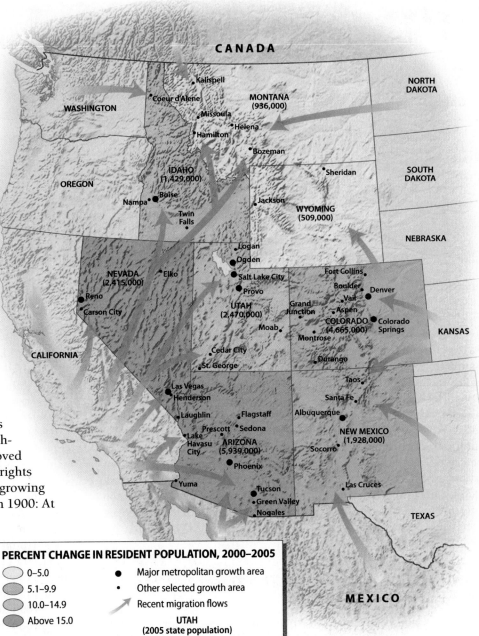

PERCENT CHANGE IN RESIDENT POPULATION, 2000–2005
- 0–5.0
- 5.1–9.9
- 10.0–14.9
- Above 15.0
- ● Major metropolitan growth area
- · Other selected growth area
- ➚ Recent migration flows

UTAH
(2005 state population)

saw accelerating out-migration as well as home values being cut in half between 2007 and 2009 (Figure 3.12).

**Black Exodus from the South** African Americans have generated distinctive patterns of interregional migration (Figure 3.13). Most blacks remained economically tied to the rural South after the Civil War. Conditions changed, however, in the early 20th century. Many African Americans migrated because of declining demands for labor in the agricultural South and growing industrial opportunities in the North and West. Migrants ended up in cities where jobs were located. Boston, New York, Philadelphia, Detroit, Chicago, Los Angeles, and Oakland became key destinations for southern blacks. Since 1970, however, more blacks have moved from North to South. Sun Belt jobs and federal civil rights guarantees now attract many northern urban blacks to growing southern cities. The net result is still a major change from 1900: At the beginning of the 20th century, more than 90 percent of African Americans lived in the South, while today only about half of the nation's 40 million blacks reside within the region.

**Rural-to-Urban Migration** Another continuing trend in North American migration has taken people from the country to the city. Two centuries ago, only 5 percent of North Americans lived in urban areas (cities of more than 2,500 people), whereas today more than 75 percent of the North American population is urban. Shifting economic opportunities account for much of the transformation: As mechanization on the farm reduced the demand for labor, many young people left for new employment opportunities in the city.

**Growth of the Sun Belt South** Particularly since the 1970s, southern states from the Carolinas to Texas have grown much more rapidly than states in the Northeast and Midwest. During the 1990s, Georgia, Florida, Texas, and North Carolina each grew by more than 20 percent. The South's expanding economy, modest living costs, adoption of air conditioning, attractive recreational opportunities, and appeal to snow-weary retirees have all contributed to its growth. Dallas–Fort Worth's bustling metropolitan area (6.0 million) is now larger than that of Boston (4.5 million) (Figure 3.14).

**FIGURE 3.12 Land of Tarnished Dreams** California's dynamic population is often on the move. This neighborhood in the Golden State reflects the combined effects of population turnover and a tough real estate market.

**FIGURE 3.13 The Block II (1972)**    Romare Bearden's collage of Harlem life is based on an actual neighborhood scene. Bearden's work as a black artist drew upon his own diverse American experiences in both the Northeast and South.

**Patterns of Counterurbanization**    During the 1970s, certain nonmetropolitan areas in North America saw significant population gains, including many rural settings that had previously lost population. Selectively, that pattern of **counterurbanization**, in which people leave large cities and move to smaller towns and rural areas, continues today. Migrants find employment in affordable smaller cities and rural settings that are pleasant to live in and often removed from the perceived problems of urban America. Recent nonmetropolitan population growth has exceeded metropolitan growth in most western states. Other smaller communities outside the West, such as Mason City, Iowa; Mankato, Minnesota; and Traverse City, Michigan, also are seen as desirable destinations for migrants interested in leaving behind the problems of metropolitan life.

## Settlement Geographies: The Decentralized Metropolis

Settlement landscapes of North American cities are characterized by **urban decentralization**, in which metropolitan areas sprawl in all directions, and suburbs take on many of the characteristics of traditional downtowns. Although both Canadian and U.S. cities have experienced decentralization, the impact has been particularly profound in the United States, where inner-city problems, poor public transportation, widespread automobile ownership, and fewer regional-scale planning initiatives have encouraged middle-class urban residents to move beyond the central city.

**Historical Evolution of the City in the United States**    Changing transportation technologies decisively shaped the evolution of the city in the United States (Figure 3.15). The pedestrian/horsecar city (pre-1888) was compact, essentially limiting urban growth to a 3- or 4-mile-diameter ring around downtown. The invention of the electric trolley in 1888 expanded the urbanized landscape farther into new "streetcar suburbs," often 5 or 10 miles from the city center. A star-shaped urban pattern resulted, with growth extending outward along and near the streetcar lines. The biggest technological revolution came after 1920, with the widespread adoption of the automobile. The automobile city (1920–1945) promoted the growth of middle-class suburbs beyond the reach of the streetcar and added even more distant settlement in the surrounding countryside. Following World War II, growth in the outer city (1945 to the present) promoted more decentralized settlement along commuter routes as built-up areas appeared 40 to 60 miles from downtown.

Urban decentralization also reconfigured land-use patterns in the city, producing metropolitan areas today that are strikingly different

**FIGURE 3.14 Downtown Dallas, Texas**    Sun Belt cities such as Dallas experienced rapid job creation between 1970 and 2005. Healthy growth in office space, specialty retailing, and entertainment districts fueled the expansion of downtown Dallas and reshaped the look of the central-city skyline.

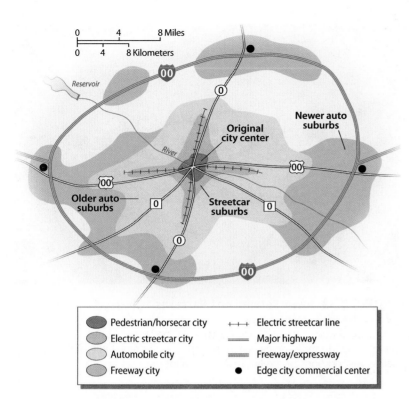

**FIGURE 3.16 Tysons Corner, Virginia**    North America's edge-city landscape is nicely illustrated by Tysons Corner, Virginia. Far from a traditional metropolitan downtown, this sprawling complex of suburban offices and commercial activities reveals how and where many North Americans will live their lives in the 21st century.

**FIGURE 3.15 Growth of the American City**    Many U.S. cities became increasingly decentralized as they moved through the pedestrian/horsecar, electric streetcar, automobile, and freeway eras. Each era left a distinctive mark on metropolitan America, including the recent growth of edge cities on the urban periphery.

from their counterparts of the early 20th century. In the city of the early 20th century, idealized in the **concentric zone model**, urban land uses were neatly organized in rings around a highly focused central business district (CBD) that contained much of the city's retailing and office functions. Residential districts beyond the CBD were added as the city expanded, with higher-income groups seeking more desirable locations on the outside edge of the urbanized area.

Today's **urban realms model** recognizes these new suburbs characterized by a mix of peripheral retailing, industrial parks, office complexes, and entertainment facilities. These areas of activity, often called "edge cities," have fewer functional connections with the central city than they have with other suburban centers. For most residents of an edge city, jobs, friends, and entertainment are located in other suburbs rather than in the old downtown. Tysons Corner, Virginia, located west of Washington, DC, is an excellent example of the edge-city landscape on the expanding periphery of a North American metropolis (Figure 3.16).

**The Consequences of Sprawl**    As suburbanization increased in the 1960s and 1970s, many inner cities, especially in the Northeast and Midwest, suffered absolute losses in population, increased levels of crime and social disruption, and a shrinking tax base. Poverty rates average almost three times those of nearby suburbs. Unemployment rates remain above the national average. Central cities in the United States also remain places of racial tension, the product of decades of discrimination, segregation, and poverty.

Even with these challenges, inner-city landscapes are also enjoying selective improvement. Referred to as **gentrification**, the process

involves the displacement of lower-income residents of central-city neighborhoods with higher-income residents, the improvement of deteriorated inner-city landscapes, and the construction of new shopping complexes, entertainment attractions, or convention centers in selected downtown locations. In some renovated neighborhoods, such as Fall Creek Place in Indianapolis, a pleasing mix of older homes and new traditional-style houses creates a new, more livable residential landscape (Figure 3.17). In Pennsylvania, Pittsburgh's urban renaissance offers an affordable housing market, an older highly skilled workforce, and mixed-use neighborhoods such as the SouthSide Works development, a 34-acre assortment of

**FIGURE 3.17 Fall Creek Place, Indianapolis**    This renovated inner-city neighborhood features a pleasing mix of century-old homes and new traditional-style houses designed to offer residents affordable housing.

residences, offices, and stores created on the site of an old steel plant (Figure 3.18).

The suburbs are also changing. Construction of new corporate office centers, fashion malls, and industrial facilities has created true "suburban downtowns" in suburbs that are no longer simply bedroom communities for central-city workers. Indeed, such localities have been growing players in the continent's globalization process. For example, many of North America's key internationally connected corporate offices (IBM, Microsoft), industrial facilities (Boeing, Oracle), and entertainment complexes (Disneyland, Walt Disney World, and the Las Vegas Strip) are now in such settings, and they are intimately tied to global information, technology, capital, and migration flows.

## Settlement Geographies: Rural North America

Rural North American landscapes trace their origins to early European settlement. Over time, these immigrants from Europe showed a clear preference for a dispersed rural settlement pattern as they created new farms on the North American landscape. In portions of the United States settled after 1785, the federal government surveyed and sold much of the rural landscape. Surveys were organized around the simple, rectangular pattern of the federal government's township-and-range survey system, which offered a convenient method of dividing and selling the public domain in 6-mile-square townships (Figure 3.19). Canada developed a similar system of regular surveys that stamped much of southern Ontario and the western provinces with a strikingly rectilinear character.

Commercial farming and technological changes further transformed the settlement landscape. Railroads opened corridors of development, provided access to markets for commercial crops, and helped to establish towns. By 1900, several transcontinental lines spanned North America, radically transforming the farm economy and the pace of rural life. After 1920, however, even greater change accompa-

**FIGURE 3.19 Iowa Settlement Patterns**   The regular rectangular look of this Iowa town and the nearby rural setting are common cultural landscape features across North America. In the United States, the township-and-range survey system stamped such predictable patterns across vast portions of the North American interior.

nied the arrival of the automobile, farm mechanization, and better rural road networks. The need for farm labor declined with mechanization, and many smaller market centers became unnecessary as farmers equipped with automobiles and trucks could travel farther and faster to larger, more diverse towns.

Today, many areas of rural North America face population declines, as they adjust to the changing conditions of modern agriculture. Both U.S. and Canadian farm populations fell by more than two-thirds during the last half of the 20th century. Typically, a smaller number of farms (but larger in acreage) dot the modern rural scene, and many young people leave the land to obtain employment elsewhere. The visual record of abandonment offers a painful reminder of the economic and social adjustments that come from population losses. Weed-choked driveways, empty farmhouses, roofless barns, and the empty marquees of small-town movie houses tell the story more powerfully than any census or government report.

Elsewhere, rural settings show signs of growth. Some places begin to experience the effects of expanding edge cities. Other growing rural settings lie beyond direct metropolitan influence but are seeing new populations that seek amenity-rich environments, removed from city pressures. These trends are shaping the settlement landscape from British Columbia's Vancouver Island to Michigan's Upper Peninsula. Newly subdivided land, numerous real estate offices, and the presence of espresso bars are all signs of growth in such surroundings.

**FIGURE 3.18 Pittsburgh's SouthSide Works Neighborhood**   New investments have transformed Pittsburgh's SouthSide Works Neighborhood into an upscale office, shopping, and entertainment district.

## Cultural Coherence and Diversity: Shifting Patterns of Pluralism

North America's cultural geography exerts global influence. At the same time, it is internally diverse. On one hand, history and technology have produced a contemporary North American cultural force that is second to none in the world. Yet North America is also a collection of different peoples who retain part of their traditional cultural identities. In fact, North Americans celebrate their varied roots and acknowledge the region's multicultural character.

## The Roots of a Cultural Identity

Powerful historical forces formed a common dominant culture within North America. While both the United States (1776) and Canada (1867) became independent from Great Britain, the two countries remained closely tied to their Anglo roots. Key Anglo legal and social institutions solidified the common set of core values that many North Americans shared with Britain and, eventually, with one another. Traditional Anglo beliefs emphasized representative government, separation of church and state, liberal individualism, privacy, pragmatism, and social mobility. From those shared foundations, particularly within the United States, consumer culture blossomed after 1920, producing a common set of experiences oriented around convenience, consumption, and the mass media.

But North America's cultural unity coexists with pluralism, the persistence and assertion of distinctive cultural identities. Closely related is the concept of **ethnicity**, in which a group of people with a common background and history identify with one another, often as a minority group within a larger society. For Canada, the early and enduring French colonization of Quebec complicates its modern cultural geography. Canadians face the challenge of creating a truly bicultural society where issues of language and political representation are central concerns. Within the United States, given its unique immigration history, a greater diversity of ethnic groups exists, and differences in cultural geography are often found on both local and regional scales.

## Peopling North America

North America is a region of immigrants. Quite literally, global-scale migrations to the region created its unique cultural geography. Decisively displacing Native Americans in most portions of the region, immigrant populations created a new cultural geography of ethnic groups, languages, and religions. Early migrants had considerable cultural influence, even though their numbers were very small. Over time, immigrant groups and their changing destinations produced a varied cultural geography across North America. Also varying between groups was the pace and degree of **cultural assimilation**, the process in which immigrants were absorbed by the larger host society.

**Migration to the United States**   In the United States, variations in the number and source regions of migrants produced five distinctive chapters in the country's history (Figure 3.20). In Phase 1 (prior to 1820), English and African influences dominated. Slaves, mostly from

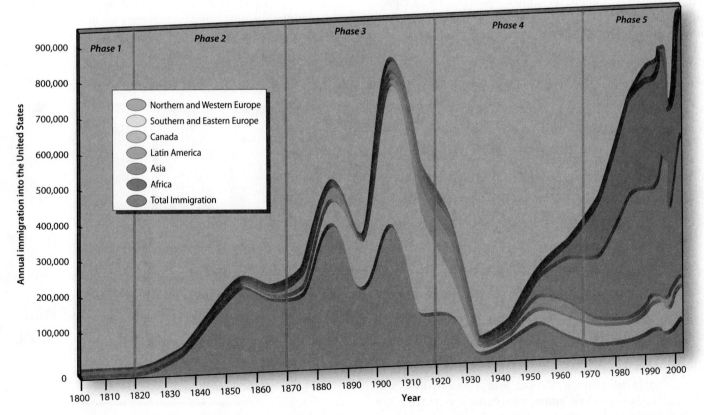

**FIGURE 3.20  U.S. Immigration, by Year and Group**   Annual immigration rates peaked around 1900, declined in the early 20th century, and then surged again, particularly beginning in 1970. The source areas of these migrants have also shifted. Note the decreased role Europeans currently play versus the growing importance of Asians and Latin Americans. *(Modified from Rubenstein, 2005,* An Introduction to Human Geography, *Upper Saddle River, NJ: Prentice Hall)*

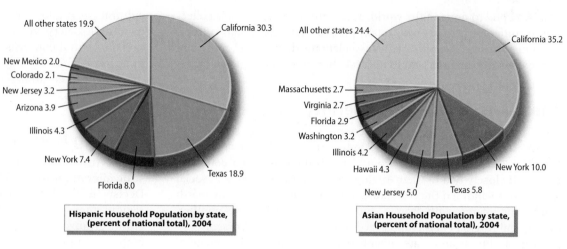

**FIGURE 3.21 Distribution of U.S. Hispanic and Asian Household Populations, by State, 2004** California and Texas claim about half of the nation's Hispanic population, but a growing number of Hispanics are locating elsewhere. California alone is still home to more than one-third of the country's Asian population. *(U.S. Census, American Community Survey Reports, issued February 2007)*

West Africa, contributed additional cultural influences in the South. Northwest Europe served as the main source region of immigrants between 1820 and 1870 (Phase 2). The emphasis, however, shifted away from English migrants. Instead, Irish and Germans dominated the flow and provided more cultural variety.

As Figure 3.20 shows, immigration reached a much higher peak around 1900, when almost 1 million foreigners entered the United States *annually*. During Phase 3 (1870–1920), the majority of immigrants were southern and eastern Europeans. Political strife and poor economies in Europe existed during this period. News of available land and expanding industrialization in the United States offered an escape from such difficult conditions. By 1910, almost 14 percent of the nation was foreign born. Very few of these immigrants, however, targeted the job-poor U.S. South, creating a cultural divergence that still exists.

Between 1920 and 1970 (Phase 4), more immigrants came to the United States from neighboring Canada and Latin America, but overall totals fell sharply, a function of more restrictive federal immigration policies (the Quota Act of 1921 and the National Origins Act of 1924), the Great Depression, and the disruption caused by World War II. After 1970 (Phase 5), the number of immigrants increased, particularly from Latin America and Asia, and total numbers matched those of the early 20th century. Since 2008, however, the pace of illegal immigration has slowed appreciably, mostly because of declining job opportunities in the United States as well an increased number of U.S. border patrol agents. Today, about 12 million unauthorized immigrants live in the United States.

The nation's Hispanic population continues to grow. An estimated 12 million Mexican-born residents (more than 10 percent of Mexico's population) now live in the United States. Mexicans make up about 64 percent of the nation's Hispanic population, but those numbers are changing. In the next 25 years, most of the projected increase in the U.S. Hispanic population will be fueled by births within the country versus new immigrants. While half of U.S. Hispanics live in California or Texas, they are increasingly moving to other areas (Figure 3.21). States such as Wisconsin, Georgia, Kansas, and Arkansas have witnessed dramatic increases in their immigrant Hispanic populations, a trend likely to continue in the early 21st century.

In percentage terms, migrants from Asia constitute the fastest-growing immigrant group, and various Asian ethnicities, both native and foreign born, account for 5 percent of the U.S. population. Chinese is now the third most commonly spoken language in the United

States (behind English and Spanish). California remains a key entry point for new migrants and is home to more than one-third of the nation's overall Asian population (Figure 3.21). Diverse cultures make up the nation's Asian population. The largest groups represented include Chinese (23 percent), Asian Indian (19 percent), Filipino (18 percent), Vietnamese (11 percent), and Korean (10 percent).

The future cultural geography of the United States will be dramatically redefined by these recent immigration patterns. The increasing diversity of the country is an expression of the globalization process and the powerful pull of North America's economy and political stability. By 2070, Asians may total more than 10 percent of the U.S. population, and almost one American in three will be Hispanic. Indeed, it is likely that the U.S. non-Hispanic white population will achieve minority status by that date (Figure 3.22).

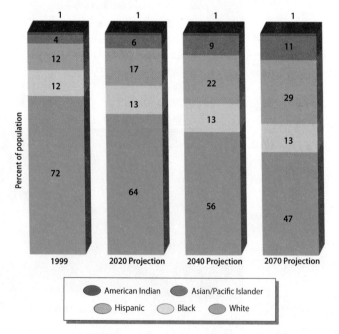

**FIGURE 3.22 Projected U.S. Ethnic Composition, 1999 to 2070** By late in the 21st century, almost one in three Americans will be Hispanic, and non-Hispanic whites will achieve minority status amid an increasingly diverse U.S. population. *(Modified from the U.S. Census, Census of Population, National Projections, 2020–2070)*

**The Canadian Pattern**  The peopling of Canada included early French arrivals that concentrated in the St. Lawrence Valley. After 1765, many migrants came from Britain, Ireland, and the United States. Canada then experienced the same surge and reorientation in migration flows seen in the United States around 1900. Between 1900 and 1920, more than 3 million foreigners ventured to Canada, an immigration rate far higher than for the United States, given Canada's much smaller population. Eastern Europeans, Italians, Ukrainians, and Russians were the most important nationalities in these later movements. Today, about 60 percent of Canada's recent immigrants are Asians, and its 17 percent foreign-born population is among the highest in the developed world. In Toronto, the city's 44 percent foreign-born population reveals a slight bias toward European backgrounds, while the west-coast metropolis of Vancouver (38 percent foreign born) is clearly dominated by Asian populations, particularly Chinese.

## Culture and Place in North America

Cultural and ethnic identity is often strongly tied to place. North America's cultural diversity is expressed geographically in two ways. First, people with similar backgrounds congregate near one another and derive meaning from the territories they occupy together. Second, these distinctive cultures leave their mark on the everyday scene: The landscape is filled with the artifacts, habits, language, and values of different groups. Boston's Italian North End simply looks and smells different from nearby Chinatown.

**Persisting Cultural Homelands**  French-Canadian Quebec is an excellent example of a cultural homeland: It is a culturally distinctive nucleus of settlement in a well-defined geographic area, and its ethnicity has survived over time, stamping the cultural landscape with an enduring personality (Figure 3.23). More than 80 percent

**FIGURE 3.23 Selected Cultural Regions of North America**  From northern Canada's Nunavut to the Southwest's Hispanic Borderlands, different North American cultural groups strongly identify with traditional local and regional homelands. *(Modified from Jordan, Domosh, and Rowntree, 1998,* The Human Mosaic, *Upper Saddle River, NJ: Prentice Hall)*

of the population of Quebec speaks French, and language remains the "cultural glue" that holds the homeland together. Indeed, policies adopted after 1976 strengthened the French language within the province by requiring French instruction in the schools and national bilingual programming by the Canadian Broadcasting Corporation (CBC). Many Quebeçois feel that the greatest cultural threat may come not from Anglo-Canadians but rather from recent immigrants to the province. Southern Europeans or Asians in Montreal, for example, show little desire to learn French, preferring instead to put their children in English-speaking private schools.

Another well-defined cultural homeland is the Hispanic Borderlands (Figure 3.23). It is similar in size to French-Canadian Quebec, significantly larger in total population, but more diffuse in its cultural and political expression. Historical roots of the homeland are deep, extending back to the 16th century, when Spaniards opened the region to the European world. A rich legacy of Spanish place names, earth-toned Catholic churches, and traditional Hispanic settlements dot the rolling highlands of northern New Mexico and southern Colorado. From California to Texas, other historical sites and place names also reflect the Hispanic legacy.

Unlike Quebec, however, large 20th-century migrations from Latin America brought an entirely new wave of Hispanic settlement to the Southwest. About 45 million Hispanics now live in the United States, with more than half in California, Texas, New Mexico, and Arizona combined. Indeed, by 2015, Hispanics will likely outnumber non-Hispanic whites in California. Cities such as San Antonio, Phoenix, and Los Angeles play leading roles in revealing the Hispanic presence in the Southwest. New York City, Chicago, and Cuban South Florida serve as key points of Hispanic influence beyond the homeland.

African Americans also retain a cultural homeland, but it has become less important because of out-migration (Figure 3.23). Reaching from coastal Virginia and North Carolina to east Texas, a zone of enduring African-American population remains a legacy of the Cotton South, when a vast majority of U.S. blacks resided as slaves in the region. Dozens of rural counties in the region still have large black majorities. More broadly, the South is home to many black folk traditions, including music such as black spirituals and the blues, which have now become popular far beyond their rural origins.

A second rural homeland in the South is Acadiana, a zone of persisting Cajun culture in southwestern Louisiana (Figure 3.23). This homeland was created in the 18th century, when French settlers were expelled from eastern Canada (an area known as "Acadia") and relocated to Louisiana. Nationally popularized today through their food and music, the Cajuns still have a long-lasting attachment to the bayous and swamps of southern Louisiana.

Native American populations are also strongly tied to their homelands. About 5.2 million Indians, Inuits, and Aleuts live in North America, and they claim allegiance to more than 1,100 tribal bands. Particularly in the American West and the Canadian and Alaskan North, native peoples control sizable reservations, including the 16-million-acre (6.5-million-hectare) Navajo Reservation in the Southwest, as well as self-governing Nunavut in the Canadian North (Figure 3.23). Although these homelands preserve traditional ties to the land, they have also been settings for pervasive poverty and increasing cultural tensions (Figure 3.24). Within the United States,

**FIGURE 3.24 Native American Poverty**   Navajo youngsters enjoy a game of basketball on the reservation. Poor housing, low incomes, and persistent unemployment plague many Native American settings across the rural West.

many Native American groups have taken advantage of the special legal status of their reservations and have built gambling casinos and tourist facilities that bring in much-needed capital but also challenge traditional lifeways.

**A Mosaic of Ethnic Neighborhoods**   North America's cultural mosaic is characterized by smaller-scale ethnic signatures that shape both rural and urban landscapes. When much of the agricultural interior was settled, immigrants often established close-knit communities. Among others, German, Scandinavian, Slavic, Dutch, and Finnish neighborhoods took shape, held together by common origins, languages, and religions. Although many of these ties weakened over time, rural landscapes of Wisconsin, Minnesota, the Dakotas, and the Canadian prairies still display these cultural imprints. Folk architecture, distinctive settlement patterns, ethnic place names, and the simple elegance of rural churches survive as signatures of cultural diversity on the visible scene of rural North America.

Ethnic neighborhoods are also a part of the urban landscape and reflect both global-scale and internal North American migration patterns. The ethnic geography of Los Angeles is an example of both economic and cultural forces at work (Figure 3.25). Because most of its economic expansion took place during the 20th century, the city's ethnic patterns reflect the movements of more recent migrants. African-American communities on the city's south side (Compton and Inglewood) represent the legacy of black population movements out of the South. Hispanic (East Los Angeles) and Asian (Alhambra and Monterey Park) neighborhoods are a reminder that about 40 percent of the city's population is foreign born.

Particularly in the United States, ethnic concentrations of non-white populations increased in many cities during the 20th century as whites left for the perceived safety of the suburbs. In terms of central-city population, African Americans make up more than 60 percent of Atlanta, while Los Angeles is now more than 40 percent Hispanic. Often these central-city ethnic neighborhoods are further isolated economically from most urban residents by high levels of poverty and unemployment. Ethnic concentrations are also growing in the suburbs of some U.S. cities: Southern California's Monterey Park has been called the "first suburban Chinatown," and growing numbers of middle-class African Americans and Hispanics shape suburban neighborhoods in metropolitan settings such as Atlanta and San Antonio.

## The Globalization of American Culture

Simply put, North America's cultural geography is becoming more global at the same time that global cultures are becoming more North American (influenced particularly by the United States). But the process of cultural globalization is becoming more complex. No longer can we think of simple flows of foreign influences into North America or of U.S. cultural dominance invading every traditional corner of the globe. In the 21st century, the story of cultural globalization will increasingly feature a mix of influences that flow in many directions at once and that feature new hybrid cultural creations.

### North Americans: Living Globally

More than ever before, North Americans in their everyday lives are exposed to people from beyond the region. With 43 million foreign-born migrants living across the region, diverse global influences are free to mingle in new ways. In 2006, the United States recorded more than 50 million international visitors arriving in the country. At U.S. colleges and universities, several hundred thousand international students also add a global flavor to the curriculum.

Globalization presents cultural challenges for North Americans. In the United States, one key issue revolves around the English language, which some have described as the "social glue" that holds the nation together. Since 1980, the continuing flood of non-English-speaking immigrants into the country has sharpened the debate over the role English should play in U.S. culture. Many in the United States argue that English should be the country's only officially recognized language, while some immigrants suggest that they need to maintain their traditional languages both to function within their ethnic communities and to preserve their cultural heritage. The evidence suggests that North America's immigrants are learning English more rapidly than ever before, seeing it as a powerful tool to accelerate their economic opportunities, both in the United States and Canada. A 2006 poll of Hispanic residents revealed that almost 60 percent of Hispanics agreed that "immigrants have to speak English to say that they are part of American society." The growing popularity of **Spanglish**, a hybrid combination of English and Spanish spoken by Hispanic Americans, also illustrates the complexities of North American globalization. Spanglish includes interesting hybrids such as *chatear*, which means to have an online conversation.

**FIGURE 3.25 Ethnicity in Los Angeles** Economic opportunities have attracted a wide variety of migrants to North American cities, producing an ethnic mixture of distinctive neighborhoods and communities. In Los Angeles, several cycles of economic expansion have attracted a diverse collection of residents from around the globe. *(Reprinted from Rubenstein, 2008,* An Introduction to Human Geography, *Upper Saddle River, NJ: Prentice Hall)*

North Americans are going global in other ways. By 2010, the vast majority of Americans and Canadians had Internet access, opening the door for far-reaching journeys in cyberspace. North Americans also travel much more widely than ever before. Within North America, the popularity of ethnic restaurants has peppered the realm with a bewildering variety of Cuban, Ethiopian, Basque, and Pakistani eateries. The growing affinity for foreign beverages mirrors the pattern; imported beer sales in the United States are expanding much more rapidly than domestic sales. American consumers now drink more than 800 million gallons of imported beer annually (Figure 3.26). Americans also have rapidly Europeanized their coffee-drinking habits. In fashion, *Gucci*, *Armani*, and *Benetton* are household words for millions who keep their eyes on European styles. The beat of German techno bands, Gaelic instrumentals, and Latin rhythms has also become an increasingly seamless part of daily life in the region. Indeed, from acupuncture and massage therapy to soccer and New Age religions, North Americans are tirelessly borrowing, adapting, and absorbing the larger world around them.

### The Global Diffusion of U.S. Culture

In parallel fashion, U.S. culture has forever changed the lives of billions of people beyond the region. Although the economic and military power of the United States was notable by 1900, it was not until after World War II that the country's popular culture reshaped global human geographies in fundamental ways. The Marshall Plan and Peace Corps initiatives exemplified the growing presence of the United States on the world stage, even as European colonialism waned. Rapid improvements in global transportation and information technologies, many of them engineered in the United States, also brought the world more surely under the region's spell. Perhaps most critical was the marriage between growing global demands for consumer goods and the rise of the multinational corporation, which was superbly

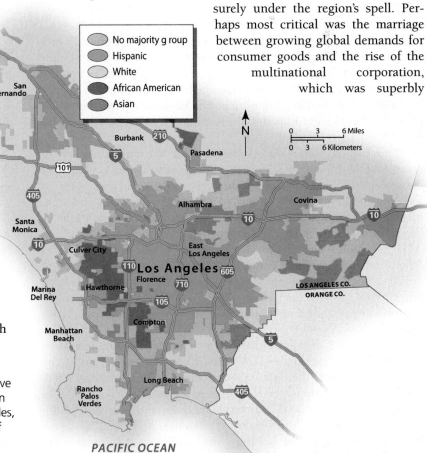

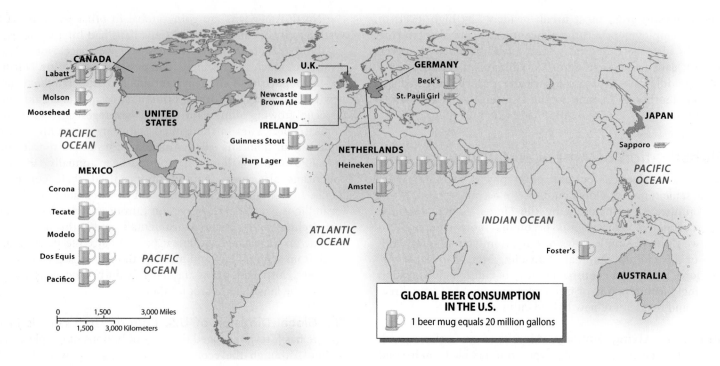

**FIGURE 3.26 Annual Beer Imports to the United States, 2002** Whether they are aware of it or not, North Americans are increasingly eating and drinking globally. Rising beer imports, including many more expensive foreign brands, exemplify the pattern. The nation's beer drinkers know no bounds to their thirsts, preferring a wide variety of Asian, Australian, European, and Latin American producers. *(Data from* Modern Brewery Age, *2003)*

structured to meet and cultivate those needs. The results of these connections are not a simple Americanization of traditional cultures or a single, synthesized global culture shaped in the U.S. image. Still, millions of people, particularly the young, are strongly attracted by the North American emphasis on individualism, consumption, youth, and mobility. The wide popularity of English-language teaching programs in places from China to Cuba is testimony to the cultural power of the United States.

The United States shapes the popular cultural landscape of every corner of the globe. Global corporate advertising, distribution networks, and mass consumption bring Cokes and Big Macs to Moscow and Beijing, golf courses to Thai jungles, Mickey and Minnie Mouse to Tokyo and Paris, and Avon cosmetics to millions of beauty-conscious Chinese. Western-style business suits have become the professional uniform of choice, while T-shirts and jeans offer standardized global comfort on days away from work. In the built landscape, central-city skylines become indistinguishable from one another, suburban apartment blocks take on a global sameness, and one airport hotel looks the same as another eight time zones away.

But U.S. cultural control has not gone unchallenged, illustrating the varied consequences of globalization. As worldwide use of the Internet has grown, the online dominance of English-speaking users has dramatically declined globally from more than 71 percent in 1998 to only 22 percent in 2005. Not surprisingly, given the rapid diffusion of the Internet and China's growing influence, Mandarin may have surpassed English as the leading global language of Internet users

by 2010. Active resistance to U.S. cultural influence is also notable. For example, Canadian government agencies routinely chastise their radio, television, and film industries for letting in too much U.S. cultural influence. The French also criticize U.S. dominance in such media as the Internet. Elsewhere, Iran has banned satellite dishes and many U.S. films, although illegal copies of top box-office hits often find their way through national borders.

# Geopolitical Framework: Patterns of Dominance and Division

North America is home to two of the world's largest states. The creation of these states, however, was neither simple nor preordained but rather the result of historical processes that might have created quite a different North American map. Once established, these two states have coexisted in a close relationship of mutual economic and political interdependence.

## Creating Political Space

The United States and Canada have very different political roots. The United States broke cleanly and violently from Great Britain. Canada, in contrast, was a country of convenience, born from a peaceful separation from Britain and then assembled as a collection of distinctive regional societies that only gradually acknowledged their common political destiny.

In the case of the future United States, Europe imposed its own political boundaries across the region. The 13 English colonies, sensing their common destiny after 1750, finally united two decades later in the Revolutionary War. By the 1790s, the young nation's territorial claims had reached the Mississippi River. Soon the Louisiana Purchase (1803) nearly doubled the national domain. By mid-century, Texas had been annexed, treaties with Britain had secured the Pacific Northwest, and an aggressive war with Mexico had captured much of the Southwest. The acquisition of Alaska (1867) and Hawaii (1898) rounded out what became the 50 states.

Canada was created under quite different circumstances. After the American Revolution, England's remaining territories in the region were controlled by administrators in British North America. In 1867 the British North America Act united the provinces of Ontario, Quebec, Nova Scotia, and New Brunswick in an independent Canadian Confederation. Within a decade, the Northwest Territories (1870), Manitoba (1870), British Columbia (1871), and Prince Edward Island (1873) joined Canada, and the continental dimensions of the country took shape. Soon the Yukon Territory (1898) separated from the Northwest Territories; Alberta and Saskatchewan gained provincial status (1905); and Manitoba, Ontario, and Quebec were enlarged (1912) north to Hudson Bay. Newfoundland finally joined in 1949. The addition of Nunavut Territory (1999) represents the latest change in Canada's political geography.

## Continental Neighbors

Geopolitical relationships between Canada and the United States have always been close: Their common 5,525-mile (8,900-kilometer) boundary requires both nations to pay close attention to one another. During the 20th century, the two countries have lived largely in political harmony with one another. In 1909, the Boundary Waters Treaty created the International Joint Commission, an early step in the common regulation of cross-boundary issues involving water resources, transportation, and environmental quality. The St. Lawrence Seaway (1959) opened the Great Lakes region to better global trade connections. The two nations also joined in cleaning up Great Lakes pollution and in making plans to reduce acid rain in eastern North America.

It has been in the area of trade relations that the close political ties between these neighbors have mattered most. The United States receives almost 90 percent of Canada's exports and supplies more than two-thirds of its imports. Conversely, Canada is the most important trading partner of the United States, accounting for roughly 20 percent of its exports and imports. One landmark agreement reached in 1989 was the signing of the bilateral (two-way) Free Trade Agreement (FTA). Five years later, the larger **North American Free Trade Agreement (NAFTA)** extended the alliance to Mexico. Paralleling the success of the European Union (EU), NAFTA has forged the world's largest trading bloc, including more than 400 million consumers and a huge free-trade zone that stretches from beyond the Arctic Circle to Latin America.

Political conflicts still divide the two countries (Figure 3.27). Environmental issues produce cross-border tensions, especially when environmental degradation in one nation affects the other. For example, Montana's North Flathead River flows out of British Columbia, where Canadian logging and mining operations periodically threaten fisheries and recreational lands south of the border. Agricultural and natural resource competition also cause occasional controversy between the two neighbors. The appearance of mad cow disease in Canadian livestock curtailed exports to the United States and elsewhere. Furthermore, when the disease appeared in U.S. cattle in 2003, Canadian sources were suspected, raising tensions between the two nations. Problems recently developed when Canadian wheat and potato growers were accused of dumping their products into U.S. markets, thus depressing prices and profits for U.S. farmers. Similar issues have arisen in the logging industry, although a 2006 agreement signed between the two countries has lessened tensions over that issue.

## The Legacy of Federalism

The United States and Canada are **federal states** in that both nations allocate considerable political power to units of government beneath the national level. Other nations, such as France, have traditionally been **unitary states**, in which power is centralized at the national level. Federalism leaves many political decisions to local and regional governments and often allows distinctive cultural and political groups to be recognized within a country. The U.S. Constitution (1787) limited centralized authority, giving all unspecified powers to the states or the people. In contrast, the Canadian Constitution (1867), which created a federal state under a parliamentary system, reserved most powers to central authorities. Ironically, the evolution of the United States produced an increasingly powerful central government, while Canada's geopolitical balance of power shifted toward more provincial autonomy and a relatively weak national government.

**Quebec's Challenge** The political status of Quebec remains a major issue in Canada (see Figure 3.27). Economic disparities between the Anglo and French populations have reinforced cultural differences between the two groups, with the French Canadians often suffering when compared with their wealthier neighbors in Ontario. Beginning in the 1960s, a separatist political party in Quebec (the Parti Quebecois) increasingly voiced French-Canadian concerns. When the party won provincial elections in 1976, it declared French the official language of Quebec. Formal provincial votes over the question of remaining within Canada were held in 1980 and 1995. Both measures failed. Since then, support for separation has ebbed in favor of a more modest strategy of increased "autonomy" within Canada. In elections held in 2007, for example, the Parti Quebecois ran a poor third in Quebec, suggesting that its separatist rhetoric is no longer as popular within the province.

**Native Peoples and National Politics** Another challenge to federal political power has come from North American Indian and Inuit populations, in both Canada and the United States. Within the United States, Native Americans asserted their political power in the 1960s, and this marked a decisive turn away from earlier policies of assimilation. Since passage of the Indian Self-Determination and Education Assistance Act of 1975, the trend has been toward increased Native American control of their economic and political destiny. The Indian Gaming Regulatory Act (1988) offered potential economic independence for many tribes. By 2006, Indian gaming operations nationally netted tribes almost $25 billion annually. In the western American

0    250    500 Miles
0    250    500 Kilometers

RUSSIA

An unstable
Russian realm

ARCTIC OCEAN

GREENLAND
(DENMARK)

Baffin
Bay

ALASKA
(U.S.)

New Canadian Territory
of Nunavut

**Salmon Wars.** *Recent negotiations between the United States and Canada have eased tensions between the two countries over contested fisheries in the North Pacific. Still, potential exists for more trouble in the future.*

C  A  N  A  D  A

Hudson
Bay

Cooperation
in Great Lakes
commerce and
environmental
cleanup

An
independent
Quebec?

**Commercial Lumbering.** *U.S. loggers have protested British Columbia timber operations, arguing that the provincial government has given Canadian lumber companies special privileges and an edge in competing with their southern neighbors.*

Dumping of
Canadian wheat
on U.S. markets

ATLANTIC
OCEAN

U  N  I  T  E  D

**Terrorist attacks.** *The hijacking of commercial airliners and the terrorist attacks on the World Trade Center and Pentagon in 2001 ushered in a new era of political uncertainty for the United States.*

S  T  A  T  E  S

PACIFIC
OCEAN

Illegal immigration
along U.S./Mexico
border

**Illegal Immigration.** *Will NAFTA's success lead to more or less immigration across the U.S./Mexico border? As trade linkages grow closer, so do the destinies of people north and south of the international border.*

Gulf of Mexico

MEXICO

Cuba's uncertain
political future

CUBA

Puerto Rico

HAITI   DOMINICAN
REPUBLIC

**FIGURE 3.27 Geopolitical Issues in North America**    Although Canada and the United States share a long and peaceful border, many political issues still divide the two countries. In addition, internal political conflicts, particularly in bicultural Canada, cause tensions.

interior, where Indians control roughly 20 percent of the land, tribes are developing their natural resources and buying up additional acreage. In Alaska, native peoples gained ownership to 44 million acres (18 million hectares) of land in 1971 under the Alaska Native Claims Settlement Act.

In Canada, even more ambitious challenges by native peoples to a weaker centralized government have yielded dramatic results. Canada established the Native Claims Office in 1975. Agreements with native peoples in Quebec, Yukon, and British Columbia turned over millions of acres of land to aboriginal control and increased

native participation in managing remaining public lands. By far the most ambitious agreement has been to create the territory of Nunavut out of the eastern portion of the Northwest Territories in 1999 (Figure 3.27). Nunavut is home to 30,000 people (85 percent Inuit) and is the largest territorial/ provincial unit in Canada (Figure 3.28). Its creation represents a new level of native self-government in North America. Recently, agreements between the federal Parliament and British Columbia tribes (the Nisga'a) have made a similar move toward more native self-government in that western province (see Figure 3.23).

**FIGURE 3.28  Life in Nunavut**   This scene on Baffin Island is from Iqaluit, the largest urban center in the Canadian province of Nunavut.

## The Politics of U.S. Immigration

Immigration policies are hotly contested in the United States. Public protests on various sides of the dilemma have helped highlight the conflict (Figure 3.29). Three key issues remain at the center of the debate. First, there are ongoing disagreements concerning the overall

**FIGURE 3.29  Immigration Protests**    Growing political disagreements over U.S. immigration policy have sparked large public protests over the issue in many American cities.

numbers of legal immigrants that should be allowed into the country. Some suggest that sharply reduced numbers of immigrants would protect American jobs and allow for a more gradual assimilation of existing foreigners, but others propose to loosen existing restrictions on immigrants to spur economic growth and business expansion. A second major issue, particularly along the border with Mexico, is how to tighten up on the daily illegal flow of immigrants. Many have argued that the country's wide-open southern border is a national security issue. Finally, there is no political consensus on a fair policy to deal with existing undocumented workers in the United States. Some policymakers advocate strict penalties for illegal immigrants, while others propose loosening requirements for citizenship or some form of amnesty to enable illegal immigrants to more easily enter the mainstream of American society. Suggested political compromises have included expanding the country's guest worker program or requiring illegal immigrants to return to their home countries ("touching base") before they can reapply for legal status in the United States.

## A Global Reach

The geopolitical reach of the United States, in particular, has taken its influence far beyond the borders of the region. The Monroe Doctrine (1824) asserted that U.S. interests extended throughout the Western Hemisphere and went beyond national boundaries. In the Pacific, the United States claimed the Philippines as a prize of the Spanish-American War (1898), and the further annexation of Hawaii (1898) contributed to the country's 20th-century dominance of the region. In Central America and the Caribbean, the growing role of the American military between 1898 and 1916 shaped politics in Cuba, Puerto Rico, Panama, Nicaragua, Haiti, Mexico, and elsewhere.

World War II and its aftermath forever redefined the role of the United States in world affairs. Postwar America emerged from the conflict as the world's dominant political power. The United States also developed multinational political and military agreements, such as those establishing the North Atlantic Treaty Organization (NATO) and the Organization of American States (OAS). Violent conflicts in Korea (1950–1953) and Vietnam (1961–1975) pitted U.S. political interests against communist attempts to extend their control beyond the Soviet Union and China. Tensions also ran high in Europe, as the Berlin Wall crisis (1961) and nuclear weapons deployments by NATO- and Soviet-backed forces brought the world closer to another global war. The Cuban missile crisis (1962) reminded Americans that traditional political boundaries provided little defense in a world uneasily brought closer together by technologies of potential mass destruction.

Even as the Cold War gradually faded during the late 1980s, the global political reach of the United States continued to expand. Policies in Central America favored governments friendly to the United States. President Carter's successful Middle East Peace Treaty between Israel and Egypt (1979) guaranteed a continuing diplomatic and military presence in the eastern Mediterranean. In the late 1990s, Serbian aggression within Kosovo prompted an American- and NATO-led intervention, which included major air attacks on the Serbian capital of Belgrade (1999) and a peacekeeping presence (with the United Nations) in the disputed area of Kosovo. The recent wars in Afghanistan (from 2001) and Iraq (from 2003) offered other examples of America's global political presence.

# Economic and Social Development: Geographies of Abundance and Affluence

North America possesses the world's most powerful economy and its wealthiest population. Its 340 million people consume huge quantities of global resources but also produce some of the world's most sought-after manufactured goods and services. The region's human capital—the skills and diversity of its population—has enabled North Americans to achieve high levels of economic development (Table 3.2). Even so, residents of the region suffered in the global recession of the late 2000s as household incomes fell and unemployment levels rose. Manufacturing states such as Michigan saw already high jobless rates rise even higher. But the impacts were widespread: Some of the worst-hit states, such as California and Arizona, included areas that had witnessed some of the highest growth—particularly in home prices—earlier in the decade.

## An Abundant Resource Base

North America is blessed with varied natural resources, which have provided diverse raw materials for development. Indeed, the direct extraction of natural resources still makes up 3 percent of the U.S. economy and more than 6 percent of the Canadian economy. Some of these North American resources are then exported to global markets, while other raw materials are imported to the region.

**Opportunities for Agriculture**   North Americans have created one of the most efficient food-producing systems in the world, and agriculture remains a dominant land use across much of the region (Figure 3.30). Farmers practice highly commercialized, mechanized, and specialized agriculture. The system emphasizes the importance of efficient transportation, global markets, and large capital investments in farm machinery. Today, agriculture employs only a small percentage of the labor force in both the United States (1 percent) and Canada (2 percent). At the same time, changes in farm ownership have sharply reduced the number of operating units, while average farm sizes have steadily risen.

The geography of North American farming represents the combined impacts of (1) diverse environments, (2) varied continental and global markets for food, and (3) historical patterns of settlement and agricultural evolution. In the Northeast, dairy operations and truck farms take advantage of their nearness to major cities in Megalopolis and southern Canada. Corn and soybeans dominate the Midwest and western Ontario, where a tradition of mixed farming combines the growing of feed grains with the production and fattening of livestock. To the south, only remnants of the old Cotton Belt remain, largely replaced by varied subtropical specialty crops; poultry, catfish, and livestock production; and commercial logging. Farther west, extensive, highly mechanized commercial grain-growing operations stretch from Kansas to Saskatchewan and Alberta. Depending on surface and groundwater resources, irrigated agriculture across western North America also offers opportunities for farming. Indeed, California's agricultural output, nourished by large agribusiness operations in the irrigated Central Valley, accounts for more than 10 percent of the nation's farm economy.

**Energy and Industrial Raw Materials**   North Americans produce and consume huge quantities of other natural resources. The U.S. produces about 12 percent of the world's oil but consumes about 25 percent. As a result, the United States imports more than half of its oil. Within the region, the major areas of oil and gas production are the Gulf Coast, the Central Interior, Alaska's North Slope, and Central Canada (especially Alberta). The most abundant fossil fuel in the United States is coal (27 percent of the world's total), but its relative importance in the overall energy economy declined in the 20th century as industrial technologies changed and environmental concerns grew. Still, the country's 400-year supply of coal reserves will be a major energy resource for the future. In addition, wind, solar, nuclear, and biofuels energy sources appear likely to make up larger portions of the region's future energy budget as the region gradually migrates away from its dependence on fossil fuels (Figure 3.31).

## Creating a Continental Economy

The timing of European settlement in North America was critical in its rapid economic transformation. The region's abundant resources came under the control of Europeans possessing new technologies that reshaped the landscape and reorganized its economy. By the 19th century, North Americans actively contributed to those technological changes. In addition, new natural resources were developed in the interior, and new immigrant populations arrived in large numbers. In the 20th century, although natural resources remained important, industrial innovations and more jobs in the service sector added to the economic base and extended the country's global reach.

**Connectivity and Economic Growth**   Dramatic improvements in North America's transportation and communication systems laid the foundation for urbanization, industrialization, and the commercialization of agriculture. Indeed, the region's economic

## TABLE 3.2   DEVELOPMENT INDICATORS

| Country | GNI per capita, PPP 2007 | GDP Average Annual %Growth 2000–07 | Human Development Index (2006)# | Percent Population Living Below $2 a Day | Life Expectancy* 2009 | Under Age 5 Mortality Rate 1990 | Under Age 5 Mortality Rate 2007 | Gender Equity 2007 |
|---|---|---|---|---|---|---|---|---|
| Canada | 35,500 | 2.7 | 0.967 | | 81 | 8 | 6 | 98 |
| United States | 45,840 | 2.6 | 0.950 | | 78 | 11 | 8 | 100 |

Source:  World Bank, *World Development Indicators, 2009.*

United Nations, *Human Development Index, 2008.* #

Population Reference Bureau, *World Population Data Sheet, 2009*\*

Gender Equity – Ratio of female to male enrollments in primary and secondary school.  Numbers below 100 have more males in primary/secondary school, numbers above 100 have more females in primary/secondary schools.

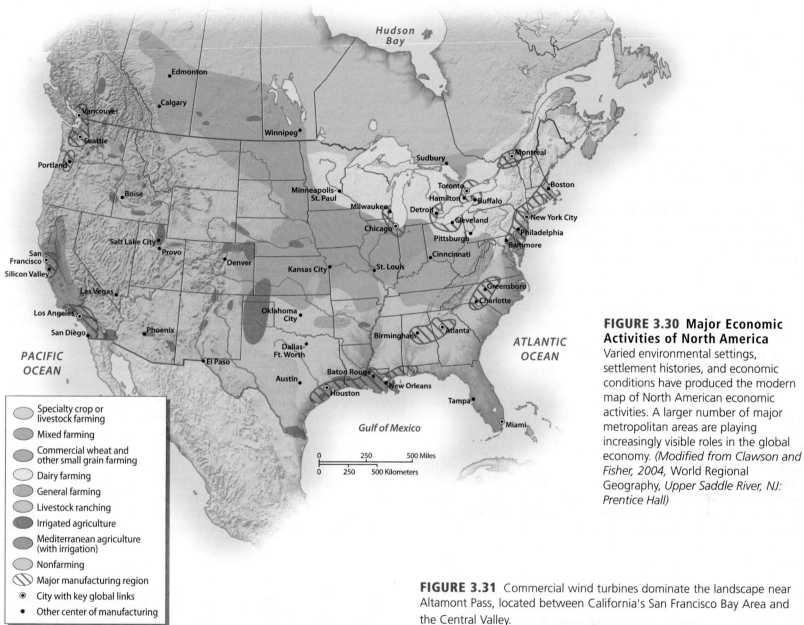

**FIGURE 3.30 Major Economic Activities of North America**
Varied environmental settings, settlement histories, and economic conditions have produced the modern map of North American economic activities. A larger number of major metropolitan areas are playing increasingly visible roles in the global economy. (*Modified from Clawson and Fisher, 2004,* World Regional Geography, *Upper Saddle River, NJ: Prentice Hall*)

Legend:
- Specialty crop or livestock farming
- Mixed farming
- Commercial wheat and other small grain farming
- Dairy farming
- General farming
- Livestock ranching
- Irrigated agriculture
- Mediterranean agriculture (with irrigation)
- Nonfarming
- Major manufacturing region
- City with key global links
- Other center of manufacturing

**FIGURE 3.31** Commercial wind turbines dominate the landscape near Altamont Pass, located between California's San Francisco Bay Area and the Central Valley.

success was a function of its **connectivity**, or how well its different locations became linked with one another through improved transportation and communications networks. Those links greatly facilitated the potential for interaction between locations and dramatically reduced the cost of moving people, products, and information over long distances.

Tremendous technological breakthroughs revolutionized North America's economic geography between 1830 and 1920. By 1860, more than 30,000 miles (48,387 kilometers) of railroad track had been laid in the United States, and the network grew to more than 250,000 miles (403,226 kilometers) by 1910. Farmers in the Midwest and Plains found ready markets for their products in cities hundreds of miles away. Industrialists collected raw materials from faraway places, processed them, and shipped manufactured goods to their final destinations. The telegraph brought similar changes to information: Long-distance messages flowed across eastern North America by the late 1840s, and 20 years later, undersea cables linked the realm to Europe, another milestone in the process of globalization.

North America's transportation and communications systems were modernized further after 1920. Automobiles, mechanized farm equipment, paved highways, commercial air links, national radio broadcasts, and dependable transcontinental telephone service reduced the cost of distance across the region. Perhaps most importantly, the region has taken the lead in the global information age, integrating computer, satellite, telecommunications, and Internet technologies in a web of connections that assists the flow of knowledge both within the region and beyond.

**The Sectoral Transformation** Changes in employment structure signaled North America's economic modernization just as surely as its increasingly interconnected society. The **sectoral transformation** refers to the evolution of a nation's labor force from one dependent on the *primary* sector (natural resource extraction) to one with more employment in the *secondary* (manufacturing or industrial), *tertiary* (services), and *quaternary* (information processing) sectors. For example, with agricultural mechanization, lower demands for primary-sector workers are replaced by new opportunities in the growing industrial sector. In the 20th century, new services (trade, retailing) and information-based activities (education, data processing, research) created other employment opportunities. Today, the tertiary and quaternary sectors employ more than 70 percent of the labor force in both Canada and the United States.

**Regional Economic Patterns** The locations of North America's industries show important regional patterns. **Location factors** are the varied influences that explain *why* an economic activity is located where it is. Many influences, both within and beyond the region, shape patterns of economic activity. Patterns of industrial location illustrate the concept (see Figure 3.30). The historical manufacturing core includes Megalopolis (Boston, New York, Philadelphia, Baltimore, and Washington, DC), southern Ontario (Toronto and Hamilton), and the industrial Midwest. The region's proximity to *natural resources* (farmland, coal, and iron ore); increasing *connectivity* (canals and railroad networks, highways, air traffic hubs, and telecommunications centers); a ready supply of *productive labor*; and a growing national, then global, *market demand* for its industrial goods encouraged continued *capital investment* within the core. Traditionally, the core has dominated in the production of steel, automobiles, machine tools, and agricultural equipment and played a critical role in producer services such as banking and insurance.

In the last half of the 20th century, industrial and service-sector growth shifted to the South and West. Cities of the South's Piedmont manufacturing belt (Greensboro to Birmingham) grew after 1960, partly because lower labor costs and Sun Belt amenities attracted new investment. By 2007, for example, the North Carolina "research triangle" area between Raleigh, Durham, and Chapel Hill emerged to become the nation's third largest biotech cluster behind California and Massachusetts. The Gulf Coast industrial region is strongly tied to nearby fossil fuels that provide raw materials for its many energy-refining and petrochemical industries (Figure 3.32). The varied West Coast industrial region stretches from Vancouver, British Columbia, to San Diego, California (and beyond into northern Mexico), and it demonstrates the increasing importance of Pacific Basin trade. Large aerospace operations in the West also suggest the role of *government spending* as a location factor. Silicon Valley is now one of North America's leading regions of manufacturing exports. Its proximity to Stanford, Berkeley, and other universities demonstrates the importance of *access to innovation and research* for many fast-changing high-technology industries. Silicon Valley's location also shows the advantages of *agglomeration economies*, in which many companies with similar and often integrated manufacturing operations locate near one another (Figure 3.33). Smaller places such as Provo, Utah, and Austin, Texas, specialize in high-technology industries and demonstrate the growing

**FIGURE 3.32 Gulf Coast Petroleum Refining** Petroleum-related manufacturing has transformed many Gulf Coast settings. Much of Houston's 20th-century growth was fueled by the dramatic expansion of oil-related industries. The port of Houston remains a major center of North America's refining and petrochemical operations.

**FIGURE 3.33 Silicon Valley** The high-technology industrial landscape of California's Silicon Valley differs from the look of traditional manufacturing centers. Here similar industries form complex links, benefiting from their proximity to one another and to nearby universities such as Stanford and Berkeley.

role of *lifestyle amenities* in shaping industrial location decisions, both for entrepreneurs and for the skilled workers who need to be attracted to such opportunities.

## Persisting Social Issues

Profound economic and social problems shape the human geography of North America. Even with its continental wealth, great differences persist between rich and poor. High median household incomes in the United States and Canada do not reveal the differences in wealth within the two countries. Broader measures of social well-being suggest that the poorer people in these countries have less access than others to quality education and health care. In addition, both nations face enduring issues related to gender inequity and social and economic problems related to aging populations.

**Wealth and Poverty**    The North American landscape displays contrasting scenes of wealth and poverty. Elite northeastern suburbs, gated California neighborhoods, upscale shopping malls, and posh alpine ski resorts are all expressions of private and exclusive landscape settings that characterize wealthier North American communities (Figure 3.34). On the other hand, substandard housing, abandoned property, aging infrastructure, and unemployed workers are visual reminders of the gap between rich and poor in the region (Figure 3.35). Specifically, in the United States, black household incomes remain only 67 percent of the national average, while Hispanic incomes fare slightly better, at 80 percent of the national average.

Poverty levels have declined in both countries since 1980, but poor populations are still grouped into a variety of geographic settings. By 2008, poverty rates in the United States had fallen to about 13 percent of the population; a similar measure of low-income Canadians

declined to around 18 percent. The problems of the rural poor remain major regional social issues in the Canadian Maritimes, Appalachia, the Deep South, and the Southwest. Most poor people in the United States, however, live in central-city locations, and links between ethnicity and poverty are strong in these communities. Nationally, 25 percent of the country's African Americans and 22 percent of its Hispanic populations live below the poverty line, and the great majority reside in poorer central-city communities. Furthermore, Hispanic workers suffered particularly large job losses in the recession of the late 2000s.

**21st-Century Challenges**    Measures of social well-being in North America compare favorably with those of most other world regions (Table 3.2). Still, many economic and social challenges exist for this region. As workers, North Americans compete globally for jobs in a fast-changing and increasingly unpredictable world economy.

Since World War II, both the United States and Canada have seen great changes in the role that women play in society, but the "gender gap" is yet to be closed when it comes to salary issues, working conditions, and political power. Women widely participate in the workforce in both countries and are as educated as men, but they still earn only about 75 cents for every dollar that men earn. Women also head the vast majority of poorer single-parent families in the United States. Canadian women, particularly single mothers who work full time, are also greatly disadvantaged, averaging only about 65 to 70 percent of the salaries of Canadian men.

Health care and aging are also key concerns in a region full of graying baby boomers. While U.S. residents spend more than

**FIGURE 3.34 Gated America**    An automatic gate protects the entrance of a gated community in Apollo Beach, near Tampa, Florida.

**FIGURE 3.35 Inner-City Neighborhood**    Many poor inner-city neighborhoods, such as this community in New York City, suffer from substandard housing and aging infrastructure. Shared ethnic bonds often supply the social support necessary to ease the challenges of high unemployment and troubled family life.

15 percent of gross domestic product on health care (Canadians spend slightly less), there are 47 million Americans without any health-care insurance. A recent report on aging in the United States predicted that 20 percent of the nation's population will be older than 65 by 2050 and that the most elderly (age 85+) are the fastest-growing part of the population. The geographic consequences of aging are already abundantly clear. Whole sections of the United States—from Florida to southern Arizona—have become increasingly oriented around retirement (Figure 3.36). Communities cater to seniors with special assisted-living arrangements, health-care facilities, and recreational opportunities.

The rising incidence of chronic diseases associated with aging (heart disease, cancer, and stroke are the three leading causes of death) will continue to pressure both health-care systems. Another cost has been the care and treatment of the region's 1 million HIV/AIDS victims. The disease hits poorer black and Hispanic populations particularly hard, where rates of new infection are still on the rise.

## North America and the Global Economy

Together with Europe, Japan, and China, North America plays a key role in the global economy. When the economy is thriving, the region benefits from global economic growth, but in periods of international instability, globalization means that the region is more vulnerable to economic downturns. In the economic slump of the late 2000s, for example, many North American job losses were directly tied to declining global demand for North American products and to lower rates of foreign investment. No community remains untouched by these links, and the consequences will shape the lives of every North American in the 21st century.

**Creating the Modern Global Economy**    The United States, with Canada's firm support, played a formative role in creating much of the new global economy and in shaping many of its key institutions. In 1944, allied nations met at Bretton Woods, New Hampshire, to discuss economic affairs. Under U.S. leadership, the group set up the International Monetary Fund (IMF) and the World Bank and gave these global organizations the responsibility for defending the world's monetary system. The United States was also the driving force for the creation (1948) of the General Agreement on Tariffs and Trade (GATT). Renamed the **World Trade Organization (WTO)** in 1995, its 153 member states are dedicated to reducing global barriers to trade. In addition, the United States and Canada participate in the **Group of Eight (G8)**, a collection of powerful countries (including, besides the United States and Canada, Japan, Germany, Great Britain, France, Italy, and Russia) that regularly meets for discussions on key global economic and political issues.

**Attracting Skilled Immigrants**    North America's dynamic role in the global economy helps attract thousands of skilled workers from other countries, adding to the region's supply of human capital. Statistics gathered by the U.S. Department of Homeland Security point to the unique contributions of highly skilled immigrants. H-1B visas are granted to special "temporary skilled workers" to encourage computer programmers, doctors, and other professionals to work in the United States. More than 400,000 such visas were issued in 2005, enabling these individuals to work within the United States. The geography of the top 20 contributing countries offers yet another snapshot of the economic impact of globalization and is a powerful example of another way in which the United States benefits in the process (Figure 3.37). While there are predictable ties to many European nations, the linkages with India (particularly through the computer, software, and electronics industries) are dramatic. Continental connections to nearby Mexico and Canada are also notable, along with growing links to Asian nations such as South Korea, Japan, the Philippines, and Taiwan. The

**FIGURE 3.36 Tomorrow's Baby Boom Landscape?**    This Arizona retirement community, its circular shape boldly stamped on the desert landscape, suggests the 21st-century destination for many aging baby boomers. As North Americans grow older, entire subregions may be devoted to age-dependent demands for housing, recreation, and health services.

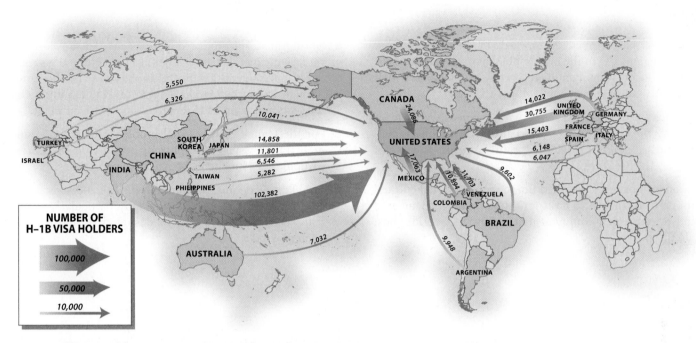

**FIGURE 3.37 Origins of Temporary Skilled Workers (H-1B Visas) in the United States, 2005 (Top 20 Contributing Countries)** India's contribution to the skilled human capital of the United States is particularly notable. Both Mexico and Canada contribute many skilled workers, as do other nations of East Asia and western Europe. (*U.S. Department of Homeland Security, 2005*)

patterns offer a powerful reminder that the economic evolution of both the United States and Canada remain intimately connected to skilled immigrant populations.

**Doing Business Globally** Patterns of capital investment and corporate power also place the North American region at the center of global money flows and economic influence. The region attracts inflows of foreign capital, both as investments in North American companies and as foreign direct investment (FDI) by international companies. Since 1980, aging U.S. baby boomers also have poured billions of pension fund and investment dollars into Japanese, European, and "emerging" stock markets. U.S. investments in foreign countries are also made directly by multinational corporations based in the United States.

But the geography of 21st-century multinational corporations is changing, illustrating three recent shifts in broader patterns of globalization. First, traditional American-based multinational corporations are adopting a new, more globally integrated model. For example, IBM now has more than 50,000 employees in India. Second, a growing array of multinational corporations based elsewhere in the world, especially in places such as China, India, Russia, and Latin America, are buying up companies and assets once controlled by North American or European capital. In 2006, for example, Brazilians

invested $26 billion overseas (especially in North America), far outpacing the $18 billion that foreign companies invested in Brazil. Third, many of these same global multinational companies are making sizable investments of their own in other portions of the less developed world, from Africa to Southeast Asia, bypassing North American control altogether. Simply put, the late-20th-century top-down model of multinational corporate control and investment, traditionally based in North America, Europe, and Japan, is being replaced by a more globally distributed model of corporate control. This new model has many origins, many destinations, and new patterns of labor, capital, production, and consumption.

North Americans have experienced direct consequences from these shifts in global capitalism. There has been a growing reaction in the United States, for example, to corporate **outsourcing**, a business practice that transfers portions of a company's production and service activities to lower-cost settings, often located overseas. In addition, millions of jobs in manufacturing, textiles, semiconductors, and electronics have effectively migrated to places such as China, India, and Mexico, as those localities offer low-cost, less regulated settings for production, for both local and foreign firms. The results are complex: North American consumers benefit from buying cheap imports, but they may find that their own jobs are threatened in the corporate restructurings that make such bargains possible.

# SUMMARY

- North Americans have reaped the rewards of the natural abundance of their region, and in the process they have transformed the environment, created a highly affluent society, and extended their global economic, cultural, and political reach.

- North America's affluence has come with a considerable price tag, and today the region faces significant environmental challenges, including soil erosion, acid rain, and air and water pollution.

- In a remarkably short time period, a unique mix of varied cultural groups from around the world has contributed to the settlement of a huge and resource-rich continent that is now the world's most urbanized region.

- North Americans produced two societies that are closely intertwined yet face distinctive national political and cultural issues. In Canada, the nation's identity remains problematic as the country works through the persisting challenges of its multicultural character and the costs and benefits of its proximity to its continental neighbor.

- For the United States, social problems linked to ethnic diversity, immigration issues, and enduring poverty remain central concerns, particularly in many of the nation's largest cities.

- The global economic downturn of the late 2000s profoundly affected North America's economic geography, particularly in many regions that were hit hardest by the housing crisis and by rising rates of unemployment.

# KEY TERMS

acid rain (p. 55)
boreal forest (p. 57)
concentric zone model (p. 63)
connectivity (p. 75)
counterurbanization (p. 62)
cultural assimilation (p. 65)

ethnicity (p. 62)
federal state (p. 71)
gentrification (p. 63)
Group of Eight (G8) (p. 78)
location factor (p. 76)
Megalopolis (p. 58)

North American Free Trade
    Agreement (NAFTA) (p. 71)
outsourcing (p. 79)
prairie (p. 57)
sectoral transformation (p. 76)
Spanglish (p. 69)

tundra (p. 57)
unitary state (p. 71)
urban decentralization (p. 62)
urban realms model (p. 63)
World Trade Organization
    (WTO) (p. 78)

# THINKING GEOGRAPHICALLY

1. Explain how "natural hazards" can be "culturally" defined. In other words, what role do humans play in shaping the distribution of hazards?

2. Summarize and map the ethnic background and migration history of your own family. Discuss how these patterns parallel or depart from larger North American trends.

3. Describe the strengths and weaknesses of federalism and cite examples from both Canada and the United States.

4. The environmental price for North American development has often been steep. Suggest why it may or may not be worth the price and defend your answer.

5. Who will America's leading trade partner be in 2050? Explain the reasons for your choice.

Log in to **www.mygeoscienceplace.com** for videos, interactive maps, RSS feeds, case studies, and self-study quizzes to enhance your study of North America.

# 4 LATIN AMERICA

## GLOBALIZATION AND DIVERSITY

Neoliberal policies have profoundly changed the way Latin American economies and societies function. Foreign investment and international trade have intensified and diversified, forging new connections with other world regions.

### ENVIRONMENTAL GEOGRAPHY

Tropical forests in Latin America, especially in the Amazon Basin, are one of the planet's greatest reserves of biological diversity. How this diversity will be managed is a critical question.

### POPULATION AND SETTLEMENT

Latin America is the most urbanized region of the developing world, with 75 percent of the population living in cities. Four megacities (10 million or more) are found there.

### CULTURAL COHERENCE AND DIVERSITY

Amerindian activism is on the rise in Latin America. Indigenous peoples from Central America to the Andes are finding their political voice and demanding cultural and territorial recognition.

### GEOPOLITICAL FRAMEWORK

Recent elections in the region have seen liberal democrats and populists gaining power. These same politicians tend to be highly skeptical of the ability of neoliberal reforms and economic globalization to reduce poverty in the region.

### ECONOMIC AND SOCIAL DEVELOPMENT

From NAFTA to Mercosur, regional economic integration is changing the way Latin Americans trade with each other and the rest of the world.

*Sprawling São Paulo is Brazil's largest city with nearly 20 million people. This Latin American megacity is part of an urban industrial core that spans from Rio de Janeiro to São Paulo.*

81

Beginning with Mexico and extending to the tip of South America, Latin America's regional unity stems largely from its shared colonial history rather than from its present-day level of development. More than 500 years ago, the Iberian countries of Spain and Portugal began their conquest of the Americas. Iberia's mark is still visible throughout the area: Officially, two-thirds of the population speaks Spanish, and the rest speaks Portuguese. Iberian architecture and town design add homogeneity to the colonial landscape. The vast majority of the population is Catholic. These European traits blended with those of different Amerindian peoples. The Indian presence remains especially strong in Bolivia, Peru, Ecuador, Guatemala, and southern Mexico, where large and diverse Amerindian populations maintain their native languages, dress, and traditions. After the initial conquest, other cultural groups were added to this mix of native and Iberian peoples.

The legacy of slavery brought several million Africans to the region, primarily on the coasts of Colombia and Venezuela and throughout Brazil. In the 19th and 20th centuries, new waves of settlers came from Spain, Italy, Germany, Japan, and Lebanon. The result is one of the world's most racially mixed regions. The modern states of Latin America are multiethnic, with distinct indigenous and immigrant profiles and very different rates of social and economic development (Figure 4.1; see also "Setting the Boundaries").

Through colonialism, immigration, and trade, the forces of globalization have been at work in Latin America for decades. The early Spanish Empire concentrated on mining precious metals, sending ships laden with silver and gold across the Atlantic. The Portuguese became important producers of natural dyes, sugar products, gold, and later coffee. In the late 19th and early 20th centuries, exports to North America and Europe fueled the region's economy. Most countries specialized in one or two products: bananas and coffee, meats and wool, wheat and corn, petroleum and copper. Such a primary export tradition led to an unhealthy economic dependence on a handful of exports. By the 1960s, some economists even argued that Latin American economies were too specialized on primary exports to foster development.

Since then, the countries of the region have industrialized and diversified their production, but they continue to be major producers of primary goods for North America, Europe, and East Asia. Today, many states have adopted neoliberal policies that encourage foreign investment, export production, and privatization. The results are mixed, with some states experiencing impressive economic growth but increased disparity between rich and poor. Intraregional trade within Latin America stimulated by Mercosur (the Southern Cone Common Market), as well as the impact of the North American Free Trade Agreement (NAFTA; see Chapter 3) and the newly formed **Central American Free Trade Association (CAFTA)**, are indicators of heightened economic integration in the western hemisphere.

Roughly equal in area to North America, Latin America has a much larger and faster-growing population of 540 million people. Its most populated state, Brazil, has 191 million people, making it the fifth largest country in the world. The next largest state, Mexico, has a population of 110 million. Despite many social and economic gains, poverty remains a major concern for this region. It is estimated that 22 percent of the people in the region live on less than $2 per day. Yet unlike most other developing areas today, Latin America is decidedly urban. Prior to World War II, most people lived in rural settings and worked as farmers. Today, three-quarters of Latin Americans are city dwellers. Even more startling is the number of **megacities**. São Paulo, Mexico City, Buenos Aires, and Rio de Janeiro all have more than 10 million residents. In addition, there are more than 40 cities of at least 1 million residents.

Despite the region's growing industrial and urban characteristics, industries focusing on natural resource extraction will continue to be

# Setting the Boundaries

The concept of Latin America as a distinct region has been commonly accepted for nearly a century. The boundaries of this region are straightforward, beginning at the Rio Grande (called the Rio Bravo in Mexico) and ending at Tierra del Fuego. French geographers are credited with coining the term "Latin America" in the 19th century to distinguish the Spanish- and Portuguese-speaking republics of the Americas plus Haiti from the English-speaking territories. There is nothing particularly "Latin" about the area, other than the prevalence of romance languages. The term stuck because it was vague enough to include areas with different colonial histories while also offering a clear cultural boundary from Anglo-America, the region referred to as North America in this text.

Latin America is one of the world regions that North Americans are most likely to visit. The trend is to visit the northern edge of this region. Tourism is strong, especially along Mexico's northern border and coastal resorts such as Cancun. Unfortunately, there is a tendency to visit one area in the region and generalize for all of it. Although it is historically sound to think of Latin America as a major world region, there are extreme variations in the physical environment, levels of social and economic development, and influences of native peoples.

Because the region is so large, geographers often divide Latin America. The continent of South America is typically distinguished from Middle America (which includes Central America, Mexico, and the Caribbean). The term "Middle America" was created to identify an area culturally distinct from North America but physically part of it because most of Mexico and all of Cuba rest on the North American Plate. Such a division has advantages, but it also separates countries with very similar histories (such as Mexico and Peru) while joining countries that have very little in common (such as El Salvador and Jamaica). In this text, the Americas are divided slightly differently. Latin America consists of the Spanish- and Portuguese-speaking countries of Central and South America, including Mexico. Chapter 5 examines the Caribbean, consisting of the islands of the Antilles, the Guianas, and Belize. So divided, the important indigenous and Iberian influences of mainland Latin America are emphasized. Similarly, the Caribbean's unique colonial and demographic history will be discussed in Chapter 5.

**LATIN AMERICA**
Political & Physical Map

⊛  ●  Over 1,000,000
⊛  ●  500,000–1,000,000 (selected cities)
⊛  ·  Selected smaller cities
- - -┐  Plate boundaries

**ELEVATION IN METERS**

4000+
2000–4000
500–2000
200–500
0–200
Sea Level
Below sea level

**FIGURE 4.1  Latin America**    Roughly equal in size to North America, Latin America supports a larger population and far greater ecological diversity. The 17 countries included in this region share a history of Iberian colonization. Seventy-four percent of the region's 540 million people live in cities, making it the most urbanized region of the developing world. It is noted for its production of primary exports and manufactured goods.

important, in part because of the area's impressive resource base. Latin America is home to Earth's largest rain forest, the greatest river by volume, and massive reserves of natural gas, oil, and copper. With its extensive territory, its tropical location, and its relatively low population density (Latin America has half the population of India in nearly seven times the area), the region is also recognized as one of the world's great reserves of biological diversity (Figure 4.2). How this diversity will be managed in the face of global demand for natural resources is an increasingly important question for the countries of this region.

**FIGURE 4.2 Tropical Wilderness** The biological diversity of Latin America—home to the world's largest rain forest—is increasingly seen as a genetic and economic asset. These forested mesas (called *tepuis*) in southern Venezuela are representative of the wild lands that many conservationists seek to protect.

# Environmental Geography: Neotropical Diversity and Degradation

Much of Latin America is characterized by its tropical nature. Travel posters of the region showcase lush forests and brightly colored parrots. The diversity and uniqueness of the **neotropics** (tropical ecosystems of the Western Hemisphere) have long been attractive to naturalists eager to understand their unique flora and fauna. It is no accident that Charles Darwin's theory of evolution was inspired by his two-year journey in tropical America. Even today, scientists throughout the region work to understand complex ecosystems, discover new species, protect them, and interpret the impact of human settlement, especially in neotropical forests. Not all of the region is tropical. Important population centers extend below the Tropic of Capricorn, most notably Buenos Aires, Argentina, and Santiago, Chile. Much of northern Mexico, including the city of Monterrey, is north of the Tropic of Cancer. Yet Latin America's tropical climate and vegetation most affect popular images of the region. Given the territory's large size and relatively low population density, Latin America has not experienced the same levels of environmental degradation witnessed in East Asia and Europe. The region's biggest environmental concerns are related to deforestation and loss of biodiversity and the livability of urban areas (Figure 4.3). Mexico City, in particular, suffers from a host of environmental problems and is a good example of the kinds of environmental challenges facing modern Latin American cities.

Huge areas of Latin America remain relatively untouched, supporting an incredible diversity of plant and animal life. Throughout the region, national parks offer some protection to unique communities of plants and animals. A growing environmental movement in countries such as Costa Rica and Brazil has yielded both popular and political support for conservation efforts. In short, Latin Americans have entered the 21st century with a real opportunity to avoid many of the environmental mistakes seen in other regions of the world. At the same time, global market forces are driving governments to exploit minerals, fossil fuels, forests, and soils. The region's biggest natural resource management challenge is to balance the economic benefits of extraction with the principles of sustainable development.

## The Destruction of Tropical Rain Forests

Perhaps the environmental issue most commonly associated with Latin America is deforestation. The Amazon Basin and portions of the eastern lowlands of Central America and Mexico still maintain unique and impressive stands of tropical forest. Other woodland areas, such as the Atlantic coastal forests of Brazil and the Pacific forests of Central America, have nearly disappeared as a result of agriculture, settlement, and ranching. In the midlatitudes, the ecologically unique evergreen rain forest of southern Chile (the Valdivian forest) is being cleared to export wood chips to Asian markets (Figure 4.4). The coniferous forests of northern Mexico are also falling, in part because of a bonanza for commercial logging stimulated by NAFTA.

The loss of tropical rain forests is most critical in terms of biological diversity. Tropical rain forests cover only 6 percent of Earth's landmass, but at least 50 percent of the world's species are found in this biome. Moreover, the Amazon contains the largest undisturbed stretches of rain forest in the world. Unlike Southeast Asian forests, where hardwood extraction drives deforestation, Latin American forests are usually seen as an agricultural frontier that state governments divide in an attempt to give land to the landless and reward political elites. Thus, forests are cut and burned, with settlers and politicians carving them up to create permanent settlements, slash-and-burn plots, or large cattle ranches. In addition, some tropical forest cutting has been motivated by the search for gold (Brazil, Venezuela, and Costa Rica) and the production of coca leaf for cocaine (Peru, Bolivia, and Colombia).

Brazil has been criticized more than other countries for its Amazon forest policies. During the past 40 years, one-fifth of the Brazilian Amazon has been deforested. In states such as Rondônia, where settlers

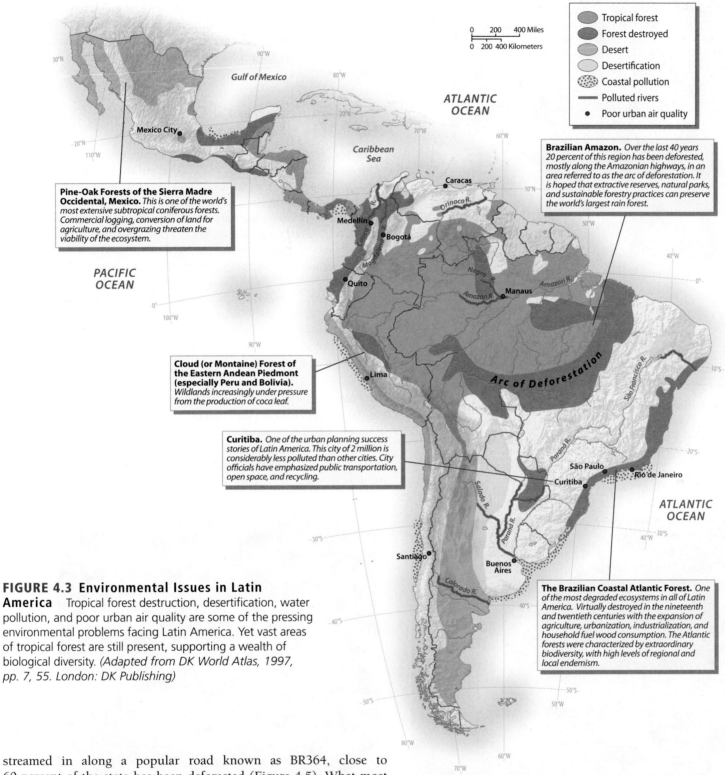

**Pine-Oak Forests of the Sierra Madre Occidental, Mexico.** *This is one of the world's most extensive subtropical coniferous forests. Commercial logging, conversion of land for agriculture, and overgrazing threaten the viability of the ecosystem.*

**Brazilian Amazon.** *Over the last 40 years 20 percent of this region has been deforested, mostly along the Amazonian highways, in an area referred to as the arc of deforestation. It is hoped that extractive reserves, natural parks, and sustainable forestry practices can preserve the world's largest rain forest.*

**Cloud (or Montaine) Forest of the Eastern Andean Piedmont (especially Peru and Bolivia).** *Wildlands increasingly under pressure from the production of coca leaf.*

**Curitiba.** *One of the urban planning success stories of Latin America. This city of 2 million is considerably less polluted than other cities. City officials have emphasized public transportation, open space, and recycling.*

**The Brazilian Coastal Atlantic Forest.** *One of the most degraded ecosystems in all of Latin America. Virtually destroyed in the nineteenth and twentieth centuries with the expansion of agriculture, urbanization, industrialization, and household fuel wood consumption. The Atlantic forests were characterized by extraordinary biodiversity, with high levels of regional and local endemism.*

Legend:
- Tropical forest
- Forest destroyed
- Desert
- Desertification
- Coastal pollution
- Polluted rivers
- Poor urban air quality

**FIGURE 4.3 Environmental Issues in Latin America** Tropical forest destruction, desertification, water pollution, and poor urban air quality are some of the pressing environmental problems facing Latin America. Yet vast areas of tropical forest are still present, supporting a wealth of biological diversity. *(Adapted from DK World Atlas, 1997, pp. 7, 55. London: DK Publishing)*

streamed in along a popular road known as BR364, close to 60 percent of the state has been deforested (Figure 4.5). What most alarmed environmentalists and forest dwellers (Indians and rubber tappers) is the dramatic increase in the rate of rain forest clearing since 2000, estimated at nearly 8,000 square miles (20,000 square kilometers) per year. The increased rates of deforestation in the Brazilian Amazon are due to the expansion of industrial mining and logging, the growth of corporate farms, the development of new road networks, the incidence of human-ignited wildfires, and continued population growth. In an effort to slow deforestation rates, the Brazilian government, under President Lula da Silva, has created new conservation

areas, many of them alongside the "arc of deforestation"—a swath of agricultural development along the southern edge of the Amazon Basin (see Figure 4.3).

The conversion of tropical forest into pasture, called **grassification**, is another practice that has contributed to deforestation. Particularly in southern Mexico, Central America, and the Brazilian Amazon, an assortment of development policies from the 1960s through the 1980s encouraged deforestation to make room for cattle in areas of frontier

**FIGURE 4.4 Chilean Wood Chips** A mountain of wood chips awaits shipment to Japanese paper mills from the southern Chilean port of Punta Arenas. The exploitation of wood products from both native and plantation forests supports Chile's booming export economy. Increasingly, these wood exports are bound for markets in East Asia.

settlement (Figure 4.6). Although many natural grasslands such as the Llanos (Venezuela and Colombia), the Chaco (Bolivia, Paraguay, and Argentina), and the Pampas (Argentina) are suitable for grazing, it was the rush to convert forest into pasture that made ranching so environmentally destructive.

## Urban Environmental Challenges: The Valley of Mexico

At the southern end of the great Central Plateau of Mexico lies the Valley of Mexico, the cradle of Aztec civilization and the site of Mexico City, a metropolitan area of approximately 18 million people. The severe problems facing this urban ecosystem underscore how environment, population, technology, and politics are interwoven in an effort to make Mexico City livable. This high-altitude basin with mild temperatures, fertile soils, and ample water surrounded by snow-capped volcanoes was an early center of Amerindian settlement and plant domestication, as well as the site of one of Spain's most important colonial cities. Given these features, it is no wonder that Mexico City has kept its primacy into the 21st century, even though the environmental setting that made it attractive centuries ago is now severely degraded. Some of the most pressing problems are air quality, adequate water, and subsidence (soil sinkage) caused by overdrawing the valley's aquifer (groundwater).

**Air Pollution**   The smog of Mexico City is so bad that most modern-day visitors have no idea that mountains surround them. Air quality has been a major issue for Mexico City since the 1960s, driven in part by the city's unusually high rate of growth. (Between 1950 and 1980, the city's annual rate of growth was 4.8 percent.) It is difficult to imagine a better setting for creating air pollution. The city sits in a

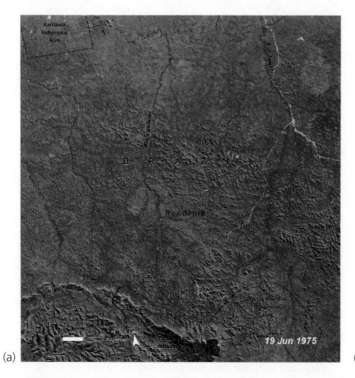

(a)

(b)

**FIGURE 4.5 Tropical Forest Settlement in the Amazon**   These satellite images of Rondônia, Brazil, illustrate the dramatic change in forest cover. In 1975 (a) there was one major road, and most of the region was tropical forest. By 2001, roads, pasture, and farmland had replaced the natural land cover.

**FIGURE 4.6 Converting Forest into Pasture** Cattle graze in northern Guatemala's Peten region. Clearing of this tropical forest lowland began in the 1960s and continues today. Ranching is a status-conferring occupation in Latin America with serious ecological costs. The beef produced from this region is for domestic and export markets.

bowl 7,400 feet (2,250 meters) above sea level, and a layer of warm air traps a layer of cold air near the surface (called a *thermal inversion*) that is filled with exhaust, industrial smoke, garbage, and fecal matter (Figure 4.7). During the worst pollution emergencies in the winter, when a layer of warm air traps pollutants, schoolchildren are required to stay indoors, and high-polluting vehicles are banned from the streets. Steps were finally taken in the late 1980s to reduce emissions from factories and cars. Unleaded gas is now widely available for the 3 million cars in the metropolitan area, and some of the worst polluting factories in the Valley of Mexico have closed. In fact, a study in 2001 suggested that the high levels of lead, carbon dioxide, and sulfur

dioxide in the atmosphere were starting to decline. Nonetheless, the health costs of breathing such contaminated air are real, as elevated death rates due to heart disease, influenza, and pneumonia suggest.

**Water Resources** One of Mexico City's most significant environmental problems is water. When Vicente Fox was president of Mexico, he declared water (both scarcity and quality) a national security issue, not just for the capital but for the entire country. Ironically, it was the abundance of water that made this site attractive for settlement. Large shallow lakes once filled the valley, but over the centuries, most were drained to expand agricultural land. As surface water became scarce, wells were dug to tap the basin's massive freshwater aquifer. Today, approximately 70 percent of the water used in the metropolitan area is drawn from the valley's aquifer. There is troubling evidence that the aquifer is being overdrawn and at risk of contamination, especially in areas where unlined drainage canals can leak pollutants into the surrounding soil, which then leach into the aquifer. To reduce reliance on the aquifer, the city now pumps water nearly a mile uphill from more than 100 miles (160 kilometers) away.

**A Sinking Land** A not-well-known but worrisome problem is that Mexico City is sinking. As the metropolis grows and pumps more water from its aquifer, subsidence worsens. Mexico City sank 30 feet (9 meters) during the 20th century. By comparison, Venice, an Italian city known for its subsidence problems, sank only 9 inches (23 centimeters) during the same period. Subsidence is a huge problem that will not go away because the city is still so dependent on groundwater. Although the amount of subsidence varies across the metropolitan area, its impact is similar throughout the city. Building foundations are destroyed, and water and sewer lines rupture. Because of subsidence, groundwater is no longer pumped from the city center, and sinking has slowed to 1 inch (2.5 centimeters) per year.

These serious urban environmental problems are made worse by poverty and governmental inaction. For decades, politicians and industrialists denied that there were problems. For most of the twentieth century,

**FIGURE 4.7 Air Pollution in Mexico City** Air pollution blankets Mexico City against a backdrop of the Ixtaccihuatl volcano. Mexico City is notorious for its smog. Its high elevation and immense size make air quality management difficult. While lead levels have declined with the introduction of unleaded gasoline, respiratory illnesses are on the rise.

one-party rule in Mexico reduced the likelihood of meaningful environmental reforms. When steps were finally taken to introduce unleaded gasoline and catalytic converters, open dumps, aging cars and minibuses, and unregulated factories continued to spew pollutants into the air. Mexico City's poorest citizens suffer the most from urban contamination. For the vast majority of Latin American urban poor, the lack of reliable water and clean air is the most pressing environmental problem.

**Modern Urban Challenges**   For most Latin Americans, air pollution, water availability and quality, and garbage removal are the pressing environmental problems of everyday life. Consequently, many environmental activists from the region focus their efforts on making urban environments cleaner by introducing "green" legislation and calling people to action. In this most urbanized region of the developing world, city dwellers do have better access to water, sewers, and electricity than their counterparts in Asia and Africa. Moreover, the density of urban settlement seems to encourage the widespread use of mass transportation; both public and private bus and van routes make getting around cities fairly easy. Yet the usual environmental problems that come from dense urban settings ultimately require expensive remedies such as new power plants and better sewer and water lines. The money for such projects is never enough, due to currency devaluation, inflation, and foreign debt. Because many urban dwellers tend to reside in unplanned squatter settlements, servicing these communities with utilities after they are built is difficult and costly. These settlements are especially vulnerable to natural hazards, especially catastrophic landslides such as the ones that buried communities around Caracas, Venezuela, in December 1999.

There are, however, examples of sustainable cities in Latin America. Curitiba is the celebrated "green city" of Brazil because of some relatively simple yet progressive planning decisions. More than 2 million people live in this industrial and commercial center, yet it is significantly less polluted than similar-sized cities. Because the city's location was vulnerable to flooding, city planners built drainage canals and set aside the remaining natural drainage areas as parks in the 1960s, well before explosive growth would have made such a policy difficult. This action added green space and reduced the negative impacts of flooding. Next, public transportation became a top priority. An extensive bus system that featured rapid loading and unloading of passengers made Curitiba a model for transportation in the developing world. Finally, a low-tech but effective recycling program has greatly reduced solid waste. Cities such as Curitiba demonstrate that designing with nature makes sense both ecologically and economically. To appreciate the diversity of environments Latin Americans must cope with when making development decisions, an understanding of the region's physical geography is important.

## Western Mountains and Eastern Shields

Latin America is a region of diverse landforms, including high mountains and extensive upland plateaus. The movement of tectonic plates explains much of the region's basic topography, including the formation of its geologically young western mountain ranges, such as the Andes and the Volcanic Axis of Central America (see Figure 4.3). This area is also geologically active, especially with regard to earthquakes that threaten people and damage property. In January 2001, for example, a major earthquake struck El Salvador, killing nearly 900 people and destroying more than 100,000 homes. In contrast, the Atlantic side of South America is characterized by humid lowlands interspersed with large upland plateaus called **shields**. The Brazilian Shield is the largest, followed by the Patagonian and Guiana shields. Across these lowlands meander some of the great rivers of the world, including the Amazon.

Historically, the most important areas of settlement in tropical Latin America were not along the region's major rivers but across its shields, plateaus, and fertile mountain valleys. In these places, the combination of arable land, mild climate, and sufficient rainfall produced the region's most productive agricultural areas and its densest settlement. The Mexican Plateau, for example, is a massive upland area ringed by the Sierra Madre mountains. The southern end of the plateau is where the Valley of Mexico is located. Similarly, the elevated and well-watered basins of Brazil's southern mountains provide an ideal setting for agriculture. These especially fertile areas are able to support high population densities, so it is not surprising that the region's two largest cities, Mexico City and São Paulo, emerged in these settings. The Latin American highlands also lend a special character to the region. Lush tropical valleys nestled below snow-covered mountains hint at the diversity of ecosystems found near one another. The most dramatic of these highland areas, the Andes, runs like a spine down the length of the South American continent.

**The Andes**   Beginning in northwestern Venezuela and ending at Tierra del Fuego, the Andes are relatively young mountains that extend nearly 5,000 miles (8,000 kilometers). They are an ecologically and geologically diverse mountain chain with some 30 peaks higher than 20,000 feet (6,000 meters). Many rich veins of precious metals and minerals are found in these mountains. In fact, the initial economic wealth of many Andean countries came from mining silver, gold, tin, copper, and iron.

Given the length of the Andes, the mountain chain is typically divided into northern, central, and southern components. In Colombia the northern Andes actually split into three distinct mountain ranges before merging near the border with Ecuador. High-altitude plateaus and snow-covered peaks distinguish the central Andes of Ecuador,

**FIGURE 4.8 Bolivian Altiplano**   The Altiplano is an elevated plateau straddling the Bolivian and Peruvian Andes. One of its striking features is beautiful Lake Titicaca, at an elevation of 12,500 feet (3,810 meters). Amerindians inhabit this stark and windswept land, which is one of the poorer areas of the Andes.

Peru, and Bolivia. The Andes reach their greatest width here. Of special interest is the treeless high plain of Peru and Bolivia, called the **Altiplano**. The floor of this elevated plateau ranges from 11,800 feet (3,600 meters) to 13,000 feet (4,000 meters) in altitude, and it has limited usefulness for grazing. Two high-altitude lakes, Titicaca on the Peruvian and Bolivian border and the smaller Poopó in Bolivia, are located in the Altiplano, as are many mining sites (Figure 4.8). The highest peaks are found in the southern Andes, shared by Chile and Argentina, including the highest peak in the Western Hemisphere, Aconcagua, at almost 23,000 feet (6,958 meters).

**The Uplands of Mexico and Central America** The Mexican Plateau and the Volcanic Axis of Central America are the most important Latin American uplands in terms of settlement. Most major cities of Mexico and Central America are found here. The southern end of the Mexican Plateau, the Mesa Central, contains a number of flat-bottomed basins interspersed with volcanic peaks. Mexico's megalopolis—a concentration of the largest population centers such as Mexico City, Guadalajara, and Puebla—is in the Mesa Central. (The Valley of Mexico, discussed earlier, is one of the basins of the Mesa Central.)

Along the Pacific Coast of Central America lies a chain of volcanoes that stretches from Guatemala to Costa Rica. The Volcanic Axis of Central America is a handsome landscape of rolling green hills, elevated basins with sparkling lakes, and volcanic peaks. More than 40 volcanoes are found here, many of them still active, which have produced a rich volcanic soil that yields a wide variety of domestic and export crops. Most of Central America's population is also concentrated in this zone, in the capital cities or the surrounding rural villages. The bulk of

**FIGURE 4.10 Patagonian Wildlife** Guanacos, native to South America, thrive on the steppe vegetation found throughout Patagonia. The numbers of guanaco fell dramatically due to hunting and competition with introduced livestock.

the agricultural land is tied up in large holdings that produce beef, cotton, and coffee for export. Yet most of the farms are small subsistence properties that produce corn, beans, squash, and assorted fruits.

**The Shields** As mentioned earlier, South America has three major shields—large upland areas of exposed crystalline rock that are similar to upland plateaus found in Africa and Australia. (The Guiana Shield will be discussed in Chapter 5.) The Brazilian and Patagonian shields vary in elevation between 600 and 5,000 feet (200 and 1,500 meters). The Brazilian Shield is the larger and more important in terms of natural resources and settlement. Far from a uniform land surface, the Brazilian Shield covers much of Brazil from the Amazon Basin in the north to the Plata Basin in the south. In the southeast corner of the plateau is the city of São Paulo, the largest urban conglomeration in South America. The other major population centers are on the coastal edge of the plateau, where large protected bays made the sites of Rio de Janeiro and Salvador attractive to Portuguese colonists. Finally, the Paraná basalt plateau, located on the southern end of the Brazilian Shield, is famous for its fertile red soils (*terra roxa*), which yield coffee, oranges, and soybeans. So fertile is this area that the economic rise of São Paulo is attributed to the expansion of commercial agriculture, especially coffee, into this area (Figure 4.9).

The Patagonian Shield lies in the southern tip of South America. Beginning south of Bahia Blanca and extending to Tierra del Fuego, the region to this day is sparsely settled and hauntingly beautiful. It is treeless, covered by scrubby steppe vegetation, and home to wildlife such as the guanaco (Figure 4.10). Sheep were introduced to Patagonia in the late 19th century, spurring a wool boom. More recently, offshore oil production has renewed the economic importance of Patagonia.

## River Basins and Lowlands

Three great river basins drain the Atlantic lowlands of South America: the Amazon, Plata, and Orinoco (Figure 4.11). Within these basins are vast interior lowlands, less than 600 feet (200 meters) in elevation. Most of these lowlands are sparsely settled and offer limited agricultural

**FIGURE 4.9 Brazilian Oranges** Most estate-grown oranges in Brazil are processed into frozen concentrate and exported. São Paulo and Paraná have some of the finest soils in Brazil. In addition to oranges, coffee and soybeans are widely cultivated.

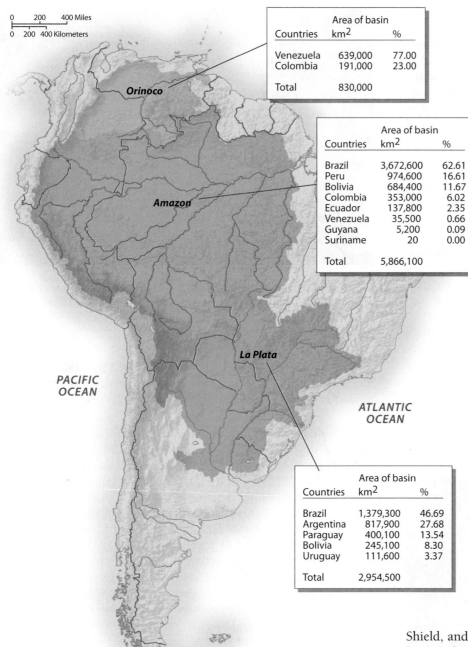

| | Area of basin | |
|---|---|---|
| Countries | km² | % |
| Venezuela | 639,000 | 77.00 |
| Colombia | 191,000 | 23.00 |
| Total | 830,000 | |

| | Area of basin | |
|---|---|---|
| Countries | km² | % |
| Brazil | 3,672,600 | 62.61 |
| Peru | 974,600 | 16.61 |
| Bolivia | 684,400 | 11.67 |
| Colombia | 353,000 | 6.02 |
| Ecuador | 137,800 | 2.35 |
| Venezuela | 35,500 | 0.66 |
| Guyana | 5,200 | 0.09 |
| Suriname | 20 | 0.00 |
| Total | 5,866,100 | |

| | Area of basin | |
|---|---|---|
| Countries | km² | % |
| Brazil | 1,379,300 | 46.69 |
| Argentina | 817,900 | 27.68 |
| Paraguay | 400,100 | 13.54 |
| Bolivia | 245,100 | 8.30 |
| Uruguay | 111,600 | 3.37 |
| Total | 2,954,500 | |

**FIGURE 4.11 South American River Basins** The three great river basins of the region are the Amazon, Plata, and Orinoco. The Amazon Basin covers 6 million square kilometers, including portions of eight countries, but the majority of the basin is within Brazil. The Amazon is the largest river system in the world in terms of volume of water and area. The Plata Basin drains nearly 3 million square kilometers across five countries and is intensely farmed and used for hydroelectricity. The Orinoco Basin, shared by Venezuela and Colombia, covers nearly 1 million square kilometers.

definitely drier and wetter times of year, with August and September being the driest months. In the basin, rainfall is likely most days, but showers often pass quickly, leaving bright blue skies. The extent of this watershed and its hydrologic cycle is highlighted by the fact that 20 percent of all freshwater discharged into the oceans comes from the Amazon.

Because the Amazon Basin draws from nine countries, it would seem that this watershed would be an ideal network to integrate the northern half of South America. Ironically, compared with the other great rivers of the world, settlement in the basin continues to be sparse, in large part due to the poor quality of the forest soils. Active colonization of the Brazilian portion of the Amazon since the 1960s has led to significant population growth. Roughly 15 million people live in the Brazilian Amazon, which is just 8 percent of the country's total population. Still, the population in the Amazonian states increases at nearly 4 percent a year.

**The Plata Basin** The region's second largest watershed begins in the tropics and discharges into the Atlantic in the midlatitudes. Three major rivers make up this system: the Paraná, the Paraguay, and the Uruguay. The Paraguay River and its tributaries drain the eastern Andes of Bolivia, the Brazilian Shield, and the Chaco. The Paraná primarily drains the Brazilian uplands before the Paraguay River joins it in northern Argentina.

Unlike the Amazon, much of the Plata Basin is now economically productive through large-scale mechanized agriculture, especially soybean production. Arid areas such as the Chaco and seasonally flooded lowlands such as the Pantanal support livestock. The Plata Basin contains several major dams, including the region's largest hydroelectric plant, the Itaipú on the Paraná, which generates all of Paraguay's electricity and much of southern Brazil's. As agricultural output in this watershed grows, sections of the Paraná River are being dredged to improve the river's capacity for barge and boat traffic.

potential except for grazing livestock. Yet the pressure to open new areas for settlement and to develop natural resources has created pockets of intense economic activity in the lowlands. Areas within the Amazon and Plata basins have experienced significant increases in resource extraction, farming, and settlement since the 1970s.

**The Amazon Basin** The Amazon drains an area of roughly 2.4 million square miles (6.1 million square kilometers), making it the largest river system in the world by volume and area and the second longest by length. Everywhere in the basin, rainfall is more than 60 inches (150 centimeters) a year and in many places more than 80 inches (200 centimeters). The basin's largest city, Belém, averages close to 100 inches (250 centimeters) a year. Although there is no real dry season, there are

## Climate and Climate Change in Latin America

In tropical Latin America, average monthly temperatures in settings such as Managua (Nicaragua), Quito (Ecuador), and Manaus (Brazil) show little variation (see Figure 4.12). Precipitation patterns, however, are variable and create distinct wet and dry seasons. In Managua, for example, January is typically a dry month and October is a wet

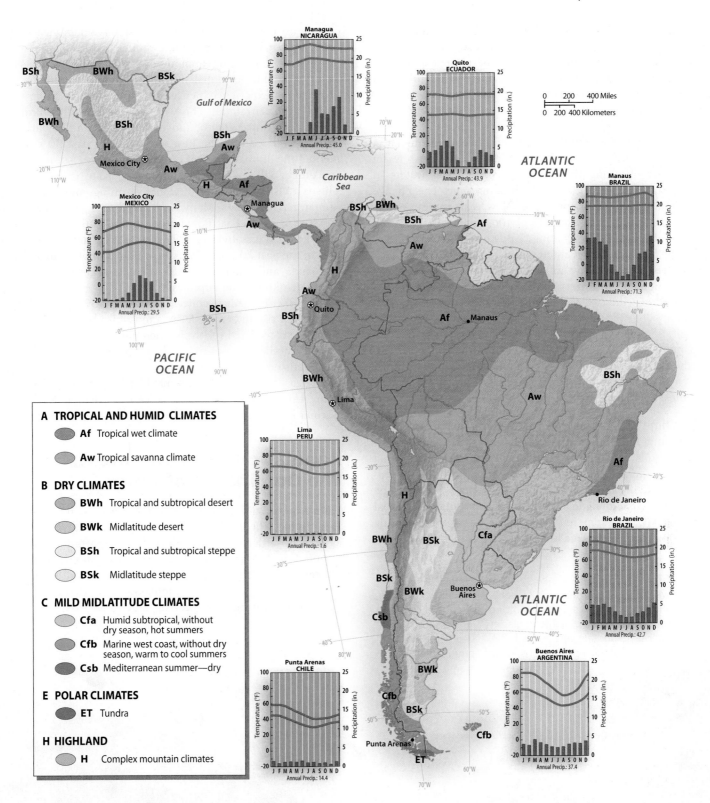

**FIGURE 4.12 Climate Map of Latin America** Latin America includes the world's largest rain forest (Af) and driest desert (BWh), as well as nearly every other climate classification. Latitude, elevation, and rainfall play important roles in determining the region's climates. Note the contrast in rainfall patterns between humid Quito and arid Lima. *(Temperature and precipitation data from Pearce and Smith, 1984, The World Weather Guide, London: Hutchinson)*

one. The tropical lowlands of Latin America, especially east of the Andes, are usually classified as tropical humid climates that support forest or savanna, depending on the amount of rainfall. The region's desert climates are found along the Pacific coasts of Peru and Chile, Patagonia, northern Mexico, and the Bahia of Brazil. Thus, a city such as Lima, Peru, which is clearly in the tropics, averages only 1.5 inches (4 centimeters) of rainfall a year due to the extreme aridity of the Peruvian coast.

Midlatitude climates, with hot summers and cold winters, prevail in Argentina, Uruguay, and parts of Paraguay and Chile. Of course, the midlatitude temperature shifts in the Southern Hemisphere are the opposite of those in the Northern Hemisphere (cold Julys and warm Januarys). In the mountain ranges, complex climate patterns result from changes in elevation.

**El Niño** One of the most studied weather phenomena in Latin America, **El Niño** (named after the Christ child) occurs when a warm Pacific current arrives along the normally cold coastal waters of Ecuador and Peru in December, around Christmastime. This change in ocean temperature, which happens every few years, produces torrential rains, signaling the arrival of an El Niño year. The 1997–1998 El Niño was especially bad. Devastating floods occurred in Peru and Ecuador. Heavy May rains in Paraguay and Argentina caused the Paraná River to rise 26 feet (8 meters) above normal.

Other than flooding, the less-talked-about result of El Niño is drought. While the Pacific Coast of South and North America experienced record rainfall in the 1997–1998 El Niño, Colombia, Venezuela, northern Brazil, Central America, and Mexico battled drought. In addition to crop and livestock losses, estimated to be in the billions of dollars, hundreds of brush and forest fires also left their mark on the region's landscape. Scientists are not yet sure whether or how global climate change will impact the frequency and strength of El Niño cycles.

**Impacts of Climate Change** Global climate change has both immediate and long-term implications for Latin America. Of greatest immediate concern is how climate change will influence agricultural productivity, water availability, changes in the composition and productivity of ecosystems, and incidence of vector-borne diseases such as malaria and, especially, dengue fever. Changes attributable to global warming are already apparent in higher-elevation regions, making these concerns more pressing. The long-term effects of global climate

change on lowland tropical forest systems is less clear; for example, some areas may experience more rainfall, others less.

Climate change research indicates that highland areas are particularly vulnerable to global warming. Tropical mountain systems are projected to experience increased temperatures of 2 to 6°F (1 to 3°C), as well as lower rainfall. This will raise the altitudinal limits of various ecosystems, impacting the range of crops and arable land available to farmers and pastoralists. Research over the past 50 years has documented the dramatic retreat of Andean glaciers; some no longer exist, and others will cease to exist in the next 10 to 15 years (Figure 4.13). This is a visible indicator of global warming, and it also has pressing human repercussions. Many Andean villages, as well as metropolitan areas such as La Paz, Bolivia, get much of their water from glacial runoff. A major Bolivian glacier, Chacaltaya, has lost 80 percent of its area in the past 20 years. Thus as average temperatures increase in the highlands and glaciers recede, there is widespread concern about future drinking water supplies.

Another immediate concern brought on by warmer temperatures is the sudden rise in dengue fever, a mosquito-borne virus. Dengue fever was once considered relatively uncommon in highland Latin America, but the number of cases has risen sharply in the past few years. Tens of thousands now suffer from its fever, headache, nausea, joint pain, and, in rare cases, external and internal bleeding that can result in death. In Latin America, the sudden rise in cases of dengue fever suggests that warmer highland temperatures have placed millions more at risk.

# Population and Settlement: The Dominance of Cities

Latin America did not have great river basin civilizations like those in Asia. In fact, the great rivers of the region are surprisingly underused as areas of settlement or corridors for transportation. While the major population clusters of Central America and Mexico are in the interior plateaus and valleys, the interior lowlands of South America are relatively empty. Historically, the highlands supported most of the region's population during the pre-Hispanic and colonial eras. In the

**FIGURE 4.13 Glacial Retreat** Research in areas such as the Peruvian Andes over the past four decades indicates a dramatic decline in the size of Andean glaciers. This is an indicator of the pace of climate change in the tropics, and highland glaciers are also sources of water for mountain villages and cities. Their demise has serious repercussions for the communities that depend on them.

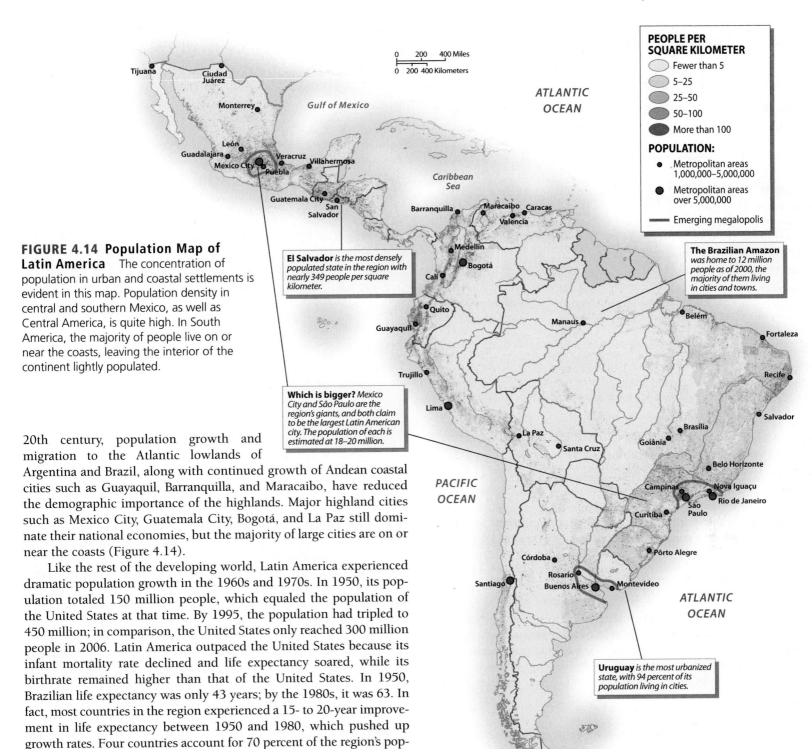

**PEOPLE PER SQUARE KILOMETER**
- Fewer than 5
- 5–25
- 25–50
- 50–100
- More than 100

**POPULATION:**
- • Metropolitan areas 1,000,000–5,000,000
- ● Metropolitan areas over 5,000,000
- ▬ Emerging megalopolis

**El Salvador** *is the most densely populated state in the region with nearly 349 people per square kilometer.*

**The Brazilian Amazon** *was home to 12 million people as of 2000, the majority of them living in cities and towns.*

**Which is bigger?** *Mexico City and São Paulo are the region's giants, and both claim to be the largest Latin American city. The population of each is estimated at 18–20 million.*

**Uruguay** *is the most urbanized state, with 94 percent of its population living in cities.*

**FIGURE 4.14 Population Map of Latin America** The concentration of population in urban and coastal settlements is evident in this map. Population density in central and southern Mexico, as well as Central America, is quite high. In South America, the majority of people live on or near the coasts, leaving the interior of the continent lightly populated.

20th century, population growth and migration to the Atlantic lowlands of Argentina and Brazil, along with continued growth of Andean coastal cities such as Guayaquil, Barranquilla, and Maracaibo, have reduced the demographic importance of the highlands. Major highland cities such as Mexico City, Guatemala City, Bogotá, and La Paz still dominate their national economies, but the majority of large cities are on or near the coasts (Figure 4.14).

Like the rest of the developing world, Latin America experienced dramatic population growth in the 1960s and 1970s. In 1950, its population totaled 150 million people, which equaled the population of the United States at that time. By 1995, the population had tripled to 450 million; in comparison, the United States only reached 300 million people in 2006. Latin America outpaced the United States because its infant mortality rate declined and life expectancy soared, while its birthrate remained higher than that of the United States. In 1950, Brazilian life expectancy was only 43 years; by the 1980s, it was 63. In fact, most countries in the region experienced a 15- to 20-year improvement in life expectancy between 1950 and 1980, which pushed up growth rates. Four countries account for 70 percent of the region's population: Brazil with 191 million, Mexico with 110 million, Colombia with 45 million, and Argentina with 40 million (Table 4.1).

## The Latin American City

A quick glance at the population map of Latin America shows a concentration of people in cities (see Figure 4.14). One of the most significant demographic shifts has been the movement out of rural areas to cities, which began in earnest in the 1950s. In 1950, just one-quarter of the region's population was urban; the rest lived in small villages and the countryside. Today, the pattern is reversed, with three-quarters of the population living in cities. In the most urbanized countries, such as Argentina, Chile, Uruguay, and Venezuela, nearly 90 percent of the

population lives in cities. This preference for urban life is attributed to cultural as well as economic factors. Under Iberian rule, people residing in cities had higher social status and greater economic opportunity. Initially, only Europeans were allowed to live in the colonial cities, but this exclusivity was not strictly enforced. Over the centuries, colonial cities became the hubs for transportation and communication, making them the primary centers for economic and social activities.

**TABLE 4.1   POPULATION INDICATORS**

| Country | Population (millions) 2009 | Population Density (per square kilometer) | Rate of Natural Increase (RNI) | Total Fertility Rate | Percent Urban | Percent <15 | Percent >65 | Net Migration (Rate per 1000) 2005–10* |
|---------|---------|---------|---------|---------|---------|---------|---------|---------|
| Argentina | 40.3 | 14 | 1.0 | 2.4 | 91 | 26 | 10 | 0.2 |
| Bolivia | 9.9 | 9 | 2.1 | 3.5 | 65 | 38 | 4 | −2.1 |
| Brazil | 191.5 | 22 | 1.0 | 2.0 | 84 | 28 | 6 | −0.2 |
| Chile | 17.0 | 22 | 1.0 | 1.9 | 87 | 25 | 8 | 0.4 |
| Colombia | 45.1 | 40 | 1.4 | 2.4 | 75 | 30 | 5 | −0.5 |
| Costa Rica | 4.5 | 88 | 1.3 | 1.9 | 59 | 27 | 6 | 1.3 |
| Ecuador | 13.6 | 48 | 1.8 | 2.8 | 63 | 33 | 6 | −5.2 |
| El Salvador | 7.3 | 349 | 1.4 | 2.5 | 60 | 34 | 5 | −9.1 |
| Guatemala | 14.0 | 129 | 2.8 | 4.4 | 47 | 43 | 4 | −3.0 |
| Honduras | 7.5 | 67 | 2.2 | 3.3 | 49 | 38 | 4 | −2.8 |
| Mexico | 109.6 | 56 | 1.6 | 2.3 | 77 | 32 | 6 | −4.5 |
| Nicaragua | 5.7 | 44 | 2.1 | 2.9 | 58 | 38 | 4 | −7.1 |
| Panama | 3.5 | 46 | 1.6 | 2.4 | 64 | 30 | 6 | 0.7 |
| Paraguay | 6.3 | 16 | 2.1 | 3.5 | 57 | 36 | 5 | −1.3 |
| Peru | 29.2 | 23 | 1.5 | 2.6 | 76 | 32 | 6 | −4.4 |
| Uruguay | 3.4 | 19 | 0.5 | 2.0 | 94 | 24 | 13 | -3.0 |
| Venezuela | 28.4 | 31 | 2.0 | 2.6 | 88 | 31 | 5 | 0.3 |

Source: Population Reference Bureau, *World Population Data Sheet, 2009.*

*Net Migration Rate from the United Nations, Population Division, *World Population Prospects: The 2008 Revision Population Database.*

Latin American cities are noted for high levels of **urban primacy**, a condition in which a country has a primate city three to four times larger than any other city in the country. Examples of primate cities are Lima, Caracas, Guatemala City, Santiago, Buenos Aires, and Mexico City. Primacy is often viewed as a problem because too many national resources are concentrated into one urban center. In three cases, the growth of urbanized regions has led to the emergence of a megalopolis: the Mexico City–Puebla–Toluca–Cuernavaca area on the Mesa Central, the Niterói–Rio de Janeiro–Santos–São Paulo–Campinas axis in southern Brazil, and the Rosario–Buenos Aires–Montevideo–San Nicolás corridor in Argentina and Uruguay's lower Rio Plata Basin (see Figure 4.14).

**Urban Form**   Latin American cities have a distinct urban form that reflects both their colonial origins and their present-day growth (Figure 4.15). Usually a clear central business district exists in the old colonial core. Radiating out from the central business district is older middle- and lower-class housing found in the zones of maturity and *in situ* accretion (an area of mixed levels of housing and services). In this model, residential quality declines as one moves from the center to the periphery. The exception is the elite spine, a newer commercial and business strip that extends from the colonial core to newer parts of the city. Along the spine one finds superior services, roads, and transportation. The city's best residential zones, as well as shopping malls, are usually on either side of the spine. Close to the elite residential sector, a limited area of middle-class tract housing is typically found. Most major urban centers also have a *periférico*

(a ring road or beltway highway) that encircles the city. Industry is located in isolated areas of the inner city and in larger industrial parks outside the ring road.

Straddling the periférico is a zone of peripheral **squatter settlements** where many of the urban poor live in self-built housing on land that does not belong to them. Services and infrastructure are extremely limited: Roads are unpaved, water is often trucked in, and sewer systems are nonexistent (Figure 4.16). The dense ring of squatter settlements that encircles Latin American cities reflects the speed and intensity with which these zones were created. In some cities, more than one-third of the population lives in these self-built homes of marginal or poor quality. These kinds of dwellings are found throughout the developing world, yet the practice of building one's home on the "urban frontier" has a longer history in Latin America than in most Asian and African cities. The combination of a rapid inflow of migrants (at times reaching 1,000 people per day), the inability of governments to meet pressing housing needs, and the eventual official recognition of many of these neighborhoods with land titles and utilities meant that this housing strategy was rarely discouraged. Each successful settlement on the urban edge encouraged more.

## Patterns of Rural Settlement

Throughout Latin America, a distinct rural lifestyle exists, especially among peasant subsistence farmers. While the majority of people live in cities, approximately 125 million people do not. In Brazil alone, at least 35 million people live in rural areas. Interestingly, the absolute

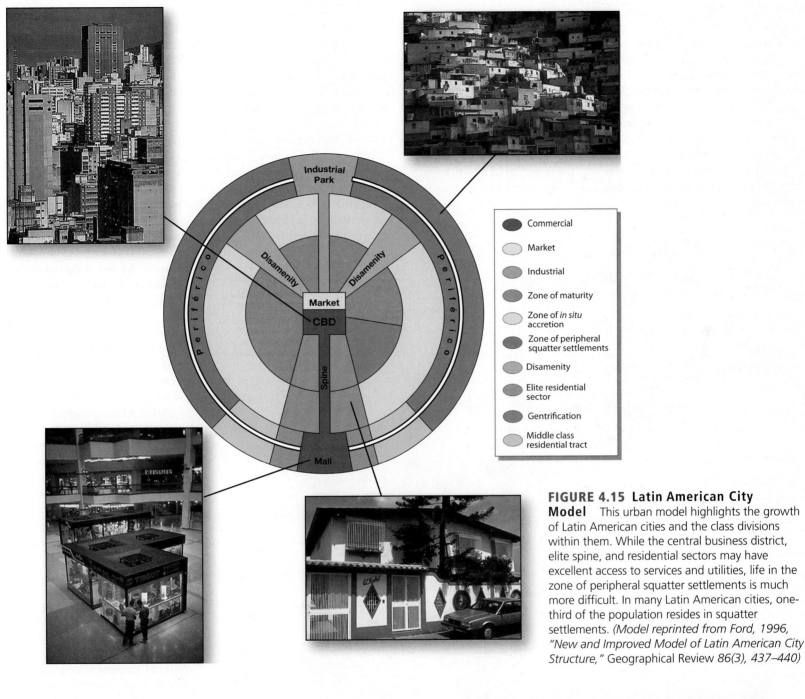

**FIGURE 4.15 Latin American City Model** This urban model highlights the growth of Latin American cities and the class divisions within them. While the central business district, elite spine, and residential sectors may have excellent access to services and utilities, life in the zone of peripheral squatter settlements is much more difficult. In many Latin American cities, one-third of the population resides in squatter settlements. *(Model reprinted from Ford, 1996, "New and Improved Model of Latin American City Structure,"* Geographical Review *86(3), 437–440)*

number of people living in rural areas today is roughly equal to the number in the 1960s. Yet rural life has definitely changed. The links between rural and urban areas are much improved, making rural people less isolated. In addition to village-based subsistence production, in most rural areas highly mechanized capital-intensive farming occurs. Much like the region's cities, the rural landscape is divided by extremes of poverty and wealth. A source of social and economic tension in the countryside is the uneven distribution of arable land.

**FIGURE 4.16 Lima Squatter Settlement** In arid Lima, squatters initially build their homes using straw mats. As settlements become more established, residents invest in adobe and cinder block to improve their homes. Life on the urban frontier is harsh. Water is trucked in, electricity is irregular, and travel to the city center is costly and slow.

**Rural Landholdings**   The control of land in Latin America was the basis for political and economic power. Historically, colonial authorities granted large tracts of land to the colonists, who were also promised the service of Indian laborers. These large estates typically took up the best lands along the valley bottoms and coastal plains. The owners were often absentee landlords, spending most of their time in the city and relying on a mixture of hired and slave labor to run their rural operations. Passed down from one generation to the next, many estates can trace their ownership back a century or more. The establishment of large blocks of estate land meant that peasants were denied access to territory of their own, so they were forced to work for the estates. This long-observed practice of maintaining large estates is called **latifundia**.

Although the pattern of estate ownership is well documented, peasants have always farmed small plots for their subsistence. This practice of **minifundia** can lead to permanent or shifting cultivation. Small farmers typically plant a mixture of crops for subsistence as well as for trade. Peasant farmers in Colombia or Costa Rica, for example, grow corn, fruits, and various vegetables alongside coffee bushes that produce beans for export. Strains on the minifundia system occur when rural populations grow and land becomes scarce, forcing farmers to divide their properties into smaller and less-productive parcels.

Much of the turmoil in 20th-century Latin America surrounded the question of land, with peasants demanding its redistribution through the process of **agrarian reform**. Governments have addressed these concerns in different ways. The Mexican Revolution in 1910 led to a system of communally held lands called *ejidos*. In the 1950s, Bolivia crafted agrarian reform policies that led to the government appropriating estate lands and redistributing them to small farmers. As part of the Sandinista revolution in Nicaragua in 1979, lands were taken from the political elite and converted into collective farms. In 2000, president Hugo Chavez ushered in a new era of agrarian reform in Venezuela, and the Bolivian president Evo Morales introduced an agrarian reform program in 2006 to give land title to indigenous communities in the eastern lowlands. Each of these programs has met with resistance and proved to be politically and economically difficult. Eventually, the path chosen by most governments has been to make frontier lands available to land-hungry peasants.

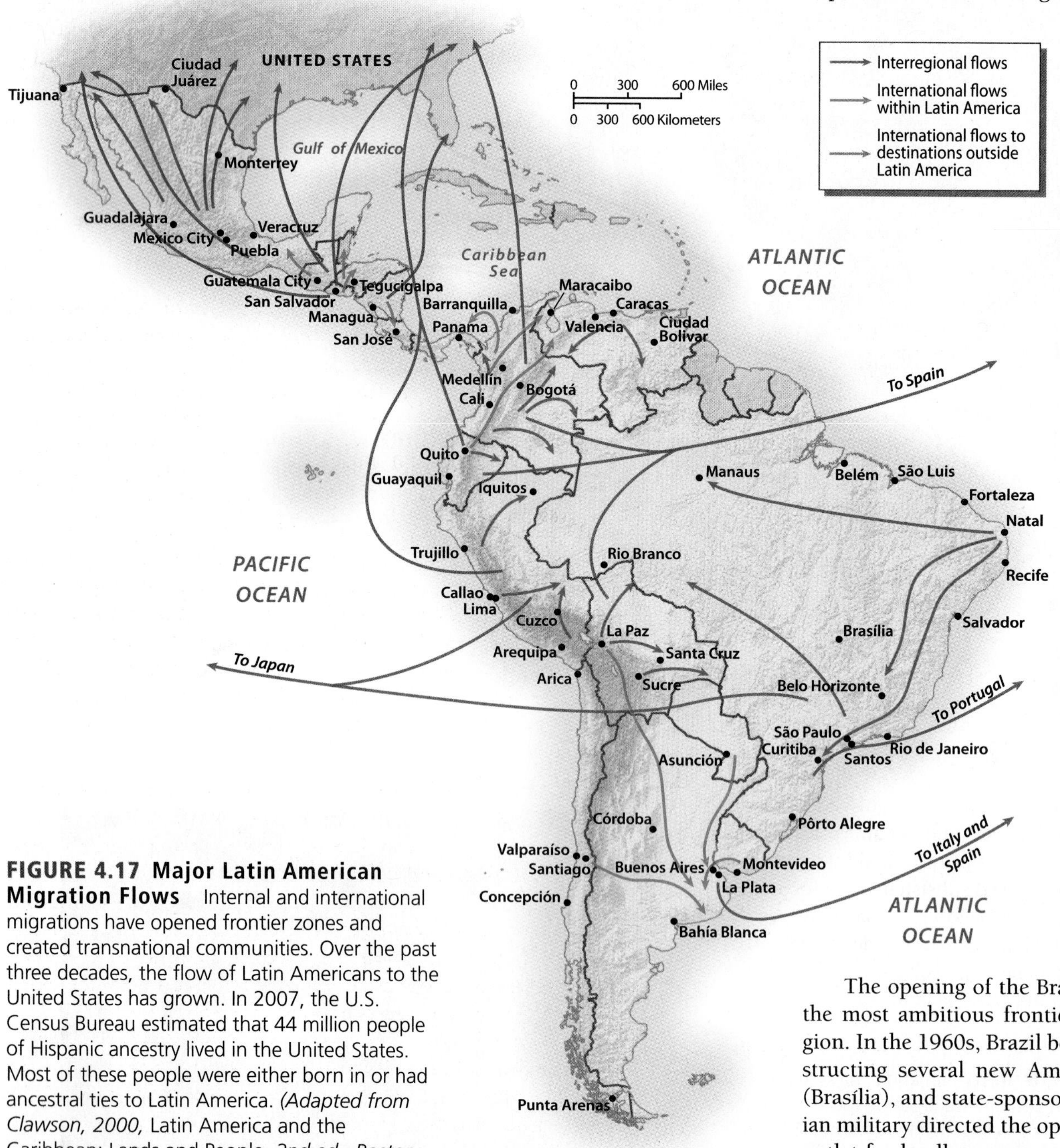

**Agricultural Frontiers**   The creation of agricultural frontiers served several purposes: providing peasants with land, tapping unused resources, and filling in blank spots on the map with settlers. Although the dominant demographic trend has been a rural-to-urban movement, an important rural-to-rural flow has changed undeveloped areas into agricultural communities (Figure 4.17).

The opening of the Brazilian Amazon for settlement was the most ambitious frontier colonization scheme in the region. In the 1960s, Brazil began its frontier expansion by constructing several new Amazonian highways, a new capital (Brasília), and state-sponsored mining operations. The Brazilian military directed the opening of the Amazon to provide an outlet for landless peasants and to extract the region's many resources. Yet the generals' plans did not deliver as intended.

**FIGURE 4.17 Major Latin American Migration Flows**   Internal and international migrations have opened frontier zones and created transnational communities. Over the past three decades, the flow of Latin Americans to the United States has grown. In 2007, the U.S. Census Bureau estimated that 44 million people of Hispanic ancestry lived in the United States. Most of these people were either born in or had ancestral ties to Latin America. *(Adapted from Clawson, 2000,* Latin America and the Caribbean: Lands and People, *2nd ed., Boston: McGraw-Hill)*

Throughout the basin, nutrient-poor tropical soils could not support permanent agricultural colonies. Government-promised land titles, agricultural subsidies, and credits were slow to reach small farmers. Instead, too much money went to subsidizing large cattle ranches through tax breaks and improvement deals in which "improved" land meant cleared land. Today, five times more people live in the Amazon than in the 1960s; thus, continued human modification of this region is inevitable.

## Population Growth and Movement

The high growth rates in Latin America throughout the 20th century are attributed to natural increases as well as immigration. The 1960s and 1970s were decades of tremendous growth, resulting from high fertility rates and increasing life expectancy. In the 1960s, for example, a typical Latin American woman had six or seven children. By the 1980s, family sizes were half as big. A number of factors explain this: more urban families, which tend to be smaller than rural ones; increased participation of women in the workforce; higher education levels of women; state support of family planning; and better access to birth control. The exceptions to this trend are the poor and more rural countries, such as Guatemala and Bolivia, where the average woman has four or five children. Cultural factors may also be at work, as Amerindian peoples in the region tend to have more children than others.

Even with family sizes shrinking and nearing replacement rates in Uruguay and Chile, there is built-in potential for continued growth because of the relative demographic youth of these countries. The average percentage of the population below age 15 is 30 percent. In North America, that same group is 20 percent of the population, and in Europe it is just 16 percent. This means that a proportionally larger segment of the population has yet to enter into its childbearing years.

Waves of immigrants into Latin America and migrant streams within Latin America have influenced population size and patterns of settlement. Beginning in the late 19th century, new immigrants from Europe and Asia added to the region's size and ethnic complexity. Important population shifts within countries have also occurred in recent decades, as witnessed by the growth of Mexican border towns and new settlements in eastern Bolivian lowlands. In an increasingly globalized economy, even more Latin Americans live and work outside the region, especially in the United States.

**European Migration** After gaining their independence from Iberia, Latin America's new leaders sought to develop their territories through immigration. Firmly believing that "to govern is to populate," many countries set up immigration offices in Europe to attract hardworking peasants to till the soils and "whiten" the *mestizo* (people of mixed European and Indian ancestry) population. The Southern Cone countries of Argentina, Chile, Uruguay, and southern Brazil were the most successful in attracting European immigrants from the 1870s until the depression of the 1930s. During this period, some 8 million Europeans arrived (more than came during the entire colonial period), with Italians, Portuguese, Spaniards, and Germans being the most numerous.

**Asian Migration** Less well-known than the European immigrants are the Asian immigrants who also arrived during the late 19th and 20th centuries. Although considerably fewer, over time they established an

**FIGURE 4.18 Japanese-Brazilians**   Japanese-Brazilians living in Japan display a banner that reads "Don't Throw Away Japanese-Brazilian Workers" during a rally in 2009.  As unemployment jumped due to the global downturn in 2009, the Japanese government starting offering cash to jobless migrants of Japanese descent, mostly from Brazil and Peru, to return to their countries of birth.

important presence in the large cities of Brazil, Peru, Argentina, and Paraguay. Beginning in the mid-19th century, the Chinese and Japanese who settled in Latin America were contracted to work on the coffee estates in southern Brazil and the sugar estates and coastal mines of Peru. In the 1990s a Japanese-Peruvian, Alberto Fujimori, was president of Peru. Between 1908 and 1978 a quarter-million Japanese immigrated to Brazil; today the country is home to 1.3 million people of Japanese descent. As a group, the Japanese have been closely associated with the expansion of soybean and orange production. Increasingly, second- and third-generation Japanese have taken professional and commercial jobs in Brazilian cities; many of them have married outside their ethnic group and are losing their fluency in Japanese. South America's economic turmoil in the 1990s encouraged many ethnic Japanese to emigrate to Japan in search of better wages. Nearly one-quarter of a million ethnic Japanese left South America in the 1990s (mostly from Brazil and Peru) and now work in Japan (Figure 4.18).

**Latino Migration and Hemispheric Change**   Movement within Latin America and between Latin America and North America has had a significant impact on both sending and receiving communities. Within Latin America, international migration is shaped by shifting economic and political realities. For example, Venezuela's oil attracted Colombian immigrants who tended to work as domestics or agricultural laborers. Argentina has long been a destination for Bolivian and Paraguayan laborers. And farmers in the United States have depended on Mexican laborers for more than a century (see Figure 4.17).

Political turmoil has also sparked waves of international migrants. The bloody civil wars in El Salvador and Guatemala, for example, sent waves of refugees into neighboring countries, such as Mexico and the United States. Because most countries in the region have democratically elected governments, most of today's immigrants are classified as economic migrants, not political refugees seeking a safe haven.

**FIGURE 4.19 Mexican–U.S. Border Crossing**   Mexican day workers cross the border into El Paso, Texas, from Ciudad Juárez. Mexicans have long used these busy border crossings to enter the United States. With the intensification of border security in the past decade, many wait hours to cross the border.

Presently, Mexico is the largest country of origin of legal immigrants to the United States (Figure 4.19). In 2007, 28 million people claimed Mexican ancestry, and of these, approximately 12 million were immigrants. Mexican labor migration to the United States dates back to the late 1800s, when relatively unskilled labor was recruited to work in agriculture, mining, and railroads. Today roughly two-thirds of the Hispanic population (both foreign born and native born) in the United States claims Mexican ancestry. Mexican immigrants are most concentrated in California and Texas, but increasingly they are found throughout the country. Although Mexicans continue to have the greatest presence among Latinos in the United States, the number of immigrants from El Salvador, Guatemala, Nicaragua, Colombia, Ecuador, and Brazil has steadily grown. The U.S. Census Bureau estimated that there were 44 million Hispanics in the United States (both foreign and native born) in 2007. Most of this population has ancestral ties with peoples from Latin America and the Caribbean (see Chapter 5 on Caribbean migration).

Today, Latin America is seen as a region of emigration rather than one of immigration. Both skilled and unskilled workers from Latin America are an important source of labor in North America, Europe, and Japan. Many of these immigrants send monthly **remittances** (monies sent back home) to sustain family members (Figure 4.20). In 2007, it was estimated that immigrants sent more than $65 billion to Latin America. Most of this money came from workers in the United States, but Latino immigrants in Spain, Japan, Canada, and Italy also sent money back to the region. Through remittances and technological advances that make communication faster and less expensive, immigrants maintain close contact with their home countries in ways that earlier generations could not (see "Exploring Global Connections: Latin Americans Bound for Europe).

# EXPLORING GLOBAL CONNECTIONS
## Latin Americans Bound for Europe

Europeans colonized Latin America, but only recently have significant numbers of Latin Americans migrated to Europe, especially Iberia (Spain and Portugal), for employment opportunities.

Several factors have converged to make Iberia a favored destination for South American immigrants, especially Ecuadorians, Colombians, Argentines, Bolivians, Peruvians, and Brazilians. Until the 1990s, most South Americans who emigrated chose the United States. Yet as both legal and illegal immigration to the United States have gotten more costly and difficult, especially in the aftermath of September 11, 2001, Spain and Portugal have increasingly become more attractive. Until recently, no visa requirement existed for Latin Americans traveling to Spain. Thus potential immigrants would enter as tourists and stay illegally as workers to serve the booming Spanish economy. The fact that they speak Spanish and understand the culture gives Latin Americans an added advantage over other immigrant groups from Africa or Asia.

Due to the onslaught of new arrivals, Spain responded by introducing new visa requirements, along with a regularization plan that legalized many foreign workers. An estimated 1.8 million Latin Americans working in Spain (mostly Madrid, Barcelona, and Valencia) sent home $5 billion in remittances in 2006. The flow to Portugal is considerably smaller, but it is the European destination of choice for Brazilians. Once in the European Union, Latin American migrants can travel and work in other countries, such as Italy, Switzerland, and the United Kingdom. Women, in particular, have a labor niche in child and elder care, while men often work jobs in construction (see Figure 4.2.1).

**FIGURE 4.2.1 Latin American Immigrants in Spain**   An immigrant family from Ecuador poses for a portrait in the streets of a small town in central Spain.

**FIGURE 4.20 Salvadorans in the Suburbs** Day laborers wait for employment in a Maryland suburb outside Washington, DC. War and economic hardship drove many Salvadorans from their country in the 1980s and 1990s. Many Salvadorans send remittances back to their country of birth, representing 17 percent of El Salvador's gross national income.

## Cultural Coherence and Diversity: Repopulating a Continent

The Iberian colonial experience brought to Latin America political and cultural unity that makes it recognizable today as a world region. Yet this was not a simple transplanting of Iberia across the Atlantic. Often a process unfolded in which European and Indian traditions blended as native groups were added into either the Spanish or the Portuguese empires. In some areas, such as southern Mexico, Guatemala, Bolivia, Ecuador, and Peru, Indian cultures have shown remarkable resilience, as evidenced by the survival of Amerindian languages. Yet the prevailing pattern is one of forced assimilation in which European religion, languages, and political organization were imposed on surviving Amerindian societies. Later, other cultures, especially more than 10 million African slaves, added to the cultural mix of Latin America, the Caribbean, and North America. The legacy of the African slave trade will be examined in greater detail in Chapters 5 and 6. For Latin America, perhaps the single most important factor in the dominance of European culture was the demographic collapse of native populations.

### The Decline of Native Populations

It is difficult to grasp the enormity of cultural change and human loss due to this encounter between two worlds (the Americas and Europe). Throughout the region, archaeological sites are reminders of the complexity of Amerindian civilizations prior to contact with Europe. Dozens of stone temples found throughout Mexico and Central America, where the Mayan and Aztec civilizations flourished, attest to the ability of these societies to thrive in the area's tropical forests and upland plateaus. The Mayan city of Tikal flourished in the lowland forests of Guatemala, supporting tens of thousands, before its mysterious collapse centuries before the arrival of Europeans (Figure 4.21). In the Andes, the complexity of Amerindian civilizations can be seen in political centers such as Cuzco and Machu Picchu. The Spanish, too,

**FIGURE 4.21 Tikal, Guatemala** This ancient Mayan city located in the lowland forests of the Petén supported up to 100,000 people before its collapse in the late 10th century. Today it is a major tourist destination.

were impressed by the sophistication and wealth they saw around them, especially in Tenochtitlán, where Mexico City sits today. Tenochtitlán was the political and ceremonial center of the Aztecs, supporting a complex metropolitan area with some 300,000 residents. The largest city in Spain at the time was considerably smaller.

**The Demographic Toll** The most telling figures of the impact of European expansion in Latin America are demographic. It is widely believed that the precontact Americas had 54 million inhabitants; by comparison, western Europe in 1500 had approximately 42 million. Of the 54 million, about 47 million were in what is now Latin America, and the rest were in North America and the Caribbean. By 1650, after a century and a half of colonization, the indigenous population was one-tenth its precontact size. The human tragedy of this population loss is difficult to comprehend. The relentless elimination of 90 percent of the native population was largely caused by epidemics of influenza and smallpox, but warfare, forced labor, and starvation due to a collapse of food production systems also contributed to the rapid population decline.

**Indian Survival** Presently, Mexico, Guatemala, Ecuador, Peru, and Bolivia have the largest indigenous populations. Not surprisingly, these areas had the densest native populations at contact. Indigenous survival also occurs in isolated settings where the workings of national and global economies are slow to break through.

In many cases Indian survival comes down to one key resource—land. Indigenous peoples who are able to maintain a territorial home, formally through land title or informally through long-term occupancy, are more likely to preserve a distinct ethnic identity. Because of this close association between identity and territory, native peoples are increasingly insisting on recognized spaces within their countries.

These efforts to define indigenous territory are seldom welcomed by the state. From Amazonia to Panama, many native groups are demanding formal political and territorial recognition as a means to address centuries of injustice.

## Patterns of Ethnicity and Culture

The Indian demographic collapse enabled Spain and Portugal to reshape Latin America into a European likeness. Yet instead of a neo-Europe rising in the tropics, a complex ethnic blend evolved. Beginning with the first years of contact, unions between European sailors and Indian women began the process of racial mixing that over time became a defining feature of the region. The courts of Spain and Portugal officially discouraged racial mixing, but such positions could not be realistically enforced in the colonial territories.

After generations of intermarriage, four broad racial categories resulted: *blanco* (European ancestry), *mestizo* (mixed ancestry), *indio* (Indian ancestry), and *negro* (African

**DOMINANT/OFFICIAL\* LANGUAGES**

- Spanish
- Portuguese

**INDIGENOUS LANGUAGES**

1. Aymara
2. Embera
3. Garifuna
4. Guaraní
5. Quechua
6. Kuna
7. Mapuche
8. Mayan
9. Miskito
10. Mixtec
11. Nawan/Spanish
12. Pemong
13. Zapotec
14. Wahiro
15. Yanomama

- Scattered indigenous language communities

\*Multiple Official Languages:
\*Bolivia: Spanish, Quechua, Aymara, Guaraní
\*Peru: Spanish, Quechua

**FIGURE 4.22 Language Map of Latin America** The dominant languages of Latin America are Spanish and Portuguese. Nevertheless, there are significant areas in which native languages still exist and, in some cases, are recognized as official languages. Smaller language groups exist in Central America, the Amazon Basin, and southern Chile. *(Adapted from the* Atlas of the World's Languages, *1994, New York: Routledge)*

ancestry). The *blancos* (or Europeans) continue to be well represented among the elites, yet the vast majority of people are of mixed racial ancestry. *Día de la Raza*, the region's observance of Columbus Day, recognizes the emergence of a new *mestizo* race as the legacy of European conquest. Throughout Latin America, more than other regions of the world, miscegenation (or racial mixing) is the norm, and it makes the process of mapping racial or ethnic groups especially difficult.

**Languages**  Roughly two-thirds of Latin Americans are Spanish speakers, and one-third speak Portuguese. These colonial languages were so widespread by the 19th century that they were the unquestioned languages of government and education for the newly independent Latin American republics. In fact, until recently, many countries actively discouraged, and even repressed, Indian tongues. It took a constitutional amendment in Bolivia in the 1990s to legalize native-language instruction in primary schools and to recognize the country's multiethnic heritage (more than half the population is Indian, and Quechua, Aymara, and Guaraní are widely spoken) (Figure 4.22).

Because Spanish and Portuguese dominate, there is a tendency to overlook the influence of indigenous languages in the region. Mapping the use of native languages, however, reveals important areas of Indian resistance and survival. In the Central Andes of Peru, Bolivia, and southern Ecuador, more than 10 million people still speak Quechua and Aymara, along with Spanish. In Paraguay and lowland Bolivia, there are 4 million Guaraní speakers, and in southern Mexico and Guatemala, at least 6 to 8 million speak Mayan languages. Small groups of native-language speakers are found scattered throughout the sparsely settled interior of South America and the more isolated forests of Central America, but many of these languages have fewer than 10,000 speakers.

**Blended Religions**  Like language, the Roman Catholic faith appears to have been imposed upon the region without challenge. Most countries report 90 percent or more of their population as Catholic. Every major city has dozens of churches, and even the smallest village maintains a graceful church on its central square (Figure 4.23). In some countries, such as El Salvador and Uruguay, a sizable portion of the population attends Protestant churches, but the Catholic core of this region is still intact.

Exactly what native peoples absorbed of the Christian faith is unclear. Throughout Latin America, **syncretic religions**, blends of different belief systems, enabled animist practices to be included in Christian worship. These blends took hold and endured, in part because Christian saints were easy replacements for pre-Christian gods and because the Catholic Church tolerated local variations in worship as long as the process of conversion was under way. The Mayan practice of paying tribute to spirits of the underworld seems to be replicated today in Mexico and Guatemala through the practice of building small cave shrines to favorite Catholic saints and leaving offerings of fresh flowers and fruits. One of the most celebrated religious symbols in Mexico is the Virgin of Guadalupe—a dark-skinned virgin seen by an Indian shepherd boy—who became the patron saint of Mexico.

Syncretic religious practices also evolved and endured among African slaves. By far the greatest concentration of slaves was in the Caribbean, where slaves were used to replace the indigenous population after it was wiped out by disease (see Chapter 5). Within Latin America, the Portuguese colony of Brazil received the most Africans. The blend of Catholicism with African traditions is most obvious in

**FIGURE 4.23  The Catholic Church**  Churches, such as the Dolores Church in Tegucigalpa, are important religious and social centers. The majority of people in Latin America define themselves as Catholic. Many churches built in the colonial era are valued as architectural treasures and are beautifully preserved.

the celebration of carnival, Brazil's most popular festival and one of the major components of Brazilian national identity. The three days of carnival known as the Reign of Momo combines pre-Lenten celebrations with African musical traditions represented by the rhythmic samba bands (Figure 4.24). Today the festival—which is most associated with Rio de Janeiro—draws thousands of participants from all over the world.

**FIGURE 4.24  Carnival in Rio de Janeiro**  A samba band marches in the streets of Rio de Janeiro. Samba, the quintessential music of carnival, draws inspiration from African rhythmic traditions.

# Geopolitical Framework: Redrawing the Map

Latin America's colonial history, more than its present condition, unifies this region geopolitically. For the first 300 years after the arrival of Columbus, Latin America was a territorial prize sought by various European countries but effectively settled by Spain and Portugal. By the 19th century, the independent states of Latin America had formed, but they continued to experience foreign influence and, at times, overt political pressure, especially from the United States. At various times a more neutral hemispheric vision of American relations and cooperation has held sway, represented by the formation of the **Organization of American States (OAS)**. The present OAS was officially formed in 1948, but its origins date to 1889. Yet there is no doubt that U.S. policies toward trade, economic assistance, political development, and at times military intervention are often seen as undermining the independence of these states.

Within Latin America there have been cycles of intraregional cooperation and antagonism. Neighboring countries have fought over territory, closed borders, imposed high tariffs, and cut off diplomatic relations. The 1990s witnessed a revival in the trade bloc concept, with the formation of **Mercosur** (the Southern Cone Common Market) in South America and NAFTA. It is possible that, as these economic ties strengthen, these trade blocks could form the basis for a new alignment of political and economic interests in the region.

## Iberian Conquest and Territorial Division

When Christopher Columbus claimed the Americas for Spain, the Spanish became the first active colonial agents in the Western Hemisphere. In contrast, the Portuguese presence in the Americas was the result of the **Treaty of Tordesillas** in 1493–1494. By that time, Portuguese navigators had charted much of the coast of Africa in an attempt to find a water route to the Spice Islands (Moluccas) in Southeast Asia. With the help of Christopher Columbus, Spain sought a western route to the Far East. When Columbus discovered the Americas, Spain and Portugal asked the pope to settle how these new territories should be divided. Without consulting other European powers, the pope divided the Atlantic world in half: The eastern half containing the African continent was awarded to Portugal, and the western half with most of the Americas was given to Spain. The line of division established by the treaty actually cut through the eastern part of South America, placing it under Portuguese rule. The treaty was never recognized by the French, English, or Dutch, who also claimed territory in the Americas, but it did provide the legal justification for the creation of Portuguese Brazil. This state would later become the largest and most populous in Latin America (Figure 4.25).

Six years after the treaty was signed, Portuguese navigator Alvares Cabral accidentally reached the coast of Brazil on a voyage to southern Africa. The Portuguese soon realized that this territory was on their side of the Tordesillas line. Initially they were unimpressed by what Brazil had to offer; there were no spices or major native settlements. Over time, they came to appreciate the utility of the coast as a provisioning site as well as a source for brazilwood, used to produce a valuable dye. Portuguese interest in the territory intensified in the late 16th century, with the development of sugar estates and the expansion of the slave trade, and in the 17th century, with the discovery of gold in the Brazilian interior.

Spain, in contrast, aggressively pursued the conquest and settlement of its new American territories from the very start. After discovering little gold in the Caribbean, by the mid-16th century, Spain's energy was directed toward developing the silver resources of Central Mexico and the Central Andes (most notably Potosí in Bolivia). Gradually, the economy diversified to include some agricultural exports, such as cacao (for chocolate) and sugar, as well as a variety of livestock. In terms of foodstuffs, the colonies were virtually self-sufficient. Manufacturing, however, was forbidden in the Spanish-American colonies in order to keep them dependent on Spain.

**Revolution and Independence** It was not until the rise of revolutionary movements between 1810 and 1826 that Spanish authority on the mainland was challenged. Ultimately, European elites born in the Americas gained control, displacing leaders loyal to the crown. In Brazil, the evolution from Portuguese colony to independent republic was a slower and less violent process that spanned eight decades (1808–1889). In the 19th century, Brazil was declared a separate kingdom from Portugal, with its own king, and later became a republic.

Later, the territorial division of Spanish and Portuguese America into administrative units provided the legal basis for the modern states of Latin America (see Figure 4.25). The Spanish colonies were first divided into two viceroyalties (the administrative units of New Spain and Peru) and within these were various subdivisions that later became the basis for the modern states. (In the 18th century, the Viceroyalty of Peru, which included all of Spanish South America, was divided to form three viceroyalties: La Plata, Peru, and New Granada.) Unlike Brazil, which evolved from a colony into a single republic, the former Spanish colonies experienced fragmentation in the 19th century.

Today the former Spanish mainland colonies include 16 states (plus 3 Caribbean islands), with a total population of over 360 million. If the Spanish colonial territory had remained a unified political unit, it would now have the third largest population in the world, following China and India.

**Persistent Border Conflicts** As the colonial administrative units turned into states, it became clear that the territories were not clearly demarcated, especially the borders that stretched into the sparsely populated interior of South America. This would later become a source of conflict, as new states struggled to define their territorial boundaries. Numerous border wars erupted in the 19th and 20th centuries, and the map of Latin America had to be redrawn many times. Some of the most notable conflicts were the War of the Pacific (1879–1882), in which Chile expanded to the north and Bolivia lost its access to the Pacific; warfare between Mexico and the United States in the 1840s, which resulted in the present border under the Treaty of Hidalgo (1848); and the War of the Triple Alliance (1864–1870), the bloodiest war of the postcolonial period, which occurred when Argentina, Brazil, and Uruguay allied themselves to defeat Paraguay in its claim to control the upper Paraná River Basin. In the 1980s, Argentina lost a war with Great Britain over control of the Falklands, or Malvinas, Islands in the South Atlantic. And as recently as 1998, Peru and Ecuador fought over a disputed boundary in the Amazon Basin.

**The Trend Toward Democracy** Early in the 21st century, most of the 17 countries in Latin America will celebrate their bicentennials. Compared with most of the rest of the developing world, Latin Americans have been independent for a long time. Yet political stability is not a characteristic of the region. Among the countries in the region, some 250 constitutions have been written since independence, and military

**FIGURE 4.25 Shifting Political Boundaries** The evolution of political boundaries in Latin America began with the 1494 Treaty of Tordesillas, which gave much of the Americas to Spain and a slice of South America to Portugal. The larger Spanish territory was gradually divided into viceroyalties and audiencias, which formed the basis for many modern national boundaries. The 1830 borders of these newly independent states were far from fixed. Bolivia would lose its access to the coast; Peru would gain much of Ecuador's Amazon; and Mexico would be stripped of its northern territory by the United States. *(From Lombardi, Cathryn L., and John V. Lombardi,* Latin American History: A Teaching Atlas, *© 1993, reprinted by permission of the University of Wisconsin Press)*

brand of democracy could make their lives better. The political left, however, has yet to produce an alternative to privatization and market-driven policies. For now, the status quo continues, although popular frustration with falling incomes, rising violence, and continual underemployment are a recipe for political change (Figure 4.26).

takeovers have been alarmingly frequent. Since the 1980s, however, the trend has been toward democratically elected governments, the opening of markets, and broader public participation in the political process. Where dictators once outnumbered elected leaders, by the 1990s each country in the region had a democratically elected president. (Cuba, the one exception, will be discussed in Chapter 5.)

Democracy may not be enough for the millions frustrated by the slow pace of political and economic reform. In survey after survey, Latin Americans reveal their dissatisfaction with politicians and governments. Most of the newly elected democratic leaders are also free-market reformers who are quick to eliminate state-backed social safety nets, such as food subsidies, government jobs, and pensions. Many of the poor and middle class have grown doubtful about whether this

## Regional Organizations

At the same time that democratically elected leaders struggle to address the pressing needs of their countries, political developments at the supranational and subnational levels pose new challenges to their authority. The most discussed **supranational organizations** (governing bodies that include several states) are the trade blocs. **Subnational organizations** (groups that represent areas or people within the state) often form along ethnic or ideological lines and can provoke serious internal divisions. Native groups seeking territorial recognition (such as the Kuna in Panama) and insurgent groups (such as the FARC [Revolutionary Armed Forces of Colombia]) have challenged the authority of the states. Finally, the financial and political force of drug cartels (mafia-like organizations in charge of the drug trade) can

**FIGURE 4.26 Protests Against Tortilla Prices**   A peasant woman joined protestors in Mexico City in 2007 as Mexicans marched against the sudden increase in corn and tortilla prices, the staple foods of the country. People carried banners proclaiming, "without corn, there is no country."

undermine judicial systems and influence areas beyond state boundaries.

**Trade Blocs**   Beginning in the 1960s, regional trade alliances were formed in an effort to promote internal markets and reduce trade barriers. The Latin American Integration Association (formerly LAFTA), the Central American Common Market (CACM), and the Andean Group have existed for decades, but their ability to influence economic trade and growth has been limited at best. In the 1990s, Mercosur and NAFTA emerged as supranational structures that could influence development (Figure 4.27). For Latin America, the lessons of Mercosur and NAFTA are causing politicians to rethink the value of regional trade.

Mercosur was formed in 1991 with Brazil and Argentina, the two largest economies in South America, and the smaller states of Uruguay and Paraguay as members. Since its formation, trade among these countries has grown tremendously, so much so that Chile, Bolivia, Peru, Ecuador, and Colombia have joined the group as associate members, and Venezuela is awaiting ratification as a full member. This is significant in two ways: It reflects the growth of these economies and the willingness to put aside old rivalries (especially long-standing antagonisms between Argentina and Brazil) for the economic benefits of cooperation. The size and productivity of this market have not gone unnoticed. Mercosur countries are the EU's largest trading partners in Latin America, and there have been renewed efforts to form a free trade agreement between Mercosur and the EU.

NAFTA took effect in 1994 as a free trade area that would gradually eliminate tariffs and ease the movement of goods among the member countries (Mexico, the United States, and Canada). NAFTA has increased intraregional trade, but there is considerable controversy about costs to the environment and to employment (see Chapter 3). NAFTA did prove, however, that a free trade area combining industrialized and developing states was possible. In 2004, the United States, five Central American countries—Guatemala, El Salvador, Nicaragua, Honduras, and Costa Rica—and the Dominican Republic signed CAFTA. CAFTA, like NAFTA, aims to increase trade and reduce tariffs between member countries. The treaty still awaits full ratification from the legislatures of member states, and much debate surrounds whether such a treaty would lead to more economic development in Central America.

**Insurgencies and Drug Trafficking**   Guerrilla groups such as the FARC in Colombia have controlled large territories of their countries through the support of those loyal to the cause, along with theft, kidnapping, and violence. The FARC, along with the ELN (National Liberation Army), gained wealth and weapons through the drug trade. The level of violence in Colombia has escalated further with the rise of paramilitary groups—armed private groups that terrorize those sympathetic to insurgency. The paramilitary groups are blamed for hundreds of politically motivated murders each year. As many as 2.5 million Colombians have been internally displaced by violence since the late 1980s, most fleeing rural areas for towns and cities. Fortunately, the situation has improved considerably since 2003. Under President Uribe, the police presence has increased throughout the state, and negotiations with insurgencies have stopped, ultimately reducing their power. Yet drug cartels and gangs in states as diverse as Mexico, El Salvador, and Brazil have been blamed for increases in violence and lawlessness. In 2009, the Mexican government was forced to bring in the army to quell the levels of violence associated with Mexican narco-traffickers in Ciudad Juarez.

Initially, most Latin governments cared little about controlling the drug trade as it brought in much-needed hard currency. Within the region, drug consumption was scarcely a problem. Some drug lords even became popular folk heroes, spending large amounts of money on housing, parks, and schools for their communities. The social costs of the drug trade to Latin America became evident by the late 1980s, when the region was crippled by a badly damaged judicial system. By paying off police, the military, judges, and politicians, the drug syndicates exert incredible political power that threatens the social order of the states in which they work. Years of counternarcotics work have done little to reduce the overall flow of drugs to North America and Europe. Since 1990, the United States has invested between $3 and $4 billion to reduce coca and cocaine production in the Andes. Yet, as Figure 4.28 shows, the overall area of coca production did not fall significantly between 1996 and 2005. The main difference is that more coca is now grown in Colombia than in Peru and Bolivia. Mexico's role in the hemispheric drug trade has also steadily increased. It is both a transshipment area for cocaine and a production region of marijuana, heroin, and methamphetamine, almost all of which is bound for consumers in the United States.

# Economic and Social Development: From Dependency to Neoliberalism

Most Latin American economies fit into the broad middle-income category set by the World Bank. Clearly part of the developing world, Latin American people are much better off than those in Sub-Saharan Africa, South Asia, and China. Still, the economic contrasts are sharp,

**FIGURE 4.27 Geopolitics and Trade Blocs in Latin America** Of the four economic trade blocs shown, Mercosur and NAFTA are the most dynamic. In fact, several members of the Andean Group have joined Mercosur. Meanwhile, Central American states signed an agreement in 2004 to form CAFTA (Central American Free Trade Association), which also includes the Dominican Republic.

both between states and within them. Generally, the Southern Cone states (including southern Brazil and excluding Paraguay) and Mexico are the richest. The poorest countries in terms of per capita PPP are Nicaragua, Bolivia, and Honduras. While per capita incomes in Latin America are well below levels of developed countries, the region has witnessed steady improvements in various social indicators, such as life expectancy, child mortality, and literacy. And some small states, such as Costa Rica, do very well in the human development index.

Even with the region's middle-income status, extreme poverty is evident throughout Latin America as more than one in five people live on less than $2 a day (Table 4.2).

The economic engines of Latin America are the two largest countries, Brazil and Mexico. According to the World Bank in 2007, Brazil's economy was the 10th largest in the world, and Mexico's was the 13th largest based on gross national income. Brazil also maintains the region's worst income disparity: 10 percent of the country's richest

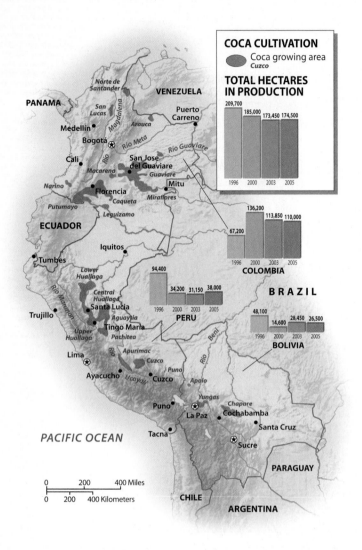

**COCA CULTIVATION**

Coca growing area
*Cuzco*

**TOTAL HECTARES IN PRODUCTION**

**FIGURE 4.28 Coca-Growing Areas in the Andes**    Although the oldest coca-growing regions are in Bolivia and Peru, Colombia traffickers turned the processing and distribution of cocaine into an international narcotics trade. By the late 1990s, the bulk of coca production had shifted out of Bolivia and Peru to Colombia. *(From U.S. Government, 2002, 2003, 2006, "Latin American Narcotics Cultivation and Production Estimates," Washington, DC: CIA Crime and Narcotics Center)*

people control nearly half the country's wealth, while the bottom 40 percent control less than one-tenth.

## Development Strategies

The Latin American reality today was not the future envisioned for the region in the mid-1960s, when Brazil, Mexico, and Argentina all seemed poised to enter the ranks of the industrialized world. Multilateral agencies such as the World Bank and the Inter-American Development Bank loaned money for big development projects: continental highways, dams, mechanized agriculture, and power plants. All sectors of the economy were radically transformed. Agricultural production increased with the application of "green revolution" technology and mechanization. State-run industries reduced the need for imported goods, and the service sector ballooned as a result of new

government and private-sector jobs. In the end, most countries made the transition from predominantly rural and agrarian economies dependent on one or two commodities to more economically diversified and urbanized countries with mixed levels of industrialization.

The industrial giant of Latin America is metropolitan São Paulo, Brazil. Rio de Janeiro has greater name recognition and was the capital before Brasília was built, but it does not have the economic muscle of São Paulo. This city of 18 million, which competes with Mexico City for the title of Latin America's largest, began to industrialize in the early 1900s, when the city's coffee merchants started to diversify their investments. Since then, a combination of private and state-owned industries have concentrated around São Paulo. Within a 60-mile radius of the city center, automobiles, aircraft, chemicals, processed foods, and construction materials are produced. There are also heavy industry and industrial parks. With the port of Santos nearby and the city of Rio de Janeiro a few hours away, São Paulo is the uncontested financial center of Brazil.

The modernization dreams of Latin American countries were trampled in the 1980s, when debt, currency devaluation, and falling commodity prices undermined the aspirations of the region. By the 1990s, most Latin American governments had radically changed their economic development strategies. National industries and tariffs were jettisoned for policy reforms that emphasized privatization and free trade that collectively were labeled *neoliberalism*. Through tough fiscal policy, increased trade, privatization, and reduced government spending, many countries saw their economies grow and poverty decline. Yet a series of economic downturns in the past 15 years have made these neoliberal policies highly unpopular with the masses, causing major political and economic turmoil. In particular, the value of increased trade has been criticized as benefiting a minority of the people in the region.

**Maquiladoras and Foreign Investment**    The growth in foreign investment and foreign-owned factories are examples of neoliberalism. The Mexican assembly plants that line the border with the United States, called **maquiladoras**, are characteristic of manufacturing systems in an increasingly globalized economy. More than 4,000 maquiladoras exist, employing 1 million people who assemble automobiles, consumer electronics, and apparel. Between 1994 and 2000, 3 out of every 10 new jobs in Mexico were in the maquiladoras, which account for nearly half of Mexico's exports. Maquiladora employment peaked in 2001, with 1.3 million people employed. Since 2003, China has become a favorite destination for the labor-intensive assembly work that Mexico has specialized in for the past four decades. As Mexican wages have gone up, some companies have relocated factories to East Asia. Northern Mexico is still an attractive location, but competition from China and even Central America may erode Mexico's various locational and structural advantages.

Considerable controversy surrounds this form of industrialization on both sides of the border. Organized labor in the United States complains that well-paying manufacturing jobs are being lost to low-cost competitors, while environmentalists point out serious industrial pollution resulting from lax government regulation. Mexicans worry that these plants are poorly integrated with the rest of the economy and that many of the workers are young unmarried women who are easily taken advantage of. With NAFTA, foreign-owned manufacturing plants are no longer restricted to the border zone and are increasingly being constructed near the population centers of Monterrey, Puebla, and Veracruz. Mexican workers and foreign corporations,

## TABLE 4.2   DEVELOPMENT INDICATORS

| Country | GNI per capita, PPP 2007 | GDP Average Annual %Growth 2000–07 | Human Development Index (2006)# | Percent Population Living Below $2 a Day | Life Expectancy* 2009 | Under Age 5 Mortality Rate 1990 | Under Age 5 Mortality Rate 2007 | Gender Equity 2007 |
|---|---|---|---|---|---|---|---|---|
| Argentina | 12,970 | 4.7 | 0.860 | 11.3 | 75 | 29 | 16 | 104 |
| Bolivia | 4,150 | 3.6 | 0.723 | 30.3 | 65 | 125 | 57 | 98 |
| Brazil | 9,270 | 3.3 | 0.807 | 12.7 | 73 | 58 | 22 | 103 |
| Chile | 12,330 | 4.5 | 0.874 | 2.4 | 78 | 21 | 9 | 99 |
| Colombia | 8,260 | 4.9 | 0.787 | 27.9 | 72 | 35 | 20 | 104 |
| Costa Rica | 10,510 | 5.4 | 0.847 | 8.6 | 79 | 18 | 11 | 102 |
| Ecuador | 7,110 | 5.0 | 0.807 | 12.8 | 75 | 57 | 22 | 100 |
| El Salvador | 5,640 | 2.8 | 0.747 | 20.5 | 71 | 60 | 24 | 101 |
| Guatemala | 4,520 | 3.6 | 0.696 | 24.3 | 70 | 82 | 39 | 93 |
| Honduras | 3,610 | 5.3 | 0.714 | 29.7 | 72 | 58 | 24 | 106 |
| Mexico | 13,910 | 2.6 | 0.842 | 4.8 | 75 | 52 | 35 | 99 |
| Nicaragua | 2,510 | 3.4 | 0.699 | 31.8 | 71 | 68 | 35 | 102 |
| Panama | 10,610 | 6.0 | 0.832 | 17.8 | 75 | 34 | 23 | 101 |
| Paraguay | 4,520 | 3.3 | 0.752 | 14.2 | 71 | 41 | 29 | 99 |
| Peru | 7,200 | 5.4 | 0.788 | 18.5 | 72 | 78 | 20 | 101 |
| Uruguay | 11,020 | 3.3 | 0.859 | 4.2 | 76 | 25 | 14 | 106 |
| Venezuela | 12,290 | 4.6 | 0.826 | 10.2 | 73 | 32 | 19 | 103 |

Source:  World Bank, *World Development Indicators, 2009.*

United Nations, *Human Development Index, 2008.* #

Population Reference Bureau, *World Population Data Sheet, 2009**

Gender Equity – Ratio of female to male enrollments in primary and secondary school.  Numbers below 100 have more males in primary/secondary school, numbers above 100 have more females in primary/secondary schools.

however, continue to locate in the border zone because there are unique advantages to being positioned next to the U.S. border.

Mexico's competitive advantage is twofold: its location along the U.S. border and its membership in NAFTA. However, other Latin American states are attracting foreign companies through tax incentives and low labor costs. Assembly plants in Honduras, Guatemala, and El Salvador are drawing foreign investors, especially in the apparel industry. A recent report from El Salvador claims that not one of its 229 apparel factories has a union. Making goods for American labels such as the Gap, Liz Claiborne, and Nike, many Salvadoran garment workers complain that they do not make a living wage, work 80-hour weeks, and will lose their jobs if they become pregnant. The situation in Costa Rica, which is now a major computer chip manufacturer for Intel and exported more than $2 billion in chips in 1999, is quite different (Figure 4.29). With a well-educated population, low crime rate, and stable political scene, Costa Rica is now attracting other high-tech firms. Hopeful officials claim that Costa Rica is transitioning from a banana republic (bananas and coffee were the country's long-standing exports) to a high-tech manufacturing center. As a result, the Costa Rican economy averaged 5.4 percent annual growth from 2000 to 2007.

### The Informal Sector

Even in prosperous San José, Costa Rica, a short drive to the urban periphery shows large neighborhoods of self-built housing filled with street traders and family-run workshops. Such activities make up the **informal sector**, which is the provision of

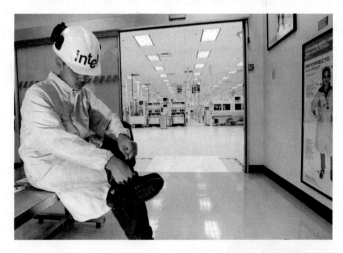

**FIGURE 4.29 High Tech in Costa Rica**   A Costa Rican worker straps on protective clothing before entering the manufacturing area of Intel's plant in Belen, not far from the capital city of San José. In 2000, Costa Rica earned more foreign exchange from exporting computer chips than it did from coffee, its traditional export.

goods and services without the benefit of government regulation, registration, or taxation. Most people in the informal economy are self-employed and receive no wages or benefits except the profits they clear. The most common informal activities are housing construction (in many cities, half of all residents live in self-built housing), manufacturing in small workshops, street vending, transportation services (messenger services, bicycle delivery, and collective taxis), garbage picking, street performing, and even paid line-waiters (Figure 4.30).

No one is sure how big this economy is, in part because it is difficult to separate formal activities from informal ones. Visit Lima, Belém, Guatemala City, or Guayaquil, and it is easy to get the impression that the informal economy *is* the economy. From self-help housing that dominates the landscape to hundreds of street vendors that crowd the sidewalks, it is impossible to avoid. There are advantages in the informal sector—hours are flexible, children can work with their parents, and there are no bosses. As important as this sector may be, however, widespread dependence on it signals Latin America's poverty, not its wealth. It reflects the inability of the formal economies of the region, especially in industry, to provide enough jobs for the many people seeking employment.

## Primary Export Dependency

Historically, Latin America's abundant natural resources were its wealth. In the colonial period, silver, gold, and sugar generated great wealth for the colonists. With independence in the 19th century, a series of export booms introduced commodities such as bananas, coffee, cacao, grains, tin, rubber, copper, wool, and petroleum to an expanding world market. One of the legacies of this export-led development was a tendency to specialize in one or two major commodities, a pattern that continued into the 1950s. During that decade, 90 percent of Costa Rica's export earnings came from bananas and coffee, 70 percent of Nicaragua's came from coffee and cotton, 85 percent of Chilean export income came from copper, and half of Uruguay's export income came from wood. Even Brazil generated 60 percent of its export earnings from coffee in 1955; by 2000, coffee accounted for less than 5 percent of the country's exports, yet Brazil remained the world leader in coffee production.

**Agricultural Production** Since the 1960s, the trend in Latin America has been to diversify and to mechanize agriculture. Nowhere is this more evident than in the Plata Basin, which includes southern Brazil, Uruguay, northern Argentina, Paraguay, and eastern Bolivia.

Soybeans, used for oil and animal feed, transformed these lowlands in the 1980s and early 1990s. Brazil is now the second largest producer of soy in the world (following the United States), and Argentina is the third largest. Between 2000 and 2003, soy production nearly doubled in both countries. In addition, there are acres of rice, cotton, and orange groves, as well as the more traditional plantings of wheat and sugar. The speed with which the shield is being converted into soy fields alarms many, but with soy prices high, the rush to plant continues (Figure 4.31).

Similar large-scale agricultural frontiers exist along the piedmont zone of the Venezuelan Llanos (mostly grains) and the Pacific slope of Central America (cotton and some tropical fruits). In northern Mexico, water supplied from dams along the Sierra Madre Occidental has turned the valleys in Sinaloa into intensive agricultural centers of fruits and vegetables for consumers in the United States. The relatively mild winters in northern Mexico allow growers to produce strawberries and tomatoes during the winter months.

In each of these cases, the agricultural sector is capital intensive. By using machinery, hybrid crops, chemical fertilizers, and pesticides, many corporate farms are extremely productive and profitable. What these operations fail to do is employ many rural people, which is especially problematic in countries where one-third or more of the population depends on agriculture for its livelihood. As industrialized agriculture becomes the norm in Latin America, subsistence peasant producers are further marginalized.

**Mining and Forestry** The exploitation of silver, zinc, copper, iron ore, bauxite, gold, oil, and gas is the economic mainstay for many countries in the region. Moreover, many commodities prices reached record levels in 2005 through 2008, boosting foreign exchange earnings. Oil-rich Venezuela, Mexico, and Ecuador are able to meet their own fuel needs and also earn vital state revenues from oil exports. Venezuela is most dependent on revenues from oil, earning up to 90 percent of its foreign exchange from crude petroleum and petroleum products; it is the fifth largest oil producer in the world (Figure 4.32). Vast oil reserves also exist in the Llanos of Colombia, but a costly and vulnerable pipeline that connects the oil fields to the coast is a regular target of guerrilla groups. Bolivia has impressive reserves of natural

**FIGURE 4.30 Peruvian Street Vendors** Street vendors selling produce in Huancayo, Peru. Street vending plays a critical role in the distribution of goods and the generation of income. Some aspects of street vending (such as access to space) are regulated by local governments and by the vendors themselves.

**FIGURE 4.31 Soy Production**   A worker sprays a field of soybeans with fertilizer in São Paulo State, Brazil. This field is an experimental field planted with genetically modified seeds. Brazil is now the second largest producer of soy in the world, following the United States.

gas in its lowlands, but it has yet to realize significant revenues from these resources.

Like agriculture, mining has become more mechanized and less labor intensive. Even Bolivia, a country dependent on tin production, cut 70 percent of its miners from state payrolls in the 1990s. This measure was part of a broad-based cutback in public spending, yet it suggests that the majority of the miners were not needed. Similarly, the vast copper mines of northern Chile are producing record amounts of copper with fewer miners. Gold mining, in contrast, continues to be labor intensive, offering employment for thousands of prospectors. Logging is another important, and controversial, resource-based activity. Several countries rely on plantation forests of introduced species of pines, teak, and eucalyptus to supply domestic fuelwood, pulp, and board lumber. These plantation forests grow single species and fall far short of the complex ecosystems occurring in natural forests. Still, growing trees for paper or fuel reduces the pressure on other forested areas. Leaders in plantation forestry are Brazil, Venezuela, Chile, and Argentina. Considered Latin America's economic star in the 1990s, Chile relied on timber and wood chips to boost its export earnings. Thousands of hectares of exotics (eucalyptus and pine) have been planted, systematically harvested, cut into boards, or chipped for wood pulp (see Figure 4.4). Japanese capital is heavily involved in this sector of the Chilean economy. The recent expansion of the wood chip business, however, led to a dramatic increase in the logging of native forests.

Due to consumer demand for certified wood, mostly in Europe, there is growing interest in certification programs that designate when a wood product has been produced sustainably. Unfortunately, such programs are small, and the lure of profit is usually stronger than the impulse to conserve for future generations.

## Latin America in the Global Economy

In order to conceptualize Latin America's place in the world economy, **dependency theory** was advanced in the 1960s by scholars from the region. The basis of the theory is that expansion of European capitalism created the region's underdevelopment. For the developed "cores" of the world to prosper, the "peripheries" became dependent and impoverished. Dependent economies, such as those in Latin America, were export oriented and vulnerable to fluctuations in the global market. Even when they experienced economic growth, it was secondary to the economic demands of the core (North America and Europe).

Economists who accepted this interpretation of Latin America's history were convinced that economic development could occur only through self-sufficiency, growth of internal markets, agrarian reform, and greater income equality. In short, they argued for strong state intervention and less trade with the economic cores of Europe and North America. Policies such as import substitution industrialization (developing a country's industrial sector by making imported products extremely expensive and domestic manufactured goods cheaper) and nationalization of key industries were partially influenced by this view. Critics of dependency theory say that in its simplest form, it becomes a means to blame forces external to Latin America for the region's problems. Unspoken in dependency theory is also the notion that the path to development taken by Europe and North America cannot be easily copied. This was a radical idea for its time.

Latin America's century-long dependence on North America, especially the United States, as its major trading partner is still evident. However, growing trade with as well as investment from Europe and East Asia suggests that a more complex and less U.S.-dependent pattern of trade is emerging. Latin America is linked to the world economy in ways other than trade. Figure 4.33 shows the changes in foreign direct investment (FDI) as a percentage of gross domestic product (GDP) from 1990 to 2004. For nearly every country in the region, the value of foreign investment in terms of percentage of GDP went up. In 1990, Brazil's FDI was less than $1 billion, and Mexico's was $2.5 billion.

**FIGURE 4.32 Oil Production**   A portable drilling platform is shown on Lake Maracaibo, Venezuela. Oil production has impacted the pace of development in countries such as Venezuela, Mexico, and Ecuador. These economies struggled in the 1990s when oil prices declined but benefited from higher oil prices in the past few years.

**FIGURE 4.33 Global Linkages: Foreign Investment and Remittances** Foreign investors and immigrants are responsible for significant increases in the amount of capital flowing into Latin America. As the map indicates, most countries saw increases in foreign direct investment between 1990 and 2004. Immigrants working abroad sent $60 billion to the region in 2006, proving much-needed capital to many poor households. *(Data from international Development Bank, Remittance Map 2006: Population Reference Bureau, World Population Data Sheet, 2006; The World Bank, World Development Indicators, 2006)*

By 2004, FDI in Brazil was valued at $18 billion, and Mexico's was $17 billion.

Remittances are another important financial flow that reflects the integration of Latin American migrants into labor markets around the world, especially in North America. Scholars debate whether this flow of capital can actually lead to sustained development or whether it is simply a survival strategy of last resort. The economic impact of remittances shown on a per capita basis is real (see Figure 4.33). El Salvador, a country of 7 million people, received over $3.3 billion in remittances in 2006, which is nearly $500 per capita.

**Neoliberalism as Globalization** By the 1990s, governments and the World Bank had become champions of neoliberalism as a sure path to economic development. **Neoliberalism** stresses privatization, export production, FDI, and few restrictions on imports. Neoliberal policies summarize the forces of globalization by turning away from policies that emphasize state intervention and self-sufficiency. Most Latin American political leaders embrace neoliberalism and the benefits that come with it, such as increased trade and more favorable terms for debt repayment. Yet there are signs of discontent with neoliberalism throughout the region. Recent protests in Peru and Bolivia reflect the popular anger against trade policies that seem to benefit only the elite.

Chile is an outspoken defender of neoliberalism. Its annual growth rate between 2000 and 2007 was 4.5 percent, one of the region's healthiest. In 1995 alone, the Chilean economy grew 10.4 percent, placing it in the same league as the Asian "tiger economies." Consequently, it is the most studied and watched country in Latin America. By the numbers, Chile's 17 million people are doing well, but the country's accomplishments are not readily transferable to other countries. To begin with, the radical move to privatize state-owned business and open the economy occurred under an oppressive military dictatorship that did not tolerate opposition. Much of Chile's export-led growth has been based on primary products: fruits, seafood, copper, and wood. Although many of these are renewable, Chile will need to develop more manu-factured goods if it hopes to be labeled "developed." Furthermore, its relatively small and homogeneous population in a resource-rich land does not have the same ethnic divisions that hinder so many other states in Latin America. Although neoliberalism has worked in the Chilean case, for many Latin America states, the social and environ-mental disruptions associated with neoliberal policies have led to political upheaval, and an active search for alternatives is under way.

**Dollarization**   As financial crises spread through Latin America in the late 1990s, governments began to consider the economic benefits of **dollarization**, a process by which a country adopts—in whole or in part—the U.S. dollar as its official currency. In a totally dollarized economy, the U.S. dollar becomes the only medium of exchange, and the country's national currency ceases to exist. This was the radical step taken by Ecuador in 2000 to address the dual problems of cur-rency devaluation and hyperinflation rates of more than 1,000 percent annually. El Salvador adopted dollarization in 2001 as a means to re-duce the cost of borrowing money. Dollarization is not a new idea; in 1904, Panama dollarized its economy, a year after it gained independ-ence from Colombia. Until 2000, however, Panama was the only fully dollarized state in Latin America.

A more common strategy in Latin America is limited dollariza-tion, in which U.S. dollars circulate and are used alongside the coun-try's national currency. Limited dollarization exists in many countries around the world but is widespread in Latin America. Because the economies of Latin America are prone to currency devaluation and hyperinflation, limited dollarization is a type of insurance. Many banks in Latin America, for example, allow customers to maintain ac-counts in dollars to avoid the problem of capital flight should a local currency be devalued.

## Social Development

Over the past three decades, Latin America has experienced marked improvements in life expectancy, child survival, and educational eq-uity. One telling indicator is the steady decline in mortality rates for children below age five between 1990 and 2007 (Table 4.2). This indi-cator is important because an increase in the number of children younger than five years surviving suggests that basic nutritional and health-care needs are being met. One can also conclude that resources are being used to sustain women and their children. Despite economic downturns, the region's social networks have been able to lessen the negative effects on children.

A combination of government policies and grassroots and nongovernmental organizations (NGOs) play a fundamental role in contributing to social well-being. Both the Brazilian and Mexican gov-ernments have launched poverty-reduction programs in an attempt to assist their poorest citizens. In states with fewer resources, interna-tional humanitarian organizations, church organizations, and commu-nity activists provide many services that state and local governments cannot. Catholic Relief Services and Caritas, for example, work with rural poor throughout the region to improve their water supplies, health care, and education. Other groups lobby local governments to build schools and recognize squatters' claims. Grassroots organizations also develop cooperatives that market everything from sweaters to cheeses. Other important indicators for social development are life expectancy, gender educational equity, and access to improved water

sources. In aggregate, 84 percent of the people in the region have ac-cess to an adequate amount of water from an improved source, slightly more girls receive secondary education than boys, and life expectancy (for both men and women) is 73 years. Masked by this combined data are extreme variations between rural and urban areas, between regions, and along racial and ethnic lines.

Within Mexico and Brazil, tremendous internal differences exist in socioeconomic indicators (Figure 4.34). The northeast part of Brazil lags behind the rest of the country in every social indicator. The country has a literacy rate of over 85 percent, but the rate in the northeast is only 60 percent. Moreover, within the northeast, literacy for city residents is 70 percent, but for rural residents it is only 40 percent. In Mexico the levels of poverty are highest in the Indian south. In contrast, Mexico City, the state of Nuevo Leon (where Monterrey is the capital), the state of Quintana Roo (home to Mexico's largest resort, Cancun), and Campeche (with oil) have the highest GDP per capita. All countries have spatial inequalities in terms of income and availability of services, but the contrasts tend to be sharper in the developing world. In the cases of Mexico and Brazil, it is difficult to ignore ethnicity and race when trying to explain these patterns.

**Race and Inequality**   There is much to admire about race relations in Latin America. The complex racial and ethnic mix that was created in Latin America encouraged tolerance for diversity. That said, Indi-ans and blacks are more likely to be counted among the poor of the re-gion. More than ever before, racial discrimination is a major political issue in Brazil. Reports of organized killings of street children, most of them Afro-Brazilian, make headlines. For decades, Brazil championed its vision of a color-blind racial democracy. True, residential segrega-tion by race is rare in Brazil, and interracial marriage is common, but certain patterns of social and economic inequality seem best ex-plained by race.

Assessing racial inequalities in Brazil is problematic. The Brazilian census asks few racial questions, and all are based on self-classification. In the 2000 census, less than 11 percent of the population called itself black. Some Brazilian sociologists, however, claim that more than half the population is of African ancestry, making Brazil the sec-ond largest "African state" after Nigeria. Racial classification is always highly subjective and relative, but there are patterns that sup-port the existence of racism. Evidence from northeastern Brazil, where Afro-Brazilians are the majority, shows death rates approach-ing those of some of the world's poorest countries. Throughout Brazil, blacks suffer higher rates of homelessness, landlessness, illit-eracy, and unemployment than others. There is even a movement in the northeastern state of Bahia to extend the same legal protection given Indians to hundreds of Afro-Brazilian villages that were once settled by runaway slaves.

In areas of Latin America where Indian cultures are strong, one also finds indicators of low socioeconomic position. In most coun-tries, areas where native languages are widely spoken regularly corre-spond with areas of persistent poverty. In Mexico, the Indian south lags behind the booming north and Mexico City. Prejudice is embed-ded in the language. To call someone an *indio* (Indian) is an insult in Mexico. It is difficult to separate status divisions based on class from those based on race. From the days of conquest, being European meant an immediate elevation in status over the Indian, African, and *mestizo* populations. But Amerindian people are asking to be heard.

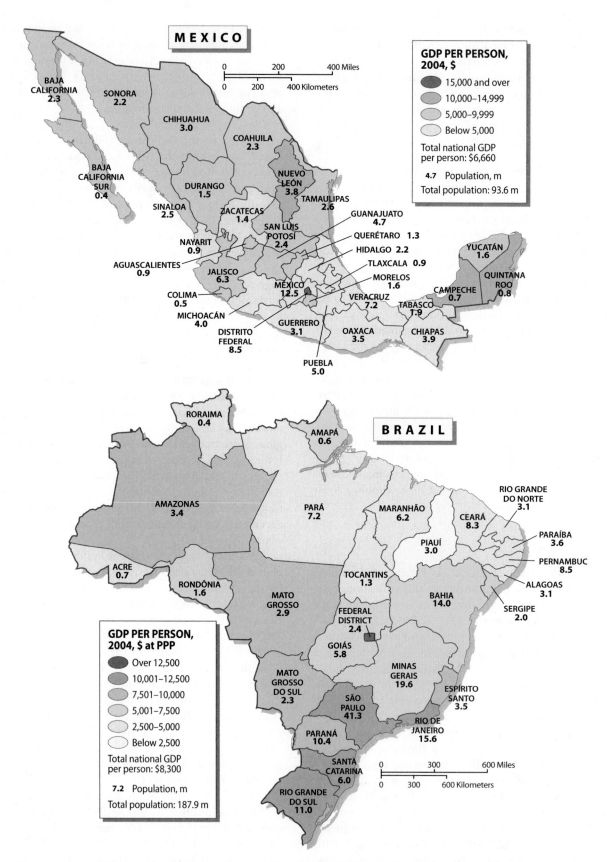

**FIGURE 4.34 Mapping Poverty and Prosperity**    Mexico and Brazil are the largest states in the region, and the levels of poverty vary greatly within each country. In Mexico, the southern states of Chiapas and Oaxaca have the lowest GDP per capita. In Brazil, the northeastern states have the country's lowest per capita GDP. *(© 2006 and 2007 The Economist Newspaper Group, Inc. Reprinted with permission. Further reproduction prohibited.)*

Evo Morales was inaugurated as president of Bolivia in 2006, making him the first Indian leader in that country.

**The Status of Women**  Many contradictions exist with regard to the status of women in Latin America. Many Latina women work outside the home. In most countries, the formal figures are between 30 and 40 percent of the workforce, not far off from many European countries but lower than in the United States. Legally speaking, women can vote, own property, and sign for loans, although they are less likely to do so than men, reflecting the patriarchal (male-dominated) tendencies in the society. Even though Latin America is predominantly Catholic, divorce is legal, and family planning is promoted. In most countries, however, abortion remains illegal.

Overall, access to education in Latin America is good compared to other developing regions, and thus illiteracy rates tend to be low. The rates of adult illiteracy are slightly higher for women than for men but usually by only a few percentage points. Throughout higher education in Latin America, male and female students are equally represented today. Consequently, women are regularly employed in the fields of education, medicine, and law.

The biggest changes for women are the trends toward smaller families, urban living, and educational parity with men. These factors have greatly improved the participation of women in the labor force. In the countryside, however, serious inequalities remain. Rural women are less likely to be educated and tend to have larger families. In addition, they are often left to care for their families alone as husbands leave in search of seasonal employment. In most cases, the conditions facing rural women have been slow to improve.

Women are increasingly playing an active role in politics. In 1990, Nicaragua elected the first woman president in Latin America, Violeta Chamorro, the owner of an opposition newspaper. Nine years later, Panamanians voted Mireya Moscoso into power. In 2005, South America elected its first woman president, when Dr. Michelle

**FIGURE 4.35 Three Latin American Presidents**  Brazilian President Luiz Inacio Lula da Silva, Argentinian President Cristina Fernandez de Kirchner and Bolivian President Evo Morales met in 2008 to discuss the development of natural gas in Bolivia. Collectively they are representative of greater diversity among political leaders in Latin America.

Bachelet, a pediatrician and single mother, took the oath of office in Chile. Two years later, Cristina Fernández de Kirchner was elected president of Argentina, following the presidency of her husband Néstor Kirchner (Figure 4.35).

Across the region, women and indigenous groups are active organizers and participants in cooperatives, small businesses, and unions. In a relatively short period, they have won a formal place in the economy and a political voice. Moreover, evidence suggests that this trend will continue.

# SUMMARY

- Latin America and the Caribbean were the first world regions to be fully colonized by Europe. In the process, perhaps 90 percent of the native population died from disease, cruelty, and forced resettlement. The slow demographic recovery of native peoples and the continual arrival of Europeans and Africans resulted in an unprecedented level of racial and cultural mixing.

- Unlike in other developing areas, most people in Latin America live in cities. This trend started early and reflects a cultural bias toward urban living with roots in the colonial past. The cities are large and combine aspects of the formal industrial economy with an informal one.

- Compared to Europe and Asia, this region is still rich in natural resources and relatively lightly populated. Yet as populations continue to grow and trade in natural resources increases, there is

growing concern for the state of the environment. Of particular concern is the relentless cutting of tropical rain forests.

- Uneven development and economic frustration have led many Latin Americans (both highly skilled and low skilled) to consider emigration as an economic option. Today, Latin America is a region of emigration, with emigrants going to work in North America, Europe, and Japan. Collectively they send billions of dollars in remittances back to Latin America each year.

- Latin American governments were early adopters of neoliberal economic policies. While some states prospered, others faltered, sparking popular protests against the negative effects of globalization. It does seem, however, that new political actors are emerging—from indigenous groups to women—who are challenging old ways of doing things.

## KEY TERMS

agrarian reform (p. 96)
Altiplano (p. 89)
Central American Free Trade
   Association (CAFTA) (p. 82)
dependency theory (p. 109)
dollarization (p. 111)
El Niño (p. 92)

grassification (p. 85)
informal sector (p. 107)
latifundia (p. 96)
maquiladora (p. 106)
megacity (p. 82)
Mercosur (p. 102)
*mestizo* (p. 100)

minifundia (p. 96)
neoliberalism (p. 110)
neotropics (p. 84)
Organization of American States
   (OAS) (p. 102)
remittance (p. 98)
shield (p. 89)

squatter settlement (p. 94)
subnational organization (p. 103)
supranational organization
   (p. 103)
syncretic religion (p. 101)
Treaty of Tordesillas (p. 102)
urban primacy (p. 94)

## THINKING GEOGRAPHICALLY

1. Discuss the processes driving tropical deforestation in Latin America and how they compare with deforestation in other areas of the world.

2. How is neoliberalism influencing the way Latin America interacts with the rest of the world? What are the social and environmental costs of neoliberalism? How is this development approach related to globalization?

3. Given the dominance of cities in this region, describe the particular urban environmental problems facing cities in the developing world. How might Latin America's megacities use their size and density to reduce the environmental problems associated with urbanization?

4. Discuss the social, environmental and economic consequences behind the modernization of agriculture. How does Latin America's experience with modern agricultural systems compare with North America's?

5. Latin America is a region of emigration. Discuss the impact mass emigration has had on this region. Why is it higher here than in other regions of the developing world? Which countries are producing the most immigrants and where are they going?

Log in to **www.mygeoscienceplace.com** for videos, interactive maps, RSS feeds, case studies, and self-study quizzes to enhance your study of Latin America.

# 5 THE CARIBBEAN

## GLOBALIZATION AND DIVERSITY

Named for some of its former inhabitants—the Carib Indians—the modern Caribbean is a culturally complex and economically peripheral world region. Settled by various colonial powers, the residents of the region are mostly a mix of African, European, and South Asian peoples. Greatly impacted by the global economy but with relatively little influence on it, the livelihoods of Caribbean peoples are filled with uncertainty.

### ENVIRONMENTAL GEOGRAPHY

Climate change threatens the Caribbean, with the potential for stronger and more frequent hurricanes, loss of territory due to sea level rising, and destruction of coral reefs.

### POPULATION AND SETTLEMENT

Large numbers of Caribbean peoples have emigrated from the region, leaving in search of economic opportunity and sending back millions of dollars.

### CULTURAL COHERENCE AND DIVERSITY

Creolization, the blending of African, European, and Amerindian elements, has resulted in many unique Caribbean expressions of culture, such as rara, reggae, and steel drum bands.

### GEOPOLITICAL FRAMEWORK

The first area of the Americas to be extensively explored and colonized by Europeans, the region has seen many rival European claims and, since the early 20th century, has experienced strong U.S. influence.

### ECONOMIC AND SOCIAL DEVELOPMENT

Environmental, locational, and economic factors make tourism a vital component of this region's economy, particularly in Puerto Rico, Cuba, the Dominican Republic, and The Bahamas.

*An aerial view of Marina Cay, British Virgin Islands in the Lesser Antilles. Many Caribbean island economies rely upon their scenic beauty and warm climate to attract foreign tourists.*

The Caribbean was the first region of the Americas to be extensively explored and colonized by Europeans. Yet its modern regional identity is unclear, often merged with Latin America but also viewed as apart from it (see "Setting the Boundaries"). Today the region is home to 43 million people scattered across 26 countries and dependent territories. They range from the small British dependency of Turks and Caicos, with fewer than 20,000 people, to the island of Hispaniola, with more than 19 million. In addition to the Caribbean Islands, Belize of Central America and the three Guianas—Guyana, Suriname, and French Guiana—of South America are included as part of the Caribbean. For historical and cultural reasons, the peoples of these mainland states identify with the island nations and are thus included in this chapter (Figure 5.1).

Historically, the Caribbean was a battleground between rival European powers competing for territorial control of these tropical lands. In the early 1900s, the United States took over as the dominant geopolitical force in the region, maintaining what some have called a neocolonial presence. As in many developing areas, external control of the Caribbean by foreign governments and companies produced highly dependent and inequitable economies. The plantation became the dominant production system, and sugar was the leading commodity. Over time, other products began to have economic importance, as did the international tourist industry. Increasingly, governments have sought to diversify their economies by expanding into banking services and manufacturing to reduce the region's dependence on agriculture and tourism.

Generally, when one thinks of the Caribbean, images of white sandy beaches and turquoise tropical waters come to mind. Tucked between the Tropic of Cancer and the equator, with year-round temperatures averaging in the high 70s, the hundreds of islands and picturesque waters of the Caribbean have often inspired comparisons to paradise. Columbus began the tradition by describing the islands of the New World as the most marvelous, beautiful, and fertile lands he had ever known, filled with flocks of parrots, exotic plants, and friendly natives. Writers today are still lured by the sea, sands, and swaying palms of the Caribbean. Since the 1960s, the Caribbean has earned much of its international reputation as a playground for northern vacationers. But there is another Caribbean that is far poorer and economically more dependent than the one portrayed on travel posters. Haiti, which has the third largest population in the region, with 9 million people, is the poorest country in the Western Hemisphere. The two largest countries—Cuba (11 million) and the Dominican Republic (10 million)—also suffer from serious economic problems.

The majority of Caribbean people are poor, living in the shadow of North America's vast wealth. The concept of **isolated proximity** has been used to explain the region's unique position in the world. The *isolation* of the Caribbean sustains the area's cultural diversity (Figure 5.2) and also explains its limited economic opportunities. Caribbean writers note that this isolation fosters a strong sense of place and a tendency to focus inward. Yet the relative *proximity* of the Caribbean to North America (and, to a lesser extent, Europe) ensures its international connections and economic dependence. For example, each year Dominican workers in the United States send nearly $3 billion to family and friends in the Dominican Republic, who rely on this money for their sustenance. Through the years, the Caribbean has evolved as a distinct but economically marginal world region. This position expresses itself today as workers flee the region in search of employment, while foreign companies are attracted to the Caribbean for its cheap labor. The economic well-being of most Caribbean countries is uncertain. Despite such uncertainty, one can see in the Caribbean an enduring cultural richness and attachment to place that may explain a growing countercurrent of immigrants back to the region.

# Setting the Boundaries

This culturally diverse area of the world does not fit neatly into current world regional divisions. In fact, in most textbooks it is scarcely mentioned, being folded into Latin America or appearing as part of Middle America (along with Mexico and Central America). By focusing on the Caribbean as a distinct world region, however, one can more clearly see the long-term processes of colonization and globalization that have created it.

For much of the colonial period, these islands were referred to as "the Indies" or "the Spanish Main," and the sea surrounding them was called *Mar del Norte* (North Sea). It was not until the late 18th century that an English cartographer first applied the name "Caribbean" to the sea (a reference to the Carib Indians who once inhabited the islands).

It took another 200 years for the term *Caribbean* to be generally applied to this region of islands and rimland.

The basis for treating the Caribbean as a distinct area lies in its particular cultural and economic history. Culturally, the islands and portions of the coast can be distinguished from the largely Iberian-influenced mainland of Latin America because of their more diverse European colonial history and much stronger African influence. Demographically, the native population of the Caribbean was virtually eliminated in the first 50 years following Columbus's arrival. Thus, unlike Latin America and North America, the Amerindians of the Caribbean islands have virtually no legacy, save a few place names. The dominance of plantation agriculture, such as sugar production, explains many of the social, economic,

and environmental patterns in the region. African slaves were introduced as replacements for lost native labor. As Caribbean forests and savannas became sugarcane fields, a pattern was set in which foreign species of plants replaced native ones. The long-term effect of plantation life is visible today in the hugely uneven distribution of land and resources among the people of the region.

Today organizations such as the Caribbean Common Market exist to promote the common interests of island and mainland states. Yet internal alliances often split along colonial lines, so that former Spanish, British, French, or Dutch colonies at times have more in common with each other than with the region as a whole. At the crossroads of the Americas, the Caribbean has a distinctive, albeit peripheral, role in the global economy.

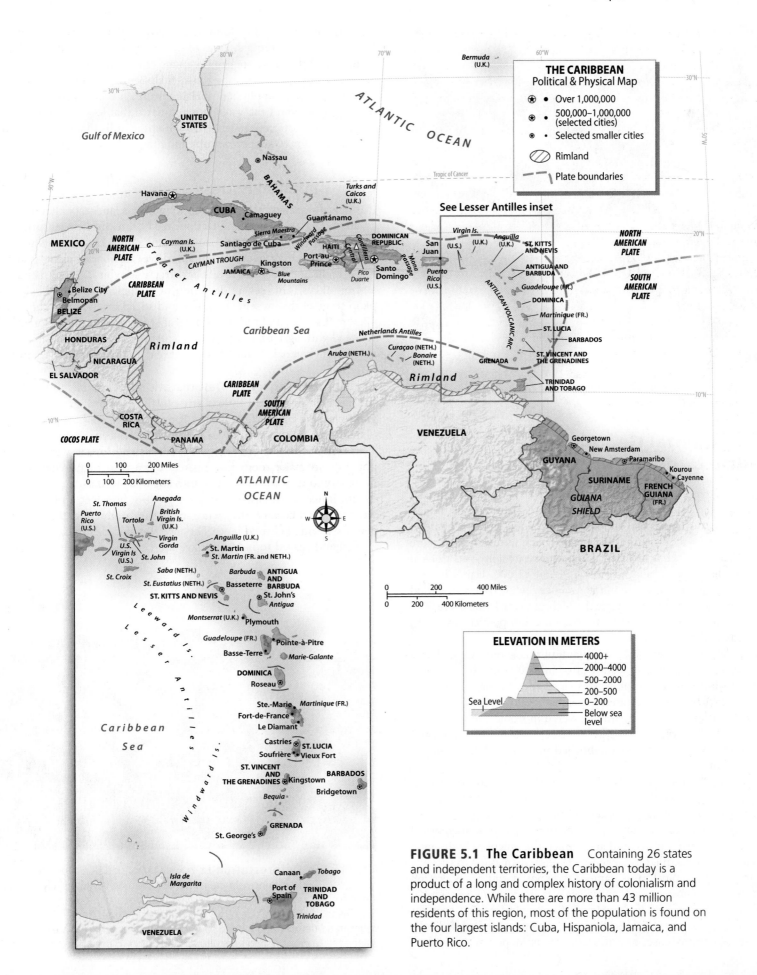

**FIGURE 5.1 The Caribbean**   Containing 26 states and independent territories, the Caribbean today is a product of a long and complex history of colonialism and independence. While there are more than 43 million residents of this region, most of the population is found on the four largest islands: Cuba, Hispaniola, Jamaica, and Puerto Rico.

**FIGURE 5.2 Carnival Dancer** A woman in elaborate costume celebrates Carnival in Port of Spain, Trinidad. Linking a Christian pre-Lenten celebration with African musical rhythms, Carnival has become an important symbol of Caribbean identity.

# Environmental Geography: Paradise Undone

The levels of environmental degradation and poverty found in Haiti are unmatched in the rest of the Western Hemisphere. Haiti's story, although extreme, illustrates how social and economic inequalities contribute to misguided choices about natural resources, which in turn lead to more poverty. What was once considered France's richest colony—its tropical jewel—now has a per capita income (PPP) of $1,000. Haiti's life expectancy (58 years) is the lowest in the Americas, while its levels of child malnutrition and infant mortality are the highest in the hemisphere. The explanations for Haiti's misfortune are complex, but political economy, demography, and natural resources are critical factors.

In the late 18th century, Haiti's plantation economy yielded vast wealth for several thousand Europeans (mostly French) who exploited the labor of half a million African slaves. During the colonial period, the lowlands were cleared to make way for sugarcane fields, but most of the mountains remained covered in tropical forest.

By the mid-20th century, a destructive cycle of environmental and economic impoverishment was established in Haiti. While under the rule of corrupt dictators from 1957 to 1986, Haiti's elite benefited as the conditions for the country's poor majority worsened. More than two-thirds of Haiti's people are peasants who work small hillside plots and seasonally labor on large estates. As the population grew, people sought more land. They cleared the remaining hillsides, subdivided their plots into smaller units, and abandoned the practice of fallowing land (leaving land untilled for several seasons) in an effort to survive. When the heavy tropical rains came, the exposed and easily eroded mountain soils were washed away, leaving rocky and semiarable surfaces for the next season. In May 2004, torrential rains brought nearly 5 feet of water in 7 days. Rain raced down the denuded hillsides, the silty rivers rose, and towns such as Mapou were washed away in the floods that followed. Nearly 3,000 people perished in Haiti and the bordering villages of the Dominican Republic; in Mapou alone, 1,600 Haitians perished. The state of the environment was further aggravated by dependence on wood fuel. Because of their poverty and limited electricity supplies, most Haitians use charcoal (made from trees) to cook meals and heat water. It is estimated that less than 3 percent of Haiti remains forested. In less than a lifetime, hills that were once covered in forest now support shrubs and grasses (Figure 5.3).

## Environmental Issues

Ecologically speaking, it is difficult to picture a landscape that has been so completely altered as the Caribbean. For nearly five centuries, the destruction of forests and the unrelenting cultivation of soils resulted in the extinction of many endemic (native) Caribbean plants and animals, including various shrubs and trees, songbirds, large mammals, and monkeys. This severe depletion of biological resources helps explain some of the present economic and social instability of the region (Figure 5.4). Most of the environmental problems in the region are associated with agricultural practices, soil erosion, excessive reliance on wood and charcoal for fuel, and the threat of global climate change.

**Agriculture's Legacy of Deforestation** Much of the Caribbean was covered in tropical forests prior to the arrival of Europeans. The great clearing of the Caribbean's forests began on European-owned plantations on the smaller islands of the eastern Caribbean in the 17th century and spread westward. The island forests were removed not only to make room for sugarcane but also to provide the fuel necessary to turn the cane juice into sugar, as well as to provide lumber for housing, fences, and ships. Primarily, however, tropical forests were removed because they were seen as unproductive; the European colonists valued cleared land. The newly exposed tropical soils easily eroded and ceased to be productive after several harvests, a situation

**FIGURE 5.3 Deforestation in Haiti** Denuded slopes on the outskirts of Port-au-Prince are indicative of how much Haitians rely on their island for fuel wood and food. Poverty and land scarcity force many Haitians to farm the hillsides. The search for wood fuels has led to widespread deforestation. With so many trees removed, soils are more vulnerable to erosion, and agricultural yields tend to decline.

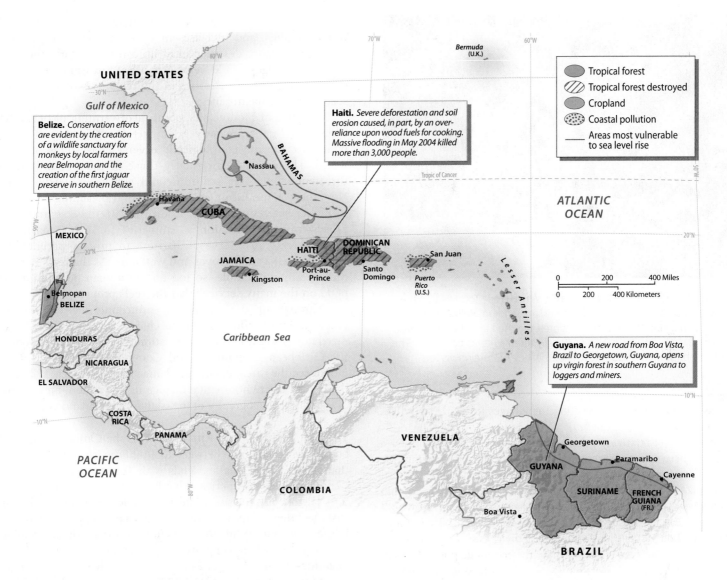

**FIGURE 5.4 Environmental Issues in the Caribbean**    It is difficult to imagine a region in which the environment has been so completely altered. Most of the island forests were removed long ago for agriculture or fuel wood, and soil erosion is a chronic problem. Coastal pollution is serious around the largest cities and industrial zones. The forest cover of the rimland states, however, is largely intact. As tourism becomes increasingly important for the Caribbean, efforts to protect the beaches and reefs, along with the fauna and flora, are growing. *(Adapted from* DK World Atlas, *1997, pp. 7, 55, London: DK Publishing)*

that led to two distinct land-use strategies. On the larger islands of Cuba and Hispaniola and on the mainland, new lands were constantly cleared and older ones abandoned or fallowed in an effort to keep up sugar production. On the smaller islands, where land was limited, such as Barbados and Antigua, labor-intensive efforts to conserve soil and maintain fertility were employed. In either case, the island forests were replaced by a landscape devoted to crops for world markets.

While Haiti has lost most of its forest cover, on Jamaica and the Dominican Republic nearly 30 percent of the land is still forested. On Puerto Rico roughly one-quarter of the land has forest cover, whereas on Cuba the figure has dropped to 20 percent. Cuba has experienced a surge in charcoal production brought on by the economic and energy crises that began in 1990. Because much of the country's electricity is generated by imported fuel, locals have turned to Cuba's forests to make charcoal for domestic energy needs.

**Managing the Rimland Forests**    The Caribbean **rimland** is the coastal zone of the mainland, beginning with Belize and extending along the coast of Central America to northern South America. In general, the biological diversity and stability of the rimland states are less threatened than in the rest of the Caribbean. Thus, current conservation efforts could produce important results. Even though much of Belize was selectively logged for mahogany in the 19th and 20th centuries, healthy forest cover still supports a diversity of mammals, birds, reptiles, and plants. Public awareness of the negative consequences of deforestation is also greater now. Many protected areas have been established in Belize. In the mid-1980s, villagers in Bermudian Landing, Belize, established a community-run sanctuary for black howler monkeys (locally referred to as baboons). The villagers banded together to maintain habitat for the monkeys and commit to land management practices that accommodate this gregarious species. The success of the

**FIGURE 5.5 Protecting Habitat and Wildlife** Tourists visit the Community "Baboon" Sanctuary in Bermudian Landing, Belize. The sanctuary is a community-run project to preserve the habitat and increase the number of black howler monkeys (locally referred to as baboons). The sanctuary, established in 1985, attracts domestic and foreign visitors.

project has resulted in tourists visiting the villages to see these indigenous primates up close (Figure 5.5).

In Guyana, the relatively pristine interior forests are becoming a battleground between conservationists and developers. During the 1990s, the Guyanese government prioritized the wood processing industry by encouraging private investment and negotiating with companies from Malaysia and China about forest concessions and sawmill operations. A new dry-season highway traverses the length of the country, connecting Boa Vista, Brazil, with Georgetown, Guyana. While governments in both states are encouraged by the economic possibilities of this road, an alliance of conservationists (local and foreign) and indigenous peoples is attempting to establish a vast national park in part of this region, where no commercial logging or mining could occur.

## The Caribbean and Climate Change

Of all the environmental issues facing the Caribbean, one of the most troubling is climate change. The effects of climate change on the Caribbean region include sea level rise, increased intensity of storms, variable rainfall leading to both floods and droughts, and loss of biodiversity (both in forests and coral reefs). The scientific consensus is that global warming could promote a sea level rise of 3 to 10 feet (1 to 3 meters) in the 21st century. In terms of land loss due to inundation, the low-lying Bahamas would be the most impacted country—losing nearly 30 percent of its land with a 10-foot (3-meter) sea level rise. In terms of people affected by inundation, Suriname, French Guiana, Guyana, Belize, and The Bahamas would be the most severely impacted: Just a 3-foot (1-meter) sea level rise would be devastating because most of the population lives near the coast. With a sea level rise of 10 feet (3 meters), 30 percent of Suriname's population and 25 percent of Guyana's would be displaced.

The Caribbean has not been a major contributor of greenhouse gases, but this maritime region is extremely vulnerable to the negative impacts of climate change. In addition to land loss and population displacement due to sea level rise, other concerns include changes in

rainfall patterns, leading to declines in agricultural yields and freshwater supplies, and increases in storm intensity—especially hurricanes—that cause destruction of infrastructure and other problems. All these changes would negatively affect tourism and thus the gross domestic income of countries in the region. Some of the worst-case scenarios are catastrophic.

In terms of biodiversity, continued warming of ocean temperatures will further negatively impact the Caribbean's coral reefs, which are the most biologically diverse ecosystems of the marine world. These reefs, particularly those of the rimland, are already threatened by water pollution and subsistence fishing practices. Now there is mounting evidence of coral bleaching and die-off due to higher sea temperatures. Coral reefs are diverse and productive ecosystems that function as nurseries for many marine species. Healthy reefs also serve as barriers to protect populated coastal zones as well as mangroves and wetlands. As the reefs become more ecologically vulnerable, so too do the human populations that depend on the many services that the reefs provide.

Throughout the Caribbean, protecting the environment and preparing for the effects of climate change are increasingly being recognized not as a luxury but as a question of economic livelihood. In fact, the Caribbean Community and Common Market (CARICOM) has monitored the threat of climate change for over a decade. To appreciate the impact of environmental change in this region, an understanding of the physical geography of the islands and rimland is essential.

**The Sea, Islands, and Rimland** It is the Caribbean Sea itself—the body of water enclosed between the *Antillean* islands (the arc of islands that begins with Cuba and ends with Trinidad) and the mainland of Central and South America—that links the states of the region (Figure 5.4). Historically, the sea connected people through its trade routes and sustained them with its marine resources of fish, green turtle, manatee, lobster, and crab. While the sea is noted for its clarity and biological diversity, it has never supported large commercial fishing because the quantities of any one species are not great. The surface temperature of the sea ranges from 73° to 84°F (23° to 29°C), over which forms a warm tropical marine air mass that influences daily weather patterns. This warm water and tropical setting continue to be key resources for the region, as millions of tourists visit the Caribbean each year (Figure 5.6).

The arc of islands that stretches across the Caribbean Sea is its most distinguishing physical feature. The Antillean islands are divided into two groups: the Greater and Lesser Antilles. The majority of the region's population live on the islands. The rimland (the Caribbean coastal zone of the mainland) includes Belize and the Guianas, as well as the Caribbean coast of Central and South America. In contrast to the islands, the rimland has low population densities (refer to Figure 5.1).

**Greater Antilles** The four large islands Cuba, Jamaica, Hispaniola (shared by Haiti and the Dominican Republic), and Puerto Rico make up the **Greater Antilles**. On these islands are found the bulk of the region's population, arable lands, and large mountain ranges. Many people are knowledgeable about the Caribbean coasts but are surprised to learn that Pico Duarte in the Cordillera Central of the Dominican Republic is more than 10,000 feet (3,000 meters) tall, Jamaica's Blue Mountains top 7,000 feet (2,100 meters), and Cuba's Sierra Maestra is more than 6,000 feet tall (1,800 meters). Historically, the mountains of the Greater Antilles were of little economic interest because plantation owners preferred the coastal plains and valleys. Yet the mountains were

**FIGURE 5.6 Caribbean Sea**   Noted for its calm turquoise waters, steady breezes, and treacherous shallows, the Caribbean Sea has both sheltered and challenged sailors for centuries. This aerial photograph shows the southern Caribbean islands of Los Roques, off the Venezuelan coast.

an important refuge for runaway slaves and subsistence farmers and thus are important in the cultural history of the region.

**Lesser Antilles**   The **Lesser Antilles** form a double arc of small islands stretching from the Virgin Islands to Trinidad. Smaller in size and population than the Greater Antilles, they were important early footholds for rival European colonial powers. The islands from St. Kitts to Grenada form the inner arc of the Lesser Antilles. These mountainous islands, with peaks ranging from 4,000 to 5,000 feet (1,200 to 1,500 meters), have volcanic origins. Erosion of the island peaks and accumulation of ash from volcanic eruptions have created small pockets of arable soils, although the steepness of the terrain places limits on agricultural development.

Just east of this volcanic arc are the low-lying islands of Barbados, Antigua, Barbuda, and the eastern half of Guadeloupe. Covered in limestone that overlays volcanic rock, these lands were much more inviting for agriculture. In particular, such soils were ideal for growing sugarcane, causing these islands to become important early settings for the plantation economy that diffused throughout the region.

**Rimland States**   Unlike the rest of the Caribbean, the rimland states of Belize and the Guianas still contain significant amounts of forest cover. As on the islands, agriculture in these states is closely tied to local geology and soils. Much of low-lying Belize is limestone. Sugarcane is the dominant crop in the drier north, while citrus is produced in the wetter central portion of the state. The Guianas are characterized by the rolling hills of the Guiana Shield, whose crystalline rock is responsible for the area's overall poor soil quality. Thus, most agriculture in the Guianas occurs on the narrow coastal plain, with sugar and rice as the primary crops. French Guiana, which is an overseas territory of France, relies mostly on French subsidies but exports shrimp and timber. It is also home to the European Space Center at Kourou (Figure 5.7).

## Climate and Vegetation

Coconut palms framed by a blue sky evoke the postcard image of the Caribbean. In this tropical region, it is warm all year, and rainfall is

abundant. Much of the Antillean islands and rimland receive more than 80 inches (200 centi-meters) of rainfall annually and can support tropical forests. Amid the forests are pockets of naturally occurring grasslands in parts of Cuba, Hispaniola, and southern Guyana. Distinctly dry areas exist, such as the rain-shadow basin in western Hispaniola. As explained earlier, much of the natural vegetation on the islands has been removed to accommodate agriculture and fuel needs, and only small forest fragments remain today. Tropical ecosystems in the rimland are largely intact.

As in many other tropical lowlands, seasonality in the Caribbean is defined by changes in rainfall more than temperature. Although some rain falls throughout the year, the rainy season is from July to October (Figure 5.8). This is also when unstable atmospheric conditions sometimes cause the formation of hurricanes. During the

**FIGURE 5.7 Kourou, French Guiana**   The European Space Agency regularly launches rockets from its center in Kourou, French Guiana. This French territory, near the equator and on the coast, makes an ideal launch site. In this photo, the *Ariane 5* rocket is being moved to the launch pad.

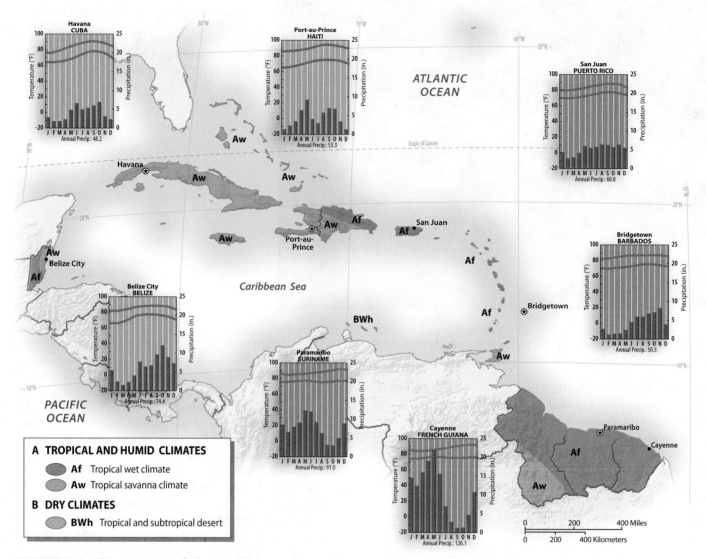

**FIGURE 5.8 Climate Map of the Caribbean**    Most of the region is classified as having either a tropical wet (Af) or tropical savanna (Aw) climate. Temperature varies little across the region, with highs slightly above 80°F (27°C) and lows around 70°F (21°C). Important differences in total rainfall and the timing of the dry season distinguish different places. In the Guianas, for example, the dry season is September through October, whereas the drier months for the islands are December through March. *(Temperature and precipitation data from Pearce and Smith, 1984,* The World Weather Guide, *London: Hutchinson)*

slightly cooler months of December through March, rainfall declines. This time of year corresponds with the peak tourist season.

**Hurricanes**    Each year several **hurricanes** pound the Caribbean as well as Central and North America with heavy rains and fierce winds. Beginning in July, westward-moving low-pressure disturbances form off the coast of West Africa, picking up moisture and speed as they move across the Atlantic. Usually no more than 100 miles across, these disturbances achieve hurricane status when wind speeds reach 74 miles per hour. Hurricanes may take several paths through the region, but they typically enter through the Lesser Antilles. They then curve north or northwest and collide with the Greater Antilles, Central America, Mexico, or southern North America before moving to the northeast and breaking up in the Atlantic Ocean. The hurricane zone lies just north of the equator on both the Pacific and Atlantic sides of

the Americas. Typically a half dozen to a dozen hurricanes form each season and move through the region, causing limited damage.

There are, of course, exceptions, and most longtime residents of the Caribbean have felt the full force of at least one major hurricane in their lifetime. The destruction caused by these storms is not just from the high winds but also from the heavy downpours that can cause severe flooding and deadly coastal storm surges. Modern tracking equipment has improved hurricane forecasting and reduced the number of fatalities, primarily by allowing early evacuation of areas in a hurricane's path. Improved forecasting saves lives, but it cannot reduce the damage to crops, forests, or infrastructure. In 2004, the island nation of Grenada was pummeled by Hurricane Ivan; although only a few people were killed, the island's infrastructure suffered 85 percent devastation (Figure 5.9). Fewer than half of the country's hotels were opened by January 2005, and exports of nutmeg—the island's principal

**FIGURE 5.9 Grenada after Hurricane Ivan** A resident of St. Georges in Grenada sits by the remains of his home after Hurricane Ivan destroyed it in 2004. One of the most destructive tropical storms to hit the Antilles in years, Ivan damaged most of Grenada's infrastructure.

cash crop—were down 90 percent. Barbados, Tobago, the Netherlands Antilles, Cuba, and Jamaica also suffered damage from Ivan. In total, about 70 lives and some $2 billion in insurance losses were claimed.

## Population and Settlement: Densely Settled Islands and Rimland Frontiers

In the Caribbean, the population density is generally quite high and, as in neighboring Latin America, increasingly urban. Eighty-seven percent of the region's population is concentrated on the four islands of the Greater Antilles (Figure 5.10). Add to this Trinidad's 1.3 million and Guyana's 800,000, and most of the population of the Caribbean is accounted for by six countries and one U.S. territory (Puerto Rico).

In terms of total population, few people inhabit the Lesser Antilles; nevertheless, some of these island states are densely settled. The small island of Barbados is an extreme example. With only 166 square miles (430 square kilometers) of territory, it has nearly 1700 people per square mile (650 per square kilometer). Bermuda, which is one-third of the size of the District of Columbia, has more than 3,000 people per

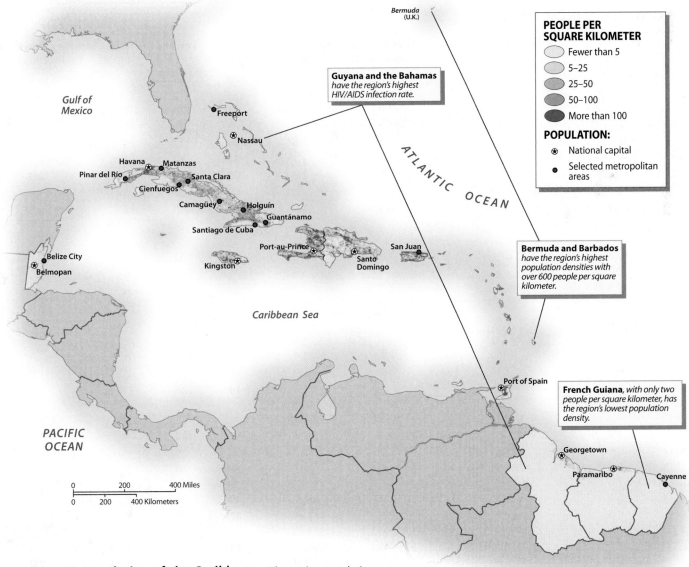

**FIGURE 5.10 Population of the Caribbean** The major population centers are on the islands of the Greater Antilles. The pattern here, as in the rest of Latin America, is a tendency toward greater urbanism. The largest city of the region is Santo Domingo, followed by Havana. In comparison, the rimland states are very lightly settled.

square mile (1,200 per square kilometer). Population densities on St. Vincent, Martinique, and Grenada, while not as high, are still more than 700 people per square mile (270 people per square kilometer). Because arable land is scarce on some of these islands, access to land is a basic resource problem for many island states. The growth in the region's population coupled with the scarcity of land has forced many people into cities or abroad.

In contrast to the islands, the mainland territories of Belize and the Guianas are lightly populated; Guyana averages 10 people per square mile (4 per square kilometer), Suriname only 8 (3 per square kilometer), and Belize 36 (14 per square kilometer). These areas are sparsely settled in part because the relatively poor quality and accessibility of arable land made them less attractive to colonial enterprises.

## Demographic Trends

During the years of sugar production based on slave labor, mortality rates were extremely high due to disease, inhumane treatment, and malnutrition. Consequently, the only way population levels could be maintained was by continuing to import African slaves. With the end of slavery in the mid- to late-19th century and the gradual improvement of health and sanitary conditions on the islands, populations began to increase naturally. In the 1950s and 1960s, many states achieved peak growth rates of 3.0 or higher, causing population totals and densities to soar. Over the past 20 years, however, growth rates have come down or stabilized. As noted earlier, the current population of the Caribbean is 43 million. The population is now growing at an annual rate of 1.2 percent, and projected population in 2025 is 48 million (see Table 5.1).

### TABLE 5.1 POPULATION INDICATORS

| Country | Population (millions) 2009 | Population Density (per square kilometer) | Rate of Natural Increase (RNI) | Total Fertility Rate | Percent Urban | Percent <15 | Percent >65 | Net Migration (Rate per 1,000) 2005–2010[*] |
|---|---|---|---|---|---|---|---|---|
| Anguilla | 0.01 | 159 | 2.3 | 1.8 | 100 | 25 | 8 | 14.0 |
| Antigua and Barbuda | 0.1 | 199 | 1.0 | 2.1 | 31 | 28 | 7 | |
| Bahamas | 0.3 | 25 | 1.1 | 1.9 | 83 | 28 | 6 | 1.2 |
| Barbados | 0.3 | 653 | 0.5 | 1.8 | 38 | 22 | 12 | −1.0 |
| Belize | 0.3 | 14 | 2.3 | 3.1 | 51 | 39 | 5 | −0.7 |
| Bermuda | 0.07 | 1256 | 0.7 | 2.0 | 100 | 18 | 15 | 2.2 |
| Cayman | 0.05 | 186 | 2.4 | 1.88 | 100 | 20 | 9 | 16.5 |
| Cuba | 11.2 | 101 | 0.3 | 1.6 | 76 | 18 | 12 | 3.5 |
| Dominica | 0.1 | 96 | 0.7 | 2.3 | 73 | 29 | 10 | |
| Dominican Republic | 10.1 | 207 | 1.8 | 2.8 | 64 | 33 | 6 | −2.8 |
| French Guiana | 0.2 | 3 | 2.8 | 3.9 | 76 | 35 | 4 | 5.5 |
| Grenada | 0.1 | 308 | 1.2 | 2.4 | 31 | 29 | 6 | −9.7 |
| Guadeloupe | 0.4 | 240 | 1.0 | 2.4 | 100 | 23 | 12 | −1.5 |
| Guyana | 0.8 | 4 | 1.2 | 2.5 | 28 | 32 | 5 | −10.5 |
| Haiti | 9.2 | 333 | 2.1 | 4.0 | 43 | 38 | 4 | −2.9 |
| Jamaica | 2.7 | 246 | 1.1 | 2.4 | 52 | 30 | 8 | −7.4 |
| Martinique | 0.4 | 368 | 0.7 | 1.9 | 98 | 22 | 12 | −1.0 |
| Montserrat | 0.005 | 50 | | 1.2 | 14 | 28 | 7 | |
| Netherlands Antilles | 0.2 | 254 | 0.5 | 1.9 | 92 | 22 | 10 | 8.7 |
| Puerto Rico | 4.0 | 447 | 0.4 | 1.7 | 94 | 20 | 14 | −1.1 |
| St. Kitts and Nevis | 0.05 | 191 | 1.0 | 2.3 | 32 | 28 | 8 | |
| St. Lucia | 0.2 | 319 | 0.8 | 1.7 | 28 | 28 | 7 | −1.2 |
| St. Vincent and the Grenadines | 0.1 | 283 | 0.9 | 2.1 | 40 | 29 | 7 | −9.2 |
| Suriname | 0.5 | 3 | 1.1 | 2.4 | 67 | 30 | 7 | −2.0 |
| Trinidad and Tobago | 1.3 | 260 | 0.6 | 1.6 | 12 | 24 | 7 | −3.0 |
| Turks and Caicos | 0.02 | 24 | 2.6 | 3.0 | 92 | 30 | 4 | 9.0 |

Source: Population Reference Bureau, *World Population Data Sheet, 2009.*

*Net Migration Rate from the United Nations, Population Division, *World Population Prospects: The 2008 Revision Population Database.*

**FIGURE 5.11 Smaller Caribbean Families**   A family in Bermuda enjoys a day at the beach. The average Caribbean family is far smaller now than 30 years ago. Higher levels of education, improved availability of contraception, an increase in urban living, and the larger percentage of women in the labor force contribute to slower population growth rates in many countries.

**Fertility Decline**   The most significant demographic trend in the Caribbean is the decline in fertility (Figure 5.11). Cuba has the region's lowest rate of natural increase (0.3), followed by Puerto Rico (0.4). In socialist Cuba, due to the education of women combined with the availability of birth control and abortion, the average woman has 1.6 children (compared to 2.1 in the United States). Yet in capitalist Puerto Rico, similarly low rates of natural increase (0.4) have also been achieved, along with a total fertility rate of 1.7. In general, educational improvements, urbanization, and a preference for smaller families have contributed to slower growth rates. Even states with relatively high total fertility rates, such as Haiti, have seen a decline in family size. Haiti's total fertility rate fell from 6.0 in 1980 to 4.0 in 2009.

**The Rise of HIV/AIDS**   The rate of HIV/AIDS infection in the Caribbean has come down in the past few years, but it is still twice that of North America, making the disease an important regional issue. Although nowhere near the infection rates in Sub-Saharan Africa (see Chapter 6), slightly more than 1 percent of the Caribbean population between ages 15 and 49 had HIV/AIDS in 2008. In Haiti, one of the earliest locations where AIDS was detected, 2.2 percent of the population between ages 15 and 49 is infected with the virus. The infection rate in Guyana is 2.5 percent, Belize is 2.1 percent, and The Bahamas is 3 percent. AIDS is already the largest single cause of death among young men in the English-speaking Caribbean.

Officials in The Bahamas have introduced drug treatments to help prevent mother-to-child transmission. And nearly every country has launched educational campaigns to bring infection rates down. Cuba, which witnessed a surge in both tourism and prostitution in the 1990s, has a very low infection rate of only 0.1 percent among its 15- to 49-year-old population. Education programs and an effective screening and reporting system for the disease have kept Cuba's infection rate down.

**Emigration**   Driven by the region's limited economic opportunities, a pattern of emigration to other Caribbean islands, North America, and Europe began in the 1950s. For more than 50 years, a **Caribbean diaspora**—the economic flight of Caribbean peoples across the globe—has become a way of life for much of the region (Figure 5.12). Barbadians generally choose to move to England. In contrast, one out of every three Surinamese has moved to the Netherlands, with most residing in Amsterdam. As for Puerto Ricans, only slightly more live on the island than reside on the U.S. mainland. In the 1980s, roughly 10 percent of Jamaica's population legally immigrated to North America (some 200,000 to the United States and 35,000 to Canada). Cubans have made the city of Miami their destination of choice since the 1960s. Today they are a large percentage of that city's population.

As a region, the Caribbean has one of the highest negative rates of net migration in the world, at –3.0. That means for every 1,000 people in the region, 3.0 annually leave. Individual countries have much higher rates, such as Guyana and Grenada at –10 per 1,000, and Jamaica at –7 per 1,000. The economic implications of this labor-related migration are significant and will be discussed later.

## The Rural–Urban Continuum

Initially, plantation agriculture and subsistence farming shaped Caribbean settlement patterns. Low-lying arable lands were dedicated to export agriculture and controlled by the colonial elite. Only small amounts of land were set aside for subsistence production. Over time, villages of freed or runaway slaves were established, especially in remote island interiors. However, the vast majority of people lived on estates as owners, managers, or slaves. Cities were formed to serve the administrative and social needs of the colonizers, but most were small, containing a small fraction of a colony's population. The colonists who linked the Caribbean to the world economy saw no need to develop major urban centers.

Even today, the structure of Caribbean communities reflects the plantation legacy. Many of the region's subsistence farmers are descendents of former slaves who continue to farm their small plots and work part time as wage laborers on estates. The social and economic patterns generated by slavery still mark the landscape. Rural communities tend to be loosely organized, labor is temporary, and

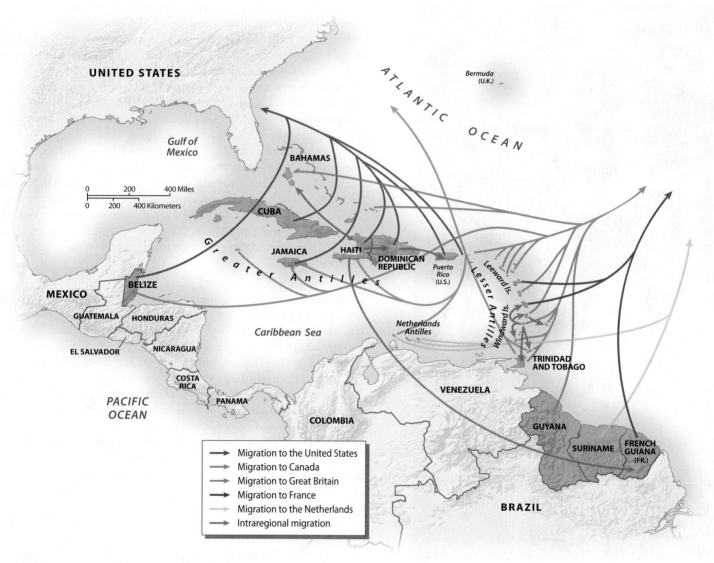

**FIGURE 5.12 Caribbean Diaspora** Emigration has long been a way of life for Caribbean peoples. With relatively high education levels but limited professional opportunities, the region commonly loses residents to North America, the United Kingdom, France, and the Netherlands. Migration within the region also occurs: Haitians work in the Dominican Republic and French Guiana, and Dominicans often travel to Puerto Rico. *(Data from Levin, 1987,* Caribbean Exodus, *New York: Praeger Publishers)*

small farms are scattered on available pockets of land. Because men tend to leave home for seasonal labor, female-headed households are common.

**Caribbean Cities** Since the 1960s, the mechanization of agriculture, offshore industrialization, and rapid population growth have caused a surge in rural-to-urban migration. Cities have grown accordingly, and today 64 percent of the region is classified as urban. Of the large countries, Puerto Rico is the most urban (94 percent) and Haiti the least (43 percent). Caribbean cities are not large by world standards, as only four have 1 million or more residents: Santo Domingo, Havana, Port-au-Prince, and San Juan. All but Port-au-Prince were former Spanish colonies.

Like their counterparts in Latin America, the Spanish Caribbean cities were laid out on a grid with a central plaza. Vulnerable to raids by rival European powers and pirates, these cities were usually walled and extensively fortified. The oldest continually occupied European city in the Americas is Santo Domingo in the Dominican Republic, settled in 1496, and today the metropolitan area has more than 2 million people. Havana emerged as the most important colonial city in the region, serving as a port for all incoming and outgoing Spanish ships. Strategically situated on Cuba's north coast at a narrow opening to a natural deep-water harbor, Havana became an essential city for the Spanish empire. Consequently, Havana possesses a handsome collection of colonial architecture, especially from the 18th and 19th centuries (Figure 5.13).

Other colonial powers left their mark on the region's cities. For example, Paramaribo, the capital of Suriname, has been described as a tropical, tulipless extension of Holland. In the British colonies, a preference for wooden whitewashed cottages with shutters was evident

**FIGURE 5.13 Old Havana**  A Cuban family strolls down a narrow cobblestone street near the cathedral in Old Havana, Cuba. The best examples of 18th- and 19th-century colonial architecture in the Caribbean are found in Havana. In 1982, UNESCO declared Old Havana a World Heritage Site, and new funds became available for its restoration.

(Figure 5.14). Yet the British and French colonial cities tended to be unplanned afterthoughts; these port cities were built to serve the rural estates, not the other way around. Most of them have grown dramatically over the past 40 years. These cities are no longer small ports for agricultural exports; increasingly their focus is on welcoming cruise ships and sun-seeking tourists.

Caribbean cities and towns do have their charms and reflect a blend of cultural influences. Throughout the region, houses are often simple structures (made of wood, brick, or stucco), raised off the ground a few feet to avoid flooding, and painted in pastels. Most people still get around by foot, bicycle, or public transportation; neighborhoods are filled with small shops and services that are within easy walking distance. Streets are narrow, and the pace of life is markedly slower than in North America and Europe. Even when space is tight in town, most settlements are close to the sea and its cooling breezes. An afternoon or evening stroll along the waterfront is a common activity.

**Housing**  The sudden surge in urbanization in the Caribbean is best explained by a decrease in rural jobs rather than a rise in urban

**FIGURE 5.14 Belize City Cottage**  Residents of Belize City build their wooden cottages on stilts as protection against flooding. Shuttered cottages with metal roofs are typical throughout the English Caribbean.

opportunities. Thousands poured into the cities as economic migrants and erected shantytowns. Squatter settlements in Port-au-Prince and Santo Domingo are especially bad, with residents living under miserable conditions, without the benefit of sewers and running water. Electricity is usually taken illegally from nearby power lines.

The one place that dramatically breaks from this pattern is Cuba. Forged in a socialist mode, Cubans are housed in uniform government-built apartment blocks like those seen throughout Russia and eastern Europe (Figure 5.15). These uninspired complexes contain hundreds of identical one- and two-bedroom apartments with basic modern services. Compared to other large cities of the developing world, the lack of squatter settlements makes Havana unusual.

**FIGURE 5.15 Government Housing**  Cubans in a state-built apartment block on the outskirts of greater Havana. Under socialism, thousands of standardized apartment blocks were built to ensure that all Cubans had access to basic modern housing.

# Cultural Coherence and Diversity: A Neo-Africa in the Americas

Linguistic, religious, and ethnic differences abound in the Caribbean. The presence of several former European colonies, millions of descendents of ethnically distinct Africans and indentured workers from India, and isolated Amerindian communities on the mainland challenges any notion of cultural coherence.

Common historical and cultural processes hold this region together. In particular, this chapter will focus on three cultural influences shared throughout the Caribbean: the European colonial presence, African influences, and the mix of European and African cultures referred to as *creolization*.

## The Cultural Impact of Colonialism

The arrival of Columbus in 1492 triggered a devastating chain of events that depopulated the Caribbean region within 50 years. A combination of Spanish brutality, enslavement, warfare, and disease reduced the densely settled islands, which supported up to 3 million Caribs and Arawaks, into an uninhabited territory ready for the colonizer's hand. By the mid-16th century, as rival European states competed for Caribbean territory, the lands they fought for were virtually uninhabited. In many ways, this simplified their task, as they did not have to recognize native land claims or work amid Amerindian societies. Instead, the colonizers reorganized the Caribbean territories to serve a plantation-based production system. The critical missing element was labor. Once slave labor from Africa, and later contract labor from Asia, was secured, the small Caribbean colonies became surprisingly profitable. Much of Caribbean culture and society today can be traced to the same processes that created "plantation America."

**Plantation America** Anthropologist Charles Wagley coined the term **plantation America** to designate a cultural region that extends from midway up the coast of Brazil through the Guianas and the

**FIGURE 5.16 Sugar Plantation** A historical illustration (1823) of slaves harvesting sugarcane on a plantation in Antigua. Sugar production was profitable but physically demanding work. Several million Africans were enslaved and forcibly relocated into the region.

**FIGURE 5.17 South Asian Indians in Trinidad** Trinidadians of Indian ancestry celebrate a Hindu festival by sending burning boats down the Marianne River in Trinidad. About 40 percent of Trinidad and Tobago's population is ethnically Indian and adheres to the Hindu faith.

Caribbean into the southeastern United States. Ruled by a European elite dependent on an African labor force, this society primarily existed in coastal regions and produced agricultural exports. It relied on **mono-crop production** (a single commodity, such as sugar) under a plantation system that concentrated land in the hands of elite families. Such a system created rigid class lines, as well as the formation of a multiracial society in which people with lighter skin were privileged. The term *plantation America* is not meant to describe a race-based division of the Americas but rather a production system that brought about specific ecological, social, and economic relations (Figure 5.16).

**Asian Immigration** Before detailing the pervasive influence of Africans on the Caribbean, the lesser-known Asian presence deserves mention. By the mid-19th century, most colonial governments in the Caribbean had begun to free their slaves. Fearful of labor shortages, they sought **indentured labor** (workers contracted to labor on estates for a set period of time, often several years) from South, Southeast, and East Asia.

The legacy of these indentured arrangements is clearest in Suriname, Guyana, and Trinidad and Tobago. In Suriname, a former Dutch colony, more than one-third of the population is of South Asian descent, and 16 percent is Javanese (from Indonesia). Guyana and Trinidad were British colonies, and most of their contract labor came from India. Today, half of Guyana's population and 40 percent of Trinidad and Tobago's claim South Asian ancestry. Hindu temples are found in the cities and villages, and many families speak Hindi at home (Figure 5.17).

Most of the former English colonies have Chinese populations of not more than 2 percent. Once these East Asian immigrants fulfilled their agricultural contracts, they often became merchants and small-business owners, positions they still hold in Caribbean society.

## Creating a Neo-Africa

The introduction of African slaves to the Americas began in the 16th century and continued into the 19th century. This forced migration of

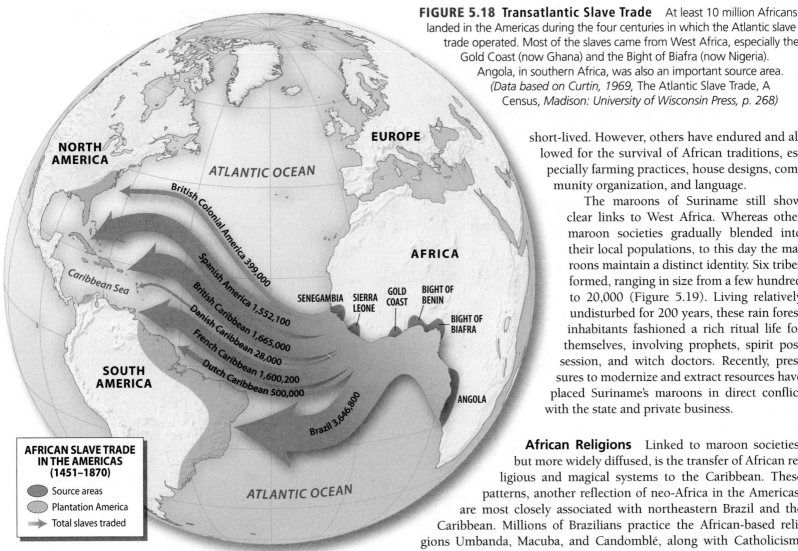

**FIGURE 5.18 Transatlantic Slave Trade** At least 10 million Africans landed in the Americas during the four centuries in which the Atlantic slave trade operated. Most of the slaves came from West Africa, especially the Gold Coast (now Ghana) and the Bight of Biafra (now Nigeria). Angola, in southern Africa, was also an important source area. *(Data based on Curtin, 1969,* The Atlantic Slave Trade, A Census, *Madison: University of Wisconsin Press, p. 268)*

short-lived. However, others have endured and allowed for the survival of African traditions, especially farming practices, house designs, community organization, and language.

The maroons of Suriname still show clear links to West Africa. Whereas other maroon societies gradually blended into their local populations, to this day the maroons maintain a distinct identity. Six tribes formed, ranging in size from a few hundred to 20,000 (Figure 5.19). Living relatively undisturbed for 200 years, these rain forest inhabitants fashioned a rich ritual life for themselves, involving prophets, spirit possession, and witch doctors. Recently, pressures to modernize and extract resources have placed Suriname's maroons in direct conflict with the state and private business.

**African Religions** Linked to maroon societies, but more widely diffused, is the transfer of African religious and magical systems to the Caribbean. These patterns, another reflection of neo-Africa in the Americas, are most closely associated with northeastern Brazil and the Caribbean. Millions of Brazilians practice the African-based religions Umbanda, Macuba, and Candomblé, along with Catholicism. Likewise, Afro-religious traditions in the Caribbean have evolved into

Africans to the Americas was only part of a much more complex **African diaspora**—the forced removal of Africans from their native area. The slave trade also crossed the Sahara to include North Africa and linked East Africa with a slave trade in the Middle East (see Chapter 6). The best-documented slave route was the transatlantic one; at least 10 million Africans landed in the Americas, and it is estimated that another 2 million died en route. More than half of these slaves were sent to the Caribbean (Figure 5.18).

This influx of slaves, combined with the elimination of nearly all the native inhabitants, remade the Caribbean as the area with the greatest concentration of relocated African people in the Americas. The African source areas extended from Senegal to Angola, and slave purchasers intentionally mixed tribal groups in order to weaken ethnic identities. Consequently, intact transfer of religion and languages into the Caribbean did not occur; instead, languages, customs, and beliefs were blended.

**Maroon Societies** Communities of runaway slaves—termed **maroons**—offer interesting examples of African cultural diffusion across the Atlantic. Hidden settlements of escaped slaves existed wherever slavery was practiced. Called maroons in English, *palenques* in Spanish, and *quilombos* in Portuguese, many of these settlements were

**FIGURE 5.19 Maroon Village in Suriname** A maroon woman carries a pail of water to her home in the village of Stonuku, Suriname. Maroon communities existed throughout the Caribbean as slaves ran away from plantations and formed villages in remote locations. The maroon communities in Suriname still retain many African traditions.

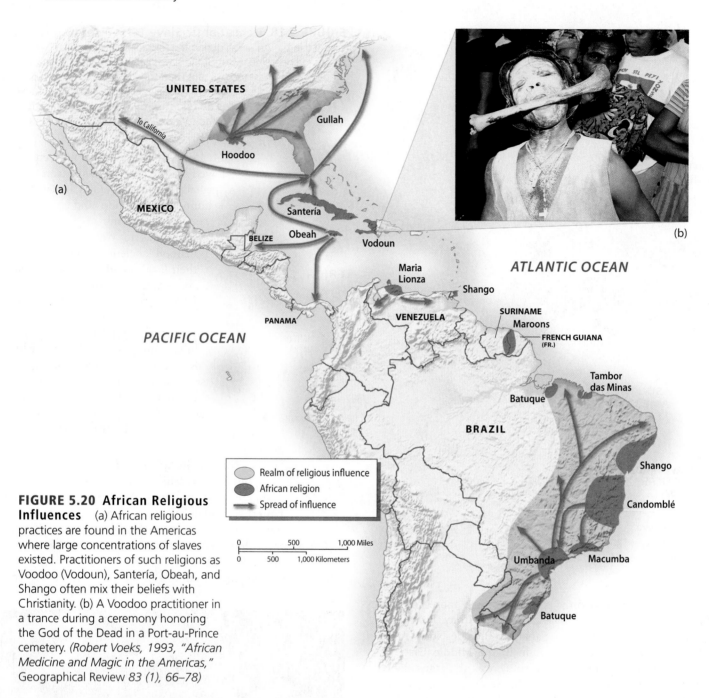

**FIGURE 5.20 African Religious Influences** (a) African religious practices are found in the Americas where large concentrations of slaves existed. Practitioners of such religions as Voodoo (Vodoun), Santería, Obeah, and Shango often mix their beliefs with Christianity. (b) A Voodoo practitioner in a trance during a ceremony honoring the God of the Dead in a Port-au-Prince cemetery. *(Robert Voeks, 1993, "African Medicine and Magic in the Americas,"* Geographical Review *83 (1), 66–78)*

unique forms that have clear ties to West Africa. The most widely practiced are Voodoo (also Vodoun) in Haiti, Santería in Cuba, and Obeah in Jamaica. Each of these religions has its own priesthood and unique pattern of worship. Their impact is considerable; the father-and-son dictators of Haiti, the Duvaliers, were known to hire Voodoo priests to scare off government opposition (Figure 5.20).

## Creolization and Caribbean Identity

**Creolization** refers to the blending of African, European, and some Amerindian cultural elements into the unique cultural systems found in the Caribbean. The Creole identities that have formed over time are complex; they illustrate the dynamic cultural and national identities of the region. Today, Caribbean writers (V. S. Naipaul, Derek Walcott, and Jamaica Kinkaid), musicians (Bob Marley, Celia Cruz, and Juan Luís

Guerra), and artists (Trinidadian costume designer Peter Minshall) are internationally regarded. Collectively, these artists are representative of their individual islands and of Caribbean culture as a whole.

**Language** The dominant languages in the region are European: Spanish (25 million speakers), French (10 million), English (6 million), and about half a million Dutch (Figure 5.21). Yet these figures tell only part of the story. In Cuba, the Dominican Republic, and Puerto Rico, Spanish is the official language, and it is universally spoken. As for the other countries, local variants of the official language exist, especially in spoken form, that can be difficult for a nonnative speaker to understand. In some cases completely new languages have emerged. In the islands of Aruba, Bonaire, and Curaçao, Papiamento (a trading language that blends Dutch, Spanish, Portuguese, English, and African languages) is the lingua franca, with use of Dutch declining. Similarly,

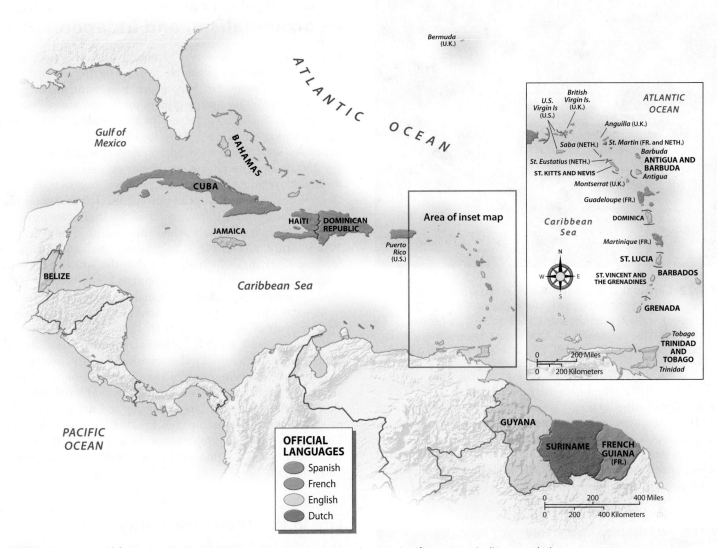

**FIGURE 5.21 Caribbean Language Map** Because this region has no significant Amerindian population (except on the mainland), the dominant languages are European: Spanish (25 million), French (10 million), English (6 million), and Dutch (0.5 million). However, many of these languages have been creolized, making it difficult for outsiders to understand them.

French Creole, or *patois*, in Haiti has been given official status as a distinct language. In practice, French is used in higher education, government, and the courts, but *patois* (with clear African influences) is the language of the street, the home, and oral tradition.

With independence in the 1960s, Creole languages became politically and culturally charged with national meaning. While most formal education is taught using standard language forms, the richness of vernacular expression and its ability to instill a sense of identity are appreciated. Locals rely on their ability to switch from standard to vernacular forms of speech. Thus, a Jamaican can converse with a tourist in standard English and then switch to a Creole variant when a friend walks by, effectively excluding the outsider from the conversation. While this ability to switch is evident in many cultures, it is widely used in the Caribbean.

**Music** The rhythmic beats of the Caribbean might be the region's best-known product. This small area is the home of reggae, calypso,

merengue, rumba, zouk, and scores of other musical forms. The roots of modern Caribbean music reflect a combination of African rhythms with European forms of melody and verse. These diverse influences coupled with a long period of relative isolation sparked distinct local sounds. As movement among the Caribbean population increased, especially during the 20th century, musical traditions were blended, but characteristic sounds remained.

The famed steel pan drums of Trinidad were created from oil drums discarded from a U.S. military base there in the 1940s. The bottoms of the cans are pounded with a sledgehammer to create a concave surface that produces different tones. During Carnival (a pre-Lenten celebration), racks of steel pans are pushed through the streets by dancers while drummers play. So skilled are these musicians that they even perform classical music, and government agencies encourage troubled teens to learn steel pan (Figure 5.22).

The distinct sound and the ingenious rhythms make Caribbean music popular. Yet it is more than good dancing music; the music is

**FIGURE 5.22 Carnival Drummer**   A steel pan drummer performs while his drum cart is pushed through the streets during Carnival. Steel drums were originally made from discarded oil cans that local peoples fashioned into drums by hammering the tops into a concave surface. Many steel drum bands perform internationally, playing everything from calypso to classical music.

closely tied to Afro-Caribbean religions and is a popular form of political protest. In Haiti, *rara* music mixes percussion instruments, saxophones, and bamboo trumpets, weaving in funk and reggae bass lines. The songs are always performed in French Creole and typically celebrate Haiti's African ancestry and the use of Voodoo. The lyrics deal with difficult issues, such as political oppression and poverty (Figure 5.23).

**FIGURE 5.23 Haiti's Rara Music**   Performed in procession, *rara* music is sung in *patois*. Considered the music of the poor, it is used to express risky social commentary. This rara band performs at a folk festival in Washington, DC.

# Geopolitical Framework: Colonialism, Neocolonialism, and Independence

Caribbean colonial history is a patchwork of competing European powers, fighting over profitable tropical territories. By the 17th century, the Caribbean had become an important proving ground for European ambitions. Spain's grip on the region was slipping, and rival European nations felt confident that they could gain territory by gradually pushing Spain out. Many territories, especially islands in the Lesser Antilles, changed European rulers several times.

Europeans viewed the Caribbean as a strategically located and profitable region in which to produce sugar, rum, and spices. Geopolitically, rival European powers also felt that their presence in the Caribbean limited Spanish authority there. Yet Europe's geopolitical dominance in the Caribbean began to diminish by the mid-19th century, just as the U.S. presence increased. Inspired by the **Monroe Doctrine**, which claimed that the United States would not tolerate European military involvement in the Western Hemisphere, the U.S. government made it clear that it considered the Caribbean to be within its sphere of influence. This view was highlighted during the Spanish–American War in 1898. Even though several English, Dutch, and French colonies remained after this date, the United States indirectly (and sometimes directly) asserted its control over the region, bringing in a period of **neocolonialism**. In an increasingly global age, however, even neocolonial interests can be short-lived or sporadic. The Caribbean has not attracted the level of private foreign investment seen by other regions. Moreover, as the Caribbean's strategic importance in a post–Cold War era fades, the leaders of the region openly worry about their area becoming more marginal.

## Life in the "American Backyard"

To this day, the United States maintains a controlling attitude toward the Caribbean, which was commonly referred to as "the American backyard" in the early 20th century. The initial foreign policy objectives were to free the region from European authority and encourage democratic governance. Yet time and again, American political and economic ambitions undermined those goals. President Theodore Roosevelt made his priorities clear with imperialistic policies that extended the influence of the United States beyond its borders. Policies and projects such as the construction of the Panama Canal and the maintenance of open sea-lanes benefited the United States but did not necessarily support social, economic, or political gains for the Caribbean people. The United States later offered benign-sounding development packages, such as the Good Neighbor Policy (1930s), the Alliance for Progress (1960s), and the Caribbean Basin Initiative (1980s). The Caribbean view of these initiatives has been wary at best. Rather than feeling liberated, many residents believe that one kind of political dependence was being traded for another—colonialism for neocolonialism.

In the early 1900s, the role of the United States in the Caribbean was overtly military and political. The Spanish–American War (1898) secured Cuba's freedom from Spain and also resulted in Spain's giving up the Philippines, Puerto Rico, and Guam to the United States; the latter two are still U.S. territories. The U.S. government also purchased the Danish Virgin Islands in 1917, renaming them the U.S. Virgin Islands and developing the harbor of St. Thomas. French, English, and Dutch colonies were tolerated, as long as these allies recognized the supremacy of the United States in the region. Outwardly against colonialism, the United States had become much like an imperial force.

One of the requirements of an empire is the ability to impose one's will, by force if necessary. When a Caribbean state refused to abide by U.S. trade rules, U.S. Navy vessels would block its ports. Marines landed, and U.S.-backed governments were installed throughout the Caribbean basin. These were not short-term engagements; U.S. troops occupied the Dominican Republic from 1916 to 1924, Haiti from 1913 to 1934, and Cuba from 1906 to 1909 and 1917 to 1922 (Figure 5.24). Even today, the United States maintains several important military bases in the region, including Guantánamo in eastern Cuba.

Many critics of U.S. policy in the Caribbean complain that business interests overwhelm democratic principles when foreign policy is determined. U.S. banana companies settled the coastal plain of the Caribbean rimland and operated as if they were independent states. Sugar and rum manufacturers from the United States bought the best lands in Cuba, Haiti, and Puerto Rico. Meanwhile, truly democratic institutions remained weak, and there was little improvement in social development. True, exports increased, railroads were built, and port facilities were improved; but levels of income, education, and health remained dreadfully low throughout the first half of the 20th century.

**The Commonwealth of Puerto Rico**    Puerto Rico is both within the Caribbean and apart from it because of its status as a commonwealth of the United States. Throughout the 20th century, various Puerto Rican independence movements sought to separate the island from the United States. Even today, residents of the island are divided about their island's political future. At the same time, Puerto Rico depends on U.S. investment and welfare programs;

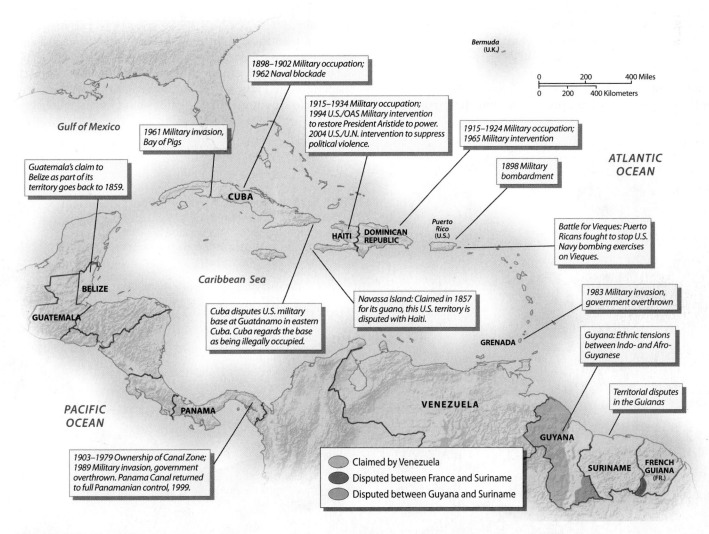

**FIGURE 5.24 Caribbean Geopolitics: U.S. Military Involvement and Regional Disputes**    The Caribbean was regarded as the geopolitical backyard of the United States, and U.S. military occupation was a common occurrence in the first half of the 20th century. Border and ethnic conflicts also exist, most notably in the Guianas. *(Data sources: Barbara Tenenbaum, ed., 1996* Encyclopedia of Latin American History and Culture, *vol. 5, p. 296, with permission of Charles Scribner's Sons; and John Allcock, 1992,* Border and Territorial Disputes, *3rd ed., Harlow, Essex, UK: Longman Group)*

U.S. food stamps are a major source of income for many Puerto Rican families. Commonwealth status also means that Puerto Ricans can freely move between the island and the U.S. mainland, a right they actively assert. In other ways, Puerto Ricans symbolically display their independence; for example, they support their own "national" sports teams and send a Miss Puerto Rico to international beauty pageants.

Puerto Rico led the Caribbean in the transition from an agrarian economy to an industrial one, beginning in the 1950s. For some U.S. officials, Puerto Rico became the model for the rest of the region. Puerto Rican President Muñoz Marín advocated an industrialization program called "Operation Bootstrap." Through tax incentives and cheap labor, hundreds of U.S. textile and clothing firms relocated to Puerto Rico. Over the next two decades, 140,000 industrial jobs were added, resulting in a marked increase in per capita gross national product. In the 1970s, when Puerto Rico faced stiff competition from Asian clothing manufacturers, the government encouraged petrochemical and pharmaceutical plants to relocate to the island (Figure 5.25). By the 1990s, Puerto Rico was one of the most industrialized places in the region, with a significantly higher per capita income than its neighbors. Yet it still showed many signs of underdevelopment, including extensive out-migration, low rates of education, and widespread poverty and crime.

**Cuba and Regional Politics** The most profound challenge to U.S. authority in the region came from Cuba and its superpower ally, the former Soviet Union. In the 1950s, a revolutionary effort began in Cuba, led by Fidel Castro against the pro-American Batista government. Cuba's economic productivity had soared, but its people were still poor, uneducated, and increasingly angry. The contrast between the lives of average sugarcane workers and the foreign elite was sharp. Castro tapped a deep vein of Cuban resentment against six decades of American neocolonialism. In 1959, Castro took power.

After Castro's government nationalized American industries and took ownership of all foreign-owned properties, the United States responded by refusing to buy Cuban sugar and ultimately ending diplomatic relations with the state. Various U.S. embargoes (laws forbidding trade with a particular country) against Cuba have existed for more than four decades. When Cuba established strong diplomatic relations with the Soviet Union in 1960, at the height of the Cold War, the island state became a geopolitical enemy of the United States. With the Soviet Union financially and militarily backing Castro, a direct U.S. invasion of Cuba was too risky. The fall of 1962 produced one of the most dangerous episodes of the Cold War, when Soviet missiles were discovered on Cuban soil. Ultimately, the Soviet Union removed its weapons; in return, the United States promised not to invade Cuba.

Even with the end of the Cold War, when Cuba lost its financial support from the Soviet Union, it managed to reinvent itself by growing its tourism sector and courting foreign investment, especially from Spain. More recently, Castro and Venezuelan President Hugo Chavez became close political allies, signing an important trade agreement in

**FIGURE 5.25 Women at Work in a Pharmaceutical Plant in Puerto Rico** Since the 1980s, U.S. pharmaceutical companies have located plants in Puerto Rico due to lower costs and tax advantages offered by the U.S. government.

2004 to exchange medical personnel (from Cuba) for petroleum (from Venezuela). Still today, the United States maintains its tough trade sanctions against Cuba and forbids U.S. tourists to visit the island.

A new political era for Cuba appears imminent. In 2008 Fidel Castro, 82 and in poor health, left office, and his younger brother, Raúl Castro, assumed the duties of president. Although many consider Raúl more willing to encourage private enterprise in Cuba, as of 2009, no major political and economic changes had occurred. (See "Exploring Global Connections: Cuba's Medical Diplomacy.")

## Independence and Integration

Given the repressive colonial history of the Caribbean, it is no wonder that the struggle for political independence began more than 200 years ago. Haiti was the second colony in the Americas to gain independence, in 1804, after the United States in 1776. However, the political independence of many states in the region has not guaranteed economic independence. Many Caribbean states struggle to meet the basic needs of their people. Today some Caribbean territories maintain their colonial status as an economic asset. For example, the French territories of Martinique, Guadeloupe, and French Guiana are overseas departments of France; residents have full French citizenship and social welfare benefits.

**Independence Movements**  Haiti's revolutionary war began in 1791 and ended in 1804. During this conflict the island's population was cut in half by casualties and emigration; ultimately, the former slaves became the rulers. Independence, however, did not allow this crown of the French Caribbean to prosper. Slowed by economic and political problems, it was ignored by the European powers and never accepted by the states of the Spanish mainland when they became independent in the 1820s.

Several revolutionary periods followed in the 19th century. In the Greater Antilles, the Dominican Republic finally gained independence in 1844, after taking control of the territory from Spain and Haiti. Cuba and Puerto Rico were freed from Spanish colonialism in 1898, but their independence was weakened by greater U.S. involvement. The British colonies also faced revolts, especially in the 1930s; it was not until the 1960s that independent states emerged from the English Caribbean. First, the larger colonies of Jamaica, Trinidad and Tobago, Guyana, and Barbados gained independence. Other British colonies followed throughout the 1970s and early 1980s. Suriname, the only Dutch colony on the rimland, became a self-governing territory in 1954 but remained part of the Netherlands until 1975, when it declared itself an independent republic.

**Regional Integration**  Perhaps the most difficult task facing the Caribbean is to increase economic integration. Scattered islands, a divided rimland, different languages, and limited economic resources hinder the formation of a meaningful regional trade bloc. It is more common to see economic cooperation between groups of islands with a shared colonial background than between, for example, former French and English colonies.

During the 1960s, the Caribbean began to experiment with regional trade associations as a means to improve its economic competitiveness. The goal of regional cooperation was to improve employment rates, increase intraregional trade, and ultimately reduce economic dependence.

The countries of the English Caribbean took the lead in this development strategy. In 1963, Guyana proposed an economic integration plan with Barbados and Antigua. In 1972, the integration process intensified, with the formation of the **Caribbean Community and Common Market (CARICOM)**. Representing the former English colonies, CARICOM proposed an ambitious regional industrialization plan and the creation of the Caribbean Development Bank to assist the poorer states. As important as this trade group is as an economic symbol of regional identity, it has produced limited improvements in intraregional trade. There are 13 full member states—all of the English Caribbean and French-speaking Haiti. Other dependencies, such as Anguilla, Turks and Caicos, Bermuda, and the British Virgin Islands, are associate members. Still, the predominance of English-speaking territories in CARICOM illustrates the deep linguistic divides in the Caribbean.

The dream of regional integration as a way to produce a more stable and self-sufficient Caribbean has never been realized. One scholar of the region argues that a limiting factor is a "small-islandist ideology." For example, islanders tend to keep their backs to the sea, oblivious to the needs of neighbors. At times, such isolationism results in suspicion, distrust, and even hostility toward nearby states. Yet economic necessity dictates engagement with partners outside the region. And so this peculiar status of isolated proximity unfolds in the Caribbean, expressing itself in uneven social and economic development trends.

# Economic and Social Development: From Cane Fields to Cruise Ships

Collectively, the population of the Caribbean, although poor by U.S. standards, is economically better off than most of Sub-Saharan Africa, South Asia, and China. Despite periods of economic stagnation in the Caribbean, social gains in education, health, and life expectancy are significant. Historically, the Caribbean's links to the world economy were through tropical agricultural exports, but several specialized industries, such as tourism, offshore banking, and assembly plants, have challenged the dominance of agriculture. These industries grew because of the region's proximity to North America and Europe, the availability of cheap labor, and the presence of policies that created a nearly tax-free environment for foreign-owned companies. Unfortunately, growth in these industries does not employ all the region's displaced rural laborers, so the lure of jobs in North America and Europe is still strong.

## From Fields to Factories and Resorts

Agriculture used to dominate the economic life of the Caribbean. Decades of unstable crop prices and decline in special trade agreements with former colonial masters have produced more hardship than prosperity. Ecologically, the soils are overworked, and there are no frontier areas to expand production, except for areas of the rimland. Moreover, agricultural prices have not kept pace with rising production costs, so wages and profits remain low. With the exception of a few mineral-rich territories, such as Trinidad, Guyana, Suriname, and Jamaica, most countries have tried to diversify their economies, relying less on their soils and more on manufacturing and services.

Comparing export figures over time demonstrates the shift away from mono-crop dependence. In 1955, Haiti earned more than

# EXPLORING GLOBAL CONNECTIONS
## Cuba's Medical Diplomacy

Cuba is widely recognized as a poor country with good health care, and it also has a plan for expanding health-care access to the developing world. Today, 30,000 Cuban doctors and health-care professionals work overseas in some 40 countries among the rural and urban poor. There are hundreds of Cuban doctors in South Africa, Gambia, and Ghana. There are thousands of Cuban doctors in Venezuela, El Salvador, Guatemala, Nicaragua, Bolivia, and Haiti. After Hurricane Mitch devastated Central America in 1998, Cuba sent medical personnel to Honduras and El Salvador. In 2005, more than 2,000 Cuban doctors traveled to northern Pakistan to attend earthquake victims. When Hurricane Katrina battered New Orleans that same year, some 1,500 medical personnel from Cuba offered to help, but the U.S. declined Cuba's support.

Cuba's health-care successes have been the pride of the revolution. In the past decade, Cuba has expanded its medical diplomacy as a means to barter for oil with Venezuela and also as an attempt to showcase something it does well. "Operation Miracle" is an outreach program that relies on Cuba's human capital and Venezuela's petro-dollars to provide free eye surgery. This joint venture, which began in 2004, has provided surgery for more than 750,000 people who could not afford it in their own countries. In Venezuela, President Chavez also began the Barrio Aldentro program to have Cuban health-care professionals live and work in Venezuela's shantytowns. Venezuelan doctors have protested this invasion of Cuban doctors, but at the same time the Cubans work in places that are typically not well served by medical professionals.

Perhaps the most lasting impact of Cuba's medical diplomacy is the creation of ELAM, the Latin American School of Medicine. ELAM students come from scores of countries, and thousands have been trained. The program selects ethnically diverse students from underserved areas and lower income groups. The students do not have to pay, but they are asked to go back to their countries and work in needy areas. As an indication of the school's reach, there are scholarships available for students from East Timor and Pakistan. The majority of students are from the Americas, including students from the United States and Puerto Rico. After completing six years of demanding training, in Spanish and English, in rural and urban areas, with limited resources, the doctors are sent back out to the world. And, in an ironic twist, the U.S. government recently recognized medical degrees from Cuba.

Adapted from Robert Huish and John M. Kirk, 2007, "Cuban Medical Internationalism and the Development of the Latin American School of Medicine," *Latin American Perspectives* 34(6): 77–92.

---

70 percent of its foreign exchange through the export of coffee; by 1990, coffee accounted for only 11 percent of its export earnings. Similarly, in 1955, the Dominican Republic earned close to 60 percent of its foreign exchange through sugar, but 35 years later, sugar earned less than 20 percent of the country's foreign exchange, and pig iron exports nearly equaled exports of sugar. The one exception to this trend is Cuba, which earned approximately 80 percent of its foreign exchange through sugar production from the 1950s to 1990. Cuba, however, was forced to diversify in the 1990s, when Russia would no longer guarantee the above-market price it paid for sugarcane. Since then, its sugar harvest has been reduced by half.

**Sugar**  The economic history of the Caribbean cannot be separated from the production of sugarcane. Even relatively small territories such as Antigua and Barbados yielded fabulous profits because there was no limit to the demand for sugar in the 18th century. Once considered a luxury crop, it became a popular necessity for European and North American laborers by the 1750s. It sweetened tea and coffee and made jams a popular spread for stale bread. In short, it made the meager and bland diets of ordinary people tolerable, and it also boosted caloric intake. Distilled into rum, sugar produced a popular intoxicant. Though it is difficult to imagine today, consumption of a pint of rum a day was not uncommon in the 1800s.

Sugarcane is still grown throughout the region for domestic consumption and export. Its economic importance has declined, however, mostly due to increased competition from corn and sugar beets grown in the midlatitudes. The Caribbean and Brazil are the world's major sugar exporters.

**The Banana Wars**  The major banana exporters are in Latin America, not the Caribbean. In fact, the success of banana plantations is mixed in this region, as banana plants are especially vulnerable to hurricanes. Still, several small states in the Lesser Antilles, most notably Dominica, St. Vincent, and St. Lucia, have become economically dependent on bananas, making as much as 60 percent of their export earnings from the yellow fruit. Bananas have not made people rich, but their production for export has led to greater economic and social development. In the eastern Caribbean, where most bananas are grown on small farms of five acres, the landowners are the laborers and thus earn two to four times more than banana plantation workers in Ecuador and Central America. Moreover, for these small states, banana exports are a link with the global economy. Yet with pressure mounting on the European Union to drop the higher prices paid to banana growers from the former colonies, the economic viability of banana production in the Caribbean is in doubt (Figure 5.26).

The case of the eastern Caribbean shows what happens to the losers in globalization. In 1996, the United States, Ecuador (the world's leading banana exporter), Mexico, Guatemala, and Honduras challenged the EU's banana trade agreement in the World Trade Organization (WTO) court. The agreement was denounced as unfair, and the EU was told to eliminate it by 1998. To make matters worse, consumer tastes had changed, with buyers preferring the uniformly large and unblemished yellow banana typical of the Latin American plantations rather than those grown in places like St. Lucia. To survive in this newly competitive environment, the growers of the eastern Caribbean will have to produce a more standardized fruit and increase

**FIGURE 5.26  Caribbean Bananas**   A laborer in St. Lucia harvests bananas on a small farm. Farmers in the eastern Caribbean have long depended on preferential markets in Europe to sell their bananas. A recent trade decision by the World Trade Organization court opens up Caribbean banana farmers to new competition, jeopardizing their economic viability.

U.S. Congress to phase out many of the tax exemptions may threaten Puerto Rico's ability to maintain its specialized industrial base.

Through the creation of **free trade zones (FTZs)**—duty-free and tax-exempt industrial parks for foreign corporations—the Caribbean is an increasingly attractive location for assembling goods for North American consumers. Manufacturing in the Dominican Republic demonstrates this production trend. The Dominican Republic took advantage of tax incentives and guaranteed access to the U.S. market offered through the Caribbean Basin Initiative. By 2001, the Dominican Republic had 16 operational FTZs, including a FTZ on the island's north shore, near the Haitian border, optimistically named "Hong Kong of the Caribbean." Plans exist to double the number of FTZs. Firms from the United States and Canada are the most frequent investors in these zones, followed by Dominican, South Korean, and Taiwanese firms. Traditional manufacturing on the island was tied to sugar refining, whereas up to 60 percent of production in the FTZs is in garments and textiles (Figure 5.27). Manufacturing is the largest economic sector in the Dominican Republic, accounting for nearly one-fifth of the country's GDP.

The growth in manufacturing depends on national and international policies that support export-led

their yield per acre. While island governments are trying to aid in this transition, local farmers have experimented with new crops, such as okra, tomatoes, avocados, and even marijuana. Reportedly, a few well-tended marijuana plants will earn 30 times more per pound than bananas. Just what will happen to family-run banana farms is hard to tell, but many Caribbean growers fear that the days of the banana economy are numbered. The banana story also reflects another major shift for the region: the decline in the economic importance of agriculture as well as the number of people employed by it.

### Assembly-Plant Industrialization
One regional strategy to deal with the unemployed agricultural workers was to invite foreign investors to set up assembly plants and thus create jobs. This was first tried successfully in Puerto Rico in the 1950s and was copied throughout the region. During Puerto Rico's Operation Bootstrap, island leaders encouraged U.S. investment by offering cheap labor, local tax breaks, and, most importantly, federal tax exemptions (something only Puerto Rico can do because of its special status as a commonwealth of the United States). Initially, the program was a tremendous success, and by 1970, nearly 40 percent of the island's gross domestic product (GDP) came from manufacturing. Today, 22 percent of the Puerto Rican labor force is employed in manufacturing, and this sector accounts for nearly half of the island's GDP. Yet competition from other states with even lower wages and the 1996 decision of the

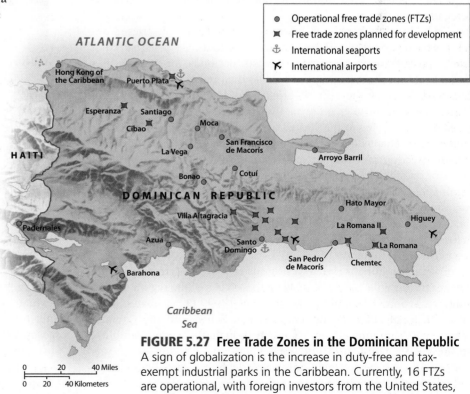

**FIGURE 5.27  Free Trade Zones in the Dominican Republic**   A sign of globalization is the increase in duty-free and tax-exempt industrial parks in the Caribbean. Currently, 16 FTZs are operational, with foreign investors from the United States, Canada, South Korea, and Taiwan. *(Source: Modified from Warf, 1995, "Information Services in the Dominican Republic," Yearbook, Conference of Latin American Geographers, p. 15)*

development through foreign investment. Certainly, new jobs are being created and national economies are diversifying in the process, but critics believe that foreign investors gain more than the host countries. Because most goods are assembled from imported materials, there is little development of national suppliers. Often higher than local averages, the wages are still low compared to those in the developed world—sometimes just $2 to $3 a day. Moreover, as other developing countries compete with the Caribbean for the establishment of FTZs, this development strategy may become less significant over time.

**Offshore Banking and Online Gambling**   The rise of offshore banking in the Caribbean is most closely associated with The Bahamas, which began this industry in the 1920s. **Offshore banking** centers appeal to foreign banks and corporations by offering specialized services that are confidential and tax exempt. Places that provide offshore banking make money through registration fees, not taxes. The Bahamas was so successful in developing this sector that by 1976 the country was the third largest banking center in the world. Its dominance began to decline due to competitors from the Caribbean, Hong Kong, and Singapore and because there was no longer a tax advantage in arranging large international loans offshore. Concerns about corruption and laundering of drug money also hurt the islands' financial status in the 1980s, and major reforms were introduced to reduce the presence of funds gained from illegal activities. By 1998, the ranking of The Bahamas in terms of global financial centers had dropped to 15th. Still, offshore banking remains an important part of the Bahamian economy. In the 1990s, the Cayman Islands emerged as the region's leader in financial services. With a population of 45,000, this crown colony of Britain has some 50,000 registered companies and the highest per capita income of the region.

Each of the offshore banking centers in the Caribbean tries to develop special financial services to attract clients, such as banking, functional operations, insurance, or trusts. Bermuda, for example, is a global leader in the reinsurance business that makes money from underwriting part of the risk of other insurance companies (Figure 5.28). The Caribbean is an attractive location for such services because of its closeness to the United States (home of many of the registered firms), client demand for these services in different countries, and the steady improvement in telecommunications that make this industry possible. The resource-poor islands of the region see providing financial services as a way to bring foreign capital to state treasuries. Envious of the economic success of The Bahamas, Bermuda, and the Cayman Islands, countries such as Antigua, Aruba, Barbados, and Belize hope they can find greater prosperity by establishing close ties to international finance.

Online gambling is the newest industry for the microstates of the Caribbean. Antigua and St. Kitts were the leaders of the region, beginning legal online gambling services in 1999. Other states soon followed; as of 2003, Dominica, Grenada, Belize, and the Cayman Islands had gambling domain sites. Although the online gambling business is illegal in the United States, nothing can stop Americans from betting in cyberspace. In only a decade, online gambling services had taken root throughout the Caribbean.

In 2007, the WTO deemed restrictions imposed on overseas Internet gambling sites by the United States were deemed illegal. The tiny nation of Antigua is currently seeking compensation from the United

**FIGURE 5.28 Financial Services in Bermuda**   Front Street in Hamilton, Bermuda, is a reflection of the territory's ties to the United Kingdom and its prosperity. Tourism and financial services in the reinsurance business explain Bermuda's wealth.

States for millions of dollars in lost revenue due to illegal restrictions placed on Antigua's business. While the WTO sided with Antigua, the future of Internet gambling remains uncertain.

**Tourism**   Environmental, locational, and economic factors converge to support tourism in the Caribbean. The earliest visitors to this tropical sea admired its clear and sparkling turquoise waters. By the 19th century, wealthy North Americans were fleeing winter to enjoy the healing warmth of the Caribbean during the dry season. Developers later realized that the simultaneous occurrence of the Caribbean dry season with the Northern Hemisphere's winter was ideal for beach resorts. By the 20th century, tourism was well established, with both destination resorts and cruise lines. By the 1950s, the leader in tourism was Cuba, and The Bahamas was a distant second. Castro's rise to power, however, eliminated this sector of the island's economy for nearly three decades and opened the door for other islands to develop tourism economies.

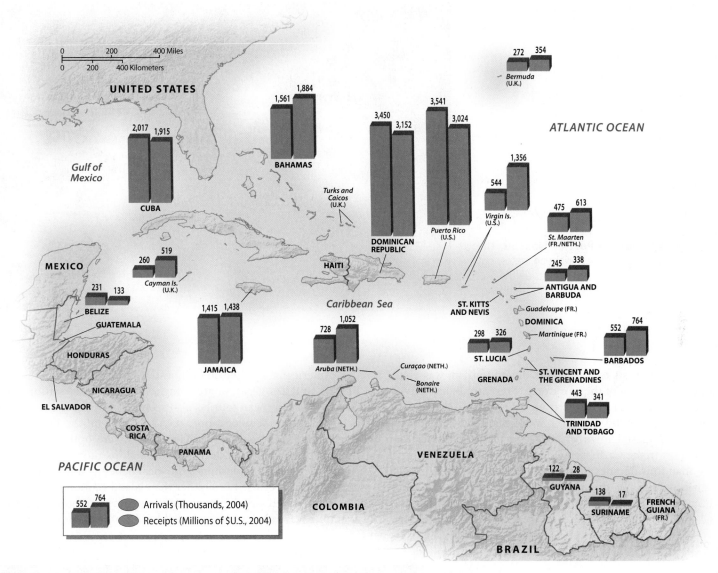

**FIGURE 5.29  Global Linkages: International Tourism in the Caribbean**   Tourism directly links the Caribbean to the global economy. Each year, 18 million tourists come to the islands, mostly from North America, Latin America, and Europe. The most popular destinations are the Dominican Republic, Puerto Rico, Cuba, The Bahamas, and Jamaica. *(From* Tourism Market Trends, *2005 edition)*

Five islands hosted two-thirds of the 18 million international tourists who came to the Caribbean in 2004: Puerto Rico, the Dominican Republic, Jamaica, The Bahamas, and Cuba (Figure 5.29). Puerto Rico saw its tourist sector begin to grow with commonwealth status in 1952. San Juan is now the largest home port for cruise lines and the second largest cruise-ship port in the world in terms of total visitors. The Bahamas attributes most of its economic development and high per capita income to tourism. With more than 1.5 million hotel guests in 2004 and 1.7 million cruise ship passengers, The Bahamas is another major hub for tourism in the region. Some 30 percent of the Bahamian population is employed in tourism, and tourism represents nearly half the country's GDP (Figure 5.30).

After years of neglect, Cuba is reviving tourism in an attempt to earn badly needed foreign currency. Tourism represented less than 1 percent of the national economy in the early 1980s. By 2004, more

than 2 million tourists (mostly Canadians and Europeans) yielded gross receipts approaching $2 billion. Conspicuous in their absence are travelers from the United States, forbidden to travel to Cuba because of the U.S.-imposed sanctions. With U.S. investors out of the picture, Spanish and other European developers are busy building up Cuba's tourist capacity, anticipating that someday the U.S. ban will be lifted.

As important as tourism is for the larger islands, it is often the principal source of income for smaller ones. The Virgin Islands, Barbados, Turks and Caicos, and, recently, Belize all greatly depend on international tourists. To show how quickly this sector can grow, consider this example: Belize began promoting tourism in the early 1980s, when just 30,000 arrivals came a year. An English-speaking country close to North America, Belize specialized in ecotourism that showcased its interior tropical forests and coastal barrier reef. By

**FIGURE 5.30 Cruise Ships in Port at Nassau, the Bahamas** Tourism is vital to many Caribbean states, but most cruise ships are owned by companies outside the region and offer relatively little employment for Caribbean workers.

1994, the number of tourists topped 300,000, and tourism was credited for employing one-fifth of the workforce. Presently, Belize City is reeling from an influx of day visitors from cruise ships. Belize City became a port of call in 2000. By 2005, more than 800,000 visitors disembarked at this coastal town of 60,000, making it the fastest-growing tourist port in the Caribbean (Figure 5.31).

For more than four decades, tourism has been the foundation of the Caribbean economy. Yet this regional industry has grown more slowly in recent years compared to other tourist destinations in the Middle East, southern Europe, and even Central America. It seems that Americans are favoring domestic destinations, such as Hawaii, Florida, and Las Vegas, or are going to more "exotic" settings, such as Costa Rica. European tourists also seem to be sticking closer to home or venturing to new locations, such as Dubai on the Persian Gulf or Goa in India. Increasingly, foreign tourists are opting to experience the Caribbean from the decks of cruise ships rather than land-based resorts. This trend undermines the local benefits of tourism, directing capital to large cruise lines rather than island economies.

Tourism-led development has detractors for other reasons. For example, it is subject to the overall health of the world economy and

**FIGURE 5.31 Resort Life** Young tourists enjoy the laid-back scene on the island of Caye Caulker, Belize. Once a village of fishermen, the main business on this small Belizean cay is serving the growing numbers of tourists coming to enjoy the barrier reef and stay in locally run hotels.

## TABLE 5.2   DEVELOPMENT INDICATORS

| Country | GNI per capita, PPP 2007 | GDP Average Annual % Growth 2000–07 | Human Development Index (2006)# | Percent Population Living Below $2 a Day | Life Expectancy* 2009 | Under Age 5 Mortality Rate 1990 | Under Age 5 Mortality Rate 2007 | Gender Equity 2007 |
|---|---|---|---|---|---|---|---|---|
| Anguilla | 8,800 | | | | 81 | | | |
| Antigua and Barbuda | 17,680 | | 0.830 | | 73 | | 11 | 95 |
| Bahamas | 28,600 | | 0.854 | | 72 | 29 | 13 | 101 |
| Barbados | 16,140 | | 0.889 | | 77 | 17 | 12 | 101 |
| Belize | 6,080 | | 0.771 | | 73 | 43 | 25 | 102 |
| Bermuda | 69,900 | | | | 80 | | | |
| Cayman | 43,800 | | | | 80 | | | 90 |
| Cuba | 9,500 | 3.4 | 0.855 | | 78 | 13 | 7 | 99 |
| Dominica | 6,930 | | 0.797 | | 75 | 18 | 11 | 101 |
| Dominican Republic | 6,350 | 4.8 | 0.768 | 15.1 | 72 | 66 | 38 | 104 |
| French Guiana | | | | | 75 | | | |
| Grenada | 5,480 | | 0.774 | | 74 | 37 | 19 | 99 |
| Guadeloupe | | | | | 79 | | | |
| Guyana | 2,580 | | 0.725 | 16.8 | 66 | 88 | 60 | 96 |
| Haiti | 1,050 | 0.2 | 0.521 | 72.1 | 58 | 152 | 76 | |
| Jamaica | 5,300 | 1.0 | 0.771 | 5.8 | 72 | 33 | 31 | 101 |
| Martinique | | | | | 80 | | | |
| Montserrat | 3,400 | | | | 73 | | | |
| Netherlands Antilles | 16,000 | | | | 76 | | | |
| Puerto Rico | 17,800 | | | | 78 | | | |
| St. Kitts and Nevis | 13,680 | | 0.830 | | 70 | 36 | 18 | 97 |
| St. Lucia | 9,240 | | 0.821 | 40.6 | 73 | 21 | 18 | 103 |
| St. Vincent and the Grenadines | 7,170 | | 0.766 | | 72 | 22 | 19 | 101 |
| Suriname | 7,640 | | 0.770 | 27.2 | 69 | 51 | 29 | 114 |
| Trinidad and Tobago | 22,420 | 8.8 | 0.833 | 13.5 | 69 | 34 | 35 | 101 |
| Turks and Caicos | 11,500 | | | | 75 | | | |

Source: World Bank, *World Development Indicators 2009.*

United Nations, *Human Development Index, 2008.*

Population Reference Bureau, *World Population Data Sheet, 2009*.*

Gender equity – Ratio of female to male enrollments in primary and secondary school. Numbers below 100 have more males in primary/secondary school, numbers above 100 have more females in primary/secondary schools.

current political affairs. Thus, if North America experiences a recession or international tourism declines due to heightened fears of terrorism, the flow of tourist dollars to the Caribbean dries up. Where tourism is on the rise, local resentment may build as residents confront the differences between their own lives and those of the tourists. There is also a serious problem of **capital leakage**, which is the huge gap between gross income and the total tourist dollars that remain in the Caribbean. Because many guests stay in hotel chains or cruise ships with corporate headquarters outside the region, leakage of profits is expected. On the plus side, tourism tends to promote stronger environmental laws and regulation. Countries quickly learn that their physical environment is the foundation for success. And while tourism does have its costs (higher energy and water consumption, as well as demand for more imports), it is environmentally less destructive than traditional export agriculture and at present more profitable.

## Social Development

While the record for economic growth in the region is inconsistent, measures of social development are stronger. With the exception of Haitians' low life expectancy of 58 years, most Caribbean peoples have an average life expectancy of 70 years or more (Table 5.2). Literacy levels are high, and there is nearly parity in terms of school enrollment by gender. Indeed, high levels of educational attainment and out-migration have contributed to a marked decline in natural increase rates over the past 30 years, so that the average Caribbean woman has two or three children (refer to Table 5.2). Despite real social gains, many inhabitants are chronically underemployed, poorly housed, and dependent on foreign remittances. For rich and poor alike, the temptation to leave the region in search of better opportunities is always present.

**Status of Women**    The matriarchal (female-dominated) basis of Caribbean households is often singled out as a distinguishing characteristic of the region. The rural custom of men leaving home for seasonal employment tends to nurture strong and self-sufficient female networks. Women typically run the local street markets. With men absent for long stretches, women tend to make household and community decisions. While giving women local power, this position does not always imply higher status. In rural areas, female status is often undermined by the relative exclusion of women from the cash economy; men earn wages, while women provide subsistence.

As Caribbean society urbanizes, more women are being employed in assembly plants (the garment industry, in particular, prefers to hire women), in data-entry firms, and in tourism. With new employment opportunities, female labor force participation has surged; in countries such as The Bahamas, Barbados, Jamaica, and Martinique, more than 40 percent of the workforce is female. Increasingly, women are the principal earners of cash, and they are more likely to complete secondary education then men. There are also signs of greater political involvement by women. In recent years Jamaica, Dominica, and Guyana have all had female prime ministers.

**Education**    Many Caribbean states have excelled in educating their citizens. Literacy is the norm, and the expectation is for most people to receive at least a high school degree. In many respects, Cuba's educational accomplishments are the most impressive, given the size of the country and its high illiteracy rates in the 1960s. Today, nearly all adults are literate. Hispaniola is the obvious contrast to Cuba's success. While the Dominican Republic has made strides in improving adult literacy (87 percent of adults are literate), half of all Haitian adults are illiterate. Political stability and economic growth have helped the Dominican Republic better its social conditions over the past decade. In fact, many Haitians have crossed the border into the Dominican Republic because conditions, although far from ideal, are much better there than in their homeland.

Education is expensive for these nations, but it is considered essential for development. Ironically, many states express frustration about training professionals for the benefit of developed countries in a phenomenon called **brain drain**. In the early 1980s, the prime minister of Jamaica complained that 60 percent of his country's newly trained workers left for the United States, Canada, and Britain, representing a subsidy to these economies far greater than the foreign aid Jamaica received from them. A 2007 World Bank study of skilled migrants revealed that 40 percent of Caribbean migrants living abroad were college educated. No other region in the world has proportionately this many educated people leaving. Brain drain occurs throughout the developing world, especially between former colonies and the mother countries. Given the small population of many Caribbean territories, each professional person lost to emigration can negatively impact local health care, education, and enterprise. While the outflow

**FIGURE 5.32 Remittance Economy**    Fernando Mateo, a native of the Dominican Republic, stands in one of his Manhattan offices, amid his many agents. Mr. Mateo's money-wiring business provides an electronic lifeline between Dominicans in New York and their families on the island. The money-wiring business has blossomed, as billions of dollars are sent to the developing world from hardworking immigrants in North America and Europe.

of professionals continues to be high, many countries are experiencing a return migration of Caribbean peoples from North America and Europe as a **brain gain**. Brain gain refers to the potential of returnees to contribute to the social and economic development of a home country with the experiences they have gained abroad.

**Labor-Related Migration**    Given the region's high educational rates and limited employment opportunities, Caribbean countries have seen their people emigrate for decades. After World War II, better transportation and political developments in the Caribbean produced a surge of migrants to North America. This trend began with Puerto Ricans going to New York in the early 1950s and intensified in the 1960s, with the arrival of nearly half a million Cubans. Since then, large numbers of Dominicans, Haitians, Jamaicans, Trinidadians, and Guyanese have also migrated to North America, typically settling in Miami, New York, Los Angeles, and Toronto.

Crucial in this exchange of labor from south to north is the flow of cash **remittances** (monies sent back home). Immigrants are expected to send something back, especially when immediate family members are left behind. Collectively, remittances add up; it is estimated that nearly $3 billion is annually sent to the Dominican Republic by immigrants in the United States, making remittance income the country's second leading source of income (Figure 5.32). Jamaicans and Haitians remit nearly $2 billion annually to their countries. Governments and individuals alike depend on these transnational family networks. Families carefully select the member most likely to succeed abroad, in the hopes that money will flow back and a base for future immigrants will be established. Labor-related migration has become a standard practice for tens of thousands of households in the region, as well as a clear expression of the globalization of labor flows.

## SUMMARY

- The Caribbean is more integrated into the global economy than most other developing nations, albeit as a dependent economic periphery. Although small in area, the region offers some of the clearest examples of the long-term efforts of globalization, from plantation agriculture to offshore banking.

- This tropical region has exploited its environment to produce export commodities such as sugar and bananas. The region's warm waters and mild climate attract millions of tourists. However, serious problems with deforestation, soil erosion, and water contamination have degraded urban and rural environments throughout the Caribbean.

- Population growth in the Caribbean has slowed over the past two decades. The average woman now has two or three children. Social measures of development such as life expectancy and literacy are very good in the region, yet countries struggle with supplying adequate employment. Emigration is a way of life in the region, as many people leave for employment opportunities abroad.

- The Caribbean was forged through European colonialism and the labor of millions of Africans. The blending of African and European elements, referred to as creolization, has resulted in many unique cultural expressions, especially in the religions, music, and languages of the region.

- Today the region contains 20 independent countries and several dependent territories. With the end of the Cold War, many microstates in the region fear that their lack of strategic significance may result in neglect by the United States and Europe, which may limit their ability to participate in world trade.

- In terms of development, the Caribbean has gradually shifted from being an exporter of primary agricultural resources (especially sugar) to a service and manufacturing economy. Employment in assembly plants, tourism, and offshore banking are steadily replacing jobs in agriculture. The region's strides in social development, especially in education, health, and the status of women, distinguish it from other developing areas.

## KEY TERMS

African diaspora (p. 129)
brain drain (p. 142)
brain gain (p. 142)
capital leakage (p. 141)
Caribbean Community and
  Common Market
  (CARICOM) (p. 135)

Caribbean diaspora (p. 125)
creolization (p. 130)
free trade zone (FTZ) (p. 137)
Greater Antilles (p. 120)
hurricane (p. 122)
indentured labor (p. 128)

isolated proximity (p. 116)
Lesser Antilles (p. 121)
maroon (p. 129)
mono-crop production (p. 128)
Monroe Doctrine (p. 132)
neocolonialism (p. 132)

offshore banking (p. 138)
plantation America (p. 128)
remittances (p. 142)
rimland (p. 119)

## THINKING GEOGRAPHICALLY

1. Contrast the historical African diaspora to the contemporary Caribbean one. What patterns are formed by these two distinct population movements? What social and economic forces are behind them? Are they comparable?

2. In the twenty-first century will agricultural exports continue to be important for Caribbean economies? What are the environmental consequences of agricultural production in this tropical region?

3. What advantages might Caribbean free trade zones retain over other competitors in Southeast and East Asia? What disadvantages might they face?

4. Are remittances a sign of the Caribbean's isolation or integration with the global economy? How might remittances be used to foster development?

5. Why have U.S. actions in the region been considered neocolonial? Are there other regions in the world where the United States exerts neocolonial tendencies?

Log in to **www.mygeoscienceplace.com** for videos, interactive maps, RSS feeds, case studies, and self-study quizzes to enhance your study of the Caribbean.

# 6 SUB-SAHARAN AFRICA

## GLOBALIZATION AND DIVERSITY

Struggling with extreme poverty, Sub-Saharan Africans have experienced few of the benefits of globalization. Yet over the past decade, foreign assistance to the region has climbed, and the explosive growth of mobile phone technology has enhanced communication.

### ENVIRONMENTAL GEOGRAPHY

Wood is a main source of energy for this region. In 2004, Kenyan Professor Wangari Maathai won the Nobel Peace Prize for her Green Belt Movement, which led to the planting of millions of trees by rural women.

### POPULATION AND SETTLEMENT

As a region, Sub-Saharan Africa has the highest infection rates of HIV/AIDS. As a result of this disease, many countries have seen life expectancy rates plummet into the 40s. Yet the population continues to grow due to high fertility rates.

### CULTURAL COHERENCE AND DIVERSITY

This is a region with large and growing numbers of Muslims, Christians, and animists. With a few exceptions, religious diversity and tolerance has been a distinctive feature of this region.

### GEOPOLITICAL FRAMEWORK

Most countries gained their independence in the 1960s. Since then, many ethnic conflicts have resulted, as governments have struggled for national unity within the boundaries drawn by European colonialists.

### ECONOMIC AND SOCIAL DEVELOPMENT

In terms of global trade, Sub-Saharan Africa's connection with the world is limited. But as global demand for natural resources grows, companies from Asia, Europe, and North America are investing in the extraction of the region's metals and fossil fuels.

*Buffalo gather at a watering hole in Zambezi National Park in Zimbabwe. The savannas of southern Africa are a noted habitat for the region's large mammals such as buffalo, elephant, zebra, and lions.*

**145**

Compared with Latin America and the Caribbean, Africa south of the Sahara is poorer and more rural, and its population is very young. More than 800 million people reside in this region, which includes 48 states and 1 territory. Demographically, this is the world's fastest-growing region (2.5 percent rate of natural increase); in most countries, nearly half the population (43 percent) is younger than 15 years. Income levels are extremely low: 73 percent of the population lives on less than $2 per day. Life expectancy is only 50 years. This part of the world is known for poverty, disease, violence, and refugees. Overlooked in the all-too-frequent negative headlines are programs by private, local, and state groups to improve the region's quality of life. Many countries have reduced infant mortality, expanded basic education, and increased food production in the past two decades, despite the economic and political crises that hinder this region's development.

Sub-Saharan Africa—that portion of the African continent lying south of the Sahara Desert—is a commonly accepted world region (Figure 6.1). The unity of this region has to do with similar livelihood systems and a shared colonial experience. No common religion, language, philosophy, or political system ever united the area. Instead, loose cultural bonds developed from a variety of lifestyles and idea systems that evolved here. The impact of outsiders also helped to determine the region's identity. Slave traders from Europe, North Africa, and Southwest Asia treated Africans as property; up until the mid-1800s, millions of Africans were taken from the region and sold into slavery. In the late 1800s, the entire African continent was divided by European colonial powers, imposing political boundaries that remain to this day. In the postcolonial period, which began in the 1960s, Sub-Saharan African countries faced many of the same economic and political challenges. These common experiences also helped unify the region (see "Setting the Boundaries").

The region is culturally complex, with dozens of languages spoken in some states. Consequently, most Africans understand and speak several languages. Ethnic identities do not follow the political divisions of Africa, sometimes resulting in bloody ethnic warfare. Nevertheless, throughout the region peaceful coexistence between distinct ethnic groups is the norm. The cultural significance of European colonizers cannot be ignored: European languages, religions, educational systems, and political ideas were adopted and modified. Yet the daily rhythms of life are far removed from the industrial world. Most Africans still engage in subsistence and cash-crop agriculture (Figure 6.2). The influence of African peoples outside the region is great, especially when one considers that human origins are traceable to this part of the world. In historic times, the legacy of the slave trade resulted in the transfer of African peoples, religious systems, and musical traditions throughout the Western Hemisphere. Even today, African-based religious systems are widely practiced in the Caribbean and Latin America, especially in Brazil.

The African economy, however, is marginal compared with the rest of the world. According to the World Bank, Sub-Saharan Africa's economic output in 2007 amounted to less than 2 percent of global output, even though the region contains 12 percent of the world's population. Moreover, the gross national income (GNI) of just one country, South Africa, accounts for more than one-third of the region's total GNI.

Many scholars feel that Sub-Saharan Africa has benefited little from its integration (both forced and voluntary) into the global economy. Slavery, colonialism, and export-oriented mining and agriculture served the needs of consumers outside the region but failed to improve domestic food supplies, infrastructure, and standards of living.

Foreign assistance in the postcolonial years initially improved agricultural and industrial output but also led to mounting foreign debt and corruption, which over time undercut the region's economic gains. Ironically, many of these same scholars and politicians worry that negative global attitudes about the region have produced a pattern of neglect. Private capital investment in Sub-Saharan Africa lags far behind that in other developing regions, but foreign investment is on the rise, especially in oil-rich states such as Angola, Chad, and Equatorial

# Setting the Boundaries

Sub-Saharan Africa includes 43 mainland states plus the island nations of Madagascar, Cape Verde, São Tomé and Principe, the Seychelles, Mauritius, and the French territory of Reunion. When setting this particular regional boundary, the major question one faces is how to treat North Africa.

Many scholars argue for keeping the African continent intact as a culture region. The Sahara has never formed a complete barrier between the Mediterranean north and the remainder of the African landmass. Moreover, the Nile River forms a several-thousand–mile-long corridor of continuous settlement linking North Africa directly to the center of the continent. There is no obvious place to divide the

watershed between North and Sub-Saharan Africa. Regional organizations such as the African Union are modern examples of the continent's unity.

The lack of a clear divide across Africa does not mean it cannot be divided into two world regions. North Africa is generally considered more closely linked, both culturally and physically, to Southwest Asia. Arabic is the dominant language and Islam the dominant religion of North Africa. Consequently, North Africans feel more closely connected to the Arab hearth in Southwest Asia than to the Sub-Saharan world.

The decision to view Sub-Saharan Africa as a world region still presents the problem of

where to divide it from Africa north of the Sahara. We preserved political boundaries when drawing the line so that the Mediterranean states of North Africa are discussed with Southwest Asia. Sudan is discussed in both Chapters 6 and 7 because it shares characteristics common to both regions. Sudan is Africa's largest state in terms of area; it is one-fourth the size of the United States. In the more populous and powerful north, Muslim leaders have crafted an Islamic state that is culturally and politically oriented toward North Africa and Southwest Asia. Southern Sudan, however, has more in common with the animist and Christian groups of the Sub-Saharan region.

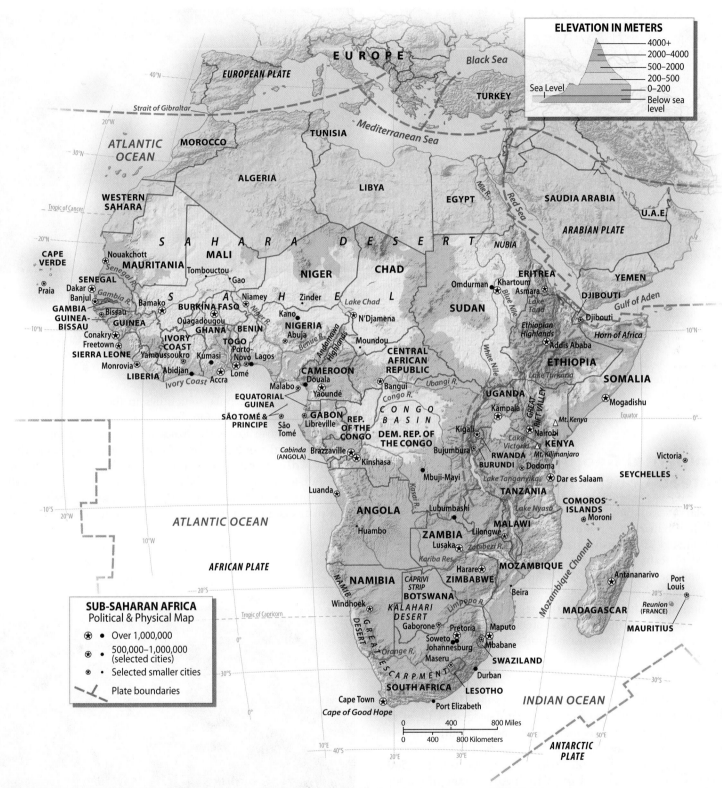

**FIGURE 6.1  Sub-Saharan Africa**    Africa south of the Sahara includes 48 states and 1 territory. This large region of rain forests, tropical savannas, and deserts is home to more than 800 million people. Much of the region consists of broad plateaus ranging from 1,600 to 6,500 feet (500 to 2,000 meters) in elevation. Although the population is growing rapidly, the overall population density of Sub-Saharan Africa is low. Considered one of the least developed regions of the world, it remains an area rich in natural resources.

**FIGURE 6.2 Women Farmers in Senegal** Senegalese women grow manioc, a drought-resistant staple introduced from the Americas. Throughout Sub-Saharan Africa, subsistence agriculture is still widely practiced, and women are often responsible for tending these crops.

Guinea. Through a combination of internal reforms, better governance, foreign assistance, and foreign investment in infrastructure and technology, many believe that social and economic gains are possible for Sub-Saharan Africa.

# Environmental Geography: The Plateau Continent

The largest landmass straddling the equator, Sub-Saharan Africa is vast in scale and remarkably beautiful. Called the plateau continent, Africa is dominated by extensive areas of geologic uplift that formed huge elevated plateaus. The highest areas are found on the eastern edge of the continent, where the Great Rift Valley forms a complex upland area of lakes, volcanoes, and deep valleys. In contrast, lowlands prevail in West Africa. Despite this region's immense biodiversity, vast water resources, and wealth of precious minerals, one finds relatively poor soils, widespread disease, and vulnerability to drought.

## Plateaus and Basins

A series of plateaus and elevated basins dominates the African interior and explain much of the region's unique physical geography (refer to Figure 6.1). Generally, elevations increase toward the south and east of the continent. Most of southern and eastern Africa lies well above 2,000 feet (600 meters), and sizable areas sit above 5,000 feet (1,500 meters). This is typically referred to as High Africa; Low Africa includes West Africa and much of Central Africa. Steep escarpments form where plateaus abruptly end, as illustrated by the majestic Victoria Falls on the Zambezi River (Figure 6.3). Much of southern Africa is rimmed by a landform called the **Great Escarpment** (a high cliff separating two comparatively level areas), which begins in southwestern Angola and ends in northeastern South Africa, creating a barrier to coastal settlement.

Though Sub-Saharan Africa is an elevated landmass, it has few significant mountain ranges. The one extensive area of mountainous

**FIGURE 6.3 Victoria Falls** The Zambezi River descends over Victoria Falls in the southern part of the region. A fault zone in the African plateau explains the existence of a 360-foot (110-meter) drop. The Zambezi has never been important for navigation, but it is a vital supply of hydroelectricity for Zimbabwe, Zambia, and Mozambique.

topography is in Ethiopia, which lies in the northern portion of the Rift Valley zone. Receiving heavy rains in the wet season, the Ethiopian Plateau forms the headwaters of several important rivers, most notably the Blue Nile, which joins the White Nile at Khartoum, Sudan. A discontinuous series of volcanic mountains, some of them quite tall, are associated with the southern half of the Rift Valley that runs through Kenya. Yet even in these areas, the dominant features are high plateaus intercut with deep valleys rather than actual mountain ranges (Figure 6.4).

**FIGURE 6.4 The Rift Valley** An aerial view of a portion of the Rift Valley in Kenya. This is one of the continent's most dramatic landforms, extending nearly 5,000 miles (8,000 kilometers) from Lake Nyasa to the Red Sea. The Rift Valley zone is a series of faults that have created volcanoes, escarpments, elongated lakes, and valleys.

**FIGURE 6.5 Congo River Fishermen** Congo River fishermen in a canoe ply the waters as a source of fish. Behind them are their village and the immense Ituri rain forest.

Sub-Saharan Africa's two largest hydroelectric installations are located on this river.

**Soils** With a few major exceptions, Sub-Saharan Africa's soils are relatively infertile. Generally speaking, fertile soils are young soils, those deposited in recent geologic time by rivers, volcanoes, glaciers, or windstorms. In older soils—especially those located in moist tropical environments—natural processes tend to wash out most plant nutrients over time. Over most of Sub-Saharan Africa, the agents of soil renewal have largely been absent.

Portions of Sub-Saharan Africa are, however, noted for their natural soil fertility, and, not surprisingly, these areas support denser settlement. Some of the most fertile soils are in the Rift Valley, made productive by the volcanic activity associated with the area. The population densities of rural Rwanda and Burundi, for example, are partially explained by the highly productive soils. The same can be said for highland Ethiopia, which supports the region's second largest population, with 30 million people. The Lake Victoria lowlands and central highlands of Kenya are also noted for their sizable populations and productive agricultural bases.

## Climate and Vegetation

Most of Sub-Saharan Africa lies in the tropical latitudes, and it is the largest tropical landmass on the planet. Only the far south of the continent extends into the subtropical and temperate belts. Much of the region averages high temperatures from 70 to 80°F (22 to 28°C) year-round (Figure 6.6). Rainfall, more than temperature, determines the different vegetation belts that characterize the region. Addis Ababa, Ethiopia, and Walvis Bay, Namibia, have similar average temperatures, but the former is in the moist highlands and receives nearly 50 inches (127 centimeters) of rainfall annually, while Walvis Bay rests on the Namibian Desert and receives less than 1 inch (2.5 centimeters).

The three main biomes of the region are tropical forests, savannas, and deserts.

**Tropical Forests** The center of Sub-Saharan Africa falls in the tropical wet climate zone. The world's second-largest expanse of humid equatorial rain forest, the Ituri, lies in the Congo Basin, extending from the Atlantic Coast of Gabon two-thirds of the way across the continent, including the northern portions of the Republic of the Congo and the Democratic Republic of the Congo (Zaire). The conditions here are constantly warm to hot, and precipitation falls year-round (see graph for Kisangani, Figure 6.6).

Commercial logging and agricultural clearing have degraded the western and southern fringes of the Ituri, but much of this vast forest is still intact. The Ituri has, so far, fared much better than Southeast Asia and Latin America in terms of tropical deforestation. Certainly

**Watersheds** Africa south of the Sahara does not have the broad, alluvial lowlands that influence patterns of settlement throughout other regions. The four major river systems are the Congo, Niger, Nile, and Zambezi (the Nile will be discussed in Chapter 7). Smaller rivers—such as the Orange in South Africa; the Senegal, which divides Mauritania and Senegal; and the Limpopo in Mozambique—are locally important but drain much smaller areas. Ironically, most people think of Africa south of the Sahara as suffering from water scarcity and tend to discount the size and importance of the watersheds (or catchment areas) drained by these river systems.

The Congo River (or Zaire) is the largest watershed in the region, both in terms of drainage area and the volume of river flow produced. It is second only to South America's Amazon River in terms of annual flow. The Congo flows across a relatively flat basin that lies more than 1,000 feet (300 meters) above sea level, meandering through Africa's largest tropical forest, the Ituri (Figure 6.5). Entry from the Atlantic into the Congo Basin is prevented by a series of rapids and falls, making the Congo River only partially navigable. Despite these limitations, the Congo River has been the major corridor for travel within the Republic of the Congo and the Democratic Republic of the Congo (formerly Zaire); the capitals of both countries, Brazzaville and Kinshasa, rest on opposite sides of the river.

The Niger River is the critical source of water for the arid countries of Mali and Niger. Beginning in the humid Guinea highlands, the Niger flows first to the northeast and then spreads out to form a huge inland delta in Mali before making a great bend southward at the margins of the Sahara near Gao. On the banks of the Niger River are the capitals of Mali (Bamako) and Niger (Niamey), as well as the historic city of Tombouctou (Timbuktu). After flowing through the desert north, the Niger River returns to the humid lowlands of Nigeria, where the Kainji Reservoir temporarily blocks its flow to produce electricity for Africa's most heavily populated state.

The considerably smaller Zambezi River begins in Angola and flows east, spilling over an escarpment at Victoria Falls, and finally reaching Mozambique and the Indian Ocean. More than other rivers in the region, the Zambezi is a major supplier of commercial energy.

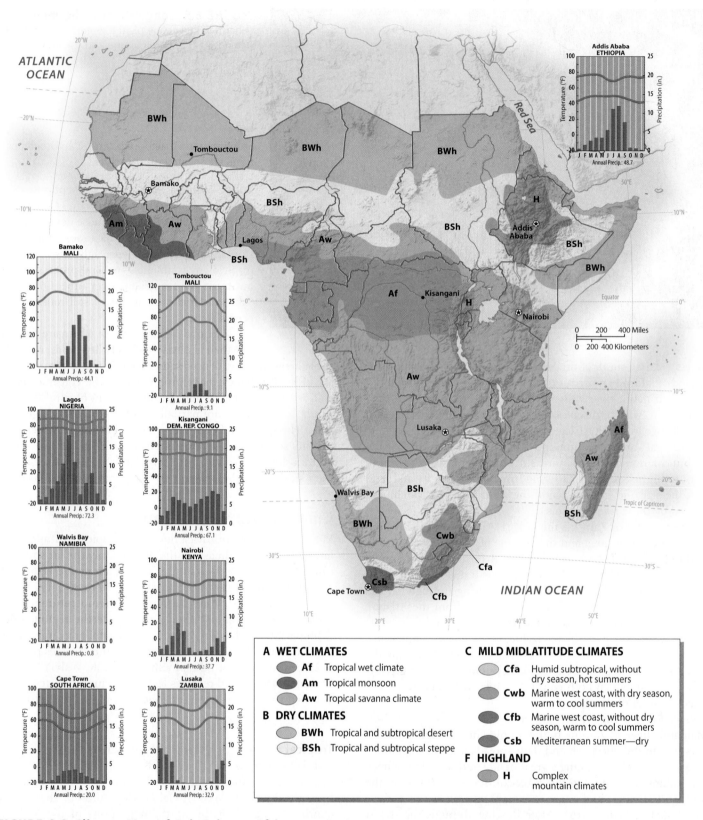

**FIGURE 6.6 Climate Map of Sub-Saharan Africa** Much of the region lies within the tropical humid and tropical dry climatic zones; thus, the seasonal temperature changes are not great. Precipitation, however, varies significantly from month to month. Compare the distinct rainy seasons in Lusaka and Lagos: Lagos is wettest in June, and Lusaka receives most of its rain in January. Although there are important tropical forests in West and Central Africa (coinciding with the tropical wet and monsoon climate zones), vegetation in much of the region is tropical savanna. *(Temperature and precipitation data from Pearce and Smith, 1984,* The World Weather Guide, *London: Hutchinson)*

**FIGURE 6.7 African Savannas**    A line of wildebeests marches across the tree-studded savannas in Masai Mara National Park, Kenya. The savannas, an essential habitat for Africa's wildlife, are steadily being converted into agricultural and pastoral lands for human needs.

the area's low population of subsistence farmers does not place much strain on the resource base. Moreover, Gabon and the Democratic Republic of the Congo export oil, making commercial logging less important to local economies. The political chaos in the Democratic Republic of the Congo has made large-scale logging a difficult proposition. In the future, however, it seems likely that Central Africa's rain forests could suffer the same kind of degradation experienced in other equatorial areas.

**Savannas**    Wrapped around the Central African rain forest belt in a great arc lie Africa's vast tropical wet and dry savannas. Savannas are dominated by a mixture of trees and tall grasses in the wetter zones immediately adjacent to the forest belt and shorter grasses with fewer trees in the drier zones (Figure 6.7). North of the equatorial belt, rain generally falls only from May to October. The farther north one travels, the less the total rainfall and the longer the dry season. Climatic conditions south of the equator are similar, only reversed, with the wet season occurring between October and May, and precipitation generally decreasing toward the south (see Lusaka in Figure 6.6). A larger area of wet savanna exists south of the equator, with extensive woodlands located in southern portions of the Democratic Republic of the Congo, Zambia, and eastern Angola. These savannas are also a critical habitat for the region's large fauna.

Elephants, zebras, rhinoceroses, and lions are found in the wooded grasslands, although their numbers have declined in most areas since the 1980s.

**Deserts**    Major deserts exist in the southern and northern boundaries of the region. The Sahara, the world's largest desert and one of its driest, crosses the landmass from the Atlantic coast of Mauritania all the way to the Red Sea coast of Sudan. A narrow belt of desert extends to the south and east of the Sahara, wrapping around the **Horn of Africa** (the northeastern corner that includes Somalia, Ethiopia, Djibouti, and Eritrea) and pushing as far as eastern and northern Kenya. An even drier zone is found in southwestern Africa. In the Namib Desert of coastal Namibia, rainfall is a rare event, although temperatures are usually mild. Inland from the Namib lies the Kalahari Desert. Because it receives slightly more than 10 inches (25 centimeters) of rain a year, most of the Kalahari is not dry enough to be classified as a true desert (Figure 6.8). Its rainy season, however, is brief, and most of the precipitation is immediately absorbed by the underlying sands.

**FIGURE 6.8 Kalahari Desert**    A San man teaches his four-year-old son to hunt with a bow and arrow. The Kalahari, though not a true desert, does not get enough rain to support agriculture. Yet this ecosystem is home to hunter-gatherers, such as the San (formerly known as the Bushmen).

Surface water is thus scarce, giving the Kalahari a desertlike appearance for most of the year.

## Africa's Environmental Issues

The prevailing perception of Africa south of the Sahara is one of environmental scarcity and degradation, no doubt encouraged by televised images of drought-ravaged regions and starving children. Single explanations such as rapid population growth or colonial exploitation cannot fully capture the complexity of Africa's environmental issues or the ways that people have adapted to living in marginal ecosystems. Because much of Sub-Saharan Africa's population is rural and poor, earning its livelihood directly from the land, sudden environmental changes have dramatic effects and can cause mass migrations, famine, and even death. As Figure 6.9 illustrates, deforestation and **desertification**, the expansion of desertlike conditions as a result of human-induced degradation, are commonplace. Sub-Saharan African is also vulnerable to drought, most notably in the Horn

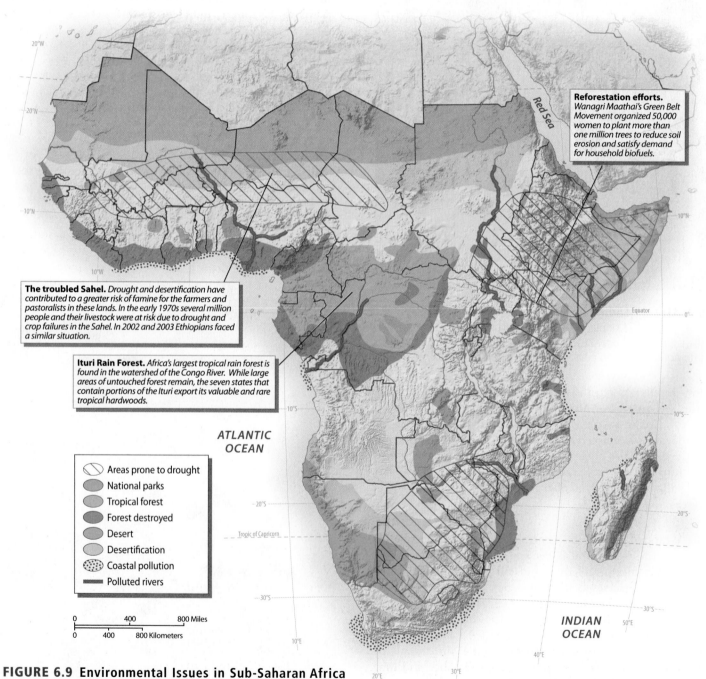

**Reforestation efforts.** *Wanagri Maathai's Green Belt Movement organized 50,000 women to plant more than one million trees to reduce soil erosion and satisfy demand for household biofuels.*

**The troubled Sahel.** *Drought and desertification have contributed to a greater risk of famine for the farmers and pastoralists in these lands. In the early 1970s several million people and their livestock were at risk due to drought and crop failures in the Sahel. In 2002 and 2003 Ethiopians faced a similar situation.*

**Ituri Rain Forest.** *Africa's largest tropical rain forest is found in the watershed of the Congo River. While large areas of untouched forest remain, the seven states that contain portions of the Ituri export its valuable and rare tropical hardwoods.*

Legend:
- Areas prone to drought
- National parks
- Tropical forest
- Forest destroyed
- Desert
- Desertification
- Coastal pollution
- Polluted rivers

0  400  800 Miles
0  400  800 Kilometers

**FIGURE 6.9 Environmental Issues in Sub-Saharan Africa**
Given the immense size of Sub-Saharan Africa, it is difficult to generalize about environmental problems. Dependence on trees for fuel places strain on forests and wooded savannas throughout the region. In semiarid regions, such as the Sahel, population pressures and land-use practices seem to have led to desertification. Yet Sub-Saharan Africa also supports the most impressive array of wildlife, especially large mammals, on Earth. (*Adapted from* DK World Atlas, *1997, p. 75, London: DK Publishing*)

**FIGURE 6.10 The Sahel in Bloom** A woman prepares millet grains grown near the city of Maradi, Niger. The soils of the Sahel are fertile and peasant farmers can produce a surplus with adequate rain. Yet in times of drought, crop failures can lead to famine in this region.

of Africa, parts of southern Africa, and the Sahel. Many scientists fear that drought will come more often and be prolonged under global climate change scenarios.

As Sub-Saharan cities grow in size and importance, urban environments increasingly face problems of air and water pollution as well as sewage and waste disposal. At the same time, wildlife tourism is an increasingly important source of income for many African states. Throughout the region, national parks have been created in an effort to strike a balance between humans' and animals' competing demands for land.

**The Sahel and Desertification** In the 1970s, the **Sahel** became a popular symbol for the dangers of unchecked population growth and human-induced environmental degradation when a relatively wet period came to an abrupt end, and six years of drought (1968–1974) ravaged the land. The Sahel is a zone of ecological transition between the Sahara to the north and wetter savannas and forest in the south (see Figure 6.9). Life in the Sahel depends on a delicate balance of limited rain, drought-resistant plants, and a pattern of animal **transhumance** (the movement of animals between wet-season and dry-season pasture). During the drought, rivers in the area diminished, and desertlike conditions began to move south. Unfortunately, tens of millions of people lived in this area, and farmers and pastoralists whose livelihoods had come to depend on the more abundant precipitation of the relatively wet period were temporarily forced out.

Considerable disagreement continues over the basic causes of desertification and drought in the Sahel. Was it a matter of too many humans degrading their environment, unsound settlement schemes encouraged by European colonizers, or a failure to understand global atmospheric cycles? Certainly, human activity in the region greatly increased in the mid-20th century, making the case for human-induced desertification more likely. But parts of the Sahel were important areas of settlement long before European colonization. What appears to be desert wasteland in April or May is transformed into productive fields of millet, sorghum, and peanuts after the drenching rains of June. Relatively free of the tropical diseases found in the wetter zones to the south, Sahelian soils are also quite fertile, which helps to explain why people continue to live there, despite the unreliable rainfall patterns of recent decades (Figure 6.10).

The main practices cited in the desertification of the Sahel are the expansion of agriculture and overgrazing, leading to the loss of natural vegetation and declines in soil fertility. Yet in the case of the Sahelian portion of Niger, local agronomists have recently documented an unanticipated increase in tree cover over the past 30 years (Figure 6.11). More interesting still, increases in tree cover have occurred in some of the most densely populated rural areas.

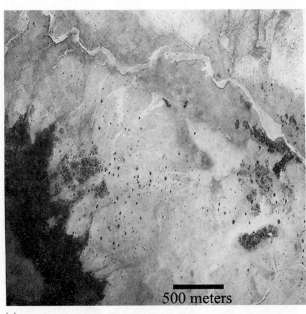
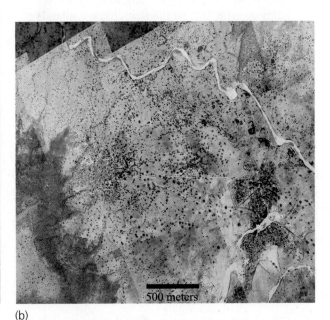

500 meters

500 meters

(a)                                                                                      (b)

**FIGURE 6.11 More Trees in Southern Niger** These two satellite images demonstrate an increase in tree cover in Niger. In 1975 (a), there were relatively few trees in this part of the Sahel. By 2003 (b), tree cover had increased substantially.

The relative greening of the Sahelian portion of Niger is due to simple actions taken by farmers, a change in government policy, and better rainfall. After a drought in 1984, farmers began to actively protect saplings instead of clearing them from their fields, including the nitrogen-fixing *goa* tree that had disappeared from many villages. During the rainy season, the goa tree loses its leaves, so it does not compete with crops for water or sun. The leaves themselves fertilize the soil. Sahelian farmers also use branches, pods, and leaves from trees for fuel and for animal fodder.

Until the 1990s, all trees were considered property of the state of Niger, thus giving farmers little incentive to protect them. Since then, the government has recognized the value of allowing individuals to own trees. Not only can farmers sell branches, pods, or fruit, they can also conserve them to ensure sustainable rural livelihoods. The villages that protect their trees are much greener than those that do not. In the village of Moussa Bara, where regeneration has been successful, not one child died of malnutrition in the famine of 2005. The Sahel is still poor and prone to drought, but, as the case of Niger shows, relatively simple conservation practices can have a positive impact.

**Deforestation** Although Sub-Saharan Africa still contains extensive forests, much of the region is either grasslands or agricultural lands that were once forest. Lush forests that existed in places such as highland Ethiopia were long ago reduced to a few remnant patches. Local populations have relied on such woodlands throughout history for their daily needs. Tropical savannas, which cover large portions of the region to the north and south of the tropical rain forest zone, are grassland areas with scattered wooded areas of trees and shrubs. The woodlands that dot this extensive African vegetation biome are few, and, unlike in tropical America, in Africa savanna deforestation is of greater local concern than the commercial logging of the rain forest. North of the equator, for example, only a few wooded areas remain in a landscape dominated by grasslands and cropland. Loss of woody vegetation has resulted in extensive hardship, especially for women and children who must spend many hours a day looking for wood (Figure 6.12). Deforestation, especially in the scattered woodlands of the savannas, aggravates problems of soil erosion and shortages of **biofuels** (wood and charcoal used for household energy needs, especially cooking).

In some countries, village women have organized into community-based nongovernmental organizations (NGOs) to plant trees and create greenbelts to meet ongoing fuel needs. One of the most successful efforts is in Kenya, under the leadership of Wangari Maathai. Maathai's Green Belt Movement has more than 50,000 members, mostly women, organized into 2,000 local community groups. Since the group's beginning in 1977, millions of trees have been successfully planted. In those areas, village women now spend less time collecting fuel, and local environments have improved. Kenya's success has drawn interest from other African countries, spurring a Pan-African Green Belt Movement largely organized through NGOs interested in biofuel generation, protection of the environment, and the empowerment of women. In 2004, Professor Maathai was awarded a Nobel Peace Prize for her contribution to sustainable development, democracy, and peace.

Destruction of tropical rain forests for logging is most evident in the fringes of Central Africa's Ituri (refer to Figure 6.9). Given the vastness of this forest and the relatively small number of people living there, however, it is less threatened than other forest areas. Two

smaller rain forests, one along the Atlantic coast from Sierra Leone to western Ghana and the other along the eastern coast of the island of Madagascar, have nearly disappeared. These rain forests have been severely degraded by commercial logging and agricultural clearing. Madagascar's eastern rain forests, as well as its western dry forests, have suffered serious degradation in the past three decades (Figure 6.13). Deforestation in Madagascar is especially worrisome because the island forms a unique environment, with a large number of endemic species (plants and animals native to a particular area).

**Wildlife Conservation** Sub-Saharan Africa is famous for its wildlife. In no other region of the world can one find such abundance and diversity of large mammals. The survival of wildlife here reflects, to some extent, the historically low human population density and the fact that sleeping sickness (transferred by the tsetse fly) and other diseases have kept people and their livestock out of many areas. In addition, many African peoples have developed various ways of successfully coexisting with wildlife.

But as is true elsewhere in the world, wildlife is quickly declining in much of Sub-Saharan Africa. The most noted wildlife reserves are in East Africa; in Kenya and Tanzania, these reserves are economically important major tourist attractions. Even there, however, population pressure, political instability, and poverty make the maintenance of large wildlife reserves difficult. Poaching is a major problem, particularly for rhinoceroses and elephants; the price of a single horn or tusk in distant markets represents several years' wages for most Africans. Ivory is sought in East Asia, especially in Japan. In China, powdered rhino horn is used as a traditional medicine, whereas in Yemen, rhino horn is prized for dagger handles. Wildlife reserves in southern Africa now seem to be the most secure. In fact, elephant populations are

**FIGURE 6.12 Wood Fuel in Ethiopia** A woman walks to Addis Ababa with a large bundle of firewood to sell in the city market. Collecting firewood is the work of women and children, both for household subsistence needs and for sale in local markets.

**FIGURE 6.13 Deforestation in Madagascar**
A man fills a bag with charcoal made from burning tropical forest. Rates of deforestation in Madagascar are high due to a reliance on slash-and-burn agriculture and biofuels. Loss of forest cover has led to serious problems with soil erosion and species extinction for this island nation.

because of the region's limited resources to both respond and adapt to environmental change. The areas most vulnerable are arid and semiarid regions such as the Sahel and the Horn of Africa, some grassland areas, and the coastal lowlands of West Africa and Angola.

Climate change models suggest that parts of highland East Africa and equatorial Central Africa may receive more rainfall in the future. Thus some lands that are currently marginal for farming might become more productive. These effects are likely to be offset, however, by the decline in agricultural productivity in the Sahelian belt as well as the grasslands of southern Africa, especially in Zambia and Zimbabwe. Drier grassland areas could deplete wildlife populations, which are a major factor behind the growing tourist economy. As in Latin America, higher temperatures in the tropics may result in the expansion of vector-borne diseases such as malaria and dengue fever into the highlands, where they have until now been relatively rare. Given the relatively high elevations in the region, the negative consequences of sea level rising would be mostly felt on the West African coast (Senegal, Gambia, Sierra Leone, Nigeria, Cameroon, and Gabon).

Even without the threat of climate change, famine stalks many areas of Africa. During 2002 and 2003, the rains did not fall in Ethiopia, and agronomists estimated that 15 million people would require food aid or face death by starvation. In this recent battle with famine, early warnings were taken seriously, and international food aid was aggressively sought and distributed. This was not the case in the early 1980s, though, when nearly 1 million Ethiopians perished, many of them needlessly because aid was not delivered to certain areas of the country due to ethnic fighting. In 2005, crops in Niger failed due to drought, and an estimated 2.5 million people were vulnerable to famine, but relief efforts resulted in relatively low mortality rates (Figure 6.14).

considered to be too large for the land to sustain in countries such as Zimbabwe. Some wildlife experts contend that herds could be reduced to prevent overgrazing and that the ivory and rhino horn should be legally sold in the international market in order to generate revenue for further conservation. Many environmentalists, not surprisingly, disagree strongly.

In 1989, a worldwide ban on the legal ivory trade was imposed as part of the Convention on International Trade in Endangered Species (CITES). While several African states, such as Kenya, lobbied hard for the ban, others, such as Zimbabwe, Namibia, and Botswana, complained that their herds were growing and the sale of ivory had helped to pay for conservation efforts. Conservationists feared that lifting the ban would bring on a new wave of poaching and illegal trade. In the late 1990s, the ban was lifted so some southern African states could sell down their inventories of ivory confiscated from poachers and continue with limited sales. The long-term effects of this policy change for elephant survival are not yet known, although political leaders from Kenya and India have voiced strong opposition to all ivory sales. The ivory controversy shows how differences in the region make it difficult to come up with a consistent conservation strategy.

## Climate Change and Vulnerability in Sub-Saharan Africa

Global climate change poses extreme risks for Sub-Saharan Africa due to its widespread poverty, recurrent droughts, and overdependence on rain-fed agriculture. Sub-Saharan Africa is the lowest emitter of greenhouse gases in the world, but it is likely to experience greater-than-average human vulnerability to global warming

**FIGURE 6.14 Food for Drought Victims** Food aid was delivered to drought victims in Central Niger in 2005. With adequate drought warming and international support, famine-related deaths can be avoided in the region.

## TABLE 6.1 POPULATION INDICATORS

| Country | Population (millions) 2009 | Population Density (per square kilometer) | Rate of Natural Increase (RNI) | Total Fertility Rate | Percent Urban | Percent <15 | Percent >65 | Net Migration (Rate per 1 000) 2005–10* |
|---|---|---|---|---|---|---|---|---|
| Angola | 17.1 | 14 | 2.7 | 6.6 | 57 | 46 | 2 | 0.9 |
| Benin | 8.9 | 79 | 3.2 | 5.7 | 41 | 44 | 3 | 1.2 |
| Botswana | 2.0 | 3 | 1.3 | 3.2 | 60 | 35 | 5 | 1.6 |
| Burkina Faso | 15.8 | 58 | 3.2 | 6.0 | 16 | 46 | 3 | −0.9 |
| Burundi | 8.3 | 298 | 2.1 | 5.4 | 10 | 41 | 3 | 8.1 |
| Cameroon | 18.9 | 40 | 2.3 | 4.7 | 57 | 42 | 4 | −0.2 |
| Cape Verde | 0.5 | 126 | 2.1 | 3.1 | 59 | 38 | 6 | −5.1 |
| Central African Republic | 4.5 | 7 | 1.9 | 5.0 | 38 | 41 | 4 | 0.2 |
| Chad | 10.3 | 8 | 2.6 | 6.3 | 27 | 46 | 3 | −1.4 |
| Comoros | 0.7 | 302 | 2.5 | 4.2 | 28 | 38 | 3 | −3.1 |
| Congo | 3.7 | 11 | 2.3 | 5.3 | 60 | 42 | 3 | −2.8 |
| Dem. Rep. of Congo | 68.7 | 29 | 3.1 | 6.5 | 33 | 47 | 3 | −0.3 |
| Djibouti | 0.9 | 37 | 1.9 | 4.2 | 87 | 37 | 3 | 0.0 |
| Equatorial Guinea | 0.7 | 24 | 2.4 | 5.4 | 39 | 41 | 3 | 3.1 |
| Eritrea | 5.1 | 43 | 2.9 | 5.3 | 21 | 42 | 2 | 2.3 |
| Ethiopia | 82.8 | 75 | 2.7 | 5.3 | 16 | 43 | 3 | −0.8 |
| Gabon | 1.5 | 6 | 1.8 | 3.6 | 84 | 37 | 4 | 0.7 |
| Gambia | 1.6 | 142 | 2.8 | 5.6 | 54 | 42 | 3 | 1.8 |
| Ghana | 23.8 | 100 | 2.1 | 4.0 | 48 | 40 | 4 | −0.4 |
| Guinea | 10.1 | 41 | 2.7 | 5.7 | 33 | 43 | 3 | −6.1 |
| Guinea-Bissau | 1.6 | 45 | 2.6 | 5.9 | 30 | 43 | 3 | −1.6 |
| Ivory Coast | 21.4 | 66 | 2.4 | 4.9 | 48 | 40 | 2 | −1.4 |
| Kenya | 39.1 | 67 | 2.7 | 4.9 | 19 | 42 | 2 | −1.0 |
| Lesotho | 2.1 | 70 | 0.2 | 3.4 | 24 | 39 | 5 | −3.5 |
| Liberia | 4.0 | 36 | 3.0 | 5.8 | 58 | 44 | 3 | 13.3 |
| Madagascar | 19.5 | 33 | 2.9 | 5.0 | 30 | 44 | 3 | −0.1 |
| Malawi | 14.2 | 120 | 3.1 | 6.3 | 17 | 46 | 3 | −0.3 |
| Mali | 13.0 | 10 | 2.8 | 6.0 | 31 | 45 | 2 | −3.2 |
| Mauritania | 3.3 | 3 | 2.5 | 5.1 | 40 | 40 | 4 | 0.6 |
| Mauritius | 1.3 | 625 | 0.7 | 1.7 | 42 | 23 | 7 | 0.0 |
| Mozambique | 22.0 | 27 | 2.4 | 5.4 | 29 | 43 | 3 | −0.2 |
| Namibia | 2.2 | 3 | 0.8 | 3.6 | 35 | 38 | 4 | −0.1 |
| Niger | 15.3 | 12 | 3.9 | 7.4 | 17 | 49 | 3 | −0.4 |
| Nigeria | 152.6 | 165 | 2.6 | 5.7 | 47 | 45 | 3 | −0.4 |
| Reunion | 0.8 | 324 | 1.3 | 2.5 | 92 | 27 | 7 | 0.0 |
| Rwanda | 9.9 | 375 | 2.5 | 5.5 | 18 | 44 | 3 | 0.3 |
| São Tomé and Principe | 0.2 | 169 | 2.6 | 4.1 | 58 | 41 | 4 | −8.8 |
| Senegal | 12.5 | 64 | 2.9 | 5.0 | 41 | 43 | 2 | −1.7 |
| Seychelles | 0.1 | 191 | 1.0 | 2.2 | 53 | 23 | 8 | |
| Sierra Leone | 5.7 | 79 | 2.0 | 5.2 | 37 | 42 | 4 | 2.2 |
| Somalia | 9.1 | 14 | 3.0 | 6.7 | 37 | 45 | 3 | −5.6 |
| South Africa | 50.7 | 42 | 0.8 | 2.7 | 59 | 32 | 5 | 2.8 |
| Sudan | 42.3 | 17 | 2.2 | 4.5 | 38 | 41 | 3 | 0.7 |
| Swaziland | 1.2 | 68 | 1.6 | 3.8 | 24 | 35 | 4 | −1.0 |
| Tanzania | 43.7 | 46 | 2.3 | 5.3 | 25 | 45 | 3 | |
| Togo | 6.6 | 117 | 2.7 | 5.1 | 40 | 41 | 3 | −0.2 |
| Uganda | 30.7 | 127 | 3.4 | 6.7 | 13 | 49 | 3 | −0.9 |
| Zambia | 12.6 | 17 | 2.9 | 6.2 | 37 | 46 | 3 | −1.4 |
| Zimbabwe | 12.5 | 32 | 1.4 | 3.8 | 37 | 40 | 4 | −11.1 |

Source: Population Reference Bureau, *World Population Data Sheet, 2009.*

*Net Migration Rate from the United Nations, Population Division, *World Population Prospects: The 2008 Revision Population Database.*

# Population and Settlement: Young and Restless

Sub-Saharan Africa's population is growing quickly. While the global population is projected to increase by nearly 40 percent by 2050, the projected population growth for Africa south of the Sahara is over 100 percent for the same time period. It is also a very young population, with 43 percent of the people younger than age 15, compared to just 17 percent for more-developed countries. Only 3 percent of the region's population is older than 65 years, whereas in Europe 16 percent is older than 65. Families tend to be large, with an average woman having five or six children (Table 6.1). Yet high child and maternal mortality rates also exist, reflecting disturbingly low access to basic health services. The most troubling indicator for the region is its low life expectancy, which dropped from 52 years in 1995 to 50 years in 2008, in part due to the AIDS epidemic. Most other developing regions have experienced gains in life expectancy. The growth of cities is also a major trend. In 1980, an estimated 23 percent of the population lived in cities; now the figure is approximately 35 percent.

Behind these demographic facts lie complex differences in settlement patterns, livelihoods, belief systems, and access to health care. Although the region is experiencing rapid population growth, Sub-Saharan Africa is not densely populated. The entire region holds some 800 million persons—roughly half the population that is crowded into the much smaller land area of South Asia. In fact, the overall population density of the region (83 people per square mile [32 people per square kilometer]) is equal to that of the United States. Just six states account for half of the region's population: Nigeria, Ethiopia, the Democratic Republic of the Congo, South Africa, Tanzania, and Sudan. Some states have very high population densities, while others are sparsely inhabited. Namibia has just 8 people per square mile (3 people per square kilometer). Many of the governments of the more densely settled territories, however, began to seriously encourage family planning policies in the 1980s.

## Population Trends and Demographic Debates

The combination of the population growth in Sub-Saharan Africa and the decline in some economic and social indicators makes demographers concerned about the region's overall well-being. The Sahel, for example, is not crowded by European or Asian standards, but it may already contain too many people for the land to support, given the unpredictability of its limited rainfall.

Some believe that Sub-Saharan Africa could support many more people than it presently does. Pessimists, however, argue that the region is a demographic time bomb and that unless fertility is quickly reduced, Sub-Saharan Africa will face massive famines in the near future. The majority of African states officially support lowering rates of natural increase and are slowly promoting modern contraception practices both to reduce family size and to protect people from sexually transmitted diseases.

**Family Size** A preference for large families is the basis for the region's demographic growth. In the 1960s, many areas in the developing world had total fertility rates (TFRs) of 6.0 or higher, but by the mid-1990s, only people in Sub-Saharan Africa and Southwest Asia continued to have such large families. For Southwest Asia, the dominance of Islam is used to explain high fertility rates. In Sub-Saharan Africa, a

**FIGURE 6.15 Large Families** This family in Senegal includes five children. Large families are still common in Sub-Saharan Africa. The average total fertility rate for the region is 5.5 children per woman.

combination of cultural practices, rural lifestyles, and economic realities encourage large families (Figure 6.15).

Throughout the region, large families guarantee a family's lineage and status. Even now, most women marry young, typically when they are teenagers, which increases their opportunity to have children. Demographers often point to the limited formal education available to women as another factor contributing to high fertility. Religious affiliation has little bearing on the region's fertility rates; Muslim, Christian, and animist communities all have similarly high birthrates.

The everyday realities of rural life make large families an asset. Children are an important source of labor; from tending crops and livestock to gathering fuelwood, they add more to the household economy than they take. Also, for the poorest places in the developing world, such as Sub-Saharan Africa, children are seen as social security: When parents' health falters, they expect their grown children to care for them.

By the 1980s, a shift in national policies had occurred. For the first time, government officials argued that smaller families and slower population growth were needed for social and economic development. Other factors are bringing down the growth rate. As African states slowly become more urban, there is a corresponding decline in family size—a pattern seen throughout the world. Tragically, declines in natural increase are also occurring as a result of AIDS.

## The Impact of AIDS on Africa

Now it its third decade, HIV/AIDS may become one of the deadliest epidemics in modern human history. As of 2007, two-thirds of the HIV/AIDS cases in the world were found in Sub-Saharan Africa, and it was estimated that in 2007 alone, there were 1.7 million AIDS-related deaths in the region. In South Africa, the most populous state in southern Africa, more than 5.5 million people (almost one in five people aged 15–49) are infected with HIV/AIDS. The rate of infection in neighboring Botswana is 24 percent (down from a staggering 36 percent in 2001) for the same age group. Although the rate of infection is highest in southern Africa, East Africa and West Africa are seeing increases. In Nigeria, it is estimated that 4 percent of the 15–49 age group has HIV or

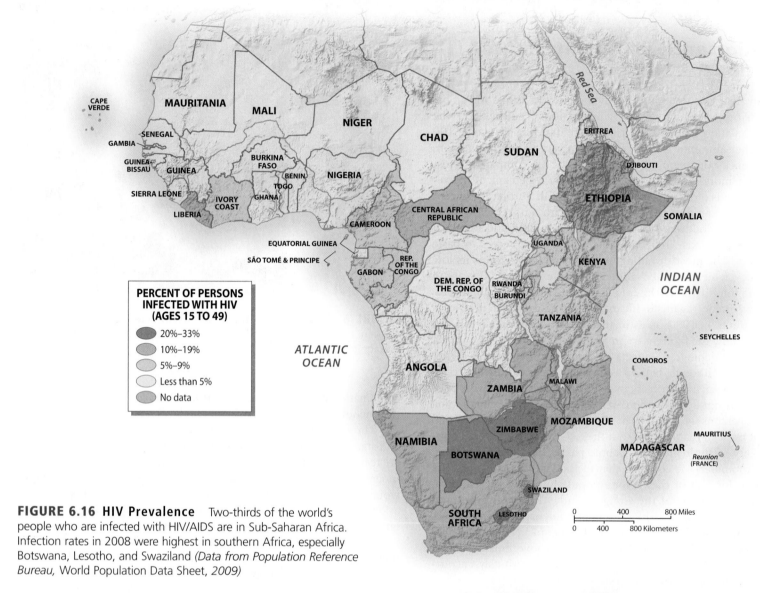

**FIGURE 6.16 HIV Prevalence** Two-thirds of the world's people who are infected with HIV/AIDS are in Sub-Saharan Africa. Infection rates in 2008 were highest in southern Africa, especially Botswana, Lesotho, and Swaziland *(Data from Population Reference Bureau, World Population Data Sheet, 2009)*

Map legend:

**PERCENT OF PERSONS INFECTED WITH HIV (AGES 15 TO 49)**
- 20%–33%
- 10%–19%
- 5%–9%
- Less than 5%
- No data

AIDS. For Kenya, the figure is 6 percent (Figure 6.16). By comparison, only 0.6 percent of the same age group in North America is infected. As of 2007, it was estimated that 22.5 million people in Sub-Saharan Africa were infected with HIV/AIDS, and it was estimated that the epidemic had already claimed 19 million lives in the region and left 12 million orphans. The virus is thought to have originated in the forests of the Congo, possibly crossing over from chimpanzees to humans sometime in the 1950s. Yet it was not until the late 1980s that the impact of the disease was widely felt in some of the most populous parts of the region.

Sadly, the social and economic implications of this epidemic are profound. Life expectancy rates have tumbled in the past decades, in a few places dropping to below 40 years. AIDS typically hits the portion of the population that is most active economically. Time lost to care for sick family members and the outlay of workers' compensation benefits has reduced economic productivity and overwhelmed public services in hard-hit areas. The disease makes no class distinctions: Countries are losing both peasant farmers and educated professionals (doctors, engineers, and teachers). Infection rates among newborns are still high, although several countries have been successful at using antiviral medicines to block transmission of HIV from mothers to children at birth. Many countries struggle to care for millions of children orphaned by AIDS.

**FIGURE 6.17 HIV/AIDS Clinic in South Africa** HIV-positive patients wait to receive doses of antiretroviral medication at a rural clinic in the Lusikisiki district, Eastern Cape. Such medications made available by the South African government can prolong the lives of HIV/AIDS patients. Programs such as these offer some hope from the ravages of this disease.

Antiretroviral drug therapies have prolonged the lives of people with HIV/AIDS for two decades in the developed world, yet governments in Sub-Saharan Africa are only beginning to disseminate these treatments. The government of South Africa insisted on its right to use cheap generic drugs to prolong the lives of those with HIV/AIDS, even though this violated patent laws. The world's major drug companies promptly sued South Africa but backed down in 2001, in the face of growing international pressure (and the realization that pharmaceutical companies in developing countries such as India would happily provide generic alternatives). Through an UN-led agreement, generic drugs are now available, and the cost of a year's supply is just $300 per person, although even this is beyond the financial reach of most Africans. In 2003, the South African government announced that it would supply antiretroviral drugs to every South African who needs them; the goal was to treat 3 million people by 2005 (Figure 6.17).

For now, the surest way to stem the epidemic is through prevention, mostly through educating people about how the virus is spread and convincing them to change their sexual behavior. In Uganda, state agencies, along with NGOs, began a national no-nonsense campaign for AIDS awareness in the 1980s that focused on the schools. President Museveni of Uganda pioneered the catchy phrase "Abstain—Be faithful—use a Condom" (ABC). As a result of explicit materials, role-playing games, and frank discussion, the prevalence of HIV among women in prenatal clinics had declined by the late 1990s.

## Patterns of Settlement and Land Use

Because of the dominance of rural settlements in Sub-Saharan Africa, people are widely scattered throughout the region (Figure 6.18). Population concentrations are the highest in West Africa, highland

**FIGURE 6.18 Population Distribution**   The majority of people in Sub-Saharan Africa live in rural areas. Some of these rural zones, however, are densely settled, such as West Africa and the East African highlands. Major urban centers, especially in South Africa and Nigeria, support millions. There is only one megacity in the region with more than 10 million residents (Lagos), but more than two dozen cities have more than 1 million residents.

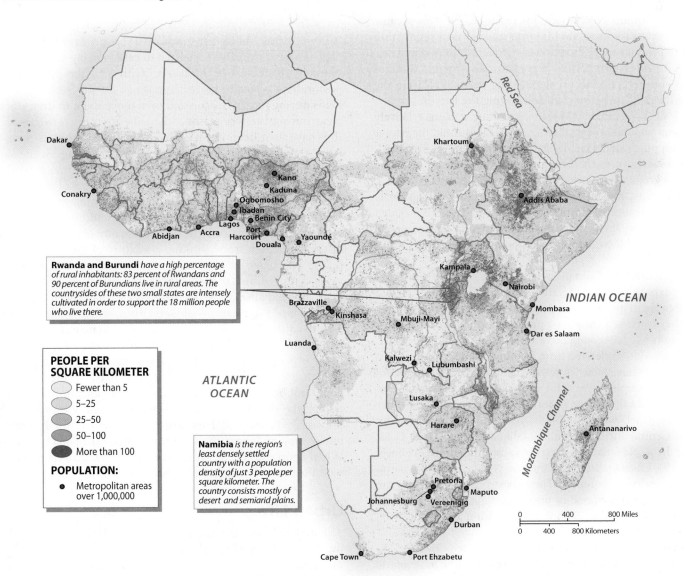

**Rwanda and Burundi** have a high percentage of rural inhabitants: 83 percent of Rwandans and 90 percent of Burundians live in rural areas. The countrysides of these two small states are intensely cultivated in order to support the 18 million people who live there.

**Namibia** is the region's least densely settled country with a population density of just 3 people per square kilometer. The country consists mostly of desert and semiarid plains.

**PEOPLE PER SQUARE KILOMETER**
- Fewer than 5
- 5–25
- 25–50
- 50–100
- More than 100

**POPULATION:**
- Metropolitan areas over 1,000,000

East Africa, and the eastern half of South Africa. The first two areas have some of the region's best soils, and native systems of permanent agriculture developed there. In South Africa, the more densely settled east is a result of an urbanized economy based on mining, as well as the forced concentration of black South Africans into eastern homelands.

As more Africans move to cities, patterns of settlement are becoming more concentrated. Towns that were once small administrative centers for colonial elites grew into major cities. The region even has its own megacity, Lagos, which recently topped 10 million residents. Throughout the continent, African cities are growing faster than rural areas. Before we examine the Sub-Saharan urban scene, a more detailed discussion of rural areas is needed.

### Agricultural Subsistence

The staple crops over most of Sub-Saharan Africa are millet, sorghum, and corn (maize), as well as a variety of tubers and root crops such as yams. Irrigated rice is widely grown in West Africa and Madagascar. Wheat and barley are grown in parts of South Africa and Ethiopia. Intermixed with subsistence foods are a variety of export crops—coffee, tea, rubber, bananas, cacao, cotton, and peanuts—that are grown in distinct ecological zones and often in some of the best soils.

In areas that support annual cropping, population densities are greater. In parts of humid West Africa, for example, the yam became the dominant subsistence crop; the Ibos' mastery of yam production allowed them to produce more food and live in denser permanent settlements in the southeastern corner of present-day Nigeria. Much of traditional Ibo culture is tied to the demanding tasks of clearing the fields, tending the delicate plants, and celebrating the harvest.

Over much of the continent, African agriculture remains relatively unproductive, and population densities tend to be low. Amid the poorer tropical soils, cropping usually entails shifting cultivation (or **swidden**). This process involves burning the natural vegetation to release fertilizing ash and planting crops such as maize, beans, sweet potatoes, banana, papaya, manioc, yams, melon, and squash. Each plot is temporarily abandoned once its source of nutrients has been exhausted. Swidden cultivation is often a very finely tuned adaptation to local environmental conditions, but it is unable to support high population densities.

### Plantation Agriculture

Plantation agriculture, designed to produce crops for export, is critical to the economies of many states. If African countries are to import the modern goods and energy resources they require, they must sell their own products on the world market. Because the region has few competitive industries, the bulk of its exports are primary products derived from farming, mining, and forestry.

A number of African countries rely heavily on one or two export crops. Coffee, for example, is vital for Ethiopia, Kenya, Rwanda, Burundi, and Tanzania. Peanuts have historically been the primary source of income in the Sahel, while cotton is tremendously important for Sudan and the Central African Republic. Ghana and the Ivory Coast have long been the world's main suppliers of cacao (the source of chocolate) (Figure 6.19); Liberia produces plantation rubber; and many farmers in Nigeria specialize in palm oil. The export of such products can bring good money when commodity prices are high, but when prices collapse, as they periodically do, economic devastation may follow.

### Herding and Livestock

Animal husbandry (the care of livestock) is extremely important in Sub-Saharan Africa, particularly in semiarid zones. Camels and goats are the principal animals in the Sahel and the

**FIGURE 6.19 Plantation Agriculture**    A worker harvests cacao on a plantation in the Ivory Coast. This area of West Africa leads the world in cacao production.

Horn of Africa, but farther south, cattle are primary. Many African peoples have traditionally specialized in cattle raising and are often tied in mutually beneficial relationships with neighboring farmers. Such **pastoralists** typically graze their stock on the stubble of harvested fields during the dry season and then move them to drier uncultivated areas during the wet season, when the pastures turn green. Farmers thus have their fields fertilized by the manure of the pastoralists' stock, while the pastoralists find good dry-season grazing. At the same time, the nomads can trade their animal products for grain and other goods of the sedentary world. Several pastoral peoples of East Africa, however, are noted for their extreme reliance on cattle and general (but never complete) independence from agriculture. The Masai of the Tanzanian–Kenyan borderlands traditionally obtain a large percentage of their nutrition from drinking a mixture of milk and blood (Figure 6.20). The blood is obtained periodically from the animal's jugular veins, a procedure that evidently causes little harm to the cattle.

Large expanses of Sub-Saharan Africa have been off-limits to cattle because of infestations of **tsetse flies**, which spread sleeping sickness to cattle, humans, and some wildlife. Where wild animals, which harbor the disease but are immune to it, were present in large numbers, especially in environments containing brush or woodland (which are necessary for the survival of tsetse flies), cattle simply could not be raised. At present, tsetse fly eradication programs are reducing the threat, and cattle raising is spreading into areas that were previously forbidden. This process is beneficial for African peoples, but it may endanger the continued survival of many wild animals. When people and their stock move into new areas in large numbers, wildlife almost inevitably declines.

## Urban Life

Although Sub-Saharan Africa is considered the least urbanized region in the developing world, with one-third of the population living in cities, most Sub-Saharan cities are growing at twice the national growth rates. If present trends continue, half of the region's population may well be living in cities by 2025. One of the consequences of this surge in city living is urban sprawl. Rural-to-urban migration, industrialization, and refugee flows are forcing the cities of the region to absorb more people and use more resources. As in Latin America, the tendency is toward urban primacy, the condition in which one major city is dominant and at least three times larger than the next largest city. Luanda, the capital of Angola, was built for half a million people but now has at least 4 million residents (many of them displaced during the country's long years of war). Most of the city's inhabitants live in the growing slums that surround the colonial core (Figure 6.21). In such rapidly growing places, city officials struggle to build enough roads and provide electricity, water, trash collection, and employment for all these people.

European colonialism greatly influenced urban form and development in the region. Although a very small percentage of the population lived in cities, Africans did have an urban tradition prior to the colonial era. Ancient cities, such as Axum in Ethiopia, thrived 2,000 years ago. Similarly, in the Sahel, major trans-Saharan trade centers, such as Timbuktu (Tombouctou) and Gao, have existed for more than a millennium. In East Africa, an urban trading culture emerged that was rooted in Islam and the Swahili language. West Africa, however, had the most developed precolonial urban network, reflecting both native and Islamic traditions. It also supports some of the region's largest cities today.

**West African Urban Traditions**   The West African coastline is dotted with cities, from Dakar, Senegal, in the far north to Lagos, Nigeria, in the east. Half of Nigerians live in cities, and in 2010, the country had seven metropolitan areas with populations of more than 1 million. Historically, the Yoruba cities in southwestern Nigeria have

**FIGURE 6.20 Masai Pastoralists**   A Masai man holds a cow steady while a woman collects blood being drained from the animal's neck. Pastoral groups such as the Masai live in the drier areas of Sub-Saharan Africa. The Masai live in Kenya and Tanzania. Other pastoral groups are found in the Sahel and the Horn of Africa.

**FIGURE 6.21 Urban Slums of Luanda, Angola**   The Roche Santiero market on the outskirts of Luanda serves the city's growing urban poor. Like many other Sub-Saharan African cities, Luanda is growing much faster than rural areas of the country.

**FIGURE 6.22 Downtown Lagos, Nigeria**
The central business district of Lagos is the country's banking and commercial center. The city streets teem with buses, cars, collective taxis, vendors, and thousands of pedestrian. A classic primate city, Lagos is well over 10 million residents. Infrastructure, however, has not kept up with the city's rapid growth.

been the best documented. Developed in the 12th century, cities such as Ibadan were walled and gated, with a palace encircled by large rectangular courtyards at the city center. An important center of trade for an extensive surrounding area, Ibadan was also a religious and political center. Lagos was also a Yoruba settlement. Founded on a coastal island on the Bight of Benin, most of the modern city has spread onto the nearby mainland. Its coastal setting and natural harbor made this relatively small native city attractive to colonial powers. When the British took control in the mid-19th century, the city's size and importance grew.

Whereas in 1960, Lagos was a city of 1 million, today, it has well over 10 million residents, making it the largest city in the region. Unable to keep up with the surge or rural migrants, Lagos's streets are clogged with traffic; for those living on the city's periphery, three-hour commutes one-way are common. For many, the informal sector (unregulated services and trade) provides employment. Crime is another major problem. The chances of being attacked on the streets or robbed in one's home are quite high, even though most windows are barred and houses are fenced. To deal with this problem, Lagos has the highest density of police in Nigeria, but the weak social bonds between urban migrants, widespread poverty, and the gap between rich and poor seem to encourage lawlessness (Figure 6.22).

**Urban Industrial South Africa**    The major cities of southern Africa, unlike those of West Africa, are colonial in origin. Most of these cities, such as Lusaka, Zambia, and Harare, Zimbabwe, grew as administrative or mining centers. The nation of South Africa is one of the most urbanized states in the entire region, and it is certainly the most industrialized. The foundations of South Africa's urban economy rest largely on its incredibly rich mineral resources (diamonds, gold, chromium, platinum, tin, uranium, coal, iron ore, and manganese). Eight of its metropolitan areas have more

than 1 million people; the largest of them are Johannesburg, Durban, and Cape Town.

The form of South African cities continues to be imprinted by the legacy of **apartheid** (an official policy of racial segregation that shaped social relations in South Africa for nearly 50 years). Even though apartheid was abolished in 1994, it is still evident in the landscape. Under apartheid rules, cities such as Cape Town were divided into residential areas according to racial categories: white, **coloured** (a South African term describing people of mixed African and European ancestry), Indian (South Asian), and African (black) (Figure 6.23). Whites

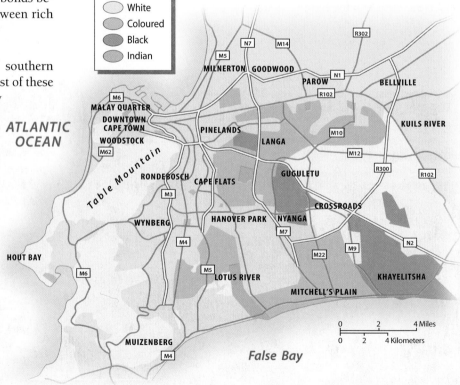

**FIGURE 6.23 Racial Segregation in Cape Town**    Under apartheid, most land in Cape Town was designated for white use. Coloureds, blacks, and Indians (South Asians) were crammed into far smaller areas on the less desirable flat and sandy soils east of downtown. *(From Christopher, 1994, The Atlas of Apartheid, p. 156, London: Routledge)*

**FIGURE 6.24 Elite Cape Town Landscape** Europeans settled in this prize farmland to the east and south of Table Mountain. Today, this area is home to South Africa's best vineyards.

occupied the largest and most desirable portions of the city, especially the scenic areas below Table Mountain (Figure 6.24). Blacks were crowded into the least desired areas, forming squatter settlements called *townships* in places such as Gugulethu and Khayelitsha. Today blacks, coloureds, and Indians are legally allowed to live anywhere they want. Yet the economic differences between racial groups as well as long-standing prejudices make residential integration difficult. South African cities such as Cape Town and Johannesburg are also home to thousands of immigrants from other African states who have settled there in the past decades in search of economic opportunities not available in their countries of origin.

## Cultural Coherence and Diversity: Unity through Adversity

No world region is culturally homogeneous, but most have been partially unified in the past by widespread systems of belief and communication. The lack of traditional cultural and political coherence across Sub-Saharan Africa is not surprising if one considers the region's huge size. Sub-Saharan Africa is more than four times larger than Europe or South Asia. Had foreign imperialism not impinged on the region, it is quite possible that West Africa and southern Africa would have developed into their own distinct world regions.

An African identity south of the Sahara was created through a common history of slavery and colonialism as well as struggles for independence and development. More telling, the people of the region often define themselves as African, especially to the outside world. That Sub-Saharan Africa is poor, no one will argue. And yet the cultural expressions of its people—its music, dance, and art—are joyous. Africans share a resilience and optimism that visitors to the region often comment on. The cultural diversity of the region is obvious, yet there is unity among the people, drawn from surviving many adversities.

## Language Patterns

In most Sub-Saharan countries, as in other former colonies, multiple languages are used that reflect tribal, colonial, and national affiliations. Indigenous languages, many from the Bantu subfamily, are often localized to relatively small rural areas. More widely spoken African trade languages, such as Swahili or Hausa, serve as a lingua franca over broader areas. Overlaying native languages are Indo-European (French and English) and Afro-Asiatic (Arabic and Somali) ones. Figure 6.25 illustrates the complex pattern of language families and major languages found in Africa today. Contrast the larger map with the inset that shows current "official" languages. A comparison of the two shows that most African countries are multilingual, which can be a source of tension within states. In Nigeria, for example, the official language is English, yet there are millions of Hausa, Yoruba, Igbo (or Ibo), Ful (or Fulani), and Efik speakers, as well as speakers of dozens of other languages.

**African Language Groups** Three of the six language groups mapped in Figure 6.25 are unique to the region (Niger-Congo, Nilo-Saharan, and Khoisan), while the other three (Afro-Asiatic, Austronesian, and Indo-European) are more closely associated with other parts of the world. Afro-Asiatic languages, especially Arabic, dominate North Africa and are understood in Islamic areas of Sub-Saharan Africa as well. Amharic in Ethiopia and Somali in Somalia are also Afro-Asiatic languages. The Malayo-Polynesian language family is limited to the island of Madagascar, which many believe was first settled by seafarers from Indonesia some 1,500 years ago. Indo-European languages, especially French, English, Portuguese, and Afrikaans, are a legacy of colonialism and are widely used today.

The Niger-Congo language group is by far the most important one in the region. This linguistic group, which originated in West Africa, includes Mandingo, Yoruba, Ful (or Fulani), and Igbo (or Ibo), among others. Around 3,000 years ago, a people of the Niger–Congo language group began to expand out of western Africa into the equatorial zone. This group, called the Bantu, began one of the most far-ranging migrations in human history, which introduced agriculture into large areas of central and southern Africa. Most individual Sub-Saharan languages are limited to relatively small areas and are significant only at the local level. One language in the Bantu subfamily, Swahili, eventually became the most widely spoken Sub-Saharan language. Swahili originated as a trade language on the East African coast, where a number of merchant colonies from Arabia were established around 1100 CE. A hybrid society grew up in a narrow coastal band of modern Kenya and Tanzania, one speaking a language of Bantu structure enriched with many Arabic words. While Swahili became the primary language only in the narrow coastal belt, it spread far into the interior as the language of trade. After independence was achieved, both Kenya and Tanzania adopted Swahili as an official language, along with English. Swahili, with some 70 million speakers, is a lingua franca for East Africa. It has generated a fairly extensive literature and is often studied in other regions of the world.

**Language and Identity** Ethnic identity as well as linguistic affiliation have historically been highly unstable over much of Sub-Saharan Africa. The tendency was for new language groups to form when people threatened by war fled to less-settled areas, where they often mixed with peoples from other places. In such situations, new languages arise quickly, and divisions between groups are blurred. Nevertheless, distinct

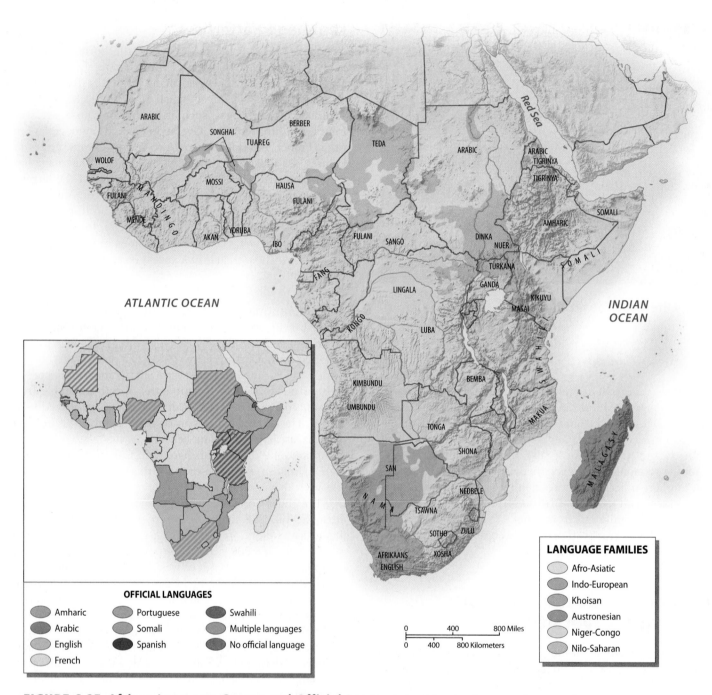

**FIGURE 6.25 African Language Groups and Official Languages** Mapping language is a complex task for Sub-Saharan Africa. There are languages with millions of speakers, such as Swahili, and there are languages spoken by a few hundred people living in isolated areas. Six language families are represented in the region. Among these families are scores of individual languages (see the labels on the map). Because most modern states have many native languages, the colonial language often became the "official" language. English and French are the most common official languages in the region (see inset).

**tribes** formed that consisted of a group of families or clans with a common kinship, language, and definable territory. European colonial administrators were eager to establish a fixed social order to better control native peoples. In this process, a flawed cultural map of Sub-Saharan Africa evolved. Some tribes were artificially divided, meaningless names were applied, and territorial boundaries were often misinterpreted.

Social boundaries between different ethnic and linguistic groups have become more stable in recent years, and a number of individual

languages have become particularly important for communication on a national scale. Wolof in Senegal; Mandingo in Mali; Mossi in Burkina Faso, Yoruba, Hausa, and Igbo in Nigeria; Kikuyu in Central Kenya; and Zulu, Xhosa, and Sotho in South Africa are all nationally significant languages spoken by millions. None, however, has the status of being the official language of any country. With the end of apartheid in South Africa, the country officially recognized 11 languages, although English is still the lingua franca of business and government. Indeed, a single language has a clear majority status in only a handful of countries. The more linguistically homogeneous states include Somalia (where virtually everyone speak Somali) and the very small states of Rwanda, Burundi, Swaziland, and Lesotho.

**European Languages**  In the colonial period, European countries used their own languages for administrative purposes in their African empires. Education in the colonial period also stressed literacy in the language of the imperial power. In the postindependence period, most Sub-Saharan African countries have continued to use the languages of their former colonizers for government and higher education. Few of these new states had a clear majority language that they could employ, and picking any minority tongue would have met with opposition from other peoples. The one exception is Ethiopia, which maintained its independence during the colonial era. The official language is Amharic, although other indigenous languages are also spoken.

Two vast blocks of European language exist in Africa today: Francophone Africa, including the former colonies of France and Belgium, where French serves as the main language of administration; and Anglophone Africa, where the use of English prevails (see the inset to Figure 6.25). Early Dutch settlement in South Africa resulted in the use of Afrikaans (a Dutch-based language) by several million South Africans. In Mauritania, Eritrea, and Sudan, Arabic serves as a main language.

## Religion

Native African religions are generally classified as animist. This is a somewhat misleading catchall term used to classify all local faiths that do not fit into one of the handful of "world religions." Most animist religions are centered on the worship of nature and ancestral spirits, but the internal diversity within the animist tradition is great. Classifying a religion as animist says more about what it is not than about what it actually is.

Both Christianity and Islam entered Sub-Saharan Africa early in their histories, but they advanced slowly for many centuries. Since the beginning of the 20th century, both religions have spread rapidly—more rapidly, in fact, than in any other part of the world. But tens of millions of Africans still follow animist beliefs, and many others combine animist practices and ideas with their observances of Christianity and Islam.

**The Introduction and Spread of Christianity**  Christianity came first to northeast Africa. Kingdoms in both Ethiopia and central Sudan were converted by 300 CE—the earliest conversions outside the Roman Empire. The peoples of northern and central Ethiopia adopted the Coptic form of Christianity and have thus historically looked to Egypt's Christian minority for their religious leadership (Figure 6.26). At present, roughly half of the population of both Ethiopia and Eritrea is Coptic Christian; most of the rest is Muslim, but there are still some animist communities, especially in Ethiopia's western lowlands.

European settlers and missionaries introduced Christianity to other parts of Sub-Saharan Africa beginning in the 1600s. The Dutch, who began to colonize South Africa at that time, brought their Calvinist Protestant faith. Later European immigrants to South Africa brought Anglicanism and other Protestant beliefs, as well as Catholicism. Most black South Africans eventually converted to one or another form of Christianity as well. In fact, churches in South Africa were instrumental in the long fight against white racial supremacy. Religious leaders such as Bishop Desmond Tutu were outspoken critics of the injustices of apartheid and worked to bring down the system.

**FIGURE 6.26 Eritrean Christians at Prayer**  Coptic Christians gather in front of St. Mary's Church in Asmara, Eritrea, on Good Friday, a holy day for Christians. Half of the populations in Eritrea and Ethiopia belong to the Coptic Church, which has ties to the Christian minority in Egypt.

Elsewhere in Africa, Christianity came with European missionaries, most of whom arrived after the mid-1800s. As was true in the rest of the world, missionaries had little success where Islam had preceded them, but they eventually made numerous conversions in animist areas. As a general rule, Protestant Christianity prevails in areas of former British colonization, while Catholicism is more important in former French, Belgian, and Portuguese territories. In the postcolonial era, African Christianity has spread out, at times taking on a life of its own, independent from foreign missionary efforts. Still active in the region are various Pentecostal, Evangelical, and Mormon missionary groups, mostly from the United States. It is difficult to map the distribution of Christianity in Africa, however, because it has spread irregularly across the non-Islamic portion of the region.

**The Introduction and Spread of Islam**  Islam began to advance into Sub-Saharan Africa 1,000 years ago (Figure 6.27). Berber traders

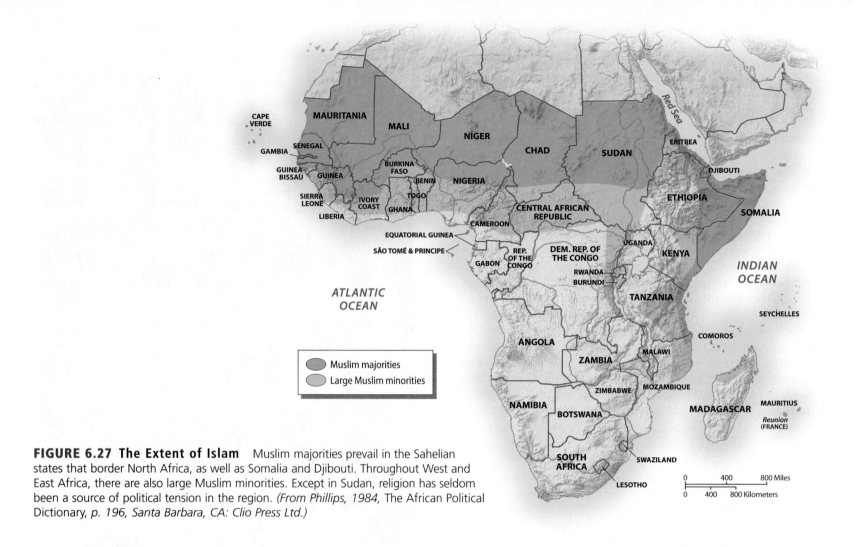

**FIGURE 6.27 The Extent of Islam** Muslim majorities prevail in the Sahelian states that border North Africa, as well as Somalia and Djibouti. Throughout West and East Africa, there are also large Muslim minorities. Except in Sudan, religion has seldom been a source of political tension in the region. *(From Phillips, 1984,* The African Political Dictionary, *p. 196, Santa Barbara, CA: Clio Press Ltd.)*

from North Africa and the Sahara introduced the religion to the Sahel, and by 1050 the Kingdom of Tokolor in modern Senegal emerged as the first Sub-Saharan Muslim state. Somewhat later, the ruling class of the powerful Mande-speaking trading empires of Ghana and Mali converted as well. In the 14th century, the emperor of Mali astounded the Muslim world when he and his royal court made the pilgrimage to Mecca, bringing with them so much gold that they set off a brief period of high inflation throughout Southwest Asia.

Mande-speaking traders, whose networks spanned the area from the Sahel to the Gulf of Guinea, gradually introduced the religion to other areas of West Africa. There, however, many peoples remained committed to animism, and Islam made slow and unsteady progress. Today, orthodox Islam prevails through most of the Sahel. Farther south, Muslims are mixed with Christians and animists, but their numbers continue to grow, and their practices tend to be orthodox as well (Figure 6.28).

**Interaction between Religious Traditions** The southward spread of Islam from the Sahel, coupled with the northward spread of Christianity from the port cities, has generated a complex religious frontier across much of West Africa. In Nigeria, the Hausa are firmly Muslim, while the southeastern Igbo are largely Christian. The Yoruba of the southwest are divided between Christian and Muslim. In the more remote parts of Nigeria, animist traditions remain strong. But despite this religious diversity, religious conflict in Nigeria has been relatively rare until recently. In 2000, seven northern Nigerian states imposed Muslim sharia laws, which has triggered intermittent violence ever

since, especially in the northern city of Kaduna. Still, most of West Africa's regional conflicts continue to be framed more along ethnic terms than religious ones.

Religious conflict historically has been far more acute in northeastern Africa, where Muslims and Christians have struggled against each other for centuries. Sudan is currently the scene of an intense conflict that is both religious and ethnic in origin. Islam was introduced to Sudan in the 1300s by an invasion of Arabic-speaking pastoralists who destroyed the indigenous Coptic Christian kingdoms of the area. Within a few hundred years, central and northern Sudan had become completely Islamic. The southern equatorial province of Sudan, however, where tropical diseases and extensive wetlands prevented Arab advances, remained animist or converted to Christianity under British colonial rule. In the 1970s, the Arabic-speaking Muslims of the north and central Sudan began to build an Islamic state. Experiencing both religious discrimination and economic exploitation, the Christian and animist peoples of the south launched a massive rebellion. Fighting since the 1980s has been intense, with the government generally controlling the main towns and roads and the rebels maintaining power in the countryside. A peace was brokered in 2003, and as part of the peace agreement southern Sudan was promised an opportunity to vote on secession from the north in 2011 (although skeptics are uncertain about whether this will actually happen). Just as peace was finally brokered in southern Sudan, a brutal ethnic conflict broke out in western Sudan in the province of Darfur, which will be discussed in detail later in the chapter.

**FIGURE 6.28 West African Muslims** Villagers leave the Larabanga mosque after Friday prayers. The Muslim village of Larabanga in Northern Ghana is home to one of the oldest mosques in West Africa and was recently declared a U.N. World Heritage Site.

forms of religious expression are also emerging. With such a diversity of faiths, it is fortunate that religion is not typically the cause of overt conflict.

**Globalization and African Culture** The slave trade that linked Africa to the Americas and Europe set in process patterns of cultural diffusion that transferred African peoples and cultures across the Atlantic. Tragically, slavery damaged the demographic and political strength of African societies, especially in West Africa, from where the most slaves were taken. An estimated 12 million Africans were shipped to the Americas as slaves from the 1500s until 1870 (Figure 6.29). Slavery impacted the entire region, sending Africans not just to the Americas, but to Europe, North Africa, and Southwest Asia. The vast majority, however, worked on plantations in the Americas.

With the exception of Sudan, a pattern of peaceful coexistence among different faiths is a distinguishing feature of Sub-Saharan Africa that deserves further study. Sub-Saharan Africa is a land of religious vitality. Both Christianity and Islam are spreading rapidly, but animism continues to hold widespread appeal. Many new and syncretic (blended)

Out of this tragic displacement of people came a blending of African cultures with Amerindian and European ones. African rhythms are at the core of various musical styles, from rumba to jazz, the blues, and rock and roll. Brazil, the largest country in Latin America, is claimed to be the

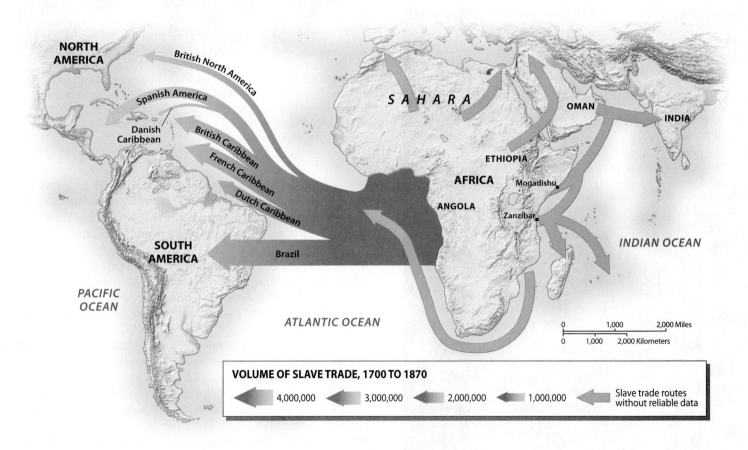

**VOLUME OF SLAVE TRADE, 1700 TO 1870**

| | | | | |
|---|---|---|---|---|
| 4,000,000 | 3,000,000 | 2,000,000 | 1,000,000 | Slave trade routes without reliable data |

**FIGURE 6.29 African Slave Trade** The slave trade had a devastating impact on Sub-Saharan societies. From ship logs, it is estimated that 12 million Africans were shipped to the Americas to work as slaves on sugar, cotton, and rice plantations; the majority went to Brazil and the Caribbean. Yet other slave routes existed, although the data on them are less reliable. Africans from south of the Sahara were used as slaves in North Africa. Others were traded across the Indian Ocean into Southwest Asia and South Asia.

second largest "African state" (after Nigeria) because of its huge Afro-Brazilian population. Thus the forced migration of Africans as slaves had a huge cultural influence on many areas of the world.

So, too, have contemporary movements of Africans influenced the cultures of many world regions. Perhaps one of the most celebrated persons of African ancestry today is President Barack Obama, the son of a Kenyan man and a woman from Kansas. Obama's heritage and upbringing embody the forces of globalization. In Kenya, he is hailed as part of the modern African diaspora—young professionals (and their offspring) who leave the continent and make their mark somewhere else. As an indicator of our global age, Kenyan President Mwai Kibaki declared a Kenyan national holiday after Obama won the U.S. election in November 2008.

Popular culture in Africa, like everywhere else in the world, is a dynamic mixture of global and local influences. Kwaito, the latest musical craze in South Africa, sounds a lot like rap from the United States. A closer listen, however, reveals an incorporation of local rhythms, lyrics in Zulu and Xhosa, and themes about life in post-apartheid townships. Lagos, Nigeria, referred to as Nollywood, has

**FIGURE 6.30 Festival in the Desert**  Every January, thousands of Touareg nomads, Malian musicians, and Western tourists gather in the oasis town of Essakane, Mali, to attend the Festival in the Desert. In this image, a Western tourist sits among the Touareg at the festival.

**FIGURE 6.31 Ethiopian Distance Runners**  Haile Gebrselassie (left) races with two other Ethiopian runners, Kenenisa Bekele (right) and Sileshi Sihine, in a 10,000-meter race in France. The sport of distance running is passionately followed in Ethiopia, and champions such as Gebrselassie are national heroes.

emerged as Africa's film capital. More than 500 films a year are made there, most shot in a few days and with budgets of $10,000. They can be seen all over English-speaking Africa, addressing such themes as religion, ethnicity, political corruption, and witchcraft.

**Music in West Africa**  Nigeria is the musical center of West Africa, with a well-developed and cosmopolitan recording industry. Modern Nigerian styles such as juju, highlife, and Afro-beat are influenced by jazz, rock, reggae, and gospel, and they are driven by an easily recognized African sound.

Further up the Niger River lies the country of Mali. Bamako, the capital, is a music center that has produced scores of recording artists. Many Malian musicians descend from a caste of musical storytellers performing on either the traditional *kora* (a cross between a harp and a lute) or the guitar. The musical style is strikingly similar to that of blues from the Mississippi Delta, so much so that Ali Farka Touré, who was one of Africa's most renowned musicians, is still referred to as the Bluesman of Africa. Each January, not far from Timbuktu, music fans from West Africa and Europe gather for the Festival in the Desert. In this remote Saharan locale, a celebration of Malian music and Touareg nomadic culture draws together western tourists, African musicians, and nomads (Figure 6.30).

**Pride in East African Runners** Ethiopia and Kenya have produced many of the world's greatest distance runners. Abebe Dikila won Ethiopia's and Africa's first Olympic gold medal, running barefoot at the Rome games in 1960. Since then, nearly every Olympic games has yielded medals for Ethiopia and Kenya. At the Beijing Olympics in 2008, Kenyan runners won 14 medals, and Ethiopians 7 (4 of them gold). These states were the top medal winners for Africa in Beijing.

Running is a national pastime in Kenya and Ethiopia, where elevation—Addis Ababa sits at 7,300 feet (2,200 meters) and Nairobi at 5,300 feet (1,600 meters)—increases oxygen-carrying capacity. Past medalists Haile Gebrselassie and Derartu Tulu are national celebrities in Ethiopia who are idealized by the country's youth. Tulu, the first black African woman to win a gold medal in distance running, is a forceful voice for women's rights in a country where women are discouraged from putting on running shorts. There is talk of a political career for Gebrselassie after he hangs up his running shoes (Figure 6.31).

## Geopolitical Framework: Legacies of Colonialism and Conflict

The duration of human settlement in Sub-Saharan Africa is unmatched by any other region. Evidence shows that humankind originated there, and many diverse ethnic groups have formed in the region over the past few thousand years. Although conflicts among these groups has existed, cooperation and coexistence among different peoples have also occurred. Some 3,000 years ago, the state of Nubia emerged in central and northern Sudan; 1,000 years later, the Kingdom at Axum arose in northern Ethiopia and Eritrea. Both of these states were strongly influenced by political models derived from Egypt and Arabia. The first wholly indigenous African states were founded in the Sahel around 700 CE. Over the next several centuries, a variety of other states emerged in West Africa. By the 1600s, the states located near the Gulf of Guinea took advantage of the opportunities presented by the slave trade—namely the selling of slaves to Europeans (Figure 6.32).

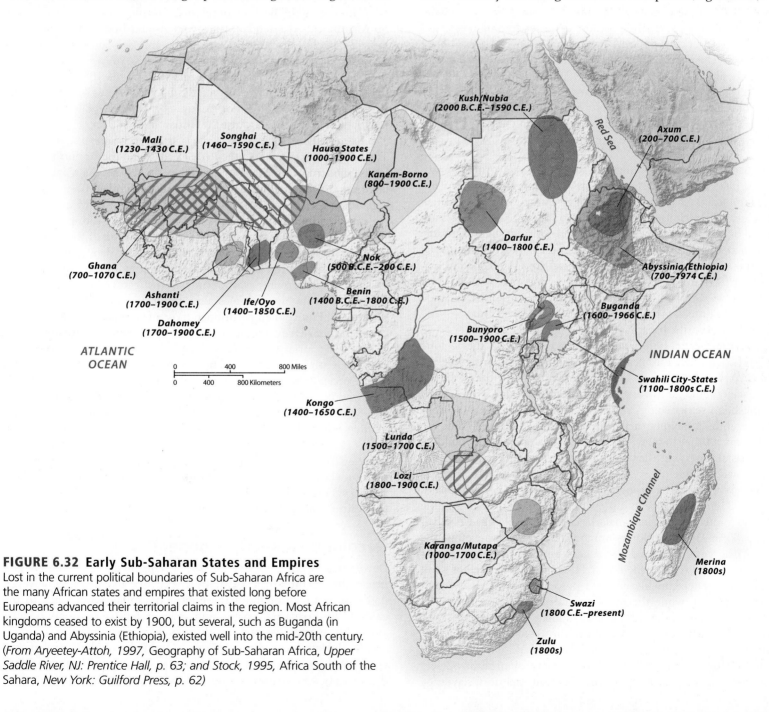

**FIGURE 6.32 Early Sub-Saharan States and Empires**
Lost in the current political boundaries of Sub-Saharan Africa are the many African states and empires that existed long before Europeans advanced their territorial claims in the region. Most African kingdoms ceased to exist by 1900, but several, such as Buganda (in Uganda) and Abyssinia (Ethiopia), existed well into the mid-20th century. (*From Aryeetey-Attoh, 1997,* Geography of Sub-Saharan Africa, *Upper Saddle River, NJ: Prentice Hall, p. 63; and Stock, 1995,* Africa South of the Sahara, *New York: Guilford Press, p. 62*)

Thus, prior to European colonization, Sub-Saharan Africa presented a complex mosaic of kingdoms, states, and tribal societies. With the arrival of Europeans, patterns of social organization and ethnic relations were changed forever. As Europeans rushed to carve up the continent to serve their imperial ambitions, they set up various administrations that heightened ethnic tensions and promoted hostility. Many of the region's modern conflicts can trace their roots back to the colonial era, especially the drawing of political boundaries.

## European Colonization

Unlike the relatively rapid colonization of the Americas, Europeans needed centuries to gain effective control of Sub-Saharan Africa. Portuguese traders arrived along the coast of West Africa in the 1400s, and by the 1500s they were well established in East Africa as well. Initially, the Portuguese made large profits, converted a few local rulers to Christianity, established several defensive trading posts, and gained control over the Swahili trading cities of the east. They stretched themselves too thin, however, and failed in many of their activities. Only along the coasts of modern Angola and Mozambique, where a sizable population of mixed African and Portuguese peoples emerged, did Portugal maintain power. Along the Swahili, or eastern, coast they were eventually expelled by Arabs from Oman, who then established their own trade-based empire in the area.

**The Disease Factor**   One of the main reasons for the Portuguese failure was the disease environment of Africa. With no resistance to malaria and other tropical diseases, roughly half of all Europeans who remained on the African mainland died within a year. Protected both by their armies and by the diseases of their native lands, African states were able to maintain an upper hand over European traders and adventurers well into the 1800s. Unlike in the Americas, where European conquest was made easy by the introduction of Old World diseases that devastated native populations (see Chapters 4 and 5), in Sub-Saharan Africa, endemic disease limited European settlement until the mid-19th century.

Colonists risked the hazards of malaria and other tropical diseases such as sleeping sickness because they were drawn to the region's wealth, and soon other European traders followed the Portuguese. By the 1600s, Dutch, British, and French firms dominated the profitable export of slaves, gold, and ivory from the Gulf of Guinea. The Dutch also established a settler colony in South Africa, safely outside the tropical disease zone, to supply their ships bound for Indonesia. For the next 200 years, European traders came and went, occasionally building fortified coastal posts, but they almost never ventured inland and seldom had any real influence on African rulers. By exporting millions of slaves, however, they had a strongly negative impact on African society.

In the 1850s the European discovery that a daily dose of quinine would offer protection against malaria completely changed the balance of power in Africa. Explorers immediately began to penetrate the interior of the continent, and merchants and expeditionary forces began to move inland from the coast. The first imperial claims soon followed. The French quickly grabbed power along the easily navigated Senegal River, while the British established protectorates over the indigenous states of the Gold Coast (modern-day Ghana).

Also in the early 1800s, two small territories were established in West Africa so that freed and runaway slaves would have a place to return to in Africa (Figure 6.33). The American Colonization Society set up a territory in 1822 to settle former African-American slaves; by 1847, it was the independent free state of Liberia. Sierra Leone served a similar function for ex-slaves from the British Caribbean, but it remained a protectorate of Britain until the 1960s. Despite the intentions behind the creation of these territories, they were colonies. Liberia, in particular, was imposed on existing indigenous groups who viewed their new "African" leaders with contempt.

**The Scramble for Africa**   In the 1880s, European colonization of the region quickly accelerated, leading to the so-called scramble for Africa. By this time, due to the invention of the machine gun, no African state could long resist European force.

As the colonization of Africa intensified, tensions among the colonizing forces of Britain, France, Belgium, Germany, Italy, Portugal, and Spain mounted. Rather than risk war, 13 countries convened in Berlin at the invitation of the German chancellor Bismarck in 1884, in a gathering known as the **Berlin Conference**. During the conference, which no African leaders attended, rules were established about what determined "effective control" of a territory, and Sub-Saharan Africa was carved up and traded like properties in a game of Monopoly (refer to Figure 6.33). Although European arms were by the 1880s far superior to anything found in Africa, several indigenous states organized effective resistance campaigns. For example, in central Sudan, an Islamic-inspired force held out against the British until 1900.

Eventually, European forces prevailed everywhere except Ethiopia. The Italians had conquered the Red Sea coast and the far northern highlands (modern Eritrea) by 1890, and they quickly set their sights on the large Ethiopian kingdom, called Abyssinia, which had been vigorously expanding for several decades. In 1896, however, Abyssinia defeated the invading Italian army, earning the respect of the European powers. In the 1930s fascist Italy launched a major invasion of the country, by this time renamed Ethiopia, to redeem its earlier defeat, and with the help of poison gas and aerial bombardment, it quickly prevailed.

Although Germany was a principal instigator of the scramble for Africa, it lost its own colonies after suffering defeat in World War I. Britain and France then partitioned most of Germany's African empire between themselves. Figure 6.33 shows the colonial status of the region in 1913, prior to Germany's territorial loss.

While the Europeans were cementing their rule over Africa, South Africa was inching toward political independence, at least for its white population. South Africa was one of the oldest colonies in Sub-Saharan Africa, and it became the first to obtain its political independence from Europe in 1910. Yet because of the formalized system of discrimination and racism, it was hardly a symbol of liberty. Ironically, as the Afrikaners tightened their political and social control over the nonwhite population through their policy of apartheid (or "separateness") introduced in 1948, the rest of the continent was preparing for political independence from Europe.

## Decolonization and Independence

Decolonization of Sub-Saharan Africa happened rather quickly and peacefully beginning in 1957. Independence movements, however, had sprung up throughout the continent, some dating back to the early 1900s. Workers' unions and independent newspapers became voices for African discontent and the hope for freedom.

By the late 1950s, Britain, France, and Belgium decided that they could no longer maintain their African empires and began to withdraw. (Italy had already lost its colonies during World War II, and

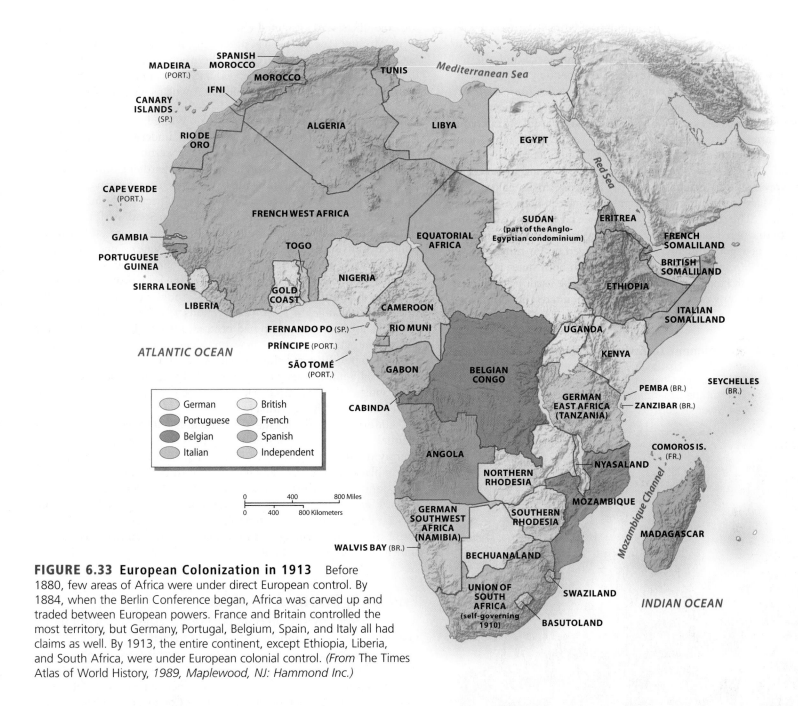

**FIGURE 6.33 European Colonization in 1913** Before 1880, few areas of Africa were under direct European control. By 1884, when the Berlin Conference began, Africa was carved up and traded between European powers. France and Britain controlled the most territory, but Germany, Portugal, Belgium, Spain, and Italy all had claims as well. By 1913, the entire continent, except Ethiopia, Liberia, and South Africa, were under European colonial control. (*From* The Times Atlas of World History, *1989, Maplewood, NJ: Hammond Inc.*)

Britain gained Somalia and Eritrea.) Once started, the decolonization process moved rapidly. By the mid-1960s, virtually the entire region had achieved independence. In most cases, the transition was relatively peaceful and smooth.

Dynamic African leaders put their mark on the region during the early decades after independence. Men such as Kenya's Jomo Kenyatta, Ivory Coast's Felix Houphuët-Boigny, and Ghana's Kwame Nkrumah became powerful father figures who molded their new nations (Figure 6.34). President Nkrumah's vision for Africa was the most expansive. After he helped to secure independence for Ghana in 1957, his ultimate aspiration was political unity of Africa. While his dream was never realized, it set the stage for the founding of the Organization of African Unity (OAU) in 1963, which was renamed the **African Union (AU)** in 2002. The AU is a continent-wide organization, whose main role has been to mediate disputes between neighbors. Certainly in the

1970s and 1980s, it was a constant voice of opposition to South Africa's minority rule, and the AU intervened in some of the most violent independence movements in southern Africa.

**Southern Africa's Independence Battles** Independence did not come easily to southern Africa. In Southern Rhodesia (modern-day Zimbabwe), the problem was the presence of some 250,000 white residents, most of whom owned large farms. Unwilling to see power pass to the country's black majority, then some 6 million strong, these settlers declared themselves the rulers of an independent, white-supremacist state in 1965. The black population continued to resist, however, and in 1978, the Rhodesian government was forced to give up power. The renamed country of Zimbabwe was henceforth ruled by the black majority, although the remaining whites still form an economically privileged community. Since the mid-1990s, disputes over government land reform (splitting up the large commercial farms mostly owned by whites and giving the land to black farmers) and President Robert Mugabe's strongman politics have resulted in serious racial and political tensions as well as the collapse of the country's economy.

In the former Portuguese colonies, independence came violently. Unlike the other imperial powers, Portugal refused to hand over its colonies in the 1960s. As a result, the people of Angola and Mozambique turned to armed resistance. The most powerful rebel movements adopted a socialist orientation and received support from the Soviet Union and Cuba. A new Portuguese government came to power in 1974, however, and it withdrew suddenly from its African colonies. At this point, Marxist regimes quickly came to power in both Angola and Mozambique. The United States, and especially South Africa, responded to this perceived threat by supplying arms to rebel groups that opposed the new governments. Fighting dragged on for three decades in Angola and Mozambique. The countryside in both states is now so heavily laden with land mines that it can hardly be used. With the end of the Cold War, however, outsiders lost their interest in continuing

**FIGURE 6.34 A Monument to Kwame Nkrumah**   Charismatic independence leader Kwame Nkrumah is remembered with this monument in Accra, Ghana. Nkrumah led Ghana to an early independence in 1957; he was also a founder of the Organization of African Unity (OAU), renamed the African Union in 2002.

these conflicts, and sustained efforts began to negotiate a peace settlement. Mozambique has been at peace since the mid-1990s. After several failed attempts at peace in Angola, the Angolan army signed a peace treaty with rebels in 2002 that ended a 27-year conflict in which more than 300,000 people died and 3 million Angolans were displaced.

**Apartheid's Demise in South Africa**   While fighting continued in the former Portuguese zone, South Africa underwent a remarkable transformation. Through the 1980s, its government had remained firmly committed to white supremacy. Under apartheid, only whites enjoyed real political freedom, while blacks were denied even citizenship in their own country—technically, they were citizens of homelands.

The first major change came in 1990, when South Africa withdrew from Namibia, which it had controlled as a protectorate (a dependent political unit under the authority of another state) since the end of World War I. South Africa now stood alone as the single white-dominated state in Africa. A few years later, the leaders of the Afrikaner-dominated political party decided they could no longer resist the pressure for change. In 1994, free elections were held in which Nelson Mandela, a black leader who had been imprisoned for 27 years by the old regime, emerged as the new president. Black and white leaders pledged to put the past behind them and work together to build a new, multiracial South Africa. Since then, orderly elections have been held, and South Africans have elected Thabo Mbeki for two terms (1999–2009) and Jacob Zuma in 2009.

Unfortunately, the legacy of apartheid is not so easily erased. Residential segregation is officially illegal, but neighborhoods are still sharply divided along race lines. Under the multiracial political system, a black middle class emerged, but most blacks remain extremely poor (and most whites remain prosperous). Violent crime has increased, and rural migrants and immigrants have poured into South African cities, producing a xenophobic anti-immigrant backlash. Because the political change was not matched by significant economic transformation, the hopes of many people have been frustrated.

## Enduring Political Conflict

Although most Sub-Saharan countries made a relatively peaceful transition to independence, virtually all of them immediately faced a difficult set of institutional and political problems. In several cases, the old authorities had done almost nothing to prepare their colonies for independence. Lacking an institutional framework for independent government, countries such as the Democratic Republic of the Congo faced a chaotic situation from the beginning. Only a handful of Congolese had received higher education, let alone been trained for administrative posts. The indigenous African political framework had been essentially destroyed by colonization, and in most cases, very little had been built in its place.

Even more problematic in the long run was the political geography of the newly independent states. Civil servants could always be trained and administrative systems built, but little could be done to rework the region's basic political map. The problem was the fact that the European colonial powers had essentially ignored indigenous cultural and political boundaries, both in dividing Africa among themselves and in creating administrative subdivisions within their own imperial territories.

**The Tyranny of the Map**   All over Africa, different ethnic groups found themselves forced into the same state with peoples of distinct linguistic and religious backgrounds, many of whom had recently been their enemies. At the same time, a number of the larger ethnic groups of the region found their territories split between two or more countries. The Hausa people of West Africa, for example, were divided between Niger (formerly French) and Nigeria (formerly British), each of which they had to share with several former enemy groups.

Given the imposed political boundaries, it is no wonder that many African countries struggle to generate a common sense of national identity or establish stable political institutions. **Tribalism**, or loyalty to the ethnic group rather than to the state, has emerged as the bane of African political life. Especially in rural areas, tribal identities are usually more important than national ones. Because nearly all of Africa's countries inherited an inappropriate set of colonial borders, one might assume that they would have been better drawing a new political map based on indigenous identities. Such a strategy was impossible, as all the leaders of the newly independent states realized. Any new territorial

divisions would have created winners and losers and thus would have resulted in more conflict. Moreover, because ethnicity in Sub-Saharan Africa was traditionally fluid, and because many ethnic groups were intermixed, it would have been difficult to generate a clear-cut system of division. Finally, most African ethnic groups were considered too small to form viable countries. With such complications in mind, the new African leaders, meeting in 1963 to form the Organization of African Unity, agreed that colonial boundaries should remain. The violation of this principle, they argued, would lead to pointless wars between states and endless civil struggles within them.

Despite the determination of Africa's leaders to build their new nations within existing political boundaries, challenges to the states began soon after independence. Figure 6.35 maps the ethnic and political conflicts that have disabled parts of Africa since 1995. The human cost of this turmoil is several million refugees and internally displaced

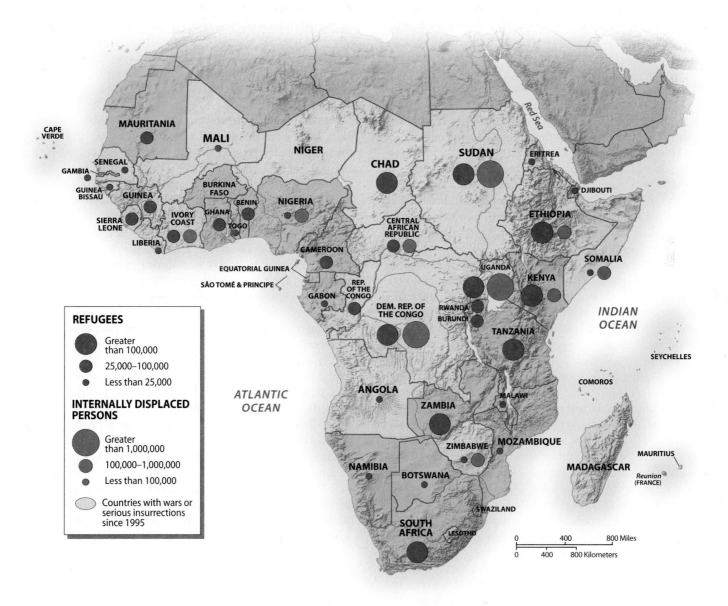

**FIGURE 6.35 Geopolitical Issues—Conflicts and Refugees** Many Sub-Saharan countries have experienced wars or serious insurrections since 1995. These same states are also likely to produce refugees (pink circles) and internally displaced persons (purple circles). As of 2009, 13 million Africans were refugees, and 3 million were internally displaced. *(Data from U.S. Committee for Refugees, 2007, World Refugee Survey)*

persons. **Refugees** are people who flee their state because of a well-founded fear of persecution based on race, ethnicity, religion, or political orientation. Nearly 3 million Africans were considered refugees in 2009. Added to this figure are another 13 million **internally displaced persons (IDPs)**. IDPs have fled from conflict but still reside in their country of origin. Sudan has the largest number of IDPs (5 to 6 million), followed by the Democratic Republic of the Congo and Uganda (over 1 million each). These populations are not technically considered refugees, making it difficult for humanitarian NGOs and the United Nations to assist them.

Over the past decade, the numbers of IDPs have risen, while the total numbers of refugees have declined. This is due, in part, to the reluctance of neighboring states to take on the burden of hosting refugees. Tanzania had more than 300,000 refugees (mostly from Burundi and the Democratic Republic of the Congo), and Chad had 300,000 (mostly from Sudan) in 2009. With international aid on the decline, it is understandable that poor African states that struggle to serve their own people are disinclined to respond to the needs of refugees. Yet the region's many ethnic conflicts continue to produce many vulnerable and needy people fleeing persecution.

**Ethnic Conflicts** As Figure 6.35 suggests, more than half of the states in the region have experienced wars or serious insurrections since 1995. Fortunately, in the past few years, peace has returned in Sierra Leon, Liberia, and Angola, states that produced large numbers of refugees in the 1990s. In the Ivory Coast, where conflict began as violence spilled over from Liberia in 2002, a peace deal was brokered in 2007 between the New Forces rebel group in the north and the government-controlled south. And, after a decade of conflict and more than 4 million deaths, the Democratic Republic of the Congo held peaceful elections in 2006. Yet as peace was advancing in Western and Central Africa, a violent ethnic conflict erupted in Darfur.

Darfur, a Sahelian province of Sudan, is the size of France. In 2003, an ethnic conflict between Arab nomads and herders and non-Arab "black" farmers ignited. Since the fighting began, 450,000 people have died and more than 2.5 million have been displaced. In 2004, the U.S. government accused the Sudanese government of **genocide**—the deliberate and systematic killing of a racial, political, or cultural group. The AU and the UN have sent peacekeeping troops to assist the displaced and reduce the violence. In 2009, the International Criminal Court issued a warrant for the arrest of Sudanese President Bashir for war crimes, but Bashir remained in office and denied the charges.

The conflict began when a rebel group (later called the Sudan Liberation Movement [SLM]) attacked a Sudanese military base. The SLM complained of economic marginalization and wanted a power-sharing arrangement with the Sudanese government. The Sudanese government responded to the rebels by encouraging Arab militiamen (called *janjaweed*—an Arabic term meaning "horse and gun") to attack non-Arab civilians in Darfur. The janjaweed storm villages on horseback and camelback, destroying grain supplies and livestock, killing villagers, and setting buildings on fire. Human rights groups are especially alarmed by the systematic use of rape by the janjaweed as a weapon of ethnic cleansing and terror. The government of Khartoum claims that it has been unable to stop the janjaweed atrocities, yet international observers continue to produce evidence that the government is supporting the Arab militiamen through aerial bombardment and attacks on civilians.

The deadliest ethnic and political conflict in the region has been in the Democratic Republic of the Congo. It is estimated that between 1998 and 2008, nearly 5 million people died, although many of the deaths were from war-induced starvation and disease rather than bullets or machetes. In 1996 and 1997, a loose alliance of armed groups from Rwanda (led by Tutsis) and Uganda joined forces with other forces in the Congo and marched their way across the country, installing Laurent Kabila as president. Under Kabila's rocky and ruthless leadership, which ended in assassination in 2001, rebel groups again invaded from Uganda and Rwanda and soon controlled the northern and eastern portions of the country, while the Kinshasa-based government loosely controlled the western and southern portions.

With Kabila's death, his son Joseph took power and signed a peace accord with the rebels in 2002. In 2003 rebel leaders were part of a transitional government, and an unsteady peace was in place, with help from the UN, the AU, and Western donors. Remarkably, when elections were held in 2006, Joseph Kabila was elected president. The country has no real experience with democracy, and it has a civil service that barely functions and virtually no roads or working infrastructure for the nearly 70 million people who live there. In 2007, foreign aid accounted for 14% of the country's GNI, down from 20 percent in 2005. For now, continuing peace and the ability of the Congo's people to improvise and cope under such adversity may gradually bring about social and economic improvements.

**Secessionist Movements** Problematic African political boundaries have occasionally led to attempts by territories to secede and form new states. The Shaba (or Katanga) province in what was then the state of Zaire tried to leave its national territory soon after independence. The rebellion was crushed a couple years after it started, with the help of France and Belgium. Zaire was unwilling to give up its copper-rich territory, and the former colonialists had economic interests in the territory worth defending. Similarly, the Igbo in oil-rich southeastern Nigeria declared an independent state of Biafra in 1967. After a short but brutal war during which Biafra was essentially starved into submission, Nigeria was reunited.

In 1991, the government of Somalia disintegrated. The territory has since been ruled by clan-based warlords and their militias, who have informally divided the country into clan-based units. **Clans** are social units that are branches of a tribe or an ethnic group larger than a family. Early in the conflict, the northern portion of the country declared its independence as a new country—Somaliland (Figure 6.36). Somaliland has a constitution, a functioning parliament, government ministries, a police force, a judiciary, and a president. The territory produces its own currency and passports. Yet no country has recognized this territory, in part because no government exists in Somalia to negotiate the secession. In 1998, neighboring Puntland also declared itself an autonomous state. Although it does not seek outright independence like Somaliland, Puntland is creating its own administration. Meanwhile, Islamic courts with their well-armed militias control the south, where Al Qaeda operatives are believed to reside, along with pirates who seize ships for ransom in the Indian Ocean and the Gulf of Aden.

Only one territory in the region has successfully seceded. In 1993, Eritrea gained independence from Ethiopia after two decades of civil conflict. This territorial secession is striking because Ethiopia gave up its access to the Red Sea, making it landlocked. Yet the creation of Eritrea still did not bring about peace. After years of fighting, the transition to Eritrean independence began remarkably well. Unfortunately, border disputes between the two countries erupted in 1998, resulting in the deaths of some 100,000 troops. In 2000, a peace

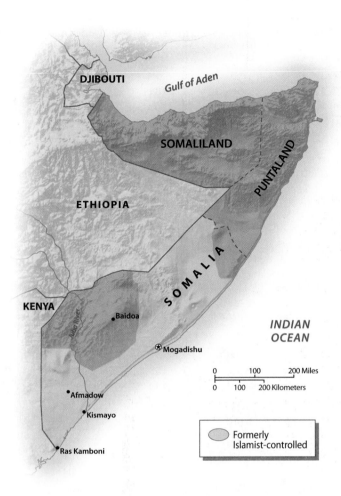

**FIGURE 6.36 Somalia Divided**    A state in name only, Somalia has had no central government since 1991. Two areas in the north seek political autonomy or independence. Meanwhile, Islamic courts have controlled areas in the south.

accord was reached, and the fighting stopped. As Ethiopia comes to accept Eritrea's independence, it might be possible for other areas torn by ethnic warfare to find peace as well. Still, major transformation of Africa's political map should not be expected.

# Economic and Social Development: The Struggle to Rebuild

By almost any measure, Sub-Saharan Africa is the poorest and least-developed world region. Whereas other regions have experienced significant improvements in life expectancy and per capita income since 1990, many Sub-Saharan African states have experienced declines in life expectancy, and income levels are either the same or lower. The average GNI per capita figure, using purchasing power parity, is about $1,869. Some states, such as Botswana, Equatorial Guinea, Mauritius, the Seychelles, and South Africa, have much higher per capita GNI–PPP (Table 6.2). Nearly all the states in the region are ranked at the bottom of most lists ranking per capita GNI or the Human Development Index.

In the 1990s, most countries in the region saw their economic productivity decline, with low or negative growth rates between 1990 and 1999. The economic and debt crisis of the 1980s and 1990s prompted the introduction of **structural adjustment programs**. Promoted by the

International Monetary Fund (IMF) and the World Bank, structural adjustment programs typically reduce government spending, cut food subsidies, and encourage private-sector initiatives. Yet these same policies have caused immediate hardships for the poor, especially women and children, and have led to social protest, most notably in cities. The idea of debt forgiveness for Africa's poorest states, argued for by the unlikely duo of economist Jeffrey Sachs and U2 rock star Bono, has gradually gained acceptance as a strategy for reducing human suffering by redirecting monies that would have gone to debt repayment to the provision of services and the building of infrastructure. Sachs argues that in order for the region to get out of the poverty trap, it will need substantial sums of new foreign aid and investment.

On the positive side, there are signs of economic growth since 2000. For most states, the average annual growth rates from 2000 to 2007 were positive. Several countries have seen average annual growth rates of 5 percent or more (see Table 6.2). While good news overall, such figures can be deceptive. Some are due to soaring oil prices (notably so for Angola, Chad, Nigeria, and Sudan), whereas others are due to countries beginning from a very low base after years of conflict (Mozambique, Sierra Leone, and Rwanda). Still, for the first time in many years, Sub-Saharan African economies are growing at a faster rate than their populations. It is unclear how much the global financial crisis of 2008-2009 will impact these trends. But as export markets weaken, economic growth could slow for the region.

## Roots of African Poverty

In the past, outside observers often attributed Africa's poverty to its environment. Favored explanations included the infertility of its soils, the erratic patterns of its rainfall, the lack of navigable rivers, and the virulence of its tropical diseases. Most contemporary scholars, however, argue that such handicaps are not prevalent everywhere throughout the region and that even where they do exist, they can be—and often have been—overcome by human labor and creativity. The favored explanations for African poverty now look much more to historical and institutional factors than to environmental circumstances.

Numerous scholars have singled out the slave trade for its debilitating effect on Sub-Saharan African economic life. Large areas of the region were depopulated, and many people were forced to flee into poor, inaccessible refuges. Colonization was another blow to Africa's economy. European powers invested little in infrastructure, education, and public health and were instead interested mainly in developing mineral and agricultural resources for their own benefit. Several plantation and mining zones did achieve some prosperity under colonial regimes, but strong national economies failed to develop. In almost all cases, the basic transport and communications systems were designed to link administration centers and zones of extraction directly to the colonial powers rather than to their own surrounding areas. As a result, after achieving independence, Sub-Saharan African countries faced economic and infrastructural challenges that were as daunting as their political problems (Figure 6.37). Today, the average amount of paved road per person in countries such as Ghana, Ethiopia, Kenya, Tanzania, and Uganda is just 33 feet (10 meters) per person. The average for the non-African developing world is 2.8 miles (4.5 kilometers) per person. South Africa is the only African state with a developed modern road network.

**Failed Development Policies**    The first decade or so of independence was a time of relative prosperity and optimism for many African

## TABLE 6.2    DEVELOPMENT INDICATORS

| Country | GNI per capita, PPP 2007 | GDP Average Annual %Growth 2000-07 | Human Development Index (2006)# | Percent Population Living Below $2 a Day | Life Expectancy* 2009 | Under Age 5 Mortality Rate 1990 | Under Age 5 Mortality Rate 2007 | Gender Equity 2007 |
|---|---|---|---|---|---|---|---|---|
| Angola | 4,270 | 12.9 | 0.484 | 70.2 | 46 | 258 | 158 | |
| Benin | 1,310 | 3.8 | 0.459 | 75.3 | 56 | 184 | 123 | 73 |
| Botswana | 12,880 | 5.3 | 0.664 | 49.4 | 49 | 57 | 40 | 101 |
| Burkina Faso | 1,120 | 5.8 | 0.372 | 81.2 | 57 | 206 | 191 | 82 |
| Burundi | 330 | 2.7 | 0.382 | 93.4 | 49 | 189 | 180 | 90 |
| Cameroon | 2,120 | 3.5 | 0.514 | 57.7 | 52 | 139 | 148 | 85 |
| Cape Verde | 2,940 | | 0.705 | 40.2 | 71 | 60 | 32 | 104 |
| Central African Republic | 710 | 0.0 | 0.352 | 81.9 | 45 | 171 | 172 | |
| Chad | 1,280 | 12.2 | 0.389 | 83.3 | 47 | 201 | 209 | 64 |
| Comoros | 1,150 | | 0.572 | 65.0 | 64 | 120 | 66 | 84 |
| Congo | 2,750 | 4.1 | 0.619 | 74.4 | 53 | 104 | 125 | 90 |
| Dem. Rep. of Congo | 290 | 5.0 | 0.361 | 79.5 | 53 | 200 | 161 | 73 |
| Djibouti | 2,260 | | 0.513 | 41.2 | 55 | 175 | 127 | 79 |
| Equatorial Guinea | 21,220 | | 0.717 | | 59 | 170 | 206 | |
| Eritrea | 620 | 1.4 | 0.442 | | 58 | 147 | 70 | 78 |
| Ethiopia | 780 | 7.5 | 0.389 | 77.5 | 53 | 204 | 119 | 83 |
| Gabon | 13,410 | 2.0 | 0.729 | 19.6 | 59 | 92 | 91 | |
| Gambia | 1,140 | 4.9 | 0.471 | 56.7 | 55 | 153 | 109 | 100 |
| Ghana | 1,320 | 5.5 | 0.533 | 53.6 | 59 | 120 | 115 | 95 |
| Guinea | 1,120 | 2.8 | 0.423 | 87.2 | 56 | 231 | 150 | 74 |
| Guinea-Bissau | 470 | 0.4 | 0.383 | 77.9 | 46 | 240 | 198 | |
| Ivory Coast | 1,620 | 0.3 | 0.431 | 46.8 | 52 | 151 | 127 | |
| Kenya | 1,550 | 4.4 | 0.532 | 39.9 | 54 | 97 | 121 | 96 |
| Lesotho | 1,940 | 3.8 | 0.496 | 62.2 | 40 | 102 | 84 | 104 |
| Liberia | 280 | −2.7 | 0.364 | 94.8 | 56 | 205 | 133 | |
| Madagascar | 930 | 3.2 | 0.533 | 89.6 | 59 | 168 | 112 | 96 |
| Malawi | 760 | 3.3 | 0.457 | 90.4 | 46 | 209 | 111 | 100 |
| Mali | 1,040 | 5.4 | 0.391 | 77.1 | 48 | 250 | 196 | 78 |
| Mauritania | 2,000 | 5.1 | 0.557 | 44.1 | 57 | 130 | 119 | 102 |
| Mauritius | 11,410 | 4.0 | 0.802 | | 72 | 24 | 15 | 102 |
| Mozambique | 730 | 8.1 | 0.366 | 90.0 | 43 | 201 | 168 | 85 |
| Namibia | 5,100 | 4.8 | 0.634 | 62.2 | 59 | 87 | 68 | 104 |
| Niger | 630 | 3.9 | 0.370 | 85.6 | 53 | 304 | 176 | 71 |
| Nigeria | 1,760 | 6.6 | 0.499 | 83.9 | 47 | 230 | 189 | 84 |
| Reunion | | | | | 76 | | | |
| Rwanda | 860 | 5.8 | 0.435 | 90.3 | 48 | 195 | 181 | 100 |
| São Tomé and Principe | 1,630 | | 0.643 | | 65 | 101 | 99 | 101 |
| Senegal | 1,650 | 4.5 | 0.502 | 60.3 | 55 | 149 | 114 | 92 |
| Seychelles | 15,440 | | 0.836 | | 73 | 19 | 13 | 106 |
| Sierra Leone | 660 | 11.2 | 0.329 | 76.1 | 48 | 290 | 262 | 86 |
| Somalia | | | | | 50 | 203 | 142 | |
| South Africa | 9,450 | 4.3 | 0.670 | 42.9 | 52 | 64 | 59 | 100 |
| Sudan | 1,880 | 7.1 | 0.526 | | 58 | 125 | 109 | 89 |
| Swaziland | 4,890 | 2.6 | 0.542 | 81.0 | 46 | 96 | 91 | 95 |
| Tanzania | 1,200 | 6.7 | 0.503 | 96.6 | 54 | 157 | 116 | |
| Togo | 770 | 2.6 | 0.479 | 69.3 | 61 | 150 | 100 | 75 |
| Uganda | 1,040 | 7.1 | 0.493 | 75.6 | 50 | 175 | 130 | 98 |
| Zambia | 1,190 | 5.1 | 0.453 | 81.5 | 43 | 163 | 170 | 96 |
| Zimbabwe | | −5.7 | | | 41 | 95 | 90 | 97 |

Source: World Bank, *World Development Indicators, 2009.*    United Nations, *Human Development Index, 2008.#*    Population Reference Bureau, *World Population Data Sheet, 2009*＊    Gender Equity – Ratio of female to male enrollments in primary and secondary school. Numbers below 100 have more males in primary/secondary school, numbers above 100 have more females in primary/secondary schools.

**FIGURE 6.37 Lack of Infrastructure**   Aid workers in the Republic of the Congo push their vehicles out of the mud along the main road from Kinkala to the capital of Brazzaville. Basic infrastructure, such as paved roads, are limited in Central Africa. During the rainy season, many dirt roads are not passable.

Many economists argue that Sub-Saharan African governments enacted counterproductive economic policies and thus brought some of their misery on themselves. Eager to build their own economies and reduce their dependency on the former colonial powers, most African countries followed a course of economic nationalism. More specifically, they set about building steel mills and other forms of heavy industry that were simply not competitive (Figure 6.38).

**Corruption**   Although prevalent through most of the world, corruption seems to have been particularly widespread in several African countries. Because civil servants are not paid a living wage, they are often forced to solicit bribes. According to a 2009 survey by an international business magazine, Somalia was ranked the most corrupt country in the world, followed by Sudan. (Skeptical observers, however, point out that several Asian nations with highly successful economies, such as China, are also noted for having high levels of corruption, so that corruption alone may not be the problem.)

countries. Most of them relied heavily on the export of mineral and agricultural products, and through the 1970s commodity prices generally remained high. Some foreign capital was attracted to the region, and in many cases, the European economic presence actually increased after decolonization.

In the 1980s, as most commodity prices began to decline, foreign debt began to weigh down many Sub-Saharan countries. By the end of the 1980s, most of the region was in serious economic decline. By the early 1990s, the region's foreign debt was around $200 billion. Although low compared to other developing regions (such as Latin America), as a percentage of its economic output, Sub-Saharan Africa's debt was the highest in the world.

With millions of dollars in loans and aid pouring into the region, officials at various levels have been tempted to take something for themselves. Some African states, such as the former Zaire, were given the name *kleptocracies*. A **kleptocracy** is a state in which corruption is so institutionalized that most politicians and government bureaucrats siphon off a huge percentage of the country's wealth. President Mobutu ruled the Democratic Republic of the Congo from 1965 until 1997 and was a legendary kleptocrat. While his country was saddled with an enormous foreign debt, he reportedly maintained several billion dollars in Belgian banks.

## Links to the World Economy

Sub-Saharan Africa's trade connection with the world is limited, accounting for less than 2 percent of global trade. The level of overall trade is low both within the region and outside it. Traditionally, most exports went to the European Union, especially the former colonial powers of England and France. The United States is the second most common destination. The trade patterns

**FIGURE 6.38 Industrialization**   Heavy industry, such as this chemical plant in Kafue, Zambia, failed to deliver Sub-Saharan Africa from poverty. In the worst cases, these industrial enterprises were unable to produce competitive products for world and domestic markets.

## FIGURE 6.39 Mobile Phones for Africa
A shopkeeper has steady customers for mobile phones in Abidjan, Ivory Coast. Mobile telephones, which do not require expensive fixed lines, are quickly becoming the preferred instrument of communication in Sub-Saharan Africa. Multinational firms are competing for the region's business.

are changing, especially as trade with China and India grows. In 2007 exports to Asia nearly equaled exports to the United States. The patterns of imports are also changing. The majority of African countries still turn to Europe for imports, but Asia (particularly China) now accounts for one-third of all imports.

By most measures of connectivity, Sub-Saharan Africa lags behind other developing regions. Fixed telephone lines are scarce; in Nigeria, there is only 1 line per 100 persons. The regional average is slightly better, at 2 lines per 100, but still far behind North Africa and Southwest Asia's 17 lines per 100. Only 18 percent of the region's households

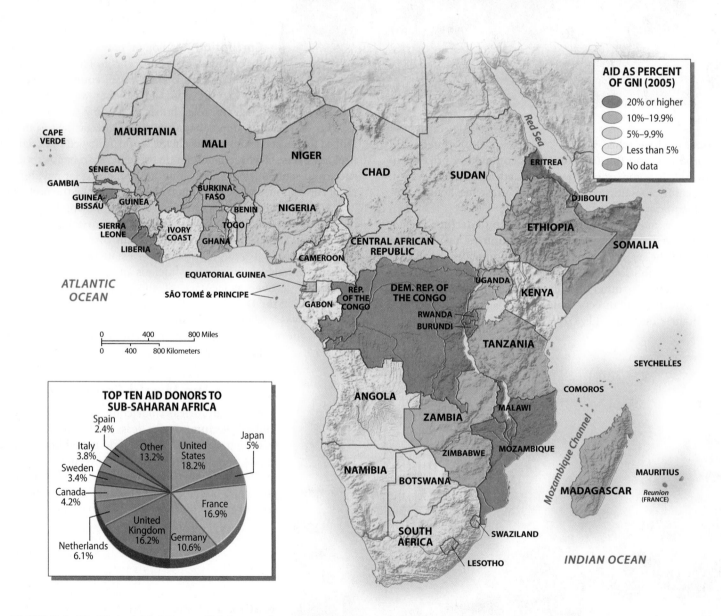

**AID AS PERCENT OF GNI (2005)**
- 20% or higher
- 10%–19.9%
- 5%–9.9%
- Less than 5%
- No data

**TOP TEN AID DONORS TO SUB-SAHARAN AFRICA**
- Spain 2.4%
- Italy 3.8%
- Sweden 3.4%
- Canada 4.2%
- Netherlands 6.1%
- United Kingdom 16.2%
- Germany 10.6%
- France 16.9%
- United States 18.2%
- Japan 5%
- Other 13.2%

## FIGURE 6.40 Global Linkages: Aid Dependency
Many states in Sub-Saharan Africa are dependent on foreign aid as their primary link to the global economy. This figure maps aid as a percentage of GNI, which ranges from less than 1 percent to nearly 50 percent in Burundi. (*From* World Development Indicators, *2006, Washington, DC: World Bank*)

have televisions. One hopeful change is the expansion of mobile telephones in Africa. No longer reliant on expensive fixed telephone lines for communication, multinational providers are now competing for mobile-phone customers. The World Bank estimated that in 2007 there were 23 cell phone subscriptions per 100 people in Sub-Saharan Africa, which is a figure comparable to that of South Asia (Figure 6.39). The number of Internet users in the region is also on the rise.

**Aid Versus Investment**   In many ways, Sub-Saharan Africa is linked to the global economy more through the flow of financial aid and loans than through the flow of goods. As Figure 6.40 shows, aid for several states accounted for more than 20 percent of the GNI, and in Burundi it was 47 percent. Most of this aid came from a handful of developed countries (see the insert in Figure 6.40). The United States provided the most aid in 2005, closely followed by France and then the United Kingdom. In contrast, net flows of private capital are extremely low. Countries such as South Africa, Equatorial Guinea, Gabon, and Angola are the exceptions because of their oil or mineral wealth.

The reasons foreign investors avoid Africa are fairly obvious. The region is generally perceived to be too poor and unstable to merit much attention, and most foreign investors eager to take advantage of low wages put their money in Asia or Latin America. Sub-Saharan Africa, some economists suggest, is therefore starved for capital. Other scholars, however, see foreign investment as more of a trap— and one that the region would be wise to avoid. Recently, in a number of African countries, a small boom has occurred in mineral exploration and production, attracting considerable overseas interest. Optimists interpret this as a sign of Africa's economic renewal, but others counter that in the past, mineral extraction has failed to lead to broad-based and lasting economic gains (see "Exploring Global Connections: China's Investment in Africa").

**Debt Relief**   The World Bank and the IMF proposed in 1996 to reduce debt levels for heavily indebted poor countries, many of which are in Sub-Saharan Africa. Most Sub-Saharan states are indebted to official creditors such as the World Bank, not to commercial banks (as is the case in Latin America and Southeast Asia). Under this World Bank/IMF program, substantial debt reduction will be given so Sub-Saharan countries that are determined to have "unsustainable" debt burdens. Mauritania, for example, spends six times more money on repaying its debts than it does on health care.

States qualify for different levels of debt relief, depending on their poverty-reduction strategies. Uganda was the first state to qualify for the program, using money it saved on debt repayment to expand primary schooling. Ghana, which qualified for debt relief in 2004, received a $3.5 billion relief package. Other countries that have benefited from debt reduction are Tanzania, Mozambique, Ethiopia, Mauritania, Mali, Niger, Nigeria, Senegal, Burkina Faso, and Benin. More Africans countries may also be able to redirect debt payments toward building infrastructure and improving basic health and educational services.

## Economic Differentiation within Africa

As in most other regions, considerable differences in levels of economic and social development persist in Sub-Saharan Africa. In many respects, the small island nations of Mauritius and the Seychelles have little in common with the mainland. With high per capita GNI, life expectancies averaging in the low 70s, and economies built on tourism, they could more easily fit into the Caribbean were it not for their Indian Ocean location. In contrast, three-quarters of the population in mainland Sub-Saharan Africa subsists on less than $2 per day. And only a few states, such as Botswana, South Africa, Gabon, Namibia, and Equatorial Guinea, have a per capita GNI–PPP of over $5,000.

## EXPLORING GLOBAL CONNECTIONS
## China's Investment in Africa

In 2006 China hosted nearly every African head of state at the China–Africa Forum. The event signaled China's growing economic investment in the region at a time when U.S. and European interests are focused on proffering aid or fighting terrorism. China wants to secure the oil and ore it needs for its massive industrial economy. In exchange, it offers African nations money for roads, railways, and schools, with relatively few strings attached. Some African leaders see China as a new kind of global partner, one that wants straight commercial relations without an ideological or political agenda. China's trade with Africa is growing faster than with any other region except Southwest Asia, increasing tenfold in the past decade. Angola, in which China has invested heavily, is now China's largest foreign supplier of oil—beating out

Saudi Arabia. Sudan, with its largely untapped oil reserves, has also received substantial Chinese investment, despite international pressure not to negotiate with the Sudanese government due to the genocide in Darfur.

China's engagement with Africa is not new. In the 1960s and 1970s China supported the socialist governments of postcolonial Africa with development assistance. In the 1970s Beijing funded and built the Tan-Zam railroad that gave landlocked Zambia an outlet to the sea via Tanzania, bypassing the need for Zambia to use apartheid South Africa's ports. China's investment today is strictly commercial and substantially larger. In addition to buying up oil, China purchases copper from Zambia and wood from the Democratic Republic of the Congo. It also sells cheap manufactured goods and clothing in record quantities.

The biggest complaint about Chinese investment in Africa comes from the West. There is frustration that China readily invests in states that have wretched human rights records (Sudan and Zimbabwe). China mostly ignores the lending standards that have been used in the region to curb corruption. Some Africans complain that China is only interested in exporting raw materials and importing manufactured goods back to Africa. Yet through its investments, China has also gained diplomatic friends, as well as access to raw materials. Interestingly, Chinese officials see themselves as the developing world's biggest beneficiary of globalization, and by partnering with the region most ignored by globalization, they will foster more development in Sub-Saharan Africa.

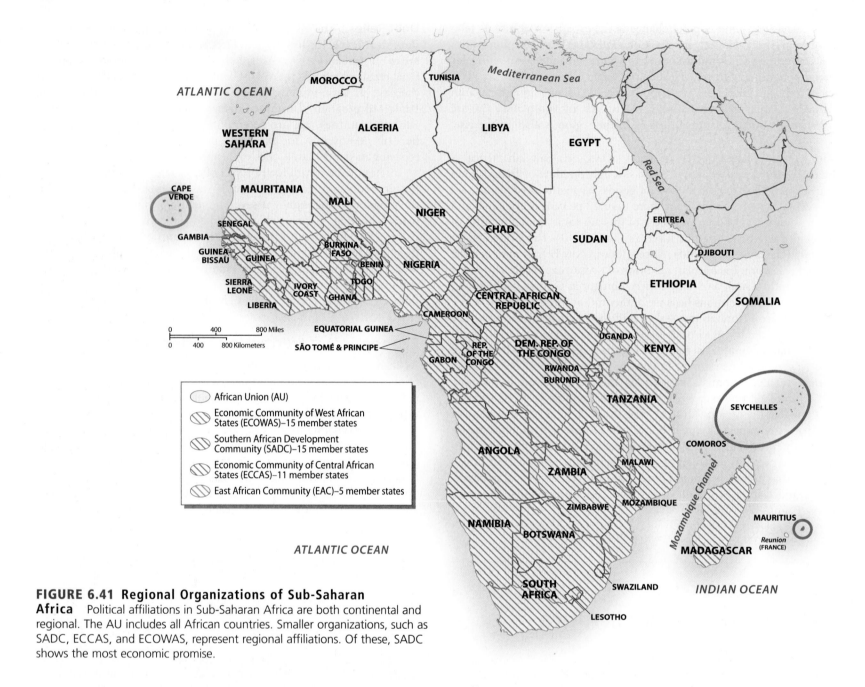

**FIGURE 6.41 Regional Organizations of Sub-Saharan Africa** Political affiliations in Sub-Saharan Africa are both continental and regional. The AU includes all African countries. Smaller organizations, such as SADC, ECCAS, and ECOWAS, represent regional affiliations. Of these, SADC shows the most economic promise.

Legend:
- African Union (AU)
- Economic Community of West African States (ECOWAS)–15 member states
- Southern African Development Community (SADC)–15 member states
- Economic Community of Central African States (ECCAS)–11 member states
- East African Community (EAC)–5 member states

Given the scale of the African continent, it is not surprising that groups of states have formed trade blocs to facilitate intraregional exchange and development. The two most active regional organizations are the Southern African Development Community (SADC) and the Economic Community of West African States (ECOWAS). Both were founded in the 1970s but became more important in the 1990s (Figure 6.41). SADC and ECOWAS are anchored by the region's two largest economies: South Africa and Nigeria. Other regional trade blocs include the Economic Community of Central African States (ECCAS) and the smaller but more effective East African Community (EAC).

**South Africa** South Africa is the unchallenged economic power-house of Sub-Saharan Africa. The per capita GNI–PPP of Nigeria, the next largest economy, is just one-fifth that of South Africa. Only South Africa has a well-developed and well-balanced industrial economy.

It also boasts a healthy agricultural sector and, more importantly, it stands as one of the world's mining superpowers. South Africa remains unchallenged in gold production and is a leader in many other minerals as well. But while South Africa is undeniably a wealthy country by African standards and while its white minority is prosperous by any standard, it is also a country beset by severe and widespread poverty. In the townships lying on the outskirts of the major cities and in the rural districts of the former homelands, employment opportunities remain limited and living standards marginal. Despite the end of apartheid, South Africa continues to suffer from one of the most unequal distributions of income in the world.

**Oil and Mineral Producers** Another group of relatively well-off Sub-Saharan countries benefits from large oil and mineral reserves and small populations (Figure 6.42). The prime example is Gabon, a country of noted oil wealth that is inhabited by 2 million people.

**FIGURE 6.42 Chad–Cameroon Pipeline** Workers unload pipe destined for the 600-mile-long pipeline that pumps oil from Chad to Cameroon's Atlantic port of Kribi. The pipeline was finished in 2003.

daily goal is to earn 75 cents, which will enable her to purchase the vegetables and ground corn (called *mealie-meal*) needed to make the evening supper. She sits at her vegetable stand among the neatly piled mounds of tomatoes for 12 hours a day. Rose earns between $12 and $18 a month from her stand. Half of her earnings go to food. She pays $2 a month for school fees, $2 a month for water, $1.50 for property tax, and 50 cents for government health insurance. Water, education, and health care were free before structural adjustment policies forced payment for these basic services. Rose used to be able to buy meat occasionally with her earnings, but not any longer because of her new expenses. There are luckless days when she returns home with no food. "You don't know what suffering is until you have watched your babies go hungry," says Rose. "I have suffered many times."

Livingstone and Maramba used to have three dozen clothing manufacturers that employed hundreds of people, including Rose's husband. The deluge of used clothing from the West killed this industry during the 1990s and plunged families like Rose's into stomach-tightening poverty. Then, Rose's husband died. With no other job prospects, she joined the ranks of the informal sector.

What begins as charity in the West—giving old clothing away—blossoms into thousands of small businesses in villages and cities across Africa. In Uganda, it's called *mivumba*, and in Zambia it's *salaula*; throughout Africa one can go to any market and find the vendors of used clothing (Figure 6.43). In a region where most people earn less than $2 per day, secondhand clothing is the norm.

Its neighbors, the Democratic Republic of the Congo and Cameroon, also benefit from oil, but both have experienced economic declines in recent years, with that of Cameroon being particularly dramatic. Farther south, Namibia and Botswana also have the advantage of small populations and abundant mineral resources, especially diamonds. Both countries have also enjoyed stable governments over the past few years and have experienced solid economic growth.

**The Leaders of ECOWAS** Nigeria has the largest oil reserves in Sub-Saharan Africa, but its huge population has kept its per capita GNI–PPP at a low $1,760. Oil money has allowed a small minority of the region's population to grow extremely wealthy more by manipulating the system than by engaging in productive activities. Most Nigerians, however, remain trapped in poverty. Oil money also led to the explosive growth of the former capital of Lagos, which by the 1980s had become one of the most expensive—and least livable—cities in the world. As a result, the Nigerian government chose to build a new capital city in Abuja, located near the country's center, a move that has proved tremendously expensive.

Ivory Coast and Senegal, formerly the core territories of the French Sub-Saharan empire, still function as commercial centers, but they also suffered economic downturns in the 1980s. In the mid-1990s, the Ivorian economy again began to grow. Supporters within the country called it an emerging "African elephant" (comparing it to the successful "economic tigers" of eastern Asia). But political turmoil in the late 1990s led to the country being divided by 2002, with the north being held by rebel forces, negatively affecting the economy and displacing half a million people. Ghana, a former British colony and an ECOWAS member, also began to see economic recovery in the 1990s. In 2001 it negotiated with the IMF and the World Bank for debt relief to reduce its nearly $6 billion foreign debt. Between 2000 and 2007, Ghana maintained an average annual growth of over 5 percent.

**Life for the Region's Poorest** Rose Shanzi, a mother of five, lives in the town of Maramba, Zambia, not far from the city of Livingstone, Zimbabwe, the site of Victoria Falls. Rose sells tomatoes in the local market, and if she sells enough, she and her children eat that day. Her

**FIGURE 6.43 Used Clothing Vendor** A vendor carries a 100-pound (45 kilos) bale of used clothing through the streets of Kampala, Uganda. The clothing is sold at local markets and is a vital part of the informal sector.

This is how a shirt sewn in a Honduras sweatshop and worn by a New Jersey teenager gets repackaged and sold by a Zambian street vendor.

Despite the resourcefulness of used clothing vendors, several African textile manufacturers are crying foul, arguing that something must be done to stop the flow of used clothing. They believe that as long as used clothes are in the markets, no textile industry can survive in the region. In some countries, most notably South Africa, the direct import of used clothing has been banned since 1999. Nigeria, Ethiopia, and Eritrea have imposed their own prohibitions. In Uganda the import duty on used clothing was raised in 2003 to appease local textile manufacturers, and also to support an industry that offers better opportunities for workers. Still, for the region's poor, used clothing and dependence on the informal sector is a way of life.

## Measuring Social Development

By world standards, measures of social development in Sub-Saharan Africa are extremely low. Yet unlike with the region's economic indicators, at least with social development there are some positive trends, especially with regard to child survival, education, and gender equity (see Table 6.2). Overall, the region needs to spend much more on education; hopeful signs of more access to news, telecommunications, and the Internet are evident in the region's cities.

**Life Expectancy** Sub-Saharan Africa's figures on life expectancy are, overall, the world's lowest. Only the poorest Asian countries, such as Afghanistan, Nepal, and Laos, stand at the average African level. Countries hard hit by HIV/AIDS or conflict have seen life expectancies tumble below 40 years; for example, the 2008 life expectancy at birth for Swaziland was a disastrous 33 years. Despite these figures, some progress has been made in enhancing life expectancy in Sub-Saharan Africa. While the region's average life expectancy of 50 years may seem incredibly short, it must be remembered that infant and childhood mortality figures depress these numbers; average life expectancy for adults alone is quite a bit higher. States such as Botswana, Kenya, and Zimbabwe established relatively high life expectancy figures in the 1980s, but the growing AIDS epidemic has detracted from those accomplishments.

The causes of low life expectancy are generally related to extreme poverty, environmental hazards (such as drought), and various environmental and infectious diseases (malaria, schistosomiasis, cholera, AIDS, and measles). Often these factors work in combination. Malaria, for example, kills a half million African children each year. The death rate is also affected by poverty, undernourished children being the most vulnerable to the effects of high fevers. Cholera outbreaks occur in crowded slums and villages where food or water is contaminated by the feces of infected persons. Such unsanitary conditions result from a lack of basic infrastructure. Tragically, diseases that are preventable, such as measles, occur when people have no access to or cannot afford vaccines.

**Meeting Educational Needs** Basic education is another obstacle confronting the region. The goal of universal access to primary education is a daunting one for a region in which 43 percent of the population is less than 15 years old. The UN estimates that 70 percent of African children are enrolled in primary school, but many of these schools are inadequately staffed and have few resources. Sub-Saharan Africa has one-sixth of the world's children under 15 but half of the world's uneducated children. Girls are still less likely than boys to attend school. In West Africa countries such as Chad, Niger, and the Ivory Coast, girls are decidedly underrepresented; in Chad, for every 100 boys there are 64 girls in school (refer to Table 6.2).

A renewed focus on education since 2000 has been attributed to the **Millennium Development Goals**, a global United Nations effort to reduce extreme poverty by focusing on basic education, health care, and access to clean water. While the region will not meet the goals set for 2015, more resources have been directed to schools from governments and from nonprofit organizations. One example of the transformative influence of better-resourced schools comes from the village of Bumwalukani, Uganda. In 2003 a new school was built there by a U.S.-based NGO that was started by a Ugandan emigrant living in the United States who grew up in this rural mountain village. Students pay a small fee to attend the school and are given hot lunches, uniforms, and an incomparable educational experience compared to the poorly funded government schools (Figure 6.44). In Uganda, students who excel receive scholarships to attend residential high schools in Kampala. Students from Bumwalukani are now able to compete for these coveted opportunities.

## Women and Development

Development gains cannot be made in Africa unless the economic contributions of African women are recognized. Officially, women are the invisible contributors to local and national economies. In agriculture, women account for 75 percent of the labor that produces more than half the food consumed in the region. Tending subsistence plots, taking in extra laundry, and selling surplus produce in local markets all contribute to household income. Yet because many of these activities are considered informal economic activities, they are not counted. For many of Africa's poorest people, however, the informal sector *is* the economy, and within this sector women dominate.

**Status of Women** The social position of women is difficult to measure for Sub-Saharan Africa. Female traders in West Africa, for example, have considerable political and economic power. By such measures as female labor force participation, many Sub-Saharan African countries show relative gender equality. And women in most Sub-Saharan societies do not suffer the kinds of traditional social restrictions encountered in much of South Asia, Southwest Asia, and North Africa; in Sub-Saharan Africa, women work outside the home, conduct business, and own property. In 2006, Ellen Johnson-Sirleaf was sworn in as Liberia's president, making her Africa's first elected female leader.

By other measures, however, such as the prevalence of polygamy, the practice of the "bride-price," and the tendency for males to inherit property over females, African women do suffer discrimination. Perhaps the most controversial issue regarding women's status is the practice of female circumcision, or genital mutilation. In Sudan, Ethiopia, Somalia, and Eritrea, as well as parts of West Africa, almost 80 percent of girls are subjected to this practice, which is extremely painful and can have serious health consequences. Yet because the practice is considered traditional, most African states are unwilling to ban it.

Regardless of their social position, most African women still live in remote villages where educational and wage-earning opportunities remain limited and bearing numerous children remains a major economic contribution to the family. As educational levels increase and urban society expands—and as reduced infant mortality provides greater security—one can expect fertility in the region to gradually

**FIGURE 6.44 Ugandan School Children**  Cheering at a school event, these children in rural Bumwalukani, Uganda, are receiving a good education as a result of efforts by a U.S.-based NGO founded by a Ugandan emigrant from this village. Educating the region's children is one of the most pressing challenges.

decrease. Governments can speed up the process by providing birth control information and cheap contraceptives—and by investing more money in health and educational efforts aimed at women. As the economic importance of women receives greater attention from national and international organizations, more programs are being directed exclusively toward them.

**Building from Within**  Major shifts in the way development agencies view women and women view themselves have the potential to transform the region. All across the continent, support groups and networks have formed, raising women's awareness, offering women micro-credit loans for small businesses, and organizing their economic power. From farm-labor groups to women's market associations, investment in the organization of women has paid off. In Kenya, for example, hundreds of women's groups organize tree plantings to prevent soil erosion and ensure future fuel supplies.

Whether inspired by feminism, African socialism, or the free market, community organizations have made a difference in meeting basic needs in sustainable ways. No doubt the majority of the groups fall short of all of their objectives. But for many people, especially women, the message of creating local networks to solve community problems is an empowering one.

# SUMMARY

- The largest landmass straddling the equator, Africa is called the plateau continent because it is dominated by extensive uplifted plains. Key environmental issues facing this tropical region are desertification, deforestation, and drought. At the same time, the region supports a tremendous diversity of wildlife, especially large mammals.

- With more than 800 million people, Sub-Saharan Africa is the fastest-growing region in terms of population. Yet it is also the poorest region, with three-quarters of its people living on less than $2 a day. In addition, it has the lowest average life expectancy, at 50 years, due to disease, especially the scourge of HIV/AIDS.

- Culturally, Sub-Saharan Africa is an extremely diverse region, where multiethnic and multi-religious societies are the norm. With a few exceptions, religious diversity and tolerance has been a distinctive feature of the region. Most states have been independent for nearly 50 years, and in that time, pluralistic but distinct national identities have been forged.

- Since 1995, there have been numerous bloody ethnic and political conflicts in the region. Fortunately, peace now exists in many conflict-ridden areas, such as Angola, Sierra Leone, Liberia, and the Democratic Republic of the Congo. However, ongoing ethnic and territorial disputes in Sudan, Democratic Republic of the Congo, and Somalia have produced millions of internally displaced persons and refugees.

- In terms of contemporary economic globalization, Sub-Saharan Africa's connections to the global economy are weak. With 12 percent of the world's population, the region accounts for only about 2 percent of the world's economic activity. Most of the region's economic ties come through international aid and loans rather than trade. Foreign direct investment, especially from China, is beginning to increase in the region.

- Poverty is the region's most pressing issue. Since 2000, there have been signs of economic growth, led in part by debt-forgiveness policies and the end of some of the longest-running conflicts in the region. It remains to be seen whether the international community is truly committed to helping millions of people in the region improve their lives.

# KEY TERMS

African Union (AU) (p. 171)
apartheid (p. 162)
Berlin Conference (p. 170)
biofuel (p. 154)
clan (p. 174)
coloured (p. 162)
desertification (p. 152)
genocide (p. 174)

Great Escarpment (p. 148)
Horn of Africa (p. 151)
internally displaced person
   (IDP) (p. 174)
kleptocracy (p. 177)
Millennium Development Goals
   (p. 182)
pastoralist (p. 160)

refugee (p. 174)
Sahel (p. 153)
structural adjustment program
   (p. 175)
swidden (p. 160)
transhumance (p. 153)

tribalism (p. 173)
tribe (p. 164)
tsetse fly (p. 160)

# THINKING GEOGRAPHICALLY

1. What factors might explain why European conquest and settlement occurred much earlier in tropical America than in tropical Africa?

2. More than any other region, Sub-Saharan Africa is noted for its wildlife, especially large mammals. What environmental and historical processes explain the existence of so much fauna? Why are there relatively fewer large mammals in other world regions?

3. Compare and contrast the role of tribalism in Sub-Saharan Africa with the role of nationalism in Europe.

4. Discuss the contrasting development models put forward by the United States and Europe with that of China. How will Chinese influence in the region alter the course of development for Sub-Saharan Africa?

5. Is desertification a natural or human-induced process? Where does it occur? How might global climate change impact desertification trends?

Log in to **www.mygeoscienceplace.com** for videos, interactive maps, RSS feeds, case studies, and self-study quizzes to enhance your study of Sub-Saharan Africa.

# 7 SOUTHWEST ASIA AND NORTH AFRICA

## GLOBALIZATION AND DIVERSITY

Globalization has played out across Southwest Asia and North Africa in complex ways. Energy-rich nations have reaped the greatest economic rewards in satisfying global demands for oil and natural gas, but varied cultural and political elements of the outside world have not been embraced by many residents of the region.

### ENVIRONMENTAL GEOGRAPHY

Water shortages are likely to increase across this arid region in the early 21st century as growing populations, rapid urbanization, and increasing demands for agricultural land strain already limited water supplies.

### POPULATION AND SETTLEMENT

Rapid population growth in North African cities such as Algiers and Cairo is far outpacing the ability of these urban places to supply adequate housing and services.

### CULTURAL COHERENCE AND DIVERSITY

The heart of the Islamic world, this region finds itself at the center of the global rise of Islamist movements that often come into conflict with Western values and traditions.

### GEOPOLITICAL FRAMEWORK

Ongoing political instability is a fact of life across much of the region. Religious and ethnic differences, shifting political allegiances, and persisting economic problems all contribute to the region's geopolitical challenges.

### ECONOMIC AND SOCIAL DEVELOPMENT

Unpredictable changes in world oil prices continue to have a tremendous economic impact on this region, which holds some two-thirds of all petroleum reserves on the planet.

*The threat of ethnic and religious violence in Southwest Asia is illustrated by the ruins of the Golden Dome Shiite shrine in Samarra, Iraq. The site was bombed by Sunni extremists in 2006.*

Climate, culture, and oil all help define the complex Southwest Asia and North Africa world region (see "Setting the Boundaries"). Located at the historic meeting ground between Europe, Asia, and Africa, the region includes thousands of miles of parched deserts, rugged plateaus, and oasis-like river valleys. It extends 4,000 miles (6,400 kilometers) between Morocco's Atlantic coastline and Iran's boundary with Pakistan. More than two dozen nations are included within its borders, with the largest populations found in Egypt, Turkey, and Iran (Figure 7.1). Climates are generally arid, although the region's diverse physical geography causes precipitation to vary greatly.

Diverse languages, religions, and ethnic identities have molded land and life within the region for centuries, strongly wedding people and place in ways that have had profound social and political implications (Figure 7.2). One traditional zone of conflict is the Middle East, where Jewish, Christian, and Islamic peoples have yet to resolve long-standing cultural tensions and political differences, particularly as they relate to the state of Israel and the pivotal Palestinian issue. Iraq also has been a setting for instability, as an American-led invasion toppled the regime of Saddam Hussein and sharpened religious and ethnic differences in the country. In addition, nearby Iran's political rhetoric and nuclear development program have increased tensions.

No world region better exemplifies the theme of globalization than Southwest Asia and North Africa. A key global **culture hearth**, the region produced many new cultural ideas that subsequently diffused widely to other portions of the world. As an early center for agriculture, several great civilizations, and three major world religions, the region has been a key human crossroads for thousands of years. Important long-distance trade routes have connected North Africa with the Mediterranean and Sub-Saharan Africa. Southwest Asia has also had historic ties to Europe, the Indian subcontinent, and Central Asia. As a result, new ideas within the region have often spread well beyond its bounds.

Particularly within the past century, processes of globalization and the region's strategic importance have made it increasingly open to outside influences (Figure 7.2). The 20th-century development of the petroleum industry, largely initiated by U.S. and European investment, had enormous consequences for economic development in the oil-rich countries of the region. Global demand for oil and natural gas has driven rapid industrial change within the region, defining its central role in world trade (Figure 7.3). Many key members of **OPEC (Organization of the Petroleum Exporting Countries)** are found within the region, and these countries greatly influence global prices and production levels for petroleum. In turn, these regional energy producers are highly vulnerable to global economic downturns, such as the recent recession of 2008–2009.

**Islamic fundamentalism** in the region has advocated a return to more traditional practices within the Islamic religion.

Fundamentalists in any religion advocate a conservative adherence to enduring beliefs within their creed, and they strongly resist change. A related political movement within Islam, known as **Islamism**, challenges the encroachment of global popular culture and blames colonial, imperial, and Western elements for many of the region's political, economic, and social problems. Islamists resent the role they claim the West has played in creating poverty in their world, and many Islamists advocate merging civil and religious authority and rejecting many characteristics of modern, Western-style consumer culture. In particular, the

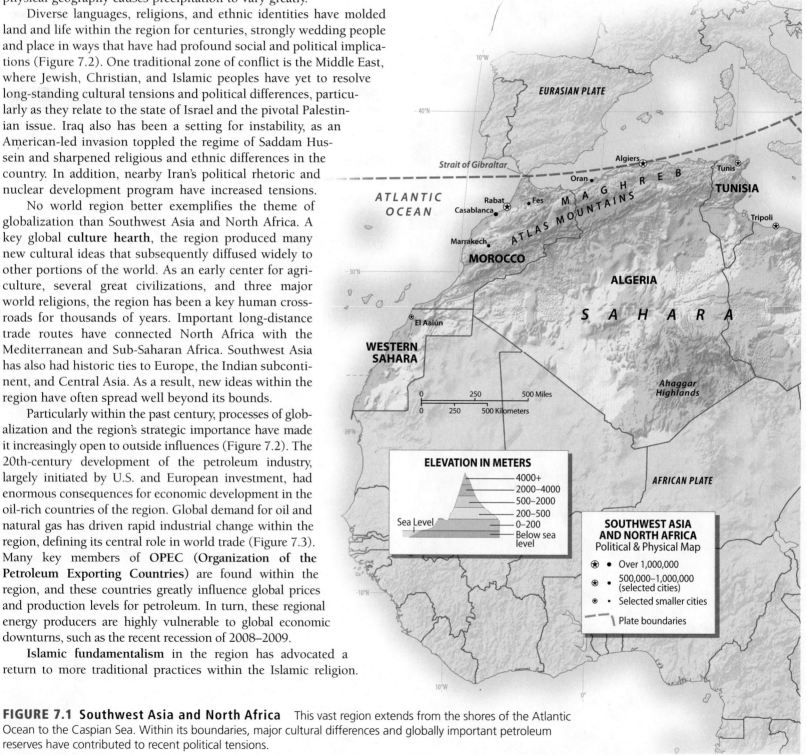

**FIGURE 7.1 Southwest Asia and North Africa**    This vast region extends from the shores of the Atlantic Ocean to the Caspian Sea. Within its boundaries, major cultural differences and globally important petroleum reserves have contributed to recent political tensions.

disruptive role played by Al Qaeda, an extremist Islamist group, particularly since the September 11, 2001, attacks against the United States, has been linked to the region in multiple ways.

The region's environment provides additional challenges. The availability of water in this largely dry portion of the world has shaped both its physical and human geographies. Biologically, plants and animals must adapt to the aridity of long dry seasons and short, often unpredictable rainy periods. Similarly, the geography of human settlement is linked to water. Whether it comes from precipitation, underground aquifers, or rivers, water has shaped patterns of human settlement and has placed severe limits on agricultural development across huge portions of the region. In the future, the region's growing population of more than 400 million people will further stress these available resources, and water shortages may increase economic and political instability.

# Environmental Geography: Life in a Fragile World

In the popular imagination, much of Southwest Asia and North Africa is a land of shifting sand dunes, searing heat, and scattered oases. Although examples of those stereotypes certainly exist, the actual physical setting is much more complex. One theme is dominant, however: A lengthy legacy of human settlement has left its mark on a fragile environment, and the entire region will face increasingly difficult ecological problems in the decades ahead.

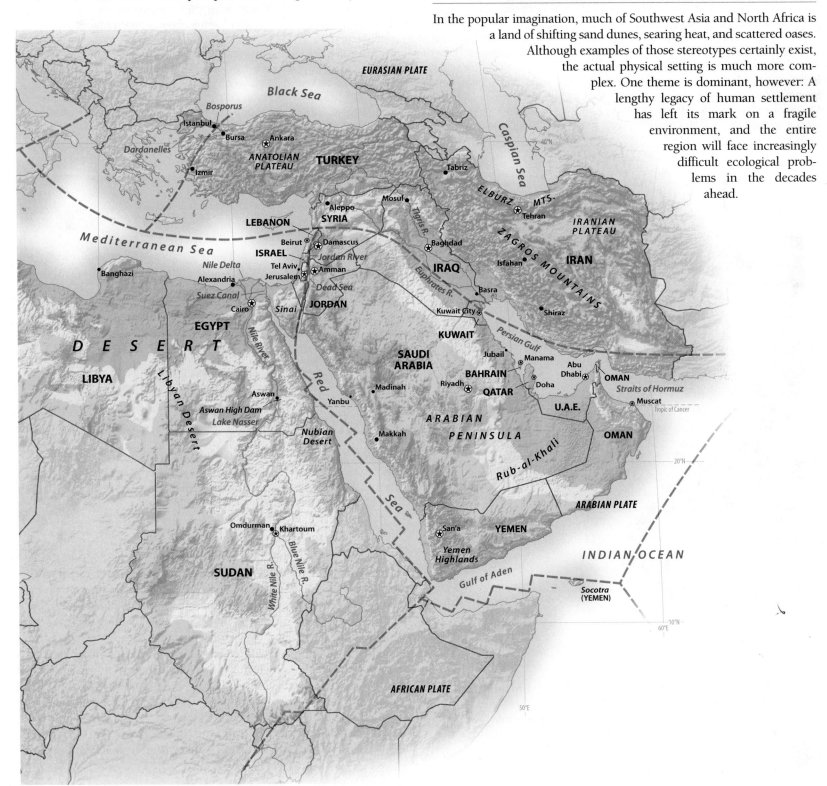

## Setting the Boundaries

"Southwest Asia and North Africa" is both an awkward term and a complex region. Often the same area is simply called the "Middle East," but some experts exclude the western parts of North Africa as well as Turkey and Iran from such a region. In addition, the "Middle East" suggests a European point of view—Lebanon is in the "middle of the east" only from the perspective of the western Europeans who colonized the region and still shape the names we give the world today. Instead, "Southwest Asia and North Africa" offers a straightforward way to describe the general limits of the region. Largely arid climates, an Islamic religious tradition, abundant fossil fuel reserves, and persistent political instability are present in many areas. Still, local variations in physical and human patterns make it difficult to generalize about the region.

There are also problems with simply defining the geographical limits of the region. In the northwest corner of Turkey, a small piece of the country actually sits west of the Bosporus Strait, generally considered to be the dividing line between Europe and Asia (see Figure 7.1). To the northeast, the largely Islamic peoples of Central Asia share many religious and linguistic ties with Turkey and Iran, but those groups are treated in a separate chapter (Chapter 10).

African borders are also problematic. The conventional regional division of "North Africa" from "Sub-Saharan Africa" cuts through the middle of modern Mauritania, Mali, Niger, Chad, and Sudan. We discuss all these transitional countries in Chapter 6, on Sub-Saharan Africa, but also include some discussion of Sudan in this chapter because of its status as an "Islamic republic," a political and cultural title that ties it strongly to the Muslim world.

**FIGURE 7.2 Man at Livestock Market, Al Ayn, United Arab Emirates** Longtime residents of Southwest Asia have witnessed incredible change in their lifetimes. In many localities, traditional lifeways have been radically transformed by urbanization, foreign investment, and external cultural influences.

**FIGURE 7.3 Saudi Arabian Oil Refinery** Eastern Saudi Arabia's Ras Tanura oil refinery links the oil-rich country to the world beyond. Huge foreign and domestic investments since 1960 have dramatically transformed many other settings in the region.

## Legacies of a Vulnerable Landscape

The environmental history of Southwest Asia and North Africa reflects both the short-sighted and resourceful practices of its human occupants. Littered with examples of environmental problems, the region reveals the hazards of lengthy human settlement in a marginal land (Figure 7.4). The island of Socotra illustrates the region's fragile and vulnerable environment and suggests how processes of globalization threaten the area's ecological health. Socotra's stony slopes rise out of the shimmering waters of the Indian Ocean southeast of Yemen. The island's unique natural and cultural history has recently caught the world's attention. Separated for millions of years from the Arabian Peninsula, Socotra's environment evolved in isolation. Often termed the "Galapagos of the Indian Ocean," the island is home to hundreds

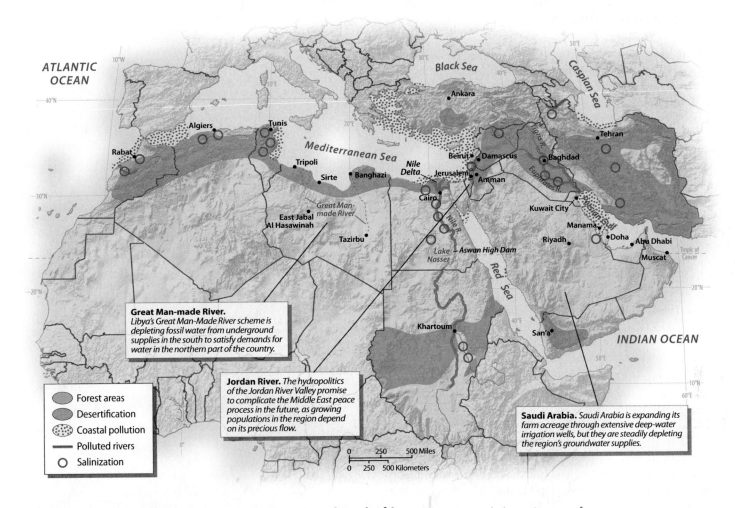

**FIGURE 7.4 Environmental Issues in Southwest Asia and North Africa** Growing populations, pressures for economic development, and widespread aridity combine to create environmental hazards across the region. A long history of human activities has contributed to deforestation, irrigation-induced salinization, and expanding desertification. Saudi Arabia's deep-water wells, Egypt's Aswan High Dam, and Libya's Great Manmade River are all recent technological attempts to expand settlement, but they may carry a high long-term environmental price tag.

Map labels and callouts:

**Great Man-made River.** *Libya's Great Man-Made River scheme is depleting fossil water from underground supplies in the south to satisfy demands for water in the northern part of the country.*

**Jordan River.** *The hydropolitics of the Jordan River Valley promise to complicate the Middle East peace process in the future, as growing populations in the region depend on its precious flow.*

**Saudi Arabia.** *Saudi Arabia is expanding its farm acreage through extensive deep-water irrigation wells, but they are steadily depleting the region's groundwater supplies.*

Legend:
- Forest areas
- Desertification
- Coastal pollution
- Polluted rivers
- Salinization

of plants that are found nowhere else on Earth. Exotic dragon's blood trees dot many of the island's dry and rocky hillsides, and dozens of other species have only recently been catalogued by botanists (Figure 7.5).

Socotra's unique environmental heritage now hangs in the balance. Participants in the 1992 environmental summit in Rio de Janeiro established a $5 million international biodiversity project on Socotra and made the island part of a global-scale research and education project known as the Darwin Initiative. In addition, Edinburgh's Royal Botanic Garden has conducted extensive studies on the island. Some of the island's unique plants have already been commercially developed as cosmetics and herbal remedies. At the same time, the Yemeni government has invited international petroleum companies to explore the island's offshore oil and gas potential, and it has considered a grand scheme for developing a series of luxury tourist hotels that would forever change the island's character. In 2003, a local group of environmentally conscious residents established the Socotra Eco-Tourism Society, dedicated to developing sustainable tourism on the island, but a 2007 environmental assessment noted that a growing network of roads on the island is fragmenting and threatening the area's environment.

**FIGURE 7.5 Socotra's Dragon's Blood Tree** Unique to Socotra, the rare dragon's blood tree reflects the island's environmental isolation. Evolving as a part of the island's ecosystem, the tree survives in the region's dry tropical climate.

**Deforestation and Overgrazing** Deforestation is an ancient problem in Southwest Asia and North Africa. Although much of the region is too dry for trees, the more humid and elevated lands that border the Mediterranean once supported heavy forests. Included in these woodlands are the cedars of Lebanon, cut down in ancient times and now reduced to a few scattered groves that survive in a largely unforested landscape.

Human activities have generally combined with natural conditions to reduce most of the region's forests to grass and scrub. Mediterranean forests often grow slowly, are highly vulnerable to fire, and usually fare poorly if subjected to heavy grazing. Browsing by sheep and goats in particular has often been blamed for much of the region's forest loss. Deforestation has resulted in a long, slow deterioration of the region's water supplies and in accelerated soil erosion. Several governments have launched reforestation drives and forest preservation efforts. Israel and Syria have expanded their forested lands since the 1980s, and more than 5 percent of Lebanon's total area is now part of the Chouf Cedar Reserve, formed in 1996 to protect old-growth cedars.

**Salinization** Salinization, or the buildup of toxic salts in the soil, is another ancient environmental issue in an arid region where irrigation has been practiced for centuries (see Figure 7.4). Hundreds of thousands of acres of once-fertile farmland within the region have been destroyed or degraded by salinization. The problem has been particularly severe in Iraq, where centuries of canal irrigation along the Tigris and Euphrates rivers have seriously degraded land quality. Similar conditions affect central Iran, Egypt, and other irrigated portions of North Africa.

**Managing Water** The people of the region have been modifying drainage systems and water flows for thousands of years. The Iranian **qanat system** of tapping into groundwater through a series of gently sloping tunnels was widely replicated on the Arabian Peninsula and in North Africa. With simple technology, farmers directed the underground flow to fields and villages, where it could be efficiently utilized. In the past half century, however, the scope of environmental change has been greatly magnified. One remarkable example is Egypt's Aswan High Dam, completed in 1970 on the Nile River south of Cairo (see Figure 7.4). Increased storage capacity in the upstream reservoir made more land and water available for agriculture. The dam also generates clean electricity for the region. But the new irrigation has increased salinization because water is not rapidly flushed from the fields. The new system has also meant increased application of costly fertilizers, the infilling of Lake Nasser behind the dam with accumulating sediments, and the collapse of the Mediterranean fishing industry near the Nile Delta, an area previously nourished by the river's silt.

Israel's "Peace Corridor" project is another massive engineering venture within the region. Designed to bring Red Sea water north into the Dead Sea, the effort received increased government backing in 2007. The ambitious plan calls for a 120-mile (200-kilometer) conduit to be built from near Aqaba to the southern end of the Dead Sea (Figure 7.6). The water could be desalinated, generate hydroelectricity, and help save the Dead Sea and its major tourist resorts from literally drying up.

Elsewhere, **fossil water**, or water supplies stored underground during earlier and wetter climatic periods, has also been put to use by

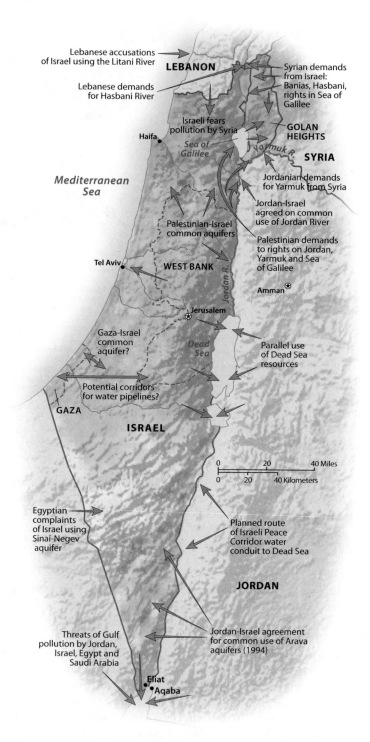

**FIGURE 7.6 Hydropolitics in the Jordan River Basin** Many water-related issues complicate the geopolitical setting in the Middle East. The Jordan River system has been a particular focus of conflict.

modern technology. Libya's "Great Manmade River" scheme taps underground water in the southern part of the country, transports it 600 miles (965 kilometers) to northern coastal zones, and uses the precious resource to expand agricultural production in one of North Africa's driest countries (see Figure 7.4). Similarly, Saudi Arabia has invested huge sums to develop deep-water wells, allowing it to greatly expand its food output (Figure 7.7). Unfortunately, these

**FIGURE 7.7 Saudi Arabian Irrigation** These irrigated fields in the Saudi Desert draw from wells more than 4,000 feet deep. While significantly expanding the country's food production, such efforts are rapidly depleting underground supplies of fossil water.

underground supplies are being depleted much more rapidly than they are recharged, thus limiting the long-term sustainability of such water projects.

Most dramatically, **hydropolitics**, or the interplay of water resource issues and politics, has raised tensions between countries that share drainage basins. For example, Turkey's growing development of the upper Tigris and Euphrates rivers (the Southeast Anatolian Project) has raised protests from Iraq and Syria. Hydropolitics has also played into negotiations between Israel, the Palestinians, and other neighboring states, particularly in the valuable Jordan River drainage, which runs through the center of the area's most hotly disputed lands (Figure 7.6). Israelis fear Palestinian and Syrian pollution, and nearby Jordanians argue for more water from Syria.

## Regional Landforms

A quick tour of Southwest Asia and North Africa reveals a surprising diversity of environmental settings and landforms (Figure 7.1). In North Africa, the **Maghreb** region (meaning "western island") includes the nations of Morocco, Algeria, and Tunisia and is dominated near the Mediterranean coastline by the Atlas Mountains. The rugged flanks of the Atlas rise like a series of islands above the narrow coastal plains to the north and the vast stretches of the lower Saharan deserts to the south (Figure 7.8). South and east of the Atlas Mountains, interior North Africa varies between rocky plateaus and extensive

**FIGURE 7.8 Atlas Mountains** The snow-capped peaks of the High Atlas Mountains near Morocco's Tizi-n-Test Pass contrast dramatically with nearby agricultural lands.

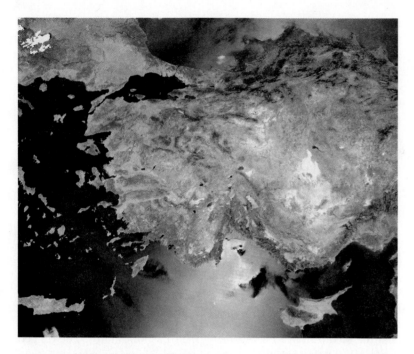

**FIGURE 7.9 Satellite View of Turkey** This satellite image of Turkey suggests the varied, quake-prone terrain encountered across the Anatolian Plateau. The Black Sea coastline is visible near the top of the image, and the island-studded Aegean Sea borders Turkey on the west.

lowlands. In northeast Africa, the Nile River dominates the scene as it flows north through Sudan and Egypt.

Southwest Asia is more mountainous than North Africa. In the **Levant**, or eastern Mediterranean region, mountains rise within 20 miles (32 kilometers) of the sea, and the highlands of Lebanon reach heights of more than 10,000 feet (3,048 meters). Farther south, the Arabian Peninsula forms a massive tilted plateau, with western highlands higher than 5,000 feet (1,524 meters) gradually sloping eastward to extensive lowlands in the Persian Gulf area. North and east of the Arabian Peninsula lie the two great upland areas of Southwest Asia: the Iranian and Anatolian plateaus (*Anatolia* refers to the large peninsula of Turkey, sometimes called Asia Minor) (Figures 7.1 and 7.9). Both of these plateaus, averaging between 3,000 and 5,000 feet (915 to 1,524 meters) in elevation, are geologically active and prone to earthquakes. One dramatic quake in western Turkey (1999) measured 7.8 on the Richter scale, killed more than 17,000 people, and left 350,000 residents homeless. Another quake near the Iranian city of Bam (2003) claimed more than 30,000 lives.

Smaller lowlands characterize other portions of Southwest Asia. Narrow coastal strips are common in the Levant, along both the southern (Mediterranean) and northern (Black Sea) Turkish coastlines, as well as north of the Iranian Elburz Mountains near the Caspian Sea. Iraq contains the most extensive alluvial lowlands in Southwest Asia, dominated by the Tigris and Euphrates rivers, which flow southeast to empty into the Persian Gulf. Although much smaller, the distinctive Jordan River Valley is also a notable lowland that straddles the strategic borderlands of Israel, Jordan, and Syria and drains southward to the Dead Sea (Figure 7.10).

## Patterns of Climate

Although Southwest Asia and North Africa is often termed the "dry world," a closer look at this region reveals a more complex climatic pattern (Figure 7.11). Both latitude and altitude come into play. Aridity dominates large portions of the region. A nearly continuous belt of desert lands stretches eastward from the Atlantic coast of southern Morocco across the continent of Africa, through the Arabian Peninsula, and into central and eastern Iran (Figure 7.12). Throughout this vast dry zone, plant and animal life have adapted to extreme conditions. Deep or extensive root systems allow desert plants to benefit from the limited moisture they receive. Similarly, animals adjust by efficiently storing water, hunting at night, or migrating seasonally to avoid the worst of the dry cycle.

Elsewhere, altitude and latitude dramatically alter the desert environment and produce a surprising amount of climatic variety. The Atlas Mountains and nearby lowlands of northern Morocco, Algeria, and Tunisia experience a Mediterranean climate, in which dry summers alternate with cooler, wet winters. In these areas, the landscape resembles that found in nearby southern Spain or Italy (Figure 7.13). A second zone of Mediterranean climate extends along the Levant coastline, into the nearby mountains, and northward across sizable portions of northern Syria, Turkey, and northwestern Iran.

## The Uncertainties of Regional Climate Change

Projected changes in global climate will aggravate existing environmental issues within North Africa and Southwest Asia. Warmer temperatures are predicted to have a greater impact on the region than changes in precipitation. Higher evaporation rates and lower soil moisture will stress crops, grasslands, and other vegetation. The region's forests (such as the cedars of Lebanon) may be particularly vulnerable to even modest changes. In irrigated zones, higher temperatures are likely to reduce yields for crops such as wheat and maize. Warmer temperatures also will likely reduce net runoff into the region's already stressed streams and rivers, possibly reducing hydroelectric potential

**FIGURE 7.10 Jordan Valley** This view of the Jordan Valley shows a fertile mix of irrigated vineyards and date palm plantations.

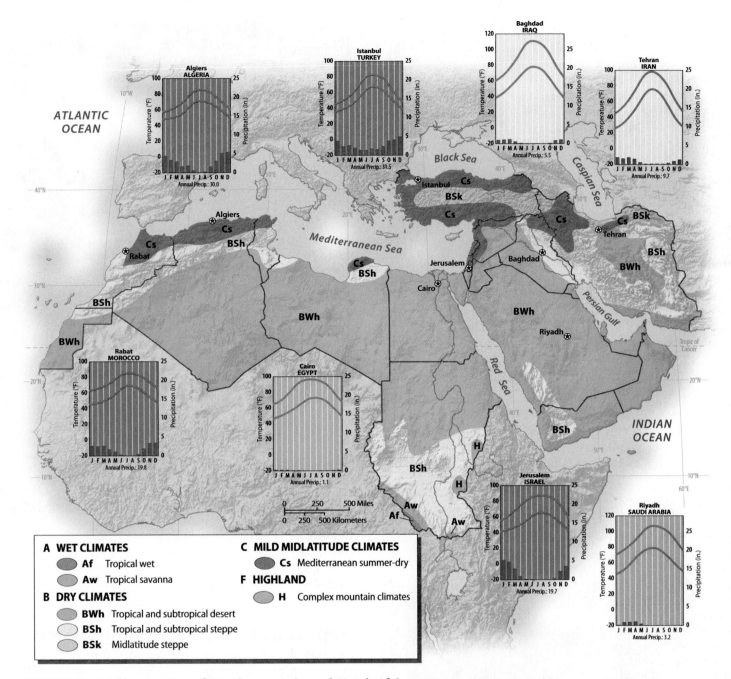

**FIGURE 7.11 Climate Map of Southwest Asia and North Africa** Dry climates dominate from western Morocco to eastern Iran. Within these zones, subtropical high-pressure systems offer only limited opportunities for precipitation. Elsewhere, mild midlatitude climates with wet winters are found near the Mediterranean basin and Black Sea. To the south, tropical savanna climates provide summer moisture to southern Sudan.

and water that is available for the region's increasingly urban population. In addition, there will be a higher likelihood of extreme weather events, such as record-setting summertime temperatures. These extreme events will lead to more heat-related deaths, as residents struggle to adapt, particularly in urban settings where many people cannot afford air conditioning.

Sea-level changes will pose special threats to the Nile Delta (Figure 7.14). This portion of northern Egypt is a vast, low-lying landscape of settlements, farms, and marshland. Studies that simulate

rising sea levels reveal that much of the region will be lost, either to inundation, erosion, or salinization. Farmland losses of more than 250,000 acres (100,000 hectares) are quite possible with even modest sea-level changes. In the Egyptian city of Alexandria, $30 billion in losses have been projected because sea-level changes will devastate the city's huge resort industry as well as nearby residential and commercial areas.

In addition, given the political instability of the Middle East, even relatively small changes in water supplies, particularly where

**FIGURE 7.12 Arid Iran**   Large portions of Iran are arid. Settlements are strongly oriented around zones of higher moisture or opportunities for irrigated agriculture. This view of a remote Iranian desert resembles the dry basin-and-range landscape of the arid American West.

they might involve several nations, could significantly add to the potential for political conflict within the region. Wealthier nations such as Israel and Saudi Arabia may also have more available resources to plan, adjust, and adapt to climate shifts and extreme events than poorer, less-developed countries such as Yemen, Syria, or Sudan.

**FIGURE 7.13 Algerian Orange Harvest**   The Mediterranean moisture in northern Algeria produces an agricultural landscape similar to that of southern Spain or Italy. Winter rains create a scene that contrasts sharply with deserts found elsewhere in the region.

**FIGURE 7.14 Nile Valley**   This satellite image of the Nile Valley dramatically reveals the impact of water on the North African desert. Cairo lies at the southern end of the delta, where it begins to widen toward the Mediterranean Sea.

# Population and Settlement: Patterns in an Arid Land

The human geography of Southwest Asia and North Africa demonstrates the intimate tie between water and life in this part of the world. The pattern is complex: Large areas of the population map remain almost devoid of permanent settlement, while lands with available moisture suffer increasingly from problems of crowding and overpopulation (Figure 7.15).

## The Geography of Population

Today, more than 400 million people live in Southwest Asia and North Africa (Table 7.1). The distribution of that population is strikingly varied (Figure 7.15). In North Africa, the moist slopes of the Atlas Mountains and nearby better-watered coastal districts have supported dense populations for centuries, a stark contrast to the almost empty lands south and east of the mountains. Egypt's zones of almost empty desert land stand in sharp contrast to crowded, irrigated locations, such as those along the Nile River. In Southwest Asia, many residents live in well-watered coastal zones, moister highland settings, and

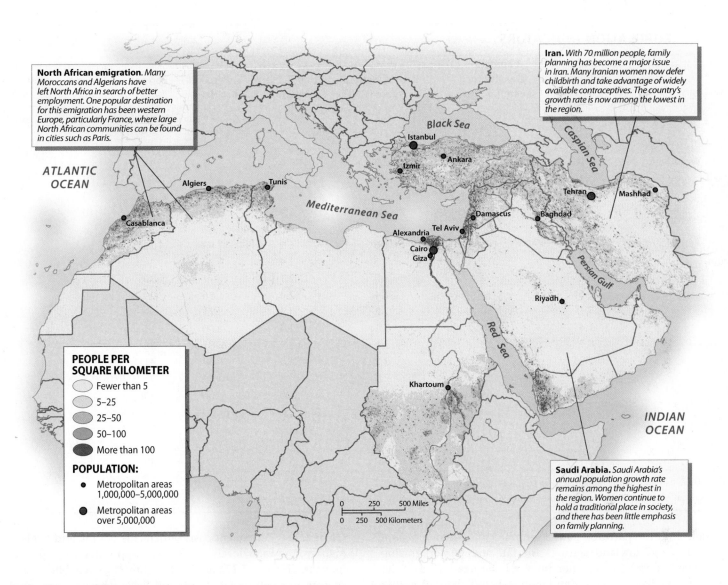

**FIGURE 7.15 Population Map of Southwest Asia and North Africa**   The striking contrasts between large, sparsely occupied desert zones and much more densely settled regions where water is available are clearly evident. The Nile Valley and the Maghreb region contain most of North Africa's people, while Southwest Asian populations cluster in the highlands and along the better-watered shores of the Mediterranean.

desert localities where water is available from nearby rivers or subsurface aquifers. High population densities are found in better-watered portions of the eastern Mediterranean (Israel, Lebanon, and Syria), Turkey, and Iran. While overall population densities in such countries appear modest, the **physiological density**, which is the number of people per unit area of arable land, is quite high by global standards. Although less than two-thirds of the overall population of the region is urban, many nations are dominated by huge cities (for example, Cairo in Egypt and Tehran in Iran) that produce the same problems of urban crowding found elsewhere in the developing world (Figure 7.16).

## Water and Life: Rural Settlement Patterns

Water and life are closely linked across the rural settlement landscapes of Southwest Asia and North Africa (Figure 7.17). Indeed, Southwest Asia is home to one of the world's earliest hearths of **domestication**, where plants and animals were purposefully selected and bred for their desirable characteristics. Beginning around 10,000 years ago, increased experimentation with wild varieties of wheat and barley led to agricultural settlements that later included domesticated animals, such as cattle, sheep, and goats. Much of the early agricultural activity focused on the **Fertile Crescent**, an ecologically diverse zone that stretches from the Levant inland through the fertile hill country of northern Syria into Iraq. Between 5,000 and 6,000 years ago, better knowledge of irrigation techniques and increasingly powerful political states encouraged the spread of agriculture into nearby lowlands such as the Tigris and Euphrates valleys (Mesopotamia) and North Africa's Nile Valley.

**Pastoral Nomadism**   In the drier portions of the region, **pastoral nomadism** is a traditional form of subsistence agriculture in which people depend on the seasonal movement of livestock for a large part of their livelihood. The settlement landscape of pastoral nomads reflects their need for mobility and flexibility as they seasonally move camels, sheep, and goats from place to place. Near highland zones such as the Atlas Mountains or the Anatolian Plateau, nomads practice **transhumance** by seasonally moving their livestock to cooler,

### TABLE 7.1 POPULATION INDICATORS

| Country | Population (millions) 2009 | Population Density (per square kilometer) | Rate of Natural Increase (RNI) | Total Fertility Rate | Percent Urban | Percent <15 | Percent >65 | Net Migration (Rate per 1,000) 2005–10* |
|---|---|---|---|---|---|---|---|---|
| Algeria | 35.4 | 15 | 1.9 | 2.3 | 63 | 28 | 5 | −0.8 |
| Bahrain | 1.2 | 1,754 | 1.3 | 2.0 | 100 | 21 | 3 | 5.2 |
| Egypt | 78.6 | 79 | 1.9 | 3.0 | 43 | 33 | 5 | −0.8 |
| Gaza and West Bank | 3.9 | 653 | 2.8 | 4.6 | 72 | 44 | 3 | −0.5 |
| Iran | 73.2 | 44 | 1.5 | 2.0 | 67 | 28 | 5 | −1.4 |
| Iraq | 30.0 | 69 | 2.3 | 4.4 | 67 | 41 | 3 | −3.9 |
| Israel | 7.6 | 345 | 1.6 | 2.9 | 92 | 28 | 10 | 2.4 |
| Jordan | 5.9 | 66 | 2.4 | 3.6 | 83 | 37 | 3 | 8.3 |
| Kuwait | 3.0 | 168 | 1.6 | 2.2 | 98 | 23 | 2 | 8.3 |
| Lebanon | 3.9 | 373 | 1.5 | 2.4 | 87 | 26 | 7 | −0.6 |
| Libya | 6.3 | 4 | 2.0 | 2.7 | 77 | 30 | 4 | 0.6 |
| Morocco | 31.5 | 71 | 1.4 | 2.4 | 56 | 29 | 6 | −2.7 |
| Oman | 3.1 | 10 | 2.2 | 3.4 | 71 | 29 | 2 | 1.7 |
| Qatar | 1.4 | 128 | 1.1 | 2.4 | 100 | 17 | 1 | 93.9 |
| Saudi Arabia | 28.7 | 13 | 2.6 | 3.9 | 81 | 38 | 2 | 1.2 |
| Sudan | 42.3 | 17 | 2.2 | 4.5 | 38 | 41 | 3 | 0.7 |
| Syria | 21.9 | 118 | 2.5 | 3.3 | 54 | 36 | 3 | 7.7 |
| Tunisia | 10.4 | 64 | 1.2 | 2.0 | 66 | 25 | 7 | −0.4 |
| Turkey | 74.8 | 95 | 1.2 | 2.1 | 63 | 27 | 6 | −0.1 |
| United Arab Emirates | 5.1 | 61 | 1.4 | 2.0 | 83 | 19 | 1 | 15.6 |
| Western Sahara | 0.5 | 2 | 1.8 | 3.0 | 81 | 31 | 2 | 19.6 |
| Yemen | 22.9 | 43 | 3.0 | 5.5 | 29 | 45 | 3 | −1.2 |

Source: Population Reference Bureau, *World Population Data Sheet, 2009.*

*Net Migration Rate from the United Nations, Population Division, *World Population Prospects: The 2008 Revision Population Database.*

greener high country pastures in the summer and then returning them to valley and lowland settings for fall and winter grazing. Elsewhere, seasonal movements often involve huge areas of desert that support small groups of a few dozen families. Fewer than 10 million pastoral nomads remain in the region today.

**FIGURE 7.16 Modern Tehran, Iran** Three Iranian teenagers ponder the latest computers at the Paytakht mall, a popular shopping spot in north Tehran.

**Oasis Life** Permanent oasis settlements dot the arid landscape where high groundwater levels or modern deep-water wells provide reliable moisture (Figures 7.17 and 7.18). Tightly clustered, often walled villages sit next to small, intensely utilized fields where underground water is carefully applied to tree and cereal crops. In more recently created oasis settlements, concrete blocks and prefabricated housing add a modern look. Traditional oasis settlements are composed of close-knit families who work their own irrigated plots or, more commonly, work for absentee landowners. Although oasis settlements are usually small, eastern Saudi Arabia's Hofuf oasis covers more than 30,000 acres (12,150 hectares). While some crops are raised for local consumption, expanding world demand for products such as figs and dates increasingly include even these remote locations in the global economy, as products end up on the tables of hungry Europeans or North Americans.

**Exotic Rivers** For centuries, the densest rural settlement of Southwest Asia and North Africa has been tied to its great river valleys and their seasonal floods of water and fertile nutrients. In such settings, **exotic rivers** transport precious water and nutrients from more humid lands into drier regions where the resources are utilized for irrigated farming. For example, some of the most efficient farms in the region are associated with Israeli **kibbutzes**, collectively worked settlements that produce grain, vegetable, and orchard crops irrigated by waters from the Jordan River and from the country's elaborate canal system. The Nile and the Tigris and Euphrates rivers offer the largest regional examples of such activity, but historically irrigation in these valleys has resulted in salinization.

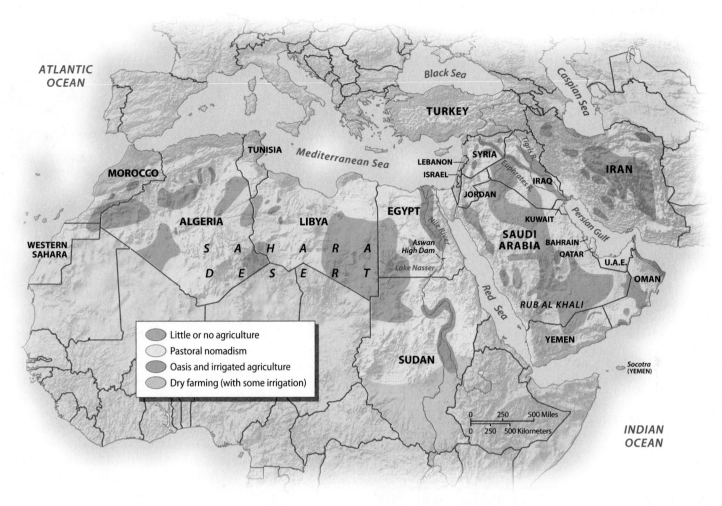

**FIGURE 7.17 Agricultural Regions of Southwest Asia and North Africa** Important agricultural zones include oases and irrigated farms where water is available. Elsewhere, dry farming supplemented with irrigation is practiced in midlatitude settings.

**FIGURE 7.18 Oasis Agriculture** The green fields and trees of Morocco's Tinghir oasis contrast dramatically with the surrounding desert landscape.

**The Challenge of Dryland Agriculture** Mediterranean climates in portions of the region allow varied forms of dryland agriculture that depend largely on seasonal moisture to support farming. These zones include the better-watered valleys and coastal lowlands of the northern Maghreb, lands along the shore of the eastern Mediterranean, and favored uplands across the Anatolian and Iranian plateaus. A mix of tree crops, grains, and livestock are raised in these settings. One farm product of growing regional and global importance is Morocco's thriving hashish crop. More than 200,000 acres (80,000 hectares) of cannabis are cultivated in the hill country near Ketama in northern Morocco, generating more than $2 billion annually in illegal exports (mostly to Europe).

## Many-Layered Landscapes: The Urban Imprint

Cities have played a key role in the human geography of Southwest Asia and North Africa. Indeed, some of the world's oldest urban places are located in the region. Today, continuing political, religious, and economic ties link the cities with the surrounding countryside.

**197**

**A Long Urban Legacy**    Cities in the region have traditionally played important roles as centers of political and religious authority, as well as focal points of local and long-distance trade. Urbanization in Mesopotamia (modern Iraq) began by 3500 BCE, and cities such as Eridu and Ur reached populations of 25,000 to 35,000 residents. Similar centers appeared in Egypt by 3000 BCE, with Memphis and Thebes assuming major importance in the middle Nile Valley. By 2000 BCE, however, a different kind of city emerged along the eastern Mediterranean and along important overland trade routes. Centers such as Beirut, Tyre, and Sidon, all in modern Lebanon, as well as Damascus in nearby Syria, exemplified the growing role of trade in shaping the urban landscape. Expanding port facilities, warehouse districts, and commercial neighborhoods suggested how trade and commerce shaped these urban settlements, and many of these early Middle Eastern trading towns survive to the present.

Islam also left an enduring mark on cities because urban centers traditionally served as places of Islamic religious power and education. By the 8th century, Baghdad emerged as a religious center, followed shortly by the appearance of Cairo as a seat of religious authority. Urban settlements from North Africa to Turkey felt the influences of Islam. Indeed, the Moors carried Islam's influence to Spain, where it shaped the architecture and culture in centers such as Córdoba and Málaga.

**FIGURE 7.20 Fes, Morocco**    These narrow streets in the old city of Fes are lined with shops and commercial stalls. The winding labyrinth of alleyways and cul-de-sacs forms an almost impenetrable maze to visitors, and it exemplifies the complex urban landscapes of the traditional Islamic city.

**FIGURE 7.19 The Islamic Landscape**    Iranian mullahs discuss the religious questions of the day beneath the minarets of the Moussavi Mosque in Qom. Islam has left a widespread mark on the region's cultural landscapes.

A characteristic Islamic cityscape exists to this day. Its traditional features include a walled urban core, or **medina**, dominated by the central mosque and its associated religious, educational, and administrative functions (Figure 7.19). A nearby bazaar, or *suq*, serves as a marketplace where products from city and countryside are traded (Figure 7.20). Housing districts feature a maze of narrow, twisting streets that maximize shade and emphasize the privacy of residents, particularly women. Houses have small windows, frequently are situated on dead-end streets, and typically open inward to private courtyards.

More recently, European colonialism added another layer of urban landscape influences in selected cities. Particularly in North Africa, colonial cities added many architectural features from Great Britain and France. Victorian building blocks, French mansard roofs, suburban housing districts, and wide, European-style commercial boulevards complicated the settlement landscapes of cities such as Algiers and Cairo. Many of these signatures remain on the modern scene.

**Signatures of Globalization**    Since 1950, cities in Southwest Asia and North Africa have become the gateways to the global economy. As the region has been opened to new investment, industrialization, and tourism, the urban landscape has reflected the fundamental changes taking place. Expanded airports, commercial and financial districts, industrial parks, and luxury tourist facilities all reveal the influence of the global economy. Many cities, such as Algiers and Istanbul, have more than doubled in population in recent years. Crowded Cairo now has more than 15 million residents. The increasing demand for homes has produced ugly, cramped, high-rise apartment houses, while elsewhere extensive squatter settlements provide little in the way of quality housing or public services.

Undoubtedly, the oil-rich states of the Persian Gulf have displayed the greatest changes in the urban landscape. Before the 20th century, urban traditions were relatively weak in the area, and even as late as 1950 only 18 percent of Saudi Arabia's population lived in cities. All that has changed, however, and today, Saudi Arabia is more urban than many industrialized nations, including the United States. Particularly since 1970, other cities, such as Abu Dhabi (United Arab Emirates), Doha (Qatar), and Kuwait City (Kuwait), have also adopted many features of modern Western urban design and futuristic architecture (Figure 7.21). The result is an urban settlement landscape in which traditional and global influences combine, producing cityscapes where domed mosques, mirrored bank buildings, and oil refineries sit beside one another beneath the dusty skies and desert sun.

## A Region on the Move

While pastoral nomads have crisscrossed Southwest Asia and North Africa for ages, new patterns of migration have been created by the global economy and recent political events. The broad rural-to-urban shift seen so widely in many other parts of the less-developed world is reworking population patterns here, too. The Saudi Arabian example is repeated in many other countries within the region. Political

**FIGURE 7.22 Labor Camp in Shariah City, United Arab Emirates** These South Asian workers live a few miles outside wealthy Dubai and enjoy a card game during their time off from work. A large majority of the country's population is foreign-born, and many immigrants work in the construction and service sectors of the economy.

**FIGURE 7.21 Modern Doha, Qatar** The rapidly changing urban landscapes of the oil-rich Persian Gulf region are illustrated in the modern architecture seen in Doha's West Bay district.

disruptions have also shaped recent patterns of migration. For example, almost 10 percent of Iraq's population left that war-torn country between 2005 and 2007. While some have returned, thousands of Iraqis remain in the nearby countries of Syria and Jordan.

In addition, changing job markets within the region are driving patterns of migration. In times of economic growth, labor-poor countries such as Saudi Arabia, Kuwait, and the United Arab Emirates (UAE) have attracted thousands of Jordanians, Palestinians, Yemenis, and Iranians to construction and service jobs within their borders. In addition, thousands of South Asians, Southeast Asians, and East Africans, many of them from Muslim nations, have taken jobs in the construction and services sectors (Figure 7.22). On the other hand, periods of recession and economic contraction produce similar waves of out-migration, as was the case in the economic downturn of 2008 and 2009.

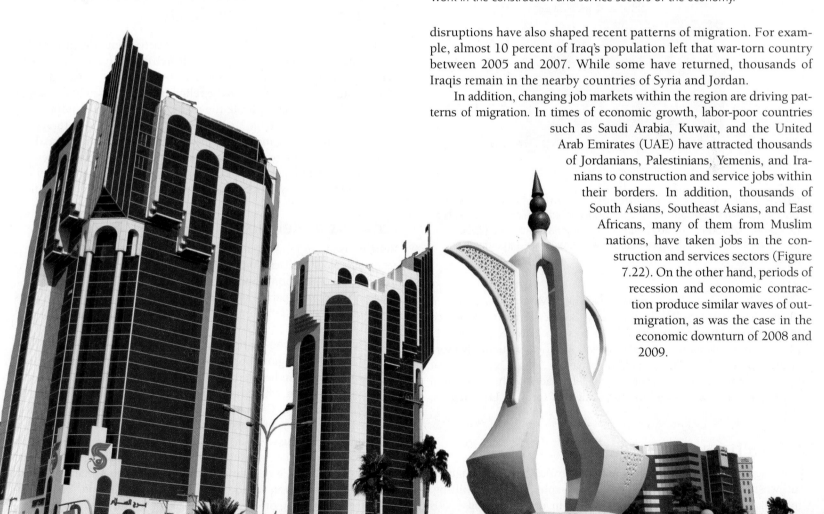

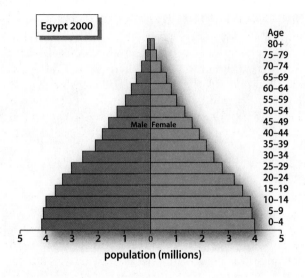

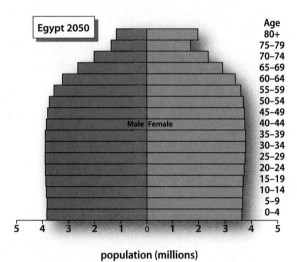

**FIGURE 7.23 Egypt's Changing Population**    While Egypt's annual population growth rates are slowing, its 2000 population pyramid reveals a large number of young people already in the population. Future (2050) implications are clear: Egypt's population will grow larger and older in the years to come.

Some regional residents migrate elsewhere. More than 2.7 million Turkish guest workers live in Germany. Both Algeria and Morocco also have seen large out-migrations to western Europe, particularly France. Thousands more journey to North America, particularly more educated professionals who seek high-paying jobs. Over the past 30 years, many wealthy residents have left places such as Lebanon and Iran, and are today living in cities such as Toronto, Los Angeles, and Paris.

## Shifting Demographic Patterns

While high population growth remains a critical issue throughout Southwest Asia and North Africa, the demographic picture is shifting. Uniformly high rates of population growth in the 1950s and 1960s have been replaced by more varied regional patterns. Many nations, such as Tunisia, Iran, and Turkey, are seeing birthrates fall fairly rapidly. Various factors help explain the changes. More urban, consumer-oriented populations have opted for fewer children. Many Arab

women are now delaying marriage into their middle 20s and even early 30s. Family planning initiatives are expanding in many countries. For example, programs in Tunisia, Egypt, and Iran have greatly increased access to contraceptive pills, IUDs, and condoms.

Still, the region faces demographic challenges. Areas such as the West Bank, Gaza, and Yemen experience rates of natural increase that are much higher than the world average. In some places, continuing patterns of poverty and traditional ways of rural life contribute to the large rates of population increase, and even in more urban and industrialized Saudi Arabia annual growth rates remain around 2 percent. The increases result from the combination of high birthrates and very low death rates. In Egypt, even though birthrates are likely to decline, the labor market will need to absorb more than 500,000 new workers annually over the next 10 to 15 years just to keep up with the country's large youthful population (Figure 7.23). As that population ages in the mid-21st century, it will be increasingly urban and demand jobs, housing, and social services. Growing populations will also increase demands on the region's already limited water resources.

# Cultural Coherence and Diversity: Signatures of Complexity

While Southwest Asia and North Africa clearly define the heart of the Islamic and Arab worlds, a surprising degree of cultural diversity characterizes the region. Muslims practice their religion in varied ways, often disagreeing strongly on basic religious views. Elsewhere, various religious minorities complicate the region's present-day cultural geography. Linguistically, Arabic languages form an important cultural core, but non-Arab peoples, including Persians, Kurds, and Turks, also dominate portions of the region. These varied patterns of cultural geography can help us understand the region's political tensions and help us appreciate why many of its residents resist processes of globalization.

## Patterns of Religion

Religion is an important part of the lives of most people in Southwest Asia and North Africa. Whether it is the quiet ritual of morning prayers or discussions about current political and social issues, religion is a central part of the daily routine of most regional residents from Casablanca to Tehran.

**Hearth of the Judeo-Christian Tradition**    Both Jews and Christians trace their religious roots to an eastern Mediterranean hearth. The roots of Judaism lie deep in the past: Abraham, an early leader in the Jewish tradition who lived some 4,000 years ago, led his people from Mesopotamia to Canaan (modern-day Israel), near the shores of the Mediterranean. Jewish history, recounted in the Old Testament of the Holy Bible, focused on a belief in one God (or **monotheism**), a strong code of ethical conduct, and a powerful ethnic identity that continues to the present. During the Roman Empire, many Jews left the eastern Mediterranean to escape Roman persecution. The resulting forced migration, or diaspora, took Jews to the far corners of Europe and North Africa. Only in the past century have many of the world's far-flung Jewish populations returned to the religion's place of origin, a process that gathered speed after the formation of the Jewish state of Israel in 1948.

Christianity also emerged in the vicinity of modern-day Israel. An outgrowth of Judaism, Christianity was based on the teachings of Jesus and his disciples, who lived and traveled in the eastern Mediterranean about 2,000 years ago. While many Christian traditions became associated with European history, some forms of early Christianity remain near the religion's hearth. For example, the Coptic Church evolved in nearby Africa, shaping the cultural geographies of places such as Egypt. In the Levant, another group of early Christians, the Maronites, also retain a separate cultural identity.

**The Emergence of Islam**  Islam originated in Southwest Asia in 622 CE, forming another cultural hearth of global significance. While Muslims can be found today from North America to the southern Philippines, the Islamic world remains centered on Southwest Asia. Most Southwest Asian and North African peoples still follow its religious teachings. Muhammad, the founder of Islam, was born in Makkah (Mecca) in 570 CE and taught in nearby Medinah (Medina) (Figure 7.24). His beliefs parallel the Judeo-Christian tradition. Muslims believe that both Moses and Jesus were prophets and that the Hebrew Bible (or Old Testament) and the Christian New Testament, while incomplete, are basically accurate. Ultimately, however, Muslims hold that the **Quran** (or Koran), a book of teachings received by Muhammad from Allah (God), represents God's highest religious and moral revelations to humankind.

The basic beliefs of Islam offer a detailed blueprint for leading an ethical and religious life. Islam literally means "submission to the will of God," and the practice of the religion rests on five essential activities: (1) repeating the basic creed ("There is no God but God, and Muhammad is his prophet"); (2) praying facing Makkah five times daily; (3) giving charitable contributions; (4) fasting between sunup and sundown during the month of Ramadan; and (5) making at least one religious pilgrimage, or **Hajj**, to Muhammad's birthplace of Makkah (Figure 7.25). Many Islamic fundamentalists also argue for a **theocratic state**, such as modern-day Iran, in which religious leaders (ayatollahs) shape many aspects of government policy.

A major religious division split Islam early on and endures today. The breakup occurred almost immediately after the death of Muhammad in 632 CE. Questions surrounded who would get religious power. One group, now called **Shiites**, favored passing on power within Muhammad's own family, specifically to Ali, his son-in-law. Most Muslims, later known as **Sunnis**, advocated passing down power through the established clergy. This group was largely victorious. Ali was killed, and his Shiite supporters went underground. Ever since, Sunni Islam has formed the mainstream branch of the religion, to which Shiite Islam has presented a recurring, and sometimes powerful, challenge.

Islam quickly spread from the western Arabian Peninsula, following camel caravan routes and Arab military campaigns as it

**FIGURE 7.24 Diffusion of Islam**  The rapid expansion of Islam that followed its birth is shown here. From Spain to Southeast Asia, Islam's legacy remains strongest nearest its Southwest Asian hearth. In some settings, its influence has faded or has come into conflict with other religions, such as Christianity, Judaism, and Hinduism.

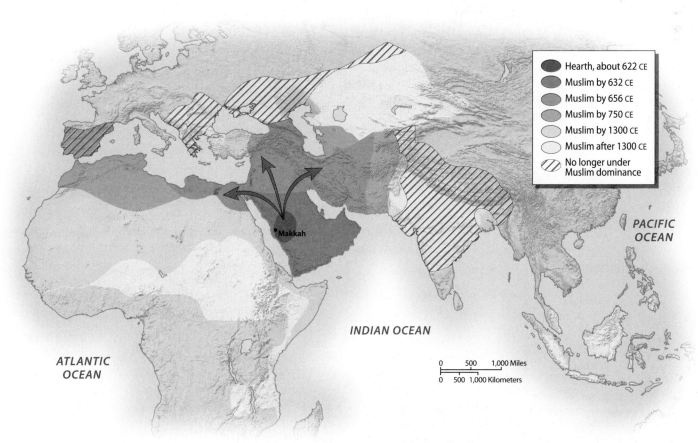

**FIGURE 7.25 Makkah** Thousands of faithful Muslims gather at the Grand Mosque in central Makkah (Mecca), part of the pilgrimage to this sacred place that draws several million visitors annually. A collection of hotels and portions of the city's commercial district can be seen in the distance.

expanded its geographic range and converted thousands to its beliefs (see Figure 7.24). By the time of Muhammad's death in 632 CE the peoples of the Arabian Peninsula were united under its banner. Shortly thereafter, the Persian Empire fell to Muslim forces and the Eastern Roman (or Byzantine) Empire lost most of its territory to Islamic influences. By 750 CE, Arab armies had swept across North Africa, conquered most of Spain and Portugal, and established footholds in Central and South Asia. By the 13th century, most people in the region were Muslims, and older religions such as Christianity and Judaism became minority faiths or disappeared altogether.

Between 1200 and 1500, Islamic influences expanded in some areas and contracted in others. The Iberian Peninsula (Spain and Portugal) returned to Christianity in 1492, although many Moorish (Islamic) cultural and architectural features remain today. At the same

**FIGURE 7.26 Modern Religions** Islam continues to be the dominant religion across the region. Most Muslims are tied to the Sunni branch, while Shiites are found in places such as Iran and southern Iraq. In some locales, however, Christianity and Judaism remain important. African animism is found in southern portions of Sudan.

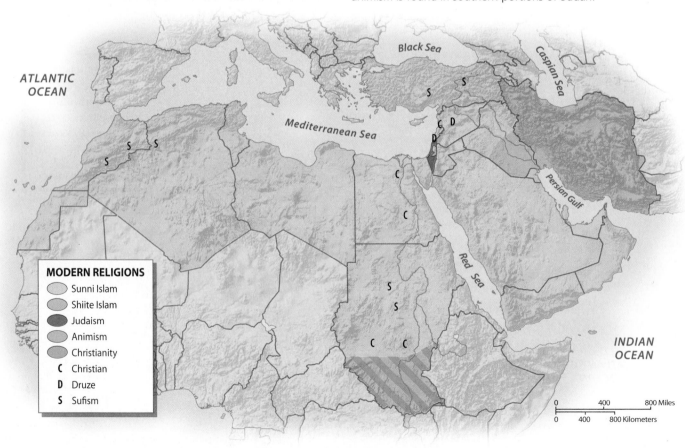

**MODERN RELIGIONS**

- Sunni Islam
- Shiite Islam
- Judaism
- Animism
- Christianity
- **C** Christian
- **D** Druze
- **S** Sufism

**FIGURE 7.27 Old Jerusalem** Jerusalem's historic center reflects its varied religious legacy. Sacred sites for Jews, Christians, and Muslims are all located within the Old City. The Western Wall, a remnant of the ancient Jewish temple, stands at the base of the Dome of the Rock and Islam's al-Aqsa Mosque.

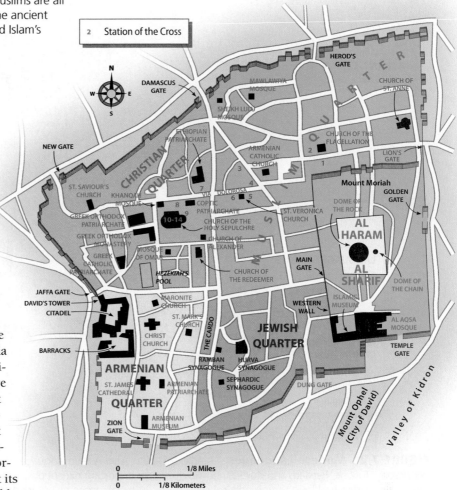

time, Muslims expanded their influence southward and eastward into Africa. In addition, Muslim Turks largely replaced Christian Greek influences in Southwest Asia after 1100. One group of Turks moved into the Anatolian Plateau and finally conquered the Byzantine Empire in 1453. These Turks soon created the huge **Ottoman Empire** (named after one of its leaders, Osman), which included southeastern Europe (including modern-day Albania, Bosnia, and Kosovo) and most of Southwest Asia and North Africa. It provided a focus of Muslim political power within the region until the Empire's disintegration in the late 19th and early 20th centuries.

**Modern Religious Diversity** Today, Muslims form the majority population in all the countries of Southwest Asia and North Africa except Israel, where Judaism is dominant (Figure 7.26). Still, divisions within Islam have created key cultural differences in the region. While most (73 percent) of the region is dominated by Sunni Muslims, the Shiites (23 percent) remain an important element in the contemporary cultural mix. In Iraq, for example, the majority Shiite population in the southern portion of the country (around Najaf, Karbala, and Basra) set its own cultural and political course following the fall of Saddam Hussein. Shiites also claim a majority in nearby Iran. In addition, they form a major religious minority in Lebanon, Bahrain, Algeria, Egypt, and Yemen. Strongly associated with the recent flowering of the Islamist movement, the Shiites also have benefited from rapid growth rates because their brand of Islam is particularly appealing to the poorer, powerless, and more rural populations of the region. While some Sunnis have been attracted to Islamist tendencies as well, many reject its more radical cultural and political precepts and argue for a more modern Islam that incorporates some accommodation with Western values and traditions.

While the Sunni–Shiite split is the great divide within the Muslim world, other variations of Islam can also be found in the region. One division separates the mystically inclined form of Islam—known as *Sufism*—from the more mainstream tradition. Sufism is especially prominent in the peripheries of the Islamic world, including the Atlas Mountains of Morocco and Algeria. The Druze of Lebanon practice yet another variant of Islam.

Southwest Asia is also home to many non-Islamic communities. Israel has a Muslim minority (15 percent) that is dominated by that nation's Jewish population (77 percent). Even within Israel's Jewish community, differences divide Jewish fundamentalists from more reform-minded Jews. In neighboring Lebanon, there was a slight Christian (Maronite and Orthodox) majority as recently as 1950. Christian out-migration and higher Islamic birthrates, however, have created a nation that today is more than 60 percent Muslim.

Jerusalem (now the Israeli capital) holds special religious significance for several groups and also stands at the core of the region's political problems (Figure 7.27). Indeed, the sacred space of this ancient Middle Eastern city remains deeply scarred and divided, as different groups argue for more control of contested neighborhoods and nearby suburbs. Jews particularly honor the city's old Western Wall (the site of a Roman-era temple); Christians honor the Church of the Holy Sepulchre (the purported burial site of Jesus); and Muslims hold sacred religious sites in the city's eastern quarter (including the place from which the prophet Muhammad supposedly ascended to heaven).

## Geographies of Language

Although the region is often referred to as the "Arab World," linguistic complexity creates many important cultural divisions across Southwest Asia and North Africa (Figure 7.28).

**Semites and Berbers** Afro-Asiatic languages dominate much of the region. Within that family, Arabic-speaking Semitic peoples are found from the Persian Gulf to the Atlantic and southward into Sudan. The Arabic language has a special religious significance for all Muslims because it was the sacred language in which God delivered his message to Muhammad. While most of the world's Muslims do not

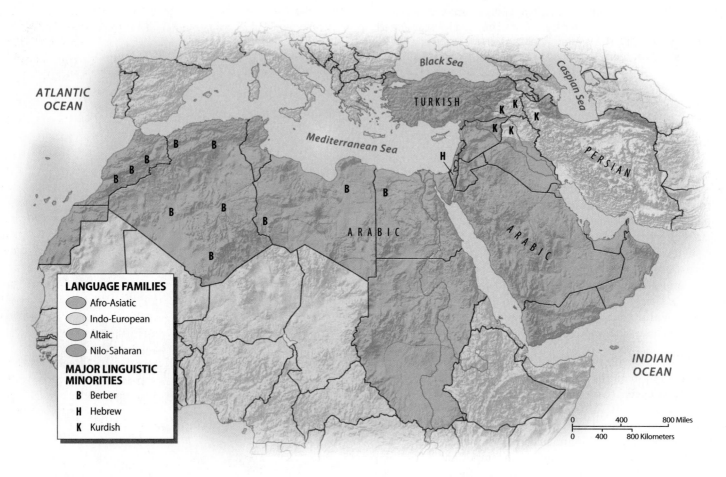

**FIGURE 7.28 Modern Languages** Arabic, a Semitic Afro-Asiatic language, dominates the region's cultural geography. Turkish, Persian, and Kurdish, however, remain important exceptions, and such differences within the region have often had long-lasting political consequences. Israel's more recent reintroduction of Hebrew further complicates the region's linguistic geography.

speak Arabic, the faithful often memorize prayers in the language, and many Arabic words have entered the other important languages of the Islamic world.

Hebrew, another Semitic language, was reintroduced into the region with the creation of Israel. This Semitic language originated in the Levant and was used by the ancient Israelites 3,000 years ago. Today, its modern version survives as the sacred tongue of the Jewish people and is the official language of Israel, although the country's non-Jewish population largely speaks Arabic.

Older Afro-Asiatic languages survive in more remote areas of the Atlas Mountains and in certain parts of the Sahara. Collectively known as Berber, these languages are related to each other but are not mutually intelligible. Most Berber languages have never been written, and none has generated a significant literature. Indeed, a Berber-language version of the Quran was not completed until 1999.

**Persians and Kurds** Although Arabic spread readily through portions of Southwest Asia, much of the Iranian Plateau and nearby mountains are dominated by older Indo-European languages. Here the principal tongue remains Persian, although, since the 10th century, the language has been enriched with Arabic words and written in the Arabic script.

Kurdish speakers of northern Iraq, northwest Iran, and eastern Turkey add further complexity to the regional pattern of languages. Kurdish, also an Indo-European language, is spoken by 10 to 15 million people in the region. The Kurds have a strong sense of shared cultural identity. Indeed, "Kurdistan" has sometimes been called the world's largest nation without its own political state. In the 2003 war in Iraq, the Kurds emerged as a cohesive group in the northern part of the country, and they have gained more political autonomy in a post–Saddam Hussein Iraq state. Nearby Kurds in eastern Turkey, however, still complain that their ethnic identity is frequently challenged by the majority Turks, leaving some to wonder if they will attempt to join forces with their Iraqi neighbors (Figure 7.29).

**The Turkish Imprint** Turkish languages provide variety across much of modern Turkey and in portions of far northern Iran. The Turkish languages are a part of the larger Altaic language family that originated in Central Asia. Turkey remains the largest nation in Southwest Asia dominated by this language family. Tens of millions of people in other countries of Southwest Asia and Central Asia speak related Altaic languages, such as Azeri, Uzbek, and Uyghur.

**FIGURE 7.29 Kurdish Family**   As an ethnic minority, these Kurdish-speaking settlers from eastern Turkey face cultural and political discrimination in their own country, as do their Kurdish neighbors in nearby nations.

## Regional Cultures in Global Context

Southwest Asian and North African peoples increasingly find themselves intertwined with cultural complexities and conflicts external to the region. Islam links the region with a Muslim population that now lives all over the world, from North American cities to western China. Islam is a truly global religion. The religion's tradition of pilgrimage ensures that Makkah will continue as a city of ongoing global significance in the 21st century.

The people of the region are also struggling to retain their traditional cultural values and also enjoy the benefits of global economic growth. Islamic fundamentalism is in many ways a reaction to the threat posed by external cultural influences, particularly those of western Europe and the United States.

Technology also contributes to cultural change. Educated Turks, Tunisians, and Egyptians have been embracing the Internet and its power to access a global wealth of information and entertainment. Mobile phone use has also increased rapidly. Widespread television viewing has transformed the region, with viewers offered everything from Islamist religious programming to American-style reality TV. Even in conservative Iran, where satellite dishes are officially banned, millions of people have access to them, beaming in multicultural programming from around the globe. Hybrid forms of popular culture

also express the shaping power of globalization. Recently, for example, both within the region and beyond, Arab hip-hop music has offered a form of artistic and political expression that represents the fusion of several cultural traditions (see "Exploring Global Connections: Arab Hip-Hop Challenges Regional Traditions").

## Geopolitical Framework: Never-Ending Tensions

Geopolitical tensions remain high in Southwest Asia and North Africa (Figure 7.30). Some tensions relate to age-old patterns of cultural geography in which different ethnic, religious, and linguistic groups struggle to live with one another in a rapidly changing world of new nation-states and political relationships. European colonialism also contributes to present difficulties because the modern boundaries of many countries were formed by colonial powers. American political power has been repeatedly exercised throughout the region, evident in the diplomatic and military roles it plays in regional conflicts. Geographies of wealth and poverty also enter the geopolitical mix: Some residents profit from petroleum resources and industrial expansion, while others struggle to feed their families. Islamist elements in many countries such as Iran, Egypt, Algeria, Turkey, Yemen, and Saudi Arabia have changed the political atmosphere dramatically in the past 25 years. The result is a political climate charged with tension, a region in which the sounds of bomb blasts and gunfire remain all-too-common characteristics of everyday life.

### The Colonial Legacy

European colonialism arrived relatively late in Southwest Asia and North Africa, but the era left an important imprint on the region's modern political geography. Between 1550 and 1850, the region was dominated by the Ottoman Empire, which expanded from its Turkish hearth to engulf much of North Africa as well as nearby areas of the Levant, the western Arabian Peninsula, and modern-day Iraq. It took a century for Ottoman influences to be replaced with largely European colonial dominance after World War I (1918).

Both France and Great Britain were major colonial players within the region. French interests in North Africa included Tunisia and Morocco. In addition, French Algeria attracted large numbers of European immigrants (Figure 7.31). Following the defeat of the Ottoman Empire after World War I, France added more colonial territories in the Levant (Syria and Lebanon). The British loosely incorporated places such as Kuwait, Bahrain, Qatar, the United Arab Emirates, and Aden (in southern Yemen) into their empire to help control sea trade between Asia and Europe. Nearby Egypt also caught Britain's attention. Once the British-engineered **Suez Canal** linked the Mediterranean and Red seas in 1869, European banks and trading companies gained more influence over the Egyptian economy. In Southwest Asia, British and Arab forces joined to force out the Turks during World War I. The Saud family convinced the British that a country (Saudi Arabia) should be established in the desert wastes of the Arabian Peninsula, and Saudi Arabia became fully independent in 1932. Britain divided its other territories into three entities: Palestine (now Israel) along the Mediterranean coast; Transjordan to the east of the Jordan River (now Jordan); and a third zone that later became Iraq.

# EXPLORING GLOBAL CONNECTIONS
## Arab Hip-Hop Challenges Regional Traditions

Tune in MTV Arabia, and you're liable to hear the newest cut from Desert Heat (United Arab Emirates) (Figure 7.1.1). These young artists represent the emergent world of Arab hip-hop music, a generation of rappers that spits out lyrics challenging many of the cultural and political stereotypes within the region. The global vibe is hard to miss: The African-American roots of rap remain a fixture in the strong beats and bops that set the tone amid the Arabic ouds (lutes), clangs of Asian pop, and soulful lyrics that might emerge in Arabic, English, or French.

The social roots of hip-hop help explain its resonance in the region. Just as urban African Americans began to use rap to vent frustration and offer commentary on their lives in the 1970s, this newer generation of Arab artists tells a similar social and cultural story. For Lebanese singer Lynn Fattouh (aka Malikah), rap has provided an opportunity to explore the painful conflict in 2006 between Israel and Hezbollah, an encounter that resulted in widespread violence within her own country. Born in France of Algerian and Lebanese parents, Fattouh also describes herself as a "rough, serious, hardcore, Arabic woman" who encourages other young women in the region "to be active in their societies; to work, study, and vote." Fattouh regularly appears in concert throughout the Arab world as well as across Europe and North America.

The pioneering venture of Salim and Abdullah Dahman (aka Desert Heat) into hip-hop within the United Arab Emirates was designed to cultivate Arab unity and to offer to the world a more positive view of the Arab experience after the painful events of September 11, 2001. The band was a recent star attraction at the Red Bull Air Race in Abu Dhabi, and its first album, *When the Desert Speaks*, was released in 2008 along with the debut video "Keep It Desert." The band's aim is clear: "We just want to fight two stereotypes. Firstly that Arabs and Muslims are not all terrorists or ignorant people, and secondly that hip hop is not all negative. Hip hop was born out of positivity; out of struggle and freedom of speech."

The Palestinian group MWR (including Mahmoud, Waseem, and Richard) has focused on describing the hardships of life in the Israeli-controlled portions of Palestine. The group's lyrics express a frustration with the history of violence in the region that has also made the group popular with many young Israelis. Authors of "Arabs in Danger," MWR raps out:

*Take care! They've entered Palestine.*
*Ruined houses, murdered people, orphans in the shadow of death.*
*Why are we quiet about these criminals?*
*The tables are turned, the world is against us.*
*A land soaked with blood, people sick with worry,*
*Yet our Arab leaders don't give a stiff.*

The new connections forged in music between younger generations of Arabs, Jews, Europeans, North Americans, and others may be powerful stimulants to redefining traditional cultural and political narratives. In this way, the global leanings and non-traditional messages of Arab hip-hop may be much more than entertainment.

**FIGURE 7.1.1 Desert Heat** Rappers Salim and Abdullah Dahman from the United Arab Emirates created the group Desert Heat and celebrate many aspects of Arab culture in their music.

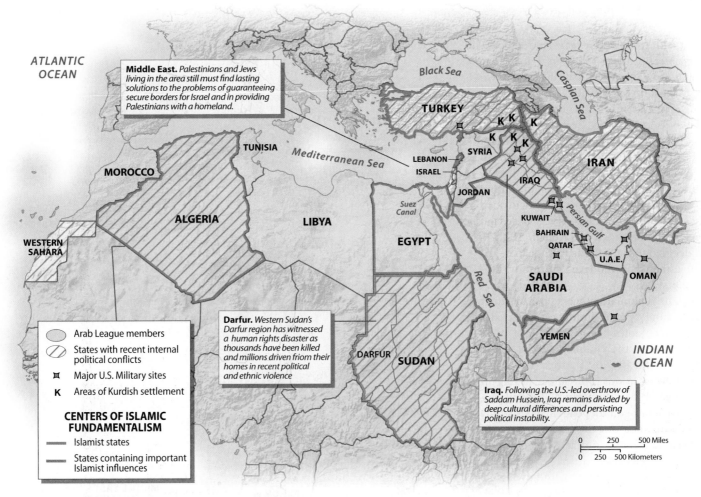

**Middle East.** *Palestinians and Jews living in the area still must find lasting solutions to the problems of guaranteeing secure borders for Israel and in providing Palestinians with a homeland.*

**Darfur.** *Western Sudan's Darfur region has witnessed a human rights disaster as thousands have been killed and millions driven from their homes in recent political and ethnic violence*

**Iraq.** *Following the U.S.-led overthrow of Saddam Hussein, Iraq remains divided by deep cultural differences and persisting political instability.*

Arab League members

States with recent internal political conflicts

Major U.S. Military sites

K    Areas of Kurdish settlement

**CENTERS OF ISLAMIC FUNDAMENTALISM**

Islamist states

States containing important Islamist influences

**FIGURE 7.30 Geopolitical Issues in Southwest Asia and North Africa**    Political tensions continue across much of the region. While the central conflict remains oriented around Israel, its neighboring states, and the rights of resident Palestinians, other regional trouble spots periodically erupt in violence. Islamic fundamentalism challenges political stability in settings from Algeria to Sudan. Elsewhere, continuing conflicts shape daily life in Iraq, Sudan, and Turkey.

**FIGURE 7.31 Algiers**    European colonial influences are still plentiful in the old French capital of Algiers in northern Algeria. The city's modern bustle continues, despite the increasing political and religious tensions that have torn the country apart since 1992.

Persia and Turkey were never directly occupied by European powers. In Persia, the British and Russians agreed to establish two spheres of economic influence in the region (the British in the south, the Russians in the north), while respecting Persian independence. In 1935, Persia's modernizing ruler, Reza Shah, changed the country's name to Iran. In nearby Turkey, European powers attempted to divide up the old core of the Ottoman Empire following World War I. The successful Turkish resistance to European control was based on new leadership provided by Kemal Ataturk. Ataturk decided to imitate the European countries and establish a modern, culturally unified, secular state.

European colonial powers began their withdrawal from several Southwest Asian and North African colonies before World War II. By the 1950s, most of the countries in the region were independent. In North Africa, Britain finally withdrew its troops from Sudan and Egypt in 1956. Libya (1951), Tunisia (1956), and Morocco (1956) achieved independence peacefully during the same era, but the French colony of Algeria became a major problem. Several million French citizens resided there, and France had no intention of simply withdrawing. A bloody war for independence began in 1954, and France finally agreed to an independent Algeria in 1962.

Southwest Asia also lost its colonial status between 1930 and 1960. While Iraq became independent from Britain in 1932, its later instability resulted in part from its imposed borders, which never recognized much of its cultural diversity. Similarly, the French division of its Levant territories into the two independent states of Syria and Lebanon (1946) greatly angered local Arab populations and set the stage for future political instability. As a favor to its Maronite Christian majority, France carved out a separate Lebanese state from largely Arab Syria, even guaranteeing the Maronites constitutional control of the national government. The action created a culturally divided Lebanon as well as a Syrian state that has repeatedly asserted its influence over its Lebanese neighbors.

## Modern Geopolitical Issues

The geopolitical instability in Southwest Asia and North Africa will continue in the 21st century. It remains difficult to predict political boundaries that seem certain to change as a result of negotiated settlements or political conflict. A quick regional transect from the shores of the Atlantic to the borders of Central Asia

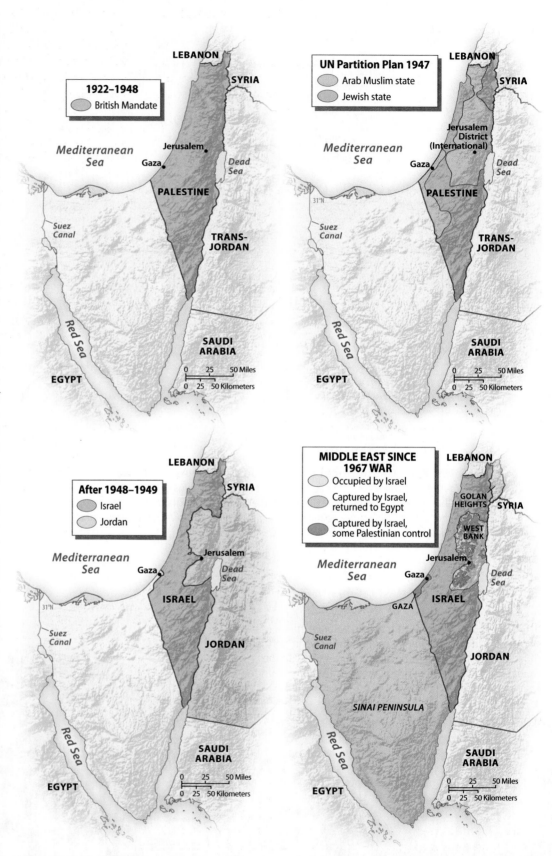

**FIGURE 7.32 Evolution of Israel**    Modern Israel's complex evolution began with an earlier British colonial presence and a United Nations partition plan in the late 1940s. Thereafter, multiple wars with nearby Arab states produced Israeli territorial victories in settings such as Gaza, the West Bank, and the Golan Heights. Each of these regions continues to be important in the country's recent relations with nearby states and with resident Palestinian populations.

suggests how these forces are playing out in different settings early in the 21st century.

**Across North Africa**  Varied North African settings threaten the region's political stability. Libya's leader, Colonel Muammar al-Qaddafi, has intermittently fomented regional tensions since he took power in 1969. Libya has financed violent political movements directed against Israel, western Europe, and the United States. Since 2004, however, Qaddafi has vowed to disarm, winning praise and growing diplomatic recognition from both the European Union and the United States.

Elsewhere in North Africa, Islamist political movements have reshaped the political landscape in several states. Most notably, Algeria was plunged into an escalating cycle of Islamist-led violence and protests for much of the 1990s, and more than 150,000 Algerians and foreign visitors were killed in the process. Since 1999, an amnesty has encouraged some rebel groups to lay down their arms, but the government still faces active opposition from Islamist extremists who are linked with Al Qaeda. Nearby Egypt has found itself ensnared in Islamist-initiated instability, and groups such as the Muslim Brotherhood have pushed for more radical political change.

Sudan faces some of the most daunting political issues in North Africa. A Sunni Islamist state since a military coup in 1989, Sudan imposed Islamic law across the country and in the process antagonized both moderate Sunni Muslims as well as the nation's large non-Muslim (mostly Christian and animist) population in the south. A long civil war between the Muslim north and the Christian and animist south produced more than 2 million casualties (mostly in the south) between 1988 and 2008. While the carnage in the south has lessened since 2001 (a tentative peace agreement was signed in 2004), a newer conflict erupted in the Darfur region in the western portion of the country (see Figure 7.30). Ethnicity, race, and control of territory seem to be at the center of the struggle in the largely Muslim region, as a well-armed Arab-led militia group (with many ties to the central government in Khartoum) has attacked hundreds of black-populated villages, killing more than 300,000 people and driving 2.5 million more from their homes.

**The Arab–Israeli Conflict**  The 1948 creation of the Jewish state of Israel produced another enduring zone of cultural and political tensions within the eastern Mediterranean. Jewish migration to Palestine increased after the defeat of the Ottoman Empire. In 1917, Britain issued the Balfour Declaration, a pledge to encourage the "establishment of Palestine as a home for the Jewish people." After World War II, the United Nations divided the region into two states: one to be predominantly Jewish, the other primarily Muslim (Figure 7.32). Indigenous Arab Palestinians rejected the partition, and war erupted. Jewish forces proved victorious, and by 1949, Israel had actually grown in size. The remainder of Palestine, including the West Bank and the Gaza Strip, passed to Jordan and Egypt, respectively. Hundreds of thousands of Palestinian refugees fled from Israel to neighboring countries, where many of them remained in makeshift camps. Under these conditions, Palestinians nurtured the idea of creating their own state on land that had become Israel.

Israel's relations with neighboring countries remained poor. Supporters of Arab unity and Muslim solidarity sympathized with the Palestinians, while their antipathy toward Israel grew. Israel fought additional wars in 1956, 1967, and 1973. In territorial terms, the Six-Day War of 1967 was the most important conflict (see Figure 7.32). In this struggle against Egypt, Syria, and Jordan, Israel occupied substantial new territories in the Sinai Peninsula, the Gaza Strip, the West Bank, and the Golan Heights. Israel annexed the eastern part of the formerly divided city of Jerusalem, arousing particular bitterness among the Palestinians, because Jerusalem is a sacred city in the Muslim tradition (it contains the Dome of the Rock and the holy al-Aqsa Mosque) (see Figure 7.27). Jerusalem is sacred in Judaism as well (it contains the Temple Mount and the holy Western Wall of the ancient Jewish temple), and Israel remains adamant in its claims to the entire city. A peace treaty with Egypt resulted in the return of the Sinai Peninsula in 1982, but tensions focused on other occupied territories that remained under Israeli control. To strengthen its geopolitical claims, Israel also built additional Jewish settlements in the West Bank and in the Golan Heights, further angering Palestinian residents.

Palestinians and Israelis began to negotiate a settlement in the 1990s. Preliminary agreements called for a quasi-independent Palestinian state in the Gaza Strip and across much of the West Bank. A tentative agreement late in 1998 strengthened the potential control of the ruling **Palestinian Authority (PA)** in the Gaza Strip and portions of the West Bank (Figure 7.33).

**FIGURE 7.33 West Bank**  Portions of the West Bank were returned to Palestinian control in the 1990s, but Israel has partially reasserted its authority in some of these regions since 2000, citing the increased violence in the region. New Israeli settlements also are scattered throughout the West Bank in areas still under their nominal control.

**FIGURE 7.34 Jewish Settlement, West Bank**　The new houses and well-planned neighborhoods of the West Bank Jewish settlement of Eli (foreground) contrast with the older Palestinian settlement in the distance. Many Palestinians resent the construction of these controversial Israeli settlements in the West Bank region.

designed to socially and economically isolate many of their settlements along the Israeli border. Nearby, Israel withdrew troops and settlers from the Gaza Strip in 2005, but political instability thereafter resulted in repeated Israeli incursions and reoccupations of the small war-torn region.

The political fragmentation of the Palestinians added further uncertainty. In 2006, control of the Palestinian government was split between the Fatah and Hamas political parties. Hamas has long been seen by many Israelis as an extremist and violent Palestinian political party, whereas the Fatah party has shown more willingness to work peacefully with the state of Israel. In 2007, the split between these rival Palestinian factions became violent, with Hamas gaining effective control of the PA within Gaza, and with Fatah maintaining its

But a new cycle of heightened violence erupted late in 2000, as Palestinian attacks against Jews increased and as the Israelis continued with the construction of new settlements in occupied lands (especially in the West Bank) (Figure 7.34). Palestinian leaders harshly criticized continuing construction of these new Jewish settlements, particularly in the vicinity of Jerusalem. Even more daunting has been construction of the Israeli security barrier, a partially completed 26-foot- (8-meter-) high wall designed to separate Israelis from Palestinians across much of the West Bank region (Figure 7.35). Israeli supporters of the barrier (to be more than 400 miles [644 kilometers] long when completed) see it as the only way they can protect their citizens from suicide bombings and more terrorist attacks. Palestinians see it as a land grab, an "apartheid wall"

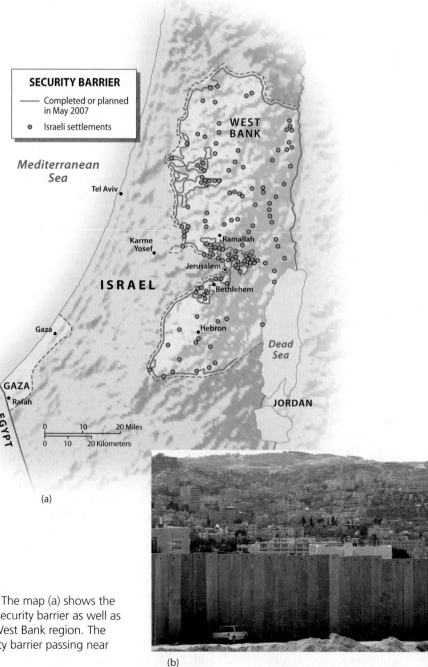

**FIGURE 7.35 Israeli Security Barrier**　The map (a) shows the completed and planned portions of the Israeli security barrier as well as many of the Israeli settlements located in the West Bank region. The photo (b) shows a segment of the Israeli security barrier passing near Bethlehem, south of Jerusalem.

greatest influence across the West Bank region. The split makes an eventual peace treaty with Israel seem even more difficult to achieve.

Instability also continues along Israel's northern border with Lebanon. In 2006, a Shiite militia group known as Hezbollah increased its rocket attacks into northern Israel, prompting an armed response by the Israelis. Along with Hamas, Hezbollah represents a radicalized Arab political element within the region that shows little inclination to negotiate with Israel (or vice versa).

One thing is certain: Geographical issues will remain at the center of the regional conflict. Palestinians hope for a land they can call their own, and Israelis continue their search for more secure borders that guarantee their political integrity in a region where they are surrounded by potentially hostile neighbors. Ultimately, the sacred geography of Jerusalem, a mere 220 acres (89 hectares) of land within the Old City, stands at the center of the conflict. Imaginative compromises in defining that political space will need to recognize its special value to both its Jewish and Palestinian residents.

**Iraq's Political Evolution**　Iraq is another nation-state born during the colonial era that has yet to escape the consequences of its geopolitical origins. When the country was carved out of the British Empire in 1932, it contained the cultural seeds of its later troubles. Iraq remains culturally complex today (Figure 7.36). Most of the country's Shiite population lives in the lower Tigris and Euphrates valleys south of Baghdad. The area around Basra contains some of the holiest Shiite shrines in the world. In northern Iraq, the culturally distinctive Kurds have their own ethnic identity and political aspirations. Many Kurds desire independence from Baghdad, and they have managed to establish a federal region that enjoys considerable autonomy from the central Iraqi government. A third major subregion is dominated by the Sunnis and includes part of the Baghdad area as well as a triangle of territory to the north and west that includes Sunni strongholds such as Fallujah and Tikrit.

Iraq created a major source of regional instability before Saddam Hussein was removed in 2003. In 1980, Hussein invaded oil-rich but politically weakened Iran to gain a better foothold in the Persian Gulf. Eight years of bloody fighting resulted in a stalemate, and the conflict left Iraq's finances in disarray. Iraq then invaded and overran Kuwait in 1990, claiming it as an Iraqi province. A U.S.-led UN coalition, receiving substantial support from Saudi Arabia, expelled Iraq from Kuwait in early 1991. Twelve years later, the 2003 American-led invasion of the country replaced one set of uncertainties with another.

When Iraqi leaders assumed control of their new state on June 28, 2004, more than 135,000 U.S. troops remained in the country. Thereafter, growing religious and ethnic violence between different Iraqi factions threw portions of the nation into civil war. Rival Sunni and Shiite groups forced many Iraqis from their communities. U.S. troops in the country in 2007 and 2008 worked with Iraqi officials to reduce the level of violence. Political debates continue concerning the future U.S. role in the region and the correct path toward solving the crisis in Iraq. Some have suggested partitioning the country into three parts (Shiite, Kurd, and Sunni), but this solution poses its own difficulties, both with complex distributions of people and resources within Iraq as well as with neighboring states that might see such divisions as further destabilizing the region. An enduring geopolitical solution has yet to take shape, but whatever it is, it will need to recognize the cultural differences that divide the country, as well as allow for a revival of the country's crippled oil industry.

**Instability in Saudi Arabia**　The region's greatest oil power is led by a conservative monarchy (the Saud family) that has been unwilling to promote many democratic reforms. On the surface, it has supported U.S. efforts in the region to provide stable flows of petroleum, but beneath the surface, certain elements of the regime may have financed radically anti-American groups such as Al Qaeda. The Saudi people themselves, largely Sunni Arabs, are torn between an allegiance to their royal family (and the economic stability it brings); the lure of a more democratic, open Saudi society; and an enduring distrust of foreigners, particularly Westerners. Furthermore, the Sunni majority includes Wahhabi sect members, whose radical Islamist philosophy has fostered anti-American sentiment. In addition, the large number of foreign laborers and the persisting U.S. military and economic presence within the country (one of the chief complaints of Al Qaeda leader Osama bin Laden) create a setting ripe for political instability. Recent economic declines and lower oil prices have made matters worse, as many Saudis have seen fewer job opportunities in their energy-dependent economy.

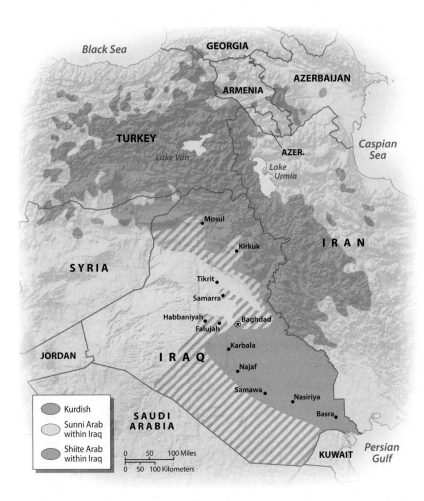

**FIGURE 7.36 Multicultural Iraq**　Iraq's complex colonial origins produced a state with varying ethnic characteristics. Shiites dominate south of Baghdad, Sunnis hold sway in the western triangle zone, and Kurds are most numerous in the north, near oil-rich Kirkuk and Mosul.

**Iranian Geopolitics**　Iran increasingly garners international attention. Islamic fundamentalism dramatically appeared on the political scene in 1978, as Shiite Muslim clerics overthrew the Shah, an authoritarian, pro-Western ruler friendly to U.S. political and economic interests. The new leaders proclaimed an Islamic republic in which religious officials ruled both clerical and political affairs.

Today, Iran supports Shiite Islamist elements throughout the region (for example, Hezbollah) and has repeatedly threatened the state of Israel. In addition, some moderate Arab states have seen the country as a growing threat. In 2009, the nearby United Arab Emirates, for example, invited the French to open their first new military base in decades on foreign soil, and French leaders pledged to defend against Iranian expansion in the region. Adding uncertainty has been Iran's ongoing nuclear development program, an initiative its government claims is solely related to the peaceful construction of power plants. Many in the West, however, remain unconvinced of the government's motives and demand that its program be stopped or opened to international inspections before greater international cooperation is given to the Iranian regime. In addition, hotly contested and controversial presidential elections in Iran in 2009 demonstrated growing frustration with the status quo within the country, suggesting that more internal political instability may be on the horizon.

# Economic and Social Development: Lands of Wealth and Poverty

Southwest Asia and North Africa is a region of both incredible wealth and discouraging poverty (Table 7.2). While some countries enjoy great prosperity, due mainly to rich reserves of petroleum and natural gas, other nations are among the least developed in the world. Overall, recent economic growth rates have fallen behind those of the more-developed world. Continuing political instability has also contributed to the region's struggling economy. Petroleum will no doubt figure significantly in the region's future economy, but many countries in the area also have focused on increasing agricultural output, investing in new industries, and promoting tourism to broaden their economic base.

## The Geography of Fossil Fuels

The striking global geographies of oil and natural gas reveal the region's persisting importance in the world oil economy, as well as the extremely uneven distribution of these resources within the region (Figure 7.37).

## TABLE 7.2  DEVELOPMENT INDICATORS

| Country | GNI per Capita, PPP 2007 | GDP Average Annual % Growth 2000–07 | Human Development Index (2006)# | Percent Population Living Below $2 a Day | Life Expectancy* 2009 | Under Age 5 Mortality Rate 1990 | Under Age 5 Mortality Rate 2007 | Gender Equity 2007 |
|---|---|---|---|---|---|---|---|---|
| Algeria | 7,640 | 4.5 | 0.748 | 23.6 | 72 | 69 | 37 | 99 |
| Bahrain | 27,210 | | 0.902 | | 75 | 19 | 10 | 101 |
| Egypt | 5,370 | 4.3 | 0.716 | 18.4 | 72 | 93 | 36 | 95 |
| Gaza and West Bank | | −0.9 | 0.731 | | 72 | 38 | 27 | 104 |
| Iran | 10,840 | 5.9 | 0.777 | 8.0 | 71 | 72 | 33 | 114 |
| Iraq | | −11.4 | | | 67 | 53 | | |
| Israel | 26,310 | 3.2 | 0.930 | | 81 | 12 | 5 | 101 |
| Jordan | 5,150 | 6.3 | 0.769 | 3.5 | 73 | 40 | 24 | 102 |
| Kuwait | 52,610 | 9.2 | 0.912 | | 78 | 15 | 11 | 100 |
| Lebanon | 10,040 | 3.3 | 0.796 | | 72 | 37 | 29 | 103 |
| Libya | 14,710 | 3.7 | 0.840 | | 73 | 41 | 18 | 105 |
| Morocco | 4,050 | 5.0 | 0.646 | 14.0 | 71 | 89 | 34 | 87 |
| Oman | 21,650 | 4.7 | 0.839 | | 72 | 32 | 12 | 99 |
| Qatar | | | 0.899 | | 76 | 26 | 15 | 99 |
| Saudi Arabia | 22,190 | 4.1 | 0.835 | | 76 | 44 | 25 | 94 |
| Sudan | 1,880 | 7.1 | 0.526 | | 58 | 125 | 109 | 89 |
| Syria | 4,430 | 4.5 | 0.736 | | 74 | 37 | 17 | 96 |
| Tunisia | 7,140 | 4.8 | 0.762 | 12.8 | 74 | 52 | 21 | 104 |
| Turkey | 12,810 | 5.9 | 0.798 | 9.0 | 72 | 82 | 23 | 90 |
| United Arab Emirates | | 7.7 | 0.903 | | 77 | 15 | 8 | 101 |
| Western Sahara | | | | | 65 | | | |
| Yemen | 2,200 | 4.0 | 0.567 | 46.6 | 63 | 127 | 73 | 66 |

Source: World Bank, *World Development Indicators 2009.*

United Nations, *Human Development Index, 2008.* #

Population Reference Bureau, *World Population Data Sheet, 2009*\*

Gender equity – Ratio of female to male enrollments in primary and secondary school. Numbers below 100 have more males in primary/secondary school, numbers above 100 have more females in primary/secondary schools.

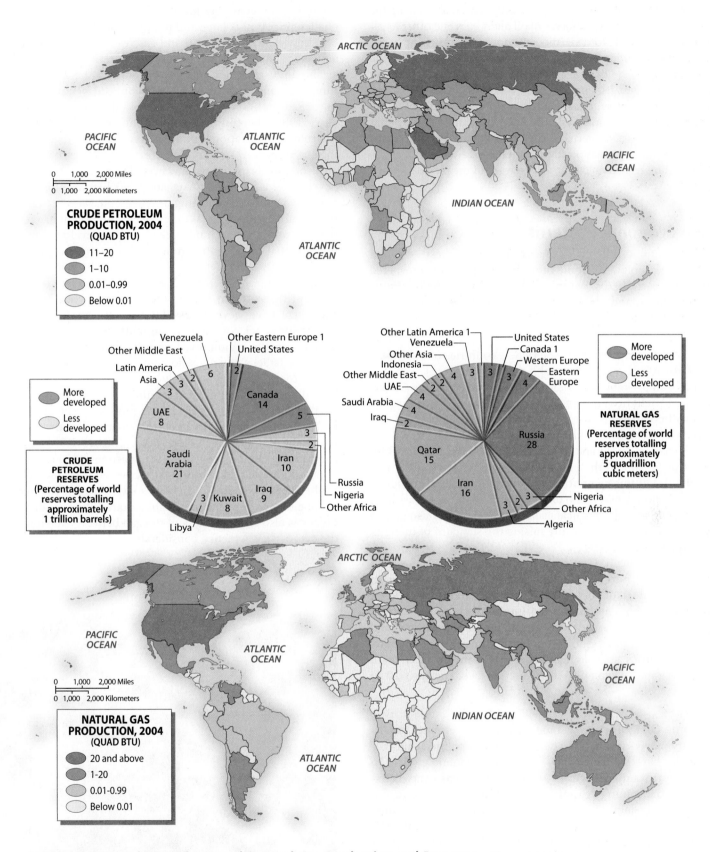

**FIGURE 7.37 Crude Petroleum and Natural Gas Production and Reserves** The region plays a pivotal role in the global geography of fossil fuels. Abundant regional reserves suggest that the pattern will continue.

North African settings (especially Algeria and Libya) as well as the Persian Gulf region have large reserves of oil and gas, while other localities (for example, Israel, Jordan, and Lebanon) lie outside zones of major fossil fuel resources. Saudi Arabia, Iran, Iraq, Kuwait, and the United Arab Emirates hold large petroleum reserves, while Iran and Qatar possess the largest regional reserves of natural gas. The distribution of fossil fuel reserves suggests that regional supplies will not be exhausted anytime soon. Overall, with only 7 percent of the world's population, the region holds a staggering 61 percent of the world's proven oil reserves. Saudi Arabia's pivotal position, both regionally and globally, is clear: Its 28 million residents live atop 20 percent of the planet's known oil supplies.

## Regional Economic Patterns

Remarkable economic differences characterize Southwest Asia and North Africa (see Table 7.2). Some oil-rich countries have prospered greatly since the early 1970s, but in many cases fluctuating oil prices, political disruptions, and rapidly growing populations have reduced the likelihood of future economic growth.

**Higher-Income Oil Exporters**   The richest countries of Southwest Asia and North Africa owe their wealth to massive oil reserves. Nations such as Saudi Arabia, Kuwait, Qatar, Bahrain, and the United Arab Emirates benefit from fossil fuel production, as well as from their relatively small populations. Large investments in transportation networks, urban centers, and other petroleum-related industries have reshaped the cultural landscape. The petroleum-processing and shipping centers of Jubail (on the Persian Gulf) and Yanbu (on the Red Sea) are examples of this commitment to expand the Saudi economic base beyond the simple extraction of crude oil. Billions of dollars have poured into new schools, medical facilities, low-cost housing, and modernized agriculture, significantly raising the standard of living in the past 40 years.

Still, problems remain. Dependence on oil and gas revenues produces economic pain in times of falling prices, such as those seen in the economic downturn of 2008–2009. Such fluctuations in world oil markets will inevitably continue in the future, disrupting construction projects, producing large layoffs of immigrant populations, and slowing investment in the region's economic and social infrastructure. In addition, countries such as Bahrain and Oman are faced with the problem of rapidly depleting their reserves over the next 20 to 30 years.

**Lower-Income Oil Exporters**   Some states in the region are important secondary players in the oil trade, but different political and economic variables have hampered their sustained economic growth. In North Africa, for example, Algerian oil and natural gas overwhelmingly dominate the country's exports, but the past 15 years have also brought political instability and increasing shortages of consumer goods. While the country contains some excellent agricultural lands in the north, the overall amount of arable land has increased little over the past 25 years, even as the country's population has grown by more than 50 percent.

In Southwest Asia, Iraq faces huge economic and political challenges. War has crippled much of Iraq's already deteriorated infrastructure, and continuing political instability has made the task of

rebuilding its economy more difficult. The situation in Iran is also challenging. Iran is large and populous, and it has a relatively diverse economy. The country's oil and gas reserves are huge and have seen active commercial development since 1912. The country also has a sizable industrial base, much of it built in the 20 years prior to the fundamentalist revolution of 1979. But today Iran is relatively poor, burdened with a stagnating, possibly declining, standard of living. Since 1980, the country's fundamentalist leaders have downplayed the role of international trade in consumer goods and services, fearing the import of unwanted cultural influences from abroad. There are economic bright spots, however: The country benefits from energy developments in Central Asia, and Iran's literacy rate (particularly for women) has risen, reflecting a new emphasis on rural education in the country. Still, political differences with potential Western trading partners are vast, and Iran's recent nuclear development program further isolated the country from many nations, including the United States.

**Prospering without Oil**   Some countries, while lacking petroleum resources, have nevertheless found paths to increasing economic prosperity. Israel, for example, supports one of the highest standards of living in the region, even with its political challenges (see Table 7.2). The Israelis and many foreigners have invested large amounts of capital to create a highly productive agricultural and industrial base. In addition, the country is emerging as a global center for high-tech computer and telecommunications products. The country is known for its fast-paced and highly entrepreneurial business culture, which resembles California's Silicon Valley (Figure 7.38). Israel also has daunting economic problems. Its persisting struggles with the Palestinians and with neighboring states have sapped much of its potential vitality. Defense spending absorbs a large share of total gross national income (GNI), necessitating high tax rates. Poverty among the Palestinians is also widespread, and the gap between rich and poor within the country has widened considerably.

**FIGURE 7.38 Israeli High-Tech Industry**   These Israeli workers are employees of the Telrad factory, a high-tech division of the Koor conglomerate.

Turkey also has a diversified economy, even though its per capita income is modest by regional standards. Lacking petroleum, Turkey produces varied agricultural and industrial goods for export. About 35 percent of the population remains employed in agriculture, and the country's principal commercial products include cotton, tobacco, wheat, and fruit. The industrial economy has grown since 1980, including exports of textiles, food, and chemicals. Turkey remains an important tourist destination in the region, as well, attracting more than 6 million visitors annually in recent years. Some Turkish leaders hope to bolster the economy by eventually joining the European Union.

**Regional Patterns of Poverty** Poorer countries of the region share the problems of much of the rest of the less-developed world. For example, Sudan, Egypt, and Yemen each face unique economic challenges. For Sudan, continuing political problems have stood in the way of progress. Civil war has resulted in major food shortages. The country's transportation and communications systems have seen little new investment, and secondary school enrollments remain very low. On the other hand, Sudan's fertile soils could support more farming, and its new oil production suggests petroleum's expanding role in the economy. Still, the country's sustained economic development appears to be delayed by continuing political instability.

Egypt's economic prospects are also unclear. On the one hand, the country experienced real economic growth during the 1990s, as President Hosni Mubarak pushed for more foreign investment, smaller government deficits, and a multibillion-dollar privatization program to put government-controlled assets under more efficient management. Even so, many Egyptians live in poverty, and the gap between rich and poor continues to widen. Illiteracy is widespread, and the country suffers from the **brain drain** phenomenon, as some of its brightest young people leave for better jobs in western Europe or the United States. Egypt's 75 million people already make it the region's most heavily populated state, and recent efforts to expand the nation's farmland have met with numerous environmental, economic, and political problems.

In Southwest Asia, Yemen remains the poorest country on the Arabian Peninsula. Positioned far from most of the region's principal oil fields, Yemen's low per capita GNI puts the country on par with many nations in impoverished Sub-Saharan Africa. The largely rural country relies mostly on marginally productive subsistence agriculture, and much of its mountain and desert interior lacks effective links to the outside world (Figure 7.39). The present state emerged in 1990, with the political union of North and South Yemen. Coffee, cotton, and fruits are commercial agricultural products, and modest oil exports bring in needed foreign currency. Overall, however, high unemployment and marginal subsistence farming remain widespread across the country. In addition, growing violence associated with Shiite rebels further destabilized the country's economy in 2008 and 2009.

Unique problems afflict the Palestinian populations of Gaza and the West Bank. Continued declines have devastated the economy as political disruptions have discouraged investment and conflicts with the Israelis have destroyed infrastructure. In many cases, when Israeli forces suspect political dissidents and terrorist elements in a given Palestinian town or urban neighborhood, they destroy the settlement, leveling houses and shops in the process. Poverty grips two-thirds of the Palestinian population, unemployment hovers above 40 percent, and the ongoing construction of the Israeli security barrier promises to further disrupt the Palestinian economy in the years to come.

## A Woman's Changing World

The role of women in largely Islamic Southwest Asia and North Africa remains a major social issue. Female labor participation rates in the workforce are among the lowest in the world, and large gaps typically exist between levels of education for males and females

**FIGURE 7.39 Rural Yemen** This isolated hilltop village in the rugged mountains of northern Yemen is surrounded by small terraced fields and livestock pastures.

**FIGURE 7.40 Algerian Women** Even within the norms of a more conservative Islamist society, Algerian women are increasingly visible and highly productive contributors to that nation's workforce.

(see Table 7.2). In conservative parts of the region few women work outside the home. Even in parts of Turkey, where Western influences are widespread, it is rare to see rural women selling goods in the marketplace or driving cars in the street. More orthodox Islamic states impose legal restrictions on the activities of women. In Saudi Arabia, for example, women are not allowed to drive, although growing protests among many younger, educated Saudi females may overturn that ban. In Iran, full veiling remains mandatory in more conservative parts of the country, and police often hassle Iranian women in more liberal dress, even in larger urban areas. Generally, Islamic women lead more private lives than men: Much of their domestic space is shielded from the world by walls and shuttered windows, and their public appearances are filtered through the use of the face veil or chador (full-body veil).

Yet in some places women's roles are changing, even within the norms of more conservative Islamist societies. Many young Algerian women demonstrate the pattern (Figure 7.40). Most studies suggest that those in the younger generation are more religious than their parents and are more likely to cover their heads and drape their bodies with traditional religious clothing. At the same time, they are more likely to be educated and employed than ever before. Today, 70 percent of Algeria's lawyers and 60 percent of its judges are women. A majority of university students are women, and women dominate the health-care field. These new social and economic roles help explain why the birthrates in such settings are declining and are likely to continue to do so. The Algerian case also demonstrates how complex social processes actually unfold, rarely following simple models for how "traditional" societies might evolve.

Women's roles are shifting throughout the region. In Sudan and Saudi Arabia, a growing number of women pursue high-level careers. Education may be segregated, but it is available. Libya's Qaddafi has singled out the modernization of women as a high priority, and today more women than men graduate from the country's university system.

In Western Sahara, Saharawi women play a leading role in the country's political fight for independence from Morocco, and their broader educational backgrounds and social freedoms (including the right to divorce their husbands) further separate them from Moroccan women. Women also have a more visible social position in Israel, except in fundamentalist Jewish communities, where conservative social customs require more traditional domestic roles.

## Global Economic Relationships

Southwest Asia and North Africa share close economic ties with the rest of the world. While oil and gas remain critical commodities that dominate international economic linkages, the growth of manufacturing and tourism are also redefining the region's role in the world.

**OPEC's Changing Fortunes** While OPEC no longer controls oil and gas prices globally, it still influences the cost and availability of these pivotal products within the developed and less-developed worlds. Western Europe, the United States, Japan, China, and many less industrialized countries depend on the region's fossil fuels. In the case of Saudi Arabia, for example, petroleum and its related products make up more than 90 percent of its exports. One recent trend evident in oil-producing countries (such as Saudi Arabia) is their willingness to form partnerships with foreign corporations, accelerating the economic integration of the region with the rest of the world. Even so, the region's major energy producers will be particularly vulnerable to global-scale recessions, such as the downturn that hit the region hard in 2008 and 2009.

Beyond key OPEC producers, other economic activities have also contributed to global economic integration. Turkey, for example, ships textiles, food products, and manufactured goods to its principal trading partners: Germany, the United States, Italy, France, and Russia. Tunisia sends more than 60 percent of its exports (mostly clothing, food products, and petroleum) to nearby France and Italy. Israeli exports emphasize the country's highly skilled workforce: Products such as cut diamonds, electronics, and machinery parts are exported to the United States, western Europe, and Japan.

**Regional and International Linkages**   Future interconnections between the global economy and Southwest Asia and North Africa may depend increasingly on cooperative economic initiatives far beyond OPEC. Relations with the European Union (EU) are critical. Since 1996, Turkey has enjoyed closer economic ties with the EU, but recent attempts at full membership in the organization have failed. Other so-called Euro-Med agreements also have been signed between the EU and countries across North Africa and Southwest Asia that border the Mediterranean Sea.

Most Arab countries, however, are wary of too much European dominance. They formed a regional political organization known as the Arab League in 1945. In 2005, seventeen league members established the Greater Arab Free Trade Area (GAFTA), designed to eliminate all intraregional trade barriers and spur economic cooperation. In addition, Saudi Arabia plays a pivotal role in regional economic development through organizations such as the Islamic Development Bank and the Arab Fund for Economic and Social Development.

Tourists are another link to the global economy. Traditional magnets such as ancient historical sites and globally significant religious localities (for a multitude of faiths) draw millions of visitors annually. As the developed world becomes wealthier, there is also a growing global demand for recreational spots that offer beaches, sunshine, and novel entertainment. Indeed, many miles of the Mediterranean, Black Sea, and Red Sea coastlines are now lined with upscale but often ticky-tack landscapes of resort hotels and condominiums dedicated to serving travelers' needs. More adventurous travelers seek ecotourist activities such as snorkeling in Naama Bay on Egypt's Sinai Coast or four-wheeling among the Berbers in the Moroccan backcountry. Endangered wildlife beckons photographers and poachers hoping to catch a glimpse of a South Arabian grey wolf, a Nubian ibex, or a darting Persian squirrel. All this activity means big business to many countries in the region, and the economic impacts of tourism seem likely to grow during the 21st century.

# SUMMARY

- Positioned at the meeting ground of Earth's largest landmasses, Southwest Asia and North Africa have played a critical role in world history and in processes of globalization that bind the planet together ever more tightly.

- In ancient times, the region was home to early examples of crop and livestock domestication, as well as some of the world's earliest urban centers. Three of the world's great religions—Judaism, Christianity, and Islam—also emerged beneath its desert skies.

- Despite the rich legacy of global influence and power, the peoples of Southwest Asia and North Africa are struggling early in the 21st century. Indeed, most countries are suffering from significant economic problems and political uncertainties. The region has faced difficulty and high costs in trying to expand its limited supplies of agricultural land and water resources amid fast-growing populations.

- Political conflicts have disrupted economic development across the region. Civil wars, conflicts between states, and regional tensions have worked against plans for greater cooperation and trade. Most importantly, the region must deal both with the basic inconsistencies between Western civilization and more fundamentalist interpretations of Islam, as well as with finding a lasting solution to the Israeli–Palestinian conflict.

- Future cultural and political change will be guided by a complex response to Western influences, a mix of fascination and suspicion that will produce its own unique regional geography. Southwest Asia and North Africa will retain its distinctive regional identity, a character defined by its environmental setting, the rich cultural legacy of its history, the selective abundance of its natural resources, and its continuing political problems.

# KEY TERMS

brain drain (p. 215)
culture hearth (p. 186)
domestication (p. 195)
exotic river (p. 196)
Fertile Crescent (p. 195)
fossil water (p. 190)
Hajj (p. 201)
hydropolitics (p. 191)

Islamic fundamentalism (p. 186)
Islamism (p. 186)
kibbutz (p. 196)
Levant (p. 192)
Maghreb (p. 191)
medina (p. 198)
monotheism (p. 200)

Organization of the Petroleum Exporting Countries (OPEC) (p. 186)
Ottoman Empire (p. 203)
Palestinian Authority (PA) (p. 209)
pastoral nomadism (p. 195)
physiological density (p. 195)

qanat system (p. 190)
Quran (p. 201)
Shiite (p. 201)
Suez Canal (p. 205)
Sunni (p. 201)
theocratic state (p. 201)
transhumance (p. 196)

# THINKING GEOGRAPHICALLY

1. How might a major project for transferring water from Turkey to the Arabian Peninsula affect the development of Saudi Arabia? What would be some of the potential political and ecological ramifications of such a project?

2. Why are birthrates declining in some countries in Southwest Asia and North Africa? Despite the cultural differences with North America, what common processes seem to be at work in both regions that have contributed to this demographic transition?

3. What economic changes could occur if Israel and the Palestinians were to reach a lasting peace? What kinds of general connections might be found between political conflict and economic conditions throughout the region?

4. Why has the idea of Arab nationalism failed to achieve any lasting geopolitical changes in the region?

5. Imagine that you are a ruler of a conservative Islamist Arab state. What might be the advantages and disadvantages of opening up your country to the Internet?

Log in to **www.mygeoscienceplace.com** for videos, interactive maps, RSS feeds, case studies, and self-study quizzes to enhance your study of Southwest Asia and North Africa.

# 8 EUROPE

## GLOBALIZATION AND DIVERSITY

**Historically, Europe was one of the world's earliest promoters of globalization as it spread its languages, laws, religion, politics, and economic systems throughout the world during the colonial period. Today, however, Europe struggles to adjust to the social and ethnic diversity sown by these earlier seeds as migrants from former colonies seek a share of Europe's high standard of living.**

### ENVIRONMENTAL GEOGRAPHY

Europe is one of the "greenest" world regions, with strong environmental laws and regulations about recycling, energy efficiency, and pollution. Europe is also a leader in addressing the many issues associated with global warming.

### POPULATION AND SETTLEMENT

With low (or no) natural population growth, immigration into Europe from other world regions is both a solution to Europe's labor needs and a troublesome social issue.

### CULTURAL COHERENCE AND DIVERSITY

Europe has a long history of internal cultural tensions linked to cultural differences in language and religion. Today, though, these internal differences have been largely resolved through the integrating force of the European Union (EU) and replaced with cultural tensions associated with immigration from other regions, such as Africa, the Middle East, and South Asia.

### GEOPOLITICAL FRAMEWORK

After half a century of the Cold War, which divided Europe into two parts, east and west, the region is now experiencing political integration of former adversaries at a national level. However, outbreaks of regionalism, with its agenda of autonomy, are still problematic in the Balkans, Spain, and the British Isles.

### ECONOMIC AND SOCIAL DEVELOPMENT

Through the past half century, the EU has achieved remarkable success in bringing peace and prosperity to this diverse region. Nonetheless, serious economic and social challenges still remain, particularly in southern and eastern Europe.

*Under the midnight sun of arctic Norway, the many troublesome issues facing Europe seem far away, at least for the moment.*

Europe is one of the most diverse regions in the world, encompassing a wide assortment of people and places, in an area considerably smaller than North America. More than half a billion people reside in this region, living in 41 countries that range in size from giant Germany to microstates such as Andorra and Monaco (Figure 8.1).

The region's remarkable cultural diversity produces a geographic mosaic of languages, religions, and landscapes. Commonly, a day's journey finds a traveler speaking two or three languages, possibly changing money several times, and sampling distinct regional food and drink.

Though a traveler may revel in Europe's cultural and environmental diversity, these regional differences are also responsible for Europe's troubled past. In the 20th century alone, Europe was the principal battleground for two world wars, followed by the 45-year **Cold War** (1945–1990) that divided the continent and the world into two hostile, highly armed camps—Europe and the United States against the former Soviet Union (now Russia).

Today, however, a spirit of cooperation prevails as Europe sets aside nationalistic pride and works toward regional economic, political, and cultural integration through the **European Union (EU)**. This supranational organization is made up of 27 countries, anchored by the western European states of Germany, France, Italy, and the United Kingdom and also including most eastern European countries. Furthermore, the geographic reach and economic policies of the EU will continue to integrate the region during the next decades, as it expands its membership (see "Setting the Boundaries"). While controversial at times, the EU has emerged as a major regional force in Europe that is unparalleled in other world regions (Figure 8.2).

## Environmental Geography: Human Transformation of a Diverse Landscape

Despite the continent's small size, Europe's environmental diversity is extraordinary. Within its borders are found a startling range of landscapes, from the Arctic tundra of northern Scandinavia to the barren hillsides of the Mediterranean islands, and from the explosive volcanoes of southern Italy to the glaciers of Iceland. Four factors explain this environmental diversity:

- The complex geology of this western extension of the Eurasian landmass has produced some of the newest, as well as the oldest, landscapes in the world.
- Europe's latitudinal extent creates opportunities for diversity because the region extends from the Arctic to the Mediterranean subtropics (Figure 8.3).
- These latitudinal controls are further modified by the moderating influence of the Atlantic Ocean and Black, Baltic, and Mediterranean seas.
- The long history of human settlement in Europe has transformed the region's natural landscapes in fundamental ways.

## Landform and Landscape Regions

European landscapes can be organized into four general topographic regions:

- The European Lowland, forming an arc from southern France to the northeast plains of Poland and also including southeastern England
- The Alpine Mountain System, extending from the Pyrenees in the west to the Balkan Mountains of southeastern Europe
- The Central Uplands, positioned between the Alps and the European Lowland
- The Western Highlands, which include mountains in Spain, portions of the British Isles, and the highlands of Scandinavia

**The European Lowland**   This lowland (also known as the North European Plain) is the unquestionable focus of western Europe, with its high population density, intensive agriculture, large cities, and major industrial regions. Though not completely flat by any means, most of the lowland lies below 500 feet (150 meters) in elevation, though it is broken in places by rolling hills, plateaus, and uplands, such as in Brittany,

## Setting the Boundaries

The European region is small compared to the United States. In fact, Europe from Iceland to the Black Sea would easily fit into the eastern two-thirds of North America. A more apt comparison would be Canada, as Europe, too, is a northern region. More than half of Europe lies north of the 49th parallel, the line of latitude forming the western border between the United States and Canada (see Figure 8.3).

Europe contains 41 countries that range in size from large countries, such as France and Germany, to microstates, such as Liechtenstein, Andorra, Monaco, and San Marino. Currently, the population of Europe totals about 531 million.

The notion that Europe is a continent with clearly defined boundaries is a mistaken belief with historical roots. The Greeks and Romans divided their worlds into the three continents of Europe, Asia, and Africa, separated by the Mediterranean Sea, the Red Sea, and the Bosporus Strait. A northward extension of the Black Sea was thought to separate Europe from Asia, and only in the 16th century was this notion proven false. Instead, explorers and cartographers discovered that the "continent" of Europe was firmly attached to the western portion of Asia.

Since that time, geographers have not agreed on the eastern boundary of Europe. During the existence of the Soviet Union, most geographers drew the line at the western boundary of the Soviet Union. However, with the disintegration of the Soviet Union in 1990, the eastern boundary of Europe became even more problematic. Now some geography textbooks extend Europe to the border with Russia, which places the two countries of Ukraine and Belarus, former Soviet republics, in eastern Europe. Though an argument can be made for an expanded definition of Europe, recent events lead us to draw our eastern border at Poland, Slovakia, and Romania, thus drawing our boundary coincidental with the eastern border of the EU. If Ukraine and Belarus join the EU, then we will reconsider how we define the boundaries of Europe.

**FIGURE 8.1 Europe** Stretching from Iceland in the Atlantic to the Black Sea, Europe includes 41 countries, ranging in size from large states, such as France and Germany, to the microstates Liechtenstein, Andorra, San Marino, and Monaco. Currently, the population of the region is about 531 million. Europe is highly urbanized and, for the most part, affluent, particularly the western portion. However, economic and social disparities between eastern and western Europe remain problematic.

**FIGURE 8.2 Anti-EU Feeling** Not everyone is happy with the European Union. Many groups feel its regulations trample on the uniqueness of individual nations. Here, a Frenchman lobbies for a "no" vote on a recent EU issue.

France, where elevations exceed 1,000 feet (300 meters). Many of Europe's major rivers, such as the Rhine, the Loire, the Thames, and the Elbe, meander across this lowland and form broad estuaries before emptying into the Atlantic. Several of Europe's great ports are located on the lowland, including London, Le Havre, Rotterdam, and Hamburg.

The Rhine River delta conveniently divides the unglaciated lowland to the south from the glaciated plain to the north, an area that was covered by a Pleistocene, or Ice Age, ice sheet until about 15,000 years ago. Because of this large continental glacier, the area of the North European Lowland that includes the Netherlands, Germany, Denmark, and Poland is far less fertile for agriculture than the unglaciated portion in Belgium and France (Figure 8.4). Rocky clay materials in Scandinavia were eroded and transported south by glaciers. As the glaciers later retreated with a warming climate, piles of glacial debris known as **moraines** were left on the plains of Germany and Poland. Elsewhere in the north, glacial meltwater created infertile outwash plains that have limited agricultural potential.

**The Alpine Mountain System** The Alpine Mountain System consists of a series of east–west-running mountains from the Atlantic to the Black Sea and the southeastern Mediterranean. Though these mountain ranges carry distinct regional names, such as the Pyrenees, Alps, Carpathians, Dinaric Alps, and Balkan Ranges, they have similar geologic traits. All were created more recently than other upland areas of Europe, and all are made up of a complex arrangement of rock types.

The Pyrenees, which form the political border between Spain and France, include the microstate of Andorra. This rugged range extends almost 300 miles (480 kilometers), stretching from the Atlantic to the Mediterranean. Within the mountains, glaciated peaks reaching to 11,000 feet (3,350 meters) alternate with broad glacier-carved valleys.

The centerpiece of this larger geologic system is the Alpine range itself, the Alps, reaching more than 500 miles (800 kilometers) from France to eastern

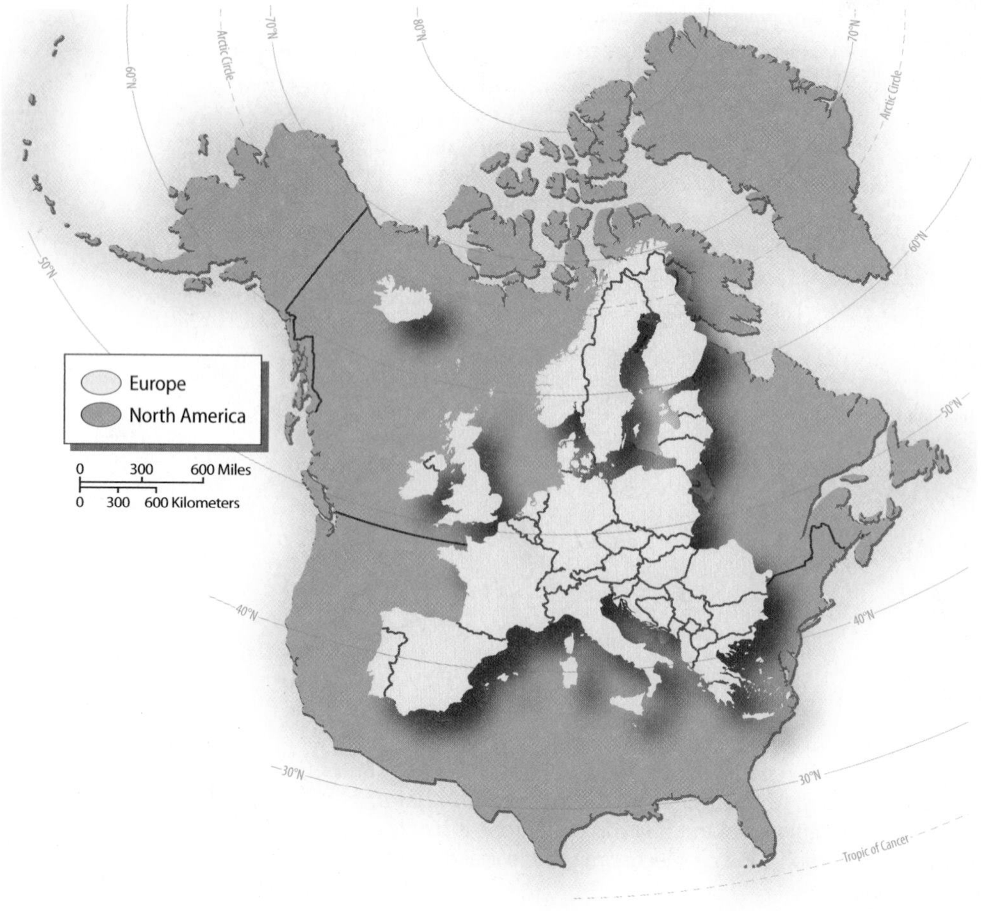

Europe
North America

0  300  600 Miles
0  300  600 Kilometers

**FIGURE 8.3 Europe: Size and Northerly Location** Europe is about two-thirds the size of North America, as shown in this cartographic comparison. Another important characteristic is the northerly location of the region, which affects its climate, vegetation, and agriculture. Much of Europe lies at the same latitude as Canada; even the Mediterranean lands are farther north than the United States–Mexico border.

**FIGURE 8.4 The European Lowland** Also known as the North European Plain, this large lowland extends from southwestern France to the plains of northern Germany and into Poland. Although this landform region has some rolling hills, most of it is less than 500 feet (150 meters) in elevation.

Austria. These impressive mountains are highest in the west, reaching more than 15,000 feet (4,575 meters) in Mt. Blanc on the French–Italian border; in Austria, to the east, few peaks exceed 10,000 feet (3,050 meters).

The Apennine Mountains, located to the south, are physically connected to the Alps by the hilly coastline of the French and Italian Riviera. Forming the mountainous spine of Italy, the Apennines are generally lower and lack the scenic glaciated peaks and valleys of the true Alps.

To the east, the Carpathian Mountains define the limits of the Alpine system in eastern Europe. They are a plow-shaped upland area

that extends from eastern Austria to the intersection of the borders of Romania and Serbia. About the same length as the main Alpine chain, the Carpathians are not nearly as high. The highest summits in Slovakia and southern Poland are less than 9,000 feet (2,780 meters).

**The Central Uplands** In western Europe, a much older highland region occupies an arc between the Alps and the European Lowland in France and Germany. These mountains are much lower in elevation than the Alpine system, with their highest peaks at 6,000 feet (1,830 meters). Their importance to western Europe is great because they contain the raw materials for Europe's industrial areas. In both Germany and France, for example, these uplands have provided the iron and coal necessary for each country's steel industry. And in the eastern part of this upland area, mineral resources have fueled major industrial areas in Germany, Poland, and the Czech Republic.

**The Western Highlands** Defining the western edge of the European subcontinent, the Western Highlands extend from Portugal in the south, through the northwest portions of the British Isles, to the highland backbone of Norway, Sweden, and Finland in the far north. These are Europe's oldest mountains, formed about 300 million years ago.

As with other upland areas that traverse many separate countries, specific place names for these mountains differ from country to country. A portion of the Western Highlands forms the highland spine of England, Wales, and Scotland, where picturesque glaciated landscapes are found at elevations of 4,000 feet (1,220 meters) or less. These U-shaped glaciated valleys are also present in Norway's uplands, where they produce a spectacular coastline of **fjords,** or flooded valley inlets similar to the coastlines of Alaska and New Zealand (Figure 8.5).

Though lower in elevation, the Fennoscandian Shield of Sweden and northern Finland is noteworthy because it is made up of some of the oldest rock formations in the world, dated conservatively at 600 million years. This **shield landscape** was eroded to bedrock by Pleistocene glaciers and, because of the cold climate and sparse vegetation, has extremely thin soils that severely limit agricultural activity.

**FIGURE 8.5 Fjord in Norway** During the Pleistocene continental ice sheets and glaciers carved deep U-shaped valleys along what is now Norway's coastline. As the ice sheets melted and sea level rose, these valleys were flooded by Atlantic waters, creating spectacular fjords. Many fjord settlements are accessible only by boat, linked to the outside world by Norway's extensive ferry system.

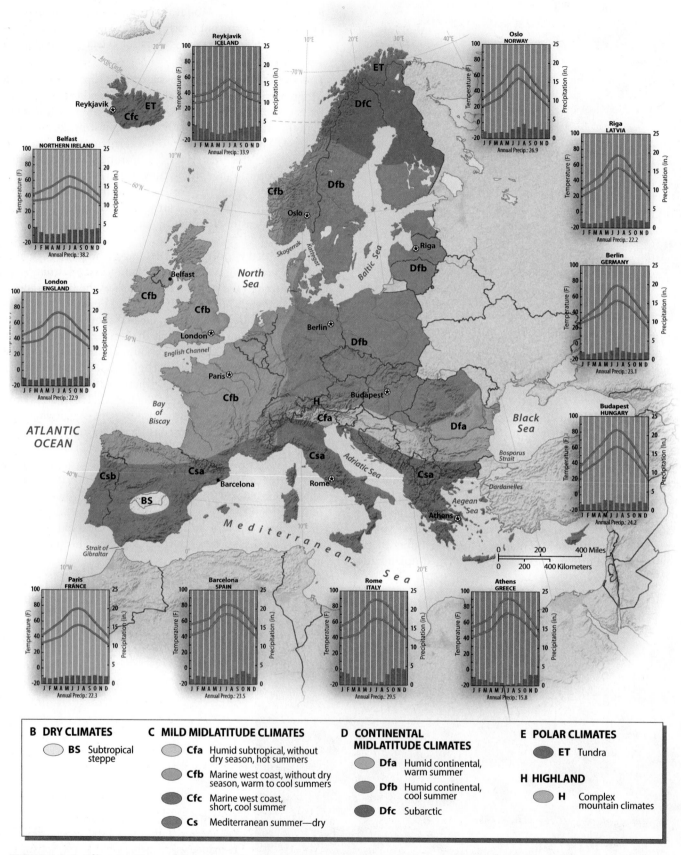

**B DRY CLIMATES**

  **BS** Subtropical steppe

**C MILD MIDLATITUDE CLIMATES**

  **Cfa** Humid subtropical, without dry season, hot summers

  **Cfb** Marine west coast, without dry season, warm to cool summers

  **Cfc** Marine west coast, short, cool summer

  **Cs** Mediterranean summer—dry

**D CONTINENTAL MIDLATITUDE CLIMATES**

  **Dfa** Humid continental, warm summer

  **Dfb** Humid continental, cool summer

  **Dfc** Subarctic

**E POLAR CLIMATES**

  **ET** Tundra

**H HIGHLAND**

  **H** Complex mountain climates

**FIGURE 8.6 Climate Map of Europe**   Three major climate zones dominate Europe. Close to the Atlantic Ocean, the marine west-coast climate has cool seasons and steady rainfall throughout the year. Farther inland, continental climates have at least one month averaging below freezing, as well as hot summers, with a precipitation maximum occurring during the warm season. Southern Europe has a dry-summer Mediterranean climate.

## Europe's Climates

Three principal climates characterize Europe (Figure 8.6). Along the Atlantic coast, a moderate and moist **maritime climate** dominates, modified by oceanic influences. Farther inland, continental climates prevail, with hotter summers and colder winters. Finally, dry-summer Mediterranean climates are found in southern Europe, from Spain to Greece.

One of the most important climate controls is the Atlantic Ocean. Though most of Europe is at a relatively high latitude (London, England, for example, is slightly farther north than Vancouver, British Columbia), the oceanic influence moderates coastal temperatures from Norway to Portugal—and even inland to the western reaches of Germany. As a result, Europe has a climate 5 to 10°F (2.8 to 5.7°C) warmer than comparable latitudes that do not have this oceanic effect. In the **marine west-coast climate** region, no winter months average below freezing, though cold rain, sleet, and occasional blizzards are common winter visitors. Summers are often cloudy and overcast, with frequent drizzle and rain. Ireland, the Emerald Isle, typifies this maritime climate.

With increasing distance from the ocean, or where a mountain chain limits the maritime influence, as in Scandinavia, landmass heating and cooling produces hotter summers and colder winters. Indeed, all **continental climates** average at least one month below freezing during the winter.

In Europe, the transition between maritime and continental climates takes place close to the Rhine River border of France and Germany. Farther north, although Sweden and other nearby countries are close to the moderating influence of the Baltic Sea, high latitude and the blocking effect of the Norwegian mountains produce cold winter temperatures characteristic of continental climates. Precipitation in continental climates comes as rain from summer storms and winter snowfall.

**FIGURE 8.7  Mediterranean Agriculture**   Because of water scarcity during the hot, dry summers in the Mediterranean climate region, local farmers have evolved agricultural strategies for making the best use of soil and water resources. Global warming is projected to make water even more scarce in the Mediterranean. In this photo from Portugal, the terraces used to prevent soil erosion on steep slopes and the mixture of tree and ground crops are evident.

The **Mediterranean climate** is characterized by a distinct dry season during the summer. While these rainless summers may attract tourists from northern Europe, the seasonal drought can be problematic for agriculture. In fact, traditional Mediterranean cultures, such as the Arab, Moorish, Greek, and Roman, all developed irrigated agriculture (Figure 8.7).

## Seas, Rivers, and Ports

Europe is a maritime region, with strong ties to its surrounding seas. Even landlocked countries such as Austria and the Czech Republic have access to the ocean through an interconnected network of navigable rivers and canals.

Europe's navigable rivers are connected by a system of canals and locks that allow inland barge travel from the Baltic and North seas to the Mediterranean and between western Europe and the Black Sea. Many rivers on the European Lowland, such as the Loire, Seine, Rhine, Elbe, and Vistula, flow into Atlantic or Baltic waters. However, the Danube and the Rhône flow in different directions. The Danube, Europe's longest river, flows east and south from Germany to the Black Sea and provides a connecting artery between central and eastern Europe. The Rhône flows southward into the Mediterranean. Both of these rivers are connected with the rivers of the European Lowland by locks and canals, allowing barge traffic to travel between all of Europe's surrounding seas and oceans.

Major ports are found at the mouths of most western European rivers, serving as transshipment points for inland waterways as well as focal points for rail and truck networks. From south to north, these ports include Bordeaux at the mouth of the Garonne, Le Havre on the Seine, London on the Thames, Rotterdam (the world's largest port in terms of tonnage) at the mouth of the Rhine, Hamburg on the Elbe River, and, to the east in Poland, Szeczin on the Oder and Gdansk on the Vistula.

## Environmental Issues: Local and Global, East and West

Because of its long history of agriculture, resource extraction, industrial manufacturing, and urbanization, Europe has its share of serious environmental problems. Compounding the situation is the fact that pollution rarely stays within political boundaries. Air pollution from England, for example, creates serious acid-rain problems in Sweden, and water pollution of the upper Rhine River by factories in Switzerland creates major problems for the Netherlands, where Rhine River water is used for urban drinking supplies. When environmental problems cross national boundaries, solutions must come from intergovernmental cooperation (Figure 8.8).

Since the 1970s, when the EU added environmental issues to its economic and political agenda, western Europe has been increasingly effective in addressing its environmental problems with regional solutions. Besides focusing on the more obvious environmental problems of air and water pollution, the EU is a world leader in recycling, waste management, reduced energy use, and sustainable resource use. As a result, western Europe is probably the "greenest" of the major world regions.

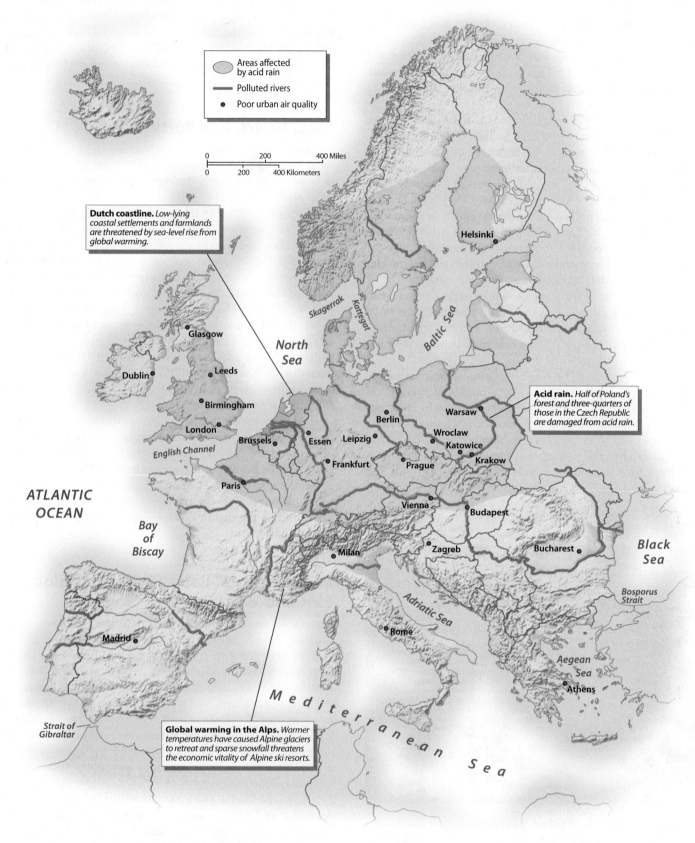

**Legend:**
- Areas affected by acid rain
- Polluted rivers
- Poor urban air quality

**Dutch coastline.** *Low-lying coastal settlements and farmlands are threatened by sea-level rise from global warming.*

**Acid rain.** *Half of Poland's forest and three-quarters of those in the Czech Republic are damaged from acid rain.*

**Global warming in the Alps.** *Warmer temperatures have caused Alpine glaciers to retreat and sparse snowfall threatens the economic vitality of Alpine ski resorts.*

**FIGURE 8.8 Environmental Issues in Europe**    While western Europe has worked energetically over the past 50 years to solve environmental problems such as air and water pollution, eastern Europe lags behind somewhat because environmental protection was not a high priority during the postwar communist period. Current efforts, though, show great promise.

**FIGURE 8.9  Acid Rain and Forest Death**   Acid precipitation has taken a devastating toll on eastern European forests, such as those shown here in Bohemia, Czech Republic. In this country, three-quarters of the forests have been damaged by acid precipitation, which is caused by industrial and auto emissions both in and outside the country.

While western Europe is successfully addressing environmental issues in that part of the region, the situation is decidedly worse in eastern Europe (Figure 8.9). During the period of Soviet economic planning (1945–1990), little attention was paid to environmental issues because of a clear emphasis on short-term industrial output. As a result, the environmental costs of that historical period have been high. Until recently, 90 percent of Poland's rivers have had no aquatic or plant life, and more than 50 percent of the country's forest trees have showed signs of damage from air pollution. Human illness is also high; one-third of Poland's population is reported to suffer from environmentally induced diseases such as cancer or respiratory illness.

The future, however, appears more positive with the expansion of EU environmental policies for new member states in eastern Europe (Figure 8.10). Even before EU membership was granted, countries such as Poland, Romania, and Slovakia had to enact environmental legislation comparable to that of western European states.

## Global Warming in Europe: Problems and Prospects

The fingerprints of global warming are everywhere in Europe, from dwindling sea ice, melting glaciers, and sparse snow cover in arctic Scandinavia to more frequent droughts in the water-starved Mediterranean (Figure 8.11). Furthermore, the projections for future climate change are ominous: World-class ski resorts in the Alps will wither away, with snowless winters, while summertime heat waves could devastate Europe's cities. Also, higher seas will threaten the Low Countries, where much settlement lies below sea level. Because of these threats, Europe has become a world leader in addressing global warming, creating numerous policies and programs to reduce greenhouse gas (GHG) emissions and adapt to changing environmental conditions.

**FIGURE 8.10  Toxic Landscape in Romania**   A troublesome legacy of Soviet communism in eastern Europe is the numerous toxic dump sites and polluted landscapes found in the region, such as this site in Romania. Money from the EU is helping to clean up these sites now that Romania is a member.

**FIGURE 8.11  Retreating Glacier in Iceland**   Glaciers throughout Europe are melting as a result of global warming. In Iceland, glacial retreat is especially complicated because subsurface volcanic activity warms the bottom layers of ice sheets.

**Europe and the Kyoto Protocol**     The EU entered the 1997 Kyoto negotiations with an innovative scheme underscoring its philosophy that regional action to solving environmental problems was superior to that of individual countries, at least in parts of the world where there was a mosaic of small states. More specifically, the EU proposed setting a collective goal of an 8 percent reduction below 1990 GHG emission levels as an umbrella for its individual member states. Under this umbrella, or EU bubble as it's called, some European states were asked to make larger reductions, while others were actually allowed an increase in GHG emissions. The point of this was to allow growth and industrial development in poorer countries, such as Spain, Greece, and Portugal, which were allowed GHG increases; at the same time, more developed states, such as Germany, Denmark, England, and France, were required to make greater reductions. As a result of this approach, Europe as a whole would see an 8 percent reduction in GHG emissions. This umbrella approach has remained a working assumption of the EU as it has grown from 15 members at Kyoto to 27 as of 2009.

**Results to Date**     Unfortunately, despite its good intentions, 2009 data show that the EU is not on target to meet its 2012 goal of an 8 percent reduction of GHG. Instead, the EU bubble has reduced emissions only 0.6 percent below the 1990 baseline. The main reason for this failure, EU experts say, is the unanticipated growth in truck transport over the past decade. In addition, Spain and other less-developed countries that were allowed GHG emission increases overshot their targets, largely because of greater-than-anticipated industrial development. Within the EU bubble, it should be noted that the major industrial powers, England and Germany, have made dramatic reductions in their emissions of over 10 percent, primarily by reducing coal usage.

To compensate for the overall slow start of the EU bubble, in early 2007, the EU agreed to set a new target of a 20 percent reduction below the 1990 baseline to be achieved by 2020. Only time will tell whether this goal is realistic and achievable (Figure 8.12).

**FIGURE 8.12 Wind Power in Northern Europe**     Not only is Europe trying to reduce its carbon dioxide emissions, but it is also a world leader in generating renewable energy from wind, sun, and biofuels. This large wind farm is in Denmark.

# Population and Settlement: Slow Growth and Rapid Migration

The map of Europe's population distribution shows that, in general, population densities are higher in the historical industrial core areas of western Europe (England, the Netherlands, northern France, northern Italy, and western Germany) than in the periphery to the east and north (Figure 8.13). While this generalization overlooks important urban clusters in Mediterranean Europe, it does express a sense of a densely settled European core set apart from a more rural, agricultural periphery.

Much of this distinctive population pattern is linked to areas of early industrialization, yet there are also many modern consequences of this core–periphery distribution. For example, economic subsidies from the wealthy, highly urbanized core to the less affluent, agricultural periphery have been an important part of the EU's development policies for several decades. Further, while the urban-industrial core is characterized by extremely low natural growth rates, it is also the target area for migrants—both legal and illegal—from Europe's peripheral countries, as well as from outside Europe.

## Natural Growth: Beyond the Demographic Transition

Probably the most striking characteristic of Europe's population is its slow natural growth (Table 8.1). More to the point, in many European countries, the death rate exceeds the birthrate, meaning that there is simply no growth at all. Instead, many countries are experiencing

**PEOPLE PER SQUARE KILOMETER**
- Fewer than 5
- 5–25
- 25–50
- 50–100
- More than 100

**POPULATION:**
- ● Metropolitan areas 1,000,000–5,000,000
- ● Metropolitan areas over 5,000,000

**Negative growth and migration.** *Negative natural growth in Germany is offset by relatively high rates of in-migration from eastern and southern Europe, former Soviet Union lands, and even Asia.*

**No growth.** *Many eastern European countries have negative natural growth. This plight is worsened by out-migration to more affluent countries of western Europe.*

**Declining population.** *Projected 2050 population for Germany is for 8 million people less than in 2009, a decline of 13%.*

**High densities.** *Dense concentrations of people in both cities and rural areas produce the highest densities in Europe—398 people per square kilometer in the Netherlands and 354 per square kilometer in Belgium.*

ATLANTIC OCEAN

North Sea

Baltic Sea

Skagerrak

Kattegat

English Channel

Mediterranean Sea

Adriatic Sea

Aegean Sea

Black Sea

Bosporus Strait

Dardanelles

Strait of Gibraltar

London · Hamburg · Berlin · Warsaw · Prague · Paris · Munich · Vienna · Budapest · Milan · Turin · Rome · Naples · Madrid · Barcelona · Belgrade · Bucharest · Sofia

**FIGURE 8.13 Population Map of Europe** The European region includes more than 531 million people, many of them clustered in large cities in both western and eastern Europe. As can be seen on this map, the most densely populated areas are in England, the Netherlands, Belgium, western Germany, northern France, and south across the Alps to northern Italy. Because most European countries have very little, if any, natural growth, in-migration is a major issue.

**TABLE 8.1   POPULATION INDICATORS**

| Country | Population (millions) 2009 | Population Density (per square kilometer) | Rate of Natural Increase (RNI) | Total Fertility Rate | Percent Urban | Percent <15 | Percent >65 | Net Migration (Rate per 1,000) 2005–2010* |
|---|---|---|---|---|---|---|---|---|
| *Western Europe* | | | | | | | | |
| Austria | 8.4 | 100 | 0.0 | 1.4 | 67 | 15 | 17 | 3.9 |
| Belgium | 10.8 | 354 | 0.2 | 1.7 | 97 | 17 | 17 | 3.8 |
| France | 62.6 | 114 | 0.4 | 2.0 | 77 | 18 | 17 | 1.6 |
| Germany | 82.0 | 230 | −0.2 | 1.3 | 73 | 14 | 20 | 1.3 |
| Liechtenstein | 0.04 | 223 | 0.4 | 1.4 | 15 | 17 | 12 | |
| Luxembourg | 0.5 | 193 | 0.4 | 1.6 | 83 | 18 | 14 | 8.4 |
| Monaco | 0.04 | 35,382 | 0.0 | | 100 | 13 | 24 | |
| Netherlands | 16.5 | 398 | 0.3 | 1.8 | 66 | 18 | 15 | 1.2 |
| Switzerland | 7.8 | 188 | 0.2 | 1.5 | 73 | 15 | 17 | 2.7 |
| United Kingdom | 61.8 | 255 | 0.4 | 1.9 | 80 | 18 | 16 | 3.1 |
| *Eastern Europe* | | | | | | | | |
| Bulgaria | 7.6 | 68 | −0.4 | 1.5 | 71 | 13 | 17 | −1.3 |
| Czech Republic | 10.5 | 133 | 0.1 | 1.5 | 74 | 14 | 15 | 4.4 |
| Hungary | 10.0 | 108 | −0.3 | 1.3 | 66 | 15 | 16 | 1.5 |
| Poland | 38.1 | 122 | 0.1 | 1.4 | 61 | 15 | 14 | −0.6 |
| Romania | 21.5 | 90 | −0.1 | 1.3 | 55 | 15 | 15 | −1.9 |
| Slovakia | 5.4 | 110 | 0.1 | 1.3 | 56 | 15 | 12 | 0.7 |
| *Southern Europe* | | | | | | | | |
| Albania | 3.2 | 111 | 0.6 | 1.3 | 49 | 25 | 9 | −4.8 |
| Bosnia and Herzegovina | 3.8 | 75 | 0.0 | 1.2 | 46 | 16 | 14 | −0.5 |
| Croatia | 4.4 | 78 | −0.2 | 1.5 | 56 | 15 | 17 | 0.5 |
| Cyprus | 1.1 | 116 | 0.5 | 1.5 | 62 | 18 | 11 | 5.8 |
| Greece | 11.3 | 85 | 0.0 | 1.4 | 60 | 14 | 19 | 2.7 |
| Italy | 60.3 | 200 | −0.0 | 1.4 | 68 | 14 | 20 | 5.6 |
| Macedonia | 2.0 | 80 | 0.2 | 1.5 | 65 | 19 | 11 | −1.0 |
| Malta | 0.4 | 1,310 | 0.2 | 1.4 | 94 | 16 | 14 | 2.5 |
| Montenegro | 0.6 | 45 | 0.3 | 1.7 | 64 | 20 | 13 | −1.6 |
| Portugal | 10.6 | 116 | −0.0 | 1.3 | 55 | 15 | 17 | 3.8 |
| San Marino | 0.03 | 515 | 0.3 | 1.2 | 84 | 15 | 16 | |
| Serbia | 7.3 | 95 | −0.5 | 1.4 | 56 | 15 | 17 | 0.0 |
| Slovenia | 2.0 | 101 | 0.2 | 1.5 | 48 | 14 | 16 | 2.2 |
| Spain | 46.9 | 93 | 0.3 | 1.5 | 77 | 14 | 17 | 7.9 |
| *Northern Europe* | | | | | | | | |
| Denmark | 5.5 | 128 | 0.2 | 1.9 | 72 | 19 | 17 | 1.1 |
| Estonia | 1.3 | 30 | −0.0 | 1.7 | 69 | 15 | 17 | 0.0 |
| Finland | 5.3 | 16 | 0.2 | 1.9 | 63 | 17 | 17 | 2.1 |
| Iceland | 0.3 | 3 | 0.9 | 2.1 | 93 | 21 | 12 | 12.8 |
| Ireland | 4.5 | 64 | 1.0 | 2.0 | 60 | 21 | 11 | 9.1 |
| Latvia | 2.3 | 35 | −0.3 | 1.4 | 68 | 14 | 17 | −0.9 |
| Lithuania | 3.3 | 51 | −0.2 | 1.5 | 67 | 15 | 16 | −6.0 |
| Norway | 4.8 | 13 | 0.4 | 2.0 | 80 | 19 | 15 | 5.7 |
| Sweden | 9.3 | 21 | 0.2 | 1.9 | 84 | 17 | 18 | 3.3 |

Source: Population Reference Bureau, *World Population Data Sheet, 2009.*

*Net Migration Rate from the United Nations, Population Division, *World Population Prospects: The 2008 Revision Population Database.*

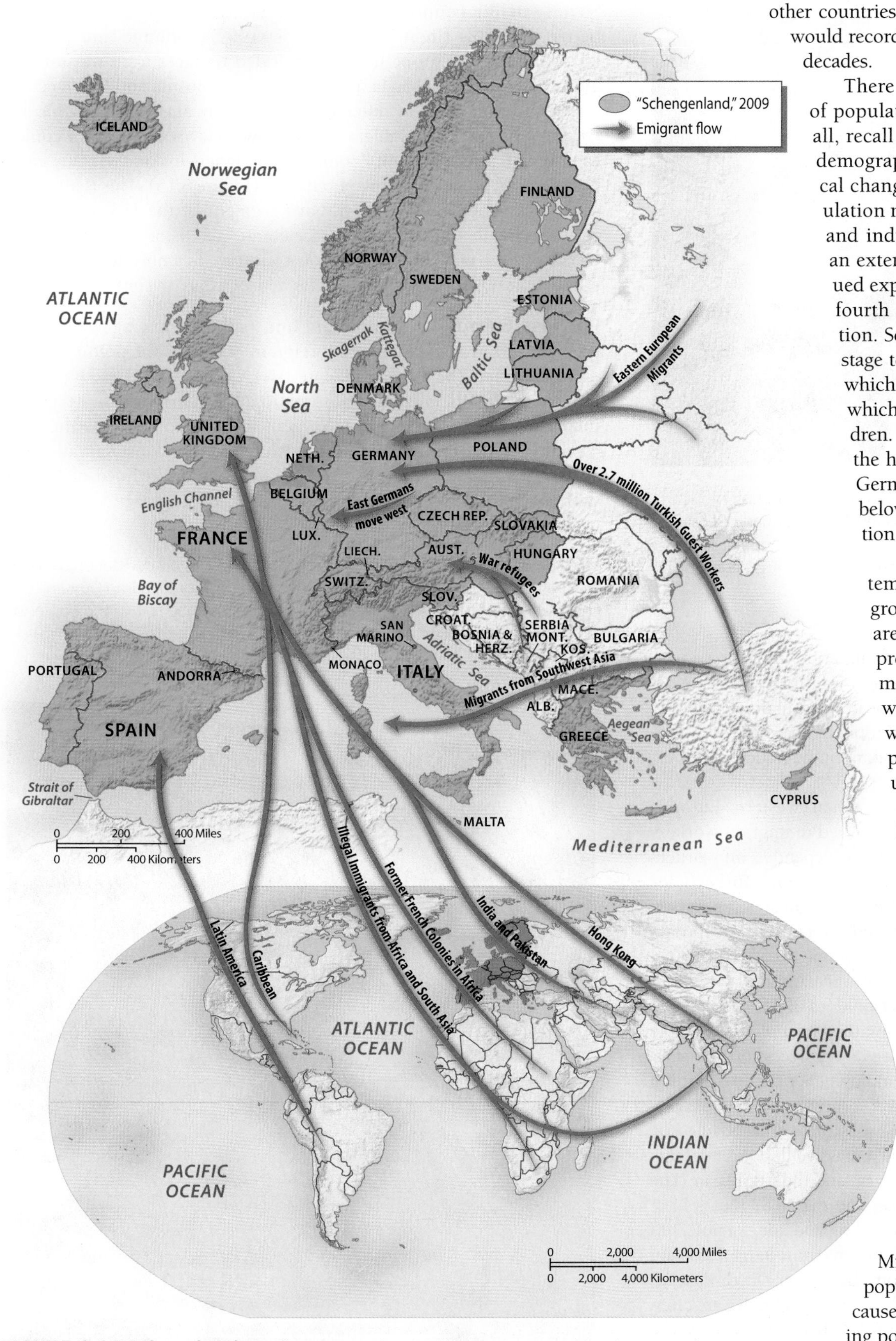

negative growth rates, and were it not for in-migration from other countries and other world regions, these countries would record declines in population over the next few decades.

There seem to be several reasons for the lack of population growth in western Europe. First of all, recall from Chapter 1 that the concept of the demographic transition was based on the historical change in European growth rates as the population moved from rural settings to more urban and industrial locations. What we see today is an extension of that model—that is, the continued expression of the low-fertility/low-mortality fourth stage of the demographic transformation. Some demo-graphers suggest adding a fifth stage to the model—a "postindustrial" phase, in which population falls below replacement levels, which means parents have fewer than two children. Evidence for this fifth stage comes from the highly urban and industrial populations of Germany, France, and England, all of which are below zero population growth when immigration is excluded.

These low-birthrate countries are attempting to increase their rates of natural growth. In Germany, for example, couples are given an outright monetary gift if they produce a child. In Austria, the government gives a modest cash award to couples when they marry, and still more follows when they give birth. France provides parents with monthly payments that last until their children reach school age.

In addition, most EU countries have liberal maternity benefits that guarantee both parents time off from their jobs without penalty. Yet despite these incentives, most European countries still have natural growth rates below replacement levels, and several, most notably in the former communist countries of eastern Europe, actually show negative growth. In these countries, the political and economic turmoil associated with the transition from communism to capitalism may explain these very low birthrates.

## Migration to and Within Europe

Migration is one of the most challenging population issues facing Europe today because the region is caught in a web of conflicting policies and values (Figure 8.14). Currently,

**FIGURE 8.14 Migration into Europe** Historically, Europe opened its doors to migrants to help solve post–World War II labor shortages, but now immigration is a contentious and controversial issue, as culturally homogeneous countries such as France, England, and Germany confront tensions resulting from large numbers of immigrants, legal and illegal.

**FIGURE 8.15 Mosque in Germany**    Girls from the Turkish Muslim community stand in front of their new mosque in Moers, Germany. Because of Germany's need for immigrant labor as the economy expanded in the 1960s, large numbers of "guest workers" settled in the country, beginning a pattern that is still evident today.

there is widespread resistance to unlimited migration into Europe, primarily because of high unemployment in the western European industrial countries. Many Europeans argue that scarce jobs should go first to European citizens, not to immigrants. This topic has become so controversial that the EU is working toward a common immigration policy for its member countries.

During the 1960s postwar recovery, when western Europe's economies were booming, many countries looked to migrant workers to ease labor shortages. Germany, for example, depended on workers from Europe's periphery—Italy, Yugoslavia, Greece, and Turkey—for industrial and service jobs. These *gastarbeiter*, or **guest workers**, arrived by the thousands. As a result, large ethnic enclaves of foreign workers became a common part of the German urban landscape. Today, there are more than 2 million Turks in Germany, most of whom are there because of the open-door worker policies of past decades.

However, these foreign workers are now the target of considerable ill will. Western Europe has suffered from economic recession and stagnation for the past decade, resulting in unemployment rates approaching 25 percent for young people. As mentioned earlier, many native Europeans protest that guest workers are taking jobs that could be filled by them. In addition, the region has witnessed a massive flood of migrants from former European colonies in Asia, Africa, and the Caribbean. The former colonial powers of England, France, and the Netherlands have been the major recipients of this immigration. England, for example, has inherited large numbers of former colonial citizens from India, Pakistan, Jamaica, and Hong Kong. Indonesians (from the former Dutch East Indies) are common in the Netherlands, while migrants in France are often from former colonies in both northern and Sub-Saharan Africa.

More recently, political and economic troubles in eastern Europe and the former Soviet Union have generated a new wave of migrants to Europe. With the total collapse of Soviet border controls in 1990,

emigrants from Poland, Bulgaria, Romania, Ukraine, and other former Soviet satellite countries poured into western Europe, looking for a better life. This flight from the post-1989 economic and political chaos of eastern Europe has also included refugees from war-torn regions of former Yugoslavia, particularly from Bosnia and Kosovo.

As a result of these different migration streams, Germany has become a reluctant land of migrants that now receives 400,000 newcomers each year. Currently, about 7.5 million foreigners live in Germany, making up about 9 percent of the population (Figure 8.15). This is about the same percentage as the foreign-born segment of the U.S. population. However, while the United States celebrates its history of immigration, Germany, like most other European countries, is ambivalent about this ethnic mixture and struggles with its new cultural diversity.

**Fortress Europe**    An important and perhaps even aggravating ingredient in Europe's issue with foreign migration is the political agreement between the heartland countries of Europe to facilitate free movement for its citizens across national borders, without passport inspections or checks. This agreement has reshaped the political geography of Europe by creating divisions between insiders and outsiders. To those on the inside, it creates a long-dreamed-of Europe without borders, while to those on the outside, it presents a "Fortress

**FIGURE 8.16 Czech Republic Border Station**    This border station between Germany and the Czech Republic was once a highly fortified Iron Curtain checkpoint. But today, with the Czech Republic's membership in the EU, most people pass through with a cursory wave after showing their passport.

**FIGURE 8.17 Urban Landscapes** This aerial view of Grosseto, Italy shows how the historic medieval city was encircled by the Renaissance-Baroque fortifications built to protect the settlement. Today, parks and public buildings are found in place of the former walls and moat.

Europe," a defensive perimeter perceived as hostile to migrants from Asia, Africa, and the Russian domain. At the so-called "hard" borders on the EU's perimeter, especially in Spain, Italy, Slovakia, and Poland, foreigners and foreign goods are subject to lengthy passport checks, visa requirements, and searches, whereas crossing a "soft" border within the European heartland involves little more than a visual check and a wave from the border police, if there is even a border station.

Taking its name from the city of Schengen, Luxembourg, where the original declaration of intent was signed in 1985, the EU's **Schengen Agreement** has as one of its goals the gradual reduction of border formalities for travelers moving between EU countries (Figure 8.16). Though reduced border formalities within western Europe seemed like reasonable and desirable goals in 1985, the situation is much more complicated today because of the large number of illegal migrants from the former Soviet lands, Africa, and South Asia.

## The Landscapes of Urban Europe

One of the major characteristics of Europe's population is its high level of urbanization. All but several Balkan countries have more than half their population in cities, while several western European countries, such as the United Kingdom and Belgium, are more than 90 percent urbanized. Once again, the gradient from the highly urbanized European heartland to the less-urbanized

**FIGURE 8.18 Historic Preservation of Cities** Most European countries have laws and regulations designed to preserve and protect their cultural heritage, as embodied in historical cities. One of the earliest historic preservation projects in Germany was the inner city of Regensburg, on the Danube River.

periphery reinforces the distinctions between the affluent, industrial core area and the more rural, less-well-developed periphery.

**The Past in the Present** Three historical eras dominate most European city landscapes. The medieval (roughly 900–1500), Renaissance–Baroque (1500–1800), and industrial (1800–present) periods have each left characteristic marks on the European urban scene. Learning to recognize these stages of historical growth provides visitors to Europe's cities with fascinating insights into both past and present landscapes (Figure 8.17).

The **medieval landscape** is one of narrow, winding streets, crowded with three- or four-story masonry buildings with little setback from the street. This is a dense landscape with few open spaces, except around churches and public buildings. Here and there, public squares or parks are clues to medieval open-air marketplaces where commerce was transacted.

As picturesque as we find medieval-era districts today, they nevertheless present challenges to modernization because of their narrow, congested streets and old housing. Often, modern plumbing and heating are lacking, and rooms and hallways are small and cramped compared to present-day standards. Because these medieval districts usually lack modern facilities, the majority of people living in them have low or fixed incomes. In many areas, this majority is made up of the elderly, university students, and ethnic migrants.

Many cities in Europe, though, are enacting legislation to upgrade and protect their historic medieval landscapes. This movement began in the late 1960s and has become increasingly popular as cultures have worked to preserve the uniqueness of special urban centers as their suburbs increasingly surrender to sprawl and shopping centers (Figure 8.18).

In contrast to the cramped and dense medieval landscape, those areas of the city built during the **Renaissance–Baroque** period produce a landscape that is much more open and spacious, with expansive ceremonial buildings and squares, monuments, ornamental gardens, and wide boulevards lined with palatial residences. During this period (1500–1800), a new artistic sense of urban planning arose in Europe that resulted in the restructuring of many European cities, particularly the large capitals.

These changes were primarily for the benefit of the new urban elite—royalty and successful merchants. City dwellers of lesser means remained in the medieval quarters, which became increasingly crowded and cramped as more city space was devoted to the ruling classes.

During this period, city fortifications limited the outward spread of these growing cities, thus increasing density and crowding within. With the advent of assault artillery, European cities were forced to build an extensive system of defensive walls. Once encircled by these walls, the cities could not expand outward. Instead, as the demand for space increased within the cities, a common solution was to add several new stories to the medieval houses.

Industrialization dramatically altered the landscape of European cities. Historically, factories clustered together in cities beginning in the early 19th century, drawn by their large markets and labor force and supplied by raw materials shipped by barge and railroad. Industrial districts of factories and worker tenements grew up around these transportation lines. In continental Europe, where many cities retained their defensive walls until the late 19th century, the new industrial districts were often located outside the former city walls, removed from the historic central city. In Paris, for example, when the railroad was constructed in the 1850s, it was not allowed to enter the city walls. As a result, terminals and train stations for the network of tracks were located beyond the original fortifications. Although the walls of Paris are long gone, this pattern of train stations lying beyond the city's historical city walls exists still today.

## Cultural Coherence and Diversity: A Mosaic of Differences

The rich cultural geography of Europe demands our attention for several reasons. First, the highly varied and fascinating mosaic of languages, customs, religions, ways of life, and landscapes that characterizes Europe has also strongly shaped the regional identities that have all too often stoked the fires of conflict.

Second, European cultures have played leading roles in processes of globalization. European colonialism has brought about changes in regional languages, religion, economies, and values in every corner of the globe. If you question this, consider cricket games in Pakistan, high tea in India, Dutch architecture in South Africa, and the millions of French-speaking inhabitants of equatorial Africa. Before modern technologies such as satellite TV, the Internet, and Hollywood films and video, European culture spread across the world, changing the speech, religion, belief systems, dress, and habits of millions of people on every continent.

Today, though, new waves of global culture spread into Europe (Figure 8.19). While some cultures embrace (or simply condone) these changes, other European cultures actively resist. France, for example, struggles against both U.S.-dominated popular culture and the multicultural influences of its large migrant population. In many ways, the same is true of Germany and England.

### Geographies of Language

Language has always been an important component of nationalism and group identity in Europe (Figure 8.20). Today, while some small ethnic groups, such as the Irish or the Bretons, work hard to preserve their local language in order to reinforce their cultural identity, millions of Europeans are also busy learning multiple languages so they can communicate across cultural and national boundaries. The EU itself, while primarily an integrating force in Europe, honors this linguistic mosaic by recognizing more than 20 official languages.

As their first language, 90 percent of Europe's population speaks Germanic, Romance, or Slavic languages, all of which are linguistic groups of the Indo-European family. Germanic and Romance speakers each number almost 200 million in the European region. There are far fewer Slavic speakers (about 80 million) when Europe's boundaries are drawn to exclude Russia, Belarus, and Ukraine.

**Germanic Languages** Germanic languages dominate Europe north of the Alps. Today, about 90 million people speak German as their first language. This is the dominant language of Germany, Austria, Liechtenstein, Luxembourg, eastern Switzerland, and several small areas in Alpine Italy. Until recently, there were also large German-speaking minorities in Romania, Hungary, and Poland, but many of these people left eastern Europe when the Iron Curtain was lifted in 1990 and resettled in Germany.

English is the second-largest Germanic language, with about 60 million speakers using it as their first language. In addition, a large number of Europeans learn English as a second language, particularly in the Netherlands and Scandinavia, where many are as fluent as native speakers. Linguistically, English is closest to the Low German spoken along

**FIGURE 8.19 Global Culture in Europe** U.S. popular culture is received with great ambivalence in Europe. While many people embrace everything from fast food to Hollywood movies, at the same time they protest the loss of Europe's traditional regional cultures. This photo was taken in Prague, Czech Republic.

**FIGURE 8.20 Language Map of Europe**   Ninety percent of
Europeans speak an Indo-European language. These languages can be grouped into the major
categories Germanic, Romance, and Slavic languages. Ninety million Europeans speak German as a first language, which places
it ahead of the 60 million who list English as their native language. However, given the large number of Europeans who speak
fluent English as a second language, one could make the case that English is the dominant language of modern Europe.

the coastline of the North Sea, which reinforces the linguistic theory that an early form of English evolved in the British Isles through contact with the coastal peoples of northern Europe. However, one of the distinctive traits of English that sets it apart from German is that almost one-third of the English vocabulary is made up of Romance words brought to England during the Norman French conquest of the 11th century.

Elsewhere in this region, Dutch (the Netherlands) and Flemish (northern Belgium) account for another 20 million people, and roughly the same number of Scandinavians speak the closely related languages Danish, Norwegian, and Swedish. Icelandic, though, is a distinct language because of its geographic separation from its Scandinavian roots.

**Romance Languages** Romance languages, such as French, Spanish, and Italian, evolved from the vulgar (or everyday) Latin used within the Roman Empire. Today, Italian is the largest of these regional dialects, with about 60 million Europeans speaking it as their first language. In addition, Italian is an official language of Switzerland and is also spoken on the French island of Corsica.

French is spoken in France, western Switzerland, and southern Belgium, where it is known as *Walloon*. Today, there are about 55 million native French speakers in Europe. As with other languages, French also has very strong regional dialects. Linguists differentiate between two forms of French in France itself: that spoken in the north (the official form because of the dominance of Paris) and the language of the south, or *langue d'oc*. This linguistic divide expresses long-standing tensions between Paris and southern France. In the past decade, the strong regional awareness of the southwest (centered on Toulouse and the Pyrenees) has led to a rebirth of its own distinct language, *Occitanian*.

Spanish also has very strong regional variations. About 25 million people speak Castilian Spanish, the country's official language, which dominates the interior and northern areas of that large country. However, the *Catalan* form, which some argue is a completely separate language, is found along the eastern coastal fringe, centered on Barcelona, Spain's major city in terms of population and the economy. This distinct language reinforces a strong sense of cultural separateness that has led to the state of Catalonia being given autonomous status within Spain.

Portuguese is spoken by another 12 million speakers in that country and in the northwestern corner of Spain, although considerably more people speak the language in Brazil, a former Portuguese colony in Latin America. Finally, Romanian represents the most eastern extent of the Romance language family; it is spoken by 24 million people in Romania. Though unquestionably a Romance language, Romanian also contains many Slavic words.

## The Slavic Language Family

Slavic is the largest European subfamily of the Indo-European languages. Traditionally, Slavic speakers are separated into northern and southern groups, divided by the non-Slavic speakers of Hungary and Romania.

To the north, Polish has 35 million speakers, and Czech and Slovakian about 14 million each. These numbers pale in comparison, however, with the number of northern Slav speakers in nearby Ukraine, Belarus, and Russia, where one can easily count more than 150 million. Southern Slav languages include three groups: 14 million Serbo-Croatian speakers (now considered separate languages because of the political troubles between Serbs and Croats), 11 million Bulgarian-Macedonian, and 2 million Slovenian.

The use of two alphabets further complicates the geography of Slavic languages (Figure 8.21). In countries with a strong Roman Catholic heritage, such as Poland and the Czech Republic, the Latin alphabet is used in writing. In contrast, countries with close ties to the Orthodox church—such as Bulgaria, Macedonia, Bosnia, and Serbia—use the Greek-derived **Cyrillic alphabet**.

## Geographies of Religion, Past and Present

Religion is an important component of the geography of cultural coherence and diversity in Europe because many of today's ethnic tensions result from historical religious events. To illustrate, strong cultural borders in the Balkans and eastern Europe are drawn based upon the 11th-century split of Christianity into eastern and western churches, or

**FIGURE 8.21 The Alphabet of Ethnic Tension** With the departure of many of Kosovo's Serbs during recent ethnic unrest, signs using the Cyrillic alphabet were taken off many stores, shops, and restaurants, as Kosovars of Albanian ethnicity restated their claims to the region. Here the Albanian owner of a restaurant in the capital city of Pristina scrapes off Cyrillic letters.

between Christianity and Islam; in Northern Ireland, blood is still shed over the tensions between the 17th-century division of Christianity into Catholicism and Protestantism; and much of the ethnic cleansing terrorism in former Yugoslavia resulted from the historical struggle between Christianity and Islam in that part of Europe. In addition, there is considerable tension regarding the large Muslim migrant populations in England, France, and Germany. Understanding these important contemporary issues thus involves taking a brief look at the historical geography of Europe's religions (Figure 8.22).

**The Schism Between Western and Eastern Christianity** In southeastern Europe, early Greek missionaries spread Christianity through the Balkans and into the lower reaches of the Danube. Progress was slower than in western Europe, perhaps because of continued invasions by peoples from the Asian steppes. There were other problems as well, primarily the refusal of these Greek missionaries to accept the control of Roman bishops from western Europe.

This tension with western Christianity led in 1054 to an official split of the eastern church from Rome. This eastern church subsequently splintered into Orthodox sects closely linked to specific nations and states. Today, for example, we find Greek Orthodox, Bulgarian Orthodox, and Russian Orthodox churches, all with different rites and rituals.

Another factor that distinguished eastern Christianity from western was the Orthodox use of the Cyrillic alphabet instead of the Latin. Because Greek missionaries were primarily responsible for the spread of early Christianity in southeastern Europe, it is not surprising that they used an alphabet based on Greek characters. More precisely, this alphabet is attributed to the missionary work of St. Cyril in the 9th century. As a result, the division between western and eastern churches, and between the two alphabets, remains one of the most problematic cultural boundaries in Europe.

**FIGURE 8.22 Religions of Europe**   This map shows the divide in western Europe between the Protestant north and the Roman Catholic south. Historically, this distinction was much more important than it is today. Note the location of the former Jewish Pale, which was devastated by the Nazis during World War II. Today, ethnic tensions with religious overtones are found primarily in the Balkans, where adherents of Roman Catholicism, Eastern Orthodoxy, and Islam are found in close proximity to one another.

**The Protestant Revolt**   Besides the division between western and eastern churches, the other great split within Christianity occurred between Catholicism and Protestantism. This division arose in Europe during the 16th century and has divided the region ever since. However, with the exception of the troubles in Northern Ireland, tensions today between these two major groups are far less troublesome than in the past.

**Conflicts with Islam**   Both eastern and western Christian churches struggled with challenges from Islamic empires to Europe's south and east. Even though historical Islam was reasonably tolerant of Christianity in its conquered lands, Christian Europe was not accepting of Muslim imperialism. The first crusade to reclaim Jerusalem from the Turks took place in 1095. After the Ottoman Turks conquered Constantinople in 1453 and gained control over the Bosporus strait and the Black

Sea, they moved rapidly to spread their Muslim empire throughout the Balkans and arrived at the gates of Vienna in the middle of the 16th century. There, Christian Europe stood firm and stopped Islam from expanding into western Europe.

Ottoman control of southeastern Europe, however, lasted until the empire's end in the early 20th century. This historical presence of Islam explains the current mosaic of religions in the Balkans, with intermixed areas of Muslims, Orthodox, and Roman Catholics.

Today, there are still tensions with Muslims as Europe struggles to address the concerns of its fast-growing immigrant population. Although many of these tensions transcend pure religious issues, there is nevertheless a strong undercurrent of concern about the varying attitudes of Muslims toward cultural assimilation in Europe. In addition, the EU must address the issue of Turkey's application to join the EU. Although Turkey is an avowed secular state, its population is predominantly Muslim, thus its membership in the EU is a contentious issue for Europe.

**A Geography of Judaism** Europe has long been a difficult homeland for Jews forced to leave Palestine during the Roman Empire. At that time, small Jewish settlements were located in cities throughout the Mediterranean. Later, by 900 CE, about 20 percent of the Jewish population was clustered in the Muslim lands of the Iberian Peninsula, where Islam showed greater tolerance for Judaism than had Christianity. Furthermore, Jews played an important role in trade activities both within and outside the Islamic lands. After the Christian reconquest of Iberia, however, Jews once more faced severe persecution and fled from Spain to more tolerant countries in western and central Europe.

One focus for migration was the area in eastern Europe that became known as the Jewish Pale. In the late Middle Ages, at the invitation of the Kingdom of Poland, Jews settled in cities and small villages in what is now eastern Poland, Belarus, western Ukraine, and northern Romania (see Figure 8.22). Jews collected in this region for several centuries, in the hope of establishing a true European homeland, despite the poor natural resources of this marshy, marginal agricultural landscape.

Until emigration to North America began in the 1890s, 90 percent of the world's Jewish population lived in Europe, and most were clustered in the Pale. Even though many emigrants to the United States and Canada came from this area, the Pale remained the largest grouping of Jews in Europe until World War II. Tragically, Nazi Germany used this ethnic clustering to its advantage by focusing its extermination activities on the Pale.

In 1939, on the eve of World War II, there were 9.5 million Jews in Europe, or about 60 percent of the world's Jewish population. During the war, German Nazis murdered some 6 million Jews in the horror of the Holocaust. Today, fewer than 2 million Jews live in Europe. Since 1990 and the lifting of quotas on Jewish emigration from Russia, Belarus, and Ukraine, more than 100,000 Jews have emigrated to Germany, giving it the fastest-growing Jewish population outside Israel (Figure 8.23).

**The Patterns of Contemporary Religion** In Europe today, there are about 250 million Roman Catholics and fewer than 100 million Protestants. Generally, Catholics are found in the southern half of the region, except for significant numbers in Ireland and Poland, while Protestants dominate in the north. In addition, since World War II, there has been a noticeable loss of interest in organized religion, mainly in western Europe, which has led to declining church attendance in

**FIGURE 8.23 Jewish Synagogue in Berlin** Before World War II, Berlin had a large and thriving Jewish population, as attested to by this synagogue built in 1866. However, during the Nazi period, the Jewish population was forcibly removed and largely exterminated. This synagogue was then used by the Nazis to store military clothing, until it was heavily damaged by Allied bombing. It was restored and opened again in 1995 as a synagogue and museum.

many areas. This trend is so marked that the term **secularization** is used, referring to the widespread movement away from the historically important organized religions of Europe.

Not to be overlooked are the 13 million Muslims in Europe. Most are migrants and their families from Africa and southwestern Europe, although some are recent converts to Islam. More than half of the Muslim population (7.5 million) is found in France, with the second largest number (around 4 million) in Germany. While most Muslims worship in mosques converted from existing buildings, several new—and rather grandiose—mosques have been built lately in France, Germany, and Britain.

Catholicism dominates the religious geography and cultural landscapes of Italy, Spain, France, Austria, Ireland, and southern Germany. In these areas, large cathedrals, monasteries, monuments to Christian saints, and religious place-names draw heavily on pre-Reformation Christian culture. Because visible beauty is part of the Catholic tradition, elaborate religious structures and monuments are more common here than in Protestant lands.

Protestantism is most widespread in northern Germany, the Scandinavian countries, and England, and it is intermixed with Catholicism in the Netherlands, Belgium, and Switzerland. Because of its reaction against ornate cathedrals and statues of the Catholic Church, the landscape of Protestantism is much more sedate and subdued.

**FIGURE 8.24 Religious Tensions in Northern Ireland**   Despite considerable and continual efforts to forge peace between Protestants and Catholics in Northern Ireland, outbreaks of violence are still common. Here, in Belfast, sectarian fighting broke out only hours after groups from both religions met to discuss ways to end the fighting.

Large cathedrals and religious monuments in Protestant countries are associated primarily with the Church of England, which has strong historical ties to Catholicism; St. Paul's Cathedral and Westminster Abbey in London are examples.

Tragically, another sort of religious landscape has emerged in Northern Ireland, where barbed-wire fences and concrete barriers separate Protestant and Catholic neighborhoods in an attempt to reduce violence between warring populations (Figure 8.24). Religious affiliation is a major social force that influences where people live, work, attend school, shop, and so on. Of the 1.6 million inhabitants of Northern Ireland, 54 percent are Protestant and 42 percent Roman Catholic. The Catholics feel that they have been discriminated against by the Protestant majority and have been treated as second-class citizens. As a result, they want a closer relationship with the Republic of Ireland across the border to the south. The Protestant reaction is to forge even stronger ties with the United Kingdom. Unfortunately, these differences have been expressed in prolonged violence that has led to about 4,000 deaths since the 1960s, largely resulting from paramilitary groups from both factions who promote their political and social agendas through terrorist activities.

## European Culture in a Global Context

Europe, like all other world regions, is currently caught up in a period of profound cultural change; in fact, many would argue that the pace of cultural change in Europe has been accelerated because of the complicated interactions between globalization and Europe's internal agenda of political and economic integration. While pundits celebrate the "New Europe" of integration and unification, other critics refer to a more tension-filled New Europe of foreign migrants and guest workers troubled by ethnic discrimination and racism.

**Globalization and Cultural Nationalism**   Since World War II, Europe has been trying to control cultural contamination from North America. Some countries are more outspoken than others. While a few large countries, such as England and Italy, seem to accept the onrush of U.S. popular culture, other countries have expressed outright indignation over the corrupting impact of U.S. popular culture on speech, music, food, and fashion. France, for example, is often thought of as the poster child for the anti-globalization struggle, as it fights to preserve its local foods and culture against the homogenizing onslaught of fast-food outlets and genetically engineered crops.

**Migrants and Culture**   Migration patterns are influencing the cultural mix in Europe. Historically, Europe spread its cultures worldwide through aggressive colonialization. Today, however, the region is experiencing a reverse flow, as millions of migrants move into Europe, bringing their own distinct cultures from the far-flung countries of Africa, Asia, and Latin America. Unfortunately, in some areas of Europe, the products of this cultural exchange are highly problematic.

Ethnic clustering leading to the formation of ghettos is now common in the cities and towns of western Europe. The high-density apartment buildings of suburban Paris, for example, are home to large numbers of French-speaking Africans and Arab Muslims caught in the crossfire of high unemployment, poverty, and racial discrimination. As a result, cultural battles have emerged in many European countries. For example, French leaders, unsettled by the country's large Muslim migrant population, attempted to speed assimilation of female high school students into French mainstream culture by banning a key symbol of conservative Muslim life: the head scarf. This rule triggered riots, demonstrations, and counterdemonstrations. As a result of these kinds of conflicts, the political landscape of many European countries now contains far-right, nationalistic parties with thinly veiled agendas of excluding migrants from their countries (Figure 8.25).

**FIGURE 8.25 Neo-Nazis in Germany**   Purporting to embrace "pure and true" Aryan values, these neo-Nazi Germans provoke police at a rally against foreigners. A rise in extreme forms of nationalism has become increasingly problematic in Germany, with anger often directed against Turks and other immigrants.

# Geopolitical Framework: A Dynamic Map

One of Europe's unique characteristics is its dense fabric of 41 independent states within a relatively small area. No other world region demonstrates such a mosaic of geopolitical division. Europe invented the nation-state. Later, these same political ideas founded in Europe fueled the flames of political independence and democracy worldwide, replacing Europe's colonial rule in Asia, Africa, and the Americas.

But Europe's diverse geopolitical landscape has been as much problem as promise. Twice in the past century, Europe shed blood to redraw its political borders. Within the past several decades, nine new states have appeared in Europe, more than half through war. Further, today's map of geopolitical troubles suggests that still more political fragmentation may take place in the near future (Figure 8.26).

## Redrawing the Map of Europe Through War

Two world wars redrew the geopolitical maps of 20th-century Europe (Figure 8.27). Because of these conflicts, empires and nation-states have appeared and disappeared within the past 100 years. By the early 20th century, Europe was divided into two opposing and highly armed camps that tested each other for a decade before the outbreak of World War I in 1914. France, Britain, and Russia were allied against the new nation-states Italy and Germany, along with the Austro-Hungarian (or Hapsburg) Empire, which controlled a complex assortment of ethnic groups in central Europe and the Balkans. While at the time World War I was referred to as the "war to end all wars," it fell far short of solving Europe's geopolitical problems. Instead, according to many experts, it made another European war unavoidable.

When Germany and Austria-Hungary surrendered in 1918, the Treaty of Versailles peace process set about redrawing the map of Europe with two goals in mind: first, to punish the losers through loss of territory and severe financial reparations and, second, to recognize the nationalistic aspirations of unrepresented peoples by creating new nation-states. As a result, the new states Czechoslovakia and Yugoslavia were born. In addition, Poland was reestablished, as were the Baltic states Finland, Estonia, Latvia, and Lithuania.

Though the goals of the treaty were admirable, few European states were satisfied with the resulting map. New states were resentful when their ethnic citizens were left outside the new borders and became minorities in other new states. This created an epidemic of **irredentism**, or state policies for reclaiming lost territory and peoples. Examples include the large German population in the western portion of the newly created state of Czechoslovakia and the Hungarians stranded in western Romania by border changes.

This imperfect geopolitical solution was greatly aggravated by the global economic depression of the 1930s that brought high unemployment, food shortages, and political unrest to Europe. Three competing ideologies promoted their own solutions to Europe's pressing problems: Western democracy (and capitalism), communism from the Soviet revolution to the east, and a fascist totalitarianism promoted by Mussolini in Italy and Hitler in Germany. With industrial unemployment commonly approaching 25 percent in western Europe, public opinion fluctuated wildly between extremist solutions of fascism and communism. In 1936, Italy and Germany once again joined forces through the Rome–Berlin "axis" agreement. As in World War I, this alignment was countered with mutual protection treaties between France, Britain, and the Soviet Union. When an imperialist Japan signed a pact with Germany, the scene was set for a second global war.

---

# EXPLORING GLOBAL CONNECTIONS
## Europe's Ties to Russian Natural Gas

The politics of heating Europe with natural gas involves complicated interactions with countries in at least three other world regions: the Russian domain, Central Asia, and Southwest Asia. Currently, one-quarter of Europe's natural gas supply comes from Russia, a fact that makes many Europeans nervous because Russia has used this resource as a political weapon by shutting off the gas and freezing out countries when political temperatures soar.

At present, two major gas pipelines transect Ukraine on their route from Russia to Europe. Ukraine, once a loyal member of the Soviet Union, has more recently become a West-leaning independent country with aspirations of joining both the European Union and NATO. Russia objects to those negotiations and has tried freezing Ukraine into a more cozy relationship by regularly shutting of its gas supply. Gazprom, the powerful Russian company that controls this resource, explains these periodic shutdowns by saying Ukraine is stealing gas from the pipelines illegally, or that it hasn't paid the bill, or, simply, that there are "technical problems" with the pipeline.

But it's not just Ukraine that suffers from these shutdowns. Europe—particularly Germany— does, too. For example, in January 2009, when political tempers flared between Russia and Ukraine and the gas was turned off, European householders froze, and the German government was forced to ration its scarce supply.

Russia's response to the Ukraine problem has been to build a new pipeline that avoids the country altogether, with a northern routing through Finland and then under the Baltic Sea to Germany. While highly controversial from an environmental perspective, this project is currently under construction, with an expected completion date of 2012. In addition, Russia is promoting a southern route that also avoids Ukraine by going under the Black Sea to Bulgaria and then on to central Europe. If this line is built, gas will flow into Europe by 2015.

However, many EU politicians are promoting a plan that doesn't rely on Russia and, instead, draws natural gas from fields in Azerbaijan, near Baku. From there, the pipeline route would go through Georgia, then southwest across Turkey to enter Europe in Bulgaria. Russia's Gazprom is outraged at this proposal and accuses the United States of advocating this plan as a way of undermining Russia.

While it's not yet clear whether Azerbaijan can actually supply the amounts of natural gas needed to make this route viable, it does seem clear that a new Cold War has begun over the issue of how to heat Europe.

**Basques.** *Basque separatists continue campaign for complete autonomy from Spain.*

**Local autonomy.** *Corsican separatists force concessions from France to increase local autonomy.*

**Continued unrest.** *There have been decades of ethnic unrest and warfare in the provinces of the former Yugoslavia.*

North Atlantic Treaty Organization (NATO) member

Former Warsaw Pact member

NATO headquarters

Note: The United States and Canada are also members of NATO.

**FIGURE 8.26 Geopolitical Issues in Europe**   While the major geopolitical issue of the early 21st century remains the integration of eastern and western Europe into the European Union, numerous issues of micro- and ethnic nationalism also engender geopolitical fragmentation. In other parts of Europe, such as Spain, France, and Great Britain, questions of local ethnic autonomy within the nation-state structure challenge central governments.

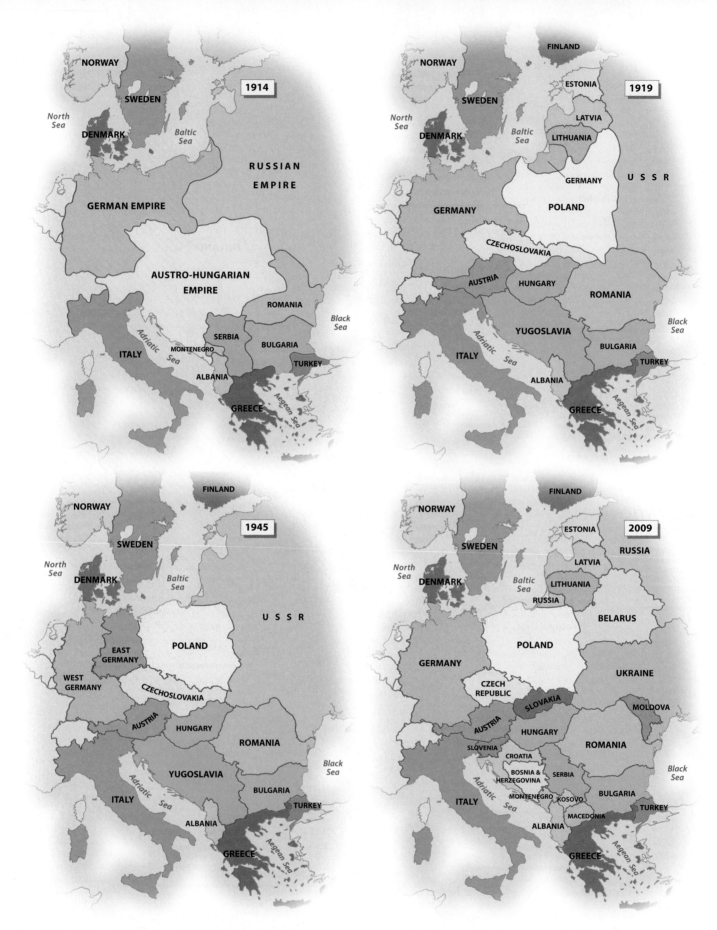

**FIGURE 8.27  A Century of Geopolitical Change**   At the outset of the 20th century, central Europe was dominated by the German, Austro-Hungarian (or Hapsburg), and Russian empires. Following World War I, these empires were largely replaced by a mosaic of nation-states. More border changes followed World War II, largely as a result of the Soviet Union's turning that area into a buffer zone between itself and western Europe. With the demise of Soviet hegemony in 1989, further political change took place.

Nazi Germany tested Western resolve in 1938 by first annexing Austria, the country of Hitler's birth, and then Czechoslovakia, under the pretense of providing protection for ethnic Germans located there. After signing a nonaggression pact with the Soviet Union, Hitler invaded Poland on September 1, 1939. Two days later, France and Britain declared war on Germany. Within a month, the Soviet Union moved into eastern Poland, the Baltic states, and Finland to reclaim territories lost through the peace treaties of World War I. Nazi Germany then moved westward and occupied Denmark, the Netherlands, Belgium, and France and began preparations to invade England.

In 1941 the war took several startling new turns. In June, Hitler broke the nonaggression pact with the Soviet Union and, catching the Red Army by surprise, took the Baltic states and drove deep into Soviet territory. When Japan attacked Pearl Harbor, Hawaii, in December, the United States entered the war in both the Pacific and Europe.

By early 1944, the Soviet army had recovered most of its territorial losses and moved against the Germans in eastern Europe, beginning the long communist domination in that region. By agreement with the Western powers, the Red Army stopped when it reached Berlin in April 1945. At that time, Allied forces crossed the Rhine River and began their occupation of Germany. With Hitler's suicide, Germany signed an unconditional surrender on May 8, 1945, ending the war in Europe. But with Soviet forces firmly entrenched in the Baltic states, Poland, Czechoslovakia, Bulgaria, Romania, Hungary, Austria, and eastern Germany, the military battles of World War II were quickly replaced by the ideological Cold War between communism and democracy that lasted 45 years, until 1990.

## A Divided Europe, East and West

From 1945 until 1990, Europe was divided into two geopolitical and economic blocs, east and west, separated by the infamous **Iron Curtain** that descended shortly after the peace agreement of World War II. East of the Iron Curtain border, the Soviet Union imposed the heavy imprint of communism on all activities—political, economic, military, and cultural. To the west, as Europe rebuilt from the destruction of the war, new alliances and institutions were created to counter the Soviet presence in Europe.

**Cold War Geography**   The seeds of the Cold War are commonly thought to have been planted at the Yalta Conference of February 1945, when Britain, the Soviet Union, and the United States met to plan the shape of postwar Europe. Because the Red Army was already in eastern Europe and moving quickly on Berlin, Britain and the United States agreed that the Soviet Union would occupy eastern Europe, and the Western allies would occupy parts of Germany.

The larger geopolitical issue, though, was the Soviet desire for a **buffer zone** between its own territory and western Europe. This buffer zone consisted of an extensive bloc of satellite countries, dominated politically and economically by the Soviet Union, that could cushion the Soviet heartland against possible attack from western Europe. In the east, the Soviet Union took control of the Baltic states, Poland, Czechoslovakia, Hungary, Bulgaria, Romania, Albania, and Yugoslavia. Austria and Germany were divided into occupied sectors by the four (former) allied powers. In both cases, the Soviet Union dominated the eastern portion of each country, areas that contained the capital cities of Berlin and Vienna. Both capital cities, in turn, were divided into French, British, U.S., and Soviet sectors.

In 1955, with the creation of an independent and neutral Austria, the Soviets withdrew from their sector, effectively moving the Iron Curtain eastward to the Hungary–Austria border. However, Germany quickly evolved into two separate states, West Germany and East Germany, that remained separate until 1990.

Along the border between east and west, two hostile military forces faced each other for almost half a century. Both sides prepared for and expected an invasion by the other across the barbed wire of a divided Europe (Figure 8.28). In the west, North Atlantic Treaty Organization (NATO) forces, including the United States, were stationed from West Germany south to Turkey. To the east, Warsaw Pact forces were anchored by the Soviets but also included small military units from satellite countries. Both NATO and the Warsaw Pact countries were armed with nuclear weapons, making Europe a tinderbox for a devastating world war.

Berlin was the flashpoint that brought these forces close to a fighting war on two occasions. In winter 1948, the Soviets imposed a blockade on the city by denying Western powers access to Berlin across its East German military sector. This attempt to starve the city into submission by blocking food shipments from western Europe was thwarted by a nonstop airlift of food and coal by NATO.

**FIGURE 8.28 The Iron Curtain**   During the Cold War, when Germany was divided into east and west, villages were often split in half by the Iron Curtain border with communist East Germany. Here, the village is truncated by the communist "kill zone" in the foreground, where those who attempted to flee East Germany were shot on sight.

(a)

(b)

**FIGURE 8.29  The Berlin Wall**   In August 1961, the East German and Soviet armies built a concrete and barbed-wire structure, known simply as "the Wall," to stem the flow of East Germans leaving the Soviet zone. It was the most visible symbol of the Cold War until November 1989, when Berliners physically dismantled it after the Soviet Union renounced its control over eastern Europe.

Then, in August 1961, the Soviets built the Berlin Wall to curb the flow of East Germans seeking political refuge in the west. The Wall became the concrete-and-mortar symbol of a firmly divided postwar Europe. For several days while the Wall was being built and the West agonized over destroying it, NATO and Warsaw Pact tanks and soldiers faced each other with loaded weapons at point-blank range. Though war was avoided, the Wall stood for 28 years, until November 1989.

**The Cold War Thaw**   The symbolic end of the Cold War in Europe came on November 9, 1989, when East and West Berliners joined forces to rip apart the Berlin Wall with jackhammers and hand tools (Figure 8.29). By October 1990, East and West Germany were officially reunified into a single nation-state. During this period, all other Soviet satellite states, from the Baltic Sea to the Black Sea, also underwent major geopolitical changes that have resulted in a mixed bag of benefits and problems.

The Cold War's end came as much from a combination of problems within the Soviet Union (discussed in Chapter 9) as from rebellion in eastern Europe. By the mid-1980s, the Soviet leadership was advocating an internal economic restructuring and also recognizing the need for a more open dialogue with the West. Financial problems from supporting a huge military establishment, along with heavy losses from an unsuccessful war in Afghanistan, lessened the Soviet appetite for occupying other countries.

In August 1989, Poland elected the first noncommunist government to lead an eastern European state since World War II. Following this, with just one exception, peaceful revolutions with free elections spread throughout eastern Europe, as communist governments renamed themselves and broke with doctrines of the past. In Romania, though, street fighting between citizens and military resulted in the violent overthrow and execution of communist dictator Nicolae Ceausescu.

As a result of the Cold War thaw, the map of Europe began changing once again. Germany reunified in 1990. Elsewhere, political separatism and ethnic nationalism, long suppressed by the Soviets, were unleashed in southeastern Europe. Over the past decades, Yugoslavia has fractured into the independent states of Slovenia, Croatia, Bosnia, Macedonia, and Kosovo. On January 1, 1993, Czechoslovakia was replaced by two separate states, the Czech Republic and Slovakia.

# Economic and Social Development: Integration and Transition

As the acknowledged birthplace of the Industrial Revolution, Europe in many ways invented the modern economic system of industrial capitalism. Though Europe was the world's industrial leader in the early 20th century, it was later eclipsed by both Japan and the United States, while it struggled to cope with the effects of two world wars, a decade of global depression, and the Cold War.

In the past 50 years, however, economic integration guided by the EU has been increasingly successful. In fact, western Europe's success at blending national economies has given the world a new model for regional cooperation, an approach that may be imitated in Latin America and Asia in the near future. Eastern Europe, however, has not fared as well. The results of four decades of Soviet economic planning were, at best, mixed. The total collapse of that system in 1990 cast eastern Europe into a period of chaotic economic, political, and social transition that has resulted in a highly differentiated pattern of rich and poor regions. While some countries, such as the Czech Republic and Slovenia, prosper, the future prospects for Albania, Slovakia, and Romania are uncertain (Table 8.2).

## TABLE 8.2    DEVELOPMENT INDICATORS

| Country | GNI per capita, PPP 2007 | GDP Average Annual % Growth 2000–07 | Human Development Index (2006)# | Percent Population Living Below $2 a Day | Life Expectancy* 2009 | Under Age 5 Mortality Rate 1990 | Under Age 5 Mortality Rate 2007 | Gender Equity 2007 |
|---|---|---|---|---|---|---|---|---|
| *Western Europe* | | | | | | | | |
| Austria | 36,750 | 2.0 | 0.951 | | 80 | 10 | 4 | 97 |
| Belgium | 35,320 | 2.0 | 0.948 | | 80 | 10 | 5 | 98 |
| France | 33,850 | 1.8 | 0.955 | | 81 | 9 | 4 | 100 |
| Germany | 34,740 | 1.0 | 0.940 | | 80 | 9 | 4 | 98 |
| Liechtenstein | | | | | 80 | 10 | 3 | 92 |
| Luxembourg | 61,860 | | 0.956 | | 80 | 9 | 3 | 103 |
| Monaco | | | | | | 9 | 4 | |
| Netherlands | 39,470 | 1.6 | 0.958 | | 80 | 9 | 5 | 98 |
| Switzerland | 44,410 | 1.8 | 0.955 | | 82 | 9 | 5 | 97 |
| United Kingdom | 34,050 | 2.6 | 0.942 | | 79 | 10 | 6 | 102 |
| *Eastern Europe* | | | | | | | | |
| Bulgaria | 11,100 | 5.7 | 0.834 | <2 | 73 | 19 | 12 | 97 |
| Czech Republic | 22,690 | 4.6 | 0.897 | <2 | 77 | 13 | 4 | 101 |
| Hungary | 17,470 | 4.0 | 0.877 | <2 | 73 | 17 | 7 | 99 |
| Poland | 15,500 | 4.1 | 0.875 | <2 | 76 | 17 | 7 | 99 |
| Romania | 12,350 | 6.1 | 0.825 | 3.4 | 72 | 32 | 15 | 100 |
| Slovakia | 19,220 | 6.0 | 0.872 | <2 | 75 | 15 | 8 | 100 |
| *Southern Europe* | | | | | | | | |
| Albania | 7,240 | 5.3 | 0.807 | 7.8 | 75 | 46 | 15 | 97 |
| Bosnia and Herzegovina | 8,020 | 5.3 | 0.802 | <2 | 75 | 22 | 14 | 99 |
| Croatia | 15,540 | 4.8 | 0.862 | <2 | 76 | 13 | 6 | 96 |
| Cyprus | 24,040 | | 0.912 | | 78 | 11 | 5 | 101 |
| Greece | 27,830 | 4.3 | 0.947 | | 80 | 11 | 4 | 98 |
| Italy | 30,190 | 1.0 | 0.945 | | 82 | 9 | 4 | 99 |
| Macedonia | 9,050 | 2.7 | 0.808 | 3.2 | 74 | 38 | 17 | 99 |
| Malta | 22,460 | | 0.894 | | 80 | 11 | 5 | 99 |
| Montenegro | 11,780 | | 0.822 | | 73 | 16 | 10 | |
| Portugal | 21,790 | 0.9 | 0.900 | | 78 | 15 | 4 | 101 |
| San Marino | | | | | 82 | 13 | 4 | |
| Serbia | 9,830 | 5.6 | 0.821 | | 73 | | 8 | 102 |
| Slovenia | 26,230 | 4.3 | 0.923 | <2 | 78 | 11 | 4 | 100 |
| Spain | 30,750 | 3.4 | 0.949 | | 81 | 9 | 4 | 103 |
| *Northern Europe* | | | | | | | | |
| Denmark | 36,800 | 1.8 | 0.952 | | 79 | 9 | 4 | 101 |
| Estonia | 18,830 | 8.1 | 0.871 | <2 | 73 | 18 | 6 | 100 |
| Finland | 34,760 | 3.0 | 0.954 | | 80 | 7 | 4 | 102 |
| Iceland | 34,070 | | 0.968 | | 81 | 7 | 3 | 101 |
| Ireland | 37,700 | 5.5 | 0.960 | | 79 | 9 | 4 | 103 |
| Latvia | 15,790 | 9.0 | 0.863 | <2 | 72 | 17 | 9 | 100 |
| Lithuania | 16,830 | 8.0 | 0.869 | <2 | 71 | 16 | 8 | 100 |
| Norway | 53,650 | 2.4 | 0.968 | | 81 | 9 | 4 | 100 |
| Sweden | 37,490 | 3.0 | 0.958 | | 81 | 7 | 3 | 100 |

Source: World Bank, *World Development Indicators, 2009.*
United Nations, *Human Development Index, 2008.* #
Population Reference Bureau, *World Population Data Sheet, 2009*\*
Gender Equity – Ratio of female to male enrollments in primary and secondary school. Numbers below 100 have more males in primary/secondary school, numbers above 100 have more females in primary/secondary schools.

Accompanying western Europe's economic boom has been an unprecedented level of social development, as measured by worker benefits, health services, education, literacy, and gender equality. Though the improved social services set an admirable standard for the world, cost-cutting politicians and businesspeople argue that these services now increase the cost of business so that European goods cannot compete in the global marketplace. As a result, and aggravated by the global economic recession of 2008–2009, many of these traditional benefits, such as job security and long vacation periods, are being eroded, causing considerable social tension in many European countries.

## Europe's Industrial Revolution

Europe is the cradle of modern industrialism. Two fundamental changes were associated with this industrial revolution: First, machines replaced human labor in many manufacturing processes, and, second, inanimate energy sources, such as water, steam, electricity, and petroleum, powered the new machines. Though we commonly apply the term *Industrial Revolution* to this transformation, implying rapid change, in reality, it took more than a century for the interdependent pieces of the industrial system to come together. This new system emerged first in England between 1730 and 1850. Later in the 19th century, this new industrialism had spread to throughout Europe and the rest of the world.

**Centers of Change**   England's textile industry, located on the flanks of the Pennine Mountains, was the center of early industrial innovation, which took place in small towns and villages, away from the rigid control of the urban guilds. The town of Yorkshire, on the eastern side of the Pennines, had been a center of woolen textile making since medieval times, drawing raw materials from the extensive sheep herds of that region and using the clean mountain waters to wash the wool before it was spun in rural cottages. By the 1730s, water wheels were used to power mechanized looms at the rapids and waterfalls of the Pennine streams (Figure 8.30). By the 1790s, the steam engine had

**FIGURE 8.30 Water Power and Textiles**   Europe's industrial revolution began on the flanks of England's Pennine Mountains, where swift-running streams and rivers were used to power large cotton and wool looms. Later, many of these textile plants switched to coal power.

become the preferred source of energy to drive the new looms. However, steam engines needed fuel, and only with the building of railroads after 1820 could coal be moved long distances at a reasonable cost.

**Development of Industrial Regions in Continental Europe**   By the 1820s, the first industrial districts had begun appearing in continental Europe. These hearth areas were, and still are, near coalfields (Figure 8.31). The first area outside Britain was the Sambre-Meuse region, named for the two river valleys straddling the French–Belgian border. Like the English Midlands, it also had a long history of cottage-based wool textile manufacturing that quickly converted to the new technology of steam-powered mechanized looms. In addition, a metalworking tradition drew on charcoal-based iron foundries in the nearby forests of the Ardenne Mountains. Coal was also found in these mountains, and in 1823, the first blast furnace outside Britain started operation in Liege, Belgium.

By the second half of the 19th century, the dominant industrial area in all Europe (including England) was the Ruhr district in northwestern Germany, near the Rhine River. Rich coal deposits close to the surface fueled the Ruhr's transformation from a small textile region to a region oriented around heavy industry, particularly iron and steel manufacturing. By the early 1900s, the Ruhr had used up its modest iron ore deposits and was importing ore from Sweden, Spain, and France. Several decades later, the Ruhr industrial region became synonymous with the industrial strength behind Nazi Germany's war machine and thus was bombed heavily in World War II (Figure 8.32).

## Rebuilding Postwar Europe: Economic Integration in the West

Europe was unquestionably the leader of the industrial world in the early 20th century. More specifically, before World War I, European industry was estimated to produce 90 percent of the world's manufactured output. However, four decades of political and economic chaos and two world wars left Europe divided and in shambles. By the mid-20th century, its cities were in ruins; industrial areas were destroyed; vast populations were dispirited, hungry, and homeless; and millions of refugees moved about Europe, looking for safety and stability.

**ECSC and EEC**   In 1950, western Europe began discussing a new form of economic integration that would avoid both the historical pattern of nationalistic independence through tariff protection and the economic inefficiencies resulting from the duplication of industrial effort. Robert Schuman, France's foreign minister, proposed that German and French coal and steel production be coordinated by a new supranational organization. In May 1952, France, Germany, Italy, the Netherlands, Belgium, and Luxembourg ratified a treaty that joined them in the European Coal and Steel Community (ECSC). Because of the immediate success of the ECSC, these six states soon agreed to work toward further integration by creating a larger European common market that would encourage the free movement of goods, labor, and capital. In March 1957, the Treaty of Rome was signed, establishing the European Economic Community (EEC), popularly called the *Common Market*.

**From the EEC to the EU**   The EEC reinvented itself in 1965, with the Brussels Treaty, which laid the groundwork for adding a political union to the successful economic community. In this "second Treaty of Rome," aspirations for more than economic integration were clearly stated, with the creation of an EEC council, court, parliament, and political commission. The EEC also changed its name to the European Community (EC).

In 1991, the EC again changed its name to the European Union (EU), and it expanded its goals once again, this time with the Treaty of Maastricht (named after the town in the Netherlands in which delegates met). While economic integration remains an underlying theme in the new constitution, particularly with its

commitment to a single currency through the European Monetary Union (discussed below), the EU has moved further into supranational affairs, with discussion of common foreign policies and mutual security agreements.

More recently, in May 2004, the EU added 10 new states, including a core group of former Soviet-controlled communist countries

**FIGURE 8.31 Industrial Regions of Europe**  From England, the Industrial Revolution spread to continental Europe, starting with the Sambre-Meuse region on the French–Belgian border and then diffusing to the Ruhr area in Germany. Readily accessible surface coal deposits powered these new industrial areas. Early on, iron ore for steel manufacture came from local deposits, but later it was imported from Sweden and other areas in the shield country of Scandinavia. Most of the newer industrial areas are closely linked to urban areas.

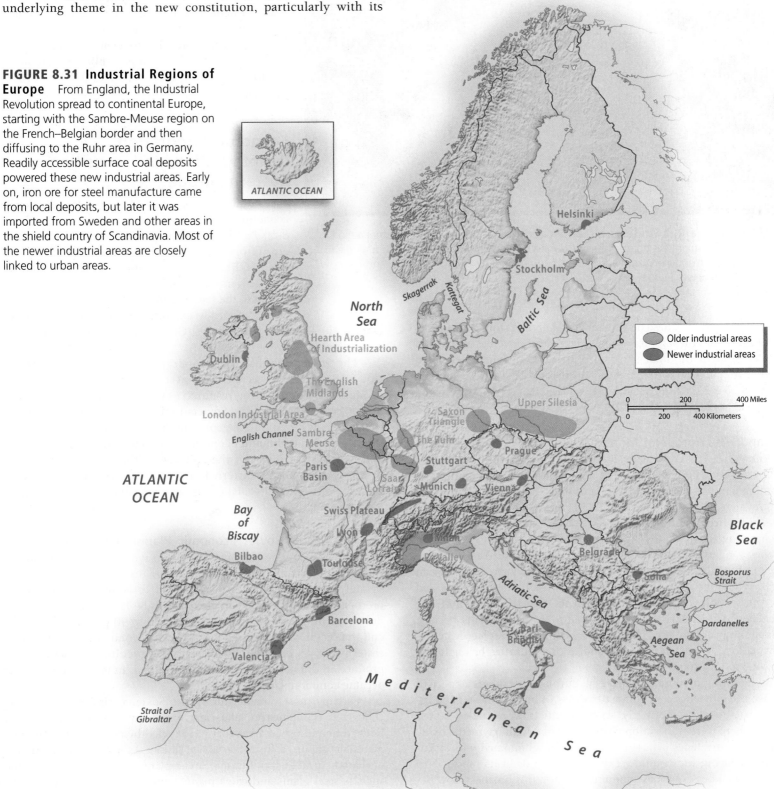

**FIGURE 8.32 The Ruhr Industrial Landscape** Long the dominant industrial region in Europe, the Ruhr region was bombed heavily during World War II because of its central role in providing heavy arms to the German military. It has been rebuilt and modernized, and it is now competitive with newer industrial regions.

However, some EU member countries, most notably the United Kingdom, have reservations about joining Euroland; as a result, membership remains a controversial political and economic topic for a large number of countries. During the economic meltdown of 2008–2009, the experiences of using a common currency were mixed. While some states in Euroland benefited from the use of a common currency, others suffered greatly because they could not adjust their national currencies to economic conditions. Greece and Spain, for example, suffered more heavily from the global recession than did France and Germany.

## Economic Integration, Disintegration, and Transition in Eastern Europe

Historically, eastern Europe has been less developed economically than its western counterpart (see Table 8.2). This is partially explained by the fact that eastern Europe is not as rich in natural resources as is western Europe, having only modest amounts of coal, less iron ore, and little oil and natural gas. In addition, the few resources found in the region have been historically exploited by outside interests, including the Ottomans, Hapsburgs, Germans, and, most recently, the Soviet Union.

from eastern Europe. These new members—Latvia, Estonia, Lithuania, Poland, Slovakia, the Czech Republic, Hungary, Slovenia, Malta, and Cyprus—brought the total to 25; Bulgaria and Romania were then admitted in January 2007, resulting in a total of 27 EU countries (Figure 8.33). As of 2009, Iceland, several Balkan countries, and Turkey had applied for EU membership.

**Euroland: The European Monetary Union** As any world traveler knows, each state usually has its own monetary system. As a result, crossing a political border usually means changing money and becoming familiar with a new system of bills and coins. In the recent past, travelers bought pounds sterling in England, marks in Germany, francs in France, lira in Italy, and so on. Today, however, much of Europe has moved from individual state monetary systems to a common currency. On January 1, 1999, in a major advance toward a united Europe, 11 of the then 15 EU member states joined to form the European Monetary Union (EMU). As of that day, cross-border business and trade transactions began taking place in the new monetary unit, the *euro*. Then, on January 1, 2002, new euro coins and bills became available for everyday use. This currency completely replaced the different national currencies of **Euroland**, those countries belonging to the EMU, in July 2002.

By adopting a common currency, Euroland members expect to increase the efficiency and competitiveness of both domestic and international business. Formerly, when products were traded across borders, there were transaction costs associated with payments made in different currencies. These have now been eliminated within Euroland. Germany, for example, exports two-thirds of its products to other EU members. With a common currency, this business now becomes essentially domestic trade, protected from the fluctuations of different currencies and not subject to transaction costs.

The Soviet-dominated economic planning of the postwar period (1945–1990) was an attempt to develop eastern Europe's economy by coordinating resource usage in a way that also served Soviet interests. When the Soviets took control of eastern Europe, their goals were complete economic, political, and social integration through a **command economy**, one that was centrally planned and controlled. However, the collapse of that centralized system in 1991 threw many eastern European countries into deep economic, political, and social chaos that has handicapped the region (Figure 8.34).

**The Results of Soviet Economic Planning** After 40 years of communist economic planning, the results were mixed within eastern Europe. In Poland and Yugoslavia, for example, many farmers strongly resisted the national ownership of agriculture; thus, most productive land remained in private hands. However, in Romania, Bulgaria, Hungary, and Czechoslovakia, 80 to 90 percent of the agricultural sector was converted to state-owned communal farms. Across eastern Europe, despite these changes in agricultural structure, food production did not increase dramatically, and, in fact, food shortages became commonplace.

Perhaps most notable during the Soviet period were dramatic changes to the industrial landscape, with many new factories built in both rural and urban areas that were fueled by cheap energy and raw materials imported from the Soviet Union. As a means to economic development in eastern Europe, Soviet planners chose heavy industry, such as steel plants and truck manufacturing, over consumer goods. As in the Soviet Union, the bare shelves of retail outlets in eastern Europe became both a sign of communist shortcomings and, perhaps

**FIGURE 8.33 The European Union** The driving force behind Europe's economic and political integration has been the European Union (EU), which was formed in the 1950s as an organization focused solely on rebuilding the region's coal and steel industries. After initial success, the Common Market (as it was then called) expanded into a wider range of economic activities. As of 2009, the EU has 27 members.

more importantly, a source of considerable public tension. As western Europeans enjoyed an increasingly high standard of living, with an abundance of consumer goods, eastern Europeans struggled to make ends meet, as the utopian vision promised by Soviet communists became increasingly elusive.

**Transition and Change Since 1991** As Soviet domination over eastern Europe collapsed in the 1990s, so did the forced economic integration of the region. In place of Soviet coordination and subsidy came a painful period of economic transition that was close to outright chaos in some eastern European countries. The causes of this economic

**FIGURE 8.34 Post-1989 Hardship** With the fall of communism in eastern Europe after 1989, economic subsidies and support from the Soviet Union ended. Accompanying this transition has been a high unemployment rate, resulting from the closure of many industries, such as this plant in Bulgaria.

pain were complex. As the Soviet Union turned its attention to its own economic and political turmoil, it stopped exporting cheap natural gas and petroleum to eastern Europe. Instead, Russia sold these fuels on the open global market to gain hard currency. Without cheap energy, many eastern European industries were unable to operate and shut down operations, laying off thousands of workers. In the first two years of the transition (1990–1992), industrial production fell 35 percent in Poland and 45 percent in Bulgaria. In addition, markets guaranteed under a command economy, many of them in the Soviet Union, evaporated. Consequently, many factories and services closed because they lacked markets for their goods.

Given these problems, eastern European countries began redirecting their economies away from Russia and toward western Europe, with the goal of joining the EU. This meant moving from a socialist-based economy of state ownership and control to a capitalist economy predicated on private ownership and free markets. To achieve this, countries such as the Czech Republic, Hungary, and Poland went through a period of **privatization**, which is the transfer to private ownership of firms and industries previously owned and operated by state governments (Figure 8.35).

**FIGURE 8.35 New Supermarket in Bulgaria** Consumers items, including food, were often in short supply in Bulgaria during the post-war Soviet period. Then, with the collapse of Soviet Russia's support in 1989, and after a decade of hardship, Bulgaria shifted its orientation toward the European Union, becoming a member in 2007. As a result, European commerce, such as this German supermarket chain, have moved into the country. Note that while Bulgaria uses the Cyrillic alphabet many of the supermarket's signs are in Roman script, illustrating yet another influence of western Europe.

GNI PER CAPITA, PPP ($2009)

- More than 20,000
- 15,001–20,000
- 7,500–15,000
- Less than 7,500

A good deal of hardship came with this transition. Price supports, tariff protection, and subsidies were removed from consumer goods as countries moved to a free market. For the first time, goods from western Europe and other parts of the world became plentiful in eastern European retail stores. The irony, though, is that this prolonged period of economic transition took a toll on many eastern European consumers because not everyone could afford these long-dreamed-of products. Unemployment was commonly in the double digits, and underemployment was widespread. Financial security was elusive, with irregular paychecks for those with jobs and uncertain welfare benefits for those without.

Nonetheless, it is quite clear that the preferred economic and political trajectory of eastern Europe is toward increased integration with western Europe and the EU—and not with Russia. More specifically, a handful of former Soviet satellite countries joined the EU in 2004 (Estonia, Latvia, Lithuania, Poland, the Czech Republic, Slovakia, and Hungary), with Bulgaria and Romania following in early 2007. While many of these countries are now successfully integrated with western European economies, there remain problems and serious uncertainties about the future of countries such as Bulgaria, Romania, Albania, Kosovo, and Macedonia (Figure 8.36).

**FIGURE 8.36 Economic Disparities in Eastern Europe**   While some countries in eastern Europe are doing fairly well economically, others are lagging behind. At the head of the pack are Slovenia and the Czech Republic, followed by Hungary and Slovakia. See Table 8.2 for more complete data.

# SUMMARY

- In terms of the environment, Europe has made great progress over the past several decades. Not only have individual countries enacted strong environmental legislation, but the EU has played an important role in working toward solutions of transboundary problems such as air and water pollution. In addition, the EU has become a strong advocate for addressing climate change issues connected with global warming.

- Europe continues to face challenges related to population and migration. The phenomena of slow and no growth has profound implications for the labor force and fiscal vitality of many countries. However, the more pressing problem is how Europe deals with immigration from Asia, Africa, Latin America, the former Soviet lands, and even its own underdeveloped regions.

- Political and cultural tensions are high regarding migration. As the bulk of Europeans meld into a common culture, the differences between Europeans and migrants from outside the region increase.

- While European countries pay lip service to building multicultural societies, the pathways to this goal are still unclear because of racism and discrimination.

- With the end of the Cold War, Europe's geopolitical issues are continually being redefined. More specifically, in the past two decades, several new countries have appeared on the European map. Another theme is the integration of the former Soviet satellites in eastern Europe into western European—and even international—geopolitics.

- Although the EU celebrated its 50th birthday in 2007 in a self-congratulatory mood, that spirit was challenged by the global economic downturn of 2008–2009 that brought many countries and their financial institutions to the brink of bankruptcy. Iceland became the unwitting poster child for that crisis, with several other eastern European and Balkan countries also brought to the brink. This crisis underscored both the strengths and weaknesses of EU policies.

# KEY TERMS

buffer zone (p. 243)
Cold War (p. 220)
command economy (p. 248)
continental climate (p. 225)
Cyrillic alphabet (p. 236)
Euroland (p. 248)

European Union (EU)
   (p. 220)
fjord (p. 223)
guest worker (p. 232)
Iron Curtain (p. 243)
irredentism (p. 240)

marine west-coast climate
   (p. 225)
maritime climate (p. 225)
medieval landscape (p. 233)
Mediterranean climate (p. 225)
moraine (p. 222)

privatization (p. 250)
Renaissance–Baroque landscape
   (p. 233)
Schengen Agreement (p. 233)
secularization (p. 238)
shield landscape (p. 223)

# THINKING GEOGRAPHICALLY

1. To compare the scale of Europe to North America, draw a circle 500 miles (800 kilometers) in diameter, which is a day's journey, around Frankfurt, Germany. Then do the same for Chicago. What does this exercise tell you about the differences in scale between the two regions?

2. Map the different rates of natural increase in Europe, noting which countries will grow and which will decline in the next 20 years. Then, link this map to a discussion of migration in Europe. Given these two factors, how might the population map change in 20 years?

3. Find good maps of several midsized European cities and deduce the historical development of the cities, based on their street patterns, arrangement of open spaces, boulevards, parks, and so on. Draw on the differences between the medieval and Renaissance–Baroque periods for your interpretation.

4. Look at the ads in a French or German news magazine and list the "globalized" English words used. Discuss your findings in terms of what sectors of society seem to use these foreign terms.

5. Research the fiscal, political, and social changes that had to take place in Hungary and Poland before these countries were admitted to the EU.

Log in to **www.mygeoscienceplace.com** for videos, interactive maps, RSS feeds, case studies, and self-study quizzes to enhance your study of Europe.

# 9 THE RUSSIAN DOMAIN

## GLOBALIZATION AND DIVERSITY

Russia's reemergence on the global stage has been obvious since 2000. Its growing role is evident in the global energy economy, in its participation in global economic meetings such as the Group of Eight (G8), and in its increasing geopolitical involvement in settings such as North Korea, Iran, and Sub-Saharan Africa.

## ENVIRONMENTAL GEOGRAPHY

Many areas within the Russian domain suffered severe environmental damage during the Soviet era (1917–1991), and today air, water, toxic chemical, and nuclear pollution plague large portions of the region.

## POPULATION AND SETTLEMENT

Unlike most other world regions, one of the key demographic challenges facing the Russian domain is its declining population, given the rising death rates and low birthrates recently experienced by the region.

## CULTURAL COHERENCE AND DIVERSITY

Although Slavic cultural influences dominate the region, many non-Slavic minorities shape the cultural and political geography of the domain, including varied indigenous peoples in Siberia and a complex collection of ethnic groups in the Caucasus Mountains.

## GEOPOLITICAL FRAMEWORK

The centralization of Russian political power has contributed to economic stability in the region, but democratic freedoms have suffered amid crackdowns on the press and personal liberties.

## ECONOMIC AND SOCIAL DEVELOPMENT

Russia's large supplies of oil and natural gas have made it a major player in the global economy, but future prosperity may increasingly hinge on unpredictable world prices for fossil fuels.

*This night view of Moscow's Red Square is centered on St. Basil's Cathedral, probably the most famous architectural landmark in the Russian domain.*

The Russian domain sprawls across the vast northern half of Eurasia and includes not only Russia itself but also the nations of Ukraine, Belarus, Moldova, Georgia, and Armenia (Figure 9.1; see also "Setting the Boundaries"). The land is rich with superlatives: endless Siberian spaces, unlimited natural resources, legends of ruthless Cossack warriors, and tales of epic wars and revolutions are all part of the region's geographic and historical mythology (Figure 9.2). Indeed, the rise of Russian civilization remarkably parallels the story of the United States. Both cultures grew from small beginnings to become imperial powers that benefited from the fur trade, gold rushes, and transcontinental railroads during the 19th century. In addition, both countries were dramatically transformed by industrialization during the 20th century.

**FIGURE 9.1  The Russian Domain**    Russia and its neighboring states of Belarus, Ukraine, Moldova, Georgia, and Armenia make up a dynamic and unpredictable world region. Sprawling from the Baltic Sea to the Pacific, the region includes huge industrial centers, vast farmlands, and almost-empty stretches of tundra.

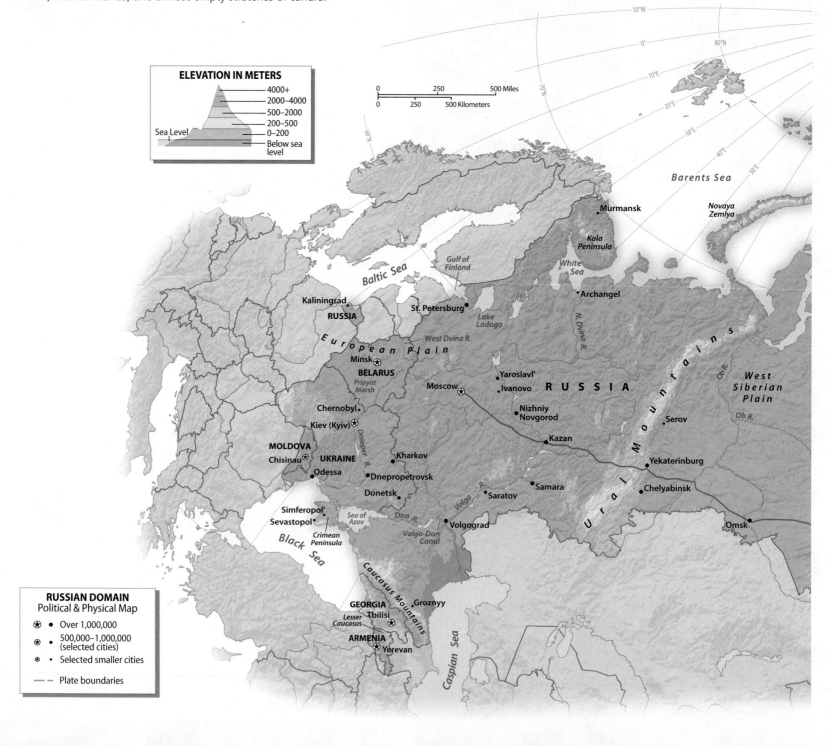

Recently, however, the Russian domain has witnessed particularly breathtaking change. With the fall of the Soviet Union in 1991, new political and economic institutions have reshaped everyday life. Economic collapse in the late 1990s produced steep declines in living standards throughout the region. Political instability also grew between neighboring states as well as within countries. After 2000, strong and increasingly centralized leadership within Russia set the region on a different course. With the help of higher energy prices (Russia is a major exporter of oil and natural gas), real economic improvements benefited most of the region, helping stabilize its economy and its larger political role in the world.

That remarkable record of recovery, however, proved vulnerable in the global economic decline of the late 2000s. Both Russia and Ukraine, for example, witnessed sharp economic contractions during

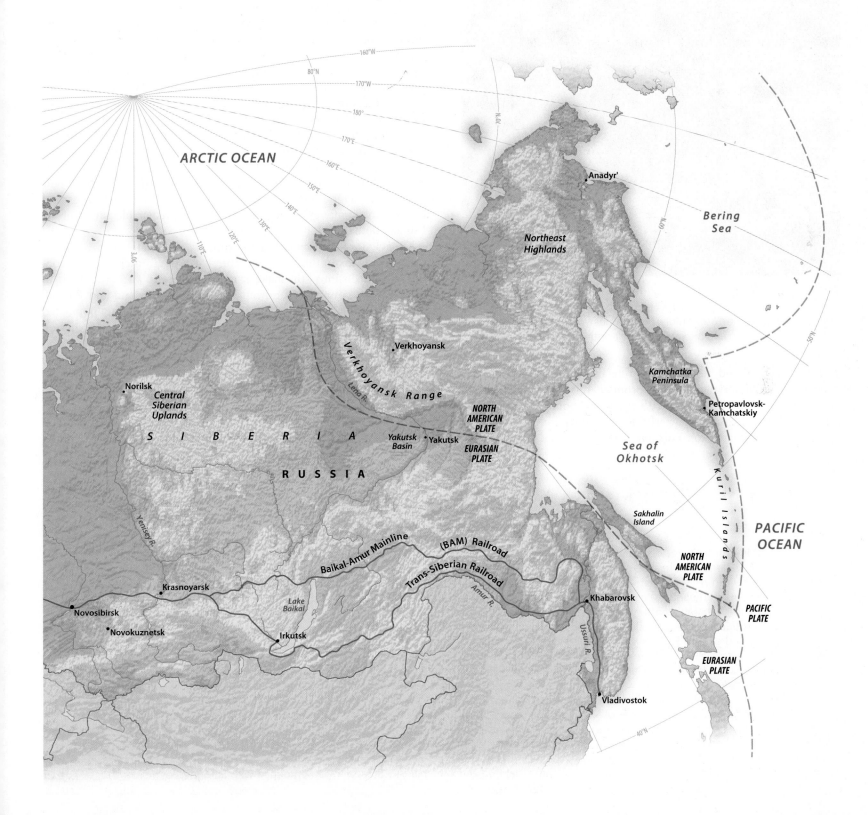

**FIGURE 9.2 Cossack**   Natives of the Ukrainian and Russian steppe, the highly mobile Cossacks played a pivotal role in aiding Russian expansion into Siberia during the 16th century. Many modern descendants retain their skills of horsemanship and are proud of their distinctive ethnic heritage.

2009, as regional and global demand declined for many of the raw materials and industrial goods they produce. In addition, many of the democratic reforms welcomed with the fall of communism in 1991 are now being eroded by a powerful Russian government committed to more direct state control. This new assertiveness on the part of the

Russian government is increasing tensions within that country as well as between Russia and its neighboring states.

Globalization is shaping the Russian domain in complex ways. The region's relationship with the rest of the world shifted during the last decade of the 20th century. Until the end of 1991, all six countries of the Russian domain belonged to the Soviet Union, the world's most powerful communist state. Under Soviet control, the region's 20th-century economy saw large increases in industrial output that made the area a major global producer of steel, weaponry, and petroleum products. The political and military reach of the Soviet Union spanned the globe, making it a superpower on par with the United States.

Suddenly, the old communist order evaporated early in the 1990s. Russia, Ukraine, Belarus, Moldova, Georgia, and Armenia had to carve out new regional and global relationships. With the breakdown of Soviet control, the region felt the growing presence of western European and American influences. The Russian domain has also been more fully exposed to both the opportunities and the competitive pressures of the global economy. The result is a world region that has seen its global linkages redefined in the recent past. Today, fluctuating world oil markets, shifting patterns of foreign investment, new patterns of migration, and the shadowy flows of illegal drugs and Russian mafia money all demonstrate the unpredictable nature of the Russian domain's global connections.

Slavic Russia (population 140 million) dominates the region. Although only about three-quarters the size of the former Soviet Union, Russia's dimensions still make it the largest state on Earth. West of Moscow, the country's European front borders Finland and Poland, while far to the east, Mongolia and China share a sparsely populated boundary with sprawling Russian Siberia. Its area of 6.6 million square miles (17 million square kilometers) dwarfs even Canada, and its 11 time zones are a reminder that dawn in Vladivostok on the Pacific Ocean is still only dinnertime in Moscow.

With the demise of the Soviet Union, the Russians ended almost 75 years of Marxist rule. After a decade of political and economic instability (1991–2000), Russia has made impressive progress in the 21st century. Russian president Vladimir Putin built a reputation for strong

# Setting the Boundaries

The boundaries of the Russian domain have shifted over time. For decades, the regional definitions were relatively easy because the highly centralized Soviet Union (or Union of Soviet Socialist Republics) dominated the region's political geography. After World War II, some geographers even included much of Soviet-dominated eastern Europe within the region in response to the country's expanded military role in nations such as East Germany, Poland, and Hungary.

The maps were suddenly redrawn late in 1991. The once-powerful Soviet state was officially dissolved, and in its place stood 15 former "republics" that had once been united under the Soviet Union. Now independent, each of these republics has tried to make its own way

in a post-Soviet world. While some geographers initially treated the region as the "former Soviet Union," it quickly became clear that diverse cultural forces, economic trends, and political orientations were taking the republics in different directions. Even so, the Russian Republic remained dominant in size and area and thus came to form the core of a new Russian domain that was considerably smaller than the Soviet Union and yet included some characteristics shared by Russia and several of its neighboring states.

The term *domain* suggests persisting Russian influence within the five other nations included in the region. Slavic Russia, Ukraine, and Belarus make up the core of the region. Nearby Moldova and Armenia also broadly

remain within Russia's geopolitical orbit, although forces within both countries are advocating closer ties with the European Union (EU). Relations between Russia and Georgia remain strained, and recent political conflicts in Georgia brought Russian troops into that small country.

Two significant areas that were once part of the Soviet Union have been eliminated from the domain. The mostly Muslim republics of Central Asia and the Caucasus (Kazakhstan, Uzbekistan, Kyrgyzstan, Turkmenistan, Tajikistan, and Azerbaijan) have become aligned with the Central Asia world region (see Chapter 10), while the Baltic republics (Estonia, Latvia, and Lithuania) are best grouped with Europe (see Chapter 8).

leadership (2000–2008), and that tradition has continued since 2008 as Putin has assumed the newly powerful role of prime minister. Much of the economic growth has come to Russian cities, where expanding middle and professional classes are enjoying better living standards. Many rural areas, however, remain deeply mired in poverty. Another concern is that Putin's desire for wealth and power has been matched by his need for more centralized political control. In addition, a 2003 agreement to form the "Common Economic Space" between Russia, Ukraine, Belarus, and Kazakhstan suggests that the Russian desire for more centralized authority within the region did not end with the demise of the Soviet Union. Such an organization may also act as a counterbalance to a rapidly expanding EU to the west.

The bordering states of Ukraine, Belarus, Moldova, Georgia, and Armenia are inevitably wed to the evolution of their giant neighbor, even as they attempt to make their own way as independent nations. Emerging from the shadows of Soviet dominance has been difficult. Ukraine, in particular, has the size, population, and resource base to become a major European nation, but it has struggled to create real political and economic change since independence. With 45 million people and a rich storehouse of resources, Ukraine's size of 233,000 square miles (604,000 square kilometers) is similar to that of France. Nearby Belarus is smaller (80,000 square miles [208,000 square kilometers]), and its population of almost 10 million is likely to remain more closely tied economically and politically to Russia. Presently, its strikingly authoritarian and antiforeign leadership reflects many aspects of the old Soviet empire.

Moldova, with 4 million people, shares many cultural links with Romania, but its economic and political connections have kept it more closely tied to the Russian domain than to nearby portions of central Europe. South of Russia and beyond the bordering Caucasus Mountains, the Transcaucasian countries of Armenia and Georgia are similar in size to Moldova. Their populations differ culturally from their Slavic neighbor to the north. In addition, these two nations face significant political challenges: Armenia shares a hostile border with Azerbaijan (see Chapter 10), and Georgia's ethnic diversity and contentious relations with neighboring Russia threaten its political stability.

## Environmental Geography: A Vast and Challenging Land

The Kamchatka Peninsula dangles dramatically into the waters of the North Pacific Ocean, far from the hectic streets of St. Petersburg or Moscow (Figure 9.1). Unlike much of the rest of the Russian domain, which has been poisoned by almost a century of reckless development and environmental exploitation, the Kamchatka region remains relatively untouched by the outside world (Figure 9.3). But that may be changing soon, and Russia's ability to preserve the region will say a great deal about that country's future commitment to preserving and restoring its environmental health.

Six native species of Pacific salmon thrive in the region, spawning in the area's free-flowing rivers. A move to preserve the salmon habitat is growing among both Russian and global environmental organizations. The salmon are at the center of a complex environmental web: They provide food for the brown bears, seals, and Stellar's sea eagles that are abundant here; they also remain part of the subsistence diet of the Koryak and other native peoples of these coastal zones; and they offer ecotourism and sports fishing opportunities found nowhere else in the world. But outside threats are rapidly appearing. Canadian, South Korean, and Russian oil companies hold drilling rights in the region. Poaching of salmon and salmon eggs is also on the rise. A concerted global effort is now under way to prevent development in seven sensitive spawning

**FIGURE 9.3 Kamchatka Peninsula**   The Kamchatka Peninsula offers many spectacular natural settings as well as a habitat for a variety of Pacific salmon.

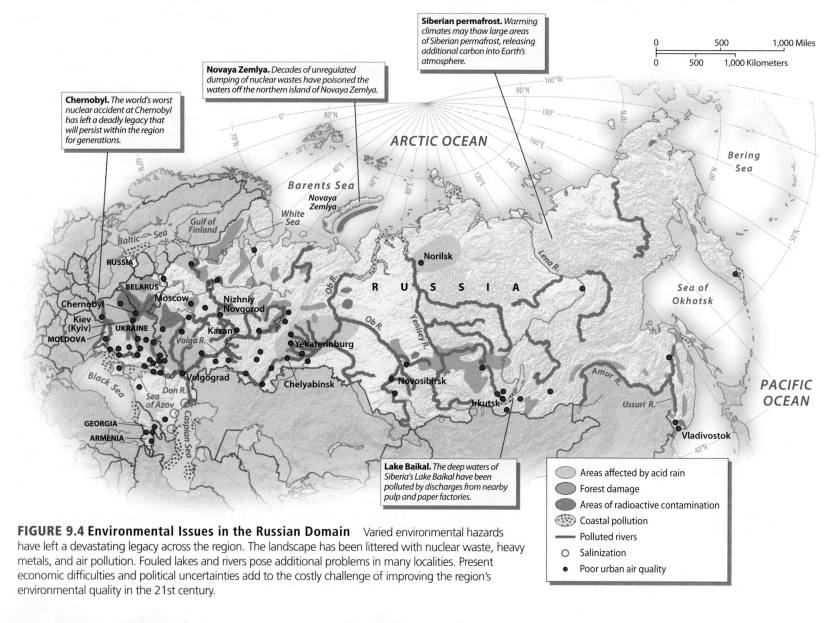

**Chernobyl.** *The world's worst nuclear accident at Chernobyl has left a deadly legacy that will persist within the region for generations.*

**Novaya Zemlya.** *Decades of unregulated dumping of nuclear wastes have poisoned the waters off the northern island of Novaya Zemlya.*

**Siberian permafrost.** *Warming climates may thaw large areas of Siberian permafrost, releasing additional carbon into Earth's atmosphere.*

**Lake Baikal.** *The deep waters of Siberia's Lake Baikal have been polluted by discharges from nearby pulp and paper factories.*

Legend:
- Areas affected by acid rain
- Forest damage
- Areas of radioactive contamination
- Coastal pollution
- Polluted rivers
- ○ Salinization
- ● Poor urban air quality

**FIGURE 9.4 Environmental Issues in the Russian Domain** Varied environmental hazards have left a devastating legacy across the region. The landscape has been littered with nuclear waste, heavy metals, and air pollution. Fouled lakes and rivers pose additional problems in many localities. Present economic difficulties and political uncertainties add to the costly challenge of improving the region's environmental quality in the 21st century.

areas of Kamchatka wilderness. The success of such initiatives would mark a new era of environmental consciousness in a part of the world that has produced some of the most toxic settings on the planet. It would also mark an important marriage between a fragile local setting and a larger collection of national and international environmental organizations dedicated to seeing the wild salmon continue to run.

## A Devastated Environment

In addition to its threatened salmon, the Russian domain faces many other enormous environmental challenges. The breakup of the Soviet Union and subsequent opening of the region to international public examination revealed some of the world's most severe environmental degradation (Figure 9.4). Seven decades of intense and rapid Soviet industrialization caused environmental problems that extend across the entire region.

**Air and Water Pollution** Poor air quality affects hundreds of cities and industrial complexes throughout the Russian domain. The Soviet policy of building large clusters of industrial processing and manufacturing plants in concentrated areas, often with minimal environmental controls, has produced a collection of polluted cities that

stretch from Belarus to Russian Siberia (Figure 9.4). A traditional reliance on abundant but low-quality coal also contributes to pollution problems. Growing rates of private automobile ownership have greatly increased automobile-related pollution, especially because many vehicles lack effective controls on emissions.

Degraded water is another hazard that residents of the region must cope with daily. Urban water supplies are constantly vulnerable to industrial pollution, flows of raw sewage, and demands that increasingly exceed capacity. Oil spills have harmed thousands of square miles in the tundra and taiga of the West Siberian Plain and along the Ob River. Water pollution has also affected the Volga River, the Black Sea, portions of the Caspian Sea shoreline, the Arctic waters off Russia's northern coast, and Siberia's Lake Baikal (the world's largest reserve of freshwater; see Figure 9.5).

**The Nuclear Threat** The nuclear era brought added dangers to the region. The Soviet Union's nuclear weapons and energy programs often ignored issues of environmental safety. Siberia, for example, suffered regular nuclear fallout when tests were conducted in the atmosphere. Nuclear explosions also were used to move earth in dam-building projects. The once-pristine Russian Arctic has been poisoned: The area around the island of Novaya

**FIGURE 9.5 Lake Baikal** Southern Siberia's Lake Baikal is one of the world's largest deep-water lakes. Industrialization devastated water quality after 1950 as pulp and paper factories poured wastes into the lake. Recent cleanup efforts have helped, but many environmental threats remain.

Zemlya served as an unregulated dumping ground for nuclear wastes in the Soviet era. Aging nuclear reactors also dot the region's landscape, often contaminating nearby rivers with plutonium leaks. Nuclear pollution is particularly pronounced in northern Ukraine, where the Chernobyl nuclear power plant suffered a catastrophic meltdown in 1986.

## A Diverse Physical Setting

The Russian domain's northern latitudinal position is an important factor in shaping its basic geographies of climate, vegetation, and agriculture (see Figure 9.1). Indeed, the Russian domain provides the world's largest example of a high-latitude continental climate, where seasonal temperature extremes and short growing seasons greatly limit opportunities for human settlement. In terms of latitude, Moscow is positioned as far north as Ketchikan, Alaska, and even the Ukrainian capital of Kiev (Kyiv) sits north of the Great Lakes in Canada. Thus, apart from a subtropical zone near the Black Sea, much of the region experiences a classic continental climate with hard, cold winters and marginal agricultural potential.

**The European West** An airplane flight over the western portions of the Russian domain would reveal a vast, barely changing landscape below. European Russia, Belarus, and Ukraine cover the eastern portions of the vast European Plain, which runs from southwest France to the Ural Mountains. One major geographic advantage of European Russia is that different river systems, all now linked by canals, flow into four separate drainages. The result is that trade goods can easily flow in many directions within and beyond the region. The Dnieper and Don rivers flow into the Black Sea; the West and North Dvina rivers drain into the Baltic and White seas, respectively; and the Volga River runs to the Caspian Sea (Figure 9.6).

Most of European Russia experiences cold winters and cool summers by North American standards (Figure 9.7). Moscow, for example, is about as cold as Minneapolis in January, but it is not nearly as warm in July. In Ukraine, Kiev is milder, however, and Simferopol, near the Black Sea, offers wintertime temperatures that average more than 20°F (11°C) warmer than those of Moscow.

Three distinctive environments shape agricultural potential in the European west (Figure 9.8). North of Moscow and St. Petersburg, poor soils and cold temperatures severely limit farming. Belarus and central portions of European Russia possess longer growing seasons, but acidic **podzol soils**, typical of northern forest environments, limit output and the ability of the region to support a productive agricultural economy. Still, diversified agriculture includes grain (rye, oats, and wheat) and potato cultivation, swine and meat production, and dairying. South of 50° latitude, agricultural conditions improve across much of southern Russia and Ukraine. Forests gradually give way to steppe environments dominated by grasslands and by fertile "black earth" **chernozem soils**, which have proven valuable for commercial wheat, corn, and sugar beet cultivation and for commercial meat production (Figure 9.9).

**The Ural Mountains and Siberia** The Ural Mountains (see Figure 9.1) mark European Russia's eastern edge, separating it from Siberia. Despite their geographic significance as the traditional division between continents, the Urals are not a particularly impressive range; several of their southern passes are less than 1,000 feet (305 meters) high. Still, the ancient rocks of these mountains contain many valuable mineral resources, and the Urals traditionally marked Russia's eastern cultural boundary.

**FIGURE 9.6 Volga Valley** This satellite image shows the lower Volga River, flowing through southwestern Russia. In the lower portion of the image, the city of Volgograd sits along the western edge of the river, where it bends sharply to the southeast on its journey to the Caspian Sea.

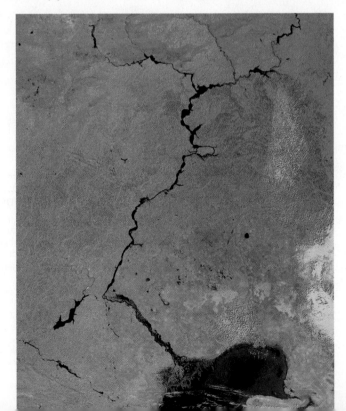

East of the Urals, Siberia unfolds across the landscape for thousands of miles. The great Arctic-bound Ob, Yenisey, and Lena rivers (see Figure 9.1) drain millions of square miles of northern country that includes the flat West Siberian Plain, the hills and plateaus of the Central Siberian Uplands, and the rugged and isolated Northeast Highlands. Along the Pacific, the Kamchatka Peninsula offers spectacular volcanic landscapes. Wintertime climatic conditions, however, are severe across the entire region.

Siberian vegetation and agriculture reflect the climatic setting. The northern portion of the region is too cold for tree growth and instead supports tundra vegetation, which is characterized by mosses, lichens, and a few ground-hugging flowering plants. Much of the tundra region

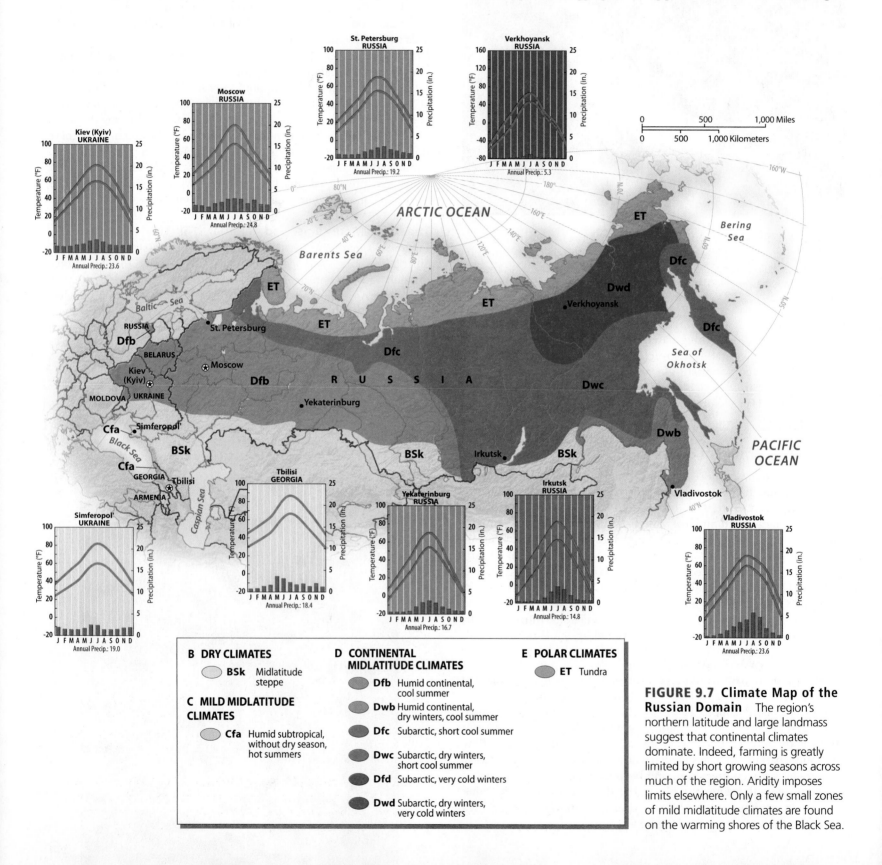

**FIGURE 9.7 Climate Map of the Russian Domain** The region's northern latitude and large landmass suggest that continental climates dominate. Indeed, farming is greatly limited by short growing seasons across much of the region. Aridity imposes limits elsewhere. Only a few small zones of mild midlatitude climates are found on the warming shores of the Black Sea.

**B DRY CLIMATES**
- BSk Midlatitude steppe

**C MILD MIDLATITUDE CLIMATES**
- Cfa Humid subtropical, without dry season, hot summers

**D CONTINENTAL MIDLATITUDE CLIMATES**
- Dfb Humid continental, cool summer
- Dwb Humid continental, dry winters, cool summer
- Dfc Subarctic, short cool summer
- Dwc Subarctic, dry winters, short cool summer
- Dfd Subarctic, very cold winters
- Dwd Subarctic, dry winters, very cold winters

**E POLAR CLIMATES**
- ET Tundra

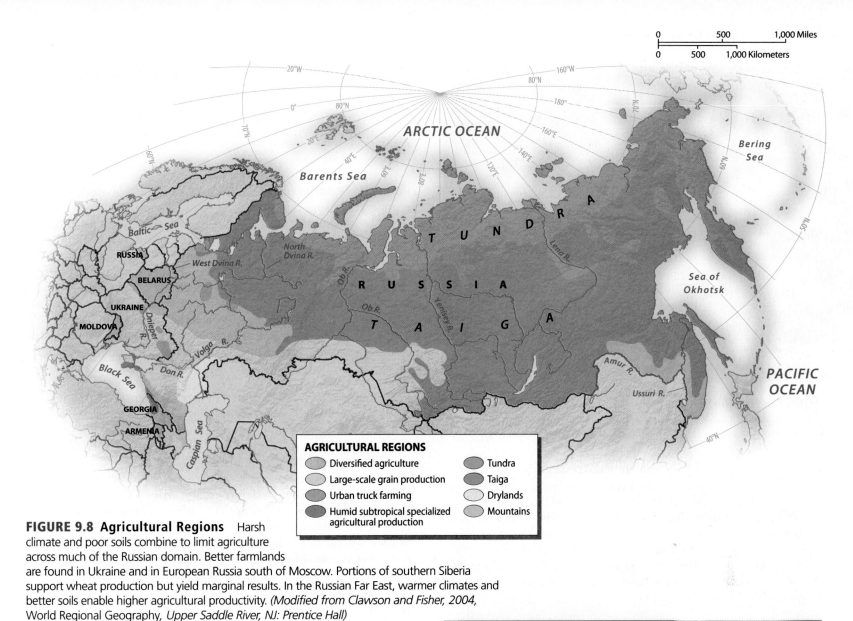

**FIGURE 9.8 Agricultural Regions** Harsh climate and poor soils combine to limit agriculture across much of the Russian domain. Better farmlands are found in Ukraine and in European Russia south of Moscow. Portions of southern Siberia support wheat production but yield marginal results. In the Russian Far East, warmer climates and better soils enable higher agricultural productivity. *(Modified from Clawson and Fisher, 2004, World Regional Geography, Upper Saddle River, NJ: Prentice Hall)*

**AGRICULTURAL REGIONS**

- Diversified agriculture
- Large-scale grain production
- Urban truck farming
- Humid subtropical specialized agricultural production
- Tundra
- Taiga
- Drylands
- Mountains

is associated with **permafrost**, a cold-climate condition of unstable, seasonally frozen ground that limits the growth of vegetation and causes problems for railroad construction. South of the tundra, the Russian **taiga**, or coniferous forest zone, dominates a large portion of the Russian interior. With huge demand for lumber from nearby Japan and China, the eastern taiga zone has been increasingly threatened by both authorized logging and illegal timber poaching (Figure 9.10).

**The Russian Far East** The Russian Far East is a distinctive subregion characterized by proximity to the Pacific Ocean, a more southerly latitude, and a pair of fertile river valleys. Located at about the same latitude as North America's New England, the region features longer growing seasons and milder climates than those found to the west or north. Here, the continental climates of the Siberian interior meet the seasonal monsoon rains of East Asia. It is a fascinating zone of ecological mixing: Conifers of the taiga mingle with Asian hardwoods, and reindeer, Siberian tigers, and leopards also find common ground.

**The Caucasus and Transcaucasia** In European Russia's extreme south, flat terrain gives way first to hills and then to the Caucasus Mountains, a large range stretching between the Black and Caspian seas (see Figure 9.1). The Caucasus Mountains mark

**FIGURE 9.9 Commercial Wheat Production** This wheat field in southern Ukraine typifies some of the region's most productive farmland. Longer growing seasons and good soils remain enduring assets in this portion of the Russian domain.

**FIGURE 9.10 Siberian Lumber**   A growing supply of Siberian lumber is flowing toward markets in both Japan and China, threatening the environmental quality and sustainability of Russia's forest resources.

Russia's southern boundary and are characterized by major earthquakes. Farther south lies Transcaucasia and the distinctive natural settings of Georgia and Armenia. Patterns of both climate and terrain in the Caucasus and Transcaucasia are very complex. Rainfall is generally higher in the western zone, while the area's eastern valleys are semiarid. In areas of adequate rainfall or where irrigation is possible, agriculture can be quite productive. Georgia in particular has long been an important producer of fruits, vegetables, flowers, and wines (Figure 9.11).

**FIGURE 9.11 Subtropical Georgia**   The moderating influences of the Black Sea and a more southern latitude produce a small zone of humid subtropical agriculture in Georgia. The fertile landscapes of these tea plantations offer a sharp contrast to the colder country found north of the Caucasus.

## Consequences of Global Climate Change

Given its latitude and continental climates, the Russian domain is often cited as a world region that would benefit from a warmer global climate. But such an interpretation oversimplifies the complex natural and human responses to global climate change, some of which are already occurring across the region.

**Potential Benefits of Global Climate Change**   Some point to the economic benefits that may result from warmer Eurasian climates. Consider agriculture (Figure 9.8). Some models predict that the northern limit of spring cereal cultivation in northwestern Russia will shift 60–90 miles (100–150 kilometers) poleward for every 1.8°F (1°C) of warming. Elsewhere, warmer springs in settings such as Belarus might allow for higher yields of maize and sunflowers. Areas once in tundra vegetation may be more suitable for boreal forest expansion. Less severe winters might make energy development in subarctic settings less costly. In the Arctic Ocean and Barents Sea, less sea ice would translate to easier navigation, more high-latitude commerce, and more ice-free days in northern Russian ports.

**Potential Hazards of Global Climate Change**   Even with such rosy scenarios, might long-term regional and global costs outweigh the benefits? Climate experts point to three particular concerns. First, rising global sea levels will hit low-lying areas of the Black and Baltic seas particularly hard. Officials in St. Petersburg, Russia's second largest city, are already contemplating significant costs associated with controlling the Baltic's rising waters. The city's natural setting is essentially in a low-lying coastal estuary. Second, changes in ecologically sensitive arctic and subarctic ecosystems are already leading to major disruptions in wildlife and indigenous human populations. For example, the shrinking volume of arctic sea ice habitat for polar bears has meant that they are forced to widen their search for food, bringing them into closer contact with villages and also disrupting traditional hunting practices. Poachers have also profited, increasing their illegal harvests.

Finally, the largest potential change within the region, with truly global implications, relates to the thawing of the Siberian permafrost. Substantial areas of northern Russia are covered with permafrost already close to thawing. Even minor increases in temperature could have large and irreversible consequences for the region. The greatest potential global impact may come with the huge release of carbon currently stored in permafrost environments. Soils frozen in permafrost contain large amounts of organic material that decomposes quickly when thawed. Most of the planet's permafrost could release its carbon reservoir within the next century, the equivalent of 80 years of burning fossil fuels. Such a contribution to the world's carbon budget, which would likely further warm Earth, is only beginning to be incorporated into models of global climate change.

## Population and Settlement: An Urban Domain

The Russian domain is home to about 200 million residents (Table 9.1). While they are widely dispersed across Eurasia, most live in cities. The region's population geography has been influenced by the distribution of natural resources and by government policies that have encouraged migration out of the traditional centers of population in the western portions of the domain.

### TABLE 9.1 POPULATION INDICATORS

| Country | Population (millions) 2009 | Population Density (per square kilometer) | Rate of Natural Increase (RNI) | Total Fertility Rate | Percent Urban | Percent <15 | Percent >65 | Net Migration (Rate per 1,000) 2005–2010* |
|---|---|---|---|---|---|---|---|---|
| Armenia | 3.1 | 104 | 0.6 | 1.7 | 64 | 20 | 10 | −4.9 |
| Belarus | 9.7 | 47 | −0.3 | 1.4 | 74 | 15 | 14 | 0.0 |
| Georgia | 4.6 | 66 | 0.2 | 1.4 | 53 | 17 | 15 | −11.5 |
| Moldova | 4.1 | 122 | −0.1 | 1.3 | 41 | 18 | 10 | |
| Russia | 141.8 | 8 | −0.3 | 1.5 | 73 | 15 | 14 | 0.4 |
| Ukraine | 46.0 | 76 | −0.5 | 1.4 | 68 | 14 | 16 | −0.3 |

*Source:* Population Reference Bureau, *World Population Data Sheet, 2009.*
* Net Migration Rate from the United Nations, Population Division, *World Population Prospects: The 2008 Revision Population Database.*

## Population Distribution

The favorable agricultural setting of the European west offered a home to more people than did the inhospitable conditions found across central and northern Siberia. Although Russian efforts over the past century have encouraged a wider dispersal of the population, it remains heavily concentrated in the west (Figure 9.12). European Russia is home to more than 100 million persons, while Siberia, although far larger, holds only some 35 million. When one adds the 60 million inhabitants of Belarus, Moldova, and Ukraine, the imbalance between east and west becomes even more striking.

**The European Core** The region's largest cities, biggest industrial complexes, and most productive farms are located in the European Core, a subregion that includes Belarus, much of Ukraine, and

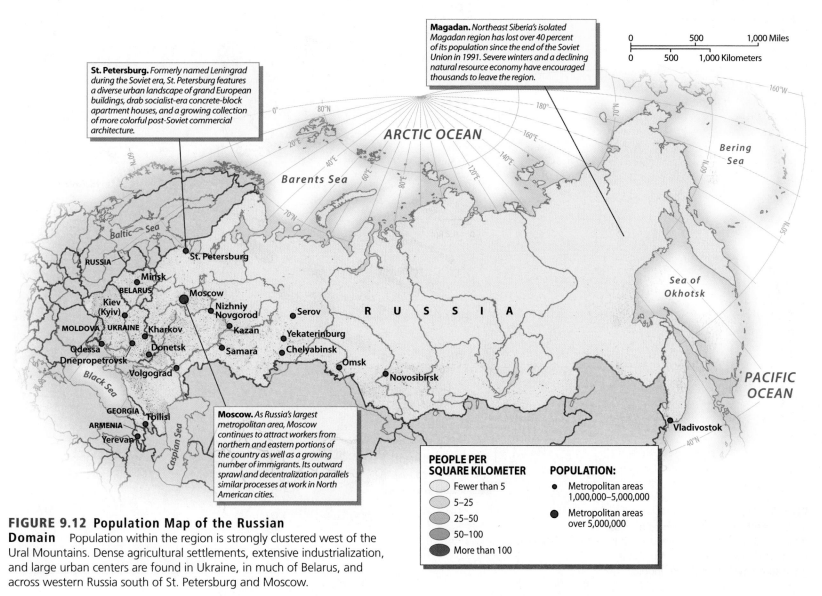

**FIGURE 9.12 Population Map of the Russian Domain** Population within the region is strongly clustered west of the Ural Mountains. Dense agricultural settlements, extensive industrialization, and large urban centers are found in Ukraine, in much of Belarus, and across western Russia south of St. Petersburg and Moscow.

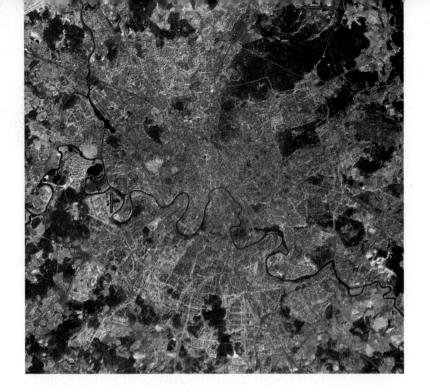

**FIGURE 9.13 Metropolitan Moscow** Sprawling Moscow extends more than 50 miles (80 kilometers) beyond the city center. The city is home to more than 10 million people, and the relative strength of its urban economy continues to attract migrants from elsewhere in the country, thus putting more pressure on its infrastructure.

Russia west of the Urals (Table 9.1). The sprawling city of Moscow and its nearby urbanized region clearly dominate the settlement landscape, with a metropolitan area of more than 10 million people (Figure 9.13).

On the shores of the Baltic Sea, St. Petersburg (Leningrad in the Soviet period, 4.7 million people) has traditionally had a great deal of contact with western Europe. Between 1712 and 1917, it served as the capital of the Russian Empire. Its rich assortment of handsome buildings, bridges, and canals gives the city an urban landscape many have compared to the great cities of western Europe (Figure 9.14).

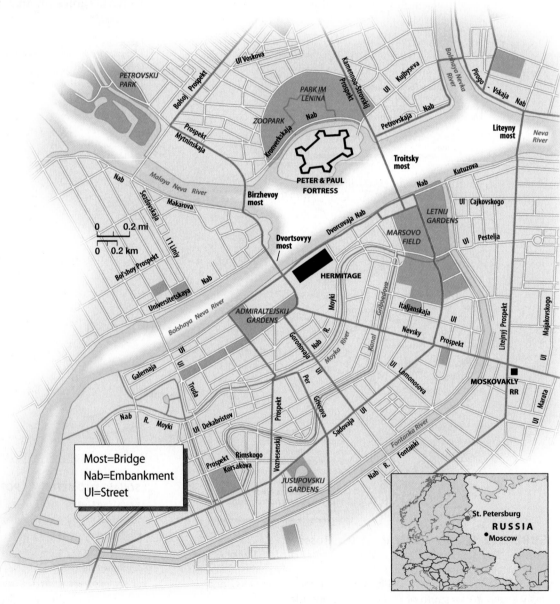

**FIGURE 9.14**
**St. Petersburg** Often beloved as Russia's most beautiful city, St. Petersburg's urban design features a varied mix of gardens, open space, waterways, and bridges.

Other urban clusters are oriented along the lower and middle stretches of the Volga River and include the cities of Kazan, Samara, and Volgograd. Industrialization in the region accelerated during World War II, as the region lay somewhat removed from German advances in the west. Today, the highly commercialized river corridor, also containing important petroleum reserves, supports a diverse industrial base strategically located to serve the large populations of the European Core. Nearby, the resource-rich Ural Mountains include the gritty industrial landscapes of Serov (1.0 million) and Chelyabinsk (1.1 million).

Beyond Russia, major population clusters within the European Core are also found in Belarus and Ukraine (Table 9.1). The Belarusian capital of Minsk (1.7 million) is the dominant urban center in that country, and its landscape recalls the drab Soviet-style architecture of an earlier era (Figure 9.15). In nearby Ukraine, the capital Kiev (Kyiv, 2.7 million) straddles the Dnieper River, and the city's old and beautiful buildings are a visual reminder of its historical role as a cultural and trading center within the European interior.

**Siberian Hinterlands**  Leaving the southern Urals city of Yekaterinburg on a Siberia-bound train, one is aware that the land ahead is ever more sparsely settled (see Figure 9.12). The distance between cities grows, and the intervening countryside reveals a landscape shifting gradually from farms to forest. The Siberian hinterland is divided into two characteristic zones of settlement, each of which can be linked to railroad lines. To the south, a collection of isolated, but sizable, urban centers follows the **Trans-Siberian Railroad**, a key railroad passage to the Pacific completed in 1904. The eastbound traveler encounters Omsk (1.1 million) as the rail line crosses the Irtysh River, Novosibirsk (1.4 million) at its junction with the Ob River, and Irkutsk (600,000) near Lake Baikal. The port city of Vladivostok (640,000) provides access to the Pacific. To the north, a thinner sprinkling of settlement appears along the more recently completed (1984) **Baikal-Amur Mainline (BAM) Railroad**, which parallels the older line but runs north of Lake Baikal to the Amur River. From the BAM line to the Arctic, the almost empty spaces of central and northern Siberia dominate the scene.

**FIGURE 9.15  Minsk**  This row of mass-produced apartment houses sits on the bank of the Svisloch River in Minsk, Belarus.

## Regional Migration Patterns

Over the past 150 years, millions of people within the Russian domain have been on the move. These major migrations, both forced and voluntary, reveal sweeping examples of human mobility that rival the great movements from Europe and Africa or the transcontinental spread of settlement across North America.

**Eastward Movement**  Just as settlers of European descent moved west across North America, exploiting natural resources and displacing native peoples, European Russians moved east across the Siberian frontier within Eurasia. Although these migrations into Siberia began several centuries earlier, the pace accelerated in the late 19th century after the Trans-Siberian Railroad was completed. Peasants were attracted to the region by its agricultural opportunities (in the south) and by greater political freedoms than they traditionally enjoyed under the **tsars** (or czars; Russian for *Caesar*), the authoritarian leaders who dominated politics during the pre-1917 Russian Empire. Almost 1 million Russian settlers moved into the Siberian hinterland between 1860 and 1914.

**Political Motives**  Political motives shaped migration patterns in the Russian domain. Particularly in the case of Russia, leaders from both the imperial and Soviet eras saw advantages in moving people to new locations. Both the tsars and the Soviet leaders saw their political power grow as Russians moved into the Eurasian interior and developed its resources. Political dissidents and troublemakers in the Soviet era were also forcibly relocated to the region's **Gulag Archipelago**, a vast collection of political prisons in which inmates often disappeared or spent years far removed from their families and communities. Until recently, political instability in Russia's internal republic of Chechnya produced large-scale, politically motivated refugee migrations into nearby regions (Figure 9.16).

**Russification**, the Soviet policy of resettling Russians into non-Russian portions of the Soviet Union, also changed the region's human geography. Millions of Russians were given economic and political incentives to move elsewhere in the Soviet Union in order to increase Russian dominance in many of the outlying portions of the country. By the end of the Soviet period, Russians made up significant minorities within former Soviet republics such as Kazakhstan (38 percent Russian), Latvia (34 percent), and Estonia (30 percent). Among its Slavic neighbors, Belarus remains 13 percent Russian and Ukraine more than 22 percent Russian, with concentrations particularly high in the eastern portions of the countries.

**New International Movements**  In the post-Soviet era, Russification has often been reversed (Figure 9.16). Several of the newly independent non-Russian countries have imposed rigid language and citizenship requirements, which have encouraged many Russian residents to leave. Often, ethnic Russians have experienced discrimination in regard to economic and social opportunities in these non-Russian countries. By 2009, more than 7 million Russians had left former Soviet republics and had returned to their homeland. Some have been attracted by a new government program that is offering relocation and employment assistance to Russian migrants willing to return.

A growing number of Russia's immigrants are from ethnically non-Slavic regions. Street cleaners from Kyrgyzstan and shopkeepers from Uzbekistan are common sights in Moscow, where almost one-third of the country's estimated 10 million illegal immigrants live (Figure 9.17).

**FIGURE 9.16 Recent Migration Flows in the Russian Domain**   Recent events are encouraging the return of ethnic Russians from former Soviet republics, while other Russians are emigrating from the domain for economic, cultural, and political reasons. Within Russia, both political and economic forces are also at work, encouraging people to be on the move.

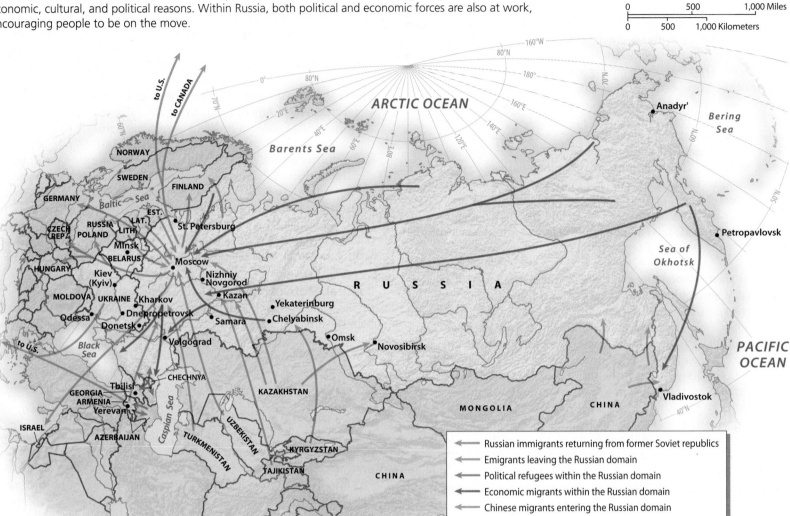

Russian immigrants returning from former Soviet republics
Emigrants leaving the Russian domain
Political refugees within the Russian domain
Economic migrants within the Russian domain
Chinese migrants entering the Russian domain

**FIGURE 9.17 Non-Russian Immigrants**   These immigrants from Tajikistan are standing in line in Moscow, applying for employment visas to work in the Russian capital city.

Elsewhere, incoming Chinese are transforming the Russian Far East, attracted by employment opportunities in that fast-changing region (see "Exploring Global Connections: Chinese Immigrants in Russia's Far East"). Future immigration flows into Russia, however, will be driven by the health of its economy. Job losses in 2009, for example, forced thousands of unemployed Tajiks in the country to return to their central Asian homeland of Tajikistan.

Today, it is also easier to leave the region (Figure 9.16). The job-related "brain drain" of young, well-educated, upwardly mobile Russians has been considerable. Sometimes, ethnic links play a part in migration patterns. For example, many Russian-born ethnic Finns have moved to nearby Finland. Russia's Jewish population also continues to fall. Furthermore, Russians have become one of the largest new immigrant groups in the United States.

## Inside the Russian City

Today, most people in the Russian domain live in cities, the product of a century of urban migration and growth (Table 9.1). Large Russian cities possess a core area, or center, that features superior transportation

# EXPLORING GLOBAL CONNECTIONS
## Chinese Immigrants in Russia's Far East

Chinese immigrants are fundamentally refashioning the economic and cultural geographies of the Russian Far East. Somewhere between 5 and 7 million immigrants (many of them illegal) may now live in the region. Much of the transformation has occurred since the early 1990s and is in response to a large, job-hungry Chinese population to the south, a shrinking supply of Russian workers to the north, and a regional economy that is increasingly being shaped by powerful forces of globalization.

Walk through the Russian cities of Vladivostok or Khabarovsk (Figure 9.1.1) and you see that Chinese immigrants are everywhere. Street signs feature both Russian and Chinese lettering. Entire neighborhoods are dominated by immigrant populations. Chinese children are learning Russian in school, and Russians find themselves working for Chinese entrepreneurs. Many Chinese arrived initially with tourist visas, while others have slipped casually over the border. Higher wages and better business opportunities have drawn people from northern China. Many have entered the construction and forestry industries, others are in retailing or in market gardening, and a growing number have come with investment capital to open manufacturing plants. At the same time, Russian populations are aging and often leaving the region, adding demographic significance to the Chinese inflow.

What are the long-term consequences of this redefinition of the Russian Far East? Many experts see the Chinese as essential ingredients in fueling continued economic growth in a Russia in which native populations are steadily shrinking. Some Chinese officials smile and remind the Russians that the area was once part of imperial China and perhaps will be once again, after several more decades of Chinese in-migration. While many Russians see the economic advantages in having hard-working Chinese living in Russia, some fear the growing foreign population in their country. A significant nationalist backlash has occurred: Chinese have been attacked by gangs in cities, Chinese shopkeepers complain of being rousted by Russian police, and recent legislation has made it more difficult for Chinese to operate businesses in the region. On the other hand, many younger Russians have welcomed their Chinese counterparts, and there are a growing number of joint Russian–Chinese companies in the region, as well as more intermarriage between the two groups. Thirty years from now, the "Russian" Far East is likely to be far more culturally diverse than it was during most of the 20th century, and the region's successful integration with the global economy will in no small part be shaped by the new human geography taking shape there today.

**FIGURE 9.1.1 Chinese Immigrants in the Russian Far East** Many Chinese immigrants in the Russian Far East have become small-scale entrepreneurs in the region's commercial economy.

connections; well-stocked, upscale department stores and shops; desirable housing; and important offices (both governmental and private). In the largest urban centers, cities such as Moscow and St. Petersburg also feature extensive public spaces and examples of monumental architecture at the city center.

Within the city, there is usually a distinctive pattern of circular land-use zones, each of which was built at a later date moving outward from the center. Such a ringlike urban morphology is not unique. However, as a result of the extensive power of government planners during the Soviet period, this urban form is probably more highly developed here than in most other parts of the world.

At their very center, the cores of many older cities predate the Soviet Union. Pre-1900 stone buildings often dominate older city centers. Some of these are former private mansions that were turned into government offices or subdivided into apartments during the communist period but are now being privatized again. Many of these older buildings, however, are being leveled in rapidly growing urban settings such as downtown Moscow (Figure 9.18). Since 1992, urban preservation experts estimate that several thousand historic buildings have been destroyed, including many structures that had supposedly received official protection. Retailing malls replace many of these older structures in the city. Nearby, nightclubs and bars are filled with pleasure seekers, as the city's professional elite mingles with foreign visitors and tourists. The "Moscow City" business zone is also expanding into older industrial areas, offering new office space for Russian and international firms.

Farther out from the city centers are **mikrorayons**, large, Soviet-era housing projects of the 1970s and 1980s (Figure 9.19). Mikrorayons are typically composed of massed blocks of standardized apartment buildings, ranging from 9 to 24 stories in height. The largest of these super-complexes house up to 100,000 residents. While planners hoped that mikrorayons would foster a sense of community, most now serve as anonymous bedroom communities for larger metropolitan areas.

Some of Russia's most rapid urban growth has occurred on the metropolitan periphery, paralleling the North American experience. Moscow, for example, has seen its urban reach expand far beyond the city center. The surrounding administrative district (the Moscow Oblast) contains about 7 million residents and is home to more than 700 international companies. Land prices and tax rates are lower than in the central city, the bureaucracy is less onerous, and the transportation and telecommunications infrastructure are relatively new. New suburban shopping malls, paralleling the North American model, are also popping up on the urban fringe, allowing residents to shop and be entertained without having to visit the city center.

## The Demographic Crisis

Russia's Ministry of Labor and Social Development recently acknowledged that the country has faced a crisis of persistently declining populations (Table 9.1). Government studies predicted that

**FIGURE 9.19 Moscow Housing**   For many residents in larger Russian cities, home is a high-rise apartment house. Many of these satellite centers were built in the Soviet era. Poor construction and a lack of landscaping often yield a bleak suburban scene, but nearby stores, entertainment, and public transportation offer important services.

**FIGURE 9.18 Downtown Moscow**   The city's downtown landscape has been rapidly transformed in the recent economic boom. It remains a complex and fascinating mix of imperial, Soviet, and post-Soviet influences.

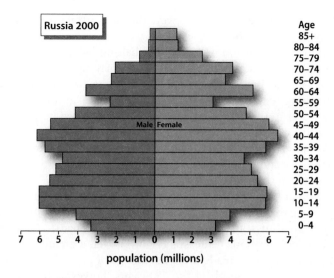

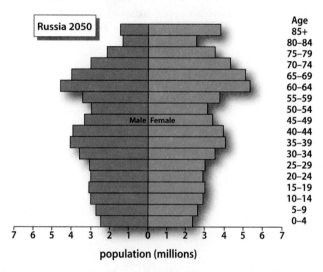

**FIGURE 9.20 Russia's Changing Population**  These two population pyramids provide a recent glimpse (2000) as well as predicted patterns (2050) of Russia's population structure. Present trends suggest that Russia's population will continue to age, with relatively fewer young people supporting a relatively large elderly population. Also note the impact of earlier wars and higher death rates on older adult Russian males. *(U.S. Census Bureau, International Database)*

Russia's population could fall by a startling 45 million by 2050 (Figure 9.20). Similar conditions are affecting the other countries in the region. Declining populations, low birthrates, and relatively high mortality, particularly among middle-aged males, are all troubling symptoms of the region's population problems.

The region's demographic crisis appears to be related to several variables. Death rates remain high, especially for Russian men. Recently, Prime Minister Putin argued that the demographic decline was Russia's "most acute problem." In 2007, he helped initiate a new multi-year program in Russia to encourage higher birthrates. Mothers of a second child receive non-taxable cash grants, extended maternity leave, and extensive day-care subsidies. Recently, there have been slight increases in Russian birthrates, perhaps a function of changing government policies. Russian birthrates are now higher than those of Germany and Japan, and there are long waiting lists for urban kindergarten spots in Moscow

(Figure 9.21). Employers are also offering more benefits to both mothers and fathers, making parenthood more appealing for some couples. But only time will tell if the region's demographic crisis is easing.

## Cultural Coherence and Diversity: The Legacy of Slavic Dominance

For hundreds of years, Slavic peoples speaking the Russian language expanded their influence from an early homeland in central European Russia. Russian cultural patterns and social institutions spread widely during this Slavic expansion, influencing many non-Russian ethnic groups that continued to live under the rule of the Russian empire. The legacy of this diffusion continues today, offering Russians a rich historical identity and suggesting how present-day Russians are dealing with forces of globalization and how non-Russian cultures have evolved in the region.

### The Heritage of the Russian Empire

The expansion of the Russian Empire paralleled similar events in western Europe. As Spain, Portugal, France, and Britain carved out overseas empires, the Russians expanded eastward and southward across Eurasia. The origin of the Russian Empire lies in the early history of the **Slavic peoples**, defined as those of a northern branch of the Indo-European language family. Slavic political power grew by 900 C.E. as these people intermarried with southward-moving warriors from Sweden known as *Varangians*, or *Rus*. Within a century, the state of Rus extended from Kiev (the capital) to near the Baltic Sea. The new Kiev-Rus state interacted with the Byzantine Empire of the Greeks, and this influence brought Christianity as well as the Cyrillic alphabet to the Russian region. Even as the Russians converted to **Eastern Orthodox Christianity**, a form of Christianity historically linked to eastern Europe and church leaders in Constantinople (modern Istanbul), their Slavic neighbors to the west

**FIGURE 9.21 Young Russian Parents**  A slight rise in Russian birthrates may help slow the country's population decline. Still, high death rates are a major issue, particularly for males throughout the region, and they are the product of many environmental and lifestyle variables.

(the Poles, Czechs, Slovaks, Slovenians, and Croatians) accepted Catholicism. This early Russian state soon faltered and split into several principalities that were then ruled by invading Mongols and Tatars (a group of Turkish-speaking peoples).

By the 14th century, however, northern Slavic peoples overthrew Tatar rule and established a new and expanding Slavic state (Figure 9.22). The core of the new Russian Empire lay near the eastern fringe of the old state of Rus. Gradually this area's language diverged from that spoken in the new Russian core, and *Ukrainians* and Russians developed into two separate peoples. A similar development took place among the northwestern Russians, who experienced several centuries of Polish rule and over time were transformed into a distinctive group known as the *Belarusians*.

The Russian Empire expanded remarkably in the 16th and 17th centuries. Former Tatar territories in the Volga Valley (near Kazan) were incorporated into the Russian state in the mid-1500s. The Russians also allied with the seminomadic **Cossacks**, Slavic-speaking Christians who had earlier migrated to the region seeking freedom in the ungoverned steppes (Figure 9.2). The alliance smoothed the way for Russian expansion into Siberia during the 17th century. Premium Siberian furs and precious metals were the chief attractions of this immense northern territory.

Westward expansion was slow and halting. When Tsar Peter the Great (1682–1725) defeated Sweden in the early 1700s, he obtained a foothold on the Baltic Sea. There he built the new capital city of St. Petersburg, designed to give the empire better access to western Europe. Later in the 18th century, Russia also defeated both the Poles and the Turks and gained all of modern-day Belarus and Ukraine. Tsarina Catherine the Great (1762–1796) was especially important in colonizing Ukraine and bringing the Russian Empire to the warm-water shores of the Black Sea.

The 19th century witnessed the Russian Empire's final expansion. Large gains were made in Central Asia, where a group of once-powerful Muslim states was no longer able to resist the Russian army. The mountainous Caucasus region proved a greater challenge, as the peoples of this area had the advantage of rugged terrain in defending their lands. South of the Caucasus, however, the Christian Armenians and Georgians accepted Russian power with little struggle because they found it preferable to rule by the Persian or Ottoman empires.

**FIGURE 9.22 Growth of the Russian Empire**     Beginning as a small principality in the vicinity of modern Moscow, the Russian Empire took shape between the 14th and 16th centuries. After 1600, Russian influence stretched from eastern Europe to the Pacific Ocean. Later, portions of the empire were added in the Far East, Central Asia, and near the Baltic and Black seas. *(Modified from Bergman and Renwick, 1999,* Introduction to Geography, *Upper Saddle River, NJ: Prentice Hall)*

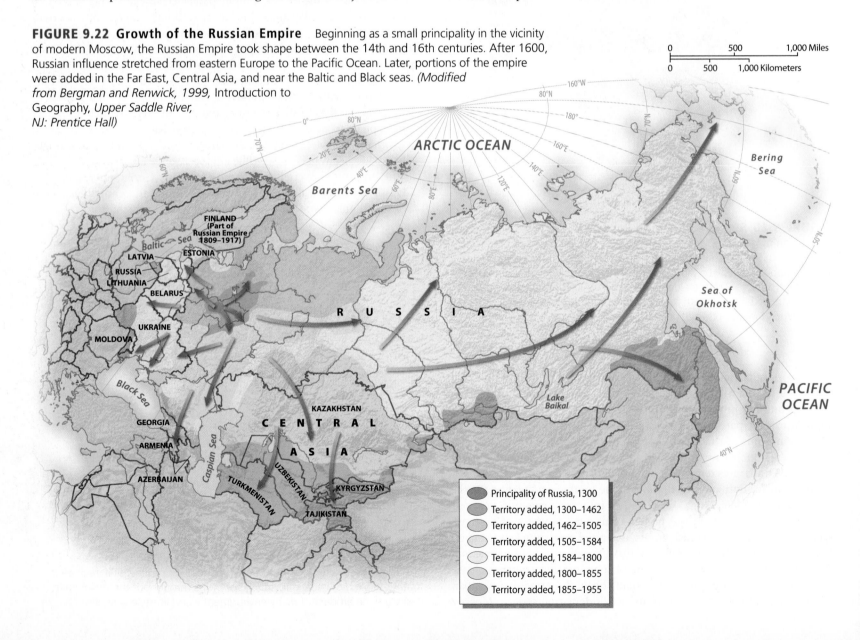

**FIGURE 9.23 Languages of the Russian Domain**   Slavic Russians dominate the region, although many linguistic minorities are present. Siberia's diverse native peoples add cultural variety in that area. To the southwest, the Caucasus Mountains and the lands beyond contain the region's most complex linguistic geography. Ukrainians and Belarusians, while sharing a Slavic heritage with their Russian neighbors, add further variety in the west.

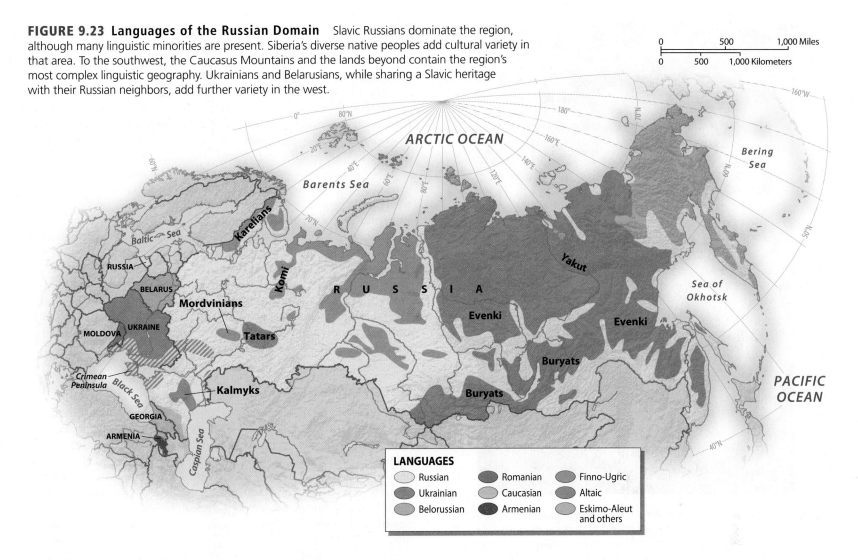

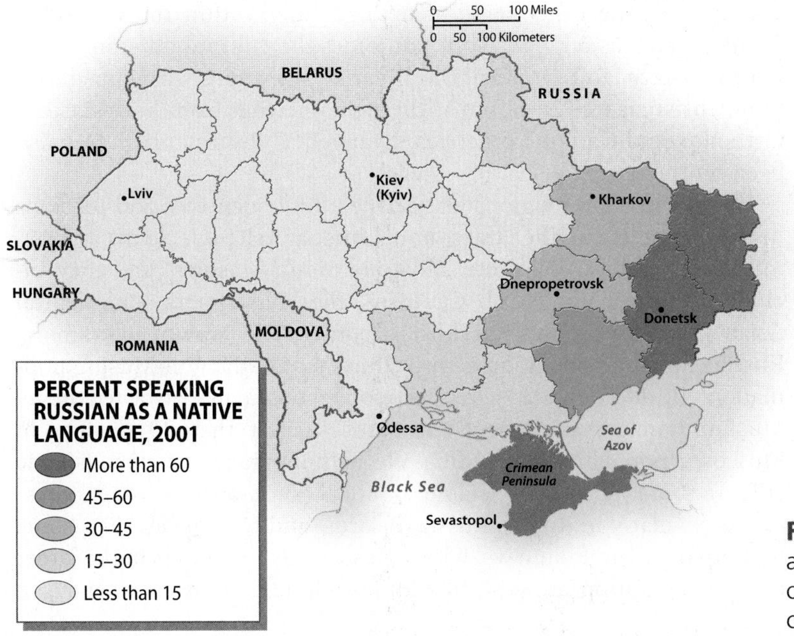

## Geographies of Language

Slavic languages dominate the region (Figure 9.23). The geographic pattern of the Belarusian people is relatively simple. The vast majority of Belarusians reside in Belarus, and most people in Belarus are Belarusians.

The situation in Ukraine, however, is more complex. Because of different historical patterns of territorial conquest, Russian speakers dominate large parts of eastern Ukraine, while they make up a much smaller portion of western Ukraine's population (Figure 9.24). Similarly, the Crimean Peninsula, now a part of Ukraine, has long had ethnic and political connections to Russia. As a result, many Ukrainian citizens in the eastern and southern parts of the country rarely speak Ukrainian. Conversely, many Ukrainians born in the west never learn Russian.

**FIGURE 9.24 The Russian Language in Ukraine**   Both eastern Ukraine and the Crimean Peninsula retain large numbers of Russian speakers, a function of long-held cultural and political ties that continue to complicate Ukraine's contemporary human geography.

**FIGURE 9.25 Minority Evenki** Russia's indigenous Evenki population speaks an Altaic language, and they share many of the economic and social problems of native peoples in North America and Australia.

In nearby Moldova, Romanian (a Romance language) speakers are dominant, although ethnic Russians and Ukrainians each make up about 13 percent of the country's population. During the Soviet period, such geographic mixing had little consequence, because the distinction between Russians, Ukrainians, and Moldavians was not viewed as important in official circles. Now that Russia, Ukraine, and Moldova are separate countries with a heightened sense of national distinction, this issue has emerged as a significant source of tension across the region.

Approximately 80 percent of Russia's population claims a Russian linguistic identity. Russians inhabit most of European Russia, but there are large groups of other peoples in the region. The Russian zone extends across southern Siberia to the Sea of Japan. In sparsely settled lands of central and northern Siberia, Russians are more numerous in many areas, but they share territory with varied native peoples.

Finno-Ugric (Finnish-speaking peoples), though small in number, dominate sizable portions of the non-Russian north. Altaic speakers also complicate the country's linguistic geography. They include the Volga Tatars, centered on the city of Kazan in the middle Volga Valley. While retaining their ethnic identity, the Turkish-speaking Tatars have extensively intermarried with and borrowed from their

Russian neighbors. Yakut and Evenki peoples of northeast Siberia also represent Turkish speakers within the Altaic family (Figure 9.25). In the east, the Buryats live in the vicinity of Lake Baikal and are closely tied to the cultures and history of Central Asia.

The plight of many native peoples in central and northern Siberia parallels the situation in the United States, Canada, and Australia. Poor, rural indigenous peoples in each of these settings remain distinct from dominant European cultures. These peoples are also internally diverse and are often divided into a number of unrelated linguistic groups. Many Siberian peoples have seen their traditional ways challenged by the pressures of Russification, just as native peoples elsewhere in the world have been subjected to similar pressures of cultural and political assimilation.

Although small in size, Transcaucasia offers a bewildering variety of languages. From Russia, along the north slopes of the Caucasus, and to Georgia and Armenia east of the Black Sea, a complex history and a complicated physical setting have combined to produce some of the most complex language patterns in the world. Several language families are spoken in a region smaller than Ohio, and many individual languages are represented by small, isolated cultural groups.

## Geographies of Religion

Most Russians, Belarusians, and Ukrainians share a religious heritage of Eastern Orthodox Christianity. For hundreds of years, Eastern Orthodoxy served as a central cultural presence in the Russian Empire (Figure 9.26). Indeed, church and state were tightly fused until the demise of the empire in 1917. Under the Soviet Union, all religion was severely discouraged and actively persecuted. Many churches were converted into museums or other kinds of public buildings. With the downfall of the Soviet Union, however, a religious revival has swept much of the Russian domain. Now an estimated 75 million Russians are members of the Orthodox Church, including almost 500 monastic orders dispersed across the country.

Other forms of Western Christianity are also present. For example, the people of western Ukraine, who experienced several hundred years of Polish rule, eventually joined the Catholic Church. Eastern Ukraine, on the other hand, remained fully within the Orthodox framework. This religious split reinforces the cultural differences between eastern and western Ukrainians. Elsewhere, Armenia has a long Christian tradition, but it differs somewhat from both Eastern Orthodox and Catholic practices. Evangelical Protestantism has also been on the rise.

Non-Christian religions also shape ethnic identities and tensions in the region. Islam, the largest non-Christian religion, claims approximately 20 million followers. Most are Sunni Muslims, and they include peoples in the North Caucasus, the Volga Tatars, and Central Asian peoples near the Kazakhstan border. The growth rate among Russia's Muslim population is three times that of the non-Muslim population. In addition, Moscow's economic boom has attracted many Muslim immigrants. Recent estimates suggest that 20 percent of Moscow's population, including illegal immigrants, is now Islamic (Figure 9.27). A growing Islamic political consciousness is also present, particularly in Russia. Russia, Belarus, and Ukraine are also home to more than 1 million Jews, who are especially numerous in the larger cities of the European west. In addition, Buddhists are represented in

arts, such as classical music and ballet, continued to receive generous state subsidies, and to this day Russian artists regularly achieve worldwide fame.

**Turn to the West**   By the 1980s it was clear that the attempt had failed to fashion a new Soviet culture based on communist ideals. The younger generation instead adopted a rebellious attitude, turning for inspiration to fashion and rock music from the West. The mass-consumer culture of the United States proved particularly attractive.

After the fall of the Soviet Union in 1991, there was an inrush of global cultural influences, particularly to the region's larger cities such as Moscow. Shops were flooded with Western books and magazines, people sought advice about home mortgages and condominium purchases, and they enjoyed the newfound pleasures of fake Chanel handbags and McDonald's hamburgers. Many people hurried to embrace the world their former leaders had warned them about for generations. Cultural influences streaming into the country were not all Western in inspiration. Films from Hong Kong and Mumbai (Bombay), as well as the televised romance novels (*telenovelas*) of Latin America, for example, proved far more popular in the Russian domain than in the United States.

**FIGURE 9.26  Russian Orthodox Church, Siberia**   A revival of interest in the Russian Orthodox Church followed the collapse of the Soviet Union in the early 1990s. The newly built Znamensky Cathedral in Kemerovo displays many of the faith's characteristic cultural landscape signatures.

the region, associated with the Kalmyk and Buryat peoples of the Russian interior.

## Russian Culture in Global Context

Russian culture has developed its own distinctive traditions and symbols, and it has also been influenced greatly by western Europe. By the 19th century, even as Russian peasants interacted rarely with the outside world, Russian high culture had become thoroughly Westernized, and Russian composers, novelists, and playwrights gained considerable fame in Europe and the United States.

**Soviet Days**   During the Soviet period, new cultural influences shaped the socialist state. Initially, European-style modern art flourished in the Soviet Union, encouraged by the radical Marxist rhetoric of the new rulers. By the late 1920s, however, Soviet leaders turned against modernism, which they viewed as the decadent expression of a declining capitalist world. Increasingly, state-sponsored Soviet artistic productions centered on **socialist realism**, a style devoted to the realistic depiction of workers heroically challenging nature or struggling against capitalism. Still, traditional high

**FIGURE 9.27  Moscow Mosque**   Moscow's growing immigrant population has included many Muslims from portions of the former Soviet Union, particularly from Central Asia. Today, their presence in the capital is an increasingly visible element in the city's cultural landscape.

**The Music Scene** Younger residents of the region have embraced the world of popular music, and their enthusiasm for U.S. and European performers, as well as their support of a budding home-grown music industry, symbolizes the changing values of an increasingly post-Soviet generation. Today, Russian MTV reaches most of the nation's younger viewers. Sony Music Entertainment established its Russian operations in 1999. Sony, BMG, and other labels have signed multiple Russian acts for domestic markets, helping to boost a native pop-music culture. Increasingly, regional talent is also going global. Ukrainian singer Ruslana won the coveted Eurovision 2004 award for her song "Wild Dance." Three years later, Moldovan singer Natalia Barbu received worldwide recognition when she appeared as a finalist in the Eurovision awards. Barbu signed a recording contract with Sony Music, and the Eurovision performance of her song "Fight" offered a jazz-folk-Gothic rock blend of traditional and contemporary music (Figure 9.28).

**FIGURE 9.28 Natalia Barbu** This young Moldovan singer has emerged as a leading European entertainer who blends traditional regional styles with a jazz-Gothic mix of contemporary music.

# Geopolitical Framework: Resurgent Global Superpower?

The geopolitical legacy of the former Soviet Union still weighs upon the Russian domain. After all, the bold lettering of the "Union of Soviet Socialist Republics" dominated the Eurasian map for much of the 20th century, and the country's global political influence affected every part of the world. Indeed, Russia's post-2000 political resurgence signals a return to its Soviet-era status of geopolitical dominance within the region and its increasing visibility in global political affairs.

## Geopolitical Structure of the Former Soviet Union

The Soviet Union rose from the ashes of the Russian Empire, which collapsed abruptly in 1917. The Russian tsars did little to modernize the country or improve the life of the peasant population. After the fall of the tsars, a government representing several political groups assumed authority. Soon, however, the **Bolsheviks**, a faction of Russian communists representing the interests of the industrial workers, seized power in the country. The leader of these Russian communists was Vladimir Ilyich Ulyanov, usually known by his self-selected name, Lenin. Lenin became the main architect of the Soviet Union, the state that replaced the failed Russian Empire. The new socialist state radically reconfigured Eurasian political and economic geography. When the Soviet Union emerged in 1917, Lenin and the other communist leaders were aware that they faced a major challenge in organizing the new state.

**Creating a Political Structure** Soviet leaders designed a geopolitical solution that maintained their country's territorial boundaries and recognized, at least theoretically, the rights of its many non-Russian citizens. Each major nationality was to receive its own "union republic," provided that it was situated on one of the nation's external borders (Figure 9.29). Eventually, 15 such republics were established, thus creating the Soviet Union. So-called **autonomous areas** within these republics gave special recognition to smaller ethnic homelands. The massive Russian Republic sprawled over roughly three-quarters of the Soviet terrain. Even with these varied internal republics and autonomous areas, the Soviet Union emerged by the late 1920s as a highly centralized state, with important decisions made in the Russian capital of Moscow.

The chief architect of this political consolidation was Soviet leader Joseph Stalin, who did everything he could to centralize power in Moscow and assert Russian authority. The Stalin period (1922–1953) also saw the enlargement of the Soviet Union. As a victorious power in World War II, the country acquired Pacific islands from Japan, the Baltic republics (Estonia, Latvia, and Lithuania), and other portions of eastern Europe. One strategic addition on the Baltic Sea was the northern portion of East Prussia (the port of Kaliningrad), previously part of Germany. It still forms a small but strategic Russian **exclave**, which is defined as a portion of a country's territory that lies outside its contiguous land area (Figure 9.30).

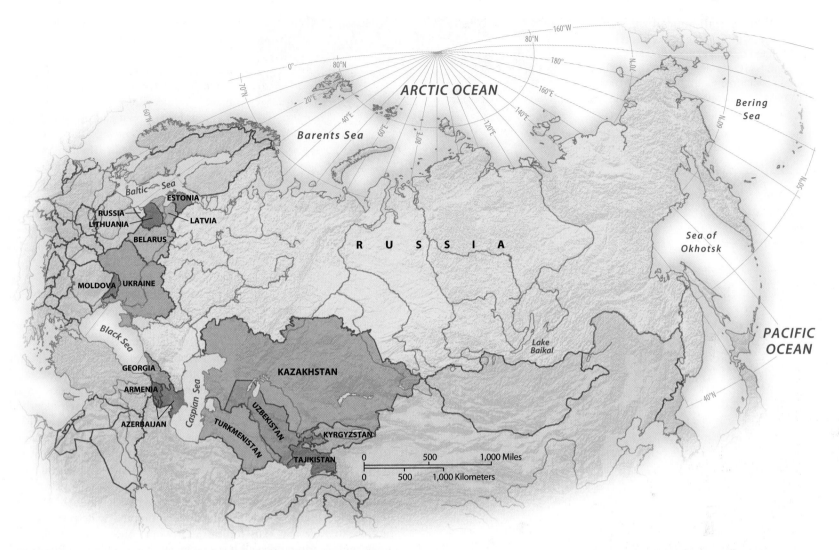

**FIGURE 9.29  Soviet Geopolitical System**  During the Soviet period, the boundaries of the country's 15 internal republics often reflected major ethnic divisions. As the Soviet empire disintegrated, the former republics became politically independent states and now form an uneasy ring of satellite nations around Russia. *(Modified from Rubenstein, 1999,* Introduction to Human Geography, *Upper Saddle River, NJ: Prentice Hall)*

After World War II, the Soviet Union actively expanded its political influence across eastern Europe. In the words of British leader Winston Churchill, the Soviets extended an **Iron Curtain** between their eastern European allies and the more democratic nations of western Europe. As eastern Europe retreated behind the Iron Curtain, the Soviet Union and the United States became antagonists in a global **Cold War** of military competition that lasted from 1948 to 1991.

**End of the Soviet System**  Ironically, Lenin's system of republics based on cultural differences sowed the seeds of the Soviet Union's demise. Even though the republics were never allowed real freedom, they provided a political framework that encouraged the survival of distinct cultural identities. Indeed, contrary to the expectations of Soviet leaders, ethnic nationalism intensified in the post–World War II era as the Soviet system grew less repressive.

When Soviet President Mikhail Gorbachev initiated his policy of **glasnost**, or greater openness, during the 1980s, several republics—most notably the Baltic states of Lithuania, Latvia, and Estonia—demanded outright independence. In addition, Gorbachev's policy of **perestroika**, or restructuring of the planned centralized economy, was an admission that domestic economic conditions increasingly lagged those of western Europe and the United States. A failed war in Afghanistan and increasing political protests in eastern Europe added to Gorbachev's problems.

By 1991, Gorbachev saw his authority slip away amid rising pressures for political decentralization and more dramatic economic reforms. During the summer, Gorbachev's regime was further endangered by the popular election of reform-minded Boris Yeltsin as the head of the Russian Republic and by a failed military takeover by communist hard-liners. By late December, all of the country's 15 constituent republics had become independent states, and the Soviet Union ceased to exist.

**FIGURE 9.30 Kaliningrad** Kaliningrad, a Russian exclave on the Baltic Sea, is now surrounded by Poland and Lithuania, both EU- and NATO-member states. The multistory Hotel Kaliningrad (right) borders the Lenin Prospekt in this downtown view of the Russian exclave.

## Current Geopolitical Setting

The political geography of post-Soviet Russia and the nearby independent republics has changed dramatically since the collapse of the Soviet Union in 1991 (Figure 9.31). All of the former republics have struggled to establish stable political relations with their neighbors.

**Russia and the Former Soviet Republics** For a time it seemed that a looser political union of most of the former republics, called the **Commonwealth of Independent States (CIS)**, would emerge from the ruins of the Soviet Union. All the former republics, with the exception of the three Baltic states, joined the CIS soon after the end of the old union (Figure 9.31). By the early 21st century, however, the CIS had developed into little more than a forum for discussion, without real economic or political power. Another complicating factor in the post-Soviet period has been the ongoing military relationships between Russia and its former republics. **Denuclearization**, the return of nuclear weapons from outlying republics to Russian control and their partial dismantling, was completed during the 1990s. The Soviet-era nuclear arsenals of Kazakhstan, Ukraine, and Belarus were removed in the process.

**Geopolitics in the South and West** While Russia shares many cultural ties with Belarus and Ukraine, growing political tensions have threatened recent relationships between these two countries. Leaders in Belarus have been extremely slow to open up their economy to outside

investment. They have also resisted too much control from nearby Russia.

Russia's rocky relations with Ukraine have transformed the geopolitical landscape in the western portion of the domain. While Russian investments in Ukraine have grown sharply since 2000, political tensions between the two countries have increased. Ukraine remains highly dependent on Russian energy supplies. In addition, major Russian pipelines to central and western Europe must pass through Ukraine. The result is that the two nations have often wrangled over energy prices and the availability of oil and natural gas supplies. Early in 2009, for example, Russia briefly shut off all natural gas supplies to Ukraine amid pricing and payment disputes between Ukraine and Gazprom, the Russian energy company.

Ukraine's internal politics have also been unpredictable, particularly since the country was so hard hit in 2009 by the global economic crisis. In 2004, Viktor Yushchenko, a reformist leader and Ukrainian nationalist, came to power after a controversial election that even included an attempt to poison him. Yushchenko often criticized Russian leaders and excessive Russian interference with Ukraine's affairs. He has also explored Ukraine's admission into both the EU and the North Atlantic Treaty Organization (NATO). On the other hand, Viktor Yanukovich, one of Ukraine's key opposition leaders, has received Moscow's political blessings. Not surprisingly, Yanukovich also received wide support from the ethnic Russian areas in southern and eastern Ukraine (Figure 9.24). Steep declines in Ukraine's economy in 2009 prompted widespread public protests in Kiev and calls for early elections, once again challenging Yushchenko's grip on power. It seems likely that neighboring Russia will continue to watch its neighbor and attempt to influence Ukrainian politics in any way it can to serve its own political and economic interests.

Tiny Moldova has also witnessed political tensions. Conflict has repeatedly flared in the Russian-dominated Transdniester region in the eastern part of the country, where Slavic separatists have pushed for independence. Others suggest that Moldova join with Romania and dump troublesome Transdniester in the process. Some argue that Moldova should seek a middle ground that would keep Moldova independent but move it out of Russia's orbit toward the potential rewards of EU, or even NATO, membership.

Transcaucasia also remains unstable. Since 2003, the Georgian government has moved toward closer ties with the United States, even suggesting that it might join NATO. At the same time, Georgia's own internal politics recently provoked an invasion from neighboring Russia. In 2008, Georgia attempted to reassert its control over Abkhazia and South Ossetia, two breakaway regions that border Russia and that have wide support in that nearby country. The Russians responded with tanks, for a time occupying considerable sections of Georgia's territory. The situation remained tense in late 2009, with Georgia still claiming control over Abkhazia and South Ossetia and with Russia formally recognizing the "independence" of these microstates.

In nearby Armenia, the territories of the Christian Armenians and the Muslim Azeris interpenetrate one other in a complex fashion. The far southwestern portion of Azerbaijan (Naxcivan) is separated from

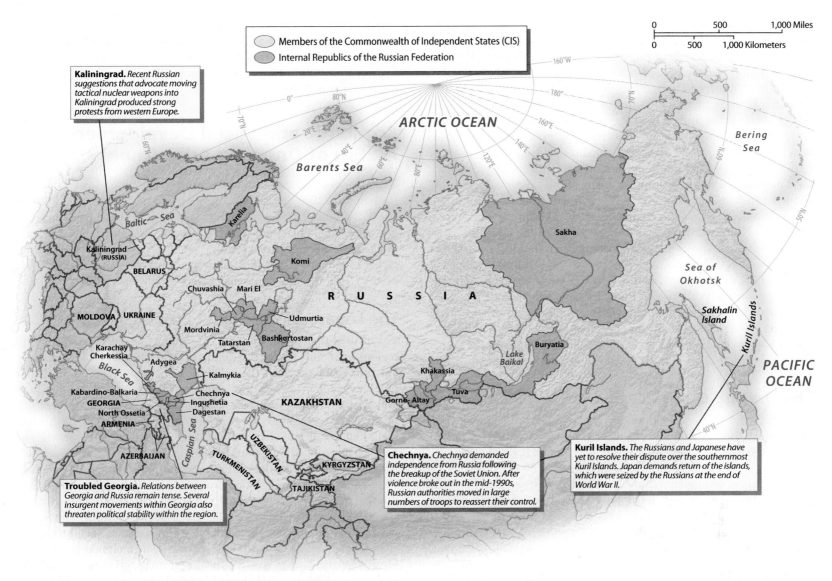

**Kaliningrad.** *Recent Russian suggestions that advocate moving tactical nuclear weapons into Kaliningrad produced strong protests from western Europe.*

Members of the Commonwealth of Independent States (CIS)
Internal Republics of the Russian Federation

**Chechnya.** *Chechnya demanded independence from Russia following the breakup of the Soviet Union. After violence broke out in the mid-1990s, Russian authorities moved in large numbers of troops to reassert their control.*

**Kuril Islands.** *The Russians and Japanese have yet to resolve their dispute over the southernmost Kuril Islands. Japan demands return of the islands, which were seized by the Russians at the end of World War II.*

**Troubled Georgia.** *Relations between Georgia and Russia remain tense. Several insurgent movements within Georgia also threaten political stability within the region.*

**FIGURE 9.31 Geopolitical Issues in the Russian Domain**   The Russian Federation Treaty of 1992 created a new internal political framework that acknowledged many of the country's ethnic minorities. Recently, however, Russian authorities have moved to centralize power and limit regional dissent. Russia's relations with several nearby states remain strained. *(Modified from Bergman and Renwick, 1999,* Introduction to Geography, *Upper Saddle River, NJ: Prentice Hall).*

the rest of the country by Armenia, while the important Armenian-speaking district of Nagorno-Karabakh is officially an autonomous portion of Azerbaijan. After Armenia successfully occupied much of Nagorno-Karabakh in 1994, fighting between the countries diminished. No final peace treaty has been signed, however, and Azerbaijan demands the return of the territory. Meanwhile, Armenia's traditionally close connections with Russia are increasingly counterbalanced with the country's interest in building ties with the United States and the EU.

**Geopolitics within Russia**   Within Russia, further pressures for devolution, or more localized political control, produced the March 1992 signing of the Russian Federation Treaty. Theoretically, the treaty granted Russia's internal autonomous republics and its lesser administrative units greater political, economic, and cultural freedoms. Defined essentially along ethnic lines, 21 regions possess status as

republics within the federation and now have constitutions that often run counter to national mandates. Scattered from central Siberia to the north-facing slopes of the Caucasus, these republics reflect much of the linguistic and religious diversity of the nation (see Figure 9.31).

Since 2002, Russian leaders, especially Vladimir Putin, have pushed for more centralized control in the country and region. Even after Putin left the presidency in 2008 (only to become prime minister thereafter), Russian President Dmitry Medvedev (a Putin ally) has maintained this strong trend toward a more centralized Russian state. Both leaders point to success in quelling the instability in the Russian republic of Chechnya. A strong hand, including multiple Russian military invasions of Chechnya, helped bring a halt to an independence movement in that largely Muslim region. In fact, in 2009, Russia proclaimed that the conflict had ended, bolstered by a Chechen government that was largely in sympathy with leaders in Moscow.

**Russian Challenge to Civil Liberties**   The move toward a more centralized Russian state has a number of troubling consequences. Citizens of the newly created Russian Republic initially enjoyed a genuine flowering of democratic freedoms during the initial post-Soviet years. Since 2002, however, many hard-won Russian civil liberties have slipped away, victims of Putin's campaign to consolidate political power, increase the authority of the central government, and silence critics who disagree with his policies. For example, since 2004, dozens of important Russian governorships and mayoral positions have been filled by appointment from Moscow leaders rather than by local popular elections. In addition, many of the country's media outlets—both television and print—have returned to state control or are now controlled by state-run companies such as Gazprom. This has silenced many critics of the government. Some of the country's most outspoken critics have been taken prisoner or mysteriously murdered. One study of global media freedoms in 2007 ranked Russia alongside regimes such as Burma and North Korea in terms of increasingly repressive policies. While some Russians have vehemently protested these losses of political liberties as well as human rights abuses, many Russians argue that such adjustments are a tolerable price to be paid for an improved economy, a better standard of living, and the country's newly elevated prestige on the global geopolitical stage.

## The Shifting Global Setting

Since the fall of the Soviet Union, regional political tensions continue to challenge the Russians in both the east and west. In East Asia, the boundary between Russia and China was imposed by the Russian Empire in 1858 and has never been fully accepted by Beijing. Millions of new Chinese immigrants in the Russian Far East serve to heighten tensions along that border. Territorial disagreements also complicate Russia's relationship with Japan in a dispute over the Kuril Islands. To the west, Russia worries about the expansion of NATO. Russian leaders strongly opposed the recent addition of the Baltic republics (Estonia, Latvia, and Lithuania) to that increasingly powerful organization.

Today, Russian leaders appear willing to reassert their nation's global political status. Its nuclear arsenal, while reduced in size, remains a powerful counterpoint to American, European, and Chinese interests. While Russia can no longer directly challenge the United States as it did in the days of the Soviet Union, it also can act as a partial counterweight to the United States in international maneuverings. Russia also retains a permanent seat on the United Nations Security Council, arguably the world's most important geopolitical body. Russia's inclusion in the G8 economic meetings also signifies its growing international clout.

# Economic and Social Development: An Era of Ongoing Adjustment

The economic future of the Russian domain remains difficult to predict. Economic declines devastated the region for much of the 1990s. But between 2002 and 2008, particularly in the case of Russia, higher oil and gas prices brought significant but selective economic improvement (Table 9.2). Recently, however, the entire region was hit hard in the global economic crisis, and significant contractions in gross domestic product, exports, and employment were widely reported in 2009.

## The Legacy of the Soviet Economy

The birth of the Soviet Union in 1917 initiated a radical change in the region's economy. Under the Russian Empire, most people were peasant farmers. Following the revolution, however, the Soviet Union quickly emerged to rival, and even surpass, many of the other most powerful economies on Earth. During that era of unmatched growth, much of the region's present economic infrastructure was established, including new urban centers and industrial developments, as well as a modern network of transportation and communication linkages.

As communist leaders such as Stalin consolidated power in the 1920s and 1930s, they nationalized Russian industries and agriculture, creating a system of **centralized economic planning** in which the state controlled production targets and industrial output. The Soviets emphasized heavy basic industries (steel, machinery, chemicals, and electricity generation), postponing demand for consumer goods to the future. By the late 1920s, Stalin also shifted agricultural land into large-scale collectives and state-controlled farms.

Much of the Russian domain's basic infrastructure—its roads, rail lines, canals, dams, and communications networks—originated during the Soviet period (Figure 9.32). Dam and canal construction, for example, turned the main rivers of European Russia into a virtual

### TABLE 9.2   DEVELOPMENT INDICATORS

| Country | GNI per capita, PPP 2007 | GDP Average Annual %Growth 2000–2007 | Human Development Index (2006)# | Percent Population Living Below $2 a Day | Life Expectancy* 2009 | Under Age 5 Mortality Rate 1990 | Under Age 5 Mortality Rate 2007 | Gender Equity 2007 |
|---|---|---|---|---|---|---|---|---|
| Armenia | 5,870 | 12.7 | 0.777 | 43.4 | 72 | 56 | 24 | 104 |
| Belarus | 10,750 | 8.3 | 0.817 | <2 | 70 | 24 | 13 | 101 |
| Georgia | 4,760 | 8.3 | 0.763 | 30.4 | 75 | 47 | 30 | 98 |
| Moldova | 2,800 | 6.5 | 0.719 | 28.9 | 69 | 37 | 18 | 102 |
| Russia | 14,330 | 6.6 | 0.806 | <2 | 68 | 27 | 15 | 99 |
| Ukraine | 6,810 | 7.6 | 0.786 | <2 | 68 | 25 | 24 | 100 |

*Source:* World Bank, *World Development Indicators, 2009.*
United Nations, *Human Development Index, 2008.*#
Population Reference Bureau, *World Population Data Sheet, 2009.**
Gender Equity – Ratio of female to male enrollments in primary and secondary school.  Numbers below 100 have more males in primary/secondary school, numbers above 100 have more females in primary/secondary schools.

network of interconnected reservoirs. Invaluable links such as the Volga-Don Canal (completed in 1952), which connected those two key river systems, have greatly eased the movement of industrial raw materials and manufactured goods within the country (Figure 9.33). The Soviets also added thousands of miles of new railroad tracks in the European west, and the Trans-Siberian line was modernized and complemented by the addition of the BAM link across central Siberia. Farther north, the Siberian Gas Pipeline was built to link the energy-rich fields of the Soviet arctic with growing demand in Europe. Overall, the postwar period produced real economic and social improvements for the Soviet people.

Despite the successes, problems increased during the 1970s and 1980s. Soviet agriculture remained inefficient, and grain imports grew. Manufacturing efficiency and quality failed to match the standards of the West, particularly in regard to consumer goods. Equally troubling was the fact that the Soviet Union was failing to participate fully in the technological revolutions that were transforming the United States, Europe, and Japan. Disparities also visibly grew between the Soviet elite and an everyday population that still enjoyed few personal freedoms. By the late 1980s, the Soviet Union had reached both an economic and a political impasse.

## The Post-Soviet Economy

Fundamental economic changes have shaped the Russian domain since the demise of the Soviet Union. Particularly within Russia itself, much of the highly centralized, state-controlled economy has been replaced by a mixed economy of state-run operations and private enterprise. The collapse of the communist state also meant that economic relationships between the former Soviet republics were no longer controlled by a single, centralized government. Fundamental problems of unstable currencies, corruption, and changing government policies plagued the system for much of the 1990s. Higher oil and natural gas prices led to an impressive recovery in the Russian economy between 2002 and 2008 (Table 9.2), followed more recently by an economic downturn throughout the region.

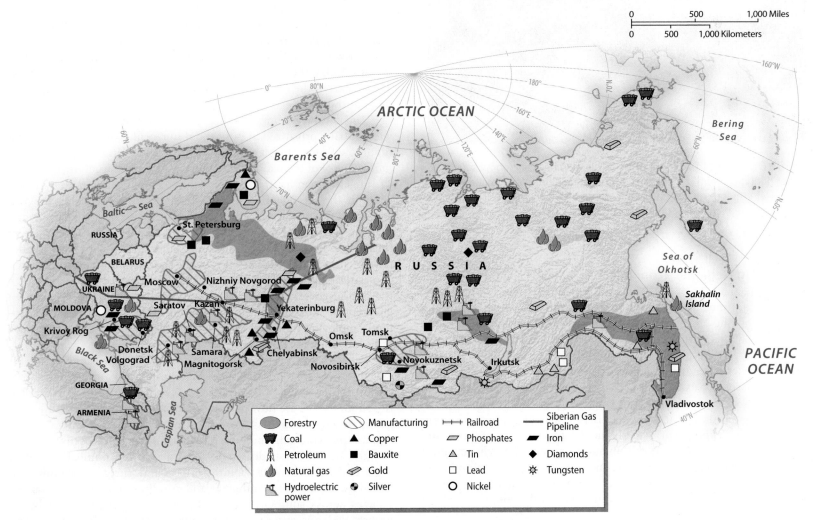

**FIGURE 9.32 Major Natural Resources and Industrial Zones** The Russian domain's varied natural resources and chief industrial zones are widely distributed. Fossil fuels are abundant, although their distance from markets often imposes special costs. In southern Siberia, rail corridors offer access to many mineral resources. In the mineral-rich Urals and eastern Ukraine, proximity to natural resources sparked industrial expansion, while Moscow's industrial might is related to its proximity to markets and capital. (*Modified from Bergman and Renwick, 1999,* Introduction to Geography, *Upper Saddle River, NJ: Prentice Hall, and Rubenstein, 2008,* Introduction to Human Geography, *Upper Saddle River, NJ: Prentice Hall*).

**FIGURE 9.33 Volga-Don Canal** This view near Volgograd suggests the enduring economic importance of the Volga-Don Canal. Built during the Soviet era, the canal remains a key commercial link that facilitates the economic integration of southern Russia.

**Redefining Regional Economic Ties** Many economic ties still link the six nations of the Russian domain. Russian dominance is likely to continue, as suggested by that country's leading role in the creation of the "Common Economic Space" with neighboring Ukraine, Belarus, and Kazakhstan. It is still unclear how effective this Russian-led response to the growth of the EU will be in the coming years. Meanwhile, other regional economic ties remain strong. Belarus, for example, is still dependent on Russian assistance, particularly in the form of cheap energy exports. Similarly, Ukraine imports a great deal of its raw materials (particularly energy) from Russia, and its exports (principally metals, food, and machinery) are led by return flows to the north. In addition, large Russian corporations have purchased many Ukrainian assets, such as aluminum smelters, television stations, and oil refineries—a new post-Soviet twist on a long-established Russian presence in the country.

**Privatization and State Control** Even as hopes initially ran high for a smooth restructuring of the post-Soviet economy, the early years were difficult. A dramatic move came in October 1993, when the government initiated a massive program to privatize the Russian economy. Millions of Russians were given the option to buy privatized agricultural lands and industrial companies. These initiatives opened the economy to more private initiative and investment. Unfortunately, the small number of legal and financial safeguards invited many abuses and often resulted in mismanagement and corruption in the new system. Elsewhere in the region, privatization proceeded more slowly. Much of the Belarusian economy, for example, remains mired in inflexible, corrupt state-controlled companies.

Almost 90 percent of the country's farmland was privatized by 2003, with many farmers forming voluntary cooperatives or joint-stock associations to work the same acreage as under the Soviet system. While crop prices have risen, costs have gone up even faster, and many farmers are not skilled in dealing with the uncertainties of a market-driven agricultural economy. Overall, agriculture employs about 10 percent of Russia's workforce, the basic distribution of crops remains little changed from the Soviet era, and the region continues to be challenged by short growing seasons, poor soil, and moisture deficiencies.

Russia has encouraged the privatization of the service sector. Thousands of privatized retailing establishments have appeared, and they now dominate that portion of the economy. In addition, the long-established "informal economy" continues to flourish. Even during the Soviet era, millions of citizens earned extra money by informally selling Western consumer goods, manufacturing food and vodka, and providing skilled services such as computer and automobile repairs. Today, these barter transactions and informal cash deals form a huge part of the regional economy that is never reported to government authorities.

The natural resource and heavy industrial sectors of the economy were initially privatized in Russia, but in recent years, under President (and now Prime Minister) Putin's control, state-run enterprises retook more control of the nation's energy assets and infrastructure. Gazprom, the huge Russian natural gas company, exemplifies the process. Privatized in the spring of 1994, about one-third of the company was auctioned off to Russian citizens, 15 percent went to managers and workers, 10 percent remained in the company's treasury, and most of the rest was kept in government hands. Since 2005, Gazprom's activities have increasingly been controlled by the state, and this company is seen as a critical part of a newly emerging "state industrial policy," in which the central government is playing a more direct role in the economy. Nicknamed "Russia, Inc.," it remains one of the world's largest companies, employing more than 300,000 people and controlling one-third of Earth's known natural gas reserves.

Particularly in Russia, the successes of the new economy are increasingly visible on the landscape, especially in the country's cities. Luxury malls, office buildings, and more fashionable housing subdivisions are now part of the urban scene as the middle class grows in settings such as Moscow. Stroll the area near that city's Red Square and you will encounter Planet Sushi; the Moscow Bentley, Ferrari, and Maserati dealership; and a growing number of trendy night clubs (Figure 9.34). On the other hand, the gap has grown between increasing urban affluence and grinding rural poverty. In many villages across southern Siberia, there are no telephones, jobs, or money. Vodka is easier to find than running water. Closed shops and a continued lack of services remain a part of everyday life in such rural settings.

**The Challenge of Corruption** Throughout the Russian domain, corruption remains widespread. Doing business often means lining the pockets of government officials, company insiders, or trade union

**FIGURE 9.34 Moscow Night Club**   Affluent young people in Moscow party late into the night as they enjoy the material fruits of selective economic growth in post-Soviet Russia.

representatives. Organized crime remains pervasive in Russia. The government's own interior ministry estimates that the Russian mafia may control about 40 percent of the private economy and 60 percent of state-run enterprises. Many ties also remain between organized crime and Russian intelligence-collecting agencies. Various local and regional crime organizations divide up much of the economy. One group might control the construction business in a Moscow suburb, whereas another syndicate oversees drug dealing and prostitution, and still another helps funnel illegal DVDs to eager consumers. The Russian mafia has also gone global; it has been implicated in huge money-laundering schemes involving Russian, British, and U.S. banks as well as the flow of International Monetary Fund investments into the region.

**Social Problems**   Many residents of the Russian domain still struggle to make ends meet. Street crime remains high in many neighborhoods. Recently, unemployment and declining social welfare expenditures have hit many families hard. Often, both husband and wife work multiple low-paying jobs, with few benefits and long hours. In many settings, women pay a particularly heavy price (Figure 9.35). Violence against women has been widely reported in the post-Soviet era. Beatings and rapes are common. A survey in Moscow suggested that one-third of divorced women had experienced domestic violence, and a women's rights group in Ukraine reported that rape was an all-too-common crime in many villages. International organizations have sharply criticized government authorities for doing little to change the situation. The region also suffers from significant health-care problems. Diseases such as tuberculosis and AIDS have been on the rise. Chronic illnesses, often related to lifestyle, produce the highest mortality rates in the developed world for cardiovascular disease, alcoholism, and smoking, particularly for Russian men.

## Growing Economic Globalization

The relationship between the Russian domain and the world beyond has shifted greatly since the end of communism. During much of the Soviet era, the region was relatively isolated from the world economic

system. But connections with the global economy have grown tremendously since the downfall of the Soviet Union in 1991.

**A New Day for Consumers**   New consumer goods now reach residents of the Russian domain, particularly those living in larger cities. All the symbols of global capitalism are visible in the heart of Moscow and, increasingly, in other settings in the region. Luxury goods from the West have also found a small but enthusiastic market among the Russian elite, a group noted for its devotion to BMW automobiles, Rolex watches, and other status symbols. Most Russians find such luxuries beyond their limited budgets, but they

**FIGURE 9.35 Life on the Street**   This young street vendor at the Moscow Food Market hopes to sell her pickled vegetables to passing consumers. Moscow's economy features both great opportunities and grinding poverty.

are interested in purchasing basic foods, inexpensive technology, and popular clothing and media from their Western neighbors or from eastern and southern Asia.

### Attracting Foreign Investment

Despite the current political and economic uncertainties, most countries of the Russian domain are successfully attracting significant inflows of foreign investment. The strongest global ties by far have been with the United States, Japan, and western Europe, particularly Germany and Great Britain. Many potential investors are still put off by continuing uncertainties with Russian legal frameworks and concerns about the ultimate political aims of Russian leaders. Large outside investments have been made in Russia's oil and gas economy, but recently some of those opportunities have cooled, amid the country's desire to limit foreign ownership of its energy resources and infrastructure.

Elsewhere in the region, connections with the global economy vary. In Belarus, a lack of economic reforms continues to slow investment. About 80 percent of the nation's industrial sector is controlled by the state and the political climate seems opposed to foreign investment. Neighboring Ukraine, however, has succeeded in attracting more investment since 2004, particularly in the western portion of the country, but it is still early in the process of fully opening its economy to outside investment. That could change dramatically in the years to come if Ukraine joins the EU.

### Globalization and Russia's Petroleum Economy

Russia's oil and gas industry remains one of the strongest economic links between the region and the global economy, and the diverse international connections it has forged suggest the increasing importance of the sector to the region's future. The statistics are impressive: Russia's energy production recently accounted for 25 percent of its entire economic output. Russia has 35 percent of the world's natural gas reserves (mostly in Siberia) and is the world's largest gas exporter. As for oil, Russia is by far the world's largest non-OPEC producer, and it is the second largest oil exporter in the world (behind Saudi Arabia). It far outpaces the United States in annual output (major oilfields are in Siberia, the Volga Valley, the Far East, and the Caspian Sea region), and it possesses more than 60 billion barrels of proven reserves.

The dynamic nature of the Russian oil and gas business exemplifies the forces of globalization and the region's changing economy. Prior to the breakup of the Soviet Union, about half of Russia's oil and gas exports went to other Soviet republics, such as Ukraine and Belarus. While these two nations still depend on Russian supplies, the primary destination for Russian petroleum products has shifted to western Europe. Russia supplies that region with more than 25 percent of its natural gas and 16 percent of its crude oil, and those linkages are likely to grow stronger. The Siberian Gas Pipeline weds distant Asian fields with western Europe via Ukraine, and those connections are being supplemented by lines through Belarus (the Yamal–Europe Pipeline) and Turkey (the Blue Stream Pipeline). Similar expansions are refashioning the geography of oil exports. The huge Druzhba Pipeline and the expanding oil port facility at Primorsk (near St. Petersburg) serve western Europe and other global markets.

Pipeline politics plays a growing role in the Russian domain (Figure 9.36). Many new and planned oil and gas pipelines pass through strategic, sometimes politically unstable, regions. For

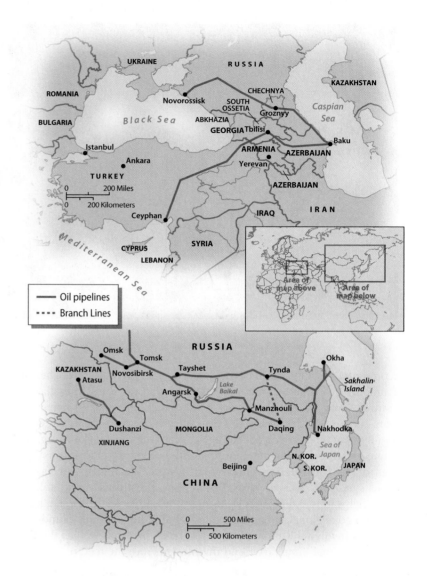

**FIGURE 9.36 Russia's Expanding Pipelines**   These two maps show new and planned oil pipelines that are designed to expand Russia's presence in the global petroleum economy. Different pipeline projects in the Russian Far East would benefit nearby China or Japan, while projects near the Caspian Sea take pipelines through politically unstable portions of the region.

example, a large export terminal opened at Novorossisk (on the Black Sea) in 2001, delivering Caspian Sea supplies to the world market via a pipeline passing through troubled Chechnya. To the south, another pipeline between Baku (on the Caspian Sea) and Ceyphan (on the Turkish Mediterranean) crosses both Azerbaijan and Georgia. In the Russian Far East, both China and Japan are lobbying hard for pipeline projects. The Russians are building a large new Siberian Pacific Pipeline to link its Siberian fields to Asian markets. The Chinese want Russian oil to flow to Daqing, where it could be refined for national and regional markets. Japan prefers a large new facility at the Pacific port of Nakhodka, well-positioned to supply Japan and offering Russia easy access to global markets via the Pacific Ocean. Other links connect the system with developments on Sakhalin Island, where several major energy projects are currently being completed (Figure 9.37).

**FIGURE 9.37 Sakhalin Island** Large-scale investments by both foreign and Russian interests have concentrated on energy-rich Sakhalin Island. The area promises to be a producer of both oil and natural gas in the years to come.

**FIGURE 9.38 Novosibirsk** The Russian city of Novosibirsk has become a major focus of investment and urban growth in Siberia since the disintegration of the Soviet Union.

While many global companies have participated in the development of Russia's energy infrastructure, recent policies aimed at centralizing the economy suggest that state-controlled Russian companies will play a larger role in the future. On Sakhalin Island, for example, Gazprom has been allowed to take control of several ventures originally dominated by companies such as Royal Dutch Shell (Figure 9.37). Ironically, Russian authorities have stripped control of these projects from foreign companies, suggesting that they failed to follow environmental regulations. Once in Russian hands, environmental concerns faded away.

**Local Impacts of Globalization** As the geography of the Russian petroleum industry suggests, the local impacts of globalization are highly selective. Obviously, portions of the region that are close to oil and gas wells, pipeline and refinery infrastructure, and key petroleum shipping points are greatly affected (both economically and environmentally) by the changing global oil economy. The same is true more broadly: Globalization has affected different locations in the Russian domain in very distinctive ways. In Russia, capitalism has brought its most dramatic, though selective, benefits to Moscow. Indeed, much of the foreign investment in the country has gone to the Moscow area. Beyond Moscow, St. Petersburg and the Siberian cities of Omsk and Novosibirsk (Figure 9.38) have also seen growing global investment, while new oil and gas prospects in Siberia, the Russian Far East (Sakhalin Island), and near the Caspian Sea have attracted other investments. In addition, port cities such as Vladivostok are well positioned to take advantage of their accessibility to nearby markets (Figure 9.39).

Elsewhere, globalization clearly has imposed penalties. Older, less globally competitive localities have been hit hard. Aging steel plants, for example, no longer have guaranteed markets for their high-cost, low-quality products, as they did in the days of the planned Soviet economy. Instead, they must compete on the global market, a market that is increasingly prone to lower prices and weakening demand for many of the industrial goods the region produces. Many of the region's extractive centers are similarly vulnerable in a global economy of rapidly changing commodity prices.

**FIGURE 9.39 Vladivostok** The busy harbor of Vladivostok remains Russia's leading trade center in the Russian Far East. With easy access to markets in Japan, China, and even the United States, this Pacific port is poised to grow as Russia's economy recovers.

# SUMMARY

- Huge environmental challenges remain for the Russian domain. The legacy of the Soviet era includes polluted rivers and coastlines, poor urban air quality, and a frightening array of toxic waste and nuclear hazards.

- Declining and aging populations are part of the sobering reality for much of the region. While some localities see modest population growth related to in-migration (mostly toward expanding urban areas), many rural areas and less competitive industrial zones are likely to see continued outflows of people and very low birthrates.

- Much of the region's underlying cultural geography was formed centuries ago, the complex product of Slavic languages, Orthodox Christianity, and numerous ethnic minorities that continue to complicate the scene today. Further changing the country are new global influences—a set of products, technologies, and attitudes that often clash with traditional cultural values.

- Much of the region's political legacy is rooted in the Russian Empire, a land-based system of colonial expansion that greatly enlarged Russian influence after 1600 and then reappeared as the Soviet Union expanded its influence. Only large remnants of that empire survive on the modern map, yet it has stamped the geopolitical character of the region in lasting ways. Beyond the region, Russia's growing visibility on the international stage signals its reemergence as a truly global political power.

- The peoples of the Russian domain have endured immense challenges since 1991. Today, growing centralized power in Moscow increasingly limits democratic reforms. In addition, the entire region has suffered in the recent global economic downturn. The region's future economic geography, particularly in Russia, remains tied to the fortunes of the unpredictable global energy economy.

# KEY TERMS

autonomous area (p. 274)
Baikal-Amur Mainline (BAM) Railroad (p. 265)
Bolshevik (p. 274)
centralized economic planning (p. 278)
chernozem soil (p. 259)

Cold War (p. 275)
Commonwealth of Independent States (CIS) (p. 276)
Cossack (p. 270)
denuclearization (p. 276)
Eastern Orthodox Christianity (p. 269)

exclave (p. 274)
glasnost (p. 275)
Gulag Archipelago (p. 265)
Iron Curtain (p. 275)
mikrorayon (p. 268)
perestroika (p. 275)
permafrost (p. 261)

podzol soil (p. 259)
Russification (p. 265)
Slavic people (p. 269)
socialist realism (p. 273)
taiga (p. 261)
Trans-Siberian Railroad (p. 265)
tsar (p. 265)

# THINKING GEOGRAPHICALLY

1. How might it be argued that Russia's natural environment is one of its greatest assets as well as one of its greatest liabilities?

2. In the future, how might the forces of capitalism and free markets reshape the landscapes and land uses of cities within the Russian domain?

3. On a base map of the Russian domain, suggest possible political boundaries 20 years from now. What forces will work for a larger Russian state? A smaller Russian state?

4. What were some of the greatest strengths and weaknesses of centralized Soviet-style planning between 1917 and 1991? How were Ukraine and Belarus impacted? Why did the system ultimately fail?

5. From the perspective of a 23-year-old Russian college student and resident of Moscow, write a short essay that suggests how the economic and political changes over the past 10 to 15 years have changed your life.

 Log in to **www.mygeoscienceplace.com** for videos, interactive maps, RSS feeds, case studies, and self-study quizzes to enhance your study of the Russian domain.

# 10 CENTRAL ASIA

## GLOBALIZATION AND DIVERSITY

For most of the past 200 years, the landlocked region of Central Asia has been geopolitically dominated by countries located in other world regions and partially cut off from the main currents of global trade. Since the downfall of the Soviet Union in 1991, however, Central Asia has emerged as a key producer of globally traded resources and as a focus of international geopolitical rivalry.

### ENVIRONMENTAL GEOGRAPHY

Intensive agriculture along the rivers that flow into the deserts of Central Asia has resulted in serious water shortages, leading to the drying up of many of the region's lakes and wetlands.

### POPULATION AND SETTLEMENT

Pastoral nomadism, the traditional way of life across much of Central Asia, is gradually disappearing as people settle in towns and cities.

### CULTURAL COHERENCE AND DIVERSITY

In much of eastern Central Asia, the growing Han Chinese population is sometimes seen as a threat to the long-term survival of the indigenous cultures of the Tibetan and Uyghur peoples.

### GEOPOLITICAL FRAMEWORK

Afghanistan and its neighbors to the north are frontline states in the struggle between radical Islamic fundamentalism and secular governments.

### ECONOMIC AND SOCIAL DEVELOPMENT

Despite its abundant resources, Central Asia remains a poor region, although much of it enjoys relatively high levels of social development.

*A buried gas pipeline in Turkmenistan has been exposed by shifting desert sands.*

285

# Setting the Boundaries

The term *Central Asia* has several different definitions. Most authorities agree that it includes five former-Soviet republics: Kazakhstan, Kyrgyzstan, Uzbekistan, Tajikistan, and Turkmenistan. This chapter adds another former Soviet state, Azerbaijan, in addition to Mongolia and Afghanistan, as well as the autonomous regions of western China (Tibet and Xinjiang). Several other provinces and regions of western China, such as Nei Mongol (Inner Mongolia) and Qinghai, are occasionally discussed.

The inclusion of these additional territories as part of Central Asia is controversial. Azerbaijan is often classified with its neighbors in the Caucasus (Georgia and Armenia); western China is obviously part of East Asia based on political criteria; and Mongolia is also often placed within East Asia because of both its location and its historical connections with China. Afghanistan is usually located within either South Asia or Southwest Asia and the Middle East.

But considering Central Asia's historical unity, its similar environmental conditions, and its recent reentry onto the stage of global geopolitics, we think that it deserves consideration in its own right. It also makes sense to define its limits rather broadly. Azerbaijan, for example, is linked by both cultural (language and religion) and economic (oil) factors more to Central Asia than it is to Armenia and Georgia.

Central Asia is not a well-defined region, nor is it politically stable. Continuing Chinese political control over, and Han Chinese migration into, southeastern Central Asia threatens whatever claims may be made for regional coherence. Central Asia itself remains deeply divided along cultural lines. Although most of its people speak Turkic languages and practice Islam, the indigenous inhabitants of both Mongolia and Tibet traditionally follow Buddhism.

**FIGURE 10.1  Central Asia**  Central Asia, an extensive region in the center of the Eurasian continent, is dominated by arid plains and basins and high mountain ranges and plateaus. Eight independent countries—Kazakhstan, Turkmenistan, Uzbekistan, Kyrgyzstan, Tajikistan, Azerbaijan, Afghanistan, and Mongolia—form Central Asia's core. The region also includes China's lightly populated far west, which has cultural and environmental similarities to the rest of Central Asia. Desertification has resulted in the spread of sand dunes in many parts of Central Asia.

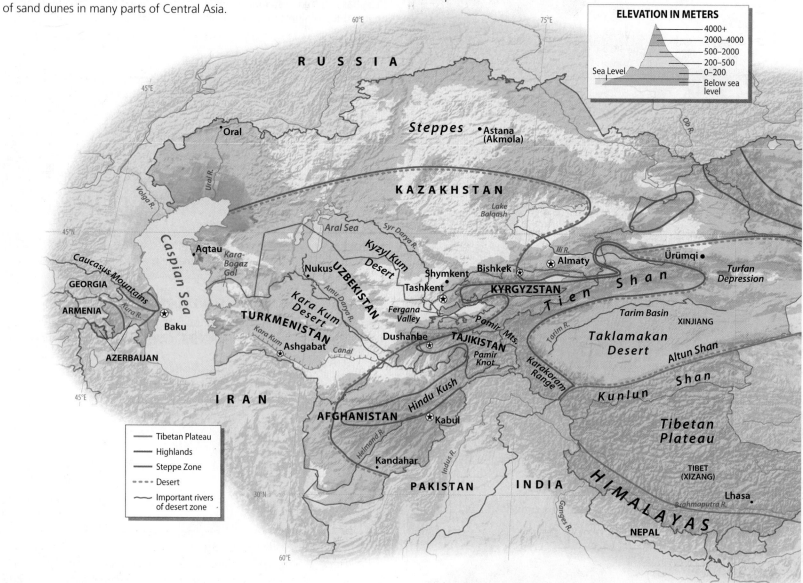

286

Central Asia does not appear in most books on world regional geography. Although it covers a larger expanse than the United States, it is a remote and lightly populated area dominated by high mountains, barren deserts, and semiarid **steppes** (grasslands). Until 1991, Central Asia contained only two independent states, Mongolia and Afghanistan. The rest of the region was divided between the Soviet Union and China (Figure 10.1).

Central Asia began to reappear in discussions of global geography following the breakup of the Soviet Union. Suddenly a handful of new countries appeared on the map, prompting scholars to pay more attention to this rapidly changing region (see "Setting the Boundaries"). Central Asia gained a more prominent place in the public imagination after September 11, 2001, when it became evident that the attack had been planned and organized by the Al Qaeda organization operating out of Afghanistan.

Until the late 20th century, Central Asia was poorly integrated into international trade networks. This situation began to change as large oil and gas reserves were found, especially in Kazakhstan, Turkmenistan, and Azerbaijan. As a result, Western oil companies began to show strong interest in Central Asia. A number of important countries, moreover, are seeking to exert influence over the region, including Iran, Pakistan, India, the United States, China, and Russia.

Central Asia forms a large region in the center of the Eurasian landmass. Alone among all the world regions, it lacks ocean access. Due to its continental position in the center of the world's largest landmass, Central Asia is noted for its severe climate. The aridity of the region has contributed to some of the most severe environmental problems in the world.

## Environmental Geography: Steppes, Deserts, and Threatened Lakes

One of the great environmental tragedies of the 20th century was the destruction of the Aral Sea, a large, saline lake (until recently, larger than Lake Michigan) located on the boundary of Kazakhstan and Uzbekistan. The Aral's only sources of water are the Amu Darya and Syr Darya rivers, which flow out of the distant Pamir Mountains. Both of these rivers have been intensively used for irrigation for thousands of years, but the scale of diversion greatly expanded after 1950. The valleys of the two rivers formed one of the southernmost farming districts of the Soviet Union and thus became critical suppliers of such warm-season crops as rice and cotton. Soviet agricultural planners favored huge engineering projects to deliver water to arid lands to "make the deserts bloom."

Unfortunately, more water delivered to produce crops meant less freshwater for the Aral Sea. As the inflow of water was reduced, the shallow Aral began to recede at a rapid rate. With less freshwater flowing into the lake, it also grew increasingly salty, destroying fish stocks. New islands began to emerge, and in 1987, the Aral Sea split into two separate lakes (Figure 10.2).

The destruction of the Aral Sea has resulted in economic and cultural damage, as well as ecological devastation. Fisheries formerly employing 40,000 workers closed down, and agriculture has suffered. The retreating lake left large salt flats on its exposed beds. Windstorms pick up the salt, along with the agricultural chemicals that have accumulated in the lake's shallows, and deposit it in nearby fields. As a result, farm yields have declined, desertification has accelerated, public health has been threatened, and the local climate has become colder in the winter and hotter in the summer.

Efforts to save what is left of the Aral are currently focused on the smaller of the two remaining lakes, located in the north. A series of dikes and dams, financed by the World Bank and the government of Kazakhstan, have managed to raise water levels over 26 feet (8 meters). Improved water quality allowed the partial revival of wildlife, and by 2009, Aral fish were once again being exported (Figure 10.3). But the southern Aral Sea, originally the larger of the two lakes, has seen no such recovery. In 2008, it was shrinking so fast that scientists began to fear that it could disappear entirely by 2050.

**CENTRAL ASIA**
Political & Physical Map
⊛ ● Over 1,000,000
⊛ ● 500,000–1,000,000 (selected cities)
⊛ ● Selected smaller cities

### Other Major Environmental Issues

Despite the tragedy of the Aral Sea, much of Central Asia has a relatively clean environment, largely due to its generally low population density. Industrial pollution, however, is a serious problem in the larger cities, such as Tashkent (in Uzbekistan) and Baku (in Azerbaijan). Elsewhere, however, the typical environmental problems of arid environments plague the region: desertification (the spread of deserts resulting from poor land-use practices), salinization (the accumulation of salt in the soil), and desiccation (the drying up of lakes and wetlands).

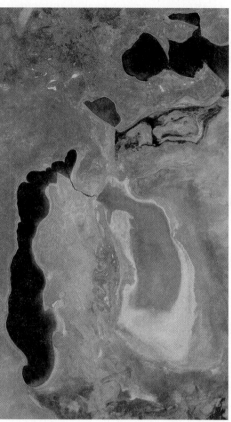

**FIGURE 10.2 The Shrinking of the Aral Sea**   The satellite image on the left, from 1989, shows the division of the Aral Sea into a larger southern lake and a much smaller northern lake. The image on the right shows the situation in 2008. Although the northern lake appears roughly the same size as it was in 1989, the southern lake had been reduced to a fraction of its former extent.

risen some 8.2 feet (2.5 meters). This enlargement, too, has caused problems, such as the flooding of some of the newly reclaimed farmlands in the Volga Delta. At present, the most serious environmental threat to the Caspian is pollution from the oil industry rather than fluctuation in size.

**Dam Building and Water Conflicts**   Roughly 80 percent of the freshwater in western Central Asia is controlled by Kyrgyzstan and Tajikistan, the two most mountainous countries of the region. Both countries are engaged in massive dam-building projects designed to control water resources and generate electricity. The Rogun Dam in Tajikistan, currently under construction with Russian aid, could be the tallest dam in the world when completed. Uzbekistan, Kazakhstan, and Turkmenistan object to the construction of these dams, fearing that Kyrgyzstan and Tajikistan will release too much water during flood periods and withhold too much during dry spells. Leaders from the five countries met in early 2009 to try to reach an agreement on water sharing but were unable to do so.

## Central Asia's Physical Regions

To understand why Central Asia experiences such great conflicts over water, it is necessary to examine the region's physical geography. In general, Central Asia is characterized by high plateaus and

**Desertification**   Desertification, caused by overgrazing and poor farming practices, is a major concern in Central Asia (Figure 10.4). In the eastern part of the region, the Gobi Desert has gradually spread southward, encroaching on densely settled lands in northeastern China proper. The Chinese have tried to prevent the march of desert with massive tree- and grass-planting campaigns, as the roots of such plants stabilize the soil and help to keep sand dunes from moving. Such efforts, however, have been only partially successful.

**Shrinking and Expanding Lakes**   Western Central Asia contains large lakes because it forms a low-lying basin, without drainage to the ocean, that is surrounded by mountains and other more humid areas. The world's largest lake, by a huge margin, is the Caspian Sea, located along the region's western boundary; the 4th largest is (or more precisely, was) the Aral Sea, situated some 300 miles (480 kilometers) to the east, and the 15th largest is Lake Balqash, found 600 miles (960 kilometers) farther east. Like the Aral Sea, Lake Balqash has become smaller and saltier over the past several decades.

The story of the Caspian is complicated. The Caspian Sea receives most of its water from the large rivers of the north, the Ural and the Volga, which drain much of European Russia. Due to extensive irrigation development in the lower Volga Basin, the volume of freshwater reaching the Caspian began to decline in the second half of the 20th century. With a reduced flow of water, the water level dropped, exposing large expanses of the former lakebed. Decreased water volume and increased salinity disrupted the ecosystem and devastated fisheries. The Russian caviar industry, centered in the northern Caspian, was particularly damaged by these changes.

The water level of the Caspian Sea reached a low point in the late 1970s. At that point, it began to rise, probably because of higher-than-normal precipitation in its drainage basin. By the late 1990s, it had

**FIGURE 10.3 The Northern Aral Sea**   Due to water diversions from the Amu Darya and Syr Darya rivers, the Aral Sea has lost most of its volume and now forms two separate lakes. The government of Kazakhstan has recently built dikes to protect the water of the smaller northern Aral, which is fed by the flow of the Syr Darya River.

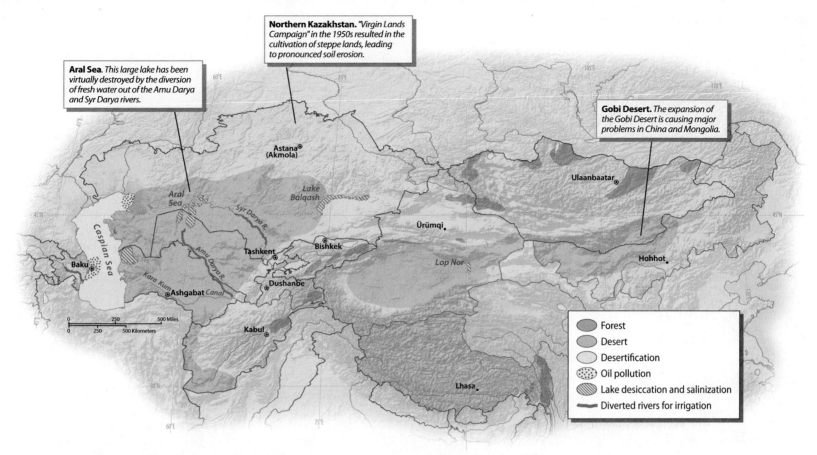

**Aral Sea.** *This large lake has been virtually destroyed by the diversion of fresh water out of the Amu Darya and Syr Darya rivers.*

**Northern Kazakhstan.** *"Virgin Lands Campaign" in the 1950s resulted in the cultivation of steppe lands, leading to pronounced soil erosion.*

**Gobi Desert.** *The expansion of the Gobi Desert is causing major problems in China and Mongolia.*

Legend:
- Forest
- Desert
- Desertification
- Oil pollution
- Lake desiccation and salinization
- Diverted rivers for irrigation

**FIGURE 10.4 Environmental Issues in Central Asia**   Desertification is perhaps more widespread in Central Asia than in any other world region. Soil erosion and overgrazing have led to the advance of desertlike conditions in much of western China and Kazakhstan. In western Central Asia, the most serious environmental problems are associated with the diversion of river-water for irrigation and the corresponding desiccation of lakes.

mountains in the south-center and southeast, grassland plains (or steppes) in the north, and desert basins in the southwestern and central areas.

**The Central Asian Highlands**   The highlands of Central Asia originated in one of the great geological events of Earth's history: the collision of the Indian subcontinent into the Asian mainland. This ongoing impact has created the world's highest mountains, the Himalayas, located along the boundary of South Asia and Central Asia. To the northwest, the Himalayas merge with the Karakoram Range

and then the Pamir Mountains. From the so-called Pamir Knot, a complex tangle of mountains located where Pakistan, Afghanistan, China, and Tajikistan meet, towering ranges spread outward in several directions. The Hindu Kush curves to the southwest through central Afghanistan, the Kunlun Shan extends to the east, and the Tien Shan swings out to the northeast into China's Xinjiang province. All these ranges have peaks higher than 20,000 feet (6,000 meters) in elevation.

Much more extensive than these mountain ranges is the Tibetan Plateau (Figure 10.5). This massive upland extends some 1,250 miles

**FIGURE 10.5 Tibetan Plateau**   Alpine grasslands and tundra interspersed with rugged mountains and saline lakes dominate the Tibetan Plateau. In summer, the sparse vegetation offers forage for the herds of nomadic Tibetan pastoralists. Much of northern Tibet, however, is too high to support pastoralism and is therefore uninhabited.

(2,000 kilometers) from east to west and 750 miles (1,200 kilometers) from north to south. Its elevation is as remarkable as its size; almost the entire area is more than 12,000 feet (3,700 meters) above sea level. Most of the Tibetan Plateau lies near the maximum elevation at which human life can exist. Rather than form a flat surface, the plateau has numerous east–west mountain ranges alternating with basins. Although the southeastern sections of the plateau receive adequate

rainfall, most of Tibet is arid (Figure 10.6). Winters on the Tibetan Plateau are cold; and while summer afternoons can be warm, summer nights remain chilly.

**The Plains and Basins**    Although the mountains of Central Asia are higher and more extensive than any others in the world, most of the region is characterized by plains and basins. This lower-lying zone

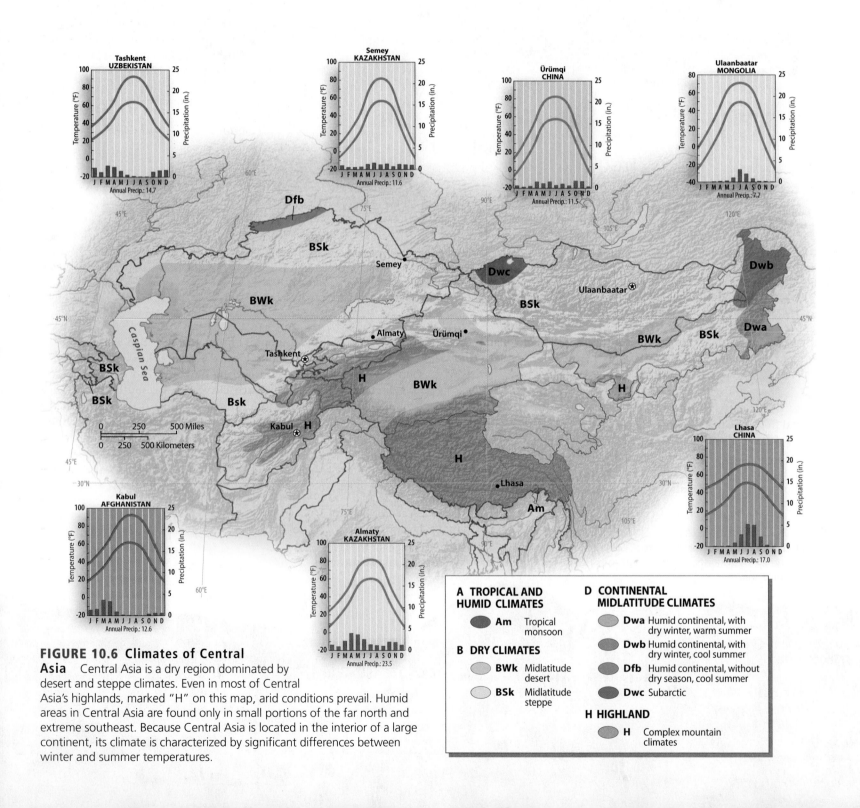

**FIGURE 10.6 Climates of Central Asia**    Central Asia is a dry region dominated by desert and steppe climates. Even in most of Central Asia's highlands, marked "H" on this map, arid conditions prevail. Humid areas in Central Asia are found only in small portions of the far north and extreme southeast. Because Central Asia is located in the interior of a large continent, its climate is characterized by significant differences between winter and summer temperatures.

**A TROPICAL AND HUMID CLIMATES**

- **Am** Tropical monsoon

**B DRY CLIMATES**

- **BWk** Midlatitude desert
- **BSk** Midlatitude steppe

**D CONTINENTAL MIDLATITUDE CLIMATES**

- **Dwa** Humid continental, with dry winter, warm summer
- **Dwb** Humid continental, with dry winter, cool summer
- **Dfb** Humid continental, without dry season, cool summer
- **Dwc** Subarctic

**H HIGHLAND**

- **H** Complex mountain climates

**FIGURE 10.7 Central Asian Desert**    Much of Central Asia is dominated by deserts and other arid lands. The Karakum Desert in Turkmenistan, seen in this photograph, is especially dry, supporting little vegetation.

can be divided into two main areas: a central belt of deserts and a northern strip of semiarid steppes.

The Tien Shan and Pamir Mountains divide Central Asia's desert belt into two separate segments. To the west lie the arid plains of the Caspian Sea and Aral Sea basins, located primarily in Turkmenistan, Uzbekistan, and southern Kazakhstan (Figure 10.7). The driest areas support little vegetation and contain extensive sand dunes. The climate of this region is continental; summers are dry and hot, while winter temperatures average well below freezing. Central Asia's eastern desert belt extends for almost 2,000 miles (3,200 kilometers) from far western China at the foot of the Pamirs to the southeastern edge of Inner Mongolia. Here one finds two major deserts: the Taklamakan, in the Tarim Basin of Xinjiang, and the Gobi, which runs along the border between Mongolia and the Chinese region of Inner Mongolia.

North of the desert zone, rainfall gradually increases, and desert eventually gives way to the grasslands, or steppes, of northern Central Asia. Near the region's northern boundary, trees begin to appear, outliers of the Siberian taiga (coniferous forest) of the north. Nearly continuous grasslands extend some 4,000 miles (6,400 kilometers) east to west across the entire region. Summers on the northern steppe are usually pleasant, but winters can be extremely cold.

## Global Warming and Central Asia

Most climate change experts expect Central Asia to be hard hit by global warming. The Tibetan Plateau has already seen marked increases in temperature, resulting in the reduction of permafrost and the rapid retreat of mountain glaciers. Eighty percent of Tibet's glaciers are presently retreating, some at rates of up to 7 percent a year, leading to predictions that the region will lose up to half of its ice fields within the next 50 years. Similar levels of glacial retreat are occurring throughout the Central Asia highlands, including the Pamir and Tien Shan mountains.

The retreat of Central Asia's glaciers is especially worrisome because of their role in providing dependable flows of water for the region's rivers. In some areas, the melting of ice has resulted in the temporary flooding of lowland basins, but the long-term result will be to reduce freshwater resources in an already dry region. Climate

change is also likely to reduce precipitation in the arid lowlands of western Central Asia. Prolonged and devastating droughts have recently struck Afghanistan, indicating a possible shift to a drier climate. As a result, the United Nations Intergovernmental Panel on Climate Change (IPCC) has predicted a 30 percent crop decline for Central Asia as a whole by the middle of the 21st century.

But as is true elsewhere in the world, global warming will not affect all parts of Central Asia in the same way. Some areas, including the Gobi Desert and the Tibetan Plateau, could possibly see increased precipitation. Some global climate models also indicate increased precipitation in the boreal forest zone of Siberia that extends into the mountains of northern Mongolia. But because the rivers coming out of these mountains generally flow north, their growth would provide few benefits for the rest of Central Asia.

# Population and Settlement: Densely Settled Oases Amid Vacant Lands

Most of Central Asia is sparsely populated (Figure 10.8). Large areas are either too arid or too high to support much human life. Even many of the most favorable areas are populated only by widely scattered groups of nomadic **pastoralists** (people who raise livestock for subsistence purposes). Mongolia, which is more than twice the size of Texas, has only 2.7 million inhabitants—fewer than live in the Dallas metropolitan area. But as is common in arid environments, those few lowland settings with good soil and dependable water supplies are thickly settled. While the nomadic pastoralists of the steppe zone have dominated much of the history of Central Asia, the settled peoples of the river valleys have always been more numerous.

## Highland Population and Subsistence Patterns

The environment of the Tibetan Plateau is particularly harsh. Only sparse grasses and herbaceous plants can survive in this high-altitude climate, making human subsistence difficult. The only feasible way of life in this area is nomadic pastoralism based on the yak, an altitude-adapted relative of the cow. Several hundred thousand people manage to make a living in such a manner, roaming with their herds over vast distances.

Although most of the Tibetan Plateau can support only nomadic pastoralism, the majority of Tibetans are sedentary farmers. Farming in Tibet is possible only in the few locations that are *relatively* low in elevation and that have good soils and either adequate rainfall or a dependable irrigation system based on local streams. The main zone of sedentary settlement lies in the far south, where protected valleys offer these favorable conditions.

Throughout Central Asia, the mountainous areas are vitally important for people living in the adjacent lowlands, whether they are migratory pastoralists or settled farmers. Many herders use the highlands for summer pasture; when the lowlands are dry and hot, the high meadows provide rich grazing. The Kyrgyz (of Kyrgyzstan) are noted for their traditional economy based on **transhumance**, moving their flocks from lowland pastures in the winter to highland meadows in the summer. The farmers of Central Asia rely on the highlands for their wood and, more importantly, for their water.

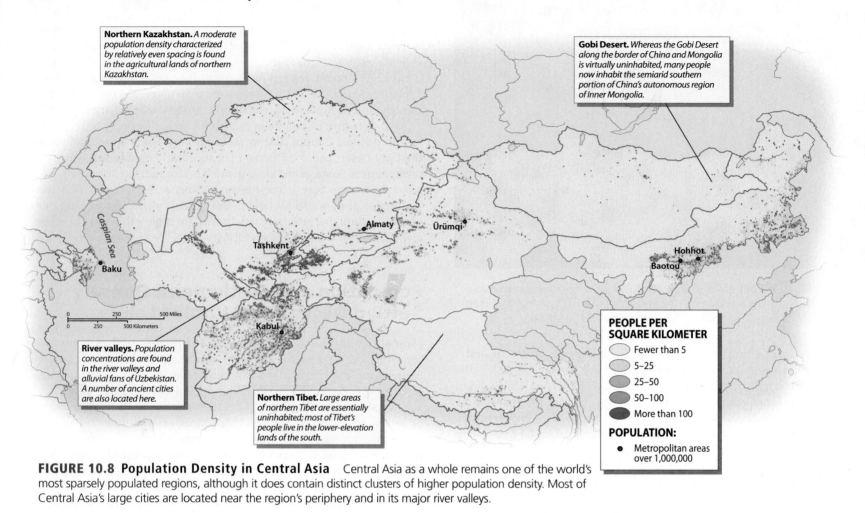

**Northern Kazakhstan.** *A moderate population density characterized by relatively even spacing is found in the agricultural lands of northern Kazakhstan.*

**Gobi Desert.** *Whereas the Gobi Desert along the border of China and Mongolia is virtually uninhabited, many people now inhabit the semiarid southern portion of China's autonomous region of Inner Mongolia.*

**River valleys.** *Population concentrations are found in the river valleys and alluvial fans of Uzbekistan. A number of ancient cities are also located here.*

**Northern Tibet.** *Large areas of northern Tibet are essentially uninhabited; most of Tibet's people live in the lower-elevation lands of the south.*

**PEOPLE PER SQUARE KILOMETER**

- Fewer than 5
- 5–25
- 25–50
- 50–100
- More than 100

**POPULATION:**
- Metropolitan areas over 1,000,000

**FIGURE 10.8 Population Density in Central Asia**   Central Asia as a whole remains one of the world's most sparsely populated regions, although it does contain distinct clusters of higher population density. Most of Central Asia's large cities are located near the region's periphery and in its major river valleys.

## Lowland Population and Subsistence Patterns

Most of the inhabitants of Central Asian deserts live in the narrow belt where the mountains meet the basins and plains. Here water supplies are adequate and soils are neither salty nor alkaline, as is often the case in the basin interiors. For example, the population distribution pattern of China's Tarim Basin forms an almost perfect ring-like structure (Figure 10.9). Streams flowing out of the mountains are diverted to irrigate fields and orchards in the fertile band along the basin's edge.

The population west of the Pamir range, in former Soviet Central Asia, is also concentrated in the transitional zone between the highlands and the plains. A series of **alluvial fans** (fan-shaped deposits of sediments dropped by streams flowing out of the mountains) have long been devoted to intensive cultivation. **Loess**, a fertile, silty soil deposited by the wind, is plentiful in this region, and in a few favored areas winter precipitation is high enough to allow for rain-fed agriculture. Several large valleys in this area also offer fertile and easily irrigated farmland (Figure 10.10). The densely populated Fergana Valley of the upper Syr Darya River is shared by three countries: Uzbekistan, Kyrgyzstan, and Tajikistan.

The steppes of northern Central Asia are the classical land of nomadic pastoralism. Until the 20th century, almost none of this area had ever been plowed and farmed. To this day, pastoralism remains a common way of life across the grasslands, particularly in Mongolia (Figure 10.11). In northwestern China and the former Soviet republics, however, many pastoral peoples have been forced to adopt sedentary lifestyles, and even in Mongolia the number of nomadic herders is steadily decreasing. In northern Kazakhstan, the Soviet regime converted the most productive pastures into wheat farms in the mid-1900s in order to increase the country's supply of grain. Consequently, northern Kazakhstan has the highest population density in the steppe region.

## Population Issues

Although Central Asia has a low population density, some portions of it are growing at a moderately rapid pace. In western China, much of the population growth over the past 30 years has stemmed from the migration of Han Chinese into the area. In 2006, China completed the construction of a railway from Beijing to the Tibetan capital of Lhasa, which has increased the flow of migrants, as well as tourists, into the region. The trains pass through such high elevations that passengers are supplied with supplemental oxygen.

**FIGURE 10.9 Population Patterns in Xinjiang's Tarim Basin** The central portion of the Tarim Basin is a nearly uninhabited expanse of sand dunes and salt flats. Along the edge of the basin, however, dense agricultural and urban settlements are located where streams running out of the surrounding mountains allow intensive irrigation. The largest of these oasis communities are found along the southwestern fringe of the basin.

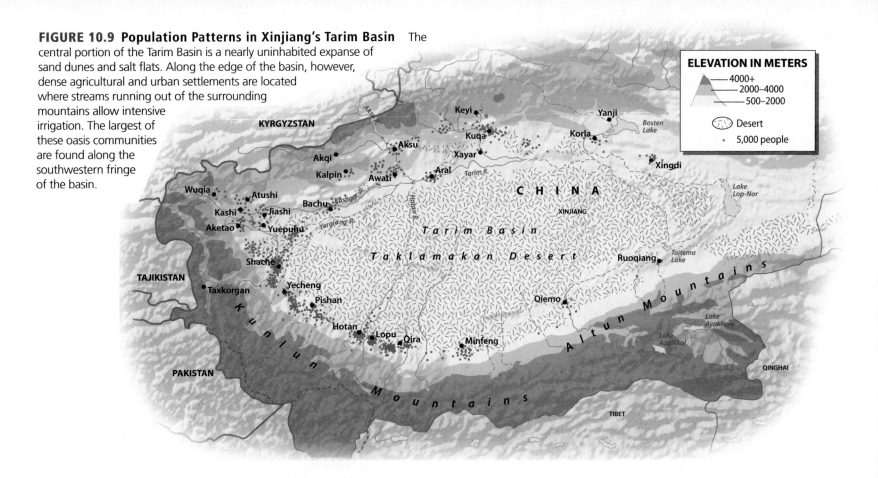

**FIGURE 10.10 Farmland in Uzbekistan** The fertile river valleys of Uzbekistan have been intensively cultivated for many centuries, producing large harvests of fruits, vegetables, and grains in addition to the cotton depicted in this photograph.

Most of the former Soviet zone of Central Asia is experiencing moderate population expansion due to its own fertility patterns. In contrast to Tibet and Xinjiang, this area has witnessed a substantial outward migration of people since 1991. At first, these emigrants were mostly ethnic Russians returning to the Russian homeland. But as the Russian economy boomed after 2000, hundreds of thousands of men from Tajikistan and other Central Asian countries sought work in Russia. The economic collapse of 2008–2009, however, forced many of these workers back home.

**FIGURE 10.11 Steppe Pastoralism** The steppes of northern and central Mongolia offer lush pastures during the summer. Mongolians, some of the world's most skilled horse riders, have traditionally followed their herds of sheep and cattle, living in collapsible, felt-covered yurts. Many Mongolians still follow this way of life.

**TABLE 10.1    POPULATION INDICATORS**

| Country | Population (millions) 2009 | Population Density (per square kilometer) | Rate of Natural Increase (RNI) | Total Fertility Rate | Percent Urban | Percent <15 | Percent >65 | Net Migration (Rate per 1,000) 2005–10* |
|---|---|---|---|---|---|---|---|---|
| Afghanistan | 28.4 | 44 | 2.1 | 5.7 | 22 | 44 | 2 | 7.5 |
| Azerbaijan | 8.8 | 101 | 1.2 | 2.3 | 52 | 23 | 7 | −1.2 |
| Kazakhstan | 15.9 | 6 | 1.3 | 2.7 | 53 | 24 | 8 | −1.3 |
| Kyrgyzstan | 5.3 | 27 | 1.6 | 2.8 | 35 | 30 | 5 | −2.8 |
| Mongolia | 2.7 | 2 | 1.8 | 2.6 | 60 | 33 | 4 | −0.8 |
| Tajikistan | 7.5 | 52 | 2.3 | 3.4 | 26 | 38 | 4 | −5.9 |
| Turkmenistan | 5.1 | 10 | 1.4 | 2.5 | 47 | 31 | 4 | −1.0 |
| Uzbekistan | 27.6 | 62 | 1.8 | 2.6 | 36 | 33 | 4 | −3.0 |

*Source:* Population Reference Bureau, *World Population Data Sheet, 2009.*
*Net Migration Rate from the United Nations, Population Division, *World Population Prospects: The 2008 Revision Population Database.*

Fertility patterns vary substantially from one part of Central Asia to another (Table 10.1). Afghanistan, the least-developed and most male-dominated country of the region, has the highest birthrate by a substantial margin. Fertility rates are in the middle range through most of the former Soviet area, but Kazakhstan's birthrate is well below the replacement level, reflecting in part the extremely low fertility level of the country's Russian speakers. According to official Chinese statistics, the fertility rate in Tibet is above the replacement level, far higher than that found in China proper.

## Urbanization in Central Asia

Although the steppes of northern Central Asia had no real cities before the modern age, the river valleys have been partially urbanized for thousands of years. Such cities as Samarkand and Bukhara in Uzbekistan were famous even in medieval Europe for their lavish architecture (Figure 10.12). These cities contrast sharply with those built during the period of Russian/Soviet rule. Tashkent in Uzbekistan, for example, is largely a Soviet creation and thus looks quite different from the more traditional Bukhara. Several major cities, such as Kazakhstan's former capital of Almaty, were minor settlements before Russian colonization. In cities throughout the former Soviet region, one can see the effects of centralized Soviet urban planning and design.

Parts of Central Asia have experienced substantial urbanization in recent decades. North-central Kazakhstan has recently seen the rise of a new major city, Astana, designated as the national capital due to its central location. Even Mongolia, long a land without many permanent settlements, now has more people living in cities than in the countryside. In the highlands of Central Asia, however, cities remain relatively few and far between. Only about one-quarter of the people of Tajikistan, for example, are urban residents. Tibet similarly remains a predominantly rural society, although change is occurring as Han Chinese migrants move into its cities.

New development projects coupled with Han migration are transforming cities throughout Chinese Central Asia. Local controversies often follow. In 2009, for example, Beijing announced that the old city of Kashgar in China's Far West would be largely demolished, due mainly to concerns about earthquake safety (Figure 10.13). Although China promised to save many of Kashgar's most historic buildings and to build new residences for its inhabitants, many local people view the plan an assault on their cultural identity.

## Cultural Coherence and Diversity: A Meeting Ground of Different Traditions

Although areas of Central Asia have a certain environmental similarity, its cultural unity is more questionable. The western half of the region is largely Muslim and is often classified as part of Southwest Asia, but in Mongolia and Tibet most people traditionally follow Tibetan Buddhism. Tibet is culturally linked to both South and East Asia, and Mongolia is historically associated with China, but neither fits easily within any other world region.

**FIGURE 10.12 Traditional Architecture in Samarkand**
Samarkand, Uzbekistan, is famous for its lavish Islamic architecture, some of it dating back to the 1400s. The city owes its rich architectural heritage in part to the fact that it was the capital of the empire created by the great medieval conqueror Tamerlane.

replace the Indo-European tongues (which were closely related to Persian) as Turkic power spread through most of Central Asia.

## Contemporary Linguistic and Ethnic Geography

Today peoples speaking Turkic and Mongolian languages inhabit most of Central Asia (Figure 10.14). A few native Indo-European languages are confined to the southwest, while Tibetan remains the main language of the plateau. Russian is also widely spoken in the west, while Chinese is increasingly important in the east.

**Tibetan**   Tibetan is usually placed in the Sino-Tibetan language family, suggesting a shared linguistic history between the Chinese and the Tibetan peoples. Some students of Tibetan, however, argue that no definite relationship between the two languages has ever been established. The Tibetan language is divided into a number of distinct dialects that are spoken over almost the entire inhabited portion of the Tibetan Plateau. Over 80 percent of the 2.8 million people who live in the Tibet Autonomous Region speak Tibetan, while another 3 million or so Tibetan speakers live in the highland portions of the Chinese provinces of Qinghai and Sichuan.

**Mongolian**   The Mongolian language includes a cluster of closely related dialects spoken by approximately 5 million people. The standard Mongolian of both the independent country of Mongolia and China's Inner Mongolia is called *Khalkha*. Mongolian has its own distinctive script, which dates back some 800 years, but Mongolia itself adopted the Cyrillic alphabet of Russia in 1941. Efforts are now being made to revive the old script.

Mongolian speakers form about 90 percent of the population of Mongolia. In China's Inner Mongolia Autonomous Region, Han Chinese who migrated into the area over the past 50 years today outnumber Mongolian speakers. Currently only about 17 percent of the 24 million residents of Inner Mongolia are Mongolian.

**Turkic Languages**   Far more Central Asians speak Turkic languages than Mongolian and Tibetan combined. The Turkic linguistic sphere extends from Azerbaijan in the west through China's Xinjiang province in the east.

Five of the six countries of the former Soviet Central Asia—Azerbaijan, Uzbekistan, Turkmenistan, Kyrgyzstan, and Kazakhstan—are named after the Turkic languages of their dominant populations. In three of these countries, these ethnic groups form a clear majority: Some 90 percent of the people of Azerbaijan speak Azeri, some 74 percent of the people of Uzbekistan speak Uzbek, and some 72 percent of the people in Turkmenistan speak Turkmen. With more than 23 million speakers, Uzbek is the most widely spoken Central Asian language. In the Amu Darya Delta in the far north of Uzbekistan, however, most people speak a different Turkish language called *Karakalpak,* while Kazak speakers are found in the sparsely populated Uzbek deserts.

The two other Turkic-speaking counties of Central Asia are more linguistically diverse. Roughly 65 percent of the inhabitants of Kyrgyzstan speak Kyrgyz as their native language, while only around 54 percent of the people of Kazakhstan speak Kazakh. In Kazakhstan the population is split between speakers of Turkic languages and European languages (Russian, Ukrainian, and German). In general, the Kazakhs and other indigenous peoples predominate in the center and south of the country, while the people of European descent live in the agricultural districts of the north as well as in the cities of the southeast.

**FIGURE 10.13  Old Kashgar before its Destruction**   A Uyghur woman is seen in an old neighborhood of Kashgar, an ancient city on the Silk Road in the Chinese region of Xinjiang. China is planning to demolish most of the old districts of Kashgar in order to modernize the city and improve its safety.

## Historical Overview: Changing Languages and Populations

The river valleys and oases of Central Asia were early sites of sedentary agricultural communities. Archaeologists have discovered abundant evidence of farming villages dating back to the Neolithic period (beginning around 8000 B.C.E.) in the Amu Darya and Syr Darya valleys. After the domestication of the horse around 4000 B.C.E., nomadic pastoralism, based on the raising of livestock, emerged in the steppe belt. Eventually, pastoral peoples gained power over the entire region.

The earliest recorded languages of Central Asia belonged to the Indo-European linguistic family. These languages were replaced on the steppe more than 1,000 years ago by languages in another major family: Altaic (which includes Turkish and Mongolian). In the river valley communities as well, Turkic languages gradually began to

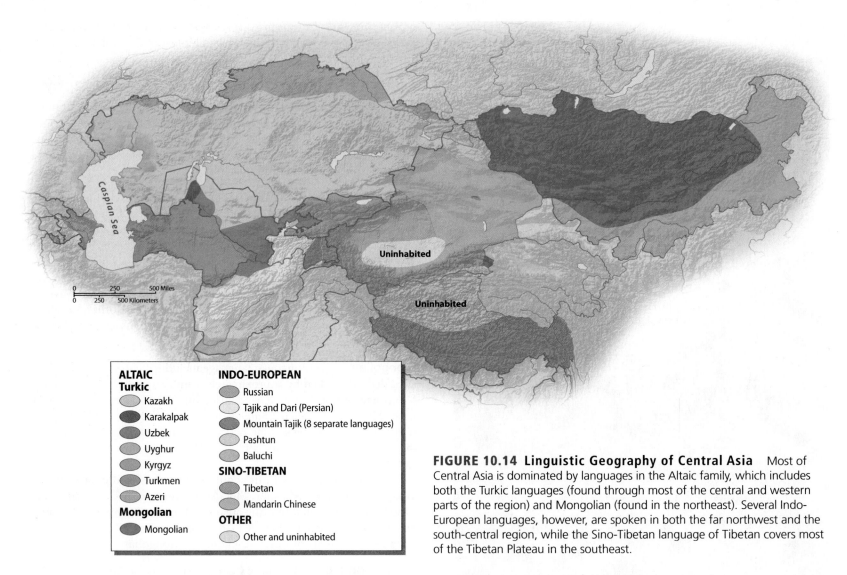

**FIGURE 10.14 Linguistic Geography of Central Asia**   Most of Central Asia is dominated by languages in the Altaic family, which includes both the Turkic languages (found through most of the central and western parts of the region) and Mongolian (found in the northeast). Several Indo-European languages, however, are spoken in both the far northwest and the south-central region, while the Sino-Tibetan language of Tibetan covers most of the Tibetan Plateau in the southeast.

Uygur, spoken in China's Xinjiang province, dates back almost 2,000 years. The Uygur people number about 11 million, most of whom live in Xinjiang (Figure 10.15). As recently as 1953, the Uygur formed about 80 percent of the population of Xinjiang, but by 2000, that number had been reduced to 45 percent, mostly due to Han Chinese migration. According to official statistics, the Han Chinese form around 40 percent of Xinjiang's population, but Uygur activists think that the number is much larger. There are also about 1 million Kazak speakers living in Xinjiang.

**Linguistic Complexity in Tajikistan**   The sixth republic of the former Soviet Central Asia, Tajikistan, is dominated by people who speak an Indo-European language rather than a Turkic language. Tajik is so closely related to Persian that it is often considered to be a Persian (or Farsi) dialect. Roughly 80 percent of the people of Tajikistan speak Tajik as their first language. In the remote mountains of eastern Tajikistan, people speak a variety of distinctive Indo-European languages, sometimes collectively referred to as "Mountain Tajik." The remainder of Tajikistan's population speaks either Uzbek or a variety of minor languages.

**Language and Ethnicity in Afghanistan**   The linguistic geography of Afghanistan is even more complex than that of Tajikistan

(Figure 10.16). Afghanistan became a country in the 1700s under the leadership of the Pashtun people. The rulers of Afghanistan, however, never tried to build a nation-state around Pashtun identity, in part because approximately half of the Pashtun people live in Pakistan. In Afghanistan itself, estimates of the amount of people speaking this language average around 40 percent. Most Pashtun speakers in Afghanistan live to the south of the Hindu Kush mountain range.

Roughly half of the people of Afghanistan speak Dari, the local form of Persian. Dari speakers are concentrated in the cities of the west, in the central mountains, and near the boundary with Tajikistan. Two separate ethnicities divide the Dari-speaking people. Those in the west and north are considered to be Tajik, whereas those in the central mountains are called *Hazaras* and are said to be descendants of Mongol conquerors who arrived in the 12th century. Finally, another 10 percent or so of the people of Afghanistan speak Turkic languages, mainly Uzbek.

## Geography of Religion

At one time, Central Asia was noted for its religious complexity. Major overland trading routes crossed the region, giving easy access to both merchants and missionaries. By 1500, however, the region was

many Pashtun rules, such as never allowing women's faces to be seen in public, are based more on ethnic rather than religious customs (Figure 10.17). The traditionally nomadic groups of the northern steppes, such as the Kazakhs and Kyrgyz, are, in contrast, considered to be rather relaxed about religious beliefs and practices. While most of the region's Muslims are Sunnis, Shiism is dominant among the Hazaras of central Afghanistan and the Azeris of Azerbaijan.

Under the communist rule of China, the Soviet Union, and Mongolia, all forms of religion were discouraged. Chinese authorities and student radicals attempted to suppress Islam in Xinjiang during the Cultural Revolution of the late 1960s and early 1970s. Chinese Muslims now enjoy basic freedom of worship, but the state still closely monitors religious expression out of fear that it will lead to resistance against Chinese rule. Periodic persecution of Muslims also occurred in Soviet Central Asia, and until the 1970s, its many observers thought that Islam might slowly disappear from the region.

Religious expression was not, however, so easily repressed. Interest in Islam began to grow in former Soviet Central Asia in the 1970s and 1980s. In the post-Soviet period, Islam continues to revive as people seek their cultural roots. In Xinjiang, Islam has served as a focal point of a social and political identity among the Uygur people. Most Uygur leaders, however, insist that their beliefs are not radically fundamentalist. Radical Islamic fundamentalism has emerged a powerful political movement only in Afghanistan, parts of Tajikistan, and the Fergana Valley (mostly in Uzbekistan).

**Tibetan Buddhism**   Mongolia and especially Tibet stand apart from the rest of Central Asia in their people's practice of Tibetan Buddhism. Buddhism entered Tibet from India many centuries ago, and it merged

**FIGURE 10.15 Uygur Mosque**   A small mosque, illustrating traditional Uygur architecture, survives amid blocks of modern apartments in Urumchi, Xinjiang. Traditional forms of housing and urban design can still be found in Uygur communities in northwestern China, but they are gradually disappearing.

divided into two spiritual camps: Islam predominated in the west and center and Tibetan Buddhism in Tibet and Mongolia. In western Central Asia, many ethnic Russians inhabitants belong to the Russian Orthodox Church.

**Islam in Central Asia**   Different Central Asian peoples are known for their different interpretations of Islam. The Pashtuns of Afghanistan are noted for their strict Islamic ideals—although critics contend that

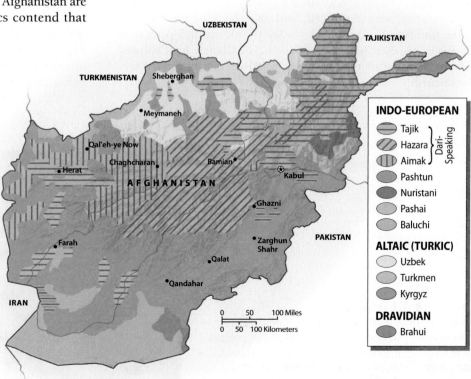

**FIGURE 10.16 Afghanistan's Ethnic Patchwork**   Afghanistan is one of the world's most ethnically complex countries. Its largest ethnic group is the Pashtuns, who live in most of the southern portion of the country as well as the adjoining borderlands of Pakistan. Northern Afghanistan, however, is mostly inhabited by Uzbeks, Tajiks, and Turkmens—whose main population centers are located in Uzbekistan, Tajikistan, and Turkmenistan, respectively. The Hazaras of Afghanistan's central mountains, like the Tajiks, speak a form of Persian. However, the Hazaras are considered to be a separate ethnic group in part because they, unlike other Afghans, follow Shiite rather than Sunni Islam.

**FIGURE 10.17 Afghan Women in Public**    Especially in the Pashtun areas of Afghanistan, women have traditionally been forced to cover their entire bodies when in public areas. In places that were controlled by the Taliban, such dress codes were strictly enforced. Extremely modest dress still prevails in Afghanistan.

with the native religion of the area, called *Bon*. The resulting mix is more oriented toward mysticism than are many other forms of Buddhism, and it is more tightly organized. Standing at the head of Lamaist society is the Dalai Lama. Until the Chinese conquest, Tibet was essentially a **theocracy** (or religious state), with the Dalai Lama enjoying political as well as religious authority (Figure 10.18).

Tibetan Buddhists suffered persecution after 1959 when China invaded Tibet. The Chinese hold on Tibet has never been as secure as that on Xinjiang, and Tibetan Buddhism has inspired many Tibetans to resist Chinese rule. The Dalai Lama, who fled Tibet for India in 1959, has been a powerful advocate for the Tibetan cause in international circles. During the 1960s and 1970s, an estimated 6,000 Tibetan Buddhist monasteries were destroyed, and thousands of monks were killed; the number of active monks today is only about 5 percent of what it was before the Chinese occupation. The Chinese have allowed many monasteries to reopen, but their activities remain limited and closely watched.

## Central Asian Culture in International and Global Context

In cultural terms, western Central Asia is closely linked to Russia, whereas eastern Central Asia is more closely tied to China. In the east, the main cultural issue is the migration of Han Chinese into the area, which has resulted in serious ethnic tensions. In the west, in contrast, the cultural influence of the Russians is gradually diminishing.

During the Soviet period, the Russian language spread widely through western Central Asia. Russian served both as a common language and as the language used in higher education. One had to be fluent in Russian in order to reach any position of responsibility. Russian speakers settled in all the major cities, and many became influential. Today, however, many Russian speakers have migrated back to Russia. And although its use is gradually declining in favor of local languages, Russian remains the common language of western Central Asia.

Although Central Asia is remote and poorly integrated into global cultures, it is hardly immune to the forces of globalization. The increased usage of English and the influence of U.S. culture throughout Central Asia show that this part of the world is not cut off from global culture. Such influences have been especially marked in the oil cities of the Caspian Basin, such as Baku, which have seen an influx of U.S. oil workers (Figure 10.19). Although their number is low, English speakers in Central Asia, especially those with computer skills, are increasingly valued as the region strives to find a place and a voice in the global community.

## Geopolitical Framework: Political Reawakening

Central Asia has played a minor role in global political affairs for the past several hundred years. Before 1991, the entire region, except Mongolia and Afghanistan, was under direct Soviet and Chinese control. Mongolia, moreover, had been a close Soviet ally, and even Afghanistan came under Soviet domination in the late 1970s. Today, of course, southeastern Central Asia is still part of China. And although the breakup of the Soviet Union saw the emergence of six new

**FIGURE 10.18 Tibetan Buddhist Monastery**    Tibet is well known for its large Buddhist monasteries, which at one time served as seats of political as well as religious authority. The Potala Palace in Lhasa, traditional seat of the Dalai Lama, is the largest, most important, and most famous of such monasteries.

**FIGURE 10.19 U.S.-Style Restaurant in Baku**   Baku, in Azerbaijan, is a vibrant and increasingly cosmopolitan city closely linked to global economic and cultural networks. This coffee shop and wine bar caters to both local and international customers.

Central Asian countries, most of them retain close political ties to Russia (Figure 10.20).

## Partitioning of the Steppes

Before 1500, Central Asia was a power center, a region whose mobile armies threatened the far more populous, sedentary states of Asia and Europe. The development of new weapons changed the balance of power, however, allowing the wealthier agricultural states to conquer the nomads. By the 1700s, the nomads' armies had been defeated and their lands taken. The winners in this struggle were the two largest states bordering the steppes: Russia and China.

By the mid-1700s, the Chinese empire (under the Manchu, or Qing, dynasty) stood at its greatest territorial extent, including Mongolia, Xinjiang, Tibet, and a slice of modern Kazakhstan. From the height of its power in the late 1700s, China declined rapidly. When the Manchu dynasty fell in 1912, Mongolia became independent, although China still ruled the extensive borderlands of Inner Mongolia (Nei Mongol). Tibet had earlier gained effective independence, although this status was not recognized by China.

Russia began to push south of the steppe zone in the mid-1800s, conquering most of western Central Asia by 1900. Russia advanced into this area in part to keep the British out. Britain attempted to conquer Afghanistan but failed to do so. Subsequently, Afghanistan's position as an independent "buffer state" between the Russian Empire (later the Soviet Union) and British India remained secure.

## Central Asia Under Communist Rule

Western Central Asia came under communist rule not long after the emergence of the Soviet Union in 1917. Mongolia followed in 1924. After the Chinese revolution of 1949, the communist system was also imposed on Xinjiang and Tibet.

**Soviet Central Asia**   The newly established Soviet Union retained all the Central Asian territories of the Russian Empire. The new regime sought to build a Soviet society that would eventually knit together all the massive territories of the Soviet Union. Central Asia's leaders were replaced by Communist Party officials loyal to the new state, Russian immigration was encouraged, and local languages had to be written in the Cyrillic (Russian) rather than the Arabic script.

Although the early Soviet leaders thought that a new Soviet nationality would eventually emerge, they realized that local ethnic diversity would not disappear overnight. They therefore divided the Soviet Union into a series of nationally defined "union republics" in which a certain degree of cultural autonomy would be allowed. In the 1920s, Kazakhstan, Kyrgyzstan, Tajikistan, Uzbekistan, Turkmenistan, and Azerbaijan assumed their present shapes as Soviet republics.

Contrary to official intentions, the new republics encouraged the development of local national identities more than a broader Soviet identity. Also undercutting Soviet unity was the fact that cultural and economic gaps separating Central Asians from Russians did not decrease as much as planned. In addition, the region's higher birthrates led many Russians to fear that the Soviet Union risked being dominated by Turkic-speaking Muslims, contributing to the breakup of the Soviet Union.

**The Chinese Geopolitical Order**   After decades of political and economic chaos, China reemerged as a united country in 1949. Its new communist government quickly reclaimed most of the Central Asian territories that had slipped out of China's grasp in the early 1900s. China's new leaders promised the non-Chinese peoples a significant degree of political self-determination and cultural autonomy and thus found much local support in Xinjiang. Gaining control over Tibet was more difficult. China occupied Tibet in 1950, but the Tibetans launched a rebellion in 1959. When this was brutally crushed, the Dalai Lama and some 100,000 followers sought refuge in India.

Loosely following the Soviet model, China established autonomous regions in areas occupied primarily by non-Han Chinese peoples, including Xinjiang, Tibet proper (called *Xizang* in Chinese), and Inner Mongolia. Such autonomy, however, often meant little, and it did not prevent the immigration of Han Chinese into these areas. Nor were all parts of Chinese Central Asia granted autonomous status. The large and historically Tibetan and Mongolian province of Qinghai, for example, remained an ordinary Chinese province.

## Current Geopolitical Tensions

Although western Central Asia made the transition to independence rather smoothly, the region still suffers from a number of conflicts. Much of China's Central Asian territory is also unsettled, but China keeps a strong political hold on the area. Afghanistan, unfortunately, continues to suffer from a particularly brutal war.

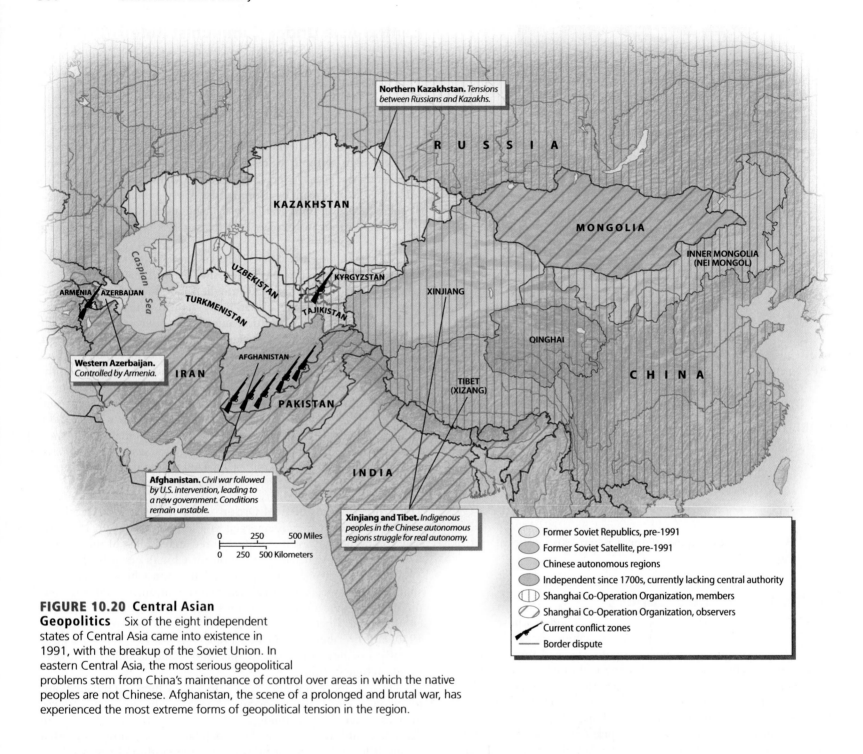

**Northern Kazakhstan.** *Tensions between Russians and Kazakhs.*

**Western Azerbaijan.** *Controlled by Armenia.*

**Afghanistan.** *Civil war followed by U.S. intervention, leading to a new government. Conditions remain unstable.*

**Xinjiang and Tibet.** *Indigenous peoples in the Chinese autonomous regions struggle for real autonomy.*

○ Former Soviet Republics, pre-1991
● Former Soviet Satellite, pre-1991
◐ Chinese autonomous regions
◑ Independent since 1700s, currently lacking central authority
⬭ Shanghai Co-Operation Organization, members
⬭ Shanghai Co-Operation Organization, observers
⚞ Current conflict zones
— Border dispute

**FIGURE 10.20 Central Asian Geopolitics** Six of the eight independent states of Central Asia came into existence in 1991, with the breakup of the Soviet Union. In eastern Central Asia, the most serious geopolitical problems stem from China's maintenance of control over areas in which the native peoples are not Chinese. Afghanistan, the scene of a prolonged and brutal war, has experienced the most extreme forms of geopolitical tension in the region.

**Independence in Former Soviet Lands** The disintegration of the Soviet Union in 1991 generally proceeded peacefully in Central Asia (Figure 10.21), although a civil war that broke out in Tajikistan lasted until 1997. The six newly independent countries, however, had been dependent on the Soviet system, and it was no simple matter for them to chart their own courses. In most cases, authoritarian rulers, rooted in the old order, retained power and undermined opposition groups. All told, democracy has made less progress in Central Asia than in other parts of the former Soviet Union.

The newly independent countries also faced a number of border conflicts, due in part to the complicated political divisions established by the Soviet Union. Such problems are particularly severe in the Fergana Valley,

a large, fertile, and densely populated lowland basin shared by Uzbekistan, Tajikistan, and Kyrgyzstan (Figure 10.22). Not only do boundaries here meander in a complex manner, but the residents of many villages have maintained citizenship in a neighboring country. In early 2009, Tajikistan began to demand that Uzbek citizens living on its side of the boundary either adopt Tajik citizenship or risk deportation to Uzbekistan.

**Strife in Western China** Many of the indigenous inhabitants of Tibet and Xinjiang in western China want independence or at least political autonomy. China maintains that all its Central Asian lands are essential parts of its national territory, and it treats separatist groups harshly. The continuing migration of Han Chinese into Tibet and

**FIGURE 10.21 The End of Soviet Rule in Central Asia** An old Tajik man looks over a toppled and decapitated statue of Lenin, founder of the Soviet Union, in Dushanbe in 1991. When the Soviet Union collapsed in 1991, emblems of the old regime were immediately destroyed in many areas.

**FIGURE 10.23 Ethnic Tension in Xinjiang** Relations between Han Chinese officials and Muslim Central Asians have grown increasingly tense in recent years. Here a Chinese official levies a trading tax on merchants in Kashgar's Sunday market.

Xinjiang has intensified anti-Chinese sentiments among the indigenous peoples while strengthening Chinese control (Figure 10.23).

In Xinjiang, a variety of groups compose the Eastern Turkistan Independence Movement. Some of these groups have a radical Islamist orientation and are classified as terrorist organizations by both China and the United States. Others are secular, basing their claims on ethnicity and territory rather than religion. Periodic violence by separatist groups typically results in quick reprisals by China. In July 2009, severe rioting broke out in Urumqi, the capital of Xinjiang, as Uygur protestors attacked Han Chinese businesses and migrants and as Han settlers responded in kind.

**FIGURE 10.22 Political Boundaries Around the Fergana Valley** Some of the world's most complex political boundaries can be found in the vicinity of the Fergana Valley. The central portion of the valley belongs to Uzbekistan, which is otherwise separated from it by high mountains. The lower valley, on the other hand, is part of Tajikistan, the core area of which is likewise separated by highlands. The Fergana's upper periphery belongs to Kyrgyzstan. Note also the small enclaves of Uzbekistan within Kyrgyzstan.

More than 180 people were killed before Chinese authorities restored order. To prevent Uygur activists from organizing, China quickly suspended Internet service and restricted cell phone coverage in the city.

Anti-China protests in Tibet have also been severely repressed, but the Tibetans have brought the attention of the world to their struggle, largely through the work of the Dalai Lama. China maintains several hundred thousand troops in the region because of both Tibetan resistance and the strategic importance of the border zone with India. In March 2008, Tibetan protestors in Lhasa rioted, burning many Chinese-owned businesses. In response, China arrested hundreds of Tibetan monks and forced others to undergo "patriotic education." Chinese officials subsequently held talks with representatives of the Dalai Lama, hoping to restore calm. The Dalai Lama now agrees that Tibet should remain part of China, but Chinese leaders reject his call for true political autonomy.

**War in Afghanistan** None of the conflicts in the former Soviet republics or western China compares in intensity to the struggle being waged in Afghanistan. Afghanistan's troubles began in 1978, when a Soviet-supported military "revolutionary council" seized power. The new Marxist government began to suppress religion, which led to widespread resistance. When the government was about to collapse, the Soviet Union responded with a massive invasion. Despite its power, the Soviet military was never able to control the more rugged parts of the country. Pakistan, Saudi Arabia, and the United States, moreover, ensured that the anti-Soviet forces remained well armed. When the exhausted Soviets finally withdrew their troops in 1989, brutal local warlords grabbed power across most of the country.

In 1995, a new movement called the Taliban arrived in Afghanistan. Founded by young Muslim religious students, the Taliban believed in the strict enforcement of Islamic law. The Taliban model attracted large numbers of soldiers, and by September 2001, only far northeastern Afghanistan lay outside Taliban power. By 2000, however, most of Afghanistan's people were turning against the group, mostly because of the severe restrictions on daily life that the Taliban imposed. These restrictions were most pronounced for women, but even men were compelled to obey the Taliban's numerous decrees. Most forms of recreation were simply outlawed, including television, films, music, and even kite flying.

The attacks of September 11, 2001, completely changed the balance of power in Afghanistan. The United States and Britain, working with Afghanistan's anti-Taliban forces, attacked the Taliban government and soon defeated it. Although a democratic Afghan government was established within a few years, peace did not return. The new government proved corrupt and ineffective, failing to establish security in most parts of the country.

By 2004, the Taliban had regrouped, operating from safe havens in Pakistan (see Chapter 12). Afghanistan's new government has had to rely on the military power of the North Atlantic Treaty Organization (NATO)-led International Security Assistance Force (ISAF). But despite its more than 58,000 troops, the ISAF has been unable to stop the Taliban resurgence (Figure 10.24). By 2008, many observers feared that Afghanistan was on the verge of becoming a failed state. In 2009, the new administration of the United States responded by sending an additional 17,000 troops to Afghanistan. U.S. forces have continued to target high-level Taliban leaders, often by bombing their compounds. This strategy has weakened the Taliban command structure but has also resulted in large numbers of civilian casualties, reducing local support for

**FIGURE 10.24 War in Afghanistan** Warfare continues in Afghanistan, as troops from the United States and other countries attempt to root out the remnants of the Taliban and maintain order. Tense encounters with local residents often result when foreign soldiers search for weapons and insurgent fighters.

the NATO war effort. Although many areas of northern Afghanistan are relatively secure, the south remains a battle-scarred war zone.

## International Dimensions of Central Asian Tension

With the collapse of the Soviet Union in 1991, Central Asia emerged as a key arena of geopolitical tension. A number of important countries, including China, Russia, Pakistan, Iran, India, and the United States, have competed for influence in the region. The political revival of Islam has also generated international geopolitical issues.

Russia's economic and military ties to Central Asia did not vanish when the Soviet Union collapsed. Russia maintains military bases in Tajikistan and Kyrgyzstan and continues to be one of the main markets for Central Asian exports. Furthermore, western Central Asia's rail links and gas and oil pipelines are mostly oriented toward Russia.

By the late 1990s, Russian and Chinese leaders became concerned about the growing influence of the United States in Central Asia, a tendency that accelerated after the September 11, 2001 attacks, when the U.S. military established bases in Uzbekistan, Kyrgyzstan, and Afghanistan. A consequence of this concern was the formation of a treaty group called the **Shanghai Cooperation Organization (SCO)**, composed of China, Russia, Kazakhstan, Kyrgyzstan, Tajikistan, and Uzbekistan (Figure 10.25). The SCO seeks cooperation on such security issues as terrorism and separatism and aims to enhance trade. In 2007, the SCO signed an agreement of cooperation with the **Collective Security Treaty Organization (CSTO)**, a Russian-led military association that includes Belarus, Armenia, Kazakhstan, Kyrgyzstan, Tajikistan, and Uzbekistan. The two organizations work together to address military threats, crime, and drug smuggling.

**FIGURE 10.25 Meeting of the Shanghai Cooperation Organization** The leaders of the six countries that make up the Shanghai Cooperation Organization have gathered together to discuss their common concerns. Pictured here are the leaders of Russia, China, Kazakhstan, Kyrgyzstan, Tajikistan, and Uzbekistan.

Turkmenistan has not joined the SCO or the CSTO. Turkmenistan's relative isolation stems from the dictatorial policies of its former president, Saparmurat Niyazov, who forced the citizens of his country to treat him as a heroic savior-figure. After Niyazov's death in 2006, Turkmenistan released political prisoners, reformed its educational system, and eased restrictions on internal travel. The United States is now vying with Russia and China to gain influence in this resource-rich country, but progress has been slow, and the new government remains repressive.

The chaotic situation in Afghanistan is Central Asia's most serious geopolitical issue, as local leaders fear that Islamic radicalism could spread from Afghanistan into their own countries. Although several

Central Asian countries have expelled U.S. military bases (see "Exploring Global Connections: Foreign Military Bases in Central Asia"), they have offered some support to the NATO forces operating in Afghanistan. India is also concerned about Afghanistan and has thus also sought to gain influence in the region. Pakistan, however, views such moves with alarm, fearing that India is trying to encircle its territory.

## Economic and Social Development: Abundant Resources, Devastated Economies

Central Asia is by most measures one of the world's poorest regions. Afghanistan in particular stands near the bottom of almost every list of economic and social indicators. In the early years of the 21st century, however, several Central Asian countries underwent an economic boom, with annual growth rates exceeding 10 percent. But the global economic crisis of 2008–2009 hit the region hard. Declining oil and mineral prices undermined Central Asia's stronger economies, while the poorer, labor-exporting countries suffered as their workers lost jobs in Russia and had to return home.

### Post-Communist Economies

During the communist period, economic planners sought to spread economic development widely across the Soviet Union. This effort required building large factories even in remote areas of Central Asia, regardless of the costs involved. Such Central Asian industries relied heavily on subsidies from the Soviet government. When those subsidies ended, the industrial base of the region collapsed, leading to a huge drop in the standard of living. As is true elsewhere in the former Soviet Union, however, certain individuals became very wealthy after the fall of communism.

Since the beginning of the 21st century, Kazakhstan, Azerbaijan, and Turkmenistan have benefited from their large supplies of energy resources. Today, Kazakhstan stands as the most developed Central Asian country. Its agricultural sector is productive, and it has the world's 11th largest reserves of both oil and natural gas. The region's oldest fossil fuel industry is located in Azerbaijan (Figure 10.26), which has attracted much international investment in recent years. Although its economy grew rapidly from 2005 to 2008, Azerbaijan remains a poor country, burdened by inadequate infrastructure. Turkmenistan, despite having the world's 12th largest reserves of natural gas, is poorer still. Turkmenistan's development has been hampered by its repressive government, which until recently discouraged foreign investments, and by its declining cotton harvests, damaged by environmental degradation.

**FIGURE 10.26 Oil Development in Azerbaijan** Although oil has brought a certain amount of wealth to Azerbaijan, it has also resulted in extensive pollution. Because most of the petroleum is located either near or under the Caspian Sea, this sea—actually the world's largest lake—is becoming increasingly polluted, and its fisheries are declining.

# EXPLORING GLOBAL CONNECTIONS
## Foreign Military Bases in Central Asia

After the downfall of the Soviet Union, Russia sought to keep its strategic influence in Central Asia's newly independent countries. Russian has thus maintained military installations in Azerbaijan and Kazakhstan, while in Tajikistan it retains a division of ground forces as well as a large airbase. After the United States built the Manas Air Base in Kyrgyzstan in 2001 to support its war in Afghanistan, Russia responded by constructing a new air force station only a few miles away from the American facility.

Outside Afghanistan, the United States has been less successful than Russia in maintaining its military influence. In 2001, the U.S. military established a sizable air base in eastern Uzbekistan, but in 2005, Uzbekistan ordered it to be shut down after the U.S. government harshly criticized that country for its violent response to an anti-government protest. In February 2009, the parliament of Kyrgyzstan voted overwhelmingly to expel U.S. forces from Manas Air Base. This move came after Russia agreed to provide Kyrgyzstan with $2 billion in new loans and $150 million in direct financial aid. Four months later, however, the government of Kyrgyzstan agreed that the United States could continue to use the facility in exchange for tripling its rent payments and referring to it as a "transit center" rather than an "air base" (Figure 10.2.1).

Pakistan's military has also long been interested in extending its power into Central Asia, leading it to support the Taliban's efforts to rule Afghanistan before September 11,

**FIGURE 10.2.1 Manas Transit Center** A soldier is seen near the entrance of Manas Transit Center, formerly called the Manas Air Base, in Kyrgyzstan.

2001. Pakistan's maneuvering, in turn, has prompted India to seek influence in the region. In 2004, the governments of India and Tajikistan ratified an agreement allowing India to construct an air base 80 miles (130 kilometers) south of Tajikistan's capital city. Fully operational since 2007, the Farkhor Air Base, India's only foreign military installation, is thought to hold a dozen MiG-29 fighter-bombers. Pakistan, not surprisingly, sees this base as part of an Indian effort to surround Pakistan with hostile forces.

The expulsion of U.S. military bases from Uzbekistan and especially Kyrgyzstan has complicated the war effort in Afghanistan, as have continuing Indian–Pakistani competition and maneuvering in the region. Historians describe competing British and Russian efforts to gain influence in Central Asia in the late 1800s as the "Great Game"; some observers now argue that a "New Great Game" in the region will result in complex geopolitical tensions for some time to come.

---

Uzbekistan's fossil fuel reserves are much smaller than those of Kazakhstan and Turkmenistan, but they are still substantial. But Uzbekistan is also a much more densely populated country, and many of its people live in overcrowded farmlands suffering from environmental stress. An authoritarian country, Uzbekistan retains many aspects of the old Soviet-style command economy (one run by governmental planners rather than by private firms responding to the market). It is, however, the world's second largest cotton exporter, and its large deposits of gold and other mineral resources help keep its economy afloat.

Central Asia's two most mountainous countries, Tajikistan and Kyrgyzstan, hold most of the region's water resources, which they are developing by building massive dams. Tajikistan, however, is burdened by its remote location, rugged topography, and lack of infrastructure. With a per capita gross national income of only $1,700, Tajikistan rates as one of the Eurasia's poorest countries, with two-thirds of its people living in poverty. In 2007, it was estimated that almost half of Tajikistan's labor force worked abroad. Kyrgyzstan also remains a very poor country, its economy heavily dependent on agriculture and gold

mining. Although Kyrgyzstan pushed through a number of market-oriented reforms in the 1990s and is the only former Soviet Central Asian country to have joined the World Trade Organization, it has experienced little sustained economic growth.

Mongolia, formerly a communist ally of the Soviet Union, also suffered an economic collapse in the 1990s, after Soviet subsidies were eliminated. Conditions worsened from 2000 until 2002, when a combination of severe winters and summer droughts devastated livestock, the traditional mainstay of the Mongolian economy. Mongolia does, however, possess vast reserves of copper, gold, and other minerals, and its economy boomed between 2004 and 2008, as foreign capital flowed in. But mineral prices dropped rapidly during the global economic crisis of 2008–2009, undermining Mongolia's recent economic progress.

**The Economy of Tibet and Xinjiang in Western China** Unlike the rest of the region, the Chinese portions of Central Asia did not experience an economic crash in the 1990s. China as a whole had one of the world's fastest-growing economies, although its centers of economic dynamism are all located in the coastal zone. But as China had been much poorer than the Soviet Union before 1991, it is not surprising that poverty in Chinese Central Asia remains widespread.

Tibet in particular remains impoverished with its economy dominated by subsistence agriculture. But most Tibetans are at least able to provide for most of their own basic needs, and their environment does not suffer from the overcrowding that strains much of China proper. China, moreover, has been rapidly building roads and railroads into Tibet, greatly expanding tourism in the process. According to official statistics, the Tibetan economy grew at a rate of 12 percent per year from 2000 to 2006. China exempts Tibet from most taxation, while heavily subsidizing its local government. Tibetan critics, however, claim that most of the region's recent economic gains have gone to Han Chinese immigrants rather than to indigenous Tibetans (Figure 10.27).

Xinjiang has tremendous mineral wealth, including China's largest oil reserves. Its agricultural sector is highly productive although limited in extent. Many of the local Muslim peoples believe that the Chinese state and Han Chinese migrants are taking the region's wealth. China is currently building major road and rail links from eastern China to Xinjiang through its official Western Development strategy. While these infrastructural projects are bringing economic benefits to the region, they are also encouraging migration from eastern China.

**The Misery of Afghanistan**    Although rich in natural gas and mineral resources, Afghanistan is a deeply impoverished country. Foreign aid and the export of hand-woven rugs have formed its legitimate economic mainstays. Official statistics indicated that the Afghan economy was finally growing at a decent pace just before the global economic crisis of 2008–2009, but war, corruption, crime, and poor infrastructure prevented most parts of the country from experiencing any real gains.

By the late 1990s, Afghanistan had emerged as the leading producer of one major commodity for the global market: narcotics. More than 90 percent of the world's opium, used to produce heroin, is grown in Afghanistan. In much of southern and western Afghanistan, opium is the main cash crop. Not only is it highly profitable and easy to transport, but it also uses relatively little water. As much as $100 million

a year in narcotics profits have supposedly gone to the Taliban, which operates mainly in the opium-growing areas of the country. As a result, both NATO and the Afghan government have tried to convince villagers to grow alternative crops. In May 2009, however, top U.S. military officials admitted that their efforts had been unsuccessful, as the opium harvest continued to expand.

In much of northern Afghanistan, opium growing has been largely eliminated. To replace lost income, many local farmers have turned to growing marijuana and making hashish. Marijuana is also grown in other parts of the country as well. In 2008, Afghan officials discovered 260 tons of hashish hidden in ditches in southern Afghanistan, in what some experts called the largest drug bust in world history.

**Central Asian Economies in Global Context**    Many foreign countries are drawn to Central Asia by its oil and natural gas deposits. Major oil companies have established significant operations in Azerbaijan and Kazakhstan, which are thought to contain some of the world's largest undeveloped fossil fuel deposits open to Western firms. But for all of the region's fossil fuel fields to be economically viable, vast pipeline systems must be constructed. The inadequate preexisting pipelines mostly pass through Russia, which charges high transit fees. Seeking an alternative route, an international consortium pushed through the massive Baku-Tbilisi-Ceyhan pipeline, which transports oil from the Caspian Basin to a Turkish port on the Mediterranean Sea. Opened in 2006, at a cost of $3.9 billion, it is the world's second longest oil pipeline (Figure 10.28).

More recently, energy development in Central Asia has focused more on natural gas pipelines. The United States and the European Union have supported the Nabucco plan that would transport natural gas from the Caspian Basin through Turkey and hence into central Europe. Construction of this $10 billion project is scheduled to begin in 2010. Russian officials fear that Nabucco is aimed at turning Central Asia away from Russia and toward the West. In response, Russia agreed in 2008 to increase its payments to Turkmenistan for the gas its transports through its own pipeline system. When natural gas prices plummeted in early 2009, the Russian gas company Gazprom therefore found itself losing money on its Central Asian ventures.

China is equally interested in Central Asia's natural resources, which is one reason it has been a strong supporter of the Shanghai Cooperation Organization. Connections to China were enhanced in 2006, with the opening of the Kazakhstan–China oil pipeline. In 2008, construction began on the ambitious Central Asia–China Gas pipeline, which will link wells in Turkmenistan, Uzbekistan, and Kazakhstan to the China market. China has also invested heavily in infrastructure, as Kazakhstan in particular hopes to replace its decaying transportation system with a "new Silk Road." In 2008, China agreed to a $2.8 billion deal with Afghanistan to expand the country's copper mines.

## Social Development in Central Asia

Social conditions in Central Asia vary more than economic conditions. Afghanistan falls at the bottom of the scale regarding almost every aspect of social development, but in the former Soviet territories, levels of health and education are still relatively high. Social conditions have improved recently in Tibet and Xinjiang, although they have probably not kept pace with the progress made in China proper.

**FIGURE 10.27 Chinese Business in Lhasa, Tibet**    An old Tibetan woman peers into a hair salon in Lhasa, Tibet, in 2006. As Han Chinese immigrants stream into Tibet's cities, rapid cultural change and economic modernization threaten to leave many Tibetans behind.

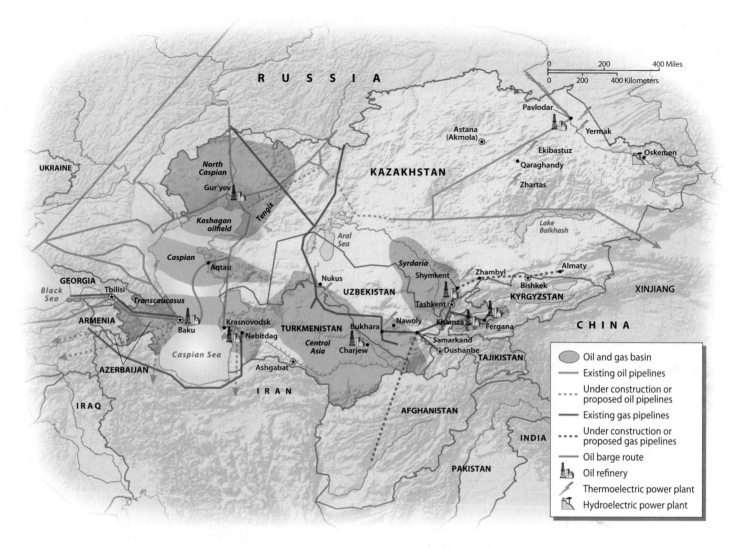

**FIGURE 10.28 Oil and Gas Pipelines** Central Asia has some of the world's largest oil and gas deposits and recently has emerged as a major center for drilling and exploration. Because of its landlocked location, Central Asia cannot easily export its petroleum products. Pipelines have been built to solve this problem, and a number of others are currently being planned. Pipeline construction is a contentious issue, however, as several of the potential pathways lie across Iran, a country that remains under U.S. sanctions.

### Social Conditions and the Status of Women in Afghanistan

Afghanistan's average life expectancy is a mere 44 years, one of the lowest figures in the world. Infant and childhood mortality levels remain extremely high. Not only does Afghanistan endure constant warfare, but its rugged topography hinders the provision of basic social and medical services. Illiteracy is commonplace, especially for women. Afghanistan's 12 percent adult female literacy rate is reportedly the lowest in the world (Table 10.2).

Women in traditional Afghan society—and especially in Pashtun society—have very little freedom. Restrictions on female activities intensified in the 1990s, when the Taliban gained power over most of Afghanistan. Taliban forces prevented women from working, from attending school, and often even from obtaining medical care. After the Taliban lost power, its fighters continued to target female education, destroying schools for girls throughout southern Afghanistan.

Under Afghanistan's new constitution, established in 2004, 25 percent of parliamentary seats are reserved for women. Much evidence, however, indicates that the social position of women has improved little outside the capital city of Kabul. In 2009, the

Afghan government pushed through a law aimed at the county's Shiite minority that would give husbands legal control over their wives' bodies. More than 300 female activists protested this new rule, facing insults as well as physical attacks (Figure 10.29). The government later changed the law, but it did so largely in answer to international objections.

### Social Conditions in the Former Soviet Area

Women in the former Soviet republics of Central Asia enjoy a much higher social position than those of Afghanistan. Traditionally, women had much more autonomy among the pastoral peoples (especially the Kazak) than among the Uzbek and Tajik farmers and city-dwellers, but under Soviet rule the position of women everywhere improved. During the Soviet period, women's educational rates were comparable to those of men, and women were well represented in the workplace. Much evidence indicates, however, that the position of women has declined since the fall of the Soviet Union.

Overall, social indicators such as literacy, average life expectancy, and levels of education remain relatively high in the former Soviet

### TABLE 10.2   DEVELOPMENT INDICATORS

| Country | GNI per capita, PPP 2007 | GDP Average Annual % Growth 2000–07 | Human Development Index (2006)# | Percent Population Living Below $2 a Day | Life Expectancy* 2009 | Under Age 5 Mortality Rate 1990 | Under Age 5 Mortality Rate 2007 | Gender Equity 2007 |
|---|---|---|---|---|---|---|---|---|
| Afghanistan | | 10.7 | | | 44 | | | |
| Azerbaijan | 6,570 | 17.6 | 0.758 | <2 | 72 | 98 | 39 | |
| Kazakhstan | 9,600 | 10.0 | 0.807 | 17.2 | 67 | 60 | 32 | 99 |
| Kyrgyzstan | 1,980 | 4.1 | 0.694 | 51.9 | 68 | 74 | 38 | 100 |
| Mongolia | 3,170 | 7.5 | 0.720 | 49.0 | 65 | 98 | 43 | 107 |
| Tajikistan | 1,710 | 8.8 | 0.684 | 50.8 | 67 | 117 | 67 | 89 |
| Turkmenistan | 4,350 | | 0.728 | 49.6 | 65 | 99 | 50 | |
| Uzbekistan | 2,430 | 6.2 | 0.701 | 76.7 | 68 | 74 | 41 | 98 |

Source:  World Bank, *World Development Indicators, 2009.*
United Nations, *Human Development Index, 2008.* #
Population Reference Bureau, *World Population Data Sheet, 2009*\*
Gender Equity – Ratio of female to male enrollments in primary and secondary school. Numbers below 100 have more males in primary/secondary school, numbers above 100 have more females in primary/secondary schools.

zone. The social successes of western Central Asia reflect investments made during the Soviet period. In much of the region, educational and health facilities have deteriorated in recent years. The situation is especially grim in Tajikistan, where classrooms typically hold 40 to 50 students, classes are held for less than 3 hours a day, and teachers earn a maximum of $52 a month (Figure 10.30).

**FIGURE 10.29  Afghan Women Protesting Proposed Law**   Afghan women in 2009 protest against a new law that would give the men of the country's Shiite minority control over their wives' bodies. The roughly 300 women who joined in the protest were pelted with stones by a group of men who were marching in favor of the law.

**FIGURE 10.30  School in Tajikistan**   Tajikistan has failed to maintain the educational system that it inherited from the Soviet Union. In this photo, students are seen wearing jackets to keep warm in an unheated classroom.

# SUMMARY

■ Central Asia, long dominated by Russia and China, has recently reappeared on the map of the world. Environmental problems, among others, have brought the region to global attention. The destruction of the Aral Sea is one of the world's worst environmental disasters. Desertification has been devastating in many areas, and the booming oil and gas industries have created environmental disasters.

■ Large movements of people in Central Asia have attracted global attention. The biggest issue at present is the migration of Han Chinese into Tibet and Xinjiang, which is threatening to turn local peoples into minority groups. The migration of Russian speakers out of the former Soviet areas of Central Asia is another important demographic issue. In Afghanistan, war and continuing chaos have generated large refugee populations.

■ Religious tension has recently become a cultural issue through much of western Central Asia. Radical Islamic fundamentalism remains a potent force in southern and western Afghanistan and, to a somewhat lesser extent, in the Fergana Valley. Although non-radical forms of Islam are prevalent through most of the region, Central Asian leaders have used the threat of religiously inspired violence to maintain repressive and anti-democratic policies.

■ While China maintains a firm grip on Tibet and Xinjiang, the rest of Central Asia has emerged as a key area of geopolitical competition. Russia, the United States, China, Iran, Pakistan, and India contend for influence. Central Asian countries have attempted to play these powers off against each other in order to maintain their power. Political structures through most of the region remain authoritarian, and in recent years a number of Central Asian countries have developed closer relations with China and Russia.

■ The economies of Central Asia are gradually opening up to global connections, largely because of the substantial fossil fuel reserves in this region. However, Central Asia is likely to face serious economic difficulties for some time, especially because the region is not a significant participant in global trade and has attracted little foreign investment outside the oil industry. Production of drugs remains a serious problem in Afghanistan.

# KEY TERMS

alluvial fan (p. 292)
Collective Security Treaty
   Organization (CSTO) (p. 302)
loess (p. 292)

pastoralist (p. 291)
Shanghai Cooperation
   Organization (SCO) (p. 302)

steppe (p. 287)
theocracy (p. 298)
transhumance (p. 291)

# THINKING GEOGRAPHICALLY

1. To what extent will Central Asia's continental location be a disadvantage in the years to come? Is coastal access truly significant in determining a country's competitive position in the global economy?

2. Which connections with the rest of the world will prove most important for Central Asia in the years to come—those based on cultural features, economic ties, or geopolitical connections?

3. Is China's control over Tibet a legitimate concern of U.S. foreign policy? Should the United States attempt to influence Chinese policy in the area?

4. How might the disappearance of Central Asia's great lakes best be addressed? Is this a topic of international concern, or should the governments of Central Asia assume full responsibility themselves?

5. What economic alternatives to drug production might be found for Afghanistan? Can the international community help the government of Afghanistan address this problem?

 Log in to **www.mygeoscienceplace.com** for videos, interactive maps, RSS feeds, case studies, and self-study quizzes to enhance your study of Central Asia.

# 11 EAST ASIA

## GLOBALIZATION AND DIVERSITY

Japan, South Korea, Taiwan, and the coastal regions of China are among the most globalized regions of the world, whereas North Korea is probably the world's most isolated country.

## ENVIRONMENTAL GEOGRAPHY

China has long experienced severe deforestation and soil erosion, and its current economic boom is generating some of the worst pollution problems in the world.

## POPULATION AND SETTLEMENT

China is currently undergoing a major transformation, as tens of millions of peasants move from impoverished villages in the interior to booming cities in the coastal region.

## CULTURAL COHERENCE AND DIVERSITY

Despite the presence of several unifying cultural features, East Asia in general and China in particular are divided along several striking cultural lines.

## GEOPOLITICAL FRAMEWORK

Both Korea and China remain divided nations, with Taiwan pitted against the People's Republic of China and South Korea set in opposition against North Korea.

## ECONOMIC AND SOCIAL DEVELOPMENT

Over the past several decades, East Asia has emerged as a core area of the world economy, with China experiencing one of the most rapid economic expansions the world has ever seen.

*A Chinese worker examines progress on the massive Three Gorges Dam built across the Yangtze River. This single hydroelectric facility provides roughly three percent of China's total electricity demand.*

East Asia, composed of China, Japan, South Korea, North Korea, and Taiwan (Figure 11.1), is the most heavily populated region of the world. China alone is inhabited by more than 1.3 billion people, more than live in any other region of the world except South Asia. Although East Asia is historically unified by cultural features, in the second half of the 20th century, it was politically divided, with the capitalist economies of Japan, South Korea, Taiwan, and Hong Kong separated from the communist bloc of China and North Korea. Distinct differences in levels of economic development also persisted. As Japan became a leader in the global economy, much of China remained extremely poor.

By the start of the 21st century, however, divisions within East Asia had been reduced. While China is still governed by the Communist Party, it is now on a path of capitalist development. Ties between its booming coastal zone and Japan, South Korea, and Taiwan have quickly strengthened. Although relations between North Korea and South Korea are still hostile, East Asia as a whole has seen a reduction in political tensions (see "Setting the Boundaries").

East Asia is one of the core areas of the world economy, and it is emerging as a center of political power as well. Japan, South Korea, and Taiwan are among the world's key trading states. Tokyo, Japan's largest city, stands alongside New York and London as one of the financial centers of the globe. China has more recently emerged as one of the key global trading powers. China's coastal zone is now closely tied to global networks of information, commerce, and entertainment.

# Environmental Geography: Resource Pressures in a Crowded Land

The most serious environmental issues in East Asia are found in China. China's environmental problems stem from its large population, its rapid industrial development, and the unique features of its physical geography (Figure 11.2). One of China's most controversial environmental actions has been the building of the Three Gorges Dam on the Yangtze River.

## Flooding and Dam Building in China

The Yangtze River (also called the Chang Jiang) is one of the most important physical features of East Asia. This river, the third largest (by volume) in the world, emerges from the Tibetan highlands onto the rolling lands of the Sichuan Basin, passes through a magnificent canyon in the Three Gorges area (Figure 11.3), and then meanders across the lowlands of central China before entering the sea in a large delta near the city of Shanghai. The Yangtze has historically been the main transportation corridor into the interior of China, and it has long been famous in Chinese literature for its beauty and power.

### The Three Gorges Controversy
The Chinese government is trying to control the Yangtze for two main reasons: to prevent flooding and to generate electricity. To do so, it built a series of large dams, the largest of which is a massive structure in the Three

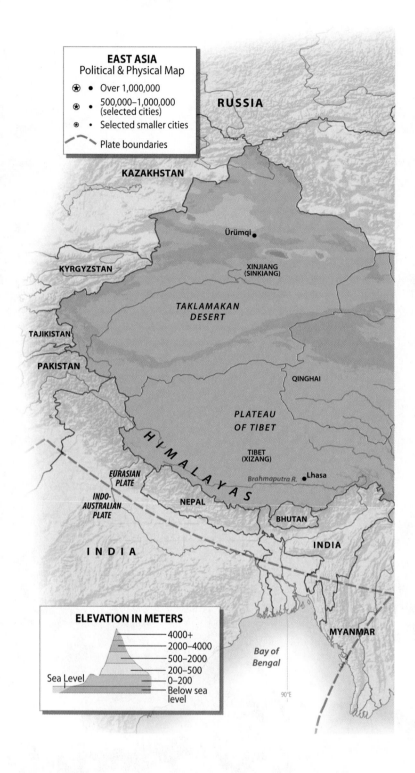

**EAST ASIA**
Political & Physical Map
- ✸ • Over 1,000,000
- ✸ • 500,000–1,000,000 (selected cities)
- ✸ • Selected smaller cities
- --- Plate boundaries

RUSSIA
KAZAKHSTAN
Ürümqi
KYRGYZSTAN
XINJIANG (SINKIANG)
TAKLAMAKAN DESERT
TAJIKISTAN
PAKISTAN
QINGHAI
PLATEAU OF TIBET
HIMALAYAS
TIBET (XIZANG)
EURASIAN PLATE
Brahmaputra R. • Lhasa
INDO-AUSTRALIAN PLATE
NEPAL
BHUTAN
INDIA
INDIA
MYANMAR
Bay of Bengal
90°E

**ELEVATION IN METERS**
- 4000+
- 2000–4000
- 500–2000
- 200–500
- 0–200
- Below sea level
Sea Level

Gorges area, completed in 2006. This $39 billion structure is the largest hydroelectric dam in the world, forming a reservoir 350 miles long. It has jeopardized several endangered species (including the Yangtze River dolphin), flooded a major scenic attraction, and displaced more than 1 million persons.

The Three Gorges Dam generates large amounts of electricity, supplying roughly 3 percent of China's demand. As China industrializes, its

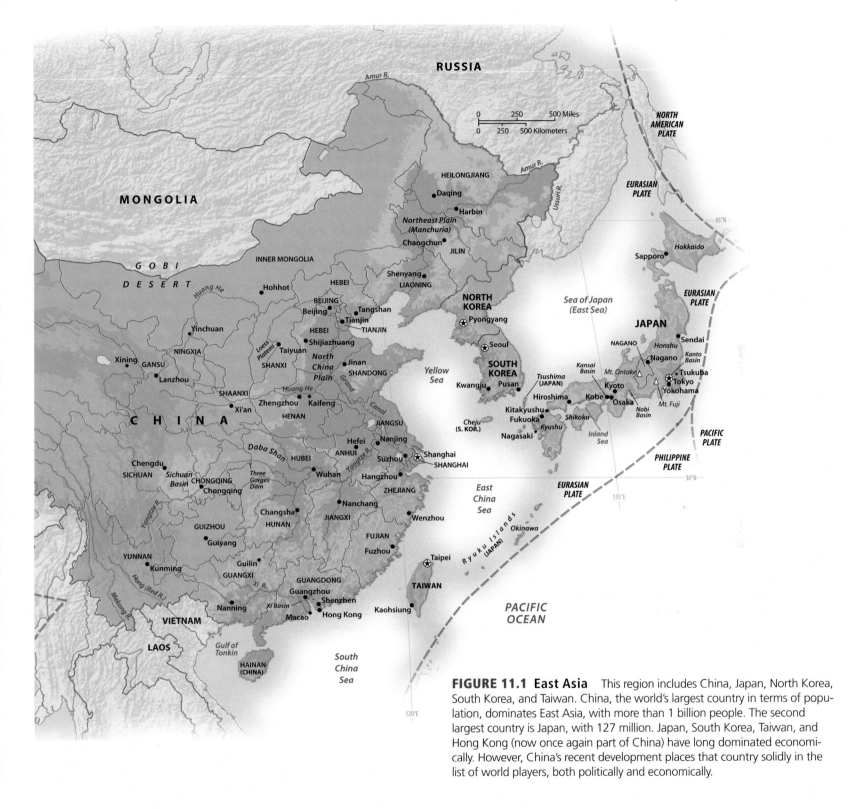

**FIGURE 11.1 East Asia**   This region includes China, Japan, North Korea, South Korea, and Taiwan. China, the world's largest country in terms of population, dominates East Asia, with more than 1 billion people. The second largest country is Japan, with 127 million. Japan, South Korea, Taiwan, and Hong Kong (now once again part of China) have long dominated economically. However, China's recent development places that country solidly in the list of world players, both politically and economically.

need for power is increasing rapidly. At present, nearly four-fifths of China's total energy supply comes from burning coal, which results in severe air pollution. But while the Three Gorges Dam may reduce air pollution somewhat, most environmentalists argue that the costs will exceed the benefits. Chinese government planners disagree, arguing that the Three Gorges Dam will both provide necessary electricity and reduce the threat of flooding that endangers the lower and middle stretches of

the Yangtze Valley. Dam-building on the Huang He River in northern China, however, has not been able to solve that region's water control problems.

**Flooding in Northern China**   The North China Plain has historically suffered from both drought and flooding. This area is dry most of the year yet often experiences heavy downpours in summer. Since

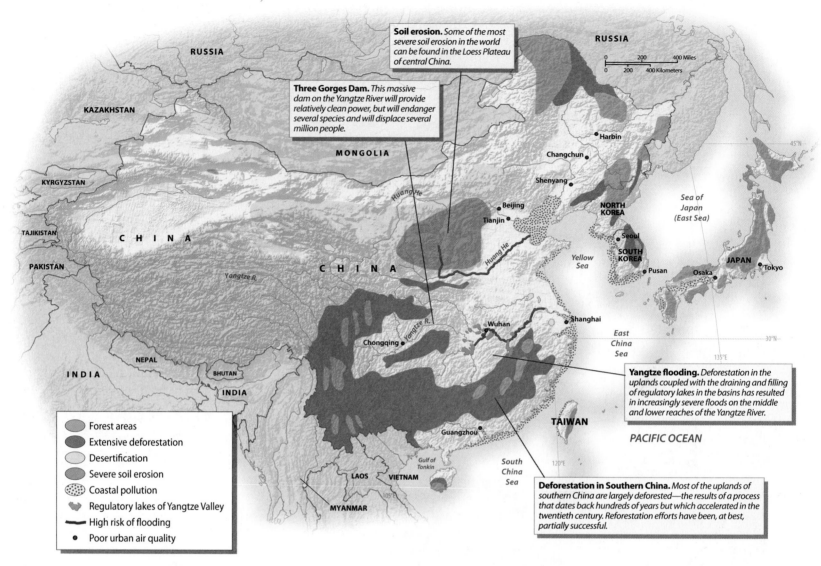

**Soil erosion.** *Some of the most severe soil erosion in the world can be found in the Loess Plateau of central China.*

**Three Gorges Dam.** *This massive dam on the Yangtze River will provide relatively clean power, but will endanger several species and will displace several million people.*

**Yangtze flooding.** *Deforestation in the uplands coupled with the draining and filling of regulatory lakes in the basins has resulted in increasingly severe floods on the middle and lower reaches of the Yangtze River.*

**Deforestation in Southern China.** *Most of the uplands of southern China are largely deforested—the results of a process that dates back hundreds of years but which accelerated in the twentieth century. Reforestation efforts have been, at best, partially successful.*

Legend:
- Forest areas
- Extensive deforestation
- Desertification
- Severe soil erosion
- Coastal pollution
- Regulatory lakes of Yangtze Valley
- High risk of flooding
- Poor urban air quality

**FIGURE 11.2 Environmental Issues in East Asia** This huge world region has been almost completely transformed from its natural state and continues to have serious environmental problems. In China, some of the most pressing environmental issues involve deforestation, flooding, water control, and soil erosion.

# Setting the Boundaries

East Asia is easily indicated on the map of the world by the territorial extent of its countries: China, Taiwan, North Korea, South Korea, and Japan (see Figure 11.1). Japan is composed of four main islands and also includes a chain of much smaller islands (the Ryukyus) that extends almost to Taiwan. Taiwan is basically a single-island country, but its political status is confusing because China claims it as part of its own territory. China itself is a vast continental state that reaches well into Central Asia. It also includes the large island of Hainan in the South China Sea. Korea forms a compact peninsula between

northern China and Japan, but it is divided into two independent states: North Korea and South Korea.

In political terms, such a straightforward definition of East Asia is appropriate. If one turns to cultural considerations, however, the issue is more complicated. The main cultural problem arises with regard to the western half of China. This is a huge but lightly populated space; some 95 percent of the residents of China live in the east (see Figure 11.15). By certain criteria, only the eastern half of China (often referred to as **China proper**) fits well into the East Asian world region. The native people of western

China are not Chinese by culture and language, and they have never accepted the religious and philosophical beliefs that have given historical unity to East Asian civilization. In the northwestern portion of China, called *Xinjiang* in Chinese, most native residents speak Turkish languages and are Muslim in religion. Tibet, in southwestern China, has a highly distinctive culture, and the Tibetans in general resent Chinese authority. Xinjiang and Tibet can thus be classified within Central Asia, which is covered in Chapter 10. In this chapter, we will examine western China only to the extent that it is politically part of China.

**FIGURE 11.3 The Three Gorges of the Yangtze** The spectacular Three Gorges landscape of the Yangtze River is the site of a controversial flood-control dam that has displaced over 1 million people. Not only are the human costs high to those displaced, but there may also be significant ecological costs to endangered aquatic species.

ancient times, levees and canals have both controlled floods and allowed irrigation. But no matter how much effort has been put into water control, disastrous flooding has never been completely prevented (Figure 11.4).

The worst floods in northern China are caused by the Huang He River, or Yellow River, which cuts across the North China Plain. As a result of upstream erosion, the Huang He carries a huge **sediment load** (the amount of suspended clay, silt, and sand in the water), making it the muddiest major river in the world. When the river enters the low-lying plain, its velocity slows, and its sediments begin to settle and accumulate in the riverbed. As a result, the level of the river gradually rises above that of the surrounding lands. Eventually, it must break free of its course to find a new route to the sea over lower-lying ground.

In the period of recorded history, the Huang He has changed course 26 times, usually causing much loss of human life. Since ancient times, the Chinese have attempted to keep the river within its banks by building ever-larger dikes. Eventually, however, the riverbed rises so high that the flow can no longer be contained. The resulting floods have been known to kill several million people in a single episode. While the river has not changed its course since the 1930s, most geographers agree that another change is inevitable.

**Erosion on the Loess Plateau** The Huang He's sediment load comes from the eroding soils of the Loess Plateau, located to the west of the North China Plain. **Loess** is a fine, windblown material that was deposited on this upland area during the last ice age. In this area, loess deposits accumulated to depths of up to several hundred feet. Loess forms fertile soil, but it washes away easily when exposed to running water. Cultivation requires plowing, which leads to soil erosion. As the population of the region gradually increased, areas of woodland and grassland diminished, leading to ever-greater rates of soil loss. As

the erosion process continued, great gullies cut across the plateau, steadily reducing the extent of productive land.

Today the Loess Plateau is one of the poorest parts of China. Population is only moderately dense by Chinese standards, but good farmland is limited, and drought is common. The Chinese government encourages the construction of terraces to conserve the soil, but such efforts have not been effective everywhere. Campaigns to plant woody vegetation to help stabilize the most severely eroded lands have been even less successful, as the young plants seldom survive the area's harsh climate.

## Other East Asian Environmental Problems

Although the problems associated with the Yangtze and Huang He rivers are among the most well-known environmental issues in East Asia, they are hardly the only ones. Pollution and deforestation also affect much of the region.

**Forests and Deforestation** Most of the uplands of China and North Korea support only grass, scrub vegetation, and stunted trees (Figure 11.5). China lacks the historical tradition of forest conservation that characterizes Japan. In much of southern China, sweet potatoes, maize, and other crops have been grown on steep and easily eroded hillsides for several hundred years. After centuries of exploitation, many upland areas have lost so much soil that they cannot easily support forests.

Although the Chinese government has started large-scale reforestation programs, few have been very successful. Today, substantial forests are found only in China's far north, where a cool climate prevents fast growth, and along the eastern slopes of the Tibetan Plateau, where rugged terrain restricts commercial forestry. As a result, China suffers a severe shortage of forest resources. As its economy continues to boom, China has become a major importer of lumber, pulp, and paper.

**FIGURE 11.4 Flooding on the North China Plain** Major floods on the Huang He River have occurred throughout history, sometimes inundating large sections of the North China Plain. At other times, severe droughts affect this region. Extensive dikes have been built along much of the river to protect the countryside from flooding, as seen in this photo taken near the historical city of Kaifeng.

**FIGURE 11.5 Denuded Hillslopes in China** Because of the need to clear forests for wood products and agricultural lands, many of China's mountain slopes have long been deforested. Without tree cover, soil erosion is a serious issue.

**Mounting Pollution** As China's industrial base expands, other environmental problems are growing increasingly severe. The burning of high-sulfur coal has resulted in serious air pollution, which is made worse by the increasing number of automobiles. According to one recent study, 16 of the world's 20 most polluted cities are located in China (Figure 11.6). Air pollution results in serious respiratory problems for many Chinese city dwellers. According to a World Health Organization (WHO) report, air pollution kills 656,000 Chinese citizens each year, while another 95,000 die from polluted drinking water.

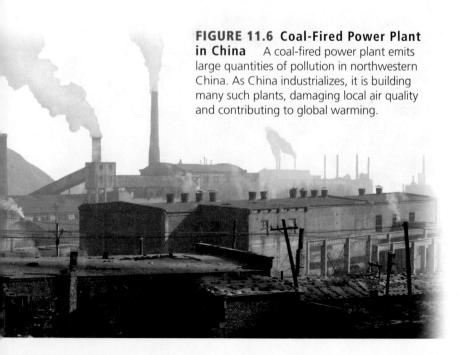

**FIGURE 11.6 Coal-Fired Power Plant in China** A coal-fired power plant emits large quantities of pollution in northwestern China. As China industrializes, it is building many such plants, damaging local air quality and contributing to global warming.

**Environmental Issues in Japan** Considering its large population and intensive industrialization, Japan's environment is relatively clean. Actually, the very density of its population gives certain environmental advantages. In crowded cities, for example, travel by private automobile is often slower than travel on subways. In the 1950s and 1960s, Japan's most intensive period of industrial growth, the country suffered from some of the world's worst water and air pollution. Soon afterward, the Japanese government passed strict environmental laws.

Japan's environmental cleanup was aided by its location because winds usually carry smog-forming chemicals out to sea. Equally important has been the phenomenon of **pollution exporting**. Because of Japan's high cost of production and its strict environmental laws, many Japanese companies have relocated their dirty factories to other areas, especially China and Southeast Asia. This practice, which is also followed by the United States and western Europe, means that Japan's pollution is partially displaced to poorer countries.

## Global Warming and East Asia

East Asia has come to occupy a central position in global warming debates, largely because of China's extremely rapid increase in carbon emissions. From levels only about half those of the United States in 2000, China's total production of greenhouse gases surpassed those of the United States in 2007. This staggering rise has been caused both by China's explosive economic growth and by the fact that it relies on burning coal to generate most of its electricity.

The potential effects of global warming in China have serious implications globally. According to one report, the country's production of wheat, corn, and rice could fall by as much as 37 percent if average temperatures increase by 3.5 to 5.5°F (2 or 3°C) over the next 50 to 80 years. Increased evaporation rates could greatly intensify the water shortages that already plague much of northern China. In the wet zones of southern China, global warming concerns center on the probability of more intense storms in this already flood-prone region.

In June 2007, China released its first national plan on climate change, which calls for a 20 percent gain in energy efficiency by 2010. China's new strategy also includes a major expansion of nuclear power, the increased use of renewable energy sources, and the continuation of ambitious reforestation efforts. While environmentalists were pleased to see China addressing these issues, most feel that the country has not gone nearly far enough to curb its extraordinarily fast rise in greenhouse gas emissions.

China's contribution to the causes of global warming has tended to overshadow those of other East Asian countries. Japan, South Korea, and Taiwan are all major emitters of greenhouse gases but all also have energy-efficient economies, releasing far less carbon than the United States on a per capita basis. Japanese high-tech companies, like those of South Korea and Taiwan, also realize that it could be very good business to develop new energy technologies to reduce the threat of climate change.

## East Asia's Physical Geography

An examination of the physical geography of East Asia helps shed light on the environmental issues it faces. East Asia is situated in the same general latitudinal range as the United States, although it extends considerably farther north and south. The northernmost tip of China lies as far north as central Quebec, while China's southernmost point is at the same latitude as Mexico City. The climate of southern China is thus roughly comparable to that of the Caribbean, while the climate of northern China is similar to that of south-central Canada (Figure 11.7).

Much of East Asia is geologically active, with earthquakes and volcanic eruptions occurring frequently. Such geological hazards are most common in Japan but are often more destructive in China, due to lax construction standards. The May 2008 Sichuan earthquake in central China resulted in roughly 70,000 deaths and left over 4 million people homeless (Figure 11.8). Many children died when poorly built schools collapsed. China's government responded by releasing $150 billion in reconstruction funds and promising to improve construction standards.

**Japan's Physical Environment**   Although slightly smaller than California, Japan is more elongated, extending farther north and south. As a result, Japan's extreme south, in southern Kyushu and the Ryukyu

**FIGURE 11.7 Climate Map of East Asia**   East Asia is located in roughly the same latitudinal zone as North America, so there are climatic parallels between the two world regions. The northernmost tip of China lies at about the same latitude as Quebec and shares a similar climate, whereas southern China approximates the climate of Florida. In Japan, maritime influences produce a milder climate.

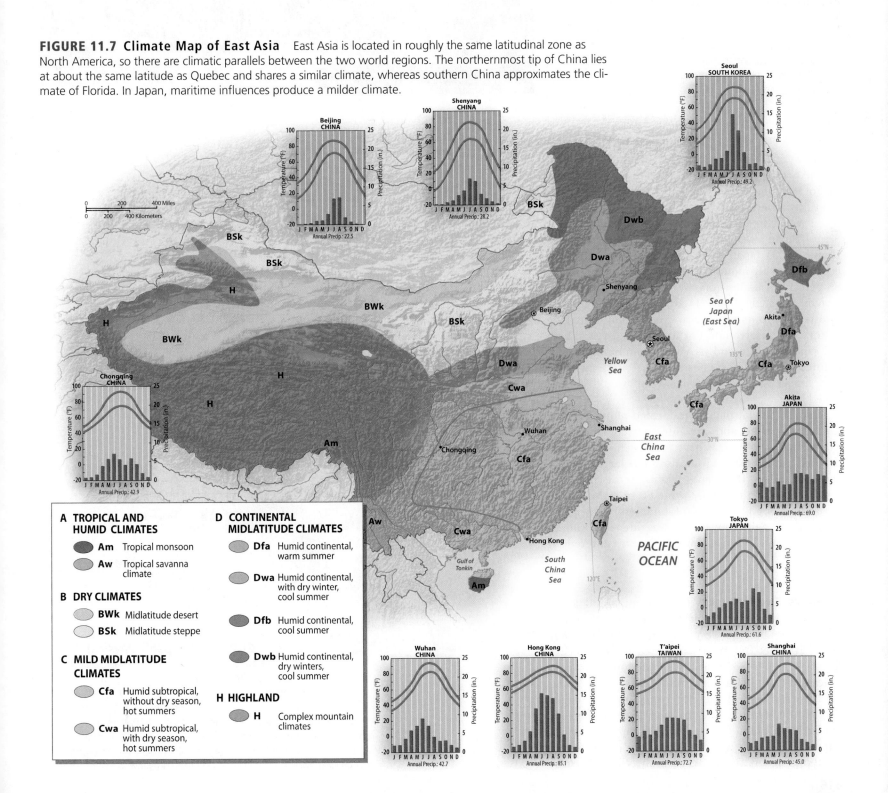

**A TROPICAL AND HUMID CLIMATES**

- **Am** Tropical monsoon
- **Aw** Tropical savanna climate

**B DRY CLIMATES**

- **BWk** Midlatitude desert
- **BSk** Midlatitude steppe

**C MILD MIDLATITUDE CLIMATES**

- **Cfa** Humid subtropical, without dry season, hot summers
- **Cwa** Humid subtropical, with dry season, hot summers

**D CONTINENTAL MIDLATITUDE CLIMATES**

- **Dfa** Humid continental, warm summer
- **Dwa** Humid continental, with dry winter, cool summer
- **Dfb** Humid continental, cool summer
- **Dwb** Humid continental, dry winters, cool summer

**H HIGHLAND**

- **H** Complex mountain climates

**FIGURE 11.8 2008 Sichuan Earthquake**
The ruins of Ying Xiu primary school, destroyed in the May, 2008 Sichuan earthquake, are visible in this photograph. According to local parents, roughly 400 of the school's 600 students perished when the buildings collapsed. Government sources claim that the casualty figures were much lower.

Archipelago, is subtropical, while northern Hokkaido is almost subarctic. Most of the country, however, is distinctly temperate. The climate of Tokyo is similar to that of Washington, DC—although Tokyo receives significantly more rain.

The Pacific coast of Japan is separated from the Sea of Japan coast by a series of mountain ranges (Figure 11.9). Japan is one of the world's most rugged countries, with mountainous terrain covering some 85 percent of its territory. Most of these uplands are heavily forested (Figure 11.10). Japan owes its lush forests both to its mild, rainy climate and to its long history of conservation. For hundreds of years, both the Japanese state and its village communities have enforced strict forest-conservation rules, ensuring that timber and firewood extraction would be balanced by tree growth.

Small alluvial plains are located along parts of Japan's coastline and are interspersed among its mountains. These areas have long been cleared and drained for intensive agriculture. The largest Japanese lowland is the Kanto Plain to the north of Tokyo, but even it is only some 80 miles wide and 100 miles long (130 by 160 kilometers). The country's other main lowland basins are the Kansai, located around Osaka, and the Nobi, centered on Nagoya.

**Taiwan's Environment** Taiwan, an island about the size of Maryland, sits at the edge of the continental landmass. To the west, the Taiwan Strait is only about 200 feet (60 meters) deep; to the east, ocean depths of many thousands of feet are found 10 to 20 miles (16 to 32 kilometers) offshore.

Taiwan's central and eastern regions are rugged and mountainous, while the west is mainly a lowland alluvial plain. Bisected by the tropic of Cancer, Taiwan has a mild winter climate, but it is often hit by typhoons in the early autumn. Unlike nearby

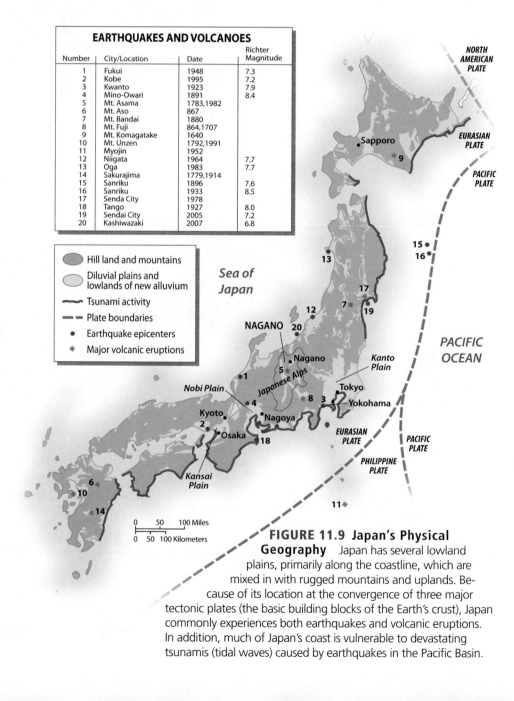

**EARTHQUAKES AND VOLCANOES**

| Number | City/Location | Date | Richter Magnitude |
|--------|---------------|------|-------------------|
| 1 | Fukui | 1948 | 7.3 |
| 2 | Kobe | 1995 | 7.2 |
| 3 | Kwanto | 1923 | 7.9 |
| 4 | Mino-Owari | 1891 | 8.4 |
| 5 | Mt. Asama | 1783,1982 | |
| 6 | Mt. Aso | 867 | |
| 7 | Mt. Bandai | 1880 | |
| 8 | Mt. Fuji | 864,1707 | |
| 9 | Mt. Komagatake | 1640 | |
| 10 | Mt. Unzen | 1792,1991 | |
| 11 | Myojin | 1952 | |
| 12 | Niigata | 1964 | 7.7 |
| 13 | Oga | 1983 | 7.7 |
| 14 | Sakurajima | 1779,1914 | |
| 15 | Sanriku | 1896 | 7.6 |
| 16 | Sanriku | 1933 | 8.5 |
| 17 | Senda City | 1978 | |
| 18 | Tango | 1927 | 8.0 |
| 19 | Sendai City | 2005 | 7.2 |
| 20 | Kashiwazaki | 2007 | 6.8 |

**FIGURE 11.9 Japan's Physical Geography** Japan has several lowland plains, primarily along the coastline, which are mixed in with rugged mountains and uplands. Because of its location at the convergence of three major tectonic plates (the basic building blocks of the Earth's crust), Japan commonly experiences both earthquakes and volcanic eruptions. In addition, much of Japan's coast is vulnerable to devastating tsunamis (tidal waves) caused by earthquakes in the Pacific Basin.

**FIGURE 11.10 Forested Landscapes of Japan** Although much of Japan is heavily forested and supports a wood products industry, the yield is not large enough to satisfy demand. As a result, Japan imports timber extensively from North America, Southeast Asia, and Latin America.

areas of China proper, Taiwan still has extensive forests, especially in its remote central and eastern uplands.

**Chinese Environments** Even if one excludes China's Central Asian provinces, China is a vast country with diverse environmental regions. For the sake of convenience, it can be divided into two main areas: one lying to the north of the Yangtze River Valley and the other including the Yangtze and all areas to the south. As Figure 11.11 shows, each of these can be subdivided into a number of distinct regions.

Southern China is a land of rugged mountains and hills interspersed with broad lowland basins. Large valleys and moderate-elevation plateaus are also found in the far south, where the climate is tropical. The coastal areas in southeast China are rugged and offer limited agricultural opportunities (Figure 11.12). North of the Yangtze Valley, the climate is both colder and drier. Summer rainfall is generally abundant, but the other seasons are usually dry. The North China Plain, a large, flat area of fertile soil crossed by the Huang He (or Yellow) River, is cold and dry in winter and hot and humid in summer. Overall precipitation is somewhat low and unpredictable, and some areas of the North China Plain

**FIGURE 11.11 Landscape Regions of China** The Yangtze Valley divides China into two general areas. Immediately to the north is the large fertile area of the North China Plain, bisected by the Huang He (or Yellow) River. To the west is the Loess Plateau, an upland area of soil derived from wind-deposited silt after the prehistoric glacial period, about 15,000 years ago.

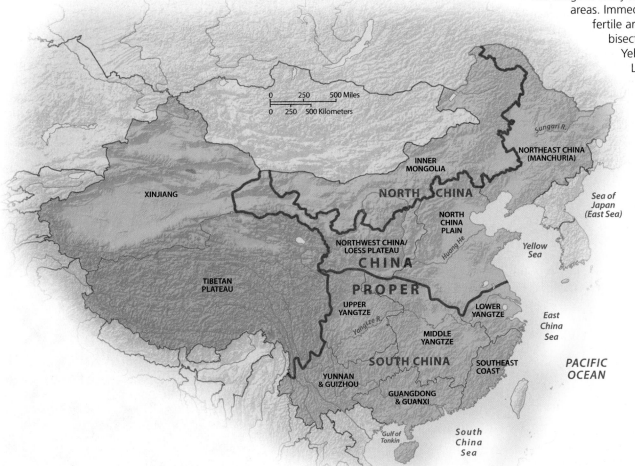

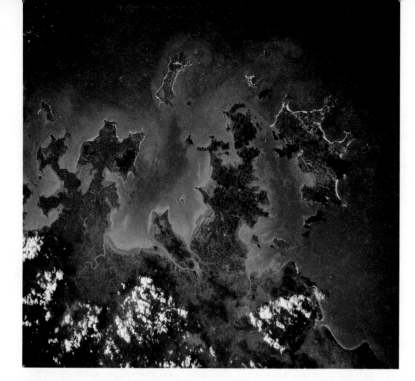

**FIGURE 11.12 The Fujian Coast of China**   In southeast China lies the rugged coastal province of Fujian. Here the coastal plain is narrow and the shoreline deeply indented, producing a charming landscape. Because of limited agricultural opportunities along this rugged coastline, many Fujianese people work in maritime activities.

are in danger of **desertification** (the spread of desert conditions). Seasonal water shortages are growing increasingly severe through much of the region, as withdrawals for irrigation and industry increase (Figure 11.13).

China's northeastern region is historically referred to as "Manchuria" in English, but it is now often called simply Northeast China. Northeast China is dominated by a broad, fertile lowland sandwiched between mountains and uplands stretching along China's borders with North Korea, Russia, and Mongolia. Although winters here can be brutally cold, summers are usually warm and moist. Manchuria's upland areas are home to some of China's best-preserved forests and wildlife refuges.

**Korean Landscapes**   Korea forms a well-defined peninsula, partially cut off from Manchuria by rugged mountains and sizable rivers. The far north, which just touches Russia's Far East, has a climate similar to that of Maine, whereas the southern tip is more similar to the Carolinas. Korea, like Japan, is a mountainous country with scattered alluvial basins. The lowlands of the southern portion of the peninsula are more extensive than those of the north, giving South Korea a distinct agricultural advantage over North Korea. The north, however, has far more abundant natural resources.

## Population and Settlement: A Realm of Crowded Lowland Basins

East Asia is one of the most densely populated regions of the world (Table 11.1). The lowlands of Japan, Korea, and China are among the most intensely used portions of Earth, containing not only the major cities but also most of the agricultural lands of these countries. Although the density of East Asia is extremely high, the region's population growth rate has declined dramatically. In Japan, the current concern is population loss. While China's population is still expanding, its rate of growth is low by global standards.

### Japanese Settlement and Agricultural Patterns

Japan is a highly urbanized country, supporting two of the largest metropolitan areas in the world: Tokyo and Osaka. It is also one of the world's most mountainous countries, and its uplands are lightly populated. Agriculture must therefore share the limited lowlands with cities and suburbs, resulting in extremely intensive farming practices.

**FIGURE 11.13 Drought in Northern China**   The Yellow River (or Huang He) is sometimes called the "cradle of Chinese civilization" owing to its historical importance.  Due to the increasing extraction of water for agriculture and industry, the river now often runs dry in its lower reaches. In this photo, a man is cycling across the dried riverbed in Shandong province.

### TABLE 11.1   POPULATION INDICATORS

| Country | Population (millions) 2009 | Population Density (per square kilometer) | Rate of Natural Increase (RNI) | Total Fertility Rate | Percent Urban | Percent <15 | Percent >65 | Net Migration (Rate per 1000) 2005-10* |
|---|---|---|---|---|---|---|---|---|
| China | 1,331.4 | 139 | 0.5 | 1.6 | 46 | 19 | 8 | −0.3 |
| Hong Kong | 7.0 | 6,403 | 0.5 | 1.1 | 100 | 13 | 13 | 3.3 |
| Japan | 127.6 | 338 | −0.0 | 1.4 | 86 | 13 | 23 | 0.2 |
| North Korea | 22.7 | 188 | 0.5 | 2.0 | 60 | 22 | 9 | 0.0 |
| South Korea | 48.7 | 490 | 0.4 | 1.2 | 82 | 17 | 10 | −0.1 |
| Taiwan | 23.1 | 641 | 0.2 | 1.0 | 78 | 17 | 10 | |

Source: Population Reference Bureau, *World Population Data Sheet, 2009*.

*Net Migration Rate from the United Nations, Population Division, *World Population Prospects: The 2008 Revision Population Database*.

**Japan's Agricultural Lands**   Japanese agriculture is largely limited to the country's coastal plains and interior basins. Rice is Japan's major crop, and irrigated rice demands flat land. Japanese rice farming has long been one of the most productive forms of agriculture in the world, helping to support a large population—some 127 million people—on a relatively small and rugged land. Although rice is grown in almost all Japanese lowlands, the country's premier rice-growing districts lie along the Sea of Japan coast of central and northern Honshu.

Vegetables are also grown intensively in all the lowland basins, even on tiny patches within urban neighborhoods (Figure 11.14). The valleys of central and northern Honshu are famous for their temperate-climate fruit, while citrus comes from the milder southwestern reaches of the country. Crops that thrive in a cooler climate, such as potatoes, are produced mainly in Hokkaido and northern Honshu.

**Settlement Patterns**   All Japanese cities—and the vast majority of the Japanese people—are located in the same lowlands that support the country's agriculture. Many of Japan's more remote areas, moreover, are currently experiencing population loss as younger people increasingly move to larger cities. Not surprisingly, the three largest metropolitan areas—Tokyo, Osaka, and Nagoya—sit near the centers of the three largest plains.

Overall, Japan is a densely populated country (Figure 11.15). The fact that its settlements are largely restricted to the lowlands means that Japan's effective population density—the actual crowding that the country experiences—is one of the highest in the world. This is especially true in the main industrial belt, which extends from Tokyo south and west through Nagoya and Osaka to the northern coast of Kyushu. Due to such space limitations, all Japanese cities are characterized by dense settlement patterns. In the major urban areas, the amount of available living space is highly restricted for all but the most affluent families. Many observers argue that Japan should allow its cities and suburbs to expand into nearby rural areas. However, because most uplands are too steep for residential use, such expansion would have to occur at the expense of agricultural land.

## Settlement and Agricultural Patterns in China, Taiwan, and Korea

Like Japan, Taiwan and Korea are essentially urban. China, however, remains largely rural, with only some 46 percent of its population living in cities. Almost all Chinese cities are growing at a rapid pace, however, despite government efforts to keep their size under control.

**China's Agricultural Regions**   A line drawn just to the north of the Yangtze Valley divides China into two main agricultural regions. To the south, rice is the dominant crop. To the north, wheat, millet, and sorghum are the most common.

In southern and central China, population is highly concentrated in the broad lowlands, which are famous for their fertile soil and intensive agriculture. More than 75 million people, for example, reside in Jiangsu, a province smaller than Ohio. Cropping occurs year-round in most of southern and central China; summer rice alternates with winter barley or vegetables, and in many areas two rice crops as well as one winter crop are grown. Southern China also produces a wide variety of tropical and subtropical crops, and moderate slopes throughout the area produce sweet potatoes, corn, and other upland crops.

In northern China, population distribution is more variable. The North China Plain has long been one of the most thoroughly **anthropogenic landscapes** in the world (that is, a landscape that has been heavily transformed by human activities). Virtually its entire extent is either cultivated or occupied by houses, factories, and

**FIGURE 11.14  Japanese Urban Farm**   Japanese landscapes often combine dense urban settlement with small patches of intensively farmed agriculture. Here a cabbage farm coexists with an urban neighborhood.

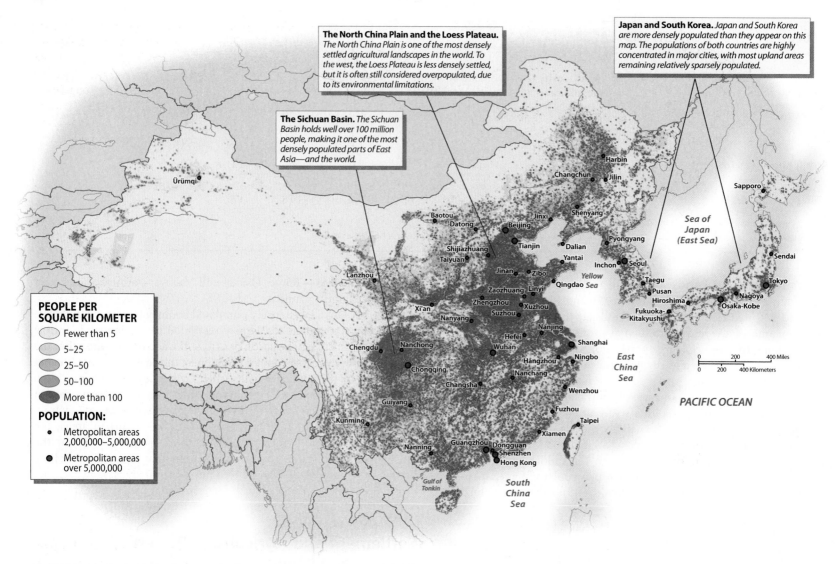

The North China Plain and the Loess Plateau. *The North China Plain is one of the most densely settled agricultural landscapes in the world. To the west, the Loess Plateau is less densely settled, but it is often still considered overpopulated, due to its environmental limitations.*

The Sichuan Basin. *The Sichuan Basin holds well over 100 million people, making it one of the most densely populated parts of East Asia—and the world.*

Japan and South Korea. *Japan and South Korea are more densely populated than they appear on this map. The populations of both countries are highly concentrated in major cities, with most upland areas remaining relatively sparsely populated.*

**PEOPLE PER SQUARE KILOMETER**
- Fewer than 5
- 5–25
- 25–50
- 50–100
- More than 100

**POPULATION:**
- Metropolitan areas 2,000,000–5,000,000
- Metropolitan areas over 5,000,000

**FIGURE 11.15 Population Map of East Asia** Parts of East Asia are very densely settled, particularly in the coastal lowlands of China and Japan. This contrasts with the sparsely settled lands of western China, North Korea, and northern Japan. Although the total population of this world region is high, as is the overall density, the rate of natural population increase has slowed rather dramatically in the past several decades because of several factors, including China's "one-child" policy.

other structures of human society. Manchuria, on the other hand, was a lightly populated frontier zone as recently as the mid-1800s. Today, with a population of more than 100 million, its central plain is thoroughly settled. The Loess Plateau is more thinly settled, supporting only some 70 million residents (Figure 11.16). But considering its aridity and widespread soil erosion, this is a high figure indeed.

**Settlement and Agricultural Patterns in Korea and Taiwan** Korea is densely populated. It contains some 71 million people (22.7 million in North Korea and 48.7 million in South Korea) in an area smaller than Minnesota. South Korea's population density is actually significantly higher than that of Japan. Most Koreans are crowded into the alluvial plains and basins of the west and south. The highland spine, extending from the far north to northeastern South Korea, remains relatively sparsely settled. South Korean agriculture is dominated by rice. North Korea relies heavily on corn and other upland crops that do not require irrigation.

Taiwan is the most densely populated state in East Asia. Roughly the size of the Netherlands, it contains more than 23 million inhabitants. Its overall population density is one of the highest in the world. Because mountains cover most of central and eastern Taiwan, virtually the entire population is concentrated in the narrow lowland belt in the north and west. In this area, large cities and numerous factories are scattered amid lush farmlands.

## East Asian Agriculture and Resources in Global Context

Although East Asian agriculture is highly productive, it cannot feed the huge number of people who live in the region. Japan, Taiwan, and South Korea are major food importers, and China has recently moved in the same direction. Other resources are also being drawn in from all quarters of the world by the powerful economies of East Asia.

**FIGURE 11.16 Loess Settlement**   This photo shows a typical subterranean dwelling carved out of the soft loess sediment in central China. This area is prone to major earthquakes that exact a high toll on the local population in terms of dwelling collapses.

Japan may be self-sufficient in rice, but it is still one of the world's largest food importers. Japan imports much of its meat and the feed used in its domestic livestock industry from the United States, Brazil, Canada, and Australia. These same countries supply soybeans and wheat. Japan has one of the highest rates of fish consumption in the world, and the Japanese fishing fleets must scour the world's oceans to meet the demand. Japan also depends on imports to supply its demand for forest resources. While its own forests produce high-quality cedar and cypress logs, Japan buys most of its construction lumber and pulp (for papermaking) from western North America and Southeast Asia. Japanese interests are increasingly turning to Latin America, Africa, and eastern Russia for lumber and pulp.

South Korea has followed Japan in obtaining food and forestry resources from abroad. In 2008, as global grain prices increased rapidly, large South Korean companies began to negotiate long-term leases for vast tracts of farmland in poor tropical countries. One deal would have given Daewoo Logistics Corporation control over 3.2 million acres of farmland in Madagascar, covering about half of the island's arable land. The project was cancelled, however, when the government of Madagascar collapsed, an event partly brought about by local opposition to the project.

Through the 1980s, China remained self-sufficient in food, despite its huge population and crowded lands. But increasing wealth, combined with changing diets and the introduction of foreign foods, has brought about an increased consumption of meat, which requires large amounts of feed grain. Economic growth has also resulted in a loss of agricultural lands to residential, commercial, and industrial development. As a result, by the end of the 20th century, China had become a net food importer.

Whether China will ever again be self-sufficient in food is an open question, but it is clear that the country's current demands for agricultural and mineral products are reordering patterns of global trade. Many Latin American, African, and Asian countries now rely heavily on the Chinese market. Although demand fell off with the global economic crisis of 2008–2009, China continues to invest heavily in infrastructure, farming, and mining projects in Africa. As a result, some observers fear that some parts of Africa are at risk of becoming virtual economic colonies of China (Figure 11.17).

## Urbanization in East Asia

China has one of the world's oldest urban foundations. In medieval and early modern times, East Asia as a whole possessed a well-developed system of cities that included some of the largest settlements on the planet. In the early 1700s, Tokyo, then called *Edo,* probably overshadowed all other cities, with a population of more than 1 million.

But despite this early urbanization, East Asia was overwhelmingly rural at the end of World War II. Some 90 percent of China's people

**FIGURE 11.17 Chinese Overseas Investments**
China is now investing vast sums of money in many of the world's underdeveloped countries. Critics contend that Chinese investment is aimed merely at securing natural resources, whereas supporters argue that it brings many local benefits, both economic and cultural. In this photograph, Senegalese and Chinese workers observe a ceremony at the site of a new national theater in Dakar, Senegal, financed by the Chinese government.

then lived in the countryside, and even Japan was only about 50 percent urbanized. But as the region's economy began to grow after the war, so did its cities. Japan, Taiwan, and South Korea are now between 78 and 86 percent urban, which is typical for advanced industrial countries. Only about 46 percent of China's people live in cities, but this figure is increasing steadily.

**Chinese Cities**    Traditional Chinese cities were clearly separated from the countryside by defensive walls. Most were planned in accordance with strict geometric principles, with straight streets meeting at right angles. The old-style Chinese city was dominated by low buildings and characterized by straight streets. Houses were typically built around courtyards, and narrow alleyways served both commercial and residential functions.

China's cities began to change as Europeans started to gain power in the region in the 1800s. A group of port cities was taken over by European interests, which proceeded to build Western-style buildings and modern business districts. By far the most important of these semicolonial cities was Shanghai, built near the mouth of the Yangtze River, the main gateway to interior China.

When the communists came to power in 1949, Shanghai was the second largest city in the world. The new authorities considered it a foreign city dominated by capitalist exploiters. They therefore refused to invest money in the city, and as a result it began to decay. Since the late 1980s, however, Shanghai has experienced a major revival. Migrants are now pouring into Shanghai, and building cranes crowd the skyline. Official statistics put the population of the metropolitan area at 19 million (Figure 11.18).

Beijing was China's capital during the Manchu period (1644–1912), a status it regained in 1949. Under communist rule, Beijing was radically transformed; old buildings were razed, and broad avenues were plowed through old neighborhoods. Crowded residential districts gave way to large blocks of apartment buildings and massive government offices. Some historically significant buildings were saved; the buildings of the Forbidden City, for example, where the Manchu rulers once lived, survived as a complex of museums.

**FIGURE 11.18 Contemporary Shanghai**    This vibrant city of almost 20 million symbolizes the new China, with its massive high-rise apartments, industrial developments, and office towers.

In the 1990s, Beijing and Shanghai vied for the first position among Chinese cities, with Tianjin, serving as Beijing's port, coming in a close third. All three of these cities have historically been removed from the regular provincial structure of the country and granted their own metropolitan governments. In 1997, another major city, Hong Kong, passed from British to Chinese control and was granted a distinctive status as a self-governing "special administrative region." The greater metropolitan area of the Xi Delta, composed of Hong Kong, Shenzhen, and Guangzhou (called *Canton* in the West), is now one of China's premier urban areas.

**City Systems of Japan and South Korea**    The urban structures of Japan and South Korea are quite different from those of China. South Korea is noted for its pronounced **urban primacy** (the concentration of total urban population in a single city), whereas Japan is the center of a new urban phenomenon—the **superconurbation**, also called a megalopolis (a huge zone of coalesced metropolitan areas) (Figure 11.19).

Seoul, the capital of South Korea, is by far the largest city in the country. The city is home to more than 10 million people, and its greater metropolitan area contains some 40 percent of South Korea's total population. All of South Korea's major governmental, economic, and cultural institutions are concentrated there. However, Seoul's explosive and generally unplanned growth has resulted in serious congestion.

Japan has traditionally been characterized by urban "bipolarity" rather than urban primacy. Until the 1960s, Tokyo, the capital and main business and educational center, together with the neighboring port of Yokohama, was balanced by the trading center of Osaka and its port, Kobe. Kyoto, the former imperial capital and the traditional center of elite culture, is also located in the Osaka region. As Japan's economy boomed in the 1960s through the 1980s, however, so did Tokyo. The capital city then outpaced all other urban areas in almost every urban function. The Greater Tokyo metropolitan area today contains 25 to 30 million persons, depending on how its boundaries are defined.

**FIGURE 11.19 Urban Concentration in Japan** The inset map shows the rapid expansion of Tokyo in the postwar decades. Today, the greater Tokyo metropolitan area is home to almost 30 million people. The larger map shows the cluster of urban settlements along Japan's southeastern coast. The major area of urban concentration is between Tokyo and Osaka, a distance of some 300 miles, known as the *Tokaido corridor.* Roughly 65 percent of Japan's population lives in this area. The Osaka–Kobe metropolitan area ranks second to Tokyo, with approximately 14 million inhabitants.

Japanese cities sometimes strike foreign visitors as rather gray and dull places, lacking historical interest (Figure 11.20). Little of the country's premodern architecture remains intact. Traditional Japanese buildings were made of wood, which survives earthquakes much better than stone or brick. Fires have therefore been a long-standing hazard, and in World War II, the U.S. Air Force fire-bombed most Japanese cities. (Hiroshima and Nagasaki were, on the other hand, completely destroyed by atomic bombs.) The one exception was Kyoto, the old imperial capital, which was spared devastation. As a result, Kyoto is famous for its beautiful temples, which ring the basin in which central Kyoto lies.

## Cultural Coherence and Diversity: A Confucian Realm?

East Asia is in some respects one of the world's most culturally unified regions. Although different parts of East Asia have their own unique cultures, the entire region shares certain historically rooted ways of life and systems of ideas. Most of these common features can be traced back to ancient Chinese civilization. Chinese culture emerged roughly 4,000 years ago, largely in isolation from the Eastern Hemisphere's other early centers of civilization, in the valleys of the Indus, Tigris-Euphrates, and Nile rivers.

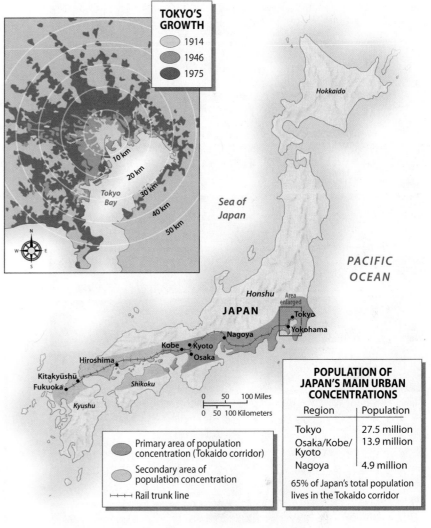

## Unifying Cultural Characteristics

The most important unifying cultural characteristics of East Asia are related to religious and philosophical beliefs. Throughout the region, Buddhism and especially Confucianism have shaped both individual beliefs and social and political structures. Although the role of traditional belief systems has been seriously challenged in recent decades, especially in China, traditional cultural patterns remain.

**The Chinese Writing System** The clearest distinction between East Asia and the world's other cultural regions is found in written language.

**FIGURE 11.20 Tokyo Apartments** The extremely high population density of Tokyo and other large Japanese cities forces most people to live in crowded apartment blocks. This photo depicts Tokyo's Danchi high-rise apartment complex.

Existing writing systems elsewhere in the world are based on the alphabetic principle, in which each symbol represents a distinct sound. East Asia evolved an entirely different system of **ideographic writing**. In ideographic writing, each symbol (or ideograph—often called *character*) usually represents an idea rather than a sound.

The East Asian writing system can be traced to the dawn of Chinese civilization. As the Chinese Empire expanded, the Chinese writing system spread. Japan, Korea, and Vietnam all came to use the same system, although in Japan it was substantially modified, while in Korea it was later largely replaced by an alphabetic system. A major disadvantage of Chinese writing is that it is difficult to learn; to be literate, a person must memorize thousands of characters. Its main benefit is that two literate persons do not have to speak the same language to be able to communicate because the written symbols they use to express their ideas are the same.

The writing system of Japan is particularly complex (Figure 11.21). Initially, the Japanese simply borrowed Chinese characters, referred to in

Japanese as *kanji*. Due to the grammatical differences between spoken Japanese and Chinese, the exclusive use of *kanji* resulted in awkward sentences. The Japanese solved this problem by developing *hiragana*, a writing system in which each symbol represents a distinct syllable, or combination of a consonant and a vowel sound. A parallel system, called *katakana*, was devised for spelling words of foreign origin. Written Japanese employs a complex mixture of all three of these symbolic systems.

**The Confucian Legacy** Just as the use of a common writing system helped build cultural linkages throughout East Asia, so too the idea system of **Confucianism** (the philosophy developed by Confucius) came to occupy a significant position in the region. Indeed, so strong is the heritage of Confucius that some writers refer to East Asia as the "Confucian world." In Japan, however, Confucianism never had the influence that it did in China and Korea.

The premier philosopher of China, Confucius (or Kung Fu Zi, in Mandarin Chinese) lived during the 6th century BCE, a period of political instability. Confucius's goal was to create a philosophy that could generate social stability. While Confucianism is sometimes considered to be a religion, Confucius himself was far more interested in how to lead a correct life and organize a proper society. He stressed obedience to

**FIGURE 11.21 Japanese Writing** The writing system of Japan was originally based on Chinese characters, known in Japan as *kanji*. Because of grammatical differences, however, the Japanese developed two unique "alphabets" of syllables, known as *katakana* and *hiragana*. *Kanji* and *katakana* symbols are visible in this photo.

**FIGURE 11.22 The Buddhist Landscape**  Mahayana Buddhism has traditionally been practiced throughout East Asia. This Golden Buddha statue is located in Baomo Park in Chi Lei Village, Guangdong Province, China.

authority but thought that those in power must act in a caring manner. Confucian philosophy also emphasizes education. The most basic level of the traditional Confucian moral order is the family unit, considered the bedrock of society. The ideal family structure is patriarchal (male-dominated), and children are told to obey and respect their parents.

The significance of Confucianism in East Asian development has long been debated. In the early 1900s, many observers believed that this conservative philosophy, based on respect for tradition and authority, was responsible for the economically backward position of China and Korea. But because East Asia has more recently enjoyed the world's fastest rates of economic growth, such a position is no longer supportable. New voices argue that Confucianism's respect for education and the social stability that it generates give East Asia an advantage in international competition. It must also be recognized, however, that Confucianism has lost much of the hold that it once had on public morality throughout East Asia.

## Religious Unity and Diversity in East Asia

Certain religious beliefs have worked alongside Confucianism to unite the East Asian region. The most important culturally unifying beliefs are associated with Mahayana Buddhism. Other religious practices, however, have tended to challenge the cultural unity of the region.

**Mahayana Buddhism**  Buddhism, a religion that stresses the quest to escape an endless cycle of rebirths and reach union with the cosmos (or nirvana), originated in India in the 6th century BCE. By the 2nd century CE, Buddhism had reached China, and within a few hundred years, it had spread throughout East Asia. Today, Buddhism remains widespread everywhere in the region, although it is far less significant here than it is in mainland Southeast Asia, Sri Lanka, and Tibet (Figure 11.22).

The variety of Buddhism practiced in East Asia—Mahayana, or Great Vehicle—is distinct from the Therevada Buddhism of Southeast Asia. Most importantly, Mahayana Buddhism simplifies the quest for nirvana, in part by putting forward the existence of beings (*boddhisatvas*) who refuse divine union for themselves in order to help others spiritually. Mahayana Buddhism also allows its followers to practice other religions. Thus many Japanese are both Buddhists and Shintoists, while many Chinese consider themselves to be both Buddhists and Taoists (as well as Confucianists).

**Shinto**  The Shinto religion is so closely bound to the idea of Japanese nationality that it is questionable whether a non-Japanese person can follow it. Shinto began as the worship of nature spirits, but it was gradually refined into a subtle set of beliefs about the harmony of nature and its connections with human existence. Shinto is still a place- and nature-centered religion. Certain mountains, particularly Mount Fuji, are considered sacred and are thus climbed by large numbers of people (Figure 11.23). Major Shinto shrines, often located in scenic places, attract numerous religious pilgrims. The most notable of these is the Ise Shrine south of Nagoya, devoted to the emperor of Japan.

**Taoism and Other Chinese Belief Systems**  Similar to Shinto, the Chinese religion Taoism (or Daoism) is rooted in nature worship. Also like Shinto, it stresses spiritual harmony. Taoism is indirectly associated with *feng shui,* commonly called **geomancy** in English, which is the Chinese and Korean practice of designing buildings in accordance with the spiritual powers that supposedly flow through the local topography. Even in hypermodern Hong Kong, skyscrapers worth millions of dollars have occasionally gone unoccupied because their construction failed to follow feng shui principles.

**Minority Religions**  Followers of virtually all world religions can be found in East Asia. More than 1 million Japanese, for example, belong to Christian churches. Christianity has spread much more extensively in South Korea, where it is now followed by between 25 and 49 percent of the population. South Korea now sends more Christian missionaries abroad than any other country except the United States. Christianity is also spreading rapidly in China, causing Beijing's communist leadership some concern. In 2008, official Chinese statistics counted some 20 million Protestants and 10 million Roman Catholics, but some independent sources argue that the Christian population could be as high as 130 million.

**FIGURE 11.23 Mt. Fuji, Japan**  This beautiful volcanic mountain, sacred to Japan's Shinto religion, is climbed by large numbers of religious pilgrims each year. In the foreground are tea fields.

China's Muslim community is much more deeply rooted than its Christian population. Roughly 10 million Chinese-speaking Muslims, called Hui, are concentrated in Gansu and Ningxia in the northwest and in Yunnan province along the south-central border. Smaller clusters of Hui, often separated in their own villages, live in almost every province of China.

**Secularism in East Asia** Despite all its varied forms of religious expression, East Asia is one of the most secular regions of the world. In Japan, while a small section of the population is highly religious, most people only occasionally observe Shinto or Buddhist rituals. Japan also has a number of "new religions," a few of which are noted for their strong beliefs. But for Japanese society as a whole, religion is not particularly important.

Chinese culture was formerly dominated by Confucianism. After the communist regime took power in 1949, all forms of religion and traditional philosophy—including Confucianism—were discouraged and sometimes severely repressed. Under the new regime, atheistic **Marxist** (or communist) philosophy became the official belief system. With the easing of Marxism during the 1980s and 1990s, however, many forms of religious expression began to return. In North Korea, however, rigid Marxist beliefs are still strictly enforced. North Korea is also noted for its official ideology of *juche,* or "self-reliance." Ironically, juche demands absolute loyalty to the country's repressive political leaders.

## Linguistic and Ethnic Diversity in East Asia

While written languages may have helped unify East Asia, the same cannot be said for spoken languages (Figure 11.24). Japanese and Mandarin Chinese may partially share a system of writing, but the two languages have no direct relationship. Like Korean, however, Japanese has adopted many words of Chinese origin.

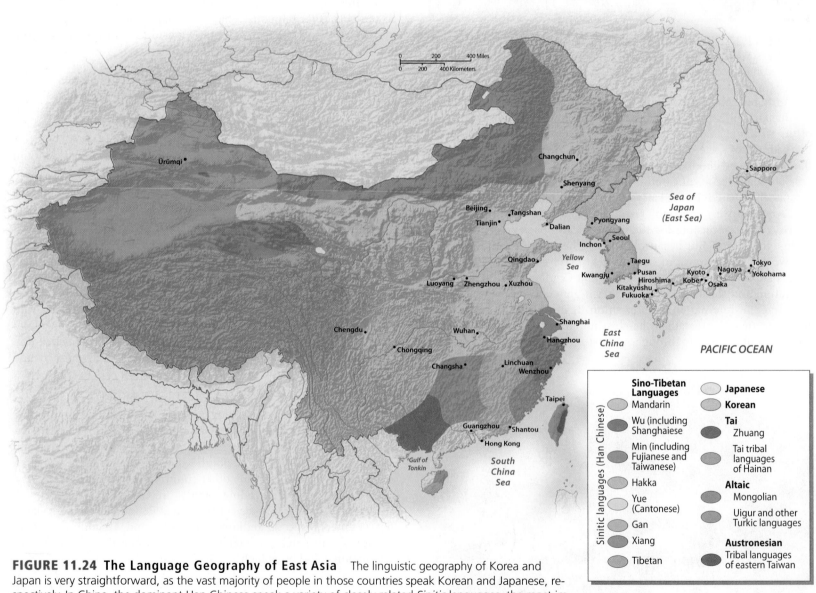

**FIGURE 11.24 The Language Geography of East Asia** The linguistic geography of Korea and Japan is very straightforward, as the vast majority of people in those countries speak Korean and Japanese, respectively. In China, the dominant Han Chinese speak a variety of closely related *Sinitic* languages, the most important of which is Mandarin Chinese. In the peripheral regions of China, a large number of languages—belonging to several different linguistic families—are spoken.

**Language and National Identity in Japan**   Japanese, according to most linguists, is not related to any other language. Korean is also usually classified as the only member of its language family. Some linguists, however, think that Japanese and Korean should be classified together because they share many basic grammatical features.

From many perspectives, the Japanese form one of the world's most homogeneous peoples. In earlier centuries, however, the Japanese islands were divided between two very different peoples: the Japanese living to the south and the Ainu inhabiting the north. The Ainu are physically distinct from the Japanese and possess their own language (Figure 11.25). For centuries, the two groups competed for land, and by the 10th century CE, the Ainu were mostly restricted to the northern island of Hokkaido. The Japanese people subsequently began to colonize Hokkaido, putting renewed pressure on the Ainu. Today, only about 24,000 Ainu remain.

The Japanese language is divided into several dialects, but only in the Ryukyu Islands does one encounter a variant of Japanese so distinct that it might be considered a separate language. Many Ryukyu people believe that they have not been considered full members of the Japanese nation, and they have suffered a certain amount of discrimination.

**FIGURE 11.25  Ainu Men**   The indigenous Ainu people of northern Japan are much reduced in population, but they still maintain a number of their cultural traditions. In this photograph, Ainu men participate in the Marimo Festival on the northern Japanese island of Hokkaido.

Approximately 600,000 persons of Korean descent living in Japan today have also felt discrimination. Most of them were born in Japan (their parents and grandparents having left Korea early in the century) and speak Japanese rather than Korean. Despite their deep bonds to Japan, however, such individuals are not easily able to obtain Japanese citizenship. Perhaps as a result of such treatment, a substantial minority of Japanese Koreans hold radical political views and have tended to support North Korea.

Starting in the 1980s, other immigrants began to arrive in Japan, mostly from the poorer countries of Asia. Because Japan severely restricts the flow of immigrants, many do not have legal status. Men from China and southern Asia typically work in the construction industry; women from Thailand and the Philippines often work as entertainers or prostitutes. Roughly 300,000 Brazilians of Japanese ancestry moved to Japan for the relatively high wages they can earn, but many returned to Brazil during the global economic crisis of 2008–2009. Overall, immigration is less pronounced in Japan than in most other wealthy countries, and relatively few migrants acquire permanent residency, let alone citizenship.

**Language and Identity in Korea**   The Koreans are also a relatively homogeneous people. The vast majority of people in both North and South Korea speak Korean and consider themselves to be members of the Korean nation. There is, however, a strong sense of regional identity, which can be traced back to the medieval period, when the peninsula was divided into three separate kingdoms.

Not all Koreans live in Korea. Several million reside directly across the border in northern China. Desperately poor North Koreans often try to sneak across the border to join these Korean-speaking Chinese communities, but the Chinese government regards such migrants as a security threat and thus returns them to North Korea when it can. A more recent Korean **diaspora** (a scattering of a particular group of people over a vast geographical area) has brought hundreds of thousands of people to the United States, Canada, Australia, New Zealand, the Philippines, and other countries.

**Language and Ethnicity Among the Han Chinese**   The geography of language and ethnicity in China is more complex than that of Korea or Japan. This is true even if one considers only the eastern half of the country, so-called *China proper*. The most important distinction is that separating the Han Chinese from the non-Han peoples. The Han, who form the vast majority, are those people who have historically been incorporated into the Chinese cultural and political systems and whose languages are expressed in the Chinese writing system. They do not, however, all speak the same language.

Northern, central, and southwestern China—a vast area extending from Manchuria through the middle and upper Yangtze Valley to the valleys of Yunnan in the far south—constitute a single linguistic zone. The spoken language here is called *Mandarin Chinese* in English. In China today, Mandarin—locally called *Putonghua,* or "common language"—is the national tongue.

In southeastern China, from the Yangtze Delta to China's border with Vietnam, a number of separate but related languages are spoken. Traveling from south to north, one encounters Cantonese (or Yue) spoken in Guangdong, Fujianese (alternatively Hokkienese, or Min locally) spoken in Fujian, and Shanghaiese (or Wu), spoken in and around Shanghai. Linguistically speaking, these are true languages because they are not mutually intelligible. They are usually

**FIGURE 11.26 Tribal Villages in South China** Non-Han people are usually classified as "tribal" in China, which assumes they have a traditional social order based on self-governing village communities. Shown are Yi people at an open-air market in the village of Xhanghe in Yunnan province.

northern Manchuria. Once Han Chinese were allowed to move to Manchuria in the 1800s, the Manchus soon found themselves vastly outnumbered. As a result, their own culture began to disappear.

Much larger communities of non-Han peoples are found in south-central China, especially in Guangxi, Guizhou, and Yunnan (in China, this area is referred to as southwestern China). Although most of the residents of Yunnan are Han Chinese, the remote areas of the province are inhabited by a wide array of indigenous peoples (Figure 11.27). Because most of the inhabitants of Guangxi's uplands and remote valleys speak languages of the Tai family, it has been designated an **autonomous region**. Critics contend, however, that very little real autonomy has ever existed. (In addition to Guangxi, there are four other autonomous regions in China. Three of these—Xizang [Tibet], Nei Mongol [Inner Mongolia], and Xinjiang—are located in Central Asia and are thus discussed at length in Chapter 10. The final autonomous region, Ningxia, located in northwestern China, is distinguished by its large concentration of Hui [Mandarin-speaking Muslims].)

**Language and Ethnicity in Taiwan** Taiwan is noted for its linguistic and ethnic complexity. In the island's mountainous eastern region, a few small groups of "tribal" peoples speak languages related to those of Indonesia (belonging to the Austronesian language family). These peoples resided throughout Taiwan before the 16th century. At that time, however, Han migrants began to arrive in large numbers. Most of the newcomers spoke Fujianese dialects, which eventually evolved into the distinctive language called Taiwanese.

Taiwan was transformed almost overnight in 1949, when China's nationalist forces, defeated by the communists, sought refuge on the

called dialects, however, because they have no distinctive written form.

Despite their many differences, all the languages of the Han Chinese are closely related to each other, belonging to the Sinitic language subfamily. Because their basic grammars and sound systems are similar, it is not difficult for a person speaking one of these languages to learn another. All Sinitic languages are **tonal** and monosyllabic; their words are all composed of a single syllable (although compound words can be formed from several syllables), and the meaning of each basic syllable changes according to the pitch in which it is uttered.

**The Non-Han Peoples** Many of the remote upland districts of China proper are inhabited by various groups of non-Han peoples speaking non-Sinitic languages. Such peoples are often classified as **tribal**, implying that they have a traditional social order based on self-governing village communities. Such a view is not entirely accurate, however, because some of these groups once had their own kingdoms and all are now subject to the Chinese state (Figure 11.26).

As many as 11 million Manchus live in Manchuria. The Manchu language is related to the languages of the tribal peoples of central Siberia. Few Manchus, however, speak their own language, having abandoned it for Mandarin Chinese. This is an ironic situation because the Manchus ruled the entire Chinese Empire from 1644 to 1912. Until the later part of this period, the Manchus prevented the Han from settling in central and

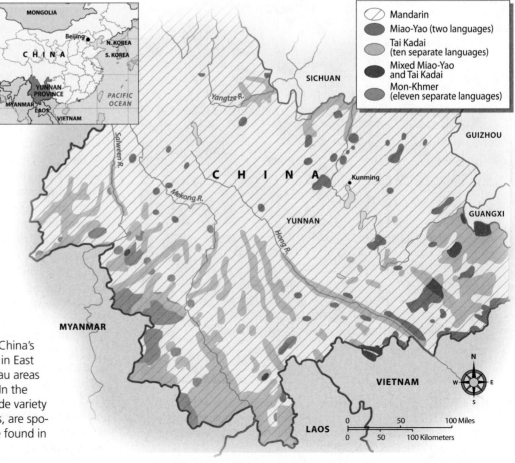

**FIGURE 11.27 Language Groups in Yunnan** China's Yunnan province is the most linguistically complex area in East Asia. In Yunnan's broad valleys and relatively level plateau areas and in its cities, most people speak Mandarin Chinese. In the hills, mountains, and steep-sided valleys, however, a wide variety of tribal languages, falling into several linguistic families, are spoken. In certain areas, several different languages can be found in very close proximity.

island. Most of the nationalist leaders spoke Mandarin, which they made the official language. Taiwan's new leadership discouraged Taiwanese, viewing it as a local dialect. As a result, considerable tension developed between the Taiwanese and the Mandarin communities. Only in the 1990s did Taiwanese speakers begin to reassert their linguistic identity. At present, many proponents of the Taiwanese language advocate formal independence from China, whereas those who favor Mandarin more often hope for eventual reunification.

## East Asian Cultures in Global Context

East Asia has long been torn between separating itself from the rest of the world and welcoming foreign influences and practices. Until the mid-1800s, all East Asian countries attempted to insulate themselves from Western cultural influences. Japan subsequently opened its doors but remained uncertain about foreign ideas. Only after its defeat in 1945 did Japan really choose to make globalization a priority. It was followed in this regard by South Korea, Taiwan, and Hong Kong (then a British colony). However, the Chinese and North Korean governments sought during the early Cold War decades to isolate themselves as much as possible from global culture. Such a stance is still maintained in North Korea.

**The Globalized Fringe**   The capitalist countries of East Asia are characterized to some extent by a cultural internationalism, especially in the large cities. Virtually all Japanese, for example, study English for 6 to 10 years, and although relatively few learn to speak it fluently, most can read and understand a good deal. Business meetings among Japanese, Chinese, and Korean firms are often conducted in English. Relatively large numbers of advanced students, moreover, study in Australia, the United States, and other English-speaking countries.

The current cultural flow is not merely from a globalist West to a previously isolated East Asia. Instead, the exchange has become reciprocal. Hong Kong's action films are popular throughout most of the world and have come to influence filmmaking techniques in Hollywood. Japan nearly dominates the world market in video games, and its *anime* style of animated film and television programming are now following karaoke bars in their overseas movement. Aspects of South Korean popular culture, including music and television shows, have recently become extremely popular through much of Asia.

**The Chinese Heartland**   In one sense, Japan is more culturally predisposed to cosmopolitanism than is China. Before the modern era, the Japanese borrowed heavily from other cultures (particularly from China itself), whereas the Chinese have historically been more self-sufficient. The southern coastal Chinese have, however, more often been oriented toward foreign lands, especially to the Chinese diaspora communities of Southeast Asia and the Pacific.

In most periods of Chinese history, the internal orientation of the center prevailed over the external orientation of the southern coast. After the communist victory of 1949, only the small British enclave of Hong Kong was able to maintain international cultural connections. In the rest of the country, a grim and puritanical cultural order was rigidly enforced. After China began to liberalize its economy and open its doors to foreign influences in the late 20th century, however, the southern coastal region suddenly assumed a new prominence. Through its doors, global cultural patterns began to penetrate the rest

**FIGURE 11.28  Chinese Theme Park**   As China's economy grows, its people are spending increasing amounts of money on entertainment. Theme parks, which now number over 2,000, are particularly popular. Happy Valley Theme Park in Beijing, shown in this photo, was China's largest when it opened in 2006.

of the country. The result has been the emergence of a vibrant and somewhat flashy urban popular culture in China that contains such global features as nightclubs, karaoke bars, fast-food franchises, and theme parks (Figure 11.28).

# Geopolitical Framework: The Imperial Legacies of China and Japan

Much of the political history of East Asia revolves around the centrality of China and the ability of Japan to remain outside China's grasp. The traditional Chinese conception of geopolitics was based on the idea of a universal empire: All territories were either supposed to be a part of the Chinese Empire, pay tribute to it and acknowledge its supremacy, or stand outside the system altogether. When China could no longer maintain its power in the face of European aggression, the East Asian political system began to fall apart. As European power declined in the 1900s, China and Japan competed for regional leadership. After World War II, East Asia was split by larger **Cold War** rivalries (Figure 11.29).

## The Evolution of China

The original core of Chinese civilization was the North China Plain and the Loess Plateau. For many centuries, periods of unification alternated with times of division into competing states. The most important episode of unification occurred in the 3rd century BCE. Once political unity was achieved, the Chinese Empire began to expand vigorously to the south of the Yangtze Valley. Subsequently, the ideal of the unity of China triumphed, helping to join the Han Chinese into a single people.

Various Chinese dynasties attempted to conquer Korea, but the Koreans resisted. Eventually, China and Korea worked out an arrangement whereby Korea paid token tribute and acknowledged the

**DIVIDED NATIONS**

- China
- Taiwan
- North Korea
- South Korea
- —— Autonomous regions

**Territorial claims.** *Japan claims the four southernmost Kuril Islands, which were annexed by Russia at the end of World War II.*

**China–India border tensions.** *The McMahon line in the east was proposed in 1913 at the main watershed in the Himalayas and is the current boundary between China and India; China has never accepted this boundary. In the west, the Aksai Chin area, formerly part of the Indian state of Kashmir, was taken over by China in 1962.*

**U.S. military bases.** *The United States has maintained several large military bases on the island of Okinawa, causing much resentment among many islanders who wish to see most, if not all, of the bases closed.*

**Spratly Islands.** *The Spratly Islands are claimed by China, Taiwan, Vietnam, Malaysia, and the Philippines. These islands, as well as the Paracel Islands, potentially hold petroleum reserves beneath the sea.*

**FIGURE 11.29 Geopolitical Issues in East Asia** East Asia remains one of the world's geopolitical hot spots. Tensions are particularly severe between capitalist, democratic South Korea and the isolated communist regime of North Korea as well as between China and Taiwan. China has had several border disputes, one of which involves a number of small islands in the South China Sea. Japan and Russia have not been able to resolve their quarrel over the southern Kuril Islands.

supremacy of the Chinese Empire and in return received trading privileges and retained independence. When foreign armies invaded Korea—as did those of Japan in the late 1500s—China sent troops to support its "vassal kingdom."

**The Manchu Qing Dynasty** The most significant conquest of China occurred in 1644, when the Manchus toppled the Ming Dynasty and replaced it with the Qing (also spelled Ch'ing) Dynasty. As earlier conquerors did, the Manchus retained the Chinese bureaucracy and made few institutional changes. Their strategy was to adapt themselves to Chinese culture yet at the same time preserve their own

identity as an elite military group. This system functioned well until the mid-19th century, when the Chinese Empire began to crumble at the hands of European and, later, Japanese power.

**The Modern Era** From its height in the 1700s, the Chinese Empire declined rapidly in the 1800s, as it failed to keep pace with the technological progress of Europe. Threats to the empire had always come from the north, and Imperial officials saw little danger from European merchants operating along their coastline. But the Europeans were distressed by the amount of silver needed to obtain Chinese silk, tea, and other products. In response to their lack of alternative goods, the

**FIGURE 11.30  Opium War**   Great Britain humiliated China in two "opium wars" in the early 1800s, forcing the much larger country to open its economy to foreign trade and to grant Europeans extraordinary privileges. This image shows the East India Company steamer *Nemesis* destroying Chinese war junks in January 1841.

British began to sell opium, which Chinese authorities rightfully viewed as a threat. When the imperial government tried to suppress the opium trade in the 1840s, Britain attacked and quickly prevailed (Figure 11.30).

This first "opium war" introduced a century of political and economic chaos in China. The British demanded and received trade privileges in selected Chinese ports. As European businesses penetrated China and weakened local economic interests, anti-Manchu rebellions broke out. At first, all such uprisings were crushed—but not before causing tremendous destruction. Meanwhile, European power continued to advance. In 1858, Russia annexed the northernmost reaches of Manchuria, and by 1900, China had been divided into separate **spheres of influence** (Figure 11.31). (In a sphere of influence, the colonial power has no formal political authority but does have informal influence and tremendous economic clout.)

A successful rebellion in 1911 finally toppled the Manchus and destroyed the empire, but subsequent efforts to establish a unified Chinese Republic were not successful. In many parts of the country, local military leaders ("warlords") grabbed power for themselves. By the 1920s, it appeared that China might be completely torn apart. The Tibetans had gained autonomy; Xinjiang was under Russian influence; and in China proper, Europeans and local warlords vied with the weak Chinese Republic for power. Japan was also increasing its demands and seeking to expand its territory.

## The Rise of Japan

Japan did not emerge as a unified state until the 7th century, some 2,000 years later than China. From its earliest days, Japan looked to China for intellectual and political models. Its offshore location, however, insulated Japan from Chinese rule. The Japanese came to view their islands as a separate empire, equal in certain respects to that of China. Between 1000 and 1580, however, Japan had no real unity, being divided into a number of small warring states.

**The Closing and Opening of Japan**   By the early 1600s, Japan had been reunited by the armies of the Tokugawa **shogunate** (a shogun is a military leader who in theory only remains under the emperor). At this time, Japan attempted to isolate itself from the rest of the world. Until the 1850s, Japan traded with China mostly through the Ryukyu islanders and with Russia through Ainu go-betweens. The only Westerners allowed to trade in Japan were the Dutch, and their activities were strictly limited.

Japan remained largely closed to foreign commerce and influence until U.S. gunboats sailed into Tokyo Bay in 1853 to demand trade access. Aware that China was losing power, Japanese leaders set about modernizing their economic, administrative, and military systems. This effort accelerated when the Tokugawa shogunate was toppled in 1868 by the Meiji Restoration. (It is called a *restoration* because it was carried out in the emperor's name, but it did not give the emperor any real power.) Unlike China, Japan successfully strengthened its government and economy.

**The Japanese Empire**   Japan's new rulers realized that their country remained threatened by European imperial powers. They decided that the only way to meet the challenge was to expand their own territory. Japan soon took control over Hokkaido and began to move farther north into the Kuril Islands and Sakhalin.

In 1895, the Japanese government tested its newly modernized army against China, winning a quick victory that gave it control of Taiwan. Tensions then mounted with Russia, as the two countries competed for power in Manchuria and Korea. The Japanese defeated the Russians in 1905, giving Japan considerable influence in northern China. With no strong rival in the area, Japan annexed Korea in 1910. Alliance with Britain, France, and the United States during World War I brought further gains, as Japan was awarded Germany's island colonies in Micronesia.

The 1930s brought a global depression, greatly reducing world trade and putting a resource-dependent Japan in a difficult situation. The country's leaders sought a military solution, and in 1931, Japan conquered Manchuria. In 1937, Japanese armies moved south, occupying the North China Plain and the coastal cities of southern China. During this period, Japan's relations with the United States deteriorated. When the United States cut off the export of scrap iron, Japan began to experience a resource crunch.

In 1941, Japan's leaders decided to destroy the American Pacific fleet in order to clear the way for the conquest of resource-rich Southeast Asia. Their grand strategy was to unite East and Southeast Asia into a "Greater East Asia Co-Prosperity Sphere." This "sphere" was to be ruled by Japan and was designed to keep the Americans and Europeans out.

## Postwar Geopolitics

With the defeat of Japan at the end of World War II in 1945, East Asia became dominated by rivalry between the United States and the Soviet Union. Initially, the American interests prevailed in Japan, South Korea, and Taiwan, while Soviet interests advanced on the mainland. Soon, however, East Asia began to experience a revival.

**Japan's Revival**   Japan lost its colonial empire when it lost World War II. Its territory was reduced to the four main islands plus the Ryukyu Archipelago. In general, the Japanese government agreed to

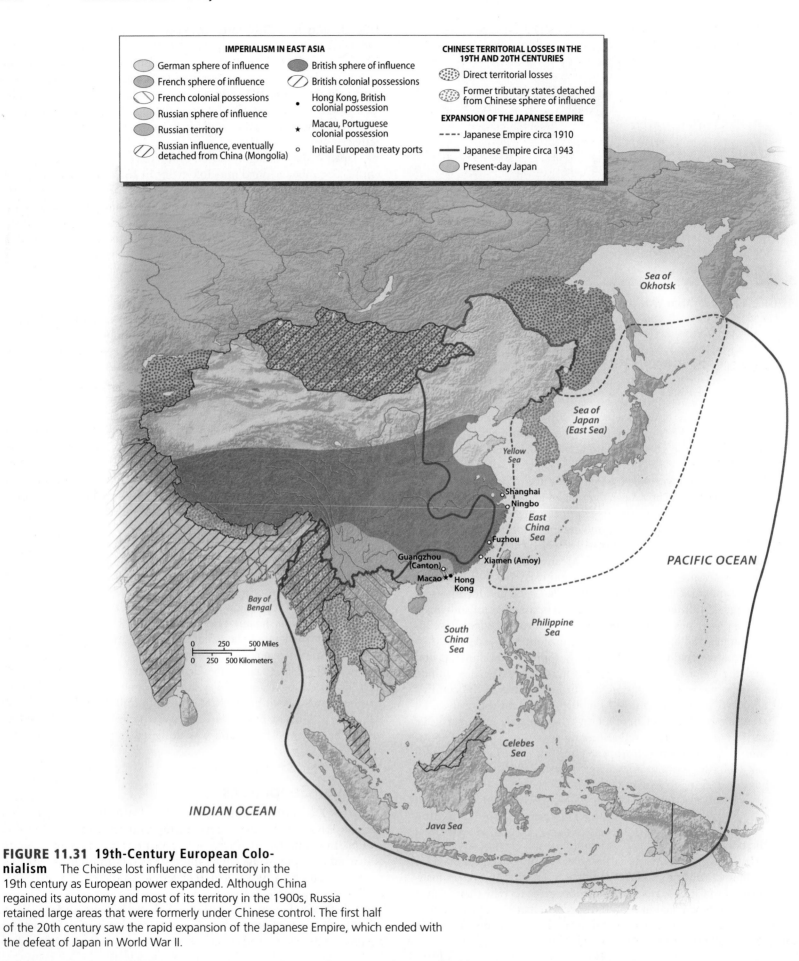

**FIGURE 11.31  19th-Century European Colonialism**  The Chinese lost influence and territory in the 19th century as European power expanded. Although China regained its autonomy and most of its territory in the 1900s, Russia retained large areas that were formerly under Chinese control. The first half of the 20th century saw the rapid expansion of the Japanese Empire, which ended with the defeat of Japan in World War II.

this loss of land. The only remaining territorial conflict concerns the four southernmost islands of the Kuril chain, which were taken by the Soviet Union in 1945. Japan still claims these islands, and Russia refuses to discuss the issue.

Japan's military power was limited by the constitution imposed on it by the United States, forcing Japan to rely in part on the U.S. military for its defense needs. The U.S. Navy patrols many of its vital sea-lanes, and U.S. armed forces maintain several bases in Japan. Many Japanese citizens, however, believe that their country ought to provide its own defense. Slowly but steadily, meanwhile, Japan's military has emerged as a strong regional force, despite the limits imposed on it. Since 2006, North Korean nuclear bomb making and missile testing have raised security concerns in Japan. Many Japanese argue that the North Korean threat shows the importance of maintaining a close alliance with the United States, whereas for others it shows the need for Japan to develop its own nuclear force.

Other East Asian countries are concerned about the potential threat posed by a remilitarized Japan. Such perceptions have been occasionally reinforced by visits of Japanese prime ministers to the Yasukuni Shrine, which contains a military cemetery in which several war criminals from World War II are buried. Anti-Japanese sentiments in China have also been intensified by the publication in Japan of textbooks that minimize Japanese atrocities during that war.

**The Division of Korea**    The end of World War II brought much greater changes to Korea than to Japan. As the end of the war approached, the Soviet Union and the United States agreed to divide the country; Soviet forces were to occupy the area north of the 38th parallel, whereas U.S. troops would occupy the south. This soon resulted in the establishment of two separate governments. In 1950, North Korea invaded South Korea, seeking to reunify the country. The United States, with support from the United Nations, supported the south, while China aided the north. The war ended in a stalemate, and Korea remained a divided country, its two governments still technically at war (Figure 11.32).

Large numbers of U.S. troops remained in South Korea after the war. South Korea in the 1960s was a poor country that could not defend itself. Subsequently, however, the south has emerged as a wealthy trading nation, while the fortunes of the north have declined. Many South Koreans came to resent the presence of U.S. forces, wanting instead to seek peace with North Korea. By the late 1990s, the South Korean government also came to favor a softer approach. In 1998, it established a "Sunshine Policy" that emphasized peaceful cooperation and reconciliation with the north. As a result, South Korean firms invested substantial funds in joint economic endeavors in North Korea, and South Korean tourists were allowed to cross the border for closely monitored trips to famous locations.

Despite these peaceful moves by South Korea, North Korea remained hostile, going so far as to detonate a small nuclear bomb in 2006. As a result of such provocations, South Korea's government essentially canceled the Sunshine Policy in 2008. This move helped convince North Korean leaders to take an even more aggressive stance. In 2009, North Korea set off another nuclear bomb and tested missiles capable of striking targets as far away as Hawaii.

The international community has made many efforts to persuade North Korea to abandon its quest for nuclear weapons. In 2007, North Korea signed an agreement with South Korea, the United

**FIGURE 11.32 The Demilitarized Zone in Korea**    North and South Korea were divided along the 38th parallel after World War II. Today, even after the conflict of the early 1950s, the demilitarized zone, or DMZ (which runs near the parallel), separates these two states. U.S. armed forces are active in patrolling the DMZ.

States, Russia, China, and Japan that required it to shut down its main nuclear reactor in exchange for economic assistance. Two years later, however, the north broke the deal and renewed its quest for a nuclear arsenal. As a result, the United Nations Security Council agreed to allow inspections of North Korean cargo ships, to place an embargo on its arms trade, and to impose financial penalties. The North Korean government responded by vowing to build more nuclear bombs.

**The Division of China**    World War II brought tremendous destruction and loss of life to China. Before the war began, China had already been engaged in a civil war between nationalist and communist forces. After Japan invaded China proper in 1937, the two camps cooperated; but as soon as Japan was defeated, China again found itself in a civil war. In 1949, the communists proved victorious, forcing the nationalists to retreat to Taiwan. The mainland was then renamed the People's Republic of China, while the nationalist government on Taiwan retained the name the Republic of China, which was originally adopted in 1911.

A dormant state of war between China and Taiwan persisted for decades after 1949. The Beijing government still claims Taiwan as an integral part of China and vows that it will eventually reclaim it. The nationalists in Taiwan long insisted that they represented the true government of China and that Taiwan was merely one province of a temporarily divided country. By the end of the 20th century, however, almost all Taiwanese had given up on the idea of taking over China itself, and many began to press openly for formal independence of the island.

The idea of Chinese unity continues to be influential both in China and abroad. In the 1950s and 1960s, the United States recognized Taiwan as the only legitimate government of China, but its policy changed after U.S. leaders decided that it would be more useful to recognize mainland China. Soon, China entered the United Nations, and Taiwan found itself diplomatically isolated from most of the world. Taiwan continues to be recognized, however, as the legitimate government of China by a number of small countries in Africa, the Americas, and the Pacific. Most of these countries receive Taiwanese economic aid in return.

The geopolitical status of Taiwan continued to be a controversial issue in Taiwan itself. When a supporter of political separation was elected president of Taiwan in 2000, China threatened to invade if the island were to declare formal independence. Military tensions remained high for several years, but during the same period, economic connections continued to strengthen. Taiwanese voters began to grow dissatisfied with an independence movement that had generated geopolitical tension without delivering substantial benefits. In the Taiwanese presidential election of 2008, the old nationalist party won a clear victory by promising better relations with mainland China. The policy of both governments then became one of "mutual non-denial," based on unofficially accepting each other's existence as separate states.

**The Chinese Territorial Domain**  Despite the fact that it has been unable to regain Taiwan, China has been successful in retaining most of the territories that the Manchus formerly controlled. In the case of Tibet, this has required considerable force. Resistance by the Tibetans compelled China to launch a full-scale invasion in 1959. The Tibetans, however, have continued to struggle for real autonomy, if not actual independence, as they fear that the Han Chinese now moving to Tibet will eventually outnumber them and undermine their culture (Figure 11.33).

The postwar Chinese government also retained control over Xinjiang in the northwest, as well as Inner Mongolia (or Nei Mongol), a vast territory stretching along the Mongolian border. The native peoples of Xinjiang prefer to call the region *Eastern Turkestan* to emphasize its Turkic heritage. The Han Chinese, however, reject this term because it challenges the unity of China. Like Tibet, Nei Mongol and Xinjiang are classified as autonomous regions. The peoples of Xinjiang are asserting their religious and ethnic identities, and separatist attitudes remain common. Most Han Chinese, however, regard Nei Mongol and Xinjiang as integral parts of their country, and they regard any talk of independence as treasonous.

One territorial issue was finally resolved in 1997, when China reclaimed Hong Kong from Britain. In the isolationist 1950s, 1960s, and 1970s, Hong Kong acted as China's window on the outside world, and it grew wealthy as a capitalist city. As Chinese relations with the outer world opened in the 1980s, Britain decided to honor its treaty provisions and return Hong Kong to China. China, in turn, promised that Hong Kong would become a **special administrative region**, retaining its fully capitalist economic system for at least 50 years. Civil liberties not enjoyed in China itself also remain protected in Hong Kong.

**FIGURE 11.33  Chinese Soldiers in Tibet**  Following a full-scale invasion of Tibet in 1959, China continues to increase its presence through its military forces, the relocation of migrants into the area from other parts of China, and rebuilding programs that mask the traditional Tibetan landscape. Here members of a Chinese paramilitary force observe a praying Tibetan while eating ice cream bars.

In 1999, Macao, the last colonial territory in East Asia, was returned to China, becoming the country's second special administrative region. This small former Portuguese enclave, located across the estuary from Hong Kong, has functioned largely as a gambling refuge. In 2008, gambling revenues in Macao surpassed those of Las Vegas, making it the world capital of commercial wagering (Figure 11.34).

## The Global Dimension of East Asian Geopolitics

In the early 1950s, East Asia was divided into two hostile Cold War camps: China and North Korea were allied with the Soviet Union, while Japan, Taiwan, and South Korea were linked to the United States. The Chinese–Soviet alliance soon deteriorated into mutual hostility, however, and in the 1970s, China and the United States found that they could work with each other, sharing as they did a common enemy in the Soviet Union.

The end of the Cold War, coupled with the rapid economic growth of China, again altered the balance of power in East Asia. The United States no longer needs China to offset the Soviet Union, and the U.S. military has become increasingly worried about the growing power of the rapidly modernizing Chinese army. Through the early years of the new century, China's military budget grew at an average annual rate of about 10 percent, creating one of the world's most powerful armed forces. Several of China's neighbors have become more concerned about its growing strength. China's relations with Russia,

**FIGURE 11.34 Casino in Macao**    Macao, reclaimed by China from Portugal in 1999, retains a unique mixture of Chinese and Portuguese cultural influences. Its economic mainstay remains gambling, which is prohibited in the rest of China.

however, have improved, and the two countries now cooperate extensively on security issues.

China is thus coming of age as a major force in global politics. Whether it is a force to be feared by other countries is a matter of considerable debate. Chinese leaders insist that they have no intention of interfering in the internal affairs of other countries. They do, however, regard concerns expressed by the United States and other countries about their human rights record, as well as their activities in Tibet, as excessive meddling in their internal affairs. In response to China's rise, the United States has sought closer military ties with both Japan and India.

## Economic and Social Development: A Core Region of the Global Economy

East Asia shows vast disparities in economic and social well-being (Table 11.2). Japan's urban belt has one of the world's greatest concentrations of wealth, whereas many interior districts of China have experienced little development. Overall, however, East Asia has experienced rapid economic growth since the 1970s. But again, growth has not been evenly distributed. North Korea, for example, has seen its living standards decline over the past two decades.

### Japan's Economy and Society

Japan was the pacesetter of the world economy in the 1960s, 1970s, and 1980s. In the early 1990s, however, the Japanese economy experienced a major setback, and growth has remained slow ever since. But despite its recent problems, Japan is still by some measurements the world's second largest economic power.

**Japan's Boom and Bust**    Although Japan's heavy industrialization began in the late 1800s, most of its people remained poor. The 1950s, however, saw the beginnings of the Japanese "economic miracle." With its empire gone, Japan was forced to export manufactured materials. Beginning with inexpensive consumer goods, Japanese industry moved to more sophisticated products, including automobiles, cameras, electronics, machine tools, and computer equipment. By the 1980s, Japan was the leader in many segments of the global high-tech economy.

The early 1990s saw the collapse of Japan's inflated real estate market, leading to a banking crisis. At the same time, many Japanese companies relocated factories to Southeast Asia and China. As a result, Japan's economy stagnated for many years. The Japanese government tried to revitalize the economy through massive state spending, resulting in large public deficits.

But despite its economic problems, Japan remains a core country of the global economic system. Its economic influence spans the globe, as Japanese multinational firms invest heavily in production

## TABLE 11.2    DEVELOPMENT INDICATORS

| Country | GNI per capita, PPP 2007 | GDP Average Annual % Growth 2000–07 | Human Development Index (2006)# | Percent Population Living Below $2 a Day | Life Expectancy* 2009 | Under Age 5 Mortality Rate 1990 | Under Age 5 Mortality Rate 2007 | Gender Equity 2007 |
|---|---|---|---|---|---|---|---|---|
| China | 5,420 | 10.3 | 0.762 | 36.3 | 73 | 45 | 22 | 100 |
| Hong Kong | 43,940 | 5.2 | 0.942 | | 82 | | | 98 |
| Japan | 34,750 | 1.7 | 0.956 | | 83 | 6 | 4 | 100 |
| North Korea | | | | 63 | 55 | 55 | | |
| South Korea | 24,840 | 4.7 | 0.928 | | 80 | 9 | 5 | 96 |
| Taiwan | | | | 78 | | | | |

Source:  World Bank, *World Development Indicators, 2009.*
United Nations, *Human Development Index, 2008.* #
Population Reference Bureau, *World Population Data Sheet,* 2009*
Gender Equity – Ratio of female to male enrollments in primary and secondary school.  Numbers below 100 have more males in primary/secondary school, numbers above 100 have more females in primary/secondary schools.

**FIGURE 11.35 Automated Japanese Auto Factory** Part of Japan's economic success has resulted from the automation of its factory assembly lines. Here Mazda automobiles are assembled in a Hiroshima plant. These cars are destined for the east coast of the United States.

facilities in North America and Europe, as well as in developing countries. Japan remains a world leader in a range of high-tech fields, including robotics, optics, and machine tools for the semiconductor industry (Figure 11.35).

By 2006, Japan's economy was finally experiencing healthy growth, largely as a result of its surging high-tech exports to China. The global economic crisis of 2008–2009, however, hit Japan hard, as export markets collapsed. Between March 2008 and March 2009, Japan's total economy contracted by 3.5 percent. Due to Japan's high level of economic globalization, recovery will depend on a revival of the world economy.

**Living Standards and Social Conditions in Japan** Despite high levels of economic development, Japanese living standards remain somewhat lower than those of the United States. Housing, food, transportation, and services are particularly expensive in Japan. Certain amenities that are standard in the United States, such as central heating, remain relatively rare in Japan.

Although the Japanese may live in cramped quarters and pay high prices for basic products, they also enjoy many benefits unknown in the United States. Unemployment remains lower than in the United States, health care is provided by the government, and crime rates are extremely low. By such social measures as literacy, infant mortality, and average life expectancy, Japan surpasses the United States by a comfortable margin. Japan also lacks the extreme poverty found in certain pockets of U.S. society.

**Women in Japanese Society** Critics often point out that Japanese women have not shared the benefits of their country's success. Advanced career opportunities remain limited for women, especially those who marry and have children. The expectation remains that mothers will devote themselves to their families and to their children's education. Japanese businessmen often work or socialize with their coworkers until late every evening and thus contribute little to child care. Japan's prolonged economic slump seems to have resulted in further reductions in career opportunities for women.

One response to the difficult conditions faced by Japanese women has been a drop in the marriage rate. Japan has seen an even more dramatic decline in its fertility rate. Whether this is due to the domestic difficulties faced by Japanese women or is merely a result of the pressures of a postindustrial society is an open question. Fertility rates have, after all, dropped even lower in many parts of Europe. But regardless of the cause, a shrinking population means an aging population, and increasing numbers of Japanese retirees will have to be supported by smaller numbers of workers. In 2005, Japan for the first time saw more deaths than births, and its population began to decline.

## The Newly Industrialized Countries

The Japanese path to development was successfully followed by its former colonies, South Korea and Taiwan. Hong Kong also emerged in the 1960s and 1970s as a newly industrialized economy, although its economic and political systems were different.

**The Rise of South Korea** The postwar rise of South Korea was even more remarkable than that of Japan. During the period of Japanese occupation, Korean industrial development was concentrated in the north, which is rich in natural resources. The south, in contrast, remained a densely populated, poor, agrarian region.

In the 1960s, the South Korean government began a program of export-led economic growth. It guided the economy with a heavy hand and denied basic political freedom to the Korean people. By the 1970s such policies had proved highly successful in the economic realm. Huge Korean industrial conglomerates, known as *chaebol* moved from exporting inexpensive consumer goods to heavy industrial products and then to high-tech equipment. As the South Korean middle class expanded, pressure for political reform increased, and in the1980s, South Korea made the transition from an authoritarian to a democratic country.

Through the 1980s, South Korean firms remained dependent on the United States and Japan for basic technology. By the 1990s, however, this was no longer the case, and South Korea emerged as one of the world's main producers of semiconductors. South Korean wages have also risen at a rapid rate. The country has invested heavily in education (by some measures, it has the world's most demanding educational system), which has served it well in the global high-tech economy. Increasingly, South Korean companies are themselves becoming **multinational**, building new factories in the low-wage countries of Southeast Asia and Latin America, as well as in the United States and Europe.

The political and social development of South Korea has not been as smooth as its economic progress. Issues of economic globalization often provoke serious political conflicts. In 2008, for example, Seoul was temporarily paralyzed when up to 40,000 demonstrators took to the streets to protest the government's decision to allow the import of beef from the United States, which had previously been banned due to concerns about mad cow disease (Figure 11.36).

Like other trade-dependent countries, South Korea experienced a major recession in 2008–2009, with its stock market declining by 40 percent and its currency dropping 26 percent. By July 2009, however, South Korean officials concluded that the economic decline had "bottomed out" and that growth would soon begin again.

**FIGURE 11.36 Protests in South Korea** Massive political protests are common in South Korea. In May 2008, some 10,000 cattle breeders took to the streets of Seoul to burn cow effigies, protesting a new policy that would allow increased beef imports from the United States to South Korea.

## Taiwan and Hong Kong

Taiwan and Hong Kong have experienced rapid economic growth since the 1960s. The Taiwanese government, like the governments of South Korea and Japan, has guided the economic development of the country. Taiwan's economy, however, is organized not around large conglomerates and linked business firms but rather around small to mid-sized family firms.

Hong Kong, unlike its neighbors, has been characterized by one of the most *laissez-faire* economic systems in the world (*laissez-faire* refers to market freedom with little governmental control). Governmental involvement has been minimal, which is one reason the city's business elite were nervous about the transition to Chinese rule. Hong Kong traditionally functioned as a trading center, but in the 1960s and 1970s, it became a major producer of textiles, toys, and other consumer goods. By the 1980s, however, such cheap products could no longer be made in such an expensive city. Hong Kong industrialists subsequently began to move their plants to southern China, while Hong Kong itself increasingly specialized in business services, banking, telecommunications, and entertainment. Fears that its economy would falter after the Chinese takeover in 1997 turned out to have been exaggerated.

Both Taiwan and Hong Kong have close overseas economic connections. Linkages are particularly tight with Chinese-owned firms located in Southeast Asia and North America. Taiwan's high-technology businesses are also intertwined with those of the United States; there is a constant back-and-forth flow of talent, technology, and money between Taipei and Silicon Valley. Hong Kong's economy is also closely bound with that of the United States (as well as those of Canada and Britain), but its closest connections are with the rest of China. Such high levels of globalization resulted in major recessions in both economies during the global economic crisis of 2008–2009.

## Chinese Development

China dwarfs all of the rest of East Asia in both physical size and population. Its economic takeoff is thus reshaping the economy of the entire world. Despite its recent growth, however, China's economy has a number of weaknesses. For example, much of the vast interior remains trapped in poverty, and many of its largest industries are not competitive. The future of the Chinese economy is thus one of the biggest uncertainties facing both East Asia and the world economy as a whole.

**China under Communism** More than a century of war, invasion, and near-chaos in China ended in 1949, when the communist forces led by Mao Zedong seized power. The new government, inheriting a weak economy, set about nationalizing private firms and building heavy industries. Their plans were most successful in Manchuria, where a large amount of heavy industrial equipment had been left by the Japanese.

In the late 1950s and 1960s, however, China experienced two economic disasters. The first, ironically called the "Great Leap Forward," entailed the idea that small-scale village workshops could produce the large quantities of iron needed for sustained industrial growth. Communist Party officials demanded that these inefficient workshops meet unreasonably high production quotas. The result was a horrific famine that may have killed 20 million persons. The early 1960s saw a return to more practical policies, but toward the end of the decade, a new wave of radicalism swept through China. This "Cultural Revolution" aimed at mobilizing young people to rid the country of "undesirable" traditional social values and replace them with communist ideology. Thousands of experienced industrial managers and college professors were expelled from their positions. Many were sent to villages to be "reeducated" through hard physical labor; others were simply killed. The economic consequences were devastating.

**Toward a Postcommunist Economy** When Mao Zedong, who had been revered as an almost superhuman being, died in 1976, China faced a crucial turning point. Its economy was nearly stagnant and its people desperately poor. However, the economy of Taiwan, its rival, was booming. This led to a political struggle between pragmatists hoping for change and dedicated communists. The pragmatists emerged victorious, and by the late 1970s it was clear that China would embark on a different economic path. The new China would seek closer connections with the world economy and take a modified capitalist road to development (Figure 11.37).

China did not, however, transform itself into a fully capitalist country. The state continued to run most heavy industries, and the Communist Party kept a monopoly on political power. Instead of suddenly abandoning the communist model, as the former Soviet Union did, China allowed cracks to appear in which capitalist businesses could take root and thrive.

**Industrial Reform** An important early industrial reform involved opening **Special Economic Zones (SEZs)**, in which foreign investment was welcome and state interference was minimal. The Shenzhen SEZ, adjacent to Hong Kong, proved particularly successful after

**FIGURE 11.37 Chinese Bank**    Although China is officially a communist country, its economy is actually highly capitalistic. As China develops, banking has become an increasingly important segment of its economy.

than most other countries. Early in the crisis, China's government announced a massive stimulus program, pledging to inject $586 billion into the economy over a two-year period. Although China's economic growth did slow down, the country did not go into recession; as of early 2009, it was still growing at an annual rate of roughly 6 percent.

China's economic expansion has created tensions with the United States. China exports far more to the United States than it imports, leading some U.S. politicians to request that China allow its currency to appreciate against the U.S. dollar (a stronger currency makes imports cheaper and exports more expensive). Other foreign critics accuse China of unfairly keeping the price of labor low in order to enhance exports. China's large and growing holdings of U.S. Treasury bonds, however, make it difficult for the United States to exert much pressure on the Chinese economy. In fact, by 2007 some experts were arguing that the two economies had become so intertwined and dependent on each other that one could describe them together as "Chimerica" (see "Exploring Global Connections: China and the Global Scrap Paper Trade").

**Social and Regional Differentiation**    The Chinese economic surge brought about by the reforms of the late 1970s and 1980s resulted in growing **social and regional differentiation**. In other words, certain groups of people—and certain portions of the country—prospered, while others did not. Despite its socialist government, the Chinese state encouraged the formation of an economic elite, having concluded that only wealthy individuals can adequately transform the economy. The least-fortunate Chinese citizens were sometimes left without work, and many millions migrated from rural villages to seek employment in the booming coastal cities. The government attempted to control the transfer of population, but with only partial success. Shantytowns and homeless populations began to emerge around some of China's cities.

Hong Kong manufacturers found it a convenient source of cheap land and labor (Figure 11.38). Additional SEZs were soon opened, mostly in the coastal region. The basic strategy was to attract foreign investment that could generate exports, the income from which could supply China with the capital it needed to build its infrastructure (roads, electrical and water systems, telephone exchanges, and the like).

Other capitalistic reforms were also enacted. Former agricultural cooperatives were allowed to produce for the market. Many of these "township and village enterprises" proved highly successful. By the early 1990s, the Chinese economy was growing at roughly 10 percent a year, perhaps the fastest rate of expansion the world has ever seen. China emerged as a major trading nation, and by the mid-1990s, it had huge trade surpluses, especially with the United States. Seeking to strengthen its connections with the global economic system, China joined the World Trade Organization (WTO), a body designed to facilitate free trade and provide ground rules for international economic exchange, in November 2001.

In 2007 China's economy was still expanding at an extremely rapid rate of about 10 percent a year. Critics argued that this growth was unsustainable and that China needed to reform its banks and fully abandon centralized planning if its economic expansion were to continue. China's leadership, however, has made it clear that economic reform is to be a gradual process. Despite China's reliance on trade, it weathered the global economic crisis of 2008–2009 better

**FIGURE 11.38 Shenzhen**    The city of Shenzhen, adjacent to Hong Kong, was one of China's first Special Economic Zones. It has recently emerged as a major city in its own right.

# EXPLORING GLOBAL CONNECTIONS
## China and the Global Scrap Paper Trade

In 2007, the richest person in China—and the wealthiest self-made woman in the world—was Zhang Yin, sometimes called the "cardboard queen" (Figure 11.2.1). Ms. Zhang founded a small scrap paper company in southern China in the late 1980s. As China's exports surged in the 1990s, so did its demand for cardboard packaging. China has few forests, and recycled paper serves as the main source material. Because the United States is the world's largest producer of scrap paper, Zhang moved to Los Angeles in 1990 to secure adequate supplies. Operating at first out of her apartment, she initially turned to garbage dumps, which were happy to sell their stocks of recycled paper.

Zhang's company, Nine Dragons Paper Holdings, Ltd., thrived through the 1990s and into the early years of the 21st century. Recycled paper in the United States was both inexpensive and of high quality. Shipping rates remained low, moreover, because hundreds of thousands of shipping containers from China arrived in the U.S. every year, but the United States exported relatively little in return to China. In 2007 alone, the United States sent 11 million tons of scrap paper to China, valued at $1.5 billion. American Chung Nam Inc., an affiliate of Nine Dragons Paper, had emerged as the largest U.S. exporter in terms of volume sent by shipping container.

With a personal fortune of $4.7 billion in 2007, Zhang Yin was able to translate her economic standing into political power, securing a place on the Chinese People's Political Consultative Conference, an important governmental advisory board. Zhang's fortune tumbled in 2008–2009, however, as the global economic crisis brought havoc to the paper recycling industry. From having expanded by more than three times between 2002 and 2008, China's exports to the United States declined by more than 50 percent between September 2008 and February 2009. The fall in China's exports undermined the demand for paper boxes, reducing the price for corrugated cardboard from $250 a ton in August 2008 to $75 a ton in October of the same year. As a result, vast quantities of scrap paper started to accumulate in warehouses in the Los Angeles region. The value of Nine Dragons stocks plummeted 80 percent from January to December 2008.

Despite the hardships imposed by the global economic crisis, Nine Dragons Paper remains optimistic about its global prospects. It is continuing to expand, building new papermaking plants in Chinese cities as well as in Vietnam and seeking alternative sources of raw material. According to the company's official Website, "The Group aims to become the world's leading containerboard product manufacturer in capacity, profitability, and efficiency." Nine Dragons is also seeking to become more environmentally responsible, both for efficiency and public relations purposes. According to the "Chairlady's Statement in the 2008/09 Interim Report," one of the firm's new slogans is "Support ND [Nine Dragons] Paper to support the environment."

**FIGURE 11.2.1  Zhang Yin, Head of Nine Dragons Paper**   One of the richest and most powerful business leaders in China, Zhang Yin is by some measures the world's wealthiest self-made woman.

---

**China's Booming Coastal Region**   Most of the benefits from China's economic transformation have flowed to the coastal region and to the capital city of Beijing. The southern provinces of Guangdong and Fujian were the first to benefit. These provinces have profited from their close connections with the overseas Chinese communities of Southeast Asia and North America. (The vast majority of overseas Chinese emigrants came from these provinces.) Their location close to Taiwan and especially Hong Kong also proved helpful.

By the 1990s, the Yangtze Delta, centered on Shanghai, reemerged as the economic core of China. The Chinese government has encouraged the development of huge industrial, commercial, and residential complexes, hoping to take advantage of the region's vitality. The Suzhou Industrial Park is now a hypermodern city of more than a half million people, thanks largely to a $20 billion investment, most of it from Singapore. Shanghai's Pudong industrial development zone has attracted $10 billion, much of it going to the construction of a new airport and subway system.

The Beijing–Tianjin region has also played a major role in China's economic boom. Its main advantage is its proximity to political power and its position as the gateway to northern China. The other coastal provinces of northern China have also done relatively well.

**Interior and Northern China**   The interior and northern parts of China, in contrast, have seen much less economic expansion than the rest of the country. Manchuria remains relatively well-off as a result of fertile soils and early industrialization, but it has not participated much in the recent boom. Many of the state-owned heavy industries of the Manchurian **rust belt** or zone of decaying factories, are relatively inefficient.

Most of the interior provinces of China have likewise missed the recent wave of growth. In many areas, rural populations are hurt by environmental degradation. One consequence is high levels of underemployment and out-migration. As a result of such regional differences, China is building roads and rail-lines and undertaking other projects to encourage development in the interior. But by most measures, poverty still increases with distance from the coast (Figure 11.39).

## Social Conditions in China

Despite its pockets of persistent poverty, China has made significant progress in social development. Since coming to power in 1949, the communist government has made large investments in medical care and education, and today, China has impressive health and longevity figures. The literacy rate remains lower than that of Japan, South

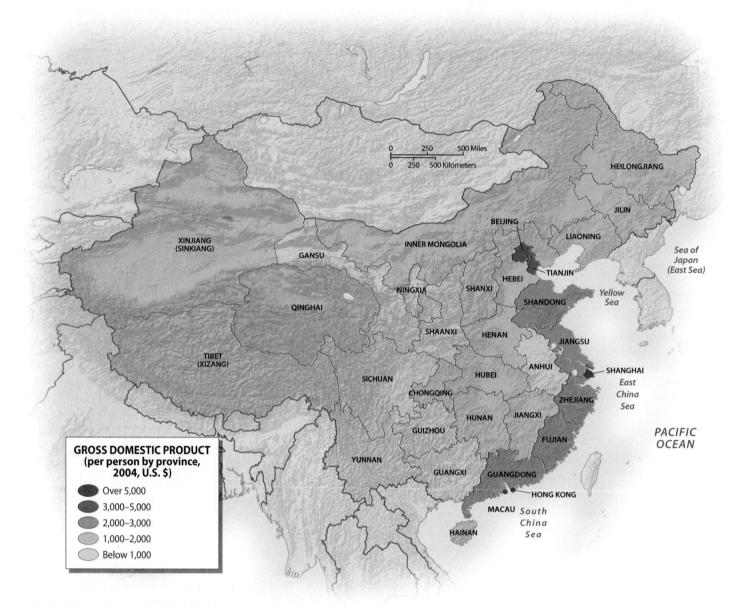

**FIGURE 11.39 Economic Differentiation in China** Although China has seen rapid economic expansion since the late 1970s, the benefits of growth have not been evenly distributed throughout the country. Economic prosperity and social development are concentrated on the coast, especially in Shanghai, Guangdong, Beijing, and Tianjin. Most of the interior remains mired in poverty. The poorest part of China is the upland region of Guizhou in the south-central part of the country. (*Benewick and Donald, 1999*, The State of China Atlas, *p. 35, New York: Penguin Reference*)

Korea, and Taiwan, but because almost all children attend elementary school, it will probably rise substantially in the coming years.

Not surprisingly, human well-being in China varies from region to region. The literacy rate remains relatively low in many of the poorer parts of China, including the uplands of Yunnan and Guizhou. Such regional disparities may increase as the gap between the wealthy and the poor grows. Some evidence also indicates that the social well-being of undocumented migrant workers in Chinese cities is also declining. Medical care is becoming more expensive, putting large burdens on China's underprivileged population.

**China's Population Quandary** Population policy remains an unsettling issue for China. With more than 1.3 billion people highly concentrated in less than half of its territory, Eastern China is very densely populated. By the 1980s, its government had become so concerned that it instituted the famous "one-child policy." Under this plan, couples in normal circumstances are expected to have only a single offspring and can suffer financial penalties if they do not comply (Figure 11.40). This strategy has been successful; the average fertility level is now only 1.6. China will, however, probably reach more than 1.5 billion people before its population stabilizes.

Fertility levels in China, as might be expected, vary from province to province. Birthrates are relatively low in most large cities, the Yangtze Delta, parts of the Sichuan Basin, and parts of the North China Plain and are relatively high in many upland areas of south China, the Loess Plateau, and northwestern China. Higher levels of fertility in these poorer and more rural areas will probably lead to increased migration to the booming coastal cities.

While China's policy has reduced its population growth rate, it has also generated social tensions and human-rights abuses. Particularly troubling is the growing gender imbalance. By 2005, some sources claimed that for every 100 births of girls, 118 boys were born. This gap reflects the practice of honoring one's ancestors; because family lines are traced through male offspring, one must produce a male heir to maintain one's lineage. Some couples decide to bear more than one child, regardless of the penalties they may face. Another option is gender-selective abortion; if ultrasound reveals a female fetus, the pregnancy is sometimes terminated. In 2004, the Chinese government banned this practice, but much evidence suggests that it continues to occur. Poor couples not uncommonly abandon baby girls, and young boys are occasionally kidnapped and sold to wealthy couples without a son.

**The Position of Women**    Women have historically had a relatively low position in Chinese society, as is true in most other civilizations. One traditional expression of this was the practice of foot binding: The feet of almost half of the girls of China in the 1800s were deformed through breaking and binding in order to produce a dainty appearance. This crippling and painful practice was eliminated only in the 20th century. In certain areas of southern China, it was also common in traditional times for girls to be married, and hence to leave their own families, when they were mere toddlers (such marriages, of course, would not be consummated for many years).

Not all women suffered such disabilities in pre-modern China. Some individuals achieved fame and fortune—a few even through military service. Both the nationalist and communist governments have, moreover, sought to begin equalizing the relations between the sexes. Many of their measures have been successful, and women now have a relatively high level of participation in the Chinese workforce. But it is still true that throughout East Asia—in Japan no less than in China—few women have achieved positions of power in either business or government. As China modernizes and its urban economy grows, the position of its women will probably improve.

**FIGURE 11.40 China's Population Policies**    One aspect of China's population policies is the expansion of child-care facilities so that mothers can be near their children while at work. This enables women to resume participating in the workforce soon after giving birth. This photo shows a typical day-care center attached to an industrial plant in Guangdong Province in coastal China.

# SUMMARY

- The economic success of East Asia has been accompanied by severe environmental degradation. Japan, South Korea, and Taiwan managed to overcome environmental crises both by enacting strict protective legislation and by moving many of their most polluting industries overseas. The major environmental issue in the region today concerns the rapid growth of the Chinese economy. At present, pollution in Chinese cities is so serious that it has had major negative effects on human health, and much of the Chinese countryside suffers from such problems as soil erosion and desertification.

- Although East Asia is a very densely populated region, it has seen its birthrates plummet in recent decades. Japan is now facing population decline, which could become quite serious within a few decades. Population decline will put severe pressure on the Japanese economy, and it is unclear how the country will meet the resulting challenges. In China, the biggest demographic challenge results from the massive movement of people from the interior to the coast and from rural villages to the rapidly expanding cities.

- East Asia is united by deep cultural and historical bonds. China has had a particularly large influence because, at one time or another, it covered nearly the entire region. Although Japan has never been under Chinese rule, it still has profound historical connections to

Chinese civilization. Overall, the most prosperous parts of East Asia have seen the striking development of cultural globalization over the past several decades. Cultural features from Europe and North America have been enthusiastically adopted, just as East Asian cultural ideas and practices have spread to many other parts of the world.

■ Geopolitically, East Asia remains a region characterized by strife. China and Korea are still suspicious of Japan, and they worry that it might rebuild a strong military force. Japan, for its part, is concerned about the growing military power of China and especially about the nuclear arms and missiles of North Korea. Relations between North and South Korea deteriorated after North Korea renewed its nuclear weapons program in 2006. North Korean weapons development, moreover, is a global concern, and it deeply involves the United States.

■ With the notable exception of North Korea, all East Asian countries have experienced major economic growth since the end of World War II. Such growth has had large global consequences, as East Asia massively exports to and imports from all other major areas of the global economy. The most important story of the global economy itself over the past two decades has probably been the rapid rise of China. That rise, however, has generated many problems, both domestically and abroad. China's relatively wealthy coastal provinces increasingly resent the control of Beijing and the flow of their tax receipts to the national government.

■ Some observers suggest that the old split between China's more capitalistic south and its more bureaucratic north, which seemed to disappear under the communist system, may be reemerging. Chinese provinces are starting to set their own economic policies rather than wait for orders from the central government. Contributing to this centrifugal process (one leading to a spreading out or a breaking apart) is the explosive growth of the private economy coupled with the near-stagnation of the state-owned sector. Some China watchers believe that this growing imbalance may eventually weaken the Chinese Communist Party and perhaps undermine the country's central authority. Others see an ever-strengthening Chinese state supported by its highly successful and globally engaged economy.

## KEY TERMS

anthropogenic landscape (p. 319)
autonomous region (p. 328)
China proper (p. 312)
Cold War (p. 329)
Confucianism (p. 324)
desertification (p. 318)
diaspora (p. 327)

geomancy (p. 325)
*hiragana* (p. 324)
ideographic writing (p. 324)
*kanji* (p. 324)
*laissez-faire* (p. 337)
loess (p. 313)
Marxism (p. 326)
multinational (p. 336)

pollution exporting (p. 314)
rust belt (p. 339)
sediment load (p. 313)
shogun, shogunate (p. 331)
social and regional differentiation (p. 338)
special administrative region (p. 334)

Special Economic Zones (SEZs) (p. 337)
sphere of influence (p. 331)
superconurbation (p. 322)
tonal (p. 328)
tribal peoples (p. 328)
urban primacy (p. 322)

## THINKING GEOGRAPHICALLY

1. What are the advantages and disadvantages of China's dam building policy? Is it wise for the Chinese government to emphasize dam construction?

2. What would be the main results, both positive and negative, if Japan were to allow the importation of rice and open its agricultural lands to urban development?

3. What might be the main consequences if China were to grant true autonomy to the Tibetans and other non-Han peoples? Is there any chance that Tibet might eventually gain independence? Could Taiwan possibly gain formal independence?

4. Discuss the potential ramifications of the United States' restricting the importation of Chinese goods in order to put pressure on the Chinese government for human-rights reforms.

5. What is the likelihood of East Asia emerging as the center of the world economy over the next 50 years? What would be some of the main consequences of such a development?

Log in to **www.mygeoscienceplace.com** for videos, interactive maps, RSS feeds, case studies, and self-study quizzes to enhance your study of East Asia.

# 12 SOUTH ASIA

## GLOBALIZATION AND DIVERSITY

Although a number of cities in western and southern India have recently emerged as centers of the global high-tech economy, much of South Asia is relatively isolated from the world economy. Resistance to globalization, moreover, remains widespread in many of the region's remote areas.

### ENVIRONMENTAL GEOGRAPHY

While the arid parts of South Asia suffer from water shortages and the salinization of the soil, the humid areas often experience devastating floods.

### POPULATION AND SETTLEMENT

South Asia will soon become the most populous region of the world. Birthrates have, however, decreased substantially in recent years.

### CULTURAL COHERENCE AND DIVERSITY

South Asia is one of the most culturally diverse regions of the world, with India alone having more than a dozen official languages as well as numerous adherents of most major religions.

### GEOPOLITICAL FRAMEWORK

South Asia is burdened not only by a number of violent secession movements but also by the struggle between India and Pakistan, which are both armed with nuclear weapons.

### ECONOMIC AND SOCIAL DEVELOPMENT

Although South Asia is one of the poorest regions of the world, certain areas are experiencing rapid economic growth and technological development.

*South Asia contains extreme poverty as well as booming high tech industries. In this photograph, the Cybergateway Building is seen rising above squatter settlements in the Indian city of Hyderabad.*

South Asia, a land of deep historical and cultural interconnections, has recently experienced intense political conflict. Since gaining independence from Britain in 1947, the two largest countries, India and Pakistan, have fought several wars and remain locked in conflict. Political tensions have reached such heights that many experts consider South Asia the leading candidate for nuclear war. Religious divisions add to the geopolitical turmoil, for India is primarily a Hindu country (with a large Muslim minority), while neighboring Pakistan and Bangladesh are both predominantly Muslim (see "Setting the Boundaries").

South Asia also has its share of economic and demographic problems. Given its current rate of growth, South Asia will soon surpass East Asia as the world's most populous region. Although agricultural production has increased slightly faster than population in recent decades, some experts think that farming improvements are approaching their limit. Compounding this serious situation is the widespread poverty of South Asia, which is home to more malnourished people than any other part of the world.

South Asia is less connected to the contemporary globalized world than are East Asia and Southeast Asia. It is, however, beginning to have a significant global impact, based on its high levels of scientific and technical talent, the international links that the region has established through migration, and the enormous size of its local markets. Several Indian cities have recently emerged as major players in the global information technology industry.

# Environmental Geography: Diverse Landscapes, from Tropical Islands to Mountain Rim

South Asia's diverse environmental geography ranges from the highest mountains in the world to densely populated delta islands barely above sea level; from some of the wettest places on Earth to scorching deserts; and from tropical rain forests to eroded scrublands (Figure 12.1). To illustrate the complexity of South Asian environmental issues, let us begin by looking at India's new "Golden Quadrilateral" highway system.

## Building the Quadrilateral Highway

India's need for improved transportation is difficult to deny. As late as 2006, the average speed for a trucker traveling between Kolkata (Calcutta) and Mumbai (Bombay) could be as low as 7 miles per hour (11 kilometers per hour). Multilane highways are rare, but checkpoints and tollbooths abound. To address the transport needs generated by

# Setting the Boundaries

South Asia forms a distinct landmass separated from the rest of the Eurasian continent by a series of sweeping mountain ranges, including the Himalayas—the highest mountains in the world. It is often called the **Indian subcontinent,** in reference to its largest country. South Asia also includes a number of islands in the Indian Ocean, including the countries of Sri Lanka and the Maldives, as well as the Indian territories of the Lakshadweep, Andaman, and Nicobar Islands.

India is by far the largest South Asian country, both in size and in population. Covering more than 1 million square miles (2.59 million square kilometers) from the Himalayan crest to the southern tip of the peninsula, India is the world's seventh largest country in terms of area and, with more than 1.1 billion inhabitants, second only to China in population. Although mostly a Hindu country, India contains tremendous religious, ethnic, linguistic, and political diversity.

Pakistan, the next largest country in South Asia, is less than one-third the size of India. Stretching from the high northern mountains to the arid coastline on the Arabian Sea, its population of 181 million is only about 15 percent of India's. Despite this imbalance, these two countries have been locked in a tense struggle, especially over the disputed territory of Kashmir. Until gaining independence in 1947, Pakistan was one portion of a larger undivided British colonial realm simply called *India.* Because of Pakistan's strong ties to Islam, however, some Pakistanis argue that their country is now more closely connected to its Muslim neighbors in Southwest Asia than it is to India and the rest of South Asia.

Bangladesh, on India's eastern shoulder, is also a largely Muslim country. Originally created as East Pakistan in the hurried division of India in 1947, it achieved independence after a brief civil war in 1971. Although a small country in area (54,000 square miles [140,000 square kilometers]), Bangladesh is one of the world's most densely populated places—and also one of the poorest—with 162 million people living in an area about the size of Wisconsin. Bangladesh has a short border with Burma (Myanmar), but it is otherwise bordered only by India.

Nepal and Bhutan are both located in the Himalayan Mountains, sandwiched between India and the Tibetan Plateau of China. Nepal, with some 27 million people, is much larger in both area and population, and is far more open to the contemporary world. Bhutan, on the other hand, has purposely disconnected itself from the global system, remaining a relatively isolated Buddhist kingdom with fewer than 1 million inhabitants.

The two island countries of Sri Lanka (formerly Ceylon) and the Maldives round out South Asia. Each of these countries faces problems that cloud the future. Sri Lanka (with a population of 20 million) experienced a civil war from 1983 to 2009 that could reignite. The predicament facing the small island nation of the Maldives is quite different: If global warming continues, and if sea levels rise as predicted, this island nation—where the highest point is only 6 feet (2 meters) above sea level—will be entirely flooded, and its 300,000 inhabitants will have to seek higher ground elsewhere.

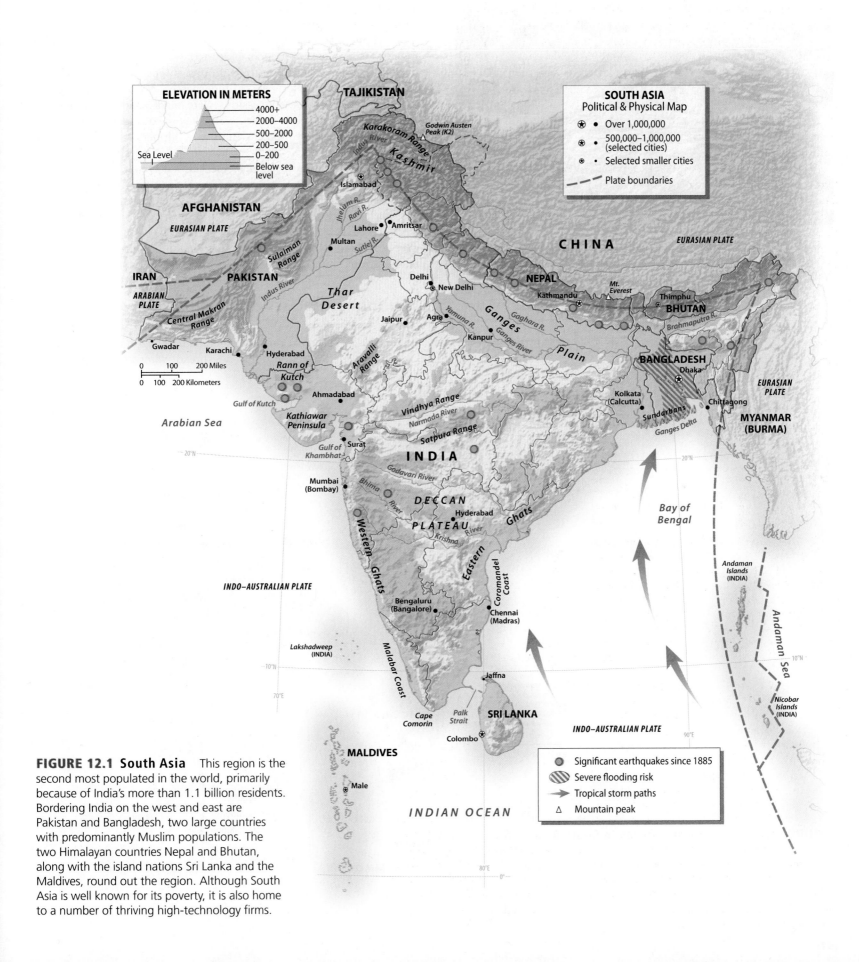

**FIGURE 12.1 South Asia**    This region is the second most populated in the world, primarily because of India's more than 1.1 billion residents. Bordering India on the west and east are Pakistan and Bangladesh, two large countries with predominantly Muslim populations. The two Himalayan countries Nepal and Bhutan, along with the island nations Sri Lanka and the Maldives, round out the region. Although South Asia is well known for its poverty, it is also home to a number of thriving high-technology firms.

**FIGURE 12.2 The Golden Quadrilateral Highway** India's infrastructure is notoriously poor, but the government is now responding with a massive highway construction program. The so-called Golden Quadrilateral highway, shown here, has generated numerous protests, as villagers object to the destruction of houses, temples, and trees that lie in its path.

its booming economy, India has recently undertaken a massive $12.2 billion road project designed to connect its four largest cities, New Delhi, Kolkata (Calcutta), Chennai (Madras), and Mumbai (Bombay), with a modern highway (Figure 12.2).

Although the new highway was largely completed in 2009, building it was not easy. Truckers protested the higher taxes and tolls needed to finance it, and citizen groups stopped construction with large protests, demanding more underpasses, overpasses, and cattle crossings. While some concessions were made, many rural Indians were infuriated as their homes were destroyed and their farms bisected by the massive project. Hindu worshippers were also angered when temples were relocated and when sacred trees were cut down.

Despite the many problems, the Golden Quadrilateral highway has generally proved an economic success. As soon as it was completed, it was carrying roughly 40 percent of India's road traffic, even though it constitutes only about 2 percent of the country's road network. As India's automobile market takes off, similar road projects may be needed elsewhere in the country, compounding the region's already serious environmental problems.

## Environmental Issues in South Asia

South Asia suffers from a number of severe natural hazards (Figure 12.3). Particular problems include flooding in the region's large river deltas, deforestation, and widespread water and air pollution.

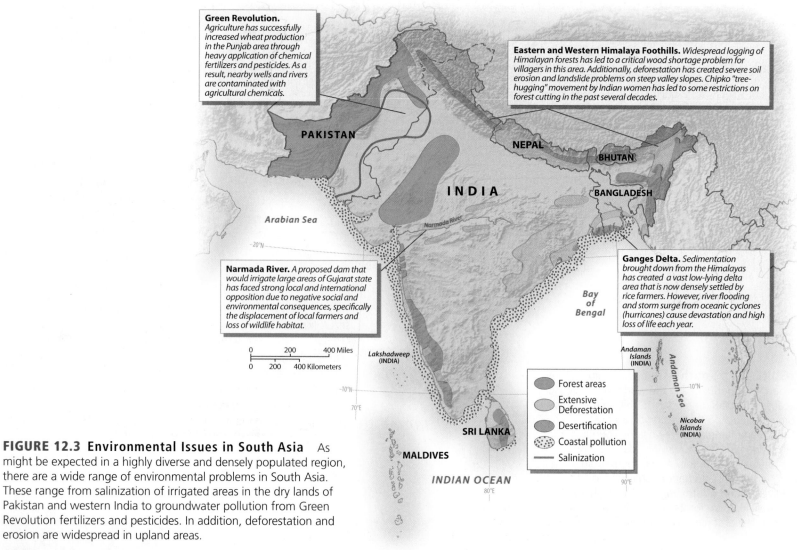

**Green Revolution.** *Agriculture has successfully increased wheat production in the Punjab area through heavy application of chemical fertilizers and pesticides. As a result, nearby wells and rivers are contaminated with agricultural chemicals.*

**Eastern and Western Himalaya Foothills.** *Widespread logging of Himalayan forests has led to a critical wood shortage problem for villagers in this area. Additionally, deforestation has created severe soil erosion and landslide problems on steep valley slopes. Chipko "tree-hugging" movement by Indian women has led to some restrictions on forest cutting in the past several decades.*

**Narmada River.** *A proposed dam that would irrigate large areas of Gujarat state has faced strong local and international opposition due to negative social and environmental consequences, specifically the displacement of local farmers and loss of wildlife habitat.*

**Ganges Delta.** *Sedimentation brought down from the Himalayas has created a vast low-lying delta area that is now densely settled by rice farmers. However, river flooding and storm surge from oceanic cyclones (hurricanes) cause devastation and high loss of life each year.*

Legend:
- Forest areas
- Extensive Deforestation
- Desertification
- Coastal pollution
- Salinization

**FIGURE 12.3 Environmental Issues in South Asia** As might be expected in a highly diverse and densely populated region, there are a wide range of environmental problems in South Asia. These range from salinization of irrigated areas in the dry lands of Pakistan and western India to groundwater pollution from Green Revolution fertilizers and pesticides. In addition, deforestation and erosion are widespread in upland areas.

**FIGURE 12.4 Flooding in Bangladesh** Devastating floods are common in the low-lying delta lands of Bangladesh. Heavy rains come with the southwest monsoon, especially to the Himalayas, and powerful cyclones often develop over the Bay of Bengal.

South Asia has suffered from some of the world's worst environmental disasters. The 1984 explosion of a fertilizer plant in Bhopal, India, for example, killed more than 2,500 persons. Compounding all these problems are the immense numbers of new people added each year through natural population growth.

**Natural Hazards in Bangladesh** The link between population pressure and environmental problems is nowhere clearer than in the delta area of Bangladesh, where the search for fertile land has driven people into hazardous areas, putting millions at risk from seasonal flooding as well as from the powerful cyclones (tropical storms) that form over the Bay of Bengal. For thousands of years, drenching monsoon rains have eroded and transported huge quantities of sediment from the Himalayan slopes to the sea by the Ganges and Brahmaputra rivers, gradually building this low-lying delta environment.

Although periodic flooding is a natural, even beneficial, phenomenon that enlarges deltas by depositing fertile river-borne sediment, flooding is a serious problem. In September 1998, for example, more than 22 million Bangladeshis were made homeless when water covered two-thirds of the country (Figure 12.4). Equally serious flooding in August 2007, however, caused much less damage, largely because of an internationally funded development program that was able to elevate hundreds of thousands of households above the 1998 high-water mark. With Bangladesh's population growing rapidly, however, there is a strong possibility that flooding will take higher tolls in the next decade, as desperate farmers continue to relocate into the hazardous lower floodplains. Continuing deforestation of the Ganges and Brahmaputra headwaters magnifies the problem.

**Forests and Deforestation** Tropical monsoon forests and savanna woodlands once covered most of the region, except for the desert areas in the northwest, but in most areas, tree cover has vanished as a result of human activities. The Ganges Valley and coastal plains of India, for example, were largely deforested hundreds of years ago to make room for agriculture. Elsewhere, forests were cleared more gradually for agricultural, urban, and industrial expansion. More recently, hillslopes in the Himalayas and elsewhere have been logged for commercial purposes. Extensive forests can still be found, however, in the far northern, southwestern, and east-central areas.

As a result of deforestation, many South Asian villages suffer from a shortage of fuelwood for household cooking, forcing people to burn dung cakes from cattle. While this low-grade fuel provides adequate heat, it prevents manure from being used as fertilizer. Where wood is available, collecting it may involve many hours of female labor because the remaining sources of wood are often far from the villages. In many areas, extensive eucalyptus stands have been planted to supply fuelwood and timber. These nonnative Australian trees support little or no wildlife, thus adding to problems of declining biodiversity throughout the region.

## South Asia's Monsoon Climates

The dominant climatic factor for most of South Asia is the **monsoon**, the seasonal change of wind direction that corresponds to wet and dry periods. During the winter, a large high-pressure system forms over the cold Asian landmass. As winds flow from high pressure to low, cold, dry winds flow outward from the continental interior across South Asia. This is the cool and dry season extending from November until February. As winter turns to spring, these winds diminish, resulting in the hot, dry season of March through May. Eventually the buildup of heat over South Asia and Southwest Asia produces a large low-pressure cell. By early June, the low-pressure cell is strong enough to cause a shift in wind direction so that warm, moist air from the Indian Ocean moves toward the continental interior. This signals the onset of the warm and rainy season of the southwest monsoon, which lasts from June through October (Figure 12.5).

**Orographic rainfall** is caused by the uplifting and cooling of moist monsoon winds over the Western Ghats and the Himalayan foothills.

**FIGURE 12.5 Monsoon Rain** During the summer monsoon, some Indian cities, such as Mumbai (Bombay), receive more than 70 inches (178 centimeters) of rain in just three months. These daily torrents cause floods, power outages, and daily inconvenience. However, these monsoon rains are crucial to India's agriculture. If the rains are late or abnormally weak, crop failure often results.

As a result, some areas receive more than 200 inches (508 centimeters) of rain during the four-month wet season (Figure 12.6). Cherrapunji, in northeastern India, is one of the world's wettest places, with an average rainfall of 450 inches (1,128 centimeters). On the Deccan Plateau, however, rainfall is dramatically reduced by a strong **rain-shadow effect**. A *rain shadow* is the area of low rainfall found on the downwind side of a mountain range. As winds move downslope, the air becomes warmer, and dry conditions usually prevail.

## Global Warming and South Asia

Due to its climate and landforms, South Asia is highly vulnerable to global warming. Even a minor rise in sea level will inundate large areas of the Ganges–Brahmaputra Delta in Bangladesh. Already, more than 18,500 acres (7,500 hectares) of swampland in the Sunderbans region have been submerged. A 2007 report from the Indian government suggests that up to 7 million people could be displaced from coastal areas by the end of the century due to a predicted 3.3-foot (1-meter) rise in sea level. If the most severe sea-level forecasts are realized, the atoll nation of the Maldives will simply vanish beneath the waves.

South Asian agriculture is likely to suffer from a number of problems linked to climate change. Most Himalayan glaciers are rapidly retreating, threatening the dry-season water supplies of the Indus–Ganges Plain, an area that already suffers from overuse of groundwater resources. Increased winter temperatures of up to 6.4°F (3°C) could destroy the vital wheat crop of Pakistan and northwestern India. In parts

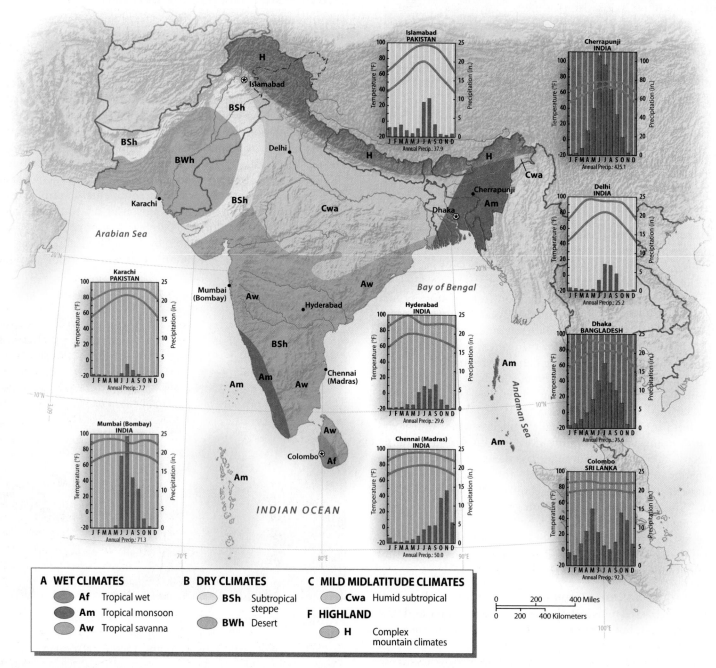

**FIGURE 12.6 Climates of South Asia** Except for the extensive Himalayas, South Asia is dominated by tropical and subtropical climates. Many of these climates show a distinct summer rainfall season that is associated with the southwest monsoon. The climographs for Mumbai (Bombay) and Delhi are excellent illustrations.

of South Asia, global warming could result in increased rainfall due to an intensification of the summer monsoon. Unfortunately, much of this rainfall would come from intense cloudbursts, and as a result, flooding and soil erosion would likely increase.

India signed the Kyoto Protocol in 2002, but as a developing country, it does not have to follow the main provisions of the treaty. With its poor and largely non-industrial economies, South Asia still has a low per capita output of greenhouse gases. But India's economy in particular is not only growing rapidly but is heavily dependent on burning coal to generate electricity. According to official estimates, if India is to maintain an economic growth rate of 8 percent over the next quarter century, it will have to triple or even quadruple its primary energy supply.

Global warming may put Pakistan in a particular bind. As much as 90 percent of the irrigation water for this mostly desert country comes from the mountains of Kashmir, a contested region divided between Pakistan and India. According to estimates of the Intergovernmental Panel on Climate Change, the glaciers of Kashmir that feed Pakistan's rivers could be mostly gone by 2035. Because Pakistan already experiences periodic water shortages, many experts believe that it must cooperate with India to develop and conserve the water resources of the Kashmir region. Considering the geopolitical tension between the two countries, which are focused on Kashmir, any such agreement seems unlikely.

## Physical Subregions of South Asia

To better understand environmental conditions in this diverse region, South Asia can be divided into four physical subregions, starting with the high mountain ranges of its northern edge and extending to the tropical islands of the far south. Lying south of the mountains are the extensive river lowlands that form the heartland of both India and Pakistan. Between river lowlands and the island countries is the vast area of peninsular India, extending more than 1,000 miles (1,600 kilometers) from north to south (Figure 12.7).

**Mountains of the North**  South Asia's northern rim of mountains is dominated by the great Himalayan Range, forming the northern borders of India, Nepal, and Bhutan. More than two dozen peaks exceed 25,000 feet (7,620 meters), including the world's highest mountain, Everest, on the Nepal–China (Tibet) border. To the east are the lower Arakan Yoma Mountains, forming the border between India and Burma (Myanmar) and separating South Asia from Southeast Asia.

These mountain ranges are a result of the dramatic collision of northward-moving peninsular India with the Asian landmass. The entire region is still geologically active, putting all of northern South Asia in serious earthquake danger. A massive earthquake in the Pakistani-controlled section of Kashmir on October 8, 2005, for example, resulted in roughly 100,000 deaths and left more than 3 million people homeless.

While most of South Asia's northern mountains are too rugged and high to support dense human settlement, major population clusters are found in the Kathmandu Valley of Nepal, situated at 4,400 feet (1,340 meters), and the Valley, or Vale, of Kashmir in northern India, at 5,200 feet (1,580 meters).

**Indus–Ganges–Brahmaputra Lowlands**  South of the northern mountains lie large lowlands created by three major river systems that have deposited sediments to build huge alluvial plains of fertile and

**FIGURE 12.7 South Asia from Space**  The four physical subregions of South Asia are clearly seen in this satellite photograph, from the snow-clad Himalayan mountains in the north to the islands of the south. The Deccan Plateau is dark, fringed by white clouds as moist air is lifted over the uplands of the Western Ghats.

easily farmed soils. These densely settled lowlands constitute the population core areas of Pakistan, India, and Bangladesh.

The Indus River, flowing from the Himalayas through Pakistan to the Arabian Sea, provides much-needed irrigation for Pakistan's southern deserts. More famous, however, is the Ganges, which flows southeasterly some 1,500 miles (2,400 kilometers) and empties into the Bay of Bengal. The Ganges has provided the fertile alluvial soil that has made northern India one of the world's most densely settled areas. Given the central role of this important river throughout Indian history, it is understandable why Hindus consider the Ganges sacred. Finally, the Brahmaputra River, which rises on the Tibetan Plateau, flows more than 1,700 miles (2,720 kilometers) before joining the Ganges in central Bangladesh and spreading out over the world's largest delta.

**Peninsular India**  Extending southward is peninsular India, made up primarily of the Deccan Plateau, which is bordered on each side by narrow coastal plains backed by north–south mountain ranges. On the west are the higher Western Ghats, which are generally about 5,000 feet in elevation (1,520 meters); to the east, the Eastern Ghats are lower and less continuous. On both coastal plains, fertile soils and an adequate water supply support population densities comparable to those of the Ganges lowland to the north.

Soil quality ranges from fair to poor over much of the Deccan Plateau, but in the state of Maharashtra, lava flows have produced particularly fertile black soils. However, much of the area does not have a reliable water supply for agriculture. The western portion of the plateau lies in the rain shadow of the Western Ghats, giving it a semi-arid climate. Small reservoirs or tanks have for centuries collected monsoon rainfall for use during the dry season. More recently, deep wells and powerful pumps have allowed groundwater development to support more widespread irrigation.

Partly because of the overuse of groundwater resources, the Indian government is building a series of large dams to provide for irrigation. Dam building, however, is controversial because the resulting reservoirs displace hundreds of thousands of rural residents. A case in point is the Sardar Sarovar Dam project on the Narmada River in the state of Madhya Pradesh, which has already dislodged more than 100,000

people and is currently being expanded. Local residents and activists throughout India have joined forces in opposition, but farmers in neighboring Gujarat, who will reap the benefits, strongly support the project.

**The Southern Islands**    At the southern tip of peninsular India lies the island country of Sri Lanka. Sri Lanka is ringed by extensive coastal plains and low hills, but mountains reaching more than 8,000 feet (2,438 meters) occupy the southern interior, providing a cool, moist climate. Because the main monsoon winds arrive from the southwest, that portion of the island is much wetter than the rain shadow areas of the north and east.

Forming a separate country are the Maldives, a chain of more than 1,200 islands stretching south to the equator some 400 miles (640 kilometers) off the southwestern tip of India. The combined land area of these islands is only about 116 square miles (290 square kilometers), and only one-quarter of the islands are inhabited. Like many islands of the South Pacific, the Maldives are low coral atolls, with a maximum elevation of just over 6 feet (2 meters) above sea level.

# Population and Settlement: The Demographic Dilemma

South Asia will soon surpass East Asia as the world's most populous region (Figure 12.8). India alone is home to more than 1.1 billion people, while Pakistan and Bangladesh, with 181 and 162 million residents, respectively, rank among the world's 10 most populous countries (Table 12.1). Furthermore, much of South Asia is still experiencing rapid population growth. Although South Asia has made remarkable agricultural gains over the past several decades, there is still widespread concern about its ability to feed

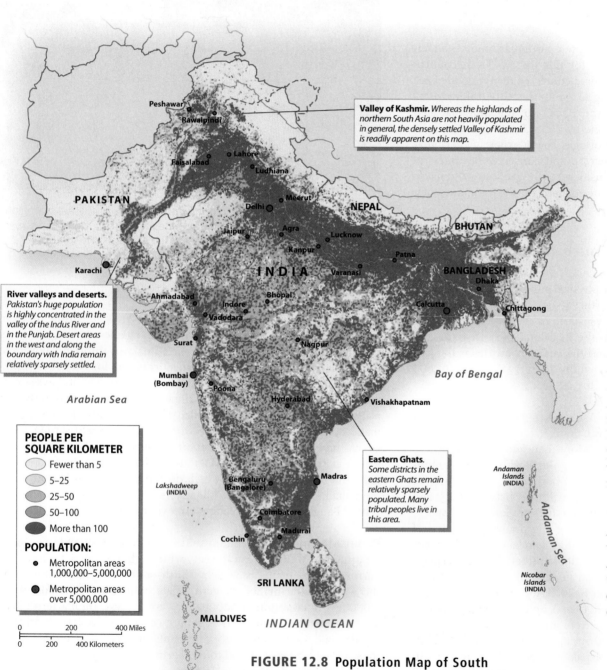

**Valley of Kashmir.** *Whereas the highlands of northern South Asia are not heavily populated in general, the densely settled Valley of Kashmir is readily apparent on this map.*

**River valleys and deserts.** *Pakistan's huge population is highly concentrated in the valley of the Indus River and in the Punjab. Desert areas in the west and along the boundary with India remain relatively sparsely settled.*

**Eastern Ghats.** *Some districts in the eastern Ghats remain relatively sparsely populated. Many tribal peoples live in this area.*

**PEOPLE PER SQUARE KILOMETER**
- Fewer than 5
- 5–25
- 25–50
- 50–100
- More than 100

**POPULATION:**
- ● Metropolitan areas 1,000,000–5,000,000
- ● Metropolitan areas over 5,000,000

**FIGURE 12.8 Population Map of South Asia**    Except for the desert areas of the west and the high mountains of the north, South Asia is a densely populated region. Particularly high densities of people are found on the fertile plains along the Indus and Ganges rivers and in India's coastal lowlands. In rural areas the population is typically clustered in villages, often located near water sources, such as streams, wells, canals, or small tanks that store water between monsoon rains.

## TABLE 12.1   POPULATION INDICATORS

| Country | Population (millions) 2009 | Population Density (per square kilometer) | Rate of Natural Increase (RNI) | Total Fertility Rate | Percent Urban | Percent <15 | Percent >65 | Net Migration (Rate per 1,000) 2005–10* |
|---|---|---|---|---|---|---|---|---|
| Bangladesh | 162.2 | 1,127 | 1.6 | 2.5 | 25 | 32 | 4 | −0.7 |
| Bhutan | 0.7 | 15 | 1.7 | 3.1 | 31 | 32 | 5 | 2.9 |
| India | 1,171.0 | 356 | 1.6 | 2.7 | 29 | 32 | 5 | −0.2 |
| Maldives | 0.3 | 1,057 | 1.8 | 2.3 | 35 | 30 | 5 | 0.0 |
| Nepal | 27.5 | 187 | 2.1 | 3.1 | 17 | 37 | 4 | −0.7 |
| Pakistan | 180.8 | 227 | 2.3 | 4.0 | 35 | 38 | 4 | −1.6 |
| Sri Lanka | 20.5 | 312 | 1.2 | 2.4 | 15 | 26 | 7 | −3.0 |

Source: Population Reference Bureau, *World Population Data Sheet, 2009.*

*Net Migration Rate from the United Nations, Population Division, *World Population*

itself. The threat of crop failure, although much reduced, remains, in part because much South Asian farming is vulnerable to the unpredictable monsoon rains.

India's total fertility rate (TFR) has dropped rapidly, falling from 6 in the 1950s to the current rate of 2.7. In western and southern India, fertility rates are now generally at or below replacement levels. In much of northern India, however, birthrates remain high; the average woman in the poor state of Bihar gives birth to 4 children. A distinct cultural preference for male children is found in most of South Asia, a tradition that further complicates family planning. Although performing sex-selective abortions is illegal in India, the practice persists. In much of northern India, only about 8 girls are born for every 10 boys. In southern India and Sri Lanka, where women generally have a higher social position, sex ratios are balanced, and birthrates are much lower.

Pakistan has seen a rapid reduction in its total fertility rate in recent years, but at 4.0, it is still well above the replacement level. As a result, Pakistan's population will probably be over 250 million by 2050, a worryingly high number, considering the country's arid environment, underdeveloped economy, and political instability. Bangladesh has been more successful than Pakistan in reducing its birthrate (Figure 12.9). As recently as 1975, its TFR was 6.3, but it dropped to 4.0 by 2009. The success of family planning can be partly attributed to strong support from Bangladesh's government, advertised through radio and billboards.

## Migration and the Settlement Landscape

South Asia is one of the least urbanized regions in the world, with fewer than one-third of its people living in cities. Most South Asians reside in compact rural villages. Rapid migration from villages to large cities, however, is occurring. This often results as much from desperate conditions in the countryside as it does from employment opportunities in the city. Increased mechanization of agriculture along with the expansion of large farms at the expense of subsistence cultivation push many people to the region's rapidly growing urban areas.

The most densely settled areas of South Asia are those with fertile soils and dependable water supplies. The highest rural population densities are found in the core area of the Ganges and Indus river valleys and on the coastal plains of India. Settlement is less dense on the Deccan Plateau and is relatively sparse in the highlands of the far north and the arid lands of the northwest.

Many South Asians have migrated in recent years from poor and densely populated areas to less densely populated or wealthier areas. Migrants are often attracted to large cities such as Mumbai (Bombay), but those from Bangladesh are settling in large numbers in rural portions of northeastern India, creating ethnic tensions. Sometimes migrants are forced out by war; a large number of both Hindus and Muslims from Kashmir, for example, have sought security away from their battle-scarred homeland.

Many experts are concerned about internal migration in South Asia, which results in huge shantytowns and soaring homeless populations in the country's largest cities. In 2009, however, the World Bank advised India to encourage relocation, noting that 60 percent of Indians live in areas with relatively stagnant economies while the more economically vibrant regions of the country sometimes experience labor shortages.

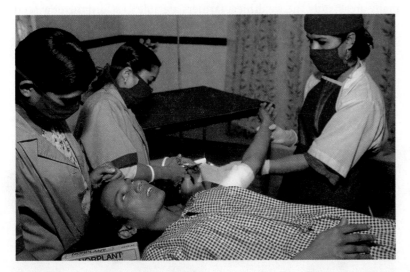

**FIGURE 12.9  Family Planning in Bangladesh**   Bangladesh has been one of the most successful nations in South Asia in reducing its fertility rate through family planning. Many women in Bangladesh use oral contraceptives. This photo shows a woman receiving a contraceptive implant.

## Agricultural Regions and Activities

South Asian agriculture has historically been relatively unproductive, especially compared with that of East Asia. Since the 1970s, however, agricultural production has grown rapidly. Many South Asian farmers are now deeply in debt, threatening future agricultural gains.

**Crop Zones**   South Asia can be divided into several distinct agricultural regions, all with different problems and potentials. These regions are based on the production of three subsistence crops—rice, wheat, and millet.

Rice is the main crop and foodstuff in the lower Ganges Valley, along the lowlands of India's eastern and western coasts, in the delta lands of Bangladesh, along Pakistan's lower Indus Valley, and in Sri Lanka (Figure 12.10). This distribution reflects the large volume of irrigation water needed to grow rice. The amount of rice grown in South Asia is impressive: India ranks behind only China in world rice production, and Bangladesh is the fourth largest producer.

Wheat is the principal crop of the northern Indus Valley and in the western half of India's Ganges Valley. South Asia's "breadbasket" is the northwestern Indian state of Punjab and adjacent areas in Pakistan. Here the so-called Green Revolution has been particularly successful in increasing grain yields. In the less fertile areas of central India, millet and sorghum are the main crops, along with root crops such as manioc. In general, wheat and rice are the preferred staples throughout South Asia, but poorer people must often subsist on millet and root crops.

**The Green Revolution**   The main reason South Asian agricultural growth has kept up with population growth is the **Green Revolution**, which originated during the 1960s in agricultural research stations established by international development agencies. By the 1970s, efforts to breed high-yield varieties of rice and wheat had reached their initial

**FIGURE 12.11  Green Revolution Farming**   Because of "miracle" wheat strains that have increased yields in the Punjab area, this region has become the breadbasket of South Asia. India has more than doubled its wheat production in the past 25 years and has moved from continual food shortages to self-sufficiency. In this photo, a man is washing down miniature greenhouses used to start seedlings that will later be transplanted.

goals. As a result, South Asia was transformed from a region of chronic food deficiency to one of self-sufficiency. India more than doubled its annual grain production between 1970 and the mid-1990s (Figure 12.11).

While the Green Revolution was an agricultural success, many experts highlight its ecological and social costs. Serious environmental problems result from the chemical dependency of the new crop strains. Not only do these crops typically need large quantities of industrial fertilizer, which is both expensive and polluting, they also require frequent pesticide applications because they lack natural resistance to plant diseases and insects.

Social problems have also followed the Green Revolution. In many areas, only the wealthiest farmers are able to afford the new seed strains, irrigation equipment, farm machinery, fertilizers, and pesticides. As a result, poorer farmers have often been forced from their

**FIGURE 12.10  Rice Cultivation**   A large amount of irrigation water is needed to grow rice, as is apparent from this photo from Sri Lanka. Rice is also the main crop in the lower Ganges Valley and Delta, along the lower Indus River of Pakistan, and in India's coastal plains.

**FIGURE 12.12 Mumbai Hutments**   Hundreds of thousands of people in Mumbai live in crude hutments, with no sanitary facilities, built on formerly busy sidewalks. Hutment construction is forbidden in many areas, but wherever it is allowed, sidewalks quickly disappear.

lands, becoming wage laborers for their more successful neighbors or migrating to crowded cities. To purchase the necessary inputs, moreover, most farmers have had to borrow large amounts of money. As lagging crop prices prevented many from repaying their debts, a wave of suicides occurred. Between 2002 and 2006, more than 85,000 Indian farmers killed themselves.

While the Green Revolution has fed South Asia's expanding population over the past several decades, it remains unclear whether it will be able to continue doing so. An alternative option is to expand water delivery systems (either through canals or wells), as many fields are not irrigated. Irrigation, however, brings its own problems. In much of Pakistan and northwestern India, where irrigation has been practiced for generations, soil **salinization**, or the buildup of salt in fields, is already a major problem. In addition, groundwater is being depleted, especially in Punjab, India's breadbasket.

## Urban South Asia

Although South Asia remains a largely rural society, many of its cities are large and growing quickly. India alone has more than 40 metropolitan areas with more than 1 million inhabitants. Because of this rapid growth, South Asian cities have serious problems with homelessness, poverty, congestion, water shortages, air pollution, and sewage disposal. Throughout South Asia, sprawling squatter settlements, or **bustees**, exist in and around urban areas, providing meager shelter for many migrants.

**Mumbai (Bombay)**   The largest city in South Asia, Mumbai (often called by its former name, Bombay) is India's financial, industrial, and commercial center. Mumbai itself contains roughly 14 million people, while its metropolitan area is home to more that 22 million. Mumbai is responsible for much of India's foreign trade, has long been a manufacturing center, and is the focus of India's film industry—the world's largest. Mumbai's economic vitality draws people from all over India, resulting in simmering ethnic tensions.

Because of the city's restricted space, most of Mumbai's growth has taken place to the north and east of the historic city. Building restrictions in the downtown area have resulted in skyrocketing commercial and residential rents, which are some of the highest in the world. Even members of the city's thriving middle class have difficulty finding adequate housing. Hundreds of thousands of less-fortunate immigrants live in "hutments," crude shelters built on formerly busy sidewalks (Figure 12.12). The least fortunate sleep on the street or in simple plastic tents, often placed along busy roadways.

Mumbai's notorious road congestion eased somewhat in 2009, after the completion of a massive eight-lane, $340 million bridge linking the central city to its northern suburbs. The Mumbai Metro, an ambitious rapid transit system, is currently under construction, promising further improvements.

**Kolkata (Calcutta)**   To many, Kolkata—more often called by its old name, Calcutta—symbolizes the problems faced by rapidly growing cities in developing countries. Approximately 1 million people here sleep on the streets every night. And with more than 15 million people in its metropolitan area, Kolkata falls far short of supplying the rest of its residents with water, power, and sewage treatment. Electrical power is woefully inadequate, and during the wet season, many streets are routinely flooded.

With rapid growth as migrants pour in from the countryside, a mixed Hindu–Muslim population that generates ethnic tension, a decayed economic base, and an overloaded infrastructure, Kolkata faces a troubled future. Yet it remains a culturally vibrant city, noted for its fine educational institutions, theaters, and publishing firms. Kolkata is currently trying to nurture an information technology industry, but it remains to be seen whether it will prove successful.

**Karachi**   Karachi, Pakistan's largest urban area and commercial core, is one of the world's fastest growing cities. Its metropolitan population, already somewhere between 12 and 18 million, is expanding at

**FIGURE 12.13 Karachi Landscape**   Karachi, Pakistan's largest city and main port, is noted for both its economic power and its ethnic violence.

about 5 percent per year (Figure 12.13). Karachi served as Pakistan's capital until 1963, when the new city of Islamabad was created in the northeast. Karachi suffered relatively little from the departure of government functions; it is still the most cosmopolitan city in Pakistan, its main streets lined with businesses and high-rise buildings.

Karachi also suffers from political and ethnic tensions that have periodically turned parts of the city into armed camps and even battlegrounds. In the early decades of Pakistan's independence, Karachi's main conflict was between the Sindis, the region's native inhabitants, and the Muhajirs, the Muslim refugees from India who settled in the city after division from India in 1947. More recently, clashes between Sunni and Shiite Muslims have intensified, as have those between Pashtun migrants from northwestern Pakistan and Afghanistan and other residents. Ethnic rioting in Karachi in April 2009 resulted in more than 30 deaths.

## Cultural Coherence and Diversity: A Common Heritage Undermined by Religious Rivalries

Historically, South Asia is a well-defined cultural region. A thousand years ago, virtually the entire area was united by the religion of Hinduism. The subsequent arrival of Islam added a new religious element but did not undermine the region's cultural unity. British imperialism later added a number of cultural features to the region, from the widespread use of English to a passion for cricket. Since the mid-20th century, however, religious strife has intensified, leading some to question whether South Asia can still be considered a culturally unified region.

India has been a secular state since its creation. Since the 1980s, this political tradition has come under pressure from the growth of **Hindu nationalism,** which promotes Hindu values as the foundation

of Indian society. Hindu nationalists have gained considerable political power through the Bharatiya Janata Party (BJP). In several high-profile instances, Hindu mobs demolished Muslim mosques that had allegedly been built on the sites of ancient Hindu temples (Figure 12.14). Since 2000, however, the Hindu nationalist movement has declined, and the more secular political parties have won most of India's recent elections.

In Pakistan, Islamic fundamentalism has been a divisive issue. Powerful fundamentalist leaders want to make Pakistan a religious state under Islamic law, a plan rejected by the country's secular intellectuals and international businesspeople. The government has attempted to mediate between the two groups, but with little success. As a result, large areas of the country have come under the control of Islamist insurgents, leading some observers to question whether Pakistan can hold together as a country.

### Origins of South Asian Civilizations

Many scholars think that the roots of South Asian culture extend back to the Indus Valley civilization, which flourished 4,500 years ago in what is now Pakistan. This remarkable urban-oriented society vanished almost entirely around 1800 BCE. By 800 BCE, however, a new focus of civilization had emerged in the middle Ganges Valley.

**Hindu Civilization**   The religion that emerged out of the early Ganges Valley civilization was Hinduism, a complicated faith that lacks a single system of belief. Certain deities are recognized, however, by all believers, as is the notion that these various gods are all expressions of a single divine entity (Figure 12.15). Hindus also share a common set of epic stories, usually written in **Sanskrit,** the sacred language of their religion. Hinduism is noted for its mystical tendencies, which have long inspired many to seek an ascetic lifestyle, renouncing property and sometimes all regular human relations. One of its hallmarks is a belief in the transmigration of souls from being to being through reincarnation. Hinduism is also associated with India's **caste system,**

**FIGURE 12.14 Destruction of the Ayodhya Mosque**   A group of Hindu nationalists are seen listening to speeches, urging them to demolish the mosque at Ayodhya, allegedly built on the site of a former Hindu temple. The mosque was later destroyed by Hindu fundamentalists.

**FIGURE 12.15 Hindu Temple**   Although India has long been noted for its ancient temples, lavish new Hindu religious complexes continue to be constructed. Visible in this photograph is the Akshardham Temple, inaugurated in New Delhi in 2005. Expected to be a major tourist attraction, Akshardham also serves as an educational center focused on the culture of India.

the strict division of society into hereditary groups that are ranked as ritually superior or inferior to each other.

**Buddhism**   Ancient India's caste system was challenged from within by Buddhism. Siddhartha Gautama, the Buddha, was born in 563 BCE in an elite caste. He rejected the life of wealth and power, however, and sought instead to attain enlightenment, or mystical union with the universe. He preached that the path to such enlightenment (or "nirvana") was open to all, regardless of social position. His followers eventually established Buddhism as a new religion. Buddhism spread through South Asia and later expanded through East, Southeast, and Central Asia. But it never fully replaced Hinduism in India, and by 500 CE, Buddhism was disappearing from most of South Asia.

**Arrival of Islam**   The next major challenge to Hindu society—Islam—came from the outside. Around the year 1000, Turkic-speaking Muslims began to invade from Central Asia. By the 1300s, most of South Asia lay under Muslim power, although Hindu kingdoms persisted in southern India. During the 16th and 17th centuries, the **Mughal (or Mogul) Empire**, the most powerful of the Muslim states, dominated much of the region from its power center in the upper Indus–Ganges Basin (Figure 12.16).

At first, Muslims formed a small ruling elite, but over time, increasing numbers of Hindus converted to the new faith. Conversions were most pronounced in the northwest and northeast, with the areas now known as Pakistan and Bangladesh becoming predominantly Muslim.

**The Caste System**   Caste is one of the historically unifying features of South Asia, as certain aspects of caste organization are even found among the Muslim and Christian populations of the region. Islam, however, has gradually reduced its significance, and even in India, caste is now being de-emphasized, especially among more educated people. But caste remains significant, especially in rural India. Marriage across caste lines is still relatively rare, and caste issues figure heavily in Indian politics. In 2006 a major controversy erupted when the Indian government proposed to increase the number of positions in public universities reserved for members of the lower castes. Proponents claimed that such a measure was necessary to address discrimination, whereas opponents claimed that it would result in lower educational standards.

*Caste* is actually a rather clumsy term that refers to the complex social order of the Hindu world. It combines two distinct local concepts: *varna* and *jati*. *Varna* refers to the ancient fourfold social hierarchy of the Hindu world, which distinguishes the Brahmins (priests), Kshatriyas (warriors), Vaishyas (merchants), and Sundras (farmers and craftsmen), in declining order of ritual purity. Standing outside this traditional order are the so-called untouchables, now usually called **Dalits**, whose ancestors held "impure" jobs, such as those associated with leather working. *Jati*, on the other hand, refers to the hundreds of local endogamous ("marrying within") groups that exist at each *varna* level. Different *jati* groups are often called *subcastes*.

## Contemporary Geographies of Religion

In the simplest terms, South Asia has a Hindu heritage overlain by a significant Muslim presence. Such a picture fails, however, to capture the enormous diversity of religion in contemporary South Asia (Figure 12.17).

**Hinduism**   Fewer than 1 percent of the people of Pakistan are Hindu, and in Bangladesh and Sri Lanka, Hinduism is a minority religion. But in India and Nepal, Hinduism is clearly the majority faith. In most of central India, more than 90 percent of the population is Hindu. Hinduism is itself a geographically complicated religion, with different aspects of faith varying across different parts of India.

**Islam**   Islam may be a minority religion for South Asia as a whole, but it is still very widespread, counting more than 450 million followers. Bangladesh and especially Pakistan are overwhelmingly Muslim. India's Muslim community, although constituting only some 15 percent of the country's population, is still roughly 150 million strong.

**FIGURE 12.16 The Red Fort**   The Red Fort of Delhi, completed in 1648, was the power center of the Mughal Empire. Today this massive fortification, one of the largest in the world, is a major tourist destination.

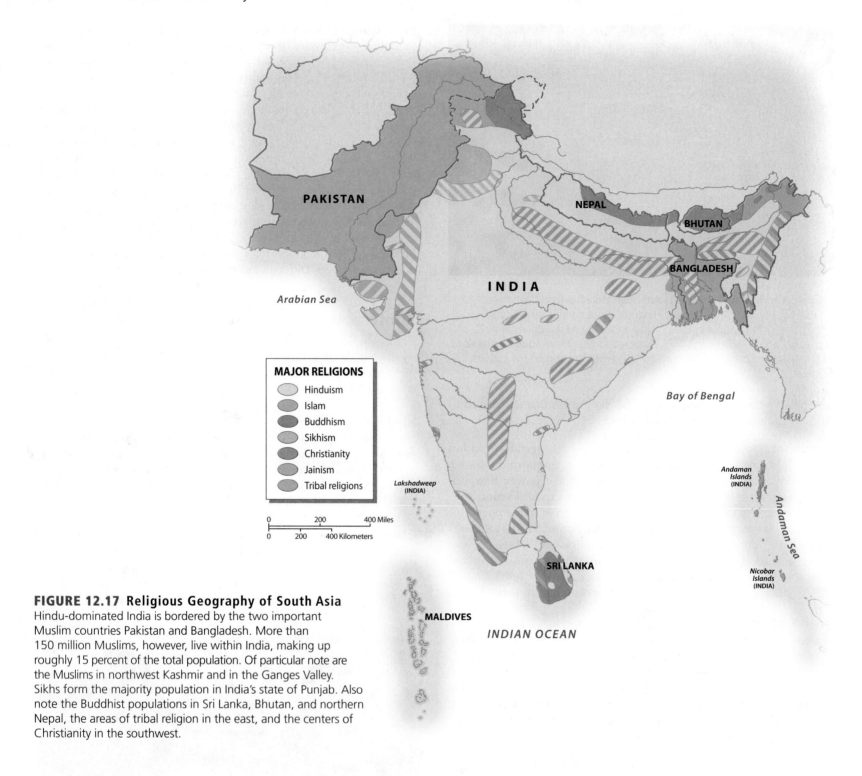

**FIGURE 12.17 Religious Geography of South Asia**
Hindu-dominated India is bordered by the two important
Muslim countries Pakistan and Bangladesh. More than
150 million Muslims, however, live within India, making up
roughly 15 percent of the total population. Of particular note are
the Muslims in northwest Kashmir and in the Ganges Valley.
Sikhs form the majority population in India's state of Punjab. Also
note the Buddhist populations in Sri Lanka, Bhutan, and northern
Nepal, the areas of tribal religion in the east, and the centers of
Christianity in the southwest.

Muslims live in almost every part of India. They are, however,
concentrated in four main areas: in most large cities; in Kashmir, in
the far north, particularly in the densely populated Vale of Kashmir
(more than 80 percent of the population here follows Islam); in the
central Ganges plain, where Muslims constitute 15 to 20 percent of
the population; and in the southwestern state of Kerala, which is approximately 25 percent Muslim.

Interestingly, Kerala was one of the few parts of India that never
experienced prolonged Muslim rule. Islam in Kerala was historically
connected to trade across the Arabian Sea. Kerala's Malabar Coast
historically supplied spices and other luxury products to Southwest
Asia, encouraging many Arab traders to settle there. Gradually, many
of Kerala's native residents converted to the new religion as well. The
same trade routes brought Islam to Sri Lanka, which is approximately
9 percent Muslim, and to the Maldives, which are almost entirely
Muslim.

**Sikhism**    The tension between Hinduism and Islam in northern South
Asia gave rise to a new religion called **Sikhism**. Sikhism originated in
the 1400s in the Punjab, near the modern boundary between India and

**FIGURE 12.18 Sikh Soldiers**    The Sikh religious minority in India has long been noted for its military traditions. In this photograph, Sikh soldiers from the Rashtriya Rifles regiment are participating in an Army Day parade in New Delhi in 2003.

Pakistan. The Punjab was the site of intense religious competition at the time; Islam was gaining converts and Hinduism was on the defensive. The new faith combined elements of both religions. Many orthodox Muslims viewed Sikhism as dangerous because it incorporated elements of their own religion in a manner contrary to accepted beliefs. Periodic persecution led the Sikhs to adopt a militantly defensive stance. Even today, many Sikh men work as soldiers and bodyguards (Figure 12.18).

At present, the Indian state of Punjab is approximately 60 percent Sikh. Small but often influential groups of Sikh are scattered across the rest of India. Devout Sikh men are immediately visible because they do not cut their hair or their beards. Instead, they wear their hair wrapped in a turban and often tie their beards close to their faces.

**Buddhism and Jainism**    Although Buddhism virtually disappeared from India in medieval times, it persisted in Sri Lanka. Among the island's dominant Sinhalese people, Theravada Buddhism developed into a national religion (Figure 12.19). In the high valleys of the Himalayas, the Tibetan form of Buddhism emerged as the majority faith. The town of Dharamsala in the northern Indian state of Himachal Pradesh is the seat of Tibet's government-in-exile and of its spiritual leader, the Dalai Lama, who fled Tibet in 1959 after an unsuccessful revolt.

At roughly the same time as the birth of Buddhism (circa 500 BCE), another religion emerged in northern India: **Jainism**. This religion also stressed nonviolence, taking this creed to its ultimate extreme. Jains are forbidden to kill any living creatures, and as a result, the most devoted members of the community wear gauze masks to prevent them from inhaling small insects. Agriculture is forbidden to Jains because plowing can kill small creatures. As a result, most members of the faith have looked to trade for their livelihoods. Today, Jains are concentrated in northwestern India.

**Other Religious Groups**    The Parsis, concentrated in Mumbai, form a small but influential religious group. Followers of Zoroastrianism, the ancient faith of Iran, Parsi refugees fled to India in the 7th century. The Parsis prospered under British rule, forming some of India's first modern, industrial companies, such as the Tata Group (see "Exploring Global Connections: The Tata Group"). Intermarriage and low fertility, however, now threaten the survival of this small community.

Indian Christians are more numerous than either Parsis or Jains. Their religion arrived some 1,700 years ago, as missionaries from Southwest Asia brought Christianity to India's southwestern coast. Today, roughly 20 percent of the people of Kerala follow Christianity. Several Christian sects are represented, but the largest are affiliated with the Syrian Christian Church of Southwest Asia. Another stronghold of Christianity is the small Indian state of Goa, a former Portuguese colony. Here Roman Catholics make up roughly half of the population.

During the colonial period, British missionaries went to great efforts to convert South Asians to Christianity. They had little success, however, in Hindu, Muslim, and Buddhist communities. The remote tribal districts of British India proved to be more receptive to missionary activity, especially those in the northeast. The Indian states of Nagaland, Meghalaya, and Mizoram now have Christian majorities, with more than 75 percent of the people of Nagaland belonging to the Baptist church.

## Geographies of Language

South Asia's linguistic diversity rivals its religious diversity. In northern South Asia, most languages belong to the Indo–European family, the world's largest. The languages of southern India, on the other hand, belong to the **Dravidian** family, which is found only in South Asia. Along the mountainous northern rim of the region, a third linguistic family, Tibeto–Burman, dominates. Within these broad divisions are many different languages, each associated with a distinct

**FIGURE 12.19 Buddhist Monastic Landscape**    The dominant Sinhalese population of Sri Lanka is noted for its devotion to Theravada Buddhism. Young Buddhist monks, like those visible in this photograph, are a common sight in the Sinhalese-speaking portions of the country.

culture. In many parts of South Asia, several languages are spoken within the same area, and the ability to speak several languages is common everywhere (Figure 12.20).

Each of the major languages of India is associated with an Indian state, as the country deliberately structured its political subdivisions along linguistic lines a decade after attaining independence. As a result, one finds Gujarati in Gujarat, Marathi in Maharashtra, Oriya in Orissa, and so on. Two of these languages, Punjabi and Bengali, extend into Pakistan and Bangladesh, respectively, as the political borders here were established on religious rather than linguistic lines. Nepali, the national language of Nepal, is also spoken in many of the mountainous areas of northern India. Minor dialects and languages abound in many of the more remote areas.

**The Indo–European North**    The most widely spoken language of South Asia is **Hindi** (not to be confused with the Hindu religion). With more than 500 million native speakers, Hindi is by some measurements the world's second most widely spoken language. It occupies a

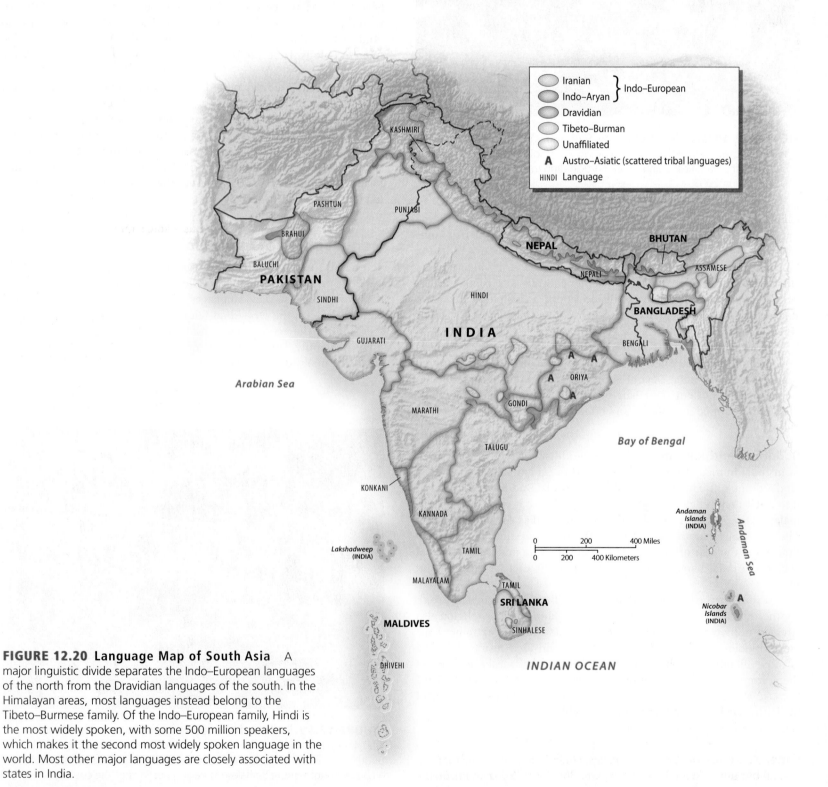

**FIGURE 12.20 Language Map of South Asia**    A major linguistic divide separates the Indo–European languages of the north from the Dravidian languages of the south. In the Himalayan areas, most languages instead belong to the Tibeto–Burmese family. Of the Indo–European family, Hindi is the most widely spoken, with some 500 million speakers, which makes it the second most widely spoken language in the world. Most other major languages are closely associated with states in India.

**FIGURE 12.21 Kolkata (Calcutta) Bookstore**    Although Kolkata (Calcutta) is noted in the West mostly for its abject poverty, the city is also known in India for its vibrant cultural and intellectual life, illustrated by its large number of bookstores, theaters, and publishing firms.

prominent role in present-day India, both because so many people speak it and because it is the main language of the Ganges Valley. Hindi is an official language in 10 Indian states, and it is widely studied by advanced students throughout the country.

Bengali, the second most widely spoken language in South Asia, is the official language of Bangladesh and the Indian state of West Bengal. Spoken by roughly 200 million people, Bengali is the world's ninth most widely spoken language. It also has an extensive literature, as West Bengal (particularly its capital city Kolkata [Calcutta]) has long been one of South Asia's leading literary and intellectual centers (Figure 12.21).

The Punjabi-speaking zone in the west was split at the time of independence between Pakistan and the Indian state of Punjab. While almost 100 million people speak Punjabi, this language does not have the significance of Bengali. Punjabi did not become the national language of Pakistan, even though it is the day-to-day language of almost half of the country's population. Instead, that position was given to Urdu.

**Urdu**, like Hindi, originated on the plains of northern India. The difference between the two was largely one of religion: Hindi was the language of the Hindu majority, Urdu that of the Muslim minority. Because of this distinction, Hindi and Urdu are written differently—the former in the Devanagari script (derived from Sanskrit) and the latter in the Arabic script. Although Urdu contains many words borrowed from Persian, its basic grammar and vocabulary are almost identical to those of Hindi. With independence in 1947, millions of Urdu-speaking Muslims from the Ganges Valley fled to Pakistan. Because Urdu had a higher status than Pakistan's native tongues, it was quickly established as the new country's official language. Although only about 8 percent of the people of Pakistan learn Urdu as their first language, more than 90 percent are able to speak and understand it.

**Languages of the South**    The four main **Dravidian languages** are confined to southern India and northern Sri Lanka. As in the north, each language is closely associated with an Indian state: Kannada in

Karnataka, Malayalam in Kerala, Telugu in Andhra Pradesh, and Tamil in Tamil Nadu. Tamil is usually considered the most important member of the family because it has the longest history and the largest literature. Tamil poetry dates back to the 1st century CE, making it one of the world's oldest written languages.

Although Tamil is spoken in northern Sri Lanka, the country's majority population, the Sinhalese, speak an Indo–European language. Apparently, the Sinhalese migrated from northern South Asia several thousand years ago, settling primarily on the island's fertile southwestern coast and central highlands. These same people also migrated to the Maldives, where the national language, Dhivehi, is essentially a Sinhalese dialect. The drier north and east of Sri Lanka, on the other hand, were settled mainly by Tamils from southern India. Some Tamils later moved to the central highlands, where they were employed as tea-pickers on British-owned estates.

**Linguistic Dilemmas**    The multilingual countries of Sri Lanka, Pakistan, and India are all troubled by linguistic conflicts. Such problems are most complex in India, simply because India is so large and has so many different languages (Figure 12.22).

Indian nationalists have long dreamed of a national language that could help unify their country. But **linguistic nationalism,** or the linking of a specific language with political goals, often faces the resistance of many people. The obvious choice for a national language would be Hindi, and Hindi was indeed declared as such in 1947. Raising Hindi to this position, however, angered many non-Hindi speakers, especially in the Dravidian south. It was eventually decided that while both Hindi and English would serve as official languages of India as a whole, each Indian state could select its own official language. As a result, 22 separate Indian languages now have such status.

Regardless of opposition, Hindi is expanding, especially in the Indo–European north. Here, local languages are closely related to Hindi, which can therefore be learned fairly easily. Hindi is spreading through education and, more significantly, through television and motion pictures. Films and television programs are made in several

**FIGURE 12.22 Multilingualism**    This four-language sign, in Malayalam, Tamil, Kannada, and English, shows the multilingual nature of contemporary South Asia. In many ways, English, the colonial language of British rule, still serves to bridge the gap between the many different languages of the region.

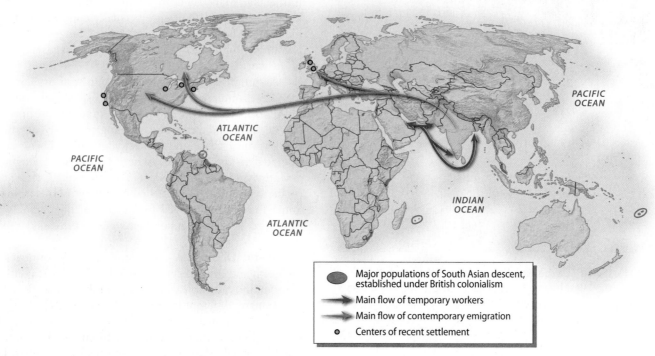

**FIGURE 12.23 The South Asian Global Diaspora** During the British imperial period, large numbers of South Asian workers settled in other colonies. Today, roughly 50 percent of the population of such places as Fiji and Mauritius are of South Asian descent. More recently, large numbers have settled, and are still settling, in Europe (particularly Britain) and North America. Large numbers of temporary workers, both laborers and professionals, are employed in the wealthy oil-producing countries of the Persian Gulf.

northern languages, but Hindi remains primary. In a poor but modernizing country such as India, where many people experience the wider world largely through moving images, the influence of a national film and television culture can be substantial.

Despite its spread, Hindi remains foreign to much of India. National-level communication is thus conducted mainly in English. While many Indians want to de-emphasize English, others advocate it as a neutral national language because all parts of the country have an equal stake in it. Furthermore, English gives substantial international benefits. English-medium schools abound throughout South Asia, and many children of the elite learn this global language well before they begin school.

## South Asia in Global Cultural Context

The widespread use of English in South Asia not only helps global culture spread through the region, but has also helped South Asians' cultural production reach a global audience. The global spread of South Asian literature, however, is nothing new. As early as the turn of the

20th century, Rabindranath Tagore gained international acclaim for his poetry and fiction, earning the Nobel Prize for Literature in 1913.

The expansion of South Asian culture abroad has been accompanied by the spread of South Asians themselves. Migration from South Asia during the time of the British Empire led to the establishment of large communities in such distant places as eastern Africa, Fiji, and the southern Caribbean (Figure 12.23). Subsequent migration

**FIGURE 12.24 Goa Beach Scene** The liberal Indian state of Goa, formerly a Portuguese colony, is now a major destination for tourists, both from within India and from Europe and Israel. European tourists come in the winter for sunbathing and for "Goan rave parties," where ecstasy and other drugs are widely available. Indian tourists typically find the scantily clad foreigners unusual, if not bizarre.

has been aimed more at the developed world; there are now several million people of South Asian origin living in Britain, and a similar number live in North America. Many present-day migrants to the United States are doctors, software engineers, and members of other professions.

In South Asia itself, the globalization of culture has brought tensions as severe as those felt anywhere else in the world. Traditional Hindu and Muslim religious norms frown on any overt display of sexuality—a staple feature of global popular culture. Religious leaders thus often criticize Western films and television shows as being immoral. Still, the pressures of internationalization are hard to resist. In the tourism-oriented Indian state of Goa, such tensions are on full display. There, German and British sun worshipers often wear nothing but skimpy clothes, whereas Indian women tourists go into the ocean fully clothed. Young Indian men, for their part, often simply walk the beach and gawk, good-naturedly, at the semi-naked foreigners (Figure 12.24).

## Geopolitical Framework: A Deeply Divided Region

Before the coming of British imperialism, South Asia had never been politically united. While a few empires at times ruled most of the subcontinent, none covered its entire extent. The British, however, brought the entire region into a single political system by the middle of the 19th century. Independence in 1947 brought the separation of Pakistan from India; in 1971, Pakistan itself was divided, with the independence of Bangladesh, formerly East Pakistan. Today, serious geopolitical issues continue to plague the region (Figure 12.25).

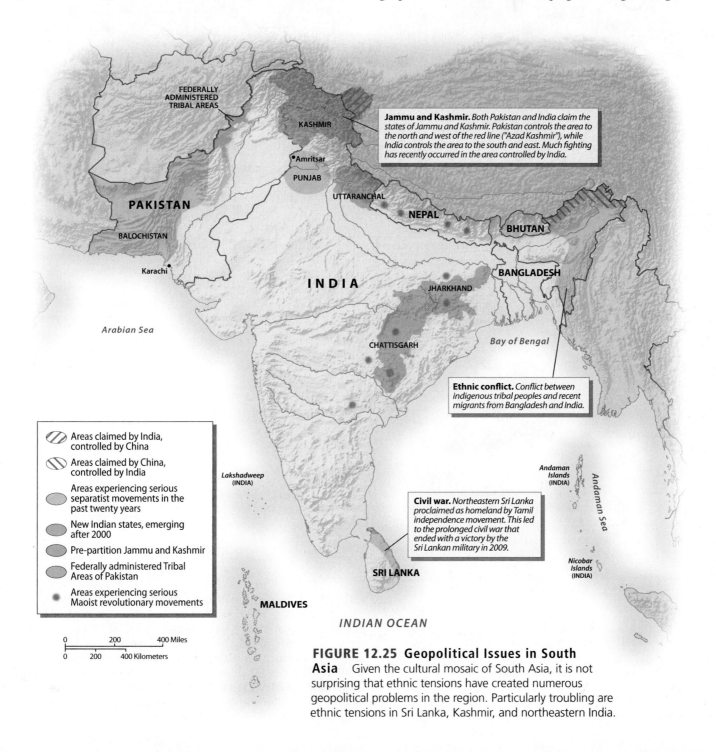

**Jammu and Kashmir.** *Both Pakistan and India claim the states of Jammu and Kashmir. Pakistan controls the area to the north and west of the red line ("Azad Kashmir"), while India controls the area to the south and east. Much fighting has recently occurred in the area controlled by India.*

**Ethnic conflict.** *Conflict between indigenous tribal peoples and recent migrants from Bangladesh and India.*

**Civil war.** *Northeastern Sri Lanka proclaimed as homeland by Tamil independence movement. This led to the prolonged civil war that ended with a victory by the Sri Lankan military in 2009.*

Legend:
- Areas claimed by India, controlled by China
- Areas claimed by China, controlled by India
- Areas experiencing serious separatist movements in the past twenty years
- New Indian states, emerging after 2000
- Pre-partition Jammu and Kashmir
- Federally administered Tribal Areas of Pakistan
- Areas experiencing serious Maoist revolutionary movements

**FIGURE 12.25 Geopolitical Issues in South Asia** Given the cultural mosaic of South Asia, it is not surprising that ethnic tensions have created numerous geopolitical problems in the region. Particularly troubling are ethnic tensions in Sri Lanka, Kashmir, and northeastern India.

## South Asia Before and After Independence in 1947

During the 1500s, when Europeans first arrived, most of northern South Asia was ruled by the Muslim Mughal Empire (Figure 12.26), while southern India remained under the control of the Hindu kingdom of *Vijayanagara*. European merchants, eager to obtain spices, textiles, and other Indian products, established a number of coastal trading posts. The Portuguese carved out an enclave in Goa, while the Dutch gained control over much of Sri Lanka, but neither was a significant threat to the Mughals. In the early 1700s, however, the Mughal Empire weakened rapidly, with a number of competing states emerging in its former territories.

**The British Conquest**  The unsettled conditions of the 1700s provided an opening for European imperialism. The British and French, having largely displaced the Dutch and Portuguese, competed for trading posts. Before the Industrial Revolution, Indian cotton textiles were considered the best in the world, and European merchants needed large quantities for their global trading networks. After

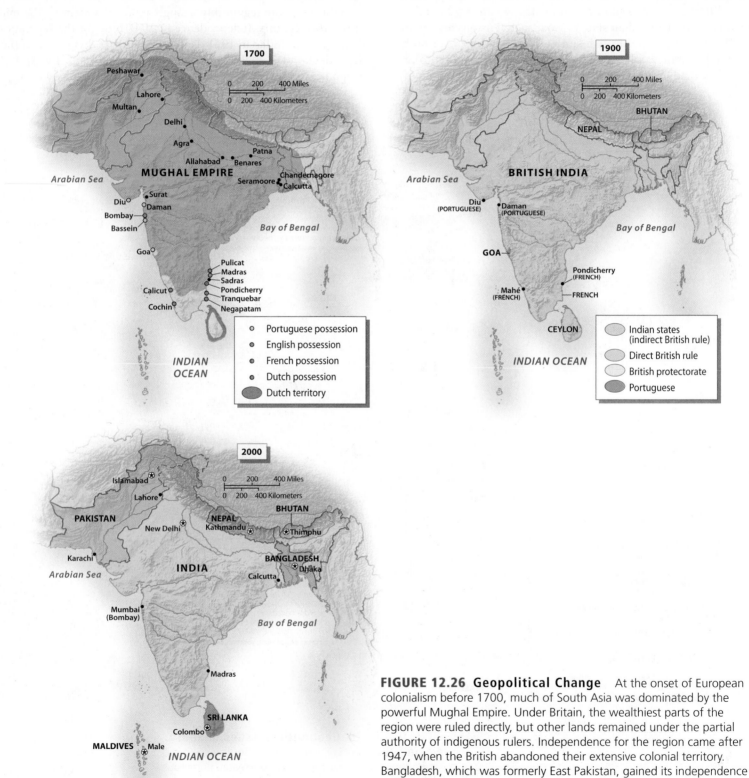

**FIGURE 12.26 Geopolitical Change**  At the onset of European colonialism before 1700, much of South Asia was dominated by the powerful Mughal Empire. Under Britain, the wealthiest parts of the region were ruled directly, but other lands remained under the partial authority of indigenous rulers. Independence for the region came after 1947, when the British abandoned their extensive colonial territory. Bangladesh, which was formerly East Pakistan, gained its independence in 1971 after a short struggle against centralized Pakistani rule from the west.

**FIGURE 12.27 Former British Hill Station**   Britain's imperial officials in India often longed for the cool weather and lush gardens of their homeland. As a result, Britain built many high-elevation "hill stations" as resorts. Today, these one-time symbols of imperialism are popular tourist destinations for the Indian elite.

Britain's victory over France in the Seven Years' War (1756–1763), the French retained only a few minor coastal cities. Elsewhere, the **British East India Company**, the private organization that acted as an arm of the British government, was free to carve out its own South Asian empire. By the 1840s, British control over South Asia was essentially completed. Valuable local allies, however, were allowed to remain in power, provided that they did not threaten British interests. The territories of these indigenous (or "princely") states were gradually reduced, while British advisors increasingly dictated their policies.

The continual expansion of British power led to a rebellion in 1856 across much of South Asia. When this uprising (often called the *Sepoy Mutiny*) was finally crushed, a new political order was implemented. South Asia was now under the authority of the British government, with the Queen of England acting as its head of state (Figure 12.27). Britain enjoyed direct control over the region's most productive and densely populated areas, including almost the entire Indus–Ganges Valley and coastal plains. In more remote areas, Britain ruled indirectly, with native "princes" continuing to occupy their thrones.

British officials were always concerned about threats to their immensely profitable Indian colony, particularly from the Russians advancing across Central Asia. In response, they attempted to secure their boundaries. In some cases, this merely required making alliances with local rulers. In such a manner, Nepal and Bhutan retained their independence. In the extreme northeast a number of small states and tribal territories, most of which had never been part of the South Asian cultural sphere, were taken over by the British Indian Empire. A similar policy was conducted on the vulnerable northwestern frontier.

**Independence and Partition**   The framework of British India began to unravel in the early 20th century, as the people of South Asia increasingly demanded independence. The British, however, were determined to stay, and by the 1920s South Asia was caught up in massive political protests.

The leaders of the rising nationalist movement faced a dilemma in attempting to organize an independent country. Many leaders, including Mohandas Gandhi—the father-figure of Indian independence—favored a unified state that would include all British territories in mainland South Asia. Most Muslim leaders, however, feared that a unified India would leave their people in a vulnerable position. They therefore argued for the division of British India into two new countries: a Hindu-majority India and a Muslim-majority Pakistan. In several parts of northern South Asia, however, Muslims and Hindus were settled in roughly equal numbers. Another problem was the location of areas of clear Muslim majority on opposite sides of the subcontinent, in present-day Pakistan and Bangladesh.

As the British finally withdrew in 1947, South Asia was indeed divided into India and Pakistan. Partition was a horrific event. Not only were some 14 million people displaced, but roughly 1 million were killed. Hindus and Sikhs fled from Pakistan, to be replaced by Muslims fleeing India (Figure 12.28).

The Pakistan that emerged from partition was for several decades a clumsy two-part country, its western section in the Indus Valley, its eastern portion in the Ganges Delta. The Bengalis, occupying the poorer eastern section, complained that they were treated as second-class citizens. In 1971 they launched a rebellion and, with the help of India, quickly prevailed. Bangladesh then emerged as a new country. This second partition did not solve Pakistan's problems, however, as the country remained politically unstable and prone to military rule. Pakistan retained the British policy of allowing almost full autonomy

**FIGURE 12.28 Partition, 1947**   Following Britain's decision to leave South Asia, violence and bloodshed broke out between Hindus and Muslims in much of the region. With the partition of British India into India and Pakistan, millions of people were forced to flee their homes, often in over-crowded trains such as those visible in this photograph.

to the Pashtun tribes living along its border with Afghanistan, a relatively lawless area marked by clan fighting. This area would later lend much support to Afghanistan's Taliban regime and to Osama bin Laden's Al Qaeda organization.

## Ethnic Conflicts in South Asia

After India and Pakistan gained independence, ethnic and religious tensions continued to plague many parts of South Asia. The region's most complex—and perilous—struggle is that in Kashmir, which involves both India and Pakistan.

**Kashmir**   Relations between India and Pakistan were hostile from the start, and the situation in Kashmir has kept the conflict burning (Figure 12.29). During the British period, Kashmir was a large princely state with a primarily Muslim core joined to a Hindu district in the south (Jammu) and a Tibetan Buddhist district in the east (Ladakh). Kashmir was ruled by a Hindu **maharaja**, a king subject to British advisors. During partition, Kashmir came under severe pressure from both India and Pakistan. After troops from Pakistan gained control of western Kashmir, the maharaja decided to join India. But neither Pakistan nor India would accept the other's control over any portion of Kashmir, and as a result, they have since fought several wars over the issue.

Although the Indo–Pakistani boundary has remained fixed, fighting in Kashmir has continued, reaching a peak in the 1990s. Many Muslim Kashmiris would like to join their homeland to Pakistan, others prefer that it remain a part of India, and a large portion would rather see it became an independent country. Indian nationalists are determined that Kashmir remain part of India. Militants from Pakistan continue to cross the border to fight the Indian army, ensuring that tensions between the two countries remain high. The result has been a low-level but periodically brutal war. The Vale of Kashmir, with its lush fields and orchards nestled among some of the world's most spectacular mountains, was once one of South Asia's premier tourist destinations; now it is a battle-scarred war zone.

**The Northeast Fringe**   A relatively obscure ethnic conflict emerged in the 1980s in the uplands of India's extreme northeast. Much of this area has never really been part of the South Asian cultural sphere, and many of its peoples want autonomy, if not actual independence. Northeastern India is still relatively lightly populated and as a result has attracted millions of migrants from Bangladesh and northern India. Many local people view this movement as a threat to their lands and culture. On several occasions, local guerillas have attacked newcomer villagers and, in turn, have suffered reprisals from the Indian military. This is a remote area, however, and relatively little information from it reaches the outside world.

Tensions in the northeast have complicated India's relations with Bangladesh. India accuses Bangladesh of allowing separatists sanctuary on its side of the border and objects as well to continuing Bangladeshi emigration. As a result, India is currently building a 2,500-mile (4,000-kilometer), $1.2 billion fence along the border between the two countries (Figure 12.30).

**Sri Lanka**   Ethnic violence in Sri Lanka has been especially severe. Here the conflict stems from both religious and linguistic differences. Northern Sri Lanka is dominated by Hindu Tamils, whereas the island's majority group is Buddhist in religion and Sinhalese in language. Relations between the two communities have historically been fairly good, but tensions mounted soon after independence (Figure 12.31). Sinhalese nationalists have favored a centralized government, some of them calling for an officially Buddhist state. Most Tamils want political and cultural autonomy, and they have accused the government of discriminating against them.

In 1983, war erupted when the rebel force known as the *Liberation Tigers of Tamil Eelam*, or "Tamil Tigers," attacked the Sri Lankan army. By the 1990s, most of northern Sri Lanka was under the control of the Tamil Tigers. A Norwegian-brokered cease-fire in 2000 brought some hope that the conflict was winding down, but fighting

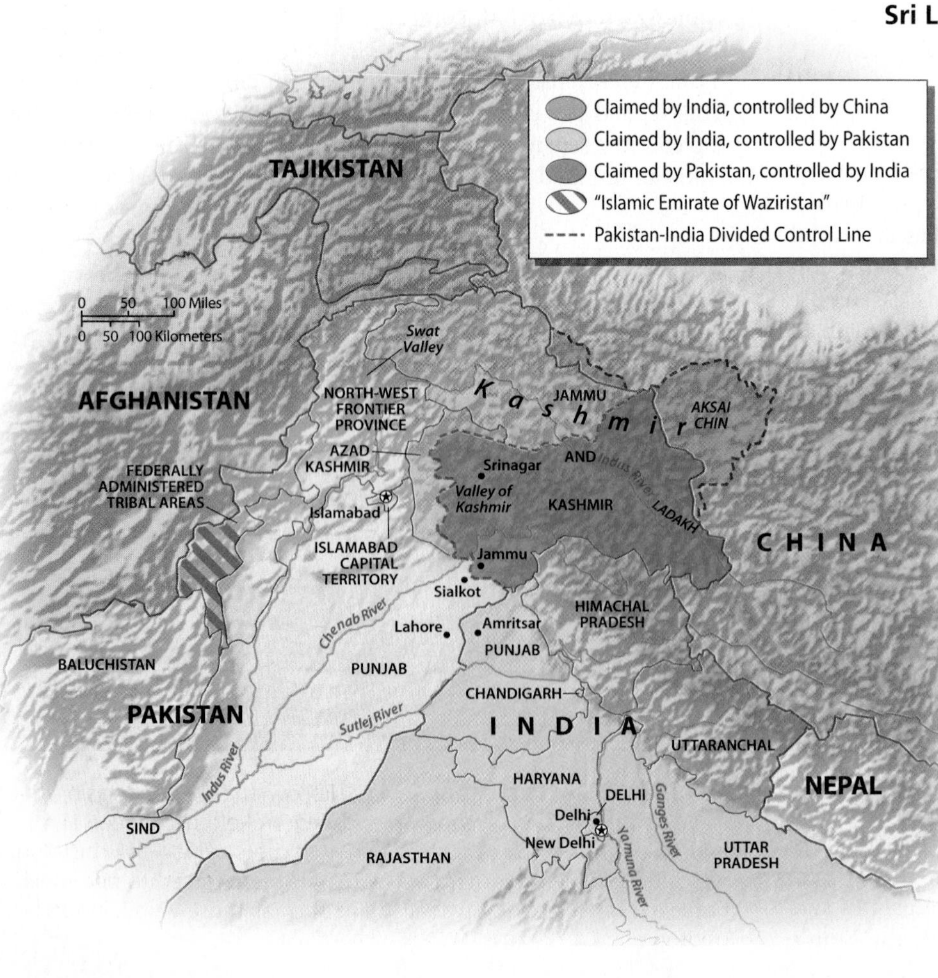

Map legend:
- Claimed by India, controlled by China
- Claimed by India, controlled by Pakistan
- Claimed by Pakistan, controlled by India
- "Islamic Emirate of Waziristan"
- Pakistan-India Divided Control Line

TAJIKISTAN · AFGHANISTAN · NORTH-WEST FRONTIER PROVINCE · Swat Valley · Kashmir · JAMMU · AKSAI CHIN · AZAD KASHMIR · FEDERALLY ADMINISTERED TRIBAL AREAS · Srinagar · AND · Valley of Kashmir · KASHMIR · LADAKH · Islamabad · ISLAMABAD CAPITAL TERRITORY · Jammu · CHINA · Sialkot · HIMACHAL PRADESH · Chenab River · Lahore · Amritsar · PUNJAB · BALUCHISTAN · PUNJAB · CHANDIGARH · PAKISTAN · Sutlej River · INDIA · UTTARANCHAL · Indus River · HARYANA · DELHI · NEPAL · SIND · Delhi · Ganges River · New Delhi · UTTAR PRADESH · RAJASTHAN · Yamuna River

**FIGURE 12.29 Conflict in Kashmir**   Unrest in Kashmir maintains hostility between the two nuclear powers of India and Pakistan. Under the British, this region of predominantly Muslim population was ruled by a Hindu maharaja, who managed to join the province to India upon partition. Today, many Kashmiris wish to join Pakistan, while many others argue for an independent state.

**FIGURE 12.30 India–Bangladesh Fence**    India began building a border fence between its territory and that of Bangladesh in 2003 in order to reduce illegal immigration and to stop the influx of militants. Members of the Indian Border Security Force visible in this photograph are patrolling a segment of the border fence.

again intensified in 2006. In 2007, the Sri Lankan government abandoned negotiations, launching instead an all-out offensive. In May 2009, government forces crushingly defeated the Tamil Tiger army, killing the organization's leaders. While Sri Lanka then entered its first period of peace in 26 years, some observers feared that Tamil diehards would adopt guerilla tactics and continue to fight.

## The Maoist Challenge

Not all of South Asia's conflicts are rooted in ethnic or religious differences. Poverty, inequality, and environmental degradation in east-central India, for example, have generated a persistent revolutionary movement that is inspired by the former Chinese communist leader, Mao Zedong. Here the fighting is low level but persistent. In July 2009, for example, Maoist rebels ambushed a detachment of Indian police in Chhattisgarh, killing 23 officers. Manmohan Singh, India's prime minister, has referred to the Maoists as "the single biggest internal security challenge ever faced by our country."

Despite Singh's warning, Maoism has been even more of a challenge to the government of Nepal. Nepalese Maoists, frustrated by the lack of development in rural areas, emerged as a significant force in the 1990s. By 2005, they controlled over 70 percent of the country. At the same time, Nepal's urban population also turned against the country's monarchy, launching massive protests. In 2006, Nepal's king agreed to restore democratic rule, while at the same time the Maoist rebels announced that they would quit fighting and enter the democratic political process. In 2008, the king stepped down, and Nepal became a republic, with the leader of the former Maoist rebels serving as prime minister.

The end of the monarchy has not brought stability to Nepal. In early 2009, the Maoist prime minister resigned after quarreling with army leaders and was soon threatening to launch demonstrations against the government. Meanwhile, the indigenous people of Nepal's southern lowlands, distressed by the migration of settlers from the more densely populated hill country, began pushing for greater representation and threatening to rebel if their demands were not met.

## International and Global Geopolitics

As the case of Kashmir shows, South Asia's major international geopolitical problem is the continuing cold war between India and Pakistan (Figure 12.32). Since India gained independence, the two countries have regarded each other as enemies. Today, the stakes are extremely high, as both India and Pakistan have nuclear weapons.

During the global Cold War, Pakistan allied itself with the United States, and India leaned slightly toward the Soviet Union. Such alliances fell apart with the end of the superpower conflict in the early 1990s. Since then, Pakistan has forged an informal alliance with China, which has long been in military competition with India. India, meanwhile, has gradually been moving into an informal alliance with the United States.

The conflict between India and Pakistan became more complex after the terrorist attacks of September 11, 2001. Until that time, Pakistan had been supporting Afghanistan's Taliban regime. After Osama bin Laden's attacks on the World Trade Center and Pentagon, the United States gave Pakistan a stark choice: Either Pakistan would assist the United States in its fight against the Taliban and receive financial aid in return, or it would lose favor with the U.S. government. Pakistan agreed to help, offering valuable intelligence to the U.S. military.

**FIGURE 12.31 Civil War in Sri Lanka**    The majority of Sri Lankans are Sinhalese Buddhist, many of whom maintain that their country should be a Buddhist state. A Tamil-speaking Hindu minority in the northeast strongly resists this idea. Tamil militants, who have waged war against the Sri Lankan government for several decades, hope to create an independent country in their northern homeland.

**FIGURE 12.32 Border Tensions** An Indian officer looks through binoculars in war-torn Kashmir. Relationships between India and Pakistan have been extremely tense since India gained independence in 1947. Moreover, with both countries now nuclear powers, the fear that border hostilities will escalate into wider warfare has become a nightmarish possibility.

Pakistan's decision to help the United States came with large risks. Both Osama bin Laden and the Taliban enjoy substantial support among the Pashtun people of Pakistan's wild North-West Frontier Province and Federally Administered Tribal Areas. After suffering several military reversals, Pakistan decided to negotiate with these radical Islamists, and on several occasions it gave them virtual control over sizable areas. From these bases, militants have launched numerous attacks on U.S. forces in Afghanistan and have attempted to gain control over broader swaths of Pakistan's territory. The United States has responded mainly by using unstaffed drones to attack insurgent leaders, a tactic that has resulted in large numbers of civilian casualties and much anti-American sentiment throughout Pakistan.

The security crisis in Pakistan intensified from 2007 to 2009. In November 2008, Pakistan's president declared a state of emergency, and in the following month Benazir Bhutto, Pakistan's former prime minister, was assassinated by radicals as she campaigned for the 2008 election. Although the elections went on relatively smoothly, the new government faced severe challenges as Islamists expanded their control over remote areas. In April 2009, Pakistan's army launched a major invasion of Swat, a former vacation zone 100 miles (160 kilometers) northwest of the capital city of Islamabad which had been turned over to Islamic militants a few months earlier. As the military advanced, hundreds of thousands of civilians fled, generating a humanitarian catastrophe. The government soon reclaimed Swat as militants retreated, but the entire area remains highly unstable.

Pakistan is a troubled country, facing not only a deep Islamist insurgency and a seemingly unsolvable geopolitical conflict with India, but also a low-level ethnic rebellion in its southwestern province (Baluchistan). Its relations with India, moreover, deteriorated in November 2008, when terrorists operating from Pakistan launched a series of coordinated attacks on tourist facilities and public places in Mumbai, killing 173 people. Pakistan responded by investigating the event and arresting a number of alleged plotters, but many Indians suspect that the attack had the support of certain elements of Pakistan's government and military. Both Indian and Pakistani leaders, however, decided that it would be in their own best interests to reduce tensions, and in July 2009, they agreed to renew their stalled peace negotiations.

# Economic and Social Development: Rapid Growth and Rampant Poverty

South Asia is one of the poorest regions of the world, yet it is also the site of great wealth. Many of South Asia's scientific and technological accomplishments are world-class, but the area also has some of the world's highest illiteracy rates. While South Asia's high-tech businesses are closely integrated with the global economy, the South Asian economy as a whole was until recently one of the world's most isolated.

## South Asian Poverty

One of the clearest measures of human well-being is nutrition, and by this measure South Asia ranks very low indeed. Nowhere else can one find so many chronically undernourished people (Figure 12.33). According to a 2007 study, 46 percent of Indian children below age three are undernourished. Roughly two-thirds of the people of India live on less than $2 a day, and Bangladesh and Nepal are poorer still (Table 12.2).

Despite such deep and widespread poverty, South Asia should not be regarded as a zone of misery. More than 300 million Indians are

**FIGURE 12.33 Poverty in India** India's rampant poverty results in a significant amount of child labor. In this photo, a 10-year-old boy is moving a large burden of plastic waste by bicycle.

### TABLE 12.2 DEVELOPMENT INDICATORS

| Country | GNI per capita, PPP 2007 | GDP Average Annual % Growth 2000–07 | Human Development Index (2006)# | Percent Population Living Below $2 a Day | Life Expectancy 2009 | Under Age 5 Mortality Rate 1990 | Under Age 5 Mortality Rate 2007 | Gender Equity 2007 |
|---|---|---|---|---|---|---|---|---|
| Bangladesh | 1,330 | 5.7 | 0.524 | 81.3 | 65 | 151 | 61 | 103 |
| Bhutan | 4,980 | | 0.613 | | 68 | 148 | 84 | 95 |
| India | 2,740 | 7.8 | 0.609 | 75.6 | 64 | 117 | 72 | 91 |
| Maldives | 4,910 | | 0.749 | | 73 | 111 | 30 | 101 |
| Nepal | 1,060 | 3.4 | 0.530 | 77.6 | 64 | 142 | 55 | 98 |
| Pakistan | 2,540 | 5.6 | 0.562 | 60.3 | 66 | 132 | 90 | 78 |
| Sri Lanka | 4,200 | 5.3 | 0.742 | 39.7 | 71 | 32 | 21 | |

Source: World Bank, *World Development Indicators 2009.*

United Nations, *Human Development Index, 2008.#*

now rated by local standards as members of the "middle class" who can buy such modern goods as televisions, motor scooters, and washing machines. This large market has begun to interest corporate executives worldwide. By the early years of the new millennium, India's economy was growing at the extremely rapid rate of 7 to 8 percent a year. The global economic crisis of 2008–2009 reduced India's economic growth rate but did not send it into recession, in part because the Indian banking sector never engaged in the risky practices that were so common in the United States and Europe.

## Geographies of Economic Development

After gaining independence, the governments of South Asia attempted to build new economic systems that would benefit their own people rather than foreign countries or corporations. Planners initially stressed heavy industry and economic self-sufficiency. While some gains were realized, the overall pace of development remained slow. Since the 1990s, however, governments in the region have gradually opened their economies to the global economic system. In the process, core areas of development and social progress have emerged, surrounded by large peripheral zones that have lagged behind.

**The Himalayan Countries** Both Nepal and Bhutan are disadvantaged by their rugged terrain and remote locations and by the fact that they have been relatively isolated from modern technology and infrastructure. But such measurements can be misleading, especially for Bhutan, because many areas in the Himalayas are still largely subsistence oriented.

Until recently, Bhutan remained purposely disconnected from the modern world economy, allowing its small population to live in a relatively pristine natural environment. While Bhutan now allows direct international flights, cable television, and the Internet, it still charts its own course, emphasizing "gross national happiness" over economic growth. Nepal, on the other hand, is more heavily populated and suffers much more severe environmental degradation. It is also more closely integrated with the Indian economy. Nepal has relied heavily on international tourism, but its tourist industry contracted sharply as its political crisis deepened after 2002 (Figure 12.34).

**Bangladesh** By several measures, Bangladesh is the poorest South Asian country. Environmental degradation and colonialism have contributed to Bangladesh's poverty, as did the partition of 1947. Most of pre-partition Bengal's businesses were located in the west, which went to India. Slow economic growth and a rising population in the first several decades after independence meant that Bangladesh remained a poor country. Almost 40 percent of Bangladeshis currently live on less than $1 a day.

Economic conditions in Bangladesh have, however, improved in recent years. Bangladesh is internationally competitive in textile and clothing manufacture, in part because its wage rate is so low. Low-interest credit provided by the internationally acclaimed Grameen

**FIGURE 12.34 Tourism in Nepal** Nepal has long been one of the world's main destinations for adventure tourism, although business has suffered greatly in recent years due to the country's Maoist insurgency. Many tourists in Nepal stay in rustic lodges such as the one shown in this photograph.

**FIGURE 12.35 Grameen Bank** This innovative institution loans money to rural women so they can buy land, purchase homes, or start cottage industries. In this photo, taken in Bangladesh, women proudly repay their loans to a bank official as testimony to their success.

Bank has given hope to many poor women in the country, allowing the emergence of a number of small-scale enterprises (Figure 12.35). Bangladesh has also discovered substantial reserves of natural gas, although it has been slow to develop them. Political instability and environmental degradation, however, cloud the country's economic future.

**Pakistan** Like Bangladesh, Pakistan suffered deeply from partition in 1947. But for several decades after independence Pakistan maintained a more productive economy than India. The country has a strong agricultural sector, as it shares the fertile Punjab with India. Pakistan also boasts a large textile industry, based on its huge cotton crop.

Pakistan's economy, however, is less dynamic than that of India, with a lower potential for growth. Pakistan is burdened by very high levels of defense spending. In addition, a small but powerful landlord class controls much of its best agricultural lands yet pays virtually no taxes. Unlike India, Pakistan has not been able to develop a successful high-tech industry.

Pakistan's economic growth did, however, accelerate after the country was granted concessions by the international community for its role in combating global terrorism. In 2008, however, its inflation rate had risen to over 20 percent, while huge budget deficits required a bailout from the International Monetary Fund. Insurgency and general political instability contribute to Pakistan's cloudy economic future.

**Sri Lanka and the Maldives** Sri Lanka's economy is by several measures the most highly developed in South Asia. The country's exports are concentrated in textiles and agricultural products such as rubber and tea. By global standards, however, Sri Lanka is still a very poor country, its progress undermined by its prolonged civil war. As the conflict comes to an end, Sri Lanka will probably benefit from its high levels of education, its abundant natural resources, and its tremendous tourist potential.

The Maldives is the most prosperous South Asian country based on per capita economic output, but its total economy, like its population, is very small. Most of the country's revenues are gained from fishing and international tourism. Critics claim that most of the revenues from the tourist economy go to a very small segment of the population.

**India's Less Developed Areas** While India's per capita gross national income (GNI) is similar to that of Pakistan, its total economy is much larger. As the region's largest country, India has far more internal variation in economic development. The most basic economic division is that between its more prosperous southern and western areas and its poorer districts in the north and east.

India's least developed and most corrupt area is Bihar, a state of 82 million people located in the lower Ganges Valley. Bihar's per capita

level of economic production is less than one-third that of India as a whole. Neighboring Uttar Pradesh, India's most populous state, is also extremely poor. It too is densely settled and has experienced little industrial development. Both states are also noted for their socially conservative outlooks and caste tensions. Other states of north-central India, such as Madhya Pradesh, Jharkhand, Chhattisgarh, and Orissa, have also experienced relatively little economic development.

**India's Centers of Economic Growth** The west-central states of Gujarat and Maharashtra are noted for their industrial and financial power, as well as for their agricultural productivity. Gujarat was one of the first parts of South Asia to experience industrialization, and its textile mills are still among the most productive in the region (Figure 12.36). Gujaratis are well known as merchants and overseas traders, and they are heavily represented in the **Indian diaspora**, the migration of Indians to foreign countries. Cash remittances sent home from these emigrants help the state's economy. Gujarat is also considered to be a well-governed state, although it also has some of the worst Hindu–Muslim relations in India.

Maharashtra is usually viewed as India's economic pacesetter. Its huge city of Mumbai (Bombay) has long been the financial center and media capital of India. Major industrial zones are located around Mumbai and in several other parts of Maharashtra. In recent years, Maharashtra's economy has grown more quickly than those of most other Indian states. Its per capita level of economic production is now roughly 50 percent greater than that of India as a whole.

In the northwestern Indian states of Punjab and Haryana, showcases of the Green Revolution, per capita levels of economic output are similar to those of Maharashtra and Gujarat. Their economies have rested largely on agriculture, but investments have been made recently in food processing and other industries. On Haryana's eastern border lies the capital district of New Delhi. India's political power and much of its wealth are concentrated here.

The center of India's fast-growing high-technology sector lies farther to the south, especially in Bangalore and Hyderabad. The Indian government selected the upland Bangalore area, which is noted for its pleasant climate, for technological investments in the 1950s. Other businesses

**FIGURE 12.36 Gujarat Factory**   A factory worker manages the flow of cotton as it blows out of the ginning equipment in a mill near Ahmedabad in the Indian state of Gujarat. Gujarat has long been a center of India's important textile industry.

soon followed. In the 1980s and 1990s, a quickly growing computer software and hardware industry emerged, earning Bangalore the label "Silicon Plateau" (Figure 12.37). By 2000 large numbers of American software, accounting, and data-processing jobs were being transferred, or "outsourced," to Bangalore and other India cities. The southern Indian states of Tamil Nadu and Kerala have also seen rapid growth in recent years, due in part to booming information technology industries.

India has proved especially competitive in software because software development does not require a sophisticated infrastructure. Computer code can be exported by wireless telecommunication systems without the use of modern roads or port facilities. What is necessary, of course, is technical talent, and this India has in great abundance. Many Indian social groups have long been highly committed to education,

and India has been a major scientific power for decades. With the growth of the software industry, India's brain power has finally begun to create economic gains (Figure 12.38). Whether such developments can spread benefits beyond the rather small high-tech areas they presently occupy remains to be seen.

## Globalization and South Asia's Economic Future

Throughout most of the second half of the 20th century, South Asia was relatively isolated from the world economy. Even today, especially compared with East Asia or Southeast Asia, the region's volume of foreign trade and its influx of foreign direct investment are still relatively small. But globalization is advancing rapidly, especially in India.

To understand South Asia's recently low level of globalization, it is necessary to examine its economic history. After independence, India's economic policy was based on widespread private ownership combined with governmental control of planning, resource allocation, and certain industries. India also established high trade barriers to protect its economy from global competition. This mixed socialist–capitalist system encouraged the development of heavy industry and allowed India to become nearly self-sufficient, even in the most technologically sophisticated goods. By the 1980s, however, problems with this model were becoming apparent. Slow economic growth meant that the percentage of Indians living in poverty remained almost constant. At the same time, countries such as China and Thailand were experiencing rapid development after opening their economies to globalization. Many Indian businesspeople also disliked the governmental regulations that made it difficult for them to expand.

In response to these difficulties India's government began to open its economy in 1991. Many regulations were eliminated, tariffs were reduced, and partial foreign ownership of local businesses was allowed. Other South Asian countries followed a somewhat similar path. Pakistan, for example, began to privatize many of its state-owned industries in 1994.

Overall, India's economic reforms have proved successful. Indian information technology firms are world class, and many have begun to expand globally. Growth has been so rapid in this sector that some companies are now having difficulty finding and retaining qualified workers; as a result, wages are increasing rapidly. Recent growth has also demonstrated the need for India to improve its infrastructure, but it is not clear how it will be able to afford the necessary investments in roads, railroads, and electricity generation and transmission facilities.

The gradual internationalization and deregulation of the Indian economy has generated substantial opposition. Foreign competitors are now seriously challenging some domestic firms. Cheap manufactured goods

**FIGURE 12.37 India's Silicon Plateau**
The Bangalore suburb of Whitefield is usually considered to be the heartland of India's high tech sector. This photograph shows Whitefield's International Technical Park.

**FIGURE 12.38 Indian Institutes of Technology** India's seven government-run institutes of technology provide world-class training for the country's top students in science and engineering, helping develop the country's globally-oriented information technology industries.

from China are seen as an especially serious threat. Hundreds of millions of Indian peasants and slum-dwellers, moreover, have seen few if any benefits from their country's rapid economic growth.

Although India gets most of the media attention, other South Asian countries have also experienced significant economic globalization in recent years. Besides exporting textiles and other consumer goods, Bangladesh, Pakistan, and Sri Lanka send large numbers of their citizens to work abroad, particularly in the Persian Gulf. Remittances from foreign workers is Bangladesh's second largest source of income, reaching almost $10 billion a year in 2009. On a per capita basis, remittances are even more important for Sri Lanka. Out of a total population of 20 million, roughly 1.5 million Sri Lankans work abroad, 90 percent of them in the Middle East.

## Social Development

South Asia has relatively low levels of health and education, which is not surprising considering its poverty. As might be expected, people in the more developed areas of western and southern India are healthier, live longer, and are better educated, on average, than people in the poorer areas, such as the lower Ganges Valley. Bihar, with a literacy rate of only about 50 percent, stands at the bottom of most social-development rankings, while the Indian states of Kerala, Punjab, Gujarat, and Maharashtra stand near the top. Several key measurements of social welfare are higher in India than in Pakistan. Pakistan has done a poor job of educating its people, which is one reason fundamentalist Islamic organizations have been able to recruit effectively in much of the country.

Several oddities stand out when one compares South Asia's map of economic development with its map of social well-being. Portions of India's extreme northeast, for example, show relatively high literacy rates despite their poverty, due largely to the educational efforts of Christian missionaries. In Mizoram, for example, the literacy rate is over 90 percent, the highest in India. In overall terms, however, southern South Asia outpaces other parts of the region in regard to social development.

**The Educated South** Southern South Asia's relatively high levels of social welfare are clearly visible when one examines Sri Lanka. Considering its meager economy and prolonged civil war, Sri Lanka must be considered a social developmental success. Sri Lanka's average life expectancy is 75 years, and its literacy rate is over 90 percent. The Sri Lankan government has achieved these results through universal primary education and inexpensive medical clinics.

On the mainland, Kerala in southwestern India has achieved even more impressive results. Kerala is extremely crowded and has long had difficulty feeding its population. Although Kerala's economy has grown strongly in recent years, its per capita economic output is only a little above average for India. Kerala's level of social development, however, is the highest in India (Figure 12.39): 90 percent of the people of Kerala are literate, the state's average life expectancy is 74 years, and a number of diseases, such as malaria, have been essentially eliminated.

Some observers attribute Kerala's social successes to its state policies. Kerala has often been led by a socialist party that has stressed education and community health care. While this has no doubt been an important factor, it does not seem to offer a complete explanation. Some researchers suggest that one of the key factors is the relatively high social position of women in Kerala.

**The Status of Women** It is often argued that South Asian women have a very low social position in both the Hindu and Muslim traditions. Throughout most of India women traditionally leave their own families shortly after puberty to join those of their husbands. As outsiders, often in distant villages, young brides have little freedom and few opportunities. In Pakistan, Bangladesh, and such Indian states as Rajasthan, Bihar, and Uttar Pradesh, female literacy lags far behind male literacy. An even more disturbing statistic is gender ratios, the relative proportion of males and females in the population. In parts of northern India, there are fewer than 850 females for every 1,000 males. An imbalance of males over females often results from differences in care. In poor families, boys typically receive better nutrition and medical care than do girls, which results in higher rates of survival. An estimated 10 million girls, moreover, have supposedly been lost in northern India due to sex-selective abortion over the past 20 years. Economics play a major role in this situation. In rural households, boys are usually viewed as a blessing because they typically remain with and work for the well-being of their families. In the poorest groups, elderly people (especially widows) subsist largely on what their sons provide. Girls, on the other hand, marry out of their families at an early age and must be provided with a dowry. They are thus seen as an economic liability.

The social position of women is improving in many parts of South Asia, especially in the more prosperous areas where employment opportunities outside the family are emerging. But even in many of the region's middle-class households, women still experience discrimination. Indeed, dowry demands are increasing in some areas, and there have been a number of well-publicized murders of young brides whose families failed to deliver an adequate supply of goods. While the social bias against women across the north is striking, it is much less evident in southern South Asia, especially Kerala and Sri Lanka.

# EXPLORING GLOBAL CONNECTIONS
## The Tata Group

Most large multinational corporations, which produce goods and services in many different countries, are based in the core areas of the world economy. Such companies generally rely on the financial and technological resources of wealthy cities yet often seek out production sites in poor countries with low wages and few regulations. As a result, multinational corporations have been criticized for their role in maintaining a stark global division between the developed and the underdeveloped worlds. As globalization proceeds, however, such a view is becoming outmoded. Increasing numbers of powerful multinational corporations are based in relatively poor areas, and several are now investing heavily in wealthy countries.

No company illustrates these trends as well as India's Tata Group. Founded in 1868 as an opium-trading firm in Bombay (Mumbai), India, the Tata Group now includes 114 companies operating in 85 countries. Despite its global reach, the company is still headquartered in Mumbai; its current chairman, Ratan Tata, is the fifth generation of the Tata family to exercise control. According to a 2009 survey, international business leaders ranked the Tata Group as the word's eleventh most reputable company.

The Tata Group is a classical conglomerate, operating businesses in the steel, motor, consultancy, electrical, chemical, medical, retail, hotel, and information technology industries. Over the past 15 years it has aggressively expanded its global reach by purchasing foreign companies. In 2008 alone, it acquired firms based in Britain, Spain, Italy, China, South Africa, Norway, and Morocco. Such well-known brands as Tetley Tea and the Taj Hotels now belong to the Tata Group.

One of the most global members of the group is Tata Motors. Although only the world's eleventh largest automaker, Tata is the second largest producer of commercial vehicles. It has reached that position in large part through joint ventures and foreign acquisitions. In 2005, for example, it purchased Daewoo Commercial Vehicle, South Korea's second largest bus manufacturer. Tata Motors's most daring international move was probably its 2008 acquisition of British Jaguar Land Rover (JLR) from Ford Motor Corporation for $2 billion. With this purchase, Tata Motors gained control of three manufacturing plants and two design centers in the United Kingdom, as well as two of the best-known luxury car brands.

While targeting high-end automobile consumers with Jaguar, Tata Motors has also taken a greater interest in the opposite end of the market. Its signal product here is the Tata Nano, a "supermini" car selling for as little as $2,500 that hit the Indian market in the summer of 2009 (Figure 12.1.1). The two-seater Nano will also be marketed in a number of African and Asian countries, and a slightly more elaborate Nano Europa will be made available in Europe.

Environmental groups have criticized Tata Motors for developing such a low-priced car, fearing that it will lead to mass automobile ownership in India and other poor countries. Tata has responded by developing low-emission vehicles. The electric Tata Indica is scheduled to be released in India and Norway by 2010. Tata is also investing in compressed air vehicles, which are potentially highly efficient. Whether any of these ventures prove successful remains to be seen, but it is clear that India's Tata Group has emerged as a global economic powerhouse.

**FIGURE 12.1.1 Tata Nano** The world's least expensive automobile, the Tata Nano averages roughly 54 miles (87 kilometers) per gallon of gasoline.

**FIGURE 12.39 Education in Kerala** India's southwestern state of Kerala, which has virtually eliminated illiteracy, is South Asia's most highly educated region. It also has the lowest fertility rate in South Asia. Because of this, many argue that women's education and empowerment is the best and most enduring form of contraception.

# SUMMARY

- South Asia, a large, complex, and densely populated area, has in many ways been overshadowed by neighboring world regions: by the uneven globalization of Southeast Asia, by the size and political weight of East Asia, and by the geopolitical tensions of Southwest Asia. Much of that is changing, however, as South Asia now figures prominently in discussions of world problems and issues.

- Environmental degradation and instability pose particular problems for South Asia. Due to its monsoon climate, both floods and droughts tend to be more problematic here than in most other world regions. Global climate change directly threatens the low-lying Maldives and may play havoc with the monsoon-dependent agricultural systems of India, Pakistan, and Bangladesh.

- Continuing population growth in this already densely populated region demands attention. Although fertility rates have declined in recent years, Pakistan, northern India, and Bangladesh cannot easily meet the demands imposed by their expanding populations. Increasing social and political instability, as well as environmental degradation, may result as cities mushroom in size and rural areas grow more crowded.

- South Asia's diverse cultural heritage, shaped by peoples speaking several dozen languages and following several major religions, makes for a particularly rich social environment. Unfortunately, cultural differences have often translated into political conflicts.

- Ethnically or religiously based separatist movements have severely challenged the governments of Pakistan, India, and Sri Lanka. In India, moreover, religious strife between Hindus and Muslims persists, whereas in Pakistan and Bangladesh, Islamic radicals clash with the state while Sunni and Shiite Muslims frequently fight against each other.

- Geopolitical tensions within South Asia are particularly severe, again demanding global attention. The long-standing feud between Pakistan and India escalated dangerously in the late 1990s, leading many observers to conclude that this was the part of the world most likely to experience a nuclear war. Although tensions between the two countries have more recently been reduced, the underlying sources of conflict—particularly the struggle in Kashmir—remain unresolved.

- Although South Asia remains one of the poorest parts of the world, much of the region has seen rapid economic expansion in recent years. Many argue that India in particular is well positioned to take advantage of economic globalization. Large segments of its huge labor force are well educated and speak excellent English, the major language of global commerce. But will these global connections help the vast numbers of India's poor or only the small number of its economic elite? Advocates of free markets and globalization tend to see a bright future, while skeptics more often see growing problems.

# KEY TERMS

British East India Company (p. 363)
bustee (p. 353)
caste system (p. 355)
Dalit (p. 355
Dravidian language (p. 359)

Green Revolution (p. 352)
Hindi (p. 358)
Hindu nationalism (p. 354)
Indian diaspora (p. 368)
Indian subcontinent (p. 344)
Jainism (p. 357)

linguistic nationalism (p. 359)
maharaja (p. 364)
monsoon (p. 347)
Mughal (or Mogul) Empire (p. 355)
orographic rainfall (p. 347)

rain-shadow effect (p. 348)
salinization (p. 353)
Sanskrit (p. 354)
Sikhism (p. 356)
Urdu (p. 359)

# THINKING GEOGRAPHICALLY

1. How might the vulnerable countries of Bangladesh and the Maldives best respond to the challenges posed by the threat of global climate change? What responsibilities do wealthy countries such as the United States have in regard to such issues?

2. What are the pros and cons of the Green Revolution as a means of increasing South Asia's food supplies? What are the advantages and disadvantages of expanding irrigated agriculture in India and Pakistan? What is the outlook for the next decade?

3. What are the drawbacks and benefits of using English as a national language in India? Might it help or hinder unity? Would this increase or decrease India's links to the contemporary world?

4. How could Pakistan best respond to the violence that currently plagues its Pashtun-speaking areas? Should the government of Pakistan try to establish control over the Federally Administered Tribal Areas?

5. Does the information technology industry offer a way for South Asia to achieve economic and social development, or are the benefits of the industry likely to help only a small segment of the population? What kinds of policies might the governments of the region enact in order to gain greater benefits from the emerging high-tech economy?

# 13 SOUTHEAST ASIA

## GLOBALIZATION AND DIVERSITY

Southeast Asia has been a key player in global commerce for hundreds of years, and it has long been open to cultural influences from other parts of the world. Today, Southeast Asian leaders are trying to cope with globalization on their own terms, working to create a more economically and politically united world region.

### ENVIRONMENTAL GEOGRAPHY

Southeast Asia's rain forests are vital centers of biological diversity, but they are rapidly disappearing due to commercial logging and agricultural expansion.

### POPULATION AND SETTLEMENT

Southeast Asia's river valleys, deltas, and areas of volcanic soil tend to be densely populated, whereas most of its upland areas are still lightly settled.

### CULTURAL COHERENCE AND DIVERSITY

Southeast Asia is noted for both its linguistic and religious diversity. Much of the region, however, is plagued by ethnic conflicts and religious tensions.

### GEOPOLITICAL FRAMEWORK

Southeast Asia is one of the most geopolitically united regions of the world, with all but one of its countries belonging to the Association of Southeast Asian Nations (ASEAN).

### ECONOMIC AND SOCIAL DEVELOPMENT

Southeast Asia contains some of the world's most globalized and dynamic economies, as well as some of the most isolated and impoverished.

*Large expanses of forest are being cleared in many parts of Southeast Asia in order to plant African oil palms. Oil palms yield large quantities of edible oil, much of which is now used for the production of bio-diesel fuel. Although bio-diesel is often considered to be an environmentally responsible form of energy, the creation of oil palm plantations of Southeast Asia releases large amounts of carbon dioxide and is thus a contributor to global climate change.*

373

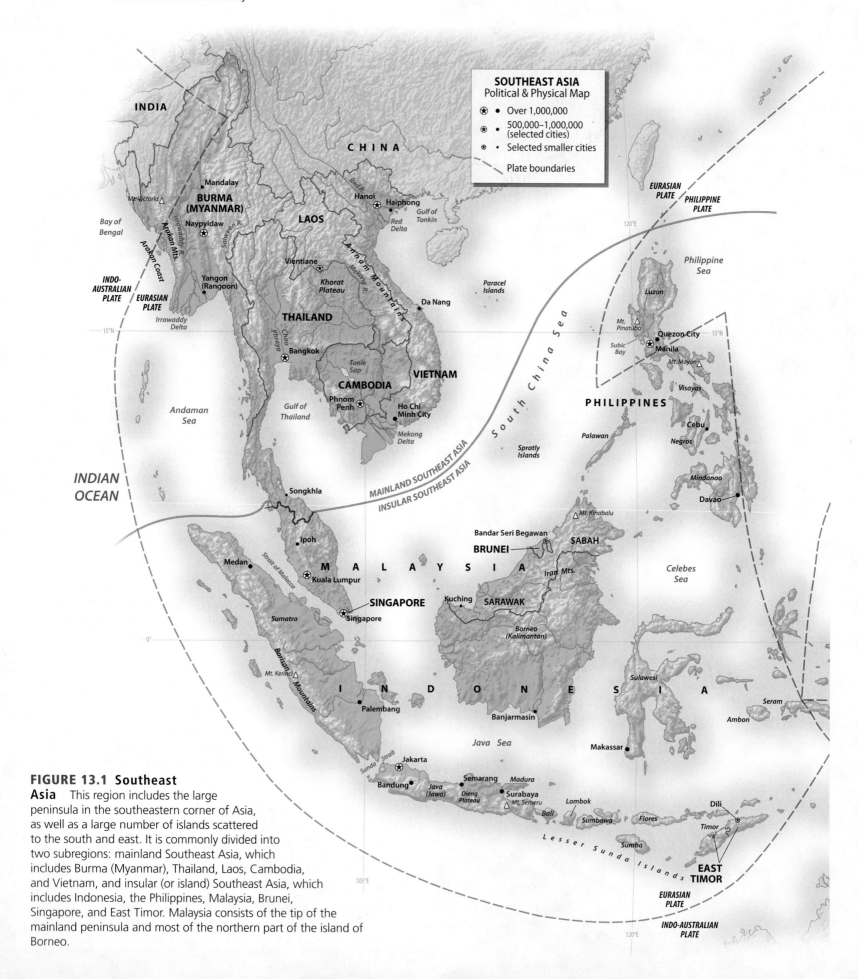

**FIGURE 13.1 Southeast Asia** This region includes the large peninsula in the southeastern corner of Asia, as well as a large number of islands scattered to the south and east. It is commonly divided into two subregions: mainland Southeast Asia, which includes Burma (Myanmar), Thailand, Laos, Cambodia, and Vietnam, and insular (or island) Southeast Asia, which includes Indonesia, the Philippines, Malaysia, Brunei, Singapore, and East Timor. Malaysia consists of the tip of the mainland peninsula and most of the northern part of the island of Borneo.

**SOUTHEAST ASIA**
Political & Physical Map
✴ ● Over 1,000,000
✴ ● 500,000–1,000,000 (selected cities)
✴ • Selected smaller cities
— — Plate boundaries

Southeast Asia demonstrates well both the promises and the perils of globalization. Over the past 30 years, the region has experienced both major economic booms and severe recessions. Since the start of the 21st century, political instability and ethnic tensions, coupled with competition from China, have presented new challenges for continuing Southeast Asian development.

Southeast Asia's involvement with the larger world is not new (Figure 13.1). Chinese and Indian connections date back many centuries. Later, commercial ties with the Middle East opened the doors to Islam, and today Indonesia is the world's most populous Muslim country. More recently came the impact of the West, as Britain, France, the Netherlands, and the United States gained control of large Southeast Asian colonies. Southeast Asia's resources and its strategic location made it a major battlefield during World War II. Yet long after world peace was restored in 1945, warfare of a different sort continued in this region. As colonial powers withdrew and were replaced by newly independent countries, Southeast Asia became a battleground for world powers and their competing economic systems. Through much of the region, communist forces, supported by China and the Soviet Union, fought for control of territory and people.

While communism eventually prevailed in Vietnam, Laos, and Cambodia, all these countries later opened their economies to the global market. Today, the struggle between capitalism and communism has taken a back seat to the economic and ethnic problems confronting Southeast Asia. Relations among the various countries of the region are now generally good. The **Association of Southeast Asian Nations (ASEAN)**, which includes every country in the region except East Timor, has created a generally effective system of regional cooperation (see "Setting the Boundaries").

# Environmental Geography: A Once-Forested Region

The mountainous area along the border between northern Thailand and Burma is a rugged place. Slopes are steep and thickly wooded, rainfall is heavy, and several strains of medicine-resistant malaria are common. These same uplands are also home to the Karen, a distinctive ethnic group numbering roughly 7 million. Unfortunately, the story of the Karen and their homeland is not a happy one. Much of their territory has been overrun by the Burmese army, and many of the Karen have been forced into refugee camps in Thailand. The struggle of these people illustrates the connections among cultural, political, economic, and environmental forces.

## The Tragedy of the Karen

The Karen were never fully included in the Burmese kingdom, which long ruled the lowlands of the country. With the beginning of British colonial rule in the 1800s, however, the Karen territory was joined to the Burmese lowlands. British and American missionaries educated many Karen and converted roughly 30 percent of them to Protestant Christianity. A number of Karen Christians obtained positions in Burma's colonial government. The Burmans of the lowlands, a strongly Buddhist people, resented this deeply, for they had long viewed the Karen as culturally inferior. (According to conventional but confusing terminology, the term *Burmese* refers to all the inhabitants of Burma, whereas *Burmans* refers only to the country's dominant, Burmese-speaking ethnic group.) After independence, the Karen lost their favored position and soon grew to resent what they saw as Burman cultural and economic control.

By the 1970s the Karen were in open rebellion, which they supported by smuggling goods between Thailand and Burma (Figure 13.2). The Burmese army, however, began to make headway against the rebels in the 1990s and by the end of the decade had overrun most of the Karen territory. Crucial to Burma's success was an agreement made with Thailand to prevent Karen soldiers from finding refuge on the Thai side of the border. This agreement was reached in part so that Thai timber interests could have access to Burma's valuable teak forests. Fighting continues, however, with the Burmese army launching a new offensive in 2008 that overran most remaining Karen bases and forced tens of

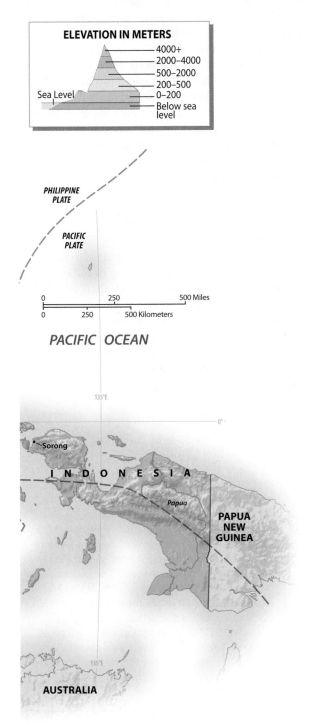

**ELEVATION IN METERS**

4000+
2000–4000
500–2000
200–500
0–200
Below sea level

Sea Level

PHILIPPINE PLATE

PACIFIC PLATE

0    250    500 Miles
0    250    500 Kilometers

PACIFIC OCEAN

135°E

0°

Sorong

INDONESIA

Papua

PAPUA NEW GUINEA

135°E

AUSTRALIA

**FIGURE 13.2 Karen Rebels** The Karen people have been in rebellion against Burma (Myanmar) since the 1970s. For several years they maintained a semi-independent state, with its own capital city and regular army. The Burmese military advanced in the 1990s and again in 2006, however, forcing the Karen into a guerrilla-style war.

thousands of civilians to flee their homes. Some 80,000 Karen now live in refugee camps in Thailand, where they suffer occasional attacks by the Burmese army.

## The Deforestation of Southeast Asia

Deforestation and related environmental problems are major issues throughout most of Southeast Asia (Figure 13.3). Although countries such as Indonesia look to their forests for agricultural expansion, population growth is not the main cause of deforestation. Most forests are cut so that the wood products can be exported to other parts of the world (Figure 13.4). Initially Japan, Europe, and the United States were the main importers, but as China has begun industrializing, its demand has grown. After China banned most logging in 1998, its imports from Southeast Asia—especially from Burma—rapidly increased. After the loggers move through, the cleared lands are often planted with oil palms and other export-oriented crops.

Malaysia has long been one of the leading exporters of tropical hardwoods from Southeast Asia. Peninsular Malaysia was largely deforested by 1985, when a cutting ban was imposed. Since then, logging has been concentrated in the states of Sarawak and Sabah on the island of Borneo. In these areas, the granting of logging concessions to Malaysian and foreign firms has caused considerable problems with local tribal people by disrupting their traditional resource base.

Thailand cut more than 50 percent of its forests between 1960 and 1980. This loss was followed by a series of logging bans that virtually eliminated commercial forestry by 1995. Damage to the landscape, however, was severe: Flooding increased in lowland areas, and erosion on hillslopes led to such problems as the accumulation of silt in irrigation works and hydroelectric facilities. Many of these cutover lands are being reforested with fast-growing Australian eucalyptus trees, which do not support local wildlife. The Thai forestry ban, moreover, has resulted in increased logging in the remote areas of Laos and Cambodia, as well as Burma, much of which is done illegally.

Indonesia, the largest country in Southeast Asia, has fully two-thirds of the region's forest area, including about 10 percent of the

## Setting the Boundaries

Southeast Asia consists of 11 countries that vary widely in spatial extent, population, cultural traits, and levels of economic and social development. Geographically, these countries are commonly divided into those on the Asian continent (mainland Southeast Asia) and those on islands (insular Southeast Asia). The mainland includes Burma (Myanmar), Thailand, Cambodia, Laos, and Vietnam. Although Burma is the largest in territory, Vietnam has the largest population of the mainland states, with 87 million people. (It is important to note that while the authoritarian government of Burma insists that the country be called Myanmar, the country's democratic

opposition favors the old name, Burma; this chapter follows the usage of the opposition.)

Insular Southeast Asia includes the large countries of Indonesia, the Philippines, and Malaysia, as well as the small countries of Singapore, Brunei, and East Timor. Although classified as part of the insular realm because of its cultural and historical background, Malaysia actually splits the difference between mainland and islands. Part of its national territory is on the mainland's Malay Peninsula, and part is on the large island of Borneo, some 300 miles (480 kilometers) distant. Borneo also includes Brunei, a small but oil-rich country of roughly 300,000 people covering an area

slightly larger than Rhode Island. Singapore is essentially a city-state, occupying a small island just to the south of the Malay Peninsula.

Indonesia is an island nation, stretching 3,000 miles (4,800 kilometers, about the distance from New York to San Francisco) from Sumatra in the west to New Guinea in the east and containing more than 13,000 separate islands. Not only does it dwarf all other Southeast Asian states in size, but it is by far the largest in population. With roughly 243 million people, it is ranked as the world's fourth most populated country. Lying north of the equator is the Philippines, a country of 92 million people spread over some 7,000 islands, both large and small.

**Mountains of northern Southeast Asia.** *Extensive forests are still found in the mountainous regions of Burma and Laos. These are increasingly threatened, however, by commercial logging and, to a lesser extent, by swidden cultivation.*

Tropical forest
Forest destroyed
Coastal pollution
• Poor urban air quality

**Kalimantan.** *Severe deforestation from commercial logging. After forests are cut, migrants from other Indonesian islands settle on small farming plots. However, soil depletion is a major problem, resulting in many abandoned farms and further environmental deterioration. Like wise, forest and field burning contribute to regional smoke pollution.*

**Java.** *Forests were cleared in most areas decades ago for rice cultivation and plantation crops. Population pressure and overfarming have resulted in serious degradation in many areas.*

**FIGURE 13.3 Environmental Issues in Southeast Asia**   Southeast Asia was once one of the most heavily forested regions of the world. Most of the tropical forests of Thailand, the Philippines, peninsular Malaysia, Sumatra, and Java, however, have been destroyed through a combination of commercial logging and agricultural settlement. The forests of Kalimantan (Borneo), Burma (Myanmar), Laos, and Vietnam, moreover, are now being rapidly cleared. Water and urban air pollution, as well as soil erosion, are also widespread in Southeast Asia.

world's true tropical rain forests. Most of Sumatra's forests have been cut, however, and those of Kalimantan (Borneo) are rapidly receding. Indonesia's last major forestry frontier is on the island of New Guinea, where forests are still extensive.

## Smoke and Air Pollution

Until recently, most of Southeast Asia's residents seemed unconcerned about the widespread air pollution created by a combination of urban smog and smoke from forest clearing. Then, late in the

1990s, the region suffered from two consecutive years of disastrous air pollution that served as a wake-up call (Figure 13.5). Because of global publicity, billions of tourism dollars were lost as potential visitors decided to travel elsewhere. Although the situation subsequently improved, Malaysian officials had to declare another emergency in 2005 because of widespread smoke from forest fires.

Several factors, both natural and economic, combined to produce the region's air pollution disaster of the late 1990s. First, large portions of insular Southeast Asia suffered a severe drought caused by El Niño (discussed in Chapter 4), turning the normally wet tropical

**377**

**FIGURE 13.4 Commercial Logging** Southeast Asia has historically been the world's most important supplier of tropical hardwoods. Unfortunately, most of the tropical forests of the Philippines and Thailand, as well as the Indonesian islands of Java and Sumatra, have been destroyed by the logging process.

forests into tinderboxes. This drought also dried out the widespread peat bogs of coastal Kalimantan, which continued to burn for months after fires began. Second, commercial forest cutting has been responsible for many fires, as the leftover slash (branches, small trees, and so forth) is often burned to clear out the land. The third factor is Southeast Asia's rapidly growing cities, where cars, trucks, and factories emit huge quantities of pollutants. One recent survey placed Vietnam's booming urban areas of Hanoi and Ho Chi Minh City among the 20 most environmentally distressed cities in the world.

## Global Warming and Southeast Asia

As most of Southeast Asia's people live in coastal environments, the region is highly vulnerable to the rise in sea level associated with global warming. Periodic flooding is already a major problem in many of the regions' low-lying cities. Southeast Asian farmland is also concentrated in delta environments and thus could suffer from saltwater intrusion and higher storm surges. It is also feared that higher temperatures could reduce rice yields through much of the region. Concerns about the region's exposure to global warming intensified in March 2009, when researchers at the International Scientific Congress on Climate Change in Copenhagen showed that global sea level would probably rise by 1 meter or more by 2100, a figure roughly double previous estimates.

Changes in precipitation across Southeast Asia brought about by global warming remain uncertain. Many experts foresee an intensification of the monsoon pattern, which could bring increased rainfall to much of the mainland. While enhanced precipitation would likely result in more destructive floods, it could bring some agricultural benefits to dry areas such as Burma's central Irrawaddy Valley. Complicating this scenario, however, is the prediction that global climate change could intensify the El Niño effect (discussed in Chapter 4), which would result in more extreme droughts, especially in the equatorial belt of Indonesia.

**FIGURE 13.5 Urban Air Pollution** Air pollution has reached a crisis stage in the rapidly industrializing cities of Southeast Asia, particularly in Bangkok, Manila, and Jakarta. People sometimes resort to using face masks to filter out soot and other forms of particulate matter. Forest fires, which often follow logging, add to the problem.

**FIGURE 13.6  Burning Peatlands**  The soil of most wetland areas in Southeast Asia is composed largely of peat, an organic substances that can burn when dry. Draining for agricultural expansion, as well as drought, often results in extensive peat fires. In this photograph, a C130 airplane is dropping water on burning peat in Lampung, Sumatra.

Most Southeast Asian countries have ratified the 1997 Kyoto Protocol. But because all—even wealthy Singapore—are officially classified as developing countries, none are obligated to reduce their emissions of greenhouse gases. Still, Southeast Asia's overall emissions from conventional sources remain low by global standards. When greenhouse gas emissions associated with deforestation are factored in, however, Southeast Asia's role in global climate change is revealed to be much larger. According to some estimates, Indonesia is the world's third largest contributor to the problem, after China and the United States.

Much of Indonesia's carbon emissions stems from the burning and oxidation of peat in deforested swamps. (Peat is the partially decayed organic matter that accumulates in saturated soils.) During dry periods, peat soils sometimes burn, releasing the stored carbon (Figure 13.6). Both Indonesia and Malaysia, moreover, are actively draining coastal wetlands to make room for agricultural expansion, a process that results in the gradual oxidation of the peat. Already some 40 percent of Southeast Asia's peatlands have been lost. Ironically, many of these swamps are being destroyed in order to produce biodiesel from oil palms, a supposedly eco-friendly form of energy.

## Patterns of Physical Geography

Two types of forest are of primary interest for commercial logging in Southeast Asia. The southern portion of the region—insular Southeast Asia—is one of the world's three main zones of tropical rain forest. The northern part of the region, or mainland Southeast Asia, is located in the tropical wet-and-dry zone that is noted for certain valuable timber species, such as teak. The distribution of these forest types is closely linked to the landforms and climates of these two distinct parts of the region.

**Mainland Environments**  Mainland Southeast Asia is an area of rugged uplands mixed together with broad lowlands associated with large rivers. The region's northern boundary lies in a cluster of mountains connected to the highlands of western Tibet and south-central China. In the far north of Burma, peaks reach 18,000 feet (5,500 meters). From this point, a series of distinct mountain ranges

spreads out, extending through western Burma, along the Burma–Thailand border, and through Laos into southern Vietnam.

Several large rivers flow southward out of Tibet and its adjacent highlands into mainland Southeast Asia. The valleys and deltas of these rivers are the centers of both population and agriculture in mainland Southeast Asia. The longest river is the Mekong, which flows through Laos and Thailand, then across Cambodia before entering the South China Sea through a large delta in southern Vietnam. Second longest is the Irrawaddy, which flows through Burma's central plain before reaching the Bay of Bengal. Two smaller rivers are equally significant: the Red River, which forms a heavily settled delta in northern Vietnam, and the Chao Phraya, which has created the fertile alluvial plain of central Thailand (Figure 13.7).

The centermost area of mainland Southeast Asia is Thailand's Khorat Plateau, which is neither a rugged upland nor a fertile river valley. This low sandstone plateau averages about 500 feet (175 meters) in height and is noted for its thin, poor soils. Water shortages and periodic droughts pose challenges throughout this extensive area.

**FIGURE 13.7  Chao Phraya Delta**    The flat, fertile, and well watered delta and lower valley of the Chao Phraya River in central Thailand are readily visible in the central portion of this satellite image. Note the rugged mountains to the west of the Chao Phraya along the border between Thailand and Burma. The drier lands of the Khorat Plateau are also visible in the right-central part of the image.

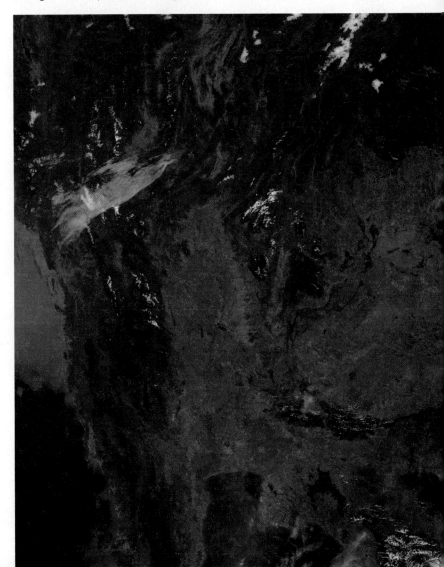

**The Influence of the Monsoon**    Almost all of mainland Southeast Asia is affected by the seasonally shifting winds known as the *monsoon*. The climate of this area is characterized by a distinct hot and rainy season from May to October. This is followed by dry but still generally hot conditions from November to April (Figure 13.8). Only the central highlands of Vietnam and a few coastal areas receive significant rainfall during this period.

Two types of tropical climates are dominant in mainland Southeast Asia. Although both are affected by the monsoon, they differ in the total amount of rainfall received during the year. Along the coasts

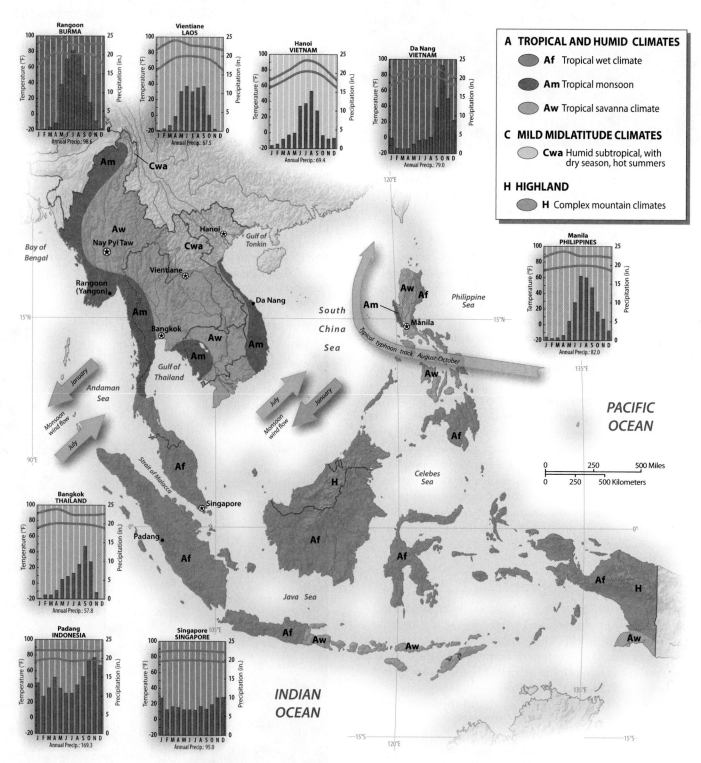

**FIGURE 13.8 Climate Map of Southeast Asia**    Most of insular Southeast Asia is characterized by the constantly hot and humid climates of the equatorial zone. Mainland Southeast Asia, on the other hand, has the seasonally wet and dry climates of the tropical monsoon and tropical savanna types. Only in the far north are subtropical climates, with relatively cool winters, encountered. The northern half of the region is strongly influenced by the seasonally shifting monsoon winds. Northeastern Southeast Asia—and especially the Philippines—often experiences typhoons from August to October.

and in the highlands, the tropical monsoon climate (Am) dominates. Rainfall totals for this climate usually average more than 100 inches (254 centimeters) each year. The greater portion of the mainland falls into the tropical savanna (Aw) climate type, where annual rainfall totals are about half those of the Am region.

**Insular Environments**  The main feature of insular Southeast Asia is its island environment. Although Indonesia contains thousands of islands, it is dominated by the four large landmasses of Sumatra, Borneo (or Kalimantan), Java, and Sulawesi. Indonesia also includes the western half of New Guinea and the Lesser Sunda Islands, which extend to the east of Java. Many of the islands in this region are mountainous. A string of active volcanoes extends the length of eastern Sumatra across Java and into the Lesser Sunda Islands. In the Philippines, the two largest and most important islands are Luzon (about the size of Ohio) in the north and Mindanao (the size of South Carolina) in the south. Sandwiched between them are the Visayan Islands, which number roughly a dozen. The topography of the Philippines includes rugged upland landscapes as well as numerous volcanoes.

Insular Southeast Asia has numerous volcanoes because it lies along the intersection of several tectonic plates. This geologically active situation generates a number of natural hazards, including earthquakes, toxic mud volcanoes, and **tsunamis** (sometimes called "tidal waves") (Figure 13.9). An undersea earthquake off the coast of Sumatra in December 2004 generated a tsunami that killed some 230,000 people in Southeast and South Asia, making it the second deadliest earthquake in recorded history.

The climates of insular Southeast Asia are more complex than those of the mainland largely because the insular belt extends across a greater span of latitude. Most of Indonesia lies in the equatorial zone, which results in high levels of precipitation evenly distributed throughout the year. Southeastern Indonesia and East Timor, however, experience a prolonged dry season from June to October. Most of the Philippines experiences dry conditions from November to April. Year-round rainfall, however, occurs

**FIGURE 13.9 Mud Volcano**  The Sidoarjo mud flow in east Java, composed of hot, toxic sludge, has been flowing out of a mud volcano since 2006. It is expected to continue to flow for another 30 years. The volcano itself was set off when a gas exploration drilling project tapped into a vent of extremely hot mud.

**FIGURE 13.10 Cyclone Nargis**  Cyclone Nargis slammed into southern Burma (Myanmar) in early May, 2008, resulting in the country's worst natural disaster in recorded history. This photograph shows devastated families waiting for relief near their destroyed homes on Haing Guy Island in southwestern Burma.

in the far eastern part of the Philippines, due to the trade winds, and in Mindanao, due to its more equatorial location.

**The Typhoon Threat**  The coastal areas of mainland Southeast Asia and the Philippines are highly vulnerable to tropical cyclones, or **typhoons**, as they are called in the western Pacific. These strong storms bring devastating winds and torrential rain. Each year a number of typhoons hit Southeast Asia with heavy damage and loss of life through flooding and landslides. Deforestation and farming on steep hillsides make the problem much worse.

In early May 2008, southern Burma was slammed by Cyclone Nargis, the worst storm in recent Southeast Asian history. Nargis caused some 150,000 deaths, left 1 million people homeless, and resulted in roughly $10 billion in damages (Figure 13.10). Casualty rates were particularly high because Burma's military government reacted very slowly to the crisis, restricting the activities of international aid agencies.

## Population and Settlement: Subsistence, Migration, and Cities

The scale of Southeast Asia's population issue is quite different from those of East Asia and South Asia. With roughly 570 million people, Southeast Asia is still *relatively* sparsely settled. Over most of the region, extensive tracts of land with infertile soil and rugged topography remain thinly inhabited. In contrast, dense populations are found in the region's deltas, coastal areas, and zones of fertile volcanic soil (Figure 13.11).

Southeast Asia experienced rapid population growth in the second half of the 20th century. More recently, birthrates have dropped quickly, especially in the wealthier countries of the region. In the poorer countries, such as Laos, Cambodia, and East Timor, however, birthrates remain high. Although birthrates have dropped in the Philippines, the country's population continues to grow at a relatively fast rate.

## Settlement and Agriculture

Much of insular Southeast Asia has relatively infertile soil that cannot easily support intensive agriculture or high rural population densities. Although the island forests are lush and biologically diverse, plant nutrients are locked up in the vegetation itself rather than being stored in the soil, where they would easily benefit agriculture. Furthermore, the constant rain of the equatorial zone tends to wash away nutrients. Agriculture must thus rely on constant field rotation or the application of heavy amounts of fertilizer.

There are, however, some notable exceptions to this generalization. Unusually rich soils connected to volcanic activity are scattered through much of the region but are particularly widespread on the island of Java. With more than 50 volcanoes, Java is a fertile island that supports a range of tropical crops and a very high population density. Some 124 million people live in Java, an island smaller than Iowa. Dense populations are also found in pockets of fertile alluvial soils along the coasts of insular Southeast Asia, where people supplement land-based farming with fishing and other commercial activities. One particularly densely settled area is the central lowlands of Luzon, near the city of Manila, the core area of the Philippines.

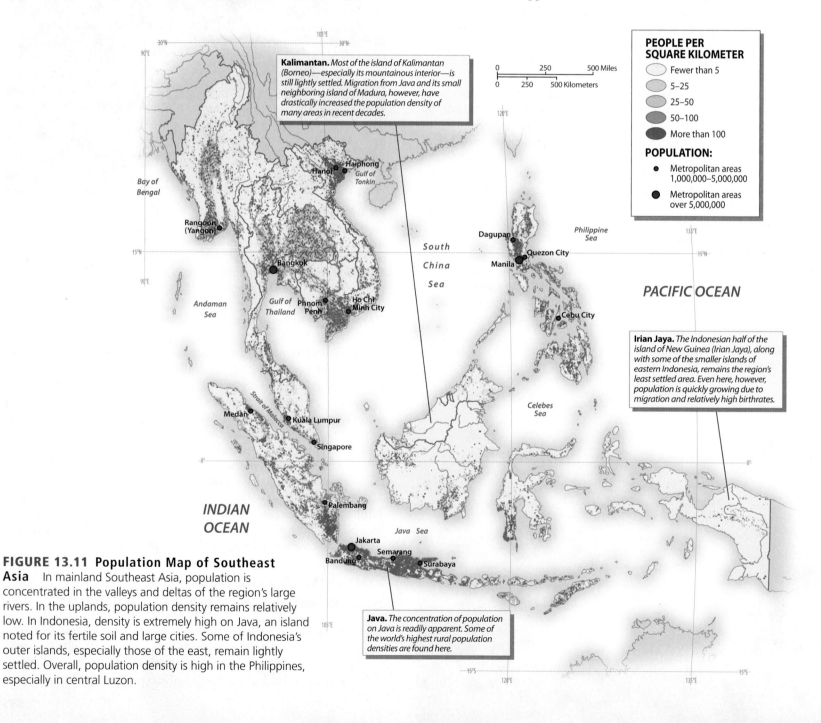

**FIGURE 13.11 Population Map of Southeast Asia**   In mainland Southeast Asia, population is concentrated in the valleys and deltas of the region's large rivers. In the uplands, population density remains relatively low. In Indonesia, density is extremely high on Java, an island noted for its fertile soil and large cities. Some of Indonesia's outer islands, especially those of the east, remain lightly settled. Overall, population density is high in the Philippines, especially in central Luzon.

**Kalimantan.** *Most of the island of Kalimantan (Borneo)—especially its mountainous interior—is still lightly settled. Migration from Java and its small neighboring island of Madura, however, have drastically increased the population density of many areas in recent decades.*

**Irian Jaya.** *The Indonesian half of the island of New Guinea (Irian Jaya), along with some of the smaller islands of eastern Indonesia, remains the region's least settled area. Even here, however, population is quickly growing due to migration and relatively high birthrates.*

**Java.** *The concentration of population on Java is readily apparent. Some of the world's highest rural population densities are found here.*

**PEOPLE PER SQUARE KILOMETER**
- Fewer than 5
- 5–25
- 25–50
- 50–100
- More than 100

**POPULATION:**
- Metropolitan areas 1,000,000–5,000,000
- Metropolitan areas over 5,000,000

**FIGURE 13.12 Swidden Agriculture** In the uplands of Southeast Asia, swidden (or "slash-and-burn") agriculture is widely practiced. When done by tribal peoples with low population densities, swidden is not environmentally harmful. When practiced by large numbers of immigrants from the lowlands, however, swidden can result in deforestation and extensive soil erosion.

In mainland Southeast Asia, population is concentrated in the agriculturally intensive valleys and deltas of the large rivers, whereas the uplands remain relatively lightly settled. The population core of Thailand is formed by the valley and delta of the Chao Phraya River, just as Burma's is focused on the Irrawaddy River. Vietnam has two distinct core areas: the Red River Delta in the far north and the Mekong Delta in the far south. In contrast to these densely settled areas, the middle reaches of the Mekong River provide only limited lowlands in Laos, which is one reason that country has such a small population.

Agricultural practices and settlement forms vary widely across the complex environments of Southeast Asia. Generally speaking, however, three farming and settlement patterns are apparent: swidden in the upland areas and both plantation agriculture and rice cultivation in the lowlands.

**Swidden in the Uplands** Also known as shifting cultivation, or "slash-and-burn" agriculture, swidden is practiced throughout the rugged uplands of Southeast Asia (Figure 13.12). In the **swidden** system, small plots of several acres of forest or brush are periodically cut by hand. The fallen vegetation is then burned to transfer nutrients to the soil before subsistence crops are planted. Yields remain high for several years and then drop off as the soil nutrients are exhausted and insect pests and plant diseases multiply. These plots are abandoned after a few years and return to woody vegetation. The cycle of cutting, burning, and planting is then moved to another small plot not far away—thus the term *shifting cultivation.*

Swidden is a sustainable form of agriculture when population densities remain relatively low. Today, however, the swidden system is increasingly threatened. With higher population densities, the rotation period must be shortened, which damages soil resources. Swidden farming is also harmed by commercial logging, which both displaces farmers and removes soil nutrients from the ecosystem as logs are exported.

When swidden can no longer support the population, upland people sometimes adapt by switching to a cash crop that will allow them to participate in the commercial economy. In the mountains of northern Southeast Asia, one of the main cash crops has historically been opium, grown by local farmers for the global drug trade. Intensive efforts by national governments and the United Nations virtually eliminated opium growing in Laos and northern Thailand by 2006, and even Burma saw a huge reduction. By early 2009, however, falling prices for alternative crops coupled with rising opium prices resulted in a return to opium poppy cultivation in many remote villages of Laos and Burma.

**Plantation Agriculture** With European colonization, Southeast Asia became a focus for plantation agriculture, growing high-value specialty crops ranging from coconuts to rubber. Even in the 19th century, Southeast Asia was linked to a globalized economy through the plantation system. Forests were cleared and swamps drained to make room for commercial farms; labor was supplied by native people or by workers brought in from India or China (Figure 13.13).

Plantations are still an important part of Southeast Asia's geography and economy. Most of the world's natural rubber is produced in Malaysia, Indonesia, and Thailand. Cane sugar has long been a major crop of the Philippines and parts of Indonesia, although it is no longer very profitable; as a result, sugar areas in the Philippines are associated with intense rural poverty. Indonesia is the region's leading producer of tea, and Vietnam dominates the production of coffee. In recent years, oil palm plantations have been spreading through much of the region,

**FIGURE 13.13 Tea Harvesting in Indonesia** Plantation crops, such as tea, are major sources of exports for several Southeast Asian countries. Coconut, rubber, palm oil, and coffee are other major cash crops. Many of these crops require large amounts of labor, particularly at harvest time.

often at the expense of tropical rain forests. Coconuts are widely grown in the Philippines, Indonesia, and elsewhere in the region.

**Rice in the Lowlands**   The lowland basins of mainland Southeast Asia are largely devoted to intensive rice cultivation. Throughout almost all of Southeast Asia, rice is the preferred staple food. Rice harvests are increasingly traded to meet the needs of expanding urban markets throughout the world. Three delta areas have been the focus for commercial rice cultivation: the Irrawaddy in Burma, the Chao Praya in Thailand, and the Mekong in Vietnam. The use of agricultural chemicals and high-yield crop varieties have allowed production to keep pace with population growth, although at the cost of significant environmental damage.

As of 2008, the world's two largest rice exporters were Thailand and Vietnam. In the same year, these two countries, along with Burma, Cambodia, and Laos, joined together to form the Organization of Rice Exporting Countries (OREC). OREC seeks to maintain high and even prices. The Philippines has strongly criticized the organization, as it is one of the world's major rice importing countries.

## Recent Demographic Change

Because Southeast Asia is not facing the same kind of population pressure as East or South Asia, a wide range of government population policies exist. While several countries are concerned about rapid growth and thus have strong family planning programs, others believe that their populations are still comparatively small. In countries facing rapid demographic expansion, internal relocation of people away from densely populated areas to outlying districts is a common outcome.

**Population Contrasts**   The Philippines, the second most populous country in Southeast Asia, has a relatively high growth rate, and effective family planning has been difficult to establish (Table 13.1).When a popular democratic government replaced a dictatorship in the 1980s, the Philippine Roman Catholic Church, which played an active role in

the peaceful revolution, pressured the new government to cut funding for family planning programs. As a result, many clinics that had given out family planning information were closed. Due to a combination of rapid growth and economic stagnation, many Filipinos have been forced to migrate either to foreign countries or to less densely settled portions of the Philippines.

The highest total fertility rate (TFR) in mainland Southeast Asia is found in Laos, a country of Buddhist religious tradition. Here the high birthrate is best explained by the country's low level of economic and social development. Thailand, which shares cultural traditions with Laos yet is considerably more developed, demonstrates the other end of the range. Here the TFR has dropped dramatically, from 5.4 in 1970 to 3.4 in 2008.

Indonesia, with the region's largest population, at 243 million, has also seen a dramatic decline in fertility in recent decades, although its fertility rate remains slightly above the replacement level. If the present trend continues, however, Indonesia will reach population stability before most other large developing countries. As with Thailand, this drop in fertility seems to have resulted from government family planning efforts coupled with urbanization and improvements in education.

The city-state of Singapore stands out on the demographic charts with a fertility rate well below replacement levels. Unless the deficit is offset by immigration or a dramatic turnabout in the birthrate, Singapore's population will soon begin to decline. Its government is concerned about this situation and is actively promoting marriage and childbearing, particularly among the most highly educated segment of its population (Figure 13.14). In 2008, Singapore's government approved a $600 per month child-care subsidy for working mothers and began to encourage pro-family environments in the workplace.

**Growth and Migration**   Indonesia has had the most explicit policy of **transmigration**, or relocation of its population from one region to another within its national territory. Primarily because of migration from densely populated Java, the population of the outer islands of Indonesia has grown rapidly since the 1970s. The

---

## TABLE 13.1   POPULATION INDICATORS

| Country | Population (millions) 2009 | Population Density (per square kilometer) | Rate of Natural Increase (RNI) | Total Fertility Rate | Percent Urban | Percent <15 | Percent >65 | Net Migration (Rate per 1000) 2005–10* |
|---|---|---|---|---|---|---|---|---|
| Burma (Myanmar) | 50.0 | 74 | 1.1 | 2.3 | 31 | 27 | 5 | −2.0 |
| Brunei | 0.4 | 66 | 1.3 | 1.7 | 72 | 26 | 4 | 1.8 |
| Cambodia | 14.8 | 82 | 1.7 | 3.0 | 1.5 | 35 | 3 | −0.1 |
| East Timor | 1.1 | 76 | 3.1 | 6.5 | 22 | 45 | 3 | 1.8 |
| Indonesia | 243.3 | 128 | 1.5 | 2.5 | 43 | 29 | 6 | −0.6 |
| Laos | 6.3 | 27 | 2.1 | 3.5 | 27 | 39 | 4 | −2.4 |
| Malaysia | 28.3 | 86 | 1.6 | 2.6 | 68 | 32 | 4 | 1.0 |
| Philippines | 92.2 | 307 | 2.1 | 3.3 | 63 | 35 | 4 | −2.0 |
| Singapore | 5.1 | 7,486 | 0.6 | 1.3 | 100 | 18 | 9 | 22.0 |
| Thailand | 67.8 | 132 | 0.6 | 1.8 | 36 | 22 | 7 | 0.9 |
| Vietnam | 87.3 | 263 | 1.2 | 2.1 | 28 | 26 | 7 | −0.5 |

*Source*: Population Reference Bureau, *World Population Data Sheet, 2009*.

*Net Migration Rate from the United Nations, Population Division, *World Population Prospects: The 2008 Revision Population Database*.

**FIGURE 13.14 Pro-Procreation Advertisement in Singapore**  Due to its extremely low birthrate, the population of Singapore will probably begin to decline quite soon. Shown here are perfumes created by local students and launched in a government-backed campaign.

province of East Kalimantan, for example, experienced a growth rate of 30 percent per year during the last two decades of the 20th century (Figure 13.15).

High social and environmental costs accompany these relocation programs. Javanese peasants, accustomed to working the fertile soils of their home island, often fail in their attempts to grow rice in the former rain forest of Kalimantan. Field abandonment is common after repeated crop failures. In some areas farmers have little choice but to adopt a semi-swidden form of cultivation, a process associated with further deforestation as well as conflicts with indigenous peoples. Partly because of these problems, the Indonesian government significantly reduced its official transmigration program in 2000. Many people, however, still borrow money or use their personal savings to move from the densely populated to the lightly settled portion of Indonesia.

## Urban Settlement

Despite the relatively high level of economic development in much of Southeast Asia, the region is not heavily urbanized; fewer than half of its people live in cities. Even Thailand's population is mostly rural, which is unusual for a country that has experienced so much industrialization. But cities are growing rapidly throughout the region, increasing the rate of urbanization.

Many Southeast Asian countries have **primate cities**, single, large urban settlements that overshadow all others. Thailand's urban system, for example, is dominated by Bangkok, just as Manila far surpasses all other cities in the Philippines. Both have grown recently

into megacities with more than 10 million residents. More than half of all city-dwellers in Thailand live in the Bangkok metropolitan area (Figure 13.16).

In both Manila and Bangkok, explosive urban growth has led to housing problems, congestion, and pollution. With its rapidly growing number of private automobiles, Bangkok suffers from some of the worst traffic in the world; the Thai government has responded with large-scale highway and mass-transit construction programs. It is estimated that more than half of the population of Manila lives in squatter settlements, usually without basic water and electricity service. Both cities suffer from a lack of parks and other public spaces, which is one reason massive shopping malls have become so popular. Bangkok's Paragon Mall has recently emerged as a major urban focus, complete with a conference center and a concert hall.

Urban primacy is less pronounced in other major Southeast Asian countries. Vietnam, for example, has two main cities, Ho Chi Minh City (formerly Saigon) in the south and the capital city of Hanoi in the north. Jakarta is the largest urban area in Indonesia, but the country has a number of other large and growing cities, including Bandung and Surabaya. Yangon (formerly Rangoon) remains the primate city of Burma, with more than 4 million residents. In Cambodia, the capital city of Phnom Penh has only about 1.3 million people, but it has been growing rapidly since 1990.

Kuala Lumpur, the largest city in Malaysia, has received heavy investments from both the national government and the global business community. This has produced a modern, forward-looking city that is largely free of most of the traffic, water, and slum problems that plague most other Southeast Asian cities. As a symbol of Kuala Lumpur's modern outlook, the Petronas Towers, owned by the country's national oil company, were the world's tallest buildings when completed in 1996.

The independent republic of Singapore is essentially a city-state of 4.8 million people on an island of 274 square miles (710 square kilometers), about three times the size of Washington, DC

**FIGURE 13.15 Migrant Settlement in Indonesia**  Migration from densely settled to sparsely settled areas of Southeast Asia has resulted in the creation of thousands of new communities. Many of these communities have minimal transportation and communication facilities and receive few governmental services. Some of them struggle to survive.

**FIGURE 13.16 Bangkok** Bangkok saw the development of an impressive skyline during its boom years from the late 1970s through the late 1990s. Unfortunately, transportation development did not keep pace with population and commercial growth, resulting in one of the most congested and polluted urban landscapes in the world.

(Figure 13.17). While space is at a premium, Singapore has been very successful at developing high-tech industries that have brought it great wealth. Unlike most other Southeast Asian cities, Singapore has no squatter settlements or slums.

# Cultural Coherence and Diversity: A Meeting Ground of World Cultures

Unlike many other world regions, Southeast Asia lacks the historical dominance of a single civilization. Instead, the region has been a meeting ground for cultural influences from South Asia, China, the Middle East, Europe, and North America. Abundant natural resources and the region's strategic location on oceanic trading routes connecting major continents have long made Southeast Asia attractive to outsiders. As a result, the cultural geography of this diverse region owes much to external influences.

## The Introduction and Spread of Major Cultural Traditions

In Southeast Asia contemporary cultural diversity is related to the historical influence of the major religions of the region: Hinduism, Buddhism, Islam, and Christianity (Figure 13.18).

**South Asian Influences** The first major external influence arrived from South Asia some 2,000 years ago, when migrants from what is now India helped establish Hindu kingdoms in coastal locations in Burma, Thailand, Cambodia, Malaysia, and western Indonesia. Although Hinduism later faded away in most locations, it is still the dominant religion on the Indonesian island of Bali. *follower*

A second wave of South Asian religious influence reached mainland Southeast Asia in the 13th century, in the form of Theravada Buddhism, which spread from Sri Lanka. Almost all the people in

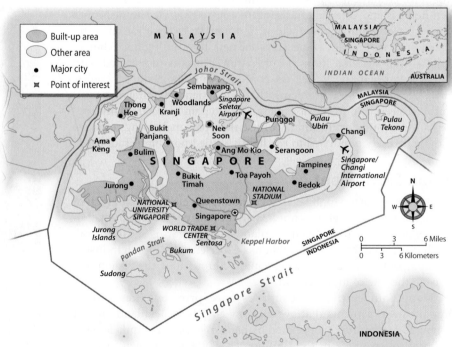

**FIGURE 13.17 Singapore** Singapore remains the economic and technological hub of Southeast Asia. It is famous for its clean, efficiently run, and very modern urban environment. Some residents complain, however, that Singapore has lost much of its charm as it developed.

lowland Burma, Thailand, Laos, and Cambodia converted to Buddhism at that time, and today it forms the foundation for their social institutions. Saffron-robed monks, for example, are a common sight, and Buddhist temples abound.

**Chinese Influences**  Unlike most other mainland peoples, the Vietnamese were not heavily influenced by South Asian civilization. Instead, their early connections were to East Asia. Vietnam was a province of China until about 1000 C.E., when the Vietnamese established a kingdom of their own. But while the Vietnamese rejected China's political rule, they retained many features of Chinese culture. The traditional religious and philosophical beliefs of Vietnam, for example, are centered on Mahayana Buddhism and Confucianism.

East Asian cultural influences in many other parts of Southeast Asia are directly linked to more recent immigration of southern Chinese. This migration reached a peak in the 19th and early 20th centuries (Figure 13.19). China was then a poor and crowded country, which made sparsely populated Southeast Asia appear to be a place of opportunity. Eventually, distinct Chinese settlements were established in every Southeast Asian country, especially in urban areas. In Malaysia the Chinese minority now constitutes roughly one-third of the population, whereas in Singapore some three-quarters of the people are of Chinese ancestry.

In many places in Southeast Asia, relationships between the Chinese minority and the native majority are strained. Even though their ancestors arrived generations ago, many Chinese are still considered resident aliens because they maintain their Chinese identities. A more significant source of tension is the fact that most Chinese communities in Southeast Asia are relatively wealthy. Many Chinese emigrants prospered as merchants, an occupation avoided by most local people. As a result, they have acquired substantial economic influence, which others often resent.

**FIGURE 13.18 Religion in Southeast Asia**  Southeast Asia is one of the world's most religiously diverse regions. Most of the mainland is Buddhist, with Theravada Buddhism dominant in Burma (Myanmar), Thailand, Laos, and Cambodia, and Mahayana Buddhism (combined with other elements of the so-called Chinese religious complex) prevailing in Vietnam. The Philippines is primarily Christian (Roman Catholic), but the rest of insular Southeast Asia is primarily Muslim. Substantial Muslim minorities are found in the Philippines, Thailand, and Burma. Animist and Christian minorities can be found in remote areas throughout Southeast Asia, especially in eastern Indonesia.

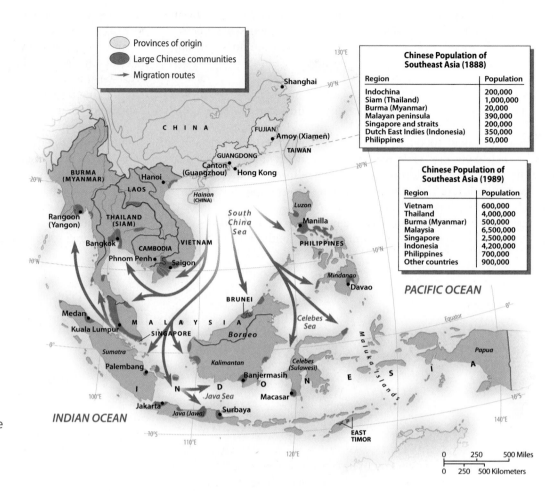

| Chinese Population of Southeast Asia (1888) | |
| --- | --- |
| Region | Population |
| Indochina | 200,000 |
| Siam (Thailand) | 1,000,000 |
| Burma (Myanmar) | 20,000 |
| Malayan peninsula | 390,000 |
| Singapore and straits | 200,000 |
| Dutch East Indies (Indonesia) | 350,000 |
| Philippines | 50,000 |

| Chinese Population of Southeast Asia (1989) | |
| --- | --- |
| Region | Population |
| Vietnam | 600,000 |
| Thailand | 4,000,000 |
| Burma (Myanmar) | 500,000 |
| Malaysia | 6,500,000 |
| Singapore | 2,500,000 |
| Indonesia | 4,200,000 |
| Philippines | 700,000 |
| Other countries | 900,000 |

**FIGURE 13.19 Chinese in Southeast Asia** People from the southern coastal region of China have been migrating to Southeast Asia for hundreds of years, a process that reached a peak in the late 1800s and early 1900s. Most Chinese migrants settled in the major urban areas, but in peninsular Malaysia large numbers were drawn to the countryside to work in the mining industry and in plantation agriculture. Today Malaysia has the largest number of people of Chinese ancestry in the region. Singapore, however, is the only Southeast Asian country with a Chinese majority.

**The Arrival of Islam** Muslim merchants from South and Southwest Asia arrived in Southeast Asia hundreds of years ago and soon began converting many of their local trading partners. From an initial focus around 1200 C.E. in northern Sumatra, Islam spread through much of insular Southeast Asia. By 1650, it had largely replaced Hinduism and Buddhism throughout Malaysia and Indonesia. The only significant holdout was the small but fertile island of Bali, where thousands of Hindu musicians and artists fled from Java, giving the island a strong artistic tradition that many international tourists appreciate today.

Some 88 percent of Indonesia's inhabitants follow Islam, making it the world's most populous Muslim country (Figure 13.20). This figure, however, hides a significant amount of internal religious diversity. In some parts of Indonesia, such as in northern Sumatra (Aceh), highly orthodox forms of Islam took root. In others, such as central and eastern Java, a more relaxed form of worship emerged that included certain Hindu and even animistic beliefs. Islamic reformers, however, have long tried to instill more mainstream forms of faith among the Javanese. Recently they have found much success, particularly among the young.

**Christianity** Islam was still spreading eastward through insular Southeast Asia when the Europeans arrived in the 16th century. When the Spanish claimed the Philippine Islands in the 1570s, they found the southwestern portion of the archipelago to be thoroughly Islamic. To this day, the southwest Philippines is still largely Muslim, although the rest of the country is mostly Roman Catholic. East Timor, long a Portuguese colony, is also a predominantly Roman Catholic country.

Christian missions spread through other parts of Southeast Asia in the late 19th and early 20th centuries, when European colonial powers controlled most of the region. While French priests converted many people in Vietnam to Catholicism, they had little influence in other lowland areas. Missionaries were more successful in the highlands inhabited by tribal peoples worshiping nature spirits and their ancestors. The general name for such religions is **animism**. While many modern hill tribes remain animist today, others were converted to Christianity. As a result, significant Christian concentrations are found in the Lake Batak area of north-central Sumatra, the mountainous borderlands of Burma, the northern peninsula of Sulawesi, and the highlands of southern Vietnam.

**Religious Persecution** Religious persecution has recently become a serious issue in parts of Southeast Asia. Vietnam's communist government is struggling against a revival of faith among the country's Buddhist majority and its 8 million Christians. Buddhist monks are sometimes harassed, and the government reserves for itself the right to appoint all religious leaders. In Burma, the authoritarian government supports Buddhism, but when Buddhist monks led massive demonstrations against the government in late 2007, it cracked down hard, killing an estimated 30 to 40 monks (Figure 13.21). Burma has also severely repressed its Rohingya Muslim minority, driving hundreds of thousands out of the country. In early 2009, Thailand's government reportedly forced hundreds of Burmese Rohingya refugees out to sea, resulting in many deaths. In the same year, prominent members of Indonesia's government sought to ban the Ahmaddiya Muslim sect, a group that many mainstream Muslims regard as heretical.

**FIGURE 13.20 Indonesian Mosque**
Indonesia is often said to be the world's largest Muslim nation, because more Muslims reside here than in any other country. Islam was first established in northern Sumatra, which is still the most devoutly Islamic part of the country. In this photo, Acehnese people are praying in front of the Baiturrahman Grand Mosque in the city of Banda Aceh.

## Geography of Language and Ethnicity

The linguistic geography of Southeast Asia is complicated (Figure 13.22). The several hundred distinct languages of the region can all be placed into four major linguistic families, discussed below.

**Austronesian Languages**   One of the world's most widespread language families is Austronesian, which extends from Madagascar to Easter Island in the eastern Pacific. Today almost all insular Southeast Asian languages belong to the Austronesian family. But despite this common linguistic grouping, more than 50 distinct languages are spoken in Indonesia alone. And in far eastern Indonesia, a variety of languages fall into the completely separate family of Papuan, closely associated with New Guinea.

The Malay language overshadows all others in insular Southeast Asia. Malay is native to the Malay Peninsula, eastern Sumatra, and coastal Borneo yet was spread historically throughout the region by merchants and seafarers. As a result, it became a common trade language, or **lingua franca**, throughout much of the insular realm. When Indonesia became an independent country in 1949, its leaders decided to use the lingua franca version of Malay as the basis for a new national language called "Bahasa Indonesia" (or simply "Indonesian"). Although Indonesian is slightly different from the Malaysian spoken in Malaysia, they form a single, mutually understandable language. Both are now written in the Roman script.

The goal of the new Indonesian government was to offer a common language that could overcome ethnic differences throughout the huge country. This policy has been generally successful, with the vast majority of Indonesians now using the language. Regionally based languages, however, such as Javanese, Balinese, and Sundanese, continue to be the primary languages of most Indonesian homes.

The people of the Philippines speak eight major languages and several dozen minor languages, all of which are closely related. Despite more than 300 years of colonialism by Spain, Spanish never became a unifying force for the islands. During the American period (1898–1946), English served as the language of government and education. After independence, Philippine nationalists selected Tagalog, the language of central and southern Luzon as well as the Metropolitan Manila areas, to

**FIGURE 13.21 Protesting Burmese Monks**   In September 2007, more than 100,000 people took to the streets of Rangoon (Yangon) to protest the Burma's repressive military government. The protest marches were led by robed monks chanting prayers for peace.

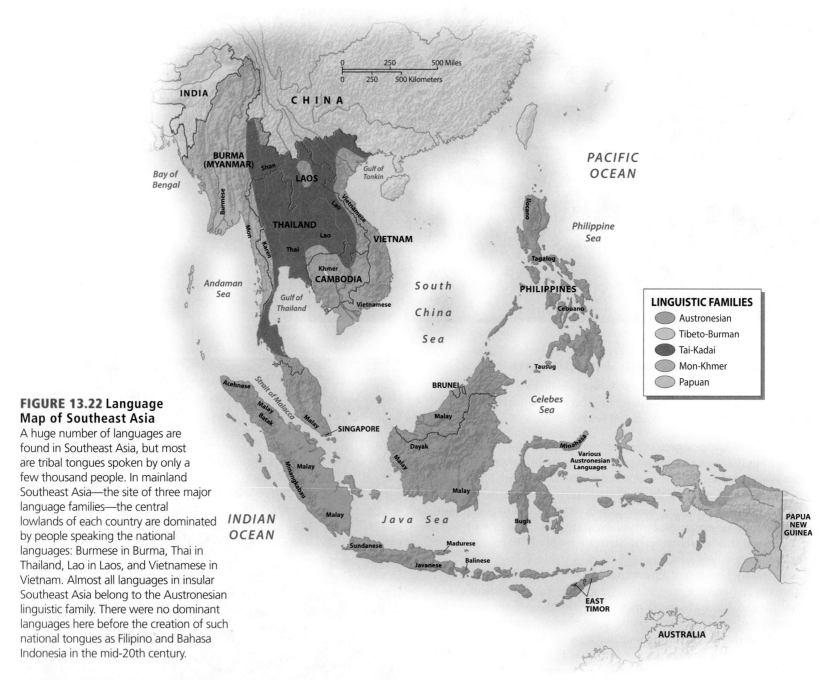

**FIGURE 13.22 Language Map of Southeast Asia**
A huge number of languages are found in Southeast Asia, but most are tribal tongues spoken by only a few thousand people. In mainland Southeast Asia—the site of three major language families—the central lowlands of each country are dominated by people speaking the national languages: Burmese in Burma, Thai in Thailand, Lao in Laos, and Vietnamese in Vietnam. Almost all languages in insular Southeast Asia belong to the Austronesian linguistic family. There were no dominant languages here before the creation of such national tongues as Filipino and Bahasa Indonesia in the mid-20th century.

replace English and help unify the new country. After Tagalog was standardized and modernized, it was renamed Filipino. Through its use in education, television, and films, Filipino has gradually emerged as the country's national language.

**Tibeto-Burman Languages**   Each country of mainland Southeast Asia is closely identified with the national language spoken in its core territory. This does not mean, however, that all the residents of these countries speak these official languages daily. In the mountains and other remote districts, other languages are commonly used. This linguistic diversity reinforces ethnic differences, often presenting challenges for programs designed to build national unity.

A good example of such linguistic challenges is Burma. Its national language is Burmese, a language that is closely related to Tibetan. Some 32 million people speak Burmese. Although the military government of Burma has sought to unify the population with one

language, a major split has developed with several non-Burman "hill tribes" that live in the rough uplands on both sides of the Burmese-speaking Irrawaddy Valley. Although most of these tribal groups speak languages in the Tibeto-Burman family, they are quite distinctive from Burmese.

**Tai-Kadai Languages**   The Tai-Kadai linguistic family probably originated in southern China and then spread into Southeast Asia starting around 1200. Today, closely related languages within the Tai subfamily are found through most of Thailand and Laos, in the uplands of northern Vietnam, and in Burma's Shan Plateau. Most Tai languages are spoken by small tribal groups. But two of them, Thai and Lao, are important national languages.

Historically, the main language of Thailand, called Siamese (just as the kingdom was called Siam), was restricted to the lower Chao Phraya valley. In the 1930s, however, the country changed its name to

Thailand to emphasize the unity of all the peoples speaking the closely related Tai languages within its territory. Siamese was similarly renamed Thai, and it has gradually become the country's unifying language. There is still much variation in dialect, however, with northern Thai sometimes considered a separate language. Somewhat more distinctive is Lao, the Tai language that became the national tongue of Laos. In Thailand's Khorat Plateau, most people speak Isan, a dialect much closer to Lao than it is to standard Thai.

**Mon-Khmer Languages**   The Mon-Khmer language family probably once covered virtually all of mainland Southeast Asia. It contains two major languages, Vietnamese and Khmer (the national language of Cambodia), as well as a host of minor languages spoken by hill peoples and a few lowland groups. Because of the historic Chinese influence in Vietnam, the Vietnamese language was written with Chinese characters until the French colonial government imposed the Roman alphabet, which remains in use today. Khmer, on the other hand, is—like Lao, Thai, and Burmese—written in its own Indian-derived script.

The most important aspect of linguistic geography in mainland Southeast Asia is the fact that in each country, the national language is spoken mainly in the core lowlands, whereas the peripheral uplands are populated by tribal peoples speaking separate languages. In Vietnam, for example, Vietnamese speakers occupy less than half of the national territory, even though they constitute a sizable majority of the country's population. Ethnic tensions here have recently mounted, as Vietnamese speakers, aided by the country's major road-building program, have begun moving into the sparsely populated highlands.

## Southeast Asian Culture in Global Context

European colonial rule initiated a new round of globalization in Southeast Asia, bringing European languages, Christianity, and new governmental, economic, and educational systems. As a result, anti-colonial movements sometimes merged with anti-globalization movements. After Burma became independent, for example, it retreated into its own form of Buddhist socialism, placing strict limits on foreign investment and tourism. Although Burma became more open to outsiders in the late 1900s, its government remains wary of foreign influences.

Other Southeast Asian countries have been receptive to foreign cultural influences. This is particularly true in the case of the Philippines, where U.S. colonialism encouraged the country to embrace many popular forms of Western culture. As a result, Filipino musicians and other entertainers are in demand elsewhere in Asia (Figure 13.23). Thailand, which was never subjected to colonial rule, is also highly open to global culture.

Cultural globalization has also been recently challenged in some Southeast Asian countries. The Malaysian government has been especially critical of American films and satellite television. Islamic revivalism in Indonesia and Malaysia also presents a challenge to cultural globalization. Islamic radicals have attacked a number of nightclubs and other tourist destination, and anti-American and anti-Western sentiments spread rapidly after the U.S.-led invasion of Iraq in 2003. More recently, however, tensions have lessened, and in the Indonesian elections of 2009, Islamist candidates fared poorly.

The use of English as a global language also causes controversy. Many conservatives oppose its use, yet it must be mastered if citizens are to participate in global business and politics. In Malaysia the widespread use of English grew increasingly controversial in the 1980s, as nationalists stressed the importance of the native tongue. This

**FIGURE 13.23 Filipino Entertainers**   The people of the Philippines have adopted popular forms of Western culture more than most other Southeast Asians, in part because of their long experience of American colonialism. As a result, Filipino performers are often in demand in other Asian countries. This photograph shows a Filipina musician singing in Hong Kong.

worried the business community, which considers English vital to Malaysia's competitive position. It also troubled the influential Chinese community, for which Malaysian is not a native language.

In Singapore, the situation is more complex. Mandarin Chinese, English, Malay, and Tamil are all official languages. Furthermore, the languages of southern China are common in home environments, because 75 percent of Singapore's population is of southern Chinese ancestry. In recent years the Singapore government has encouraged Mandarin Chinese and discouraged the southern Chinese dialects. It also launched a campaign against "Singlish," a popular form of speech based on English but employing many words from Malay and Chinese (Figure 13.24).

**FIGURE 13.24 Singlish Sign**   The Malay- and Chinese-influenced dialect of English known as "Singlish" is sometimes said to be the most important factor of cultural unity in Singapore. Singapore's government, however, discourages Singlish, viewing it as crude slang. The sign visible in this photograph ends with the common Singlish particle "la," which is used to soften a sentence and to show agreement with listeners or readers.

In the Philippines, nationalists complain about the common use of English, but widespread fluency has proved beneficial to the millions of Filipinos who work abroad or for international businesses. Many people from other Asian countries, moreover, come to the Philippines to study English; in 2008, it was estimated that almost 100,000 South Koreans were residing in the country for this purpose. Although the Philippine government has sought to gradually replace English with Filipino, English remains a widespread official language. At the same time, Filipino itself is increasingly incorporating words and phrases from English, giving rise to a hybrid dialect known as "Taglish."

## Geopolitical Framework: War, Ethnic Strife, and Regional Cooperation

Southeast Asia is sometimes defined as the geopolitical grouping of 10 different countries that have joined together as ASEAN (Figure 13.25). Although East Timor is not a member of ASEAN, its government hopes to join the organization by 2012. ASEAN has significantly reduced geopolitical problems among its member states, while giving Southeast Asia as a whole a greater degree of regional coherence. But despite ASEAN's successes, many Southeast Asian countries still experience internal ethnic conflicts as well as tensions with their neighbors.

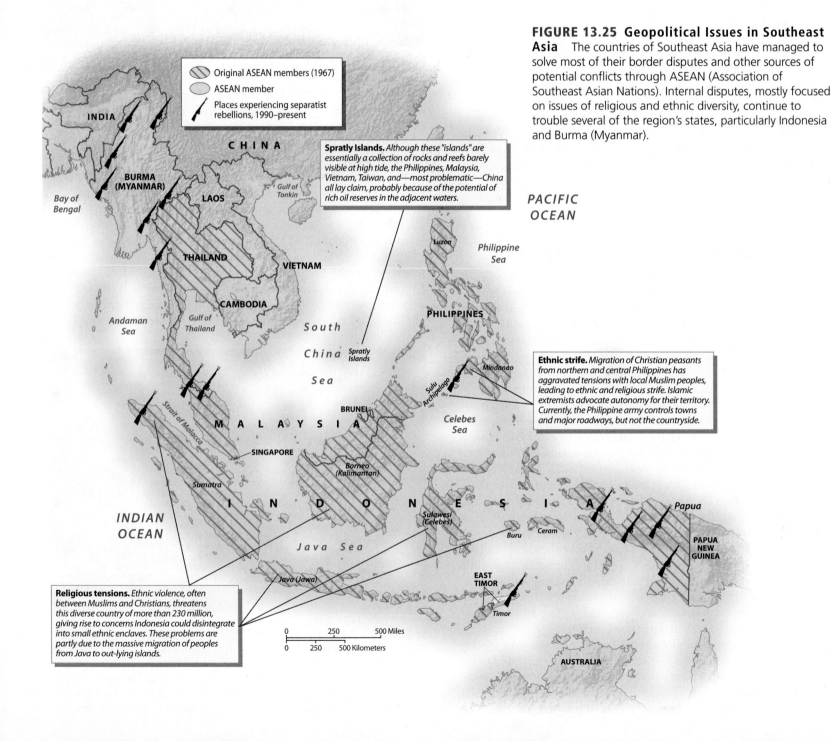

**FIGURE 13.25 Geopolitical Issues in Southeast Asia** The countries of Southeast Asia have managed to solve most of their border disputes and other sources of potential conflicts through ASEAN (Association of Southeast Asian Nations). Internal disputes, mostly focused on issues of religious and ethnic diversity, continue to trouble several of the region's states, particularly Indonesia and Burma (Myanmar).

**Spratly Islands.** *Although these "islands" are essentially a collection of rocks and reefs barely visible at high tide, the Philippines, Malaysia, Vietnam, Taiwan, and—most problematic—China all lay claim, probably because of the potential of rich oil reserves in the adjacent waters.*

**Ethnic strife.** *Migration of Christian peasants from northern and central Philippines has aggravated tensions with local Muslim peoples, leading to ethnic and religious strife. Islamic extremists advocate autonomy for their territory. Currently, the Philippine army controls towns and major roadways, but not the countryside.*

**Religious tensions.** *Ethnic violence, often between Muslims and Christians, threatens this diverse country of more than 230 million, giving rise to concerns Indonesia could disintegrate into small ethnic enclaves. These problems are partly due to the massive migration of peoples from Java to out-lying islands.*

Original ASEAN members (1967)
ASEAN member
Places experiencing separatist rebellions, 1990–present

## Before European Colonialism

The modern countries of mainland Southeast Asia all existed in one form or another as kingdoms before European colonialism. Cambodia emerged first, over 1,000 years ago, and by the 1300s, independent kingdoms had been established by the Burmese, Siamese, Lao, and Vietnamese people. All these realms were centered on major river valleys and deltas.

The situation in insular Southeast Asia was different from that of the mainland, with the premodern map being completely different from that of the modern nation-states. Many kingdoms existed on the Malay Peninsula and on the islands of Sumatra, Java, and Sulawesi, but few were territorially stable. Indonesia, the Philippines, and Malaysia thus owe their territorial shape almost completely to European colonialism (Figure 13.26).

## The Colonial Era

The Portuguese were the first Europeans to arrive in Southeast Asia (around 1500), lured mainly by the spices of the Maluku Islands in eastern Indonesia. In the late 1500s, the Spanish conquered most of the Philippines, which they used as a base for their silver trade between China and the Americas. By the 1600s, the Dutch began to establish trading bases, followed by the British. With superior naval weapons, the Europeans were able to conquer key ports and control

**FIGURE 13.26  Colonial Southeast Asia**   With the exception of Thailand, all of Southeast Asia was under Western colonial rule by the early 1900s. The Netherlands had the largest empire in the region, covering the territory that was later to become Indonesia. France maintained a large colonial realm in Vietnam, Laos, and Cambodia, as did Britain in Burma and Malaysia (including Singapore and Brunei). The Philippines was initially colonized by Spain but passed to the control of the United States in 1898.

strategic waterways. Yet for the first 200 years of colonialism, except in the Philippines, the Europeans made no major geopolitical changes.

By the 1700s, the Netherlands had become the most powerful force in the region. As a result, a Dutch empire in the "East Indies" began appearing on world maps. This empire continued to grow into the early 20th century, when it defeated its last major enemy, the Islamic state of Aceh in northern Sumatra. Later, the Netherlands divided the island of New Guinea with Germany and Britain, rounding out its Indonesian colony.

The British, preoccupied with their empire in India, concentrated their attention on the sea-lanes linking South Asia to China. As a result, they established several fortified outposts along the Strait of Malacca, the most important of which was Singapore. To avoid conflict, the British and Dutch agreed that the British would limit their attention to the Malay Peninsula and the northern portion of Borneo. The British allowed Muslim sultans to retain limited powers, much as they had done in parts of India.

In the 1800s, European colonial power spread through most of mainland Southeast Asia. The British conquered the kingdom of Burma and extended their power into the nearby highlands. During the same period, the French moved into Vietnam's Mekong Delta, gradually expanding their territorial control into Cambodia and northward to China's border. Thailand was the only country to avoid colonial rule, although it did lose territories to the British and French. The final colonial power to enter the region was the United States, which took the Philippines first from Spain and then from Filipino nationalists between 1898 and 1900. The U.S. army later conquered the Muslim areas of the southwest that had never been fully under Spanish authority.

Organized resistance to European rule began in the 1920s, but it took the Japanese occupation of World War II to show that colonial power was vulnerable. After Japan's surrender in 1945, pressure for independence intensified throughout Southeast Asia. Britain withdrew from Burma in 1948 and began to pull out of its colonies in insular Southeast Asia in the late 1950s, although Brunei did not gain independence until 1984. Singapore briefly joined Malaysia but then withdrew and became independent in 1965. In the Philippines, the United States granted long-promised independence on July 4, 1946, although it retained military bases for several decades. Although the Dutch attempted to reestablish their colonial rule after World War II, they were forced to acknowledge Indonesia's independence in 1949.

## The Vietnam War and Its Aftermath

After World War II, France was determined to regain control of its Southeast Asian colonies. Resistance to French rule was organized primarily by communist groups based mainly in northern Vietnam. Open warfare between French soldiers and the communist forces continued until 1954, when France agreed to withdraw after a major military defeat. An international peace council then divided Vietnam into a communist North Vietnam, allied with the Soviet Union and China, and a capitalist-oriented South Vietnam, with close ties to the United States.

The peace accord did not, however, end the fighting. Communist guerrillas in South Vietnam fought to overthrow the new government and unite it with the north. North Vietnam sent troops and war materials across the border to aid the rebels. Most of these supplies reached the south over the Ho Chi Minh Trail, a confusing network of forest passages through Laos and Cambodia, thus steadily drawing these two countries into the conflict. In Laos the communist Pathet Lao forces challenged the government, while in Cambodia the **Khmer Rouge** guerrillas gained considerable power.

In Washington, DC, the **domino theory** guided foreign policy. According to this notion, if Vietnam fell to the communists, then so would Laos and Cambodia; once those countries were lost, Burma, Thailand, and perhaps Malaysia and Indonesia would also become members of the communist bloc. Fearing such an outcome, the United States was drawn ever deeper into the war. By 1965 thousands of U.S. troops were fighting to support the government of South Vietnam (Figure 13.27). But despite superiority in arms and troops, U.S. forces gradually lost control over much of the countryside. As casualties mounted and the antiwar movement back home strengthened, the United States began holding secret talks in search of a negotiated settlement. U.S. troop withdrawals began in earnest in the early 1970s.

With the withdrawal of U.S. forces, the noncommunist governments began to collapse. Saigon fell in 1975, and in the following year Vietnam was officially reunited under the government of the north. Reunification was a traumatic event in southern Vietnam. Hundreds of thousands of people fled from the new regime, with many settling in the United States.

Vietnam proved fortunate compared to Cambodia. There the Khmer Rouge installed one of the most brutal regimes the world has ever seen. City-dwellers were forced into the countryside to become peasants, and most wealthy and educated people were executed. The goal of the Khmer Rouge was to create a completely new agricultural society by returning to what it called "year zero." After several years of horrific bloodshed, neighboring Vietnam invaded Cambodia and installed a far less brutal, but still repressive, regime. Fighting between different groups continued until 1991, when a comprehensive peace settlement was finally reached.

**FIGURE 13.27 U.S. Soldier and Vietcong Prisoners**   The United States maintained a substantial military presence in Vietnam in the 1960s and early 1970s. Although U.S. forces claimed many victories, they were ultimately forced to withdraw, leading to the victory of North Vietnam and the reunification of the country.

# Geopolitical Tensions in Contemporary Southeast Asia

In several parts of Southeast Asia, local ethnic groups have been struggling against national governments that inherited their territory from former colonial powers. Tensions have also emerged where tribal groups attempted to preserve their homelands from logging, mining, or migrant settlers.

**Conflicts in Indonesia**    When Indonesia gained independence in 1949, it included all of the former Dutch possessions in the region except western New Guinea (Papua). In 1962, the Netherlands organized an election to see whether the people of this area wished to join Indonesia or form an independent country. The vote went for union, but many observers believed that the election was rigged by the Indonesian government. As a result, many of the local people began to rebel. Indonesia is determined to maintain control of the region, in part because it is the site of the country's largest source of tax revenue, the Grasberg mine, run by Phoenix-based Freeport-McMoRan Corporation. Not only is Grasberg the world's largest gold mine, but it is also one of the most environmentally destructive mines.

In 2000, the Indonesian government granted partial autonomy to Papua, which reduced support for the rebellion. In early 2009, however, tensions flared after Indonesian police broke up a pro-independence rally, shooting nine protestors (Figure 13.28). The underlying source of conflict is the continuing immigration of people to Papua from central and western Indonesia. Most indigenous Papuans are either Christians or animists, but the area's population as a whole is now roughly one-third Muslim. Islamist groups are active on the island, where they frequently clash with Christian and animist leaders.

The island of Timor has also experienced political bloodshed in recent years. The eastern half of this poor and rather dry island had been a Portuguese colony and had evolved into a largely Christian society. The East Timorese expected independence when the Portuguese finally withdrew in 1975. Indonesia, however, viewed the area as its own and immediately invaded. A brutal war followed, which the Indonesian army won in part by preventing food from reaching the province.

After the economic crisis of 1997, Indonesia's power in the region slipped. A new Indonesian government promised an election in 1999 to determine whether the East Timorese still wanted independence. At the same time, however, the Indonesian army began to organize militias to intimidate the people of East Timor into voting to remain within the country. When it was clear that the vote would be for independence, the militias began rioting, looting, and killing civilians. Under international pressure, Indonesia finally withdrew its armed forces, and the East Timorese began to build a new country. The process has not been easy, in part because of continuing ethnic strife. In 2006, major rioting based on tensions between eastern East Timor and western East Timor further damaged the country, requiring Australian intervention.

The Aceh region of northern Sumatra has also experienced prolonged political violence, as local rebels have sought to create an independent Islamic state. While the Indonesian government has given Aceh "special autonomy," it has been determined to prevent actual independence. The devastation of the December 2004 tsunami seems to have generated a solid peace, as the needs of the province were so great that the separatist fighters finally agreed to lay down their weapons.

Elsewhere in Indonesia, fighting between Muslims and Christians peaked in the late 1990s but subsequently declined. In the southern

**FIGURE 13.28 Papuan Protest**    The indigenous people of Papua, the Indonesian half of the island of New Guinea, have long struggled against the domination of their homeland by the Indonesian government and the exploitation of its natural resources by foreign corporations. In this photo, Papuan protestors clash with police in Jakarta outside of the local offices of Freeport-McMoRan, a U.S.-based company that operates a huge and highly polluting mine in Papua.

Maluku Islands (especially Ambon and Seram) and in central Sulawesi, where the two religions are roughly equal in strength, however, tensions remain high, and violence is common.

**Regional Tensions in the Philippines**    The Philippines has also suffered from regional political violence. Its most persistent problem is the Islamic southwest, where rebels have long demanded independence. After successful negotiations with the main separatist group in 1989, the government created the Autonomous Region in Muslim Mindanao (ARMM), which it expanded in 2008. The more radical Islamist groups, however, rejected the settlement and continued to fight. In late 2008, the struggle intensified as rebels set off bombs and kidnapped civilians. In April 2009, the Philippine army captured two bases of the Moro Islamic Liberation Front, but critics pointed out that up to 1,000 villagers were driven from their homes as a result.

The Muslim southwest does not present the only political problem in the Philippines. A revolutionary communist group called the New People's Army operates in most parts of the country and controls many rural districts. Furthermore, the country's national government, although democratic, is far from stable, suffering from continual coup threats, corruption scandals, mass protests, and impeachment efforts.

**Burma's Many Problems**    Burma (Myanmar) has also suffered from ethnic conflict. Burma's simultaneous wars have pitted the central government, dominated by the Burmans (the Burmese-speaking ethnic group), against the country's varied non-Burman societies. Fighting intensified gradually after independence in 1948, and by the 1980s almost half of the country's territory had become a combat zone. In the 1990s, however, successful Burmese army offensives combined with cease-fire agreements resulted in a marked decrease in warfare. Sporadic fighting continues to erupt in many remote areas, however, particularly in the Karen-inhabited areas near the Thai border.

**FIGURE 13.29 Aung San Suu Kyi** The noted Burmese democratic opposition leader Aung San Suu Kyi received the Nobel Peace Prize in 1991. As of 2009, she remained under house arrest.

As long as Burma retains its repressive Burman-dominated government, social unrest and ethnic turmoil are not likely to diminish. The United States and other countries have placed trade and investment sanctions on Burma, hoping that such pressure would result in democratic change. Such policies, however, have shown few signs of success. Burma's military government still keeps the country's democratically oriented opposition leader, the 1991 Nobel Peace Prize winner Aung San Suu Kyi, under house arrest (Figure 13.29). In an unusual defensive maneuver, in 2006, Burma created a new capital city called Naypyidaw, located in a remote, forested area 200 miles (322 kilometers) north of the old capital of Yangon (Rangoon). Naypyidaw remains virtually closed off from the rest of the world.

**Thailand's Troubles** Unlike Burma, Thailand enjoys basic human freedoms and a generally free press. Until 2006, most observers thought that it had emerged as a stable democracy. In that year, however, Thai Prime Minister Thaksin Shinawatra was overthrown by a military coup that apparently had the blessing of Thailand's revered king. Thaksin, a wealthy businessman, had gained the support of Thailand's poor in part by setting up a national health-care system, but he infuriated the middle and upper classes by taking too much power into his own hands. A new national election in 2007 brought one of Thaksin's followers to power, but he was removed from office by the Thai Constitutional Court after anti-Thaksin protestors virtually shut down the government. In early 2009, Thaksin supporters responded with their own massive protests, thereby threatening the new government (Figure 13.30).

The recent political chaos in Thailand has made it difficult for the country to deal with its main political threat, that posed by the rebellion in its southernmost provinces. Far southern Thailand is relatively poor, and its people are mostly Malay in language and Muslim in religion. They have long resented being ruled by Thailand, and periodic rebellions have flared up for decades. In 2004, violence sharply increased, resulting in more than 3,000 deaths by 2009. One odd feature of the insurgency in southern Thailand is the fact that no group has emerged to claim responsibility for the anti-government violence, and hence no political demands have been made.

## International Dimensions of Southeast Asian Geopolitics

Southeast Asia was formerly troubled by a number of border disputes, but most of them have been settled or at least put on hold. With the rise of ASEAN, national leaders have concluded that friendly relations with neighbors are more important than the possible gain of additional territory. ASEAN members pledge to seek peaceable solutions to regional problems and to avoid interfering in each other's internal affairs. As a result, ASEAN leaders have been reluctant to criticize Burma over its miserable human rights record.

Despite ASEAN's general successes, border disputes continue to flare up. In late 2008 and early 2009, a conflict between Thailand and Cambodia over the area around the 11th-century Preah Vihear Temple resulted in 26 deaths. A more complicated territorial dispute centers on the Spratly and Paracel islands in the South China Sea, two groups of tiny islands and reefs that might sit over substantial oil reserves (Figure 13.31). The Philippines, Malaysia, and Vietnam have all claimed territory there, as have China and Taiwan. International tensions in the South China Sea were reduced in 2002, however, when all the interested countries agreed to seek a peaceful settlement.

Although Southeast Asian leaders have been concerned about Chinese territorial claims in the Spratly Islands, ASEAN has encouraged political cooperation with China and the rest of East Asia. It thus runs an annual conference called ASEAN + 3, where its ministers meet with leaders from China, Japan, and South Korea. A larger grouping is the ASEAN Regional Forum (ARF), in which Southeast Asian leaders meet with representatives of both East Asian and Western powers to ease tensions in the region.

One of the biggest problems faced by ASEAN leaders has been the establishment of radical Islamist networks in the region. The

**FIGURE 13.30 Anti-Government Thai Protest** Thailand has recently been plagued by massive political protests that have challenged the country's democratic institutions. In this photograph, supporters of the ousted Thai prime minister Thaksin Shinawatra protest outside of the Finance Ministry in Bangkok on April 2, 2009.

**FIGURE 13.31 The Spratly Islands** The Spratly Islands are small and barely above water at high tide, but they are geopolitically important. Oil may exist in large quantities in the surrounding areas, heightening the competition over the islands. Southeast Asian countries are especially concerned about China's military activities in the Spratlys.

largest of these is Jemaah Islamiya (JI), a militant group dedicated to establishing an Islamic state to contain all Muslim areas within Southeast Asia. JI agents are believed to have detonated bombs that killed 202 people in Bali in 2002 and to have set off major explosions in Jakarta in 2004 and Bali (again) in 2005. Indonesia responded by creating an elite counter-terrorism squad ("Detachment 88") and by establishing a "deradicalization program" aimed at convincing radical Islamists to change sides. By 2008, most observers had concluded that JI was no longer a serious threat, but periodic arrests continued to be made.

## Economic and Social Development: The Roller-Coaster Ride of Economic Growth

Over the past few decades, Southeast Asia has experienced extreme economic fluctuations (Table 13.2). Between 1980 and 1997, much of the region experienced a major economic boom. In 1997, however, Thailand's real estate market collapsed, which led to an economic crisis throughout the region. By 2000, most Southeast Asian economies were doing reasonably well, although they were no longer growing at a breakneck pace. The global economic crisis of 2008–2009, however, resulted in another blow to Southeast Asia. Singapore in particular, the region's main banking center, experienced a severe recession. In early 2009, Singaporean officials warned that the country's economy could decline by as much a 9 percent by the end of the year.

### Uneven Economic Development

While the region as a whole has experienced pronounced ups and downs, parts of Southeast Asia have done much better in the global economy than have others. Oil-rich Brunei and technologically sophisticated Singapore rank among the world's most prosperous countries, whereas Cambodia, Laos, Burma, and East Timor are among the poorest. And while Malaysia, Thailand, and Vietnam have seen tremendous economic gains over the past 50 years, the Philippines has experienced major disappointments during the same period.

**The Philippine Decline** In the 1950s, the Philippines was the most highly developed Southeast Asian country. By the late 1960s, however, Philippine development had been derailed. Through the 1980s and early 1990s, the country's economy failed to outpace its population growth, resulting in declining living standards for both the poor and the middle class. The Philippine people are still well educated and

## TABLE 13.2 DEVELOPMENT INDICATORS

| Country | GNI per capita, PPP 2007 | GDP Average Annual %Growth 2000–07 | Human Development Index (2006)# | Percent Population Living Below $2 a Day | Life Expectancy* 2009 | Under Age 5 Mortality Rate 1990 | Under Age 5 Mortality Rate 2007 | Gender Equity 2007 |
|---|---|---|---|---|---|---|---|---|
| Burma (Myanmar) | | 9.2 | 0.585 | | 61 | 130 | 103 | |
| Brunei | | | 0.919 | | 77 | 11 | 9 | 101 |
| Cambodia | 1,720 | 9.9 | 0.575 | 68.2 | 61 | 119 | 91 | 90 |
| East Timor | 3,090 | 0.9 | 0.483 | 77.5 | 61 | 184 | 97 | 95 |
| Indonesia | 3,570 | 5.1 | 0.726 | | 71 | 91 | 31 | 98 |
| Laos | 2,080 | 6.7 | 0.608 | 76.8 | 65 | 163 | 70 | 86 |
| Malaysia | 13,230 | 5.4 | 0.823 | 7.8 | 74 | 22 | 11 | 104 |
| Philippines | 3,710 | 5.1 | 0.745 | 45.0 | 69 | 62 | 28 | 102 |
| Singapore | 47,950 | 5.8 | 0.918 | | 81 | 8 | 3 | |
| Thailand | 7,880 | 5.3 | 0.786 | 11.5 | 69 | 31 | 7 | 104 |
| Vietnam | 2,530 | 7.8 | 0.718 | 48.4 | 74 | 56 | 15 | |

*Source:* World Bank, *World Development Indicators 2009.*
United Nations, *Human Development Index, 2008.* #
Population Reference Bureau, *World Population Data Sheet, 2009*\*
Gender Equity – Ratio of female to male enrollments in primary and secondary school. Numbers below 100 have more males in primary/secondary school, numbers above 100 have more females in primary/secondary schools.

reasonably healthy by world standards, but even the country's educational and health systems declined during this period.

Why did the Philippines fail despite its earlier promise? While there are no simple answers, it is clear that dictator Ferdinand Marcos (who ruled from 1965 to 1986) wasted—and perhaps even stole—billions of dollars while failing to create conditions that would lead to genuine development. The Marcos regime instituted a kind of **crony capitalism**, in which the president's friends were given huge economic favors, while those believed to be enemies had their properties taken.

An elected democratic government finally replaced the Marcos dictatorship in 1986, but corruption remains deeply entrenched, and the Philippine economy has remained unstable. Many Filipinos have responded to the economic crisis by working overseas, either in the oil-rich countries of Southwest Asia, the wealthy cities of North America and Europe, or the newly industrialized nations of East and Southeast Asia. Men primarily work in the construction industry or on ships, and women work as nurses, nannies, or domestic servants (Figure 13.32). Many suffer exploitation, both economic and sexual, in such positions.

One recent bright spot in the Philippine economy has been the expansion of business outsourcing operations, attracted by the country's educated and English-speaking population. In 2009, an estimated 225,000 Filipinos worked in international call centers, handling telephone inquiries from customers in the United States and other wealthy countries.

### The Regional Hub: Singapore

Singapore and Malaysia have been Southeast Asia's major developmental successes. Singapore has transformed itself from an **entrepôt** city, a place where goods are imported, stored, and then transshipped, to one of the world's wealthiest and most modern states. Singapore is now the communications and financial hub of Southeast Asia, as well as a thriving high-tech manufacturing center. The Singaporean government has played an active role in the development process. Singapore has encouraged investment by multinational technology companies and has itself invested heavily in housing, education, and some social services (Figure 13.33). The

**FIGURE 13.33 Housing in Singapore** Despite its free-market approach to economics, the government of Singapore has invested heavily in public housing. Most Singaporeans live in buildings similar to the ones depicted in this photograph.

Singaporean government, however, remains only partly democratic, as the ruling party maintains a firm grip on government.

### The Malaysian Boom

Although not nearly as well-off as Singapore, Malaysia has also experienced rapid economic growth. Development was initially concentrated in agriculture and natural resources, focused on tropical hardwoods, plantation products, and tin. More recently, manufacturing, especially in labor-intensive high-tech sectors, has become the main engine of growth.

The modern economy of Malaysia is not uniformly distributed across the country. One difference is geographic: Most industrial development has occurred on the west side of peninsular Malaysia. More important, however, are differences based on ethnicity. The industrial wealth generated in Malaysia has been concentrated in the Chinese community. Ethnic Malays remain less prosperous than Chinese-Malaysians, and those of South Asian descent are poorer still. Evidence suggests, moreover, that the gap between Malaysia's rich and poor has been growing in recent years.

The unbalanced wealth of the local Chinese community is a feature of most Southeast Asian countries. The problem is particularly acute in Malaysia, however, because its Chinese minority is so large. The government's response has been one of aggressive "affirmative action," by which economic power is transferred to the dominant Malay, or **Bumiputra** ("sons of the soil") community. This policy has been reasonably successful. Because the economy as a whole has expanded significantly, the Chinese community has been able to thrive even as its relative share of the country's wealth has declined. However, Malaysia's Chinese population still feels considerable resentment.

### Thailand's Ups and Downs

Thailand, like Malaysia, climbed rapidly during the 1980s and 1990s into the ranks of the world's newly industrialized countries. Japanese companies were leading players in the earlier Thai boom, attracted by Thailand's low-wage, yet reasonably well-educated, workforce. Thailand experienced a major downturn in the late 1990s, however, that undercut much of this development.

**FIGURE 13.32 Filipina Migrant Workers in Kuwait** The long period of economic stagnation in the Philippines has resulted in an outflow of workers from the country. Women from the Philippines often work as domestic servants in the Persian Gulf region and in Singapore and Hong Kong.

# EXPLORING GLOBAL CONNECTIONS
## Russian Investment in Pattaya, Thailand

Visitors to the seaside resort city of Pattaya, 100 miles (165 kilometers) southeast of Bangkok, may be surprised to find a strong Russian presence. In 2007, almost 900,000 Russian tourists visited this city of some 100,000 residents, more than from any other foreign country. According to local sources, Russians tend to stay longer and spend more money than other visitors. One recently chartered flight back to Russia supposedly could not take off because the passengers were overloaded with goods purchased in Thailand. Many wealthy Russians have invested in the city; in 2007, some 80 percent of property sales in Pattaya went to Russians, resulting in a huge condominium boom. Along with Russian money comes Russian culture. In March 2009, construction began on Pattaya's Russian Orthodox Church of All Saints, and a Russia circus operates permanently near the city's major park.

While the Russian interest has benefited the local economy, it also has troubling aspects. By early 2009, Pattaya's formerly red-hot property market was collapsing as nervous Russian investors, reeling from the global economic crisis, began withdrawing their funds. Allegations of Russian Mafia involvement are also common, contributing further to Pattaya's already seedy reputation.

Pattaya's questionable reputation, however, is one of the reasons Russians were attracted to the city in the first place. Pattaya is, simply put, the world's capital of sexual tourism, and it is no secret that Russian organized crime is heavily involved in trafficking prostitutes from Russian and other former-Soviet countries to global commercial sex centers (Figure 13.2.1). Most of the trafficked women are evidently deceived into the business, contracting with job placement agencies that offer high-paid work as dancers, waitresses, or sale representatives. Once in Thailand, they often have their passports confiscated and are essentially held in bondage. The fact that Russian prostitutes are able to charge a significantly higher price than other sex workers in Pattaya helps propel this sordid trade.

Pattaya is not the only place in Thailand infamous for commercial sex. A 2003 study claimed that the total value of the trade came in at US$4.3 billion per year, accounting for roughly 3 percent of the Thai economy. Pattaya's estimated 20,000 prostitutes actually constitute a relatively small percentage of Thailand's sex workers, but they cater largely to well-off foreigners and thus form a significant part of the business. Pattaya's sex trade has been globalized from the start. Until the late 1960s, Pattaya was little more than a sleepy fishing village. At that time, U.S. servicemen fighting in the Vietnam War discovered the village as a rest and recreation center, and commercial sexual services soon followed. The end of the Vietnam War brought economic distress, which city leaders determined to overcome by selling the city's distinctive trade to the global market.

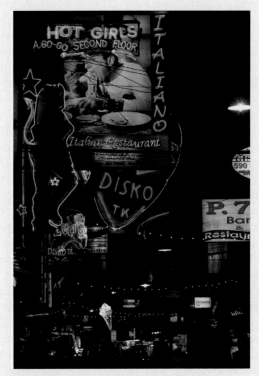

**FIGURE 13.2.1 Sex Industry in Pattaya** Hundreds of "go-go bars" and other sexually oriented businesses line the streets of Pattaya, a seaside resort community 100 miles (165 kilometers) southeast of Bangkok. Although most of the sex workers in Pattaya are Thai, Russian women form a significant presence.

---

Growth had resumed by 2000, but then Thailand's political crisis, coupled with the global economic crisis, brought about another severe recession in 2008–2009.

As is true in most other parts of the world, Thailand's economic growth over the past several decades has by no means benefited the entire country to an equal extent. Most industrial development has occurred in the historical core, especially in the city of Bangkok. Yet even in Bangkok, the blessings of progress have been mixed. As the city began to choke on its own growth, industrial growth started to spread outward. The entire Chao Phraya lowland area shares to some extent in the general prosperity because of both its proximity to Bangkok and its rich agricultural resources. Thailand's Lao-speaking northeast (the Khorat Plateau) and Malay-speaking far south remain the country's poorest regions. Because of the poverty of their homeland, northeasterners are often forced to seek employment in Bangkok. Men typically find work in the construction industry; northeastern women often make their living as sex workers.

**Unstable Economic Expansion in Indonesia**   At the time of independence (1949), Indonesia was one of the poorest countries in the world. The Indonesian economy finally began to expand in the 1970s. Oil exports fueled the early growth, as did the logging of tropical forests. But unlike most other oil exporters, Indonesia continued to grow even after oil prices plummeted in the 1980s. Like Thailand and Malaysia, Indonesia attracted multinational companies looking for low wages. Large Indonesian firms, many of them owned by local Chinese families, have also capitalized on the country's human and natural resources. The national government has attempted to build technologically oriented businesses, but their success remains uncertain.

Despite its recent economic growth, Indonesia remains a poor country. Its pace of economic expansion never matched those of Singapore and Malaysia, and it has remained much more dependent on the unsustainable exploitation of natural resources. The financial crisis of the late 1990s, moreover, hurt Indonesia more severely than it hurt any other country. By 2008, however, Indonesia's political situation was beginning to stabilize, improving its economic outlook. The global economic crisis of 2008–2009 was not as severe in Indonesia as it was in the wealthier countries of Southeast Asia.

### The Recent Rise of Vietnam and Cambodia

In 2006, 2007, and 2008, Vietnam's economy grew at a rate of more than 8 percent per year, one of the fastest rates in the world. But Vietnam is still a poor country that has a long way to go to catch up with Thailand, let alone Malaysia. From the time of the Vietnam War through the early 1990s, Vietnam experienced little economic development. Frustrated with the country's economic performance, Vietnam's leaders began to follow China by embracing market economics while retaining the political forms of a communist state (Figure 13.34).

Vietnam now welcomes multinational corporations, which are attracted by its extremely low wages and relatively well-educated workforce. Japanese and South Korean companies often favor Vietnam over other Southeast Asian countries. Vietnam's entry into the World Trade Organization in 2006 brought further economic benefits. Local businesses, however, complain of harassment by state officials, and development remains geographically uneven. Southern Vietnam is still much more entrepreneurial and capitalistic than the north, while deep and persistent poverty remains entrenched in many rural areas, particularly those in the tribal highlands.

Cambodia's recent economic history is somewhat similar to that of Vietnam, only more extreme. Long burdened by war and corruption, Cambodia was one of Asia's poorest countries, its economy focused largely on subsistence agriculture. The discovery of oil and other mineral resources after 2000, however, combined with a thriving tourist economy and large-scale international investment, resulted in a major economic boom. Until the global economic crisis of 2008–2009 hit, Cambodia's economy was growing by more than 10 percent each year.

Cambodia's recent economic expansion has resulted in problems as well as opportunities. A property boom in Phnom Penh, for example, saw thousands of poor people being forced out of their homes to make way for development projects (Figure 13.35). Tourism has also

**FIGURE 13.35 Phnom Penh** Severe conflicts have emerged in Cambodia's booming capital city, Phnom Penh, as slum areas are cleared to make way for development projects. In this photograph, a woman wearing a helmet tries to stop a bulldozer from demolishing her home while her daughter tries to pull her away from the dangerous machine.

proved to be a mixed blessing. Several Cambodian border towns, most notably Poipet, have set themselves up as gambling centers, attracting investments—and criminal activities—from Thai underworld figures. Many Cambodians are now concerned that their country is coming under the economic domination of Thailand and Vietnam.

### Persistent Poverty in Laos and East Timor

Like Cambodia, Laos has long been dominated by subsistence agriculture, which employs roughly three-quarters of its workforce. Laos has particular economic difficulties due to its rough terrain and relative isolation; outside its few cities, paved roads and reliable electricity are rare. As a result, it remains heavily dependent on foreign aid.

The Laotian government is pinning its economic hopes on hydropower development, mining, tourism, and investment from Thailand and China. Hydropower is particularly important, as the country is mountainous, with many rivers, and could therefore generate and export large quantities of electricity. Laos has also benefited from the increasing volume of barge traffic going up the Mekong River to China. This shipping route was opened in 2004, when Chinese engineers blasted away a series of rocky rapids blocking boat traffic. Environmentalists are concerned that such developments will damage the freshwater fisheries of the Mekong River, which are among the richest in the world.

The weakest economy in Southeast Asia is undoubtedly that of East Timor, which is among the world's poorest countries. East Timor has hardly begun to recover from the devastation that accompanied its independence, and it has been further weakened by the gradual withdrawal of international aid agencies. A recent agreement with Australia to share the revenues of offshore natural gas deposits, however, promises some hope for this beleaguered country.

### Burma's Troubled Economy

Burma (Myanmar) also stands near the bottom of the scale of Southeast Asian economic development. For all of its many problems, however, Burma remains a land of great

**FIGURE 13.34 Capitalism in Vietnam** Although Vietnam is a communist state, it has, like China, embraced many forms of capitalism. Private shops abound, and foreign investment is welcome. In general, the market economy is more highly developed in the south than in the north.

potential. It has abundant natural resources (including oil and other minerals, water, and timber), as well as a large expanse of fertile farmland. Its population density is moderate, and its people are reasonably well educated. But despite these advantages, Burma's economy has remained relatively stagnant since independence in 1948.

Although Burma's woes can be traced in part to the continual warfare the country has experienced, most observers primarily blame economic policy. Beginning in 1962, Burma attempted to isolate its economic system from global forces under a system of Buddhist socialism. While intentions may have been admirable, the experiment was not successful; instead of creating a self-contained economy, Burma found itself burdened by smuggling and black-market activities. More recently, Burma has opened its economy to some degree, actively trading with China and its Southeast Asian neighbors. Burma is also establishing closer economic and security ties with India, which is keenly interested in Burma's oil and gas deposits.

Despite its recent moves into the global economy, Burma remains an extremely poor country. Its crackdown on anti-government protestors in 2007 and the devastation that accompanied Typhoon Nargis in 2008 further undermined its economy. In 2007, moreover, Burma was ranked as the most corrupt country in the world, along with Somalia. Another indication of the country's economic failure is the fact that the official exchange rate, as of January 2009, was 6.43 Burmese kyat to the dollar, whereas the black-market rate was approximately 1,200 kyat per dollar!

## Globalization and the Southeast Asian Economy

As the preceding discussion shows, Southeast Asia as a whole has undergone rapid integration into the global economy. Singapore has thoroughly staked its future to the success of multinational capitalism, as have several other countries. Even communist Vietnam and once-isolationist Burma are opening their doors to the global system, although, in the case of Burma, with much hesitation.

Regardless of what happens in the coming years, global economic integration has already brought about the significant development in Singapore, Malaysia, Thailand, and even Indonesia. Outside Singapore and Malaysia, moreover, successful development has generally been heavily based on labor-intensive manufacturing in which workers are paid low wages and subjected to harsh discipline. Movements have thus begun in Europe, the United States, and elsewhere to pressure both multinational corporations and Southeast Asian governments to improve the working conditions of laborers in the export industries. Some Southeast Asian leaders, however, object, accusing Western activists of wanting to prevent Southeast Asian development under the excuse of concern over worker rights.

China figures prominently in the recent globalization of Southeast Asian economies. China has invested heavily in infrastructural projects, most notably in Laos, Burma, and Cambodia. Chinese investment in East Timor's new natural gas industry is also increasing, as is its interest in Vietnam's bauxite (aluminum ore) industry. China's most ambitious proposal involves the construction of a canal across Thailand's Kra Isthmus, which would allow oil tankers from the Persian Gulf to avoid the Strait of Malacca. Environmental groups, however, as well as the governments of Singapore and the United States, have expressed serious concerns about this $20 billion proposed project.

China is not the only foreign country to develop a new interest in Southeast Asian resources. In 2008, both Kuwait and Qatar invested heavily in Cambodia, lending money for dam building, road construction, and agricultural development. Many Cambodian farmers, however, are concerned that such projects could result in loss of their lands.

## Issues of Social Development

As might be expected, several key indicators of social development in Southeast Asia are closely linked to levels of economic development. Singapore ranks among the world leaders in regard to health and education, as does Brunei. Laos and Cambodia, not surprisingly, come out near the bottom of the chart. The people of Vietnam, however, are healthier and better educated than might be expected on the basis of their country's overall economic performance. In the early years of the new millennium, Thailand began building a comprehensive system of national health care, but recent political instability has tended to stall progress on this front.

With the exceptions of Laos, Cambodia, East Timor, and Burma, Southeast Asia has achieved relatively high levels of social welfare. In Laos and Cambodia, however, life expectancy at birth hovers around 61 to 65 years (as compared to Thailand's 69 years), and female literacy rates remain below 65 percent. But even the poorest countries of the region have made some improvements. War-torn Cambodia, however, has achieved relatively small gains, in part because most of its budget is devoted to maintaining security, with little left for social programs.

Most of the governments of Southeast Asia have placed a high priority on basic education. Literacy rates are relatively high in most countries of the region, even in the impoverished and authoritarian country of Burma. Much less success, however, has been realized in university and technical education. As Southeast Asian economies continue to grow, this educational gap is beginning to have negative consequences, forcing many advanced students to study abroad. If Southeast Asian countries other than Singapore are to become fully developed, they will probably have to invest more money in their own human resources.

# SUMMARY

- In many ways, Southeast Asia presents a prime example of diversity amid globalization. As is true elsewhere, globalization in Southeast Asia has created both challenges and opportunities. Some of the most serious problems that it has generated are environmental. Given the emphasis placed by global trade on wood products, it is perhaps understandable that Southeast Asia has

sacrificed so many of its forests to support economic development. But in most of the region, forests are now seriously depleted.

- Deforestation in Southeast Asia is also linked to domestic population growth and changes in settlement patterns. As people move from densely populated, fertile lowland areas into remote uplands, both environmental damage and cultural conflicts often follow. Population movements in Southeast Asia also have a global dimension. This is particularly true in regard to the Philippines, which has sent millions of workers to more prosperous parts of the world.

■ Southeast Asia, unlike many other world regions, has never had a single major cultural influence and is characterized today by tremendous cultural diversity. One might argue, however, that globalization has helped Southeast Asia find a new sense of regional identity, as expressed through the Association of Southeast Asian Nations (ASEAN). As a result, the historical and cultural unity that Southeast Asia has lacked may be forced upon it by 21st-century globalization.

■ The relative success of ASEAN has by no means solved all of Southeast Asia's political tensions. Many of its countries still argue about geographic, political, and economic issues, while ethnic and religious conflicts have generated major problems in Indonesia, the Philippines, and Thailand. Several of the region's countries, such as Cambodia, Laos, and especially Burma, have also been held back by repressive and corrupt governments.

■ Although ASEAN has played an economic as well as a political role, its economic successes have been limited. Most of the region's trade is still directed outward, toward the traditional centers of the global economy—North America, Europe, and East Asia. As is true in many other parts of the world, China is becoming a key trading partner. This orientation is not surprising, considering the export-focused policies of most Southeast Asian countries. A significant question for Southeast Asia's future is whether the region will develop an integrated regional economy. A more important issue is whether social and economic development will be able to lift the entire region out of poverty instead of benefiting just the more fortunate areas.

## KEY TERMS

animism (p. 388)
Association of Southeast Asian
   Nations (ASEAN) (p. 375)
Bumiputra (p. 398)

crony capitalism (p. 398)
domino theory (p. 394)
entrepôt (p. 398)
Khmer Rouge (p. 394)

lingua franca (p. 389)
primate city (p. 385)
swidden (p. 383 )
transmigration (p. 384)

tsunami (p. 395)
typhoon (p. 381)

## THINKING GEOGRAPHICALLY

1. What might the fate of animism be in the new millennium? Consider whether it is doomed to extinction before the forces of modern economics and national integration, or whether it may persist as tribal peoples struggle to retain their cultural identities.

2. What should be the position of the English language in the educational systems of Southeast Asia? What should be the position of each country's national language? What about local languages?

3. How might ethnic tensions and human rights abuses in Burma (Myanmar) be reduced? What role should the international community play?

4. Can Singapore continue to experience economic growth and technological development while limiting freedom of expression? What role might the Internet play as Singapore struggles with these issues?

5. Is the Southeast Asian economic path of integration into the global economy, marked by an openness to multinational corporations and foreign investment, going to prove wise in the long run, or do its potential hazards outweigh its benefits?

Log in to **www.mygeoscienceplace.com** for videos, interactive maps, RSS feeds, case studies, and self-study quizzes to enhance your study of Southeast Asia.

# 14 AUSTRALIA AND OCEANIA

## GLOBALIZATION AND DIVERSITY

Australia and Oceania were once isolated from the world, but globalization has now integrated them into the global community through a complex history of colonization, geopolitical arrangements, and economic ties. Traditionally oriented toward Europe because of colonial histories, the region now is linked more closely with Asia.

## ENVIRONMENTAL GEOGRAPHY

Nonnative plants and animals, including snakes, rabbits, and feral pigs, are damaging the region's unique biodiversity by preying on and replacing local flora and fauna. In addition, sea-level rise from global warming threatens many low-lying islands.

## POPULATION AND SETTLEMENT

Large areas remain virtually empty in both Australia and New Zealand. On many Pacific islands, however, high population density resulting from rapid population growth is problematic.

## CULTURAL COHERENCE AND DIVERSITY

Until 1973, Australia protected its European ethnic roots with the White Australia Policy. But recent immigration—largely from Asia—is creating a new multicultural society.

## GEOPOLITICAL FRAMEWORK

From Hawaii to Australia, native peoples are demanding ownership or, minimally, access to their ancestral lands. More often than not, these land claims are fraught with controversy and tension.

## ECONOMIC AND SOCIAL DEVELOPMENT

Increased trade linkages with China have brought economic benefits to many countries in Oceania, particularly Australia. However, those benefits also involve uncertainty as China's economy responds to global fluctuations.

*Vast expanses of ocean, dotted with thousands of small islands, captures the essence of Oceania. This photo is of the village of Taveuni, Fiji.*

This vast world region, dominated mostly by water, includes the island continent of Australia as well as **Oceania**, a collection of islands that reach from New Guinea and New Zealand to the U.S. state of Hawaii in the mid-Pacific (Figure 14.1, and "Setting the Boundaries"). Although native peoples settled the area long ago, more recent European and North American colonization began the process of globalization that now characterizes the region. Today, the region is caught up in global processes that are producing new and sometimes unsettled environmental, cultural, and political geographies.

Ongoing political and ethnic unrest in Fiji illustrates how the heat of 21st-century globalization has fired the cauldron of change (Figure 14.2). Currently, the country struggles to deal with new political and economic relationships between the two dominant ethnic

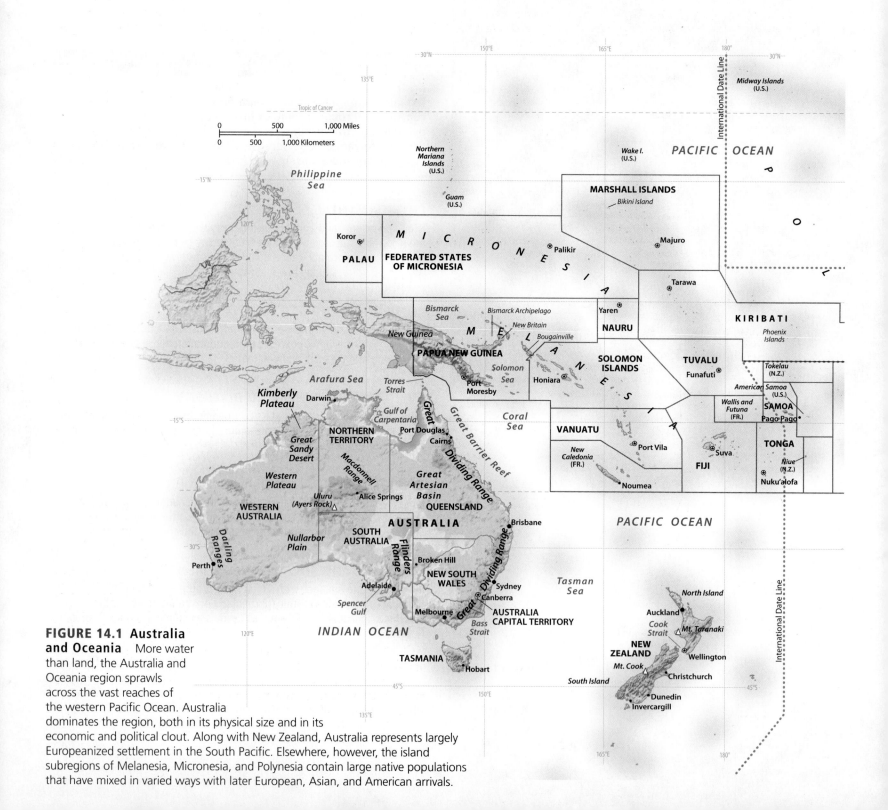

**FIGURE 14.1 Australia and Oceania** More water than land, the Australia and Oceania region sprawls across the vast reaches of the western Pacific Ocean. Australia dominates the region, both in its physical size and in its economic and political clout. Along with New Zealand, Australia represents largely Europeanized settlement in the South Pacific. Elsewhere, however, the island subregions of Melanesia, Micronesia, and Polynesia contain large native populations that have mixed in varied ways with later European, Asian, and American arrivals.

# Setting the Boundaries

While the vast distances of the Pacific stretching from New Guinea to Hawaii help define the boundaries of this region, many of the national boundaries were born from political convenience during an earlier period of colonial globalization. Australia (or "southern land"), which is often thought of as a continent, forms a coherent political unit and subregion. To the east, New Zealand, a three-hour flight from Australia, and closely linked by shared historical ties to Britain, is usually considered part of Polynesia because of the Maori, the original inhabitants of the country. These shared cultural ties also act as a bridge between many of Oceania's island nations and New Zealand.

Hawaii, 4,400 miles (7,084 kilometers) northeast of New Zealand, shares the same Polynesian heritage and is thought of as the northeastern boundary of Oceania. The southeastern boundary of the region is usually delimited by the Polynesian islands of Tahiti, 3,000 miles (4,416 kilometers) to the southeast.

Four thousand miles (6,437 kilometers) west of French Polynesia, well across the International Date Line, lies the island of New Guinea, the accepted yet sometimes confusing boundary between Oceania and Asia. Today, an arbitrary boundary line bisects the island, dividing the eastern half (Papua New Guinea, which is usually considered part of Oceania) from neighboring Irian Jaya (the western part), which, as part of Indonesia, is usually thought of as part of Southeast Asia.

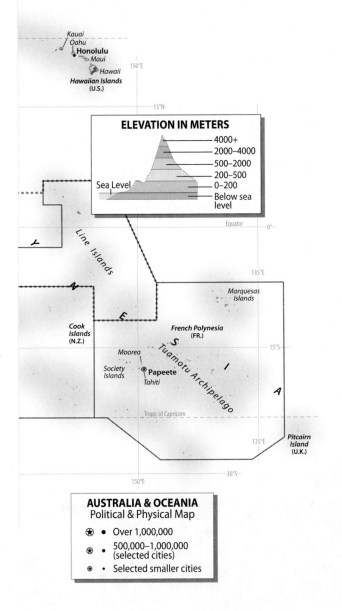

**FIGURE 14.2 Unrest in Fiji** South Asians were brought to Fiji as workers in the sugarcane fields. Today, the Indo-Fijians generally control the sugar industry, which is a source of tension with indigenous Fijians (such as this cane cutter), many of whom support politicians who advocate Fijian control of the industry.

groups, indigenous Fijians and the descendants of South Asian sugarcane workers (called Indo-Fijians) who were brought to the islands in the 19th century as a solution to labor shortages in the cane fields. Strong cultural and religious barriers exist between the two groups, resulting in extremely low rates of intermarriage. Generally speaking, the Indo-Fijians dominate the country's commercial life and are better off economically than the Fijians, even though the indigenous tribal communities own the land that produces much of the country's wealth. The fields are leased from Fijian tribes, and currently, as these leases expire, the Fijian tribes are not renewing them in hopes of improving their own economic position.

While tensions have long caused conflict between these two groups, this unrest has been heightened by modern-day globalization, as foreign investors push for economic restructuring and market development of the sugar industry. As a result, island governance has been unsettled by a series of coups and countercoups over the past two decades. In May 2000, armed Fijians took hostage the nation's first prime minister of Indian descent, creating both an internal and international crisis as other countries boycotted Fiji until an elected government was installed. Elections in 2001 and 2006 led to a Fijian-dominated government. Not surprisingly, large numbers of Indo-Fijians fled the islands, leaving the commercial and economic life of the country in shambles.

By all accounts, these current tensions are more than a continuation of historic ethnic unrest between the two groups. Instead, observers attribute much of the problem to unrest within the traditional native Fijian community as its members became empowered by the riches of economic globalization.

In many ways, the same tensions found in Fiji between indigenous peoples and "outsiders" (even if they arrived centuries ago) are also found in many other island countries of Oceania and, to a lesser degree, within Australia and New Zealand. This serves to remind us of the way environmental, settlement, cultural, geopolitical, and economic activities are inseparably linked in the contemporary world.

## Overview of the Region

Australia and New Zealand share many geographic characteristics. Major population clusters in both countries are located in the middle latitudes rather than the tropics. Australia's 21.9 million residents occupy a vast land area of 2.97 million square miles (7.69 million square kilometers), while New Zealand's combined North and South Islands (104,000 square miles, or 269,000 square kilometers) are home to 4.3 million people. Most residents of both countries live in urban settlements near the coasts (Figure 14.3). Australia's huge and dry interior, often termed the **Outback**, is as thinly settled as North Africa's Sahara Desert, and much of the New Zealand countryside is a visually spectacular but sparsely occupied collection of volcanic peaks and rugged, glaciated mountain ranges (Figure 14.4). Taken together, the land areas of these two South Pacific nations almost equal that of the United States, but their populations total less than 10 percent of their distant North Pacific neighbor. All three countries, however, share a European cultural heritage, the product of common global-scale processes that sent Europeans far from their homelands over the past several centuries. The highly Europeanized populations of both Australia and New Zealand also retain particularly close cultural links to Britain; further, they maintain relatively high levels of income and economic development within the Pacific world.

**FIGURE 14.3 Sydney, Australia** Most Australians live in cities, and the country's urban landscapes often resemble their North American counterparts. This view of Sydney features its world-famous harbor and displays the dramatic interplay of land and water.

**FIGURE 14.4  The Australian Outback**   Arid and generally treeless, the vast lands of the Australian Outback resemble some of the dry landscapes of the U.S. West. In this photo, wildflowers blossom along a dirt road near Tom Price, in the Pilbara region of Western Australia.

Punctuated with isolated chains of sand-fringed and sometimes mountainous islands, the blue waters of the tropical Pacific dominate much of the rest of the region. Three major subregions of Oceania each contain a surprising variety of human settlements and political units. Farthest west, **Melanesia** (meaning "dark islands") contains the culturally complex, generally darker-skinned peoples of New Guinea, the Solomon Islands, Vanuatu, and Fiji. The largest of these countries, Papua New Guinea (179,000 square miles, or 463,000 square kilometers), includes the eastern half of the island of New Guinea (the western half is part of Indonesia), as well as nearby portions of the northern Solomon Islands. Its population of 6.6 million people is slightly higher than that of New Zealand.

To the east, the small island groups, or **archipelagos**, of the central South Pacific are called **Polynesia** (meaning "many islands"), and this linguistically unified subregion includes French-controlled Tahiti in the Society Islands, the Hawaiian Islands, and smaller political states such as Tonga, Tuvalu, and Samoa. New Zealand is also often considered a part of Polynesia because its native peoples, known collectively as the **Maori**, share many cultural and physical characteristics with the somewhat lighter-skinned peoples of the mid-Pacific region. Finally, the more culturally diverse region of **Micronesia** (meaning "small islands") is north of Melanesia and west of Polynesia and includes microstates such as Nauru and the Marshall Islands, as well as the U.S. territory Guam.

## Environmental Geography: A Varied Natural and Human Habitat

The region's physical setting speaks to the power of space: The geology and climate of the seemingly limitless Pacific Ocean define much of the physical geography of Oceania, while the expansive interior of Australia shapes the basic physical geography of that island continent.

## Environments at Risk

Despite their relatively small populations, many areas in Australia and Oceania face significant human-induced environmental problems. Some environmental challenges are caused by natural events that increasingly affect larger and more widely distributed human populations. For instance, Pacific Rim earthquakes, periodic Australian droughts, and tropical cyclones now pose greater threats than they once did, as new settlements have made increasing populations vulnerable to these problems. Other environmental issues, however, are even more directly related to human causes (Figure 14.5). Specifically, European colonization introduced many environmental threats, and recent economic globalization has further pressured the region's natural resource base.

**Global Resource Pressures**   Globalization has exacted an environmental toll on Australia and Oceania. Specifically, the region's considerable base of natural resources has been opened to development, much of it by outside interests. While gaining from the benefits of global investment, the region has also paid a considerable price for encouraging development, and the result is an increasingly threatened environment.

Major mining operations have greatly affected Australia, Papua New Guinea, New Caledonia, and Nauru. Some of Australia's largest gold, silver, copper, and lead mines are located in sparsely settled portions of Queensland and New South Wales, putting watersheds in these semiarid regions at risk for metals pollution. In Western Australia, huge open-pit iron mines dot the landscape, unearthing ore that is usually bound for global markets, particularly China and Japan. To the north, Papua New Guinea's Bougainville copper mine has transformed the Solomon Islands, while even larger gold-mining ventures have raised increasing environmental concerns on the island of New Guinea (Figure 14.6). Elsewhere, Micronesia's tiny Nauru has been virtually turned inside out as much of the island's jungle cover has been removed to get at some of the world's richest phosphate deposits. Former Australian and New Zealand mine owners have already paid millions of dollars to settle environmental damage claims.

Deforestation is another major environmental threat across the region. Vast stretches of Australia's eucalyptus woodlands, for example, have been destroyed to create better pastures. In addition, coastal rainforests in Queensland are only a fraction of their original area, although a growing environmental movement in the region is fighting to save the remaining forest tracts. Tasmania has also been an environmental battleground, particularly given the biodiversity of its midlatitude forest landscapes. While the island's earlier European and Australian development featured many logging and pulp mill operations, more than 20 percent of the island is now protected by national parks.

Many islands in Oceania are also threatened by deforestation. With limited land areas, islands are subject to rapid tree loss, which in turn often leads to soil erosion. Although rainforests still cover 70 percent of Papua New Guinea, more than 37 million acres (15 million hectares) have been identified as suitable for logging (Figure 14.7). Some of the world's most biologically diverse environments are being threatened in these operations, but landowners see the quick cash sales to loggers as attractive, even though this nonsustainable practice is contrary to their traditional lifestyles.

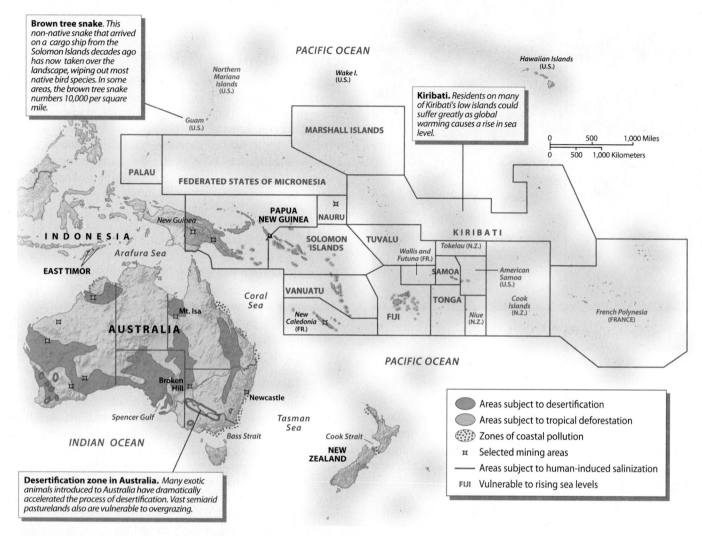

**Brown tree snake.** *This non-native snake that arrived on a cargo ship from the Solomon Islands decades ago has now taken over the landscape, wiping out most native bird species. In some areas, the brown tree snake numbers 10,000 per square mile.*

**Kiribati.** *Residents on many of Kiribati's low islands could suffer greatly as global warming causes a rise in sea level.*

**Desertification zone in Australia.** *Many exotic animals introduced to Australia have dramatically accelerated the process of desertification. Vast semiarid pasturelands also are vulnerable to overgrazing.*

Legend:
- Areas subject to desertification
- Areas subject to tropical deforestation
- Zones of coastal pollution
- ☒ Selected mining areas
- Areas subject to human-induced salinization
- FIJI Vulnerable to rising sea levels

**FIGURE 14.5 Environmental Issues in Australia and Oceania**   Modern environmental problems belie the region's myth that it is an earthly paradise. Tropical deforestation, extensive mining, and a long record of nuclear testing by colonial powers have brought varied challenges to the region. Human settlements have also extensively modified the pattern of natural vegetation. Future environmental threats loom for low-lying Pacific islands as sea levels rise from global warming.

**Global Warming in Oceania**   Even though the Pacific world contributes relatively little atmospheric pollution, the harbingers of climate change are already widespread and problematic. In New Zealand, mountain glaciers are melting away, while Australia suffers from frequent droughts and devastating wildfires. Warmer ocean waters have caused widespread bleaching of the Great Barrier Reef off Australia's coast, and rising sea levels are flooding several low-lying island nations, forcing residents to migrate to higher land (Figure 14.8). United Nations projections for the future are also highly disturbing: Stronger tropical cyclones could

**FIGURE 14.6 Mining in Papua New Guinea**   Open-pit mining for gold, silver, copper, and lead mark the landscapes of Papua New Guinea, New Caledonia, and Nauru. Although bringing some economic benefit to local peoples, these activities also cause immense environmental damage to the region. In New Guinea, for example, sediments from upland mines have severely damaged the Fly River ecosystem.

**FIGURE 14.7 Logging in Oceania** Foreign logging companies have made large investments in tropical Pacific settings such as Papua New Guinea and Samoa. While bringing new jobs, these ventures dramatically alter local environments, as precious hardwood forests are harvested for export.

devastate Pacific islands, with widespread damage to land and life; island inhabitants will suffer from reduced coastal resources as ocean waters warm; and increased wildfire will threaten population centers in southeast Australia, while agriculture will suffer from droughts and severe water shortages.

In response to these threats, the actions and policies taken by Oceania's countries vary considerably from full-on action to elusive disengagement. Until a recent change of government, Australia was the only industrial country besides the United States to not ratify the Kyoto Protocol. Perhaps the fact that Australia is the world's largest exporter of coal influenced that decision; not to be overlooked is that most of that country's emissions are produced by coal-fired power plants. However, while the former national government rejected the Kyoto Protocol and downplayed the threat of global warming, as in the United States, Australia's states and cities compensated for this inaction with plans for local and regional emission control and carbon trading schemes.

In contrast, the national government of New Zealand has taken an aggressive stance toward global warming, with specific plans for taxing carbon emissions. When New Zealand ratified the Kyoto agreement, it committed itself to a 5 percent reduction over its 1990 baseline;

**FIGURE 14.8 Sea Level Rise in Tuvalu** These photos show the effect of recent sea level rise on the small island of Tuvalu. When the house was built, it was rarely flooded at high tide, but today it is often isolated by high waters. Scenes like this contribute to the concern that this small island may soon be uninhabitable. As a result, many people are migrating to other islands, making them some of the first global warming refugees.

however, a booming economy has resulted in a 50 percent increase of emissions over the past several decades.

A major component of New Zealand's greenhouse gas pollution is methane emission from the country's large livestock population. These emissions, in fact, account for over half of the global warming pollution coming from that small country. As a result, New Zealand is discussing a much-publicized "flatulence tax" that would be levied on livestock owners (Figure 14.9).

The small, low-lying Pacific islands have emphasized adapting to climate change problems rather than curbing their miniscule carbon emissions. As noted earlier, several Pacific nations, most notably Tuvalu, Kiribati, and the Marshall Islands, are already experiencing disastrous

**FIGURE 14.9 New Zealand's Greenhouse Gas Problem** Worldwide, cattle and sheep emit the greenhouse gas methane, but only in New Zealand are these livestock emissions greater than those produced by human activities. While New Zealand has discussed a "flatulence tax" on sheep farms, the country's scientists are also working on a anti-flatulence inoculation that would reduce sheep emissions.

**FIGURE 14.10 Island Pest**    The brown tree snake, which arrived in Guam accidentally in the 1950s, has now taken over large parts of the island's forestlands and killed off most native bird species. Because these snakes, which reach 10 feet in length, climb along electrical wires, they frequently cause power outages throughout Guam.

flooding during high tides and storm surges, and, as a result, several thousand islanders have abandoned their homeland and migrated to New Zealand and Australia. In addition, fish and reef resources have reportedly been degraded because of coral bleaching and warming waters.

**Exotic Plants and Animals**    The introduction of exotic (nonnative) plants and animals has caused problems for endemic (native) species throughout the Pacific region. In Australia, for example, nonnative rabbits successfully multiplied in an environment that lacked the diseases and predators that kept their numbers in check in Europe. Before long, rabbit populations had reached plague-like proportions, stripping large sections of land of vegetation. The animals were brought under control only through the purposeful introduction of the rabbit disease myxomatosis. Introduced sheep and cattle populations have also stressed the region's environment by increasing soil erosion and contributing to desertification.

The introduction of exotic plants and animals to island environments has had similar effects. For example, many small islands possessed no native land mammals, and their native bird and plant species proved vulnerable to the ravages of introduced rats, pigs, and other animals. The larger islands of the region, such as those of New Zealand, originally supported several species of large, flightless birds that filled some of the ecological niches held by mammals on the continents. The largest of these, the moas, were substantially larger than ostriches. During the first wave of human settlement in New Zealand some 1,500 years ago, moa numbers fell rapidly, as they were hunted, their habitat burned, and their eggs consumed by invading rats. By 1800, the moas had been completely exterminated.

The spread of nonnative species continues today, perhaps at even greater pace. In Guam, the brown tree snake, which arrived accidentally by cargo ship from the Solomon Islands in the 1950s, has taken over the landscape (Figure 14.10). In some forest areas, with more than 10,000 snakes per square mile, they have wiped out nearly all the native bird species. In addition, the snakes cause frequent power outages as they crawl along electrical wires. While the brown tree snake has already done

its damage to Guam, it threatens other islands as well because it readily hides in cargo containers that are shipped to other island destinations.

## Australian and New Zealand Environments

Curiously, Australia is one of the world's most urbanized societies, yet most people associate the country with its vast and arid Outback, a sparsely settled land of sweeping distances, scrubby vegetation, and unusual animals. In contrast, the two small islands that make up New Zealand are known for their varied landscapes of rolling foothills and rugged mountains.

**Regional Landforms**    Three major landform regions dominate Australia's physical geography. The Western Plateau occupies more than half of the continent. Most of the region is a vast, irregular plateau that averages only 1,000 to 1,800 feet in height (305 to 550 meters). Further east, the Interior Lowland Basins stretch north to south for more than 1,000 miles (1,609 kilometers) from the swampy coastlands of the Gulf of Carpentaria to the Murray and Darling valleys, Australia's largest river system. Finally, more forested and mountainous country exists near Australia's Pacific coast. The Great Dividing Range extends from the Cape York Peninsula in northern Queensland to southern Victoria. Nearby, off the eastern coast of Queensland, the Great Barrier Reef offers a final dramatic subsurface feature: Over the past 10,000 years, one of the world's most spectacular examples of coral reef-building has produced a living legacy now protected by the Great Barrier Reef Marine Park (Figure 14.11).

Part of the Pacific Rim of Fire, New Zealand owes its geologic origins to volcanic mountain-building that produced two rugged and spectacular islands in the South Pacific. The North Island's active volcanic peaks, reaching heights of more than 9,100 feet (2,775 meters), and geothermal features reveal the country's fiery origins (Figure 14.12). Even higher and more rugged mountains run down the western spine of the South Island. Mantled by high mountain glaciers and surrounded by steeply sloping valleys, the Southern Alps are some of the world's most visually spectacular mountains, complete with narrow, fjordlike valleys that indent much of the South Island's isolated western coast.

**Climate**    Generally, zones of somewhat higher precipitation encircle Australia's arid center (Figure 14.13). In the tropical low-latitude

**FIGURE 14.11 The Great Barrier Reef**    Stretching along the eastern Queensland coast, the famed Great Barrier Reef is one of the world's most spectacular examples of coral reef-building. Threatened by varied forms of coastal pollution, much of the reef is now protected in a national marine park.

**FIGURE 14.12 Mt. Taranaki** New Zealand's North Island contains several volcanic peaks, including Mt. Taranaki. The 8,000-foot (2,440-meter) peak offers everything from subtropical forests to challenging ski slopes and attracts both local and international tourists.

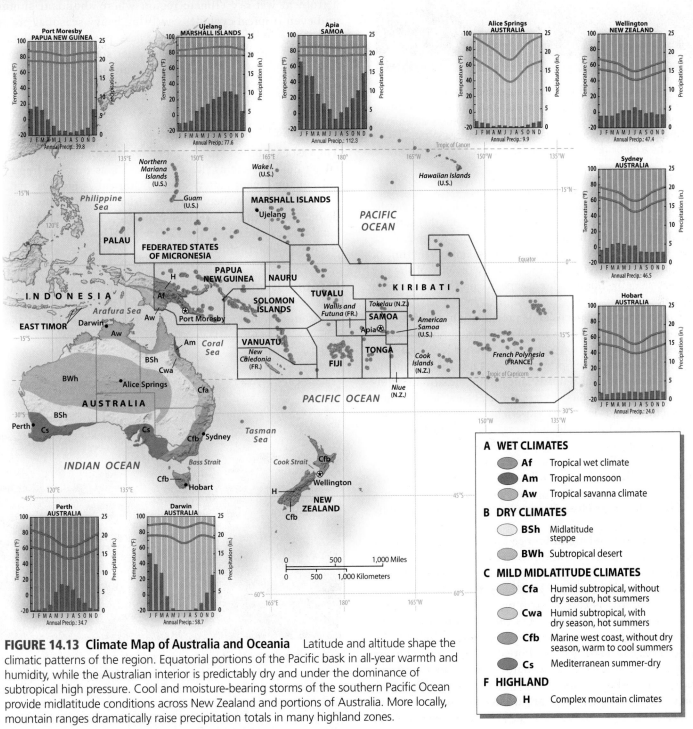

**FIGURE 14.13 Climate Map of Australia and Oceania** Latitude and altitude shape the climatic patterns of the region. Equatorial portions of the Pacific bask in all-year warmth and humidity, while the Australian interior is predictably dry and under the dominance of subtropical high pressure. Cool and moisture-bearing storms of the southern Pacific Ocean provide midlatitude conditions across New Zealand and portions of Australia. More locally, mountain ranges dramatically raise precipitation totals in many highland zones.

**A WET CLIMATES**

**Af** Tropical wet climate

**Am** Tropical monsoon

**Aw** Tropical savanna climate

**B DRY CLIMATES**

**BSh** Midlatitude steppe

**BWh** Subtropical desert

**C MILD MIDLATITUDE CLIMATES**

**Cfa** Humid subtropical, without dry season, hot summers

**Cwa** Humid subtropical, with dry season, hot summers

**Cfb** Marine west coast, without dry season, warm to cool summers

**Cs** Mediterranean summer-dry

**F HIGHLAND**

**H** Complex mountain climates

**FIGURE 14.14 Australian Wildfires**  Huge and savage dry season wildfires (known as bushfires in Australia) threaten both rural settlement and sprawling city suburbs in the southeast. This fire in February 2009, just 70 miles from the heart of Melbourne, was the worst fire disaster in 25 years and may be a harbinger of even more damaging fires accompanying global warming.

north, seasonal changes are dramatic and unpredictable. For example, Darwin can experience drenching monsoonal rains in the summer (December to March), followed by bone-dry winters (June to September). Indeed, life across the region is shaped by this annual rhythm of what is called locally "the wet" and "the dry." By the end of the dry season, wildfires usually dot the landscape of northern Australia (Figure 14.14).

Along the east coast of Queensland, precipitation remains high (60 to 100 inches, or 153 to 254 centimeters), but it diminishes rapidly as one moves into the interior. Rainfall at interior locations such as the Northern Territory's Alice Springs averages less than 10 inches (25 centimeters) annually. South of Brisbane, more midlatitude influences dominate eastern Australia's climate. Coastal New South Wales, southeastern Victoria, and Tasmania experience the country's most dependable year-round rainfall, which averages 40 to 60 inches (102 to 152 centimeters) of precipitation per year. Nearby mountains see frequent winter snows. Farther west, summers are hot and dry in much of South Australia and in the southwest corner of Western Australia, producing a distinctively Mediterranean climate. These zones of Mediterranean climate produce the **mallee** vegetation, a scrubby eucalyptus woodland.

Climates in New Zealand are influenced by latitude, the moderating effects of the Pacific Ocean, and proximity to local mountains or mountain ranges. Most of the North Island is distinctly subtropical. The coastal lowlands near Auckland are mild and wet year-round. Still, local variations can be striking, as the area's volcanic peaks create their own microclimates. On the South Island, conditions become distinctly cooler as one moves poleward. Indeed, the island's southern edge feels the seasonal breath of Antarctic chill, as it lies more than 46° south of the equator. Mountain ranges on New Zealand's South Island also display incredible local variations in precipitation: West-facing slopes are drenched with more than

100 inches (254 centimeters) of precipitation annually, while lowlands to the east average only 25 inches (64 centimeters) per year. The Otago region, inland from Dunedin, sits partially in the rain shadow of the Southern Alps, and its rolling, open landscapes resemble the semiarid expanses of North America's Intermountain West (Figure 14.15).

**Island Climates**  Many Pacific islands receive abundant precipitation, and high islands in particular are often noted for their heavy rainfall and dense tropical forests. In American Samoa, this environment has been protected in one of the nation's newest national parks. Much of the zone is located in the rainy tropics or in a tropical wet-dry climate region where abundant summer rains and even tropical cyclones can bring heavy seasonal precipitation. Low-lying atolls usually receive less precipitation than high islands and very often experience water shortages. During dry periods, the limited stores of water on these islands are quickly depleted.

## The Oceanic Realm

The vast expanse of Pacific waters reveals another set of environmental settings that are as rich and complex as they are fragile. Oceanic currents and wind patterns have historically served as natural highways of movement between these island worlds. Those same forces define the rhythm of weather patterns across the region in a broad band that extends more than 20° north (Hawaiian Islands) and south (New Caledonia) of the equator.

**Creating Island Landforms**  Much of Melanesia and Polynesia is part of the seismically active Pacific Basin. As a result, volcanic eruptions, major earthquakes, and **tsunamis**, or earthquake-induced

**FIGURE 14.15 Central Otago, South Island**  On New Zealand's South Island, the Southern Alps capture rainfall on the west coast but leave areas to the east in a drier rain shadow. As a result, the Central Otago region has a semiarid landscape resembling portions of the U.S. West.

sea waves, are not uncommon across the region, and they impose major environmental hazards on the population. For example, volcanic eruptions and earthquakes on the island of New Britain (Papua New Guinea) forced more than 100,000 people from their homes in 1994. Only four years later, a massive tsunami triggered by an offshore earthquake swept across the north coast of New Guinea, killing 3,000 residents and destroying numerous villages. Such events are unfortunately a part of life in this geologically active part of the world.

Most of the islands of Polynesia and Micronesia, however, are truly oceanic, having originated from volcanic activity on the ocean floor. The larger active and recently active volcanoes form **high islands**, which often rise to a considerable elevation and cover a large area. The island of Hawaii, the largest and youngest of the Pacific's high islands, is more than 80 miles (128 kilometers) across and rises to a height of more than 13,000 feet (3,980 meters). Indeed, the entire Hawaiian archipelago exemplifies a geological **hot spot**, where moving oceanic crust passes over a supply of magma from Earth's interior, thus creating a chain of volcanic islands. Many of the islands of French Polynesia, including Bora Bora, are smaller examples of high islands (Figure 14.16). Indeed, high islands are widely scattered throughout Micronesia and Polynesia. In tropical latitudes, most high islands are ringed by coral reefs, which quickly grow in the shallow waters near the shore.

The combination of narrow sandy islands, barrier coral reefs, and shallow central lagoons is also known as an **atoll**. The islands and reefs of an atoll characteristically form a circular or oval shape, although some are quite irregular (Figure 14.17). The world's largest atoll, Kwajalein in Micronesia's Marshall Islands, is 75 miles (120 kilometers) long and 15 miles (24 kilometers) wide. Polynesia and Micronesia are dotted with extensive atoll systems, and a number are found in Melanesia as well.

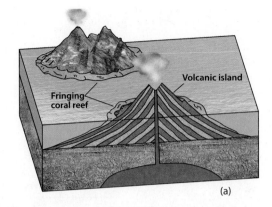

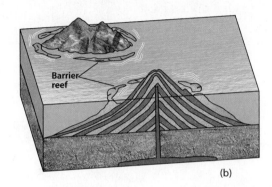

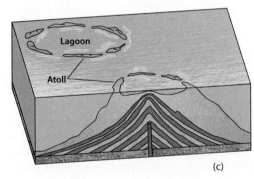

**FIGURE 14.17  Evolution of an Atoll**   Many Pacific islands begin as rugged volcanoes (a) with fringing coral reefs. However, as the extinct volcano subsides and erodes away, the coral reef expands, becoming a larger barrier reef (b). The term *barrier reef* comes from the hazards these features pose to navigation when approaching the island from sea. Finally, all that remains (c) is a coral atoll surrounding a shallow lagoon.

**FIGURE 14.16  Bora Bora**   The jewel of French Polynesia, Bora Bora displays many of the classic features of Pacific high islands. As the island's central volcanic core retreats, surrounding coral reefs produce a mix of wave-washed sandy shores and shallow lagoons.

## Population and Settlement: A Diverse Cultural Landscape

Modern population patterns across the region reflect the combined influences of indigenous and European settlement. In countries such as New Zealand, Australia, and the Hawaiian Islands, Anglo-European migration has structured the distribution and concentration of contemporary populations. In contrast, on smaller islands elsewhere in

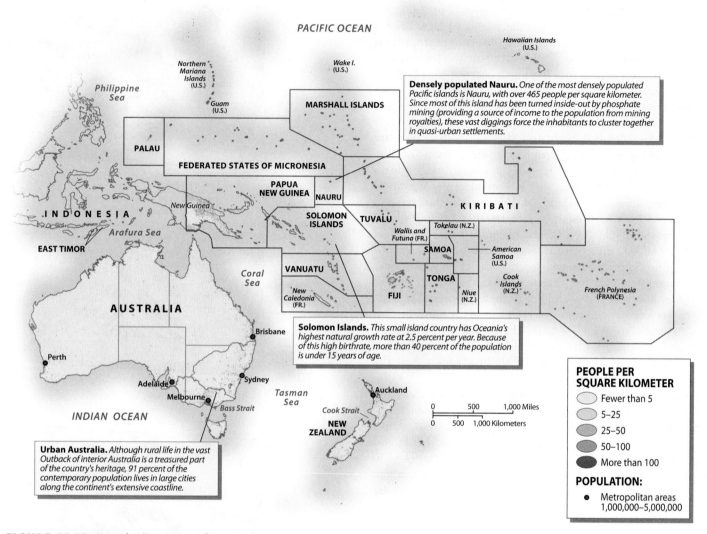

**PEOPLE PER SQUARE KILOMETER**

- Fewer than 5
- 5–25
- 25–50
- 50–100
- More than 100

**POPULATION:**

- Metropolitan areas 1,000,000–5,000,000

**Densely populated Nauru.** *One of the most densely populated Pacific islands is Nauru, with over 465 people per square kilometer. Since most of this island has been turned inside-out by phosphate mining (providing a source of income to the population from mining royalties), these vast diggings force the inhabitants to cluster together in quasi-urban settlements.*

**Solomon Islands.** *This small island country has Oceania's highest natural growth rate at 2.5 percent per year. Because of this high birthrate, more than 40 percent of the population is under 15 years of age.*

**Urban Australia.** *Although rural life in the vast Outback of interior Australia is a treasured part of the country's heritage, 91 percent of the contemporary population lives in large cities along the continent's extensive coastline.*

**FIGURE 14.18 Population Map of Australia and Oceania** About 36 million people occupy this world region. While Papua New Guinea and many Pacific islands feature mainly rural settlements, most regional residents live in the large urban areas of Australia and New Zealand. Sydney and Melbourne account for almost half of Australia's population, and most New Zealand residents live on the North Island, home to Auckland and Wellington.

Oceania, population geographies are determined by the needs of native peoples (Figure 14.18). More recently, however, migration has taken place from outlying islands to Australia and New Zealand because of a combination of push forces, including unemployment, resource depletion, and the threat of flooding associated with global warming (Table 14.1).

## Contemporary Population Patterns

Despite the popular stereotypes of life in the Outback, modern Australia has one of the most highly urbanized populations in the world. Indeed, 91 percent of the country's residents live within either the Sydney or Melbourne metropolitan areas. Australia's eastern and southern areas are home to the majority of its 21.9 million people. Inland, population densities decline as rapidly as the rainfall: Semiarid hills west of the Great Dividing Range still contain significant rural settlement, but the state's southwestern periphery

remains sparsely populated. New South Wales is the country's most heavily populated state, and its sprawling capital city of Sydney (4 million), focused around one of the world's most magnificent natural harbors, is the largest metropolitan area in the entire South Pacific In the nearby state of Victoria, Melbourne's 3.8 million residents have long competed with Sydney for status as Australia's premiere city, claiming cultural and architectural supremacy over their slightly larger neighbor (Figure 14.19). In between these two metropolitan giants, the location of the much smaller federal capital of Canberra (population 325,000) represents a classic geopolitical compromise in the same spirit that created Washington, DC, midway between the populous southern and northern portions of the United States.

Smaller clusters of population are found inland from Australia's eastern coast. More fertile farmlands in the interior of New South Wales and Victoria feature higher population densities than are found in most of the nation's arid heartland. Inland

**TABLE 14.1   POPULATION INDICATORS**

| Country | Population (millions) 2009 | Population Density (per square kilometer) | Rate of Natural Increase (RNI) | Total Fertility Rate | Percent Urban | Percent <15 | Percent >65 | Net Migration (Rate per 1,000) 2005–2010* |
|---|---|---|---|---|---|---|---|---|
| Australia | 21.9 | 3 | 0.7 | 2.0 | 83 | 19 | 13 | 4.8 |
| Fed. States of Micronesia | 0.1 | 158 | 1.9 | 3.9 | 22 | 37 | 4 | –16.3 |
| Fiji | 0.8 | 46 | 1.7 | 2.6 | 51 | 29 | 5 | –8.3 |
| French Polynesia | 0.3 | 67 | 1.3 | 2.2 | 53 | 26 | 6 | 0.0 |
| Guam | 0.2 | 332 | 1.5 | 2.6 | 93 | 28 | 7 | 0.0 |
| Kiribati | 0.1 | 136 | 1.8 | 3.5 | 44 | 36 | 4 | |
| Marshall Islands | 0.1 | 298 | 2.9 | 4.5 | 68 | 41 | 2 | |
| Nauru | 0.01 | 465 | 2.1 | 3.4 | 100 | 39 | 1 | |
| New Caledonia | 0.3 | 14 | 1.2 | 2.2 | 58 | 27 | 7 | 4.5 |
| New Zealand | 4.3 | 16 | 0.8 | 2.2 | 86 | 21 | 13 | 2.4 |
| Palau | 0.02 | 45 | 0.6 | 2.0 | 77 | 24 | 6 | |
| Papua New Guinea | 6.6 | 14 | 2.2 | 4.1 | 13 | 40 | 2 | 0.0 |
| Samoa | 0.2 | 67 | 2.0 | 4.2 | 22 | 40 | 5 | –18.4 |
| Solomon Islands | 0.5 | 18 | 2.7 | 4.6 | 17 | 41 | 3 | 0.0 |
| Tonga | 0.1 | 138 | 2.1 | 4.2 | 24 | 38 | 6 | –17.5 |
| Tuvalu | 0.01 | 427 | 1.4 | 3.7 | 47 | 32 | 6 | |
| Vanuatu | 0.2 | 20 | 2.5 | 4.0 | 21 | 41 | 3 | 0.0 |

*Source:* Population Reference Bureau, *World Population Data Sheet, 2009.*

*Net Migration Rate from the United Nations, Population Division, *World Population Prospects: The 2008 Revision Population Database.*

**FIGURE 14.19  Downtown Melbourne**   Metropolitan Melbourne lies along the Yarra River. Capital of the Australian state of Victoria, Melbourne resembles many growing North American cities, with its highrise office buildings, entertainment districts, and downtown urban redevelopment.

Aboriginal populations are widely but thinly scattered across districts such as northern Western Australia and South Australia, as well as in the Northern Territory, accounting for smaller but regionally important centers of settlement.

The population geography of the rest of Oceania shows a broad distribution of peoples, both native and European, who have clustered near favorable resource opportunities. In New Zealand, more than 70 percent of the country's 4.3 million residents live on the North Island, with the Auckland region (1.1 million) dominating the metropolitan scene in the north and the capital city of Wellington (164,000) anchoring settlement along the Cook Strait in the south. Settlement on the South Island is mostly located in the somewhat drier lowlands and coastal districts east of the mountains, with Christchurch (340,000) serving as the largest urban center. Elsewhere, rugged and mountainous terrain on both the North and South Islands feature much lower densities. Such is not the case in Papua New Guinea, where only 13 percent of the country's population is urban, with many people living in the isolated, interior highlands. The nation's largest city is the capital, Port Moresby (200,000), located along the narrow coastal lowland in the far southeastern corner of the country. The largest urban area on the northern margin of Oceania is Honolulu (1 million), on the island of Oahu, where rapid metropolitan growth since World War II has occurred because of U.S. statehood and the scenic attractions of its mid-Pacific setting.

## Historical Settlement

The historical settlement of the Pacific realm can never be precisely reconstructed, but humans in several major migrations succeeded in occupying the region over time. The region's remoteness from many of the world's early population centers meant that it often lay beyond the dominant migratory paths of earlier peoples. Even so, settlers found their way to the isolated Australian interior and the far reaches of the Pacific. Later, the pace of new in-migrations increased once Europeans identified the region and its resource potential.

**Peopling the Pacific**    The large islands of New Guinea and Australia, given their nearness to the Asian landmass, were settled much earlier than the more distant islands of the Pacific. Around 40,000 years ago, the ancestors of today's native Australian, or **Aborigine**, populations were making their way out of Southeast Asia and into Australia (Figure 14.20). The first Australians most likely arrived using some kind of watercraft. However, because such boats were probably not very seaworthy, the more distant islands remained inaccessible to

humankind for tens of thousands of years. During the last glacial period, however, sea levels were much lower than they are now, which would have allowed easier movement to Australia across relatively narrow spans of water. It is not known whether the original Australians arrived in one wave of people or in many, but the available evidence suggests that they soon occupied large portions of the continent, including Tasmania, which was at that time connected to the mainland by a land bridge.

Eastern Melanesia was settled much later than Australia and New Guinea. By 3,500 years ago, certain Pacific peoples had mastered long-distance sailing and navigation, which eventually opened the entire oceanic realm to human habitation. In that era, people gradually moved east to occupy New Caledonia, the Fiji Islands, and Samoa. From there, later movements took seafaring folk north into Micronesia, and the Marshall Islands were occupied around 2,000 years ago.

Continuing movements from Asia further complicated the story of these migrating Melanesians. Some of the migrants mixed culturally and eventually reached western Polynesia, where they formed the core population of the Polynesian people. By 800 C.E., they had reached such

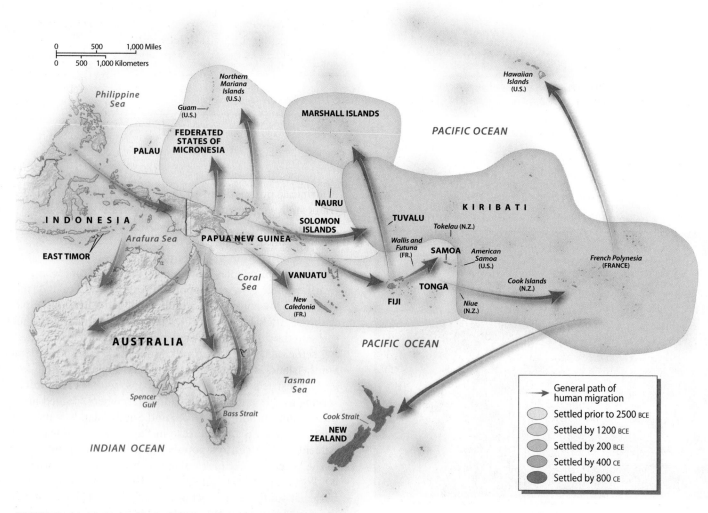

**FIGURE 14.20 Peopling the Pacific**    Ancestors of Australia's aboriginal population may have made their way into the island continent more than 60,000 years ago. Much more recent settlement of Pacific islands by Austronesian peoples from Southeast Asia shaped cultural patterns across the oceanic portions of the realm. Eastward migrations through the Solomon Islands, Fiji, and the Cook Islands were followed by late movements to the north and south.

distant places as New Zealand, Hawaii, and Easter Island. Debate has centered on whether these Polynesians purposefully set out to colonize new lands or whether they were blown off course during routine voyages, ending up on new islands. Prehistorians hypothesize that population pressures may have quickly reached crisis stage on the relatively small islands, and people would have therefore made dangerous voyages to colonize other Pacific islands. Equipped with sturdy outrigger sailing vessels and ample supplies of food, the Polynesians were quickly able to colonize most of the islands they discovered.

**European Colonization** About six centuries after the Maori brought New Zealand into the Polynesian realm, Dutch navigator Abel Tasman spotted the islands on his global exploration of 1642. Tasman's initial sighting marked the beginning of a new chapter in the human occupation of the South Pacific. Late in the following century, more lasting European contacts were made. British sea captain James Cook surveyed the shorelines of both New Zealand and Australia between 1768 and 1780. Cook and others believed that these distant lands might be worthy of European development. In addition, other expeditions were exploring the Pacific, and most of Oceania's major island groups assumed a place on European maps by the end of the 18th century.

European colonization of the region began in Australia when the British needed a remote penal colony to which convicts could be exiled. The southeastern coast of Australia was selected as an appropriate site, and in 1788 the First Fleet arrived with 750 prisoners in Botany Bay, near what is now Sydney. Other fleets and more convicts soon followed, as did boatloads of free settlers. Before long, free settlers outnumbered the convicts, who were themselves gradually gaining freedom after serving their sentences. The growing population of English-speaking people soon moved inland and also settled other favorable coastal areas. British and Irish settlers were attracted by the agricultural and stock-raising potential of the distant colony and by the lure of gold and other minerals (a major gold rush occurred in Australia during the 1850s). The British government also encouraged the emigration of its own citizens, often paying the transportation fare of those too poor to afford it themselves.

The new settlers came into conflict with the Aborigines almost immediately after arriving. No treaties were signed, however, and in most cases, Aborigines were simply expelled from their lands. In some places, most notably Tasmania, they were hunted down and killed. In mainland Australia, the Aborigines were greatly reduced in numbers by disease, removal from their lands, and pure economic hardship. By the mid-19th century, Australia was primarily an English-speaking land, as the native peoples were driven into submission.

British settlers were also attracted to the lush and fertile lands of New Zealand. European whalers and sealers arrived shortly before 1800, but more permanent agricultural settlement took shape after 1840, as the British formally declared sovereignty over the region. As new arrivals grew in number and the scope of planned settlement colonies on the North and South islands expanded, tensions with the native Maori population increased. Organized in small kingdoms, or chiefdoms, the Maori were formidable fighters. In 1845, one native group decided to resist further settlement by Europeans, leading to the widespread Maori wars that engulfed New Zealand until 1870. The British eventually prevailed, however, and the Maori lost most of their land.

The native Hawaiians also lost control of their lands to immigrants. Hawaii emerged as a united and powerful kingdom in the early 1800s, and for many years, its native rulers limited U.S. and European claims to their islands. Increasing numbers of missionaries and settlers from the United States were allowed in, however, and by the late 19th century, control of the Hawaiian economy had largely passed to foreign plantation owners. By 1898, U.S. forces were strong enough to overthrow the Hawaiian monarchy and to annex the islands to the United States.

## Modern Settlement Landscapes

The settlement geography of Australia and Oceania offers an interesting mixture of local and global influences. The contemporary cultural landscape still reflects the imprint of indigenous peoples in those settings where native populations remain numerically dominant. Elsewhere, patterns of recent colonization have produced a modern scene mainly shaped by Europeans. The result includes everything from German-owned vineyards in South Australia to houses on New Zealand's South Island that appear to be plucked directly from the British Isles. In addition, processes of economic and cultural globalization have resulted in urban forms that make cities such as Perth or Auckland look strikingly similar to such places as San Diego or Seattle.

**The Urban Transformation** Both Australia and New Zealand are highly urbanized, Westernized societies, and thus the vast majority of their populations live in urban and suburban environments. As in Europe and North America, much of this urban transformation came during the 20th century, as the rural economy became less labor intensive and as opportunities for urban manufacturing and service employment grew. As urban landscapes evolved, they took on many of the characteristics of their largely European populations, but they blended these with a strong dose of North American influences, as well as with the unique settings native to each urban place. The result is an urban landscape in which many North Americans are quite comfortable, even though the varied local accents heard on the street and many features of the metropolitan scene are reminders of the strong and lasting attachments to British traditions.

The affluent Western-style urban settings in Australia and New Zealand offer a dramatic contrast with the urban landscapes found in less-developed settings in the region. Walk the streets of Port Moresby in Papua New Guinea, and a very different urban landscape is evidence of the large gap between rich and poor within Oceania (Figure 14.21).

**FIGURE 14.21  Port Moresby, Papua New Guinea** Urban poverty and high crime haunt the city of Port Moresby, the capital of Papua New Guinea. The city's slums, many built out on the water, reflect stresses of recent urban growth as rural residents emigrate from nearby highlands.

Rapid growth in Port Moresby, the country's political capital and largest commercial center, has produced many of the classic problems of urban underdevelopment: There is a shortage of adequate housing, the building of roads and schools lags far behind the need, and street crime and alcoholism are on the rise. Elsewhere, urban centers such as Suva (Fiji), Noumea (New Caledonia), and Apia (Samoa) also reflect the economic and cultural tensions generated as local populations are exposed to Western influences. Rapid growth is a common problem in the smaller cities of Oceania because native people from rural areas and nearby islands gravitate toward the job opportunities available. In the past 50 years, the huge global growth of tourism in places such as Fiji and Samoa has also transformed the urban scene (Figure 14.22): Village life reminiscent of the 19th century has often been replaced by a landscape of souvenir shops, honking taxicabs, and crowded seaside resorts.

**The Rural Scene**    Rural landscapes across Australia and the Pacific region express a complex mosaic of cultural and economic influences.

**FIGURE 14.22 Tourism in Oceania**    On many islands, traditional ways of life are being replaced by tourist activities, both in the city (such as this scene in Fiji) as well as the countryside. While the economic benefits are welcome, the accompanying social, cultural, and political side effects of this form of globalization are not always beneficial.

In some settings, Australian Aborigines or native Papua New Guinea Highlanders can still be found in their familiar homelands, their traditional lifeways and settlements barely changed from pre-European times. Yet such settlement landscapes are becoming increasingly rare. Global influences penetrate the scene as the cash economy, foreign tourism and investment, and the currents of popular culture work their way from city to countryside.

Much of rural Australia is too dry for farming or serves as only marginally valuable agricultural land. Although modest in size, the area in crops has doubled since 1960, as increased use of fertilizers, more widespread irrigation, and more aggressive rabbit eradication efforts have opened up new areas for development. Much of the remainder of the interior, however, features range-fed livestock, areas beyond the pale of any agricultural potential, and isolated areas where Aboriginal peoples still pursue their traditional forms of hunting and gathering.

Sheep and cattle dominate rural Australia's livestock economy. Many rural landscapes in the interior of New South Wales, Western Australia, and Victoria, for example, are oriented around isolated sheep stations, ranch operations that move the flocks from one large pasture to the next. Cattle can sometimes be found in these same areas, although many of the more extensive, range-fed cattle operations are concentrated farther north, in Queensland. Croplands also vary across the region. Sometimes mingling with the sheep country, a band of commercial wheat farming includes southern Queensland; the moister interiors of New South Wales, Victoria, and South Australia; and a swath of favorable land east and north of Perth. Elsewhere, specialized sugarcane operations thrive along the narrow, warm, and humid coastal strip of Queensland. To the south and west, productive irrigated agriculture has developed in places such as the Murray River Basin, allowing for the production of orchard crops and vegetables. **Viticulture,** or grape cultivation, increasingly shapes the rural scene in places such as South Australia's Barossa Valley, the Riverina district in New South Wales, and Western Australia's Swan Valley. Indeed, the area under grape cultivation grew by 50 percent between 1991 and 1998, as the popular Chardonnay, Cabernet Sauvignon, and Shiraz varieties boosted wine production to revenues of more than $540 million per year.

**New Zealand's Landscapes**    Although much smaller in area than Australia, New Zealand's rural settlement landscape includes a variety of agricultural activities. Pastoral activities clearly dominate the New Zealand scene, with the vast majority of agricultural land devoted to livestock production, particularly sheep grazing and dairying. Commercial livestock outnumber people in New Zealand by a ratio of more than 20 to 1, and this is apparent everywhere on the rural scene. Dairy operations are present mostly in the lowlands of the north, where they sometimes mingle with suburban landscapes in the vicinity of Auckland. One of the largest zones of more specialized cropping spreads across the fertile Canterbury Plain near Christchurch (Figure 14.23). This spectacular South Island setting proved fertile ground for English settlement and continues to feature a varied landscape of pastures, grain fields, orchards, and vegetable gardens, all spread beneath the towering peaks of the Southern Alps.

**Rural Oceania**    Elsewhere in Oceania, varied influences shape the rural scene. On high islands with more water, denser populations take advantage of more diverse agricultural opportunities than are usually found on the more barren low islands, where fishing is often

**FIGURE 14.23 Canterbury Plain** The varied agricultural landscape of South Island's Canterbury Plain offers a mix of grain fields, livestock, orchard crops, and vegetable gardens. The rugged Southern Alps are a dramatic backdrop to this productive region.

more important. Several types of rural settlement can be identified across the island realm. In rural New Guinea, village-centered shifting cultivation dominates: Farmers clear a patch of forest and then, after a few years, shift to another patch, thus practicing a form of land rotation. Subsistence foods such as sweet potatoes, taro (another starchy root crop), coconut palms, bananas, and other garden crops are often found in the same field, and growing numbers of planters also include commercial crops such as coffee. In other parts of Oceania, traditional agricultural patterns are similar (Figure 14.24). Commercial plantation agriculture has also made its mark in many more accessible rural settings. In these places, settlements consist of worker housing near crops that are typically controlled by absentee landowners. For example,

**FIGURE 14.24 Yam Harvest** These farmers on the Melanesian island of Vakuta (Papua New Guinea) are harvesting yams. Traditional tropical agriculture features a mix of crops, often grown in the same field. Where possible, fields are periodically rotated to maintain productivity.

copra (coconut), cocoa, and coffee operations have transformed many agricultural settings in places such as the Solomon Islands and Vanuatu. Sugarcane plantations have reshaped other island settings, particularly in Fiji and Hawaii.

### Diverse Demographic Paths

A variety of population-related issues face residents of the region today. In Australia and New Zealand, while populations grew rapidly (mostly from natural increases) in the 20th century, today's low birthrates parallel the pattern in North America. Just as in the United States and Canada, however, significant population shifts within these countries continue to create challenges. To illustrate, the departure of farmers from Australia's wheat-growing and sheep-raising interior mirrors similar processes at work in the rural Midwest of the United States and in the Canadian prairies. Communities see many of their productive young people and professionals leave for the better employment opportunities of the city.

Different demographic challenges grip many less-developed island nations of Oceania. Population growth rates are often above 2 percent per year and are even higher in countries such as Vanuatu and the Solomon Islands. While the larger islands of Melanesia contain some room for settlement expansion, competitive pressures from commercial mining and logging operations limit the amount of new agricultural land that will probably be available in the future. On some of the smaller island groups in Micronesia and Polynesia, population growth is a more pressing problem. Tuvalu (north of Fiji), for example, has just over 10,000 inhabitants, but they are crowded onto a land area of about 10 square miles (26 square kilometers), making it one of the world's most thickly populated countries.

## Cultural Coherence and Diversity: A Global Crossroads

The Pacific world offers excellent examples of how culture is transformed as different groups migrate to a region, interact with one another, and evolve over time. As Europeans and other outsiders arrived in the region, colonization forced native peoples to adjust. Worldwide processes of globalization have also redefined the region's cultural geography, provoking fears of homogenization while at the same time promoting more cultural preservation efforts as native groups attempt to protect their heritage.

### Multicultural Australia

Australia's cultural patterns illustrate many of the fundamental processes of globalization at work. Today, while still dominated by its colonial European roots, the country's multicultural character is becoming increasingly visible as native peoples assert their cultural identity and as varied immigrant populations play larger roles in society, particularly in major metropolitan areas (Figure 14.25).

**Aboriginal Imprints** For thousands of years, Australia's Aborigines dominated the cultural geography of the continent. They never practiced agriculture, opting instead for a hunting-and-gathering way of life that persisted up to the time of the European conquest. As the

**FIGURE 14.25 Asian Immigrants in Australia** Unlike historical immigration, which was predominately from Britain and other European countries, today fully 40 percent of Australian immigrants are from Asia. As a result, neighborhoods in Sydney and other cities have a distinct Asian flavor as they serve the needs of this new population.

population consisted of foragers and hunters living in a relatively dry land, settlement densities remained low; tribal groups were often isolated from one another, and overall populations probably never numbered more than 300,000 inhabitants. To survive, Aborigines developed great adaptive skills, often subsisting in harsh environments that Europeans avoided. Because these people were clustered in many different areas, their language became fragmented. Although precise counts vary, there were probably 250 languages spoken at the time of European contact, and almost 50 indigenous languages can still be found today.

Radical cultural and geographic changes accompanied the arrival of Europeans, and Aboriginal populations were decimated in the process. The geographic results of colonization were striking, as Aboriginal settlements were relocated to the sparsely settled interior, particularly in

northern and central Australia, where fewer Europeans competed for land. In most cases, the European attitude toward the Aboriginal population was even more discriminatory than it was toward native peoples of the Americas.

Today, Aboriginal cultures persevere in Australia, and a growing native peoples movement is similar to activities in the Americas. Indigenous peoples account for approximately 2 percent (or 430,000) of Australia's population, but their geographic distribution changed dramatically in over the past century. Aborigines account for almost 30 percent of the Northern Territory's population (many of these near Darwin), and other large native reserves are located in northern Queensland and Western Australia. Most native peoples, however, live in the same urban areas that dominate the country's overall population geography. Indeed, more than 70 percent of Aborigines live in cities, and very few of them still practice traditional hunting-and-gathering lifestyles. Processes of cultural assimilation are clearly at work: Urban Aborigines are frequently employed in service occupations, Christianity has often replaced traditional animist religions, and only 13 percent of the native population still speaks a native language.

Still, forces of diversity are at work, suggesting a growing Aboriginal interest in preserving traditional cultural values. Particularly in the Outback, a handful of Aboriginal languages remain strong and have growing numbers of speakers. In addition, cultural leaders are preserving some aspects of Aboriginal spiritualism, and these religious practices often link local populations to surrounding places and natural features that are considered sacred. In fact, a growing number of these sacred locations are at the center of land-use controversies (such as mining on sacred lands) between Aboriginal populations and Australia's European majority. The future for Aboriginal cultures remains unclear: Pressures for cultural assimilation will be intense as many native peoples move to more Western-oriented urban settlements and lifestyles. At the same time, rapid rates of natural increase (almost twice the national average) and a growing cultural awareness of Aboriginal traditions will work to preserve elements of the country's indigenous cultures.

**A Land of Immigrants** Most Australians reflect the continent's more recent European-dominated migration history, but even these patterns have become more complex as a rising tide of Asian cultures becomes important. Overall, more than 70 percent of Australia's population continues to reflect a British or Irish cultural heritage. These groups dominated many of the 19th- and early 20th-century migrations into the country, and the close cultural ties to the British Isles remain strong.

A need for laborers along the fertile Queensland coast also caused European plantation owners to import inexpensive workers from the Solomons and New Hebrides. These Pacific Island laborers, known as **kanakas,** were spatially and socially segregated from their Anglo employers but further diversified the cultural mix of Queensland's "sugar coast." Historically, however, nonwhite migrations to the country were strictly limited by what is often termed the **White Australia Policy,** in which governmental guidelines promoted European and North American immigration at the expense of other groups. This remained national policy until 1973.

Recent migration trends have reversed this historical bias, and more diverse inflows of new workers and residents are adding to the country's multicultural character. Since the 1970s, the government's Migration Program has been dominated by a variety of people chosen on the basis of their educational background and potential for succeeding economically in Australian society. For example, a growing number of families have come from places such as China, India, Malaysia, and the Philippines. Smaller numbers have qualified as migrants through their New Zealand citizenship, while others have arrived as refugees from troubled parts of the world, such as Southeast Asia and the former Yugoslavia. The result is a more diverse foreign-born population. Indeed, 25 percent of Australia's people are now foreign born, reflecting the country's global popularity as a migration destination. In the early 21st century, almost 40 percent of the settlers arriving in the country have been from Asia. Major cities offer particularly attractive possibilities: Sydney's Asian population already exceeds 10 percent and is growing rapidly, while Perth's culture and economy are increasingly linked to its Asian neighbors.

## Cultural Patterns in New Zealand

New Zealand's cultural geography broadly reflects the patterns seen in Australia, although the precise cultural mix differs slightly. Native Maori populations are more numerically important and culturally visible in New Zealand than their Aboriginal counterparts in Australia. While British colonization clearly mandated the dominance of Anglo cultural traditions by the late 19th century, Maori populations survived, although they lost most of their land in the process. After the initial decline, native populations began rebounding in the 20th century, and today the Maori account for more than 8 percent of the country's 4 million residents. Geographically, the Maori remain most numerous on the North Island, including a sizable concentration in metropolitan Auckland. While urban living is on the rise, many Maori, like their Aboriginal counterparts, are also committed to preserving their religion, traditional arts, and Polynesian lifeways (Figure 14.26). In addition, Maori is now an official language in the country (along with English).

While many New Zealanders still identify with their largely British heritage, the country's cultural identity has increasingly separated from its British roots. Several processes have forged New

**FIGURE 14.26 Maori Artisans**  New Zealand's native Maori population actively preserves its cultural traditions and has recently increased its political role in national affairs. These artisans are carving decorations for a traditional Maori canoe.

Zealand's special cultural character. As Britain tightened its own links with the European continent after World War II, New Zealanders increasingly formed a more independent and diverse identity. In many ways, popular culture ties the country ever more closely to Australia, the United States, and continental Europe, a function of increasingly global mass media. A number of major movies, for example, have been filmed in New Zealand, including *The Lord of the Rings, Whale Rider,* and *The Piano.*

## The Mosaic of Pacific Cultures

Native and exotic cultural influences produce a variety of cultures across the islands of the South Pacific. In more isolated places, traditional cultures are largely insulated from outside influences. In most cases, however, modern life in the islands revolves around an intricate cultural and economic interplay of local and Western influences. One thing is certain: The relative cultural insularity of the past is gone forever, and in its place is a Pacific realm rapidly adjusting to powerful forces of colonization, global capitalism, and popular culture.

**Language Geography**  A modern language map reveals some significant cultural patterns that both unite and divide the region (see Figure 14.27). Most of the native languages of Oceania belong to the Austronesian language family, which encompasses wide expanses of the Pacific, much of insular Southeast Asia, and Madagascar. Linguists hypothesize that the first great oceanic mariners spoke Austronesian languages and thus spread them throughout this vast realm of islands and oceans. Within the broad Austronesian family, the Malayo-Polynesian subfamily includes most of the related languages of Micronesia and Polynesia, suggesting a common cultural and migratory history for these widespread peoples.

Melanesia's language geography is more complex and still incompletely understood by outside experts: While coastal peoples often speak languages brought to the region by the seafaring Austronesians, more isolated highland cultures, particularly on the island of New Guinea, speak varied Papuan languages. Indeed, the linguistic complexity of that island is so complex—more than 1,000 languages have been identified—that many experts question whether they even constitute a unified "Papuan family" of related languages. Some scholars estimate that half of New Guinea's languages are spoken by fewer than 500 persons, suggesting that the region's rugged topography plays a strong role in isolating cultural groups. These New Guinea highlands may hold some of the world's few remaining **uncontacted peoples,** cultural groups that have yet to be "discovered" by the Western world.

**Village Life**  Traditional patterns of social life are as complex and varied as the language map. In many cases, however, life revolves around predictable settings. For example, across much of Melanesia, including Papua New Guinea, most people live in small villages, often occupied by a single clan or family group. Many of these traditional villages contain fewer than 500 residents, although some larger communities may house more than 1,000 people. Life often revolves around the gathering and growing of food, annual rituals and festivals, and complex networks of kin-based social interactions.

Traditional Polynesian culture also focuses on village life (Figure 14.28), although there are often strong class-based relationships between local elites, who are often religious leaders, and

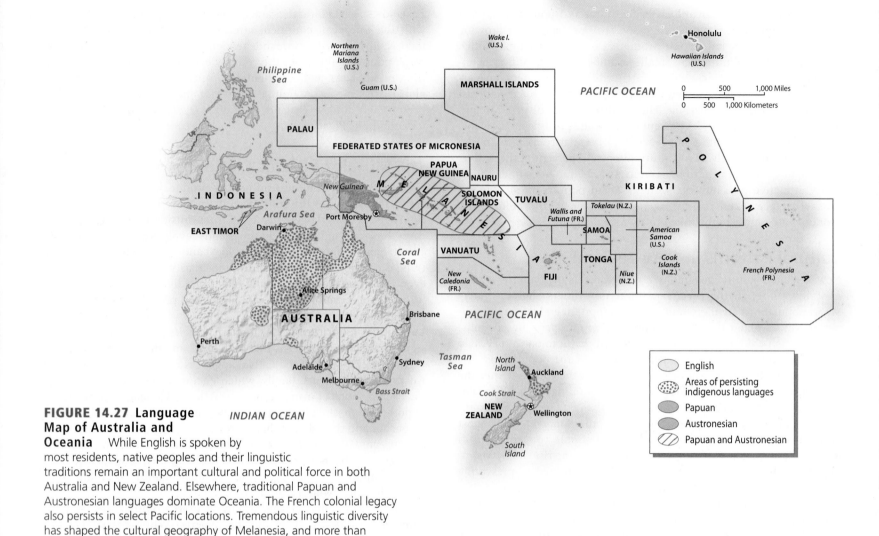

**FIGURE 14.27 Language Map of Australia and Oceania** While English is spoken by most residents, native peoples and their linguistic traditions remain an important cultural and political force in both Australia and New Zealand. Elsewhere, traditional Papuan and Austronesian languages dominate Oceania. The French colonial legacy also persists in select Pacific locations. Tremendous linguistic diversity has shaped the cultural geography of Melanesia, and more than 1,000 languages have been identified in Papua New Guinea.

**Map labels:**

Philippine Sea
Northern Mariana Islands (U.S.)
Wake I. (U.S.)
Honolulu
Hawaiian Islands (U.S.)
PACIFIC OCEAN
Guam (U.S.)
MARSHALL ISLANDS
PALAU
FEDERATED STATES OF MICRONESIA
INDONESIA
New Guinea
PAPUA NEW GUINEA
NAURU
KIRIBATI
MELANESIA
POLYNESIA
SOLOMON ISLANDS
TUVALU
Tokelau (N.Z.)
EAST TIMOR
Darwin
Arafura Sea
Port Moresby
Wallis and Futuna (FR.)
SAMOA
American Samoa (U.S.)
Coral Sea
VANUATU
TONGA
Cook Islands (N.Z.)
French Polynesia (FR.)
New Caledonia (FR.)
FIJI
Niue (N.Z.)
PACIFIC OCEAN
Alice Springs
Brisbane
AUSTRALIA
Perth
Tasman Sea
North Island
Auckland
Adelaide
Sydney
Cook Strait
Melbourne
Bass Strait
NEW ZEALAND
Wellington
INDIAN OCEAN
South Island

**Legend:**
- English
- Areas of persisting indigenous languages
- Papuan
- Austronesian
- Papuan and Austronesian

0 500 1,000 Miles
0 500 1,000 Kilometers

ordinary residents. Polynesian villages are also more likely linked to other islands by wider cultural and political ties. Despite the Western stereotype of depicting Polynesian communities in idyllic and peaceful terms, violent warfare was actually quite common across much of the region prior to European contact.

**External Cultural Influences** While traditional culture persists in some settings, most Pacific islands have witnessed tremendous cultural transformations in the past 150 years. Outsiders from Europe, the United States, and Asia brought new settlers, values, and technological innovations that have forever changed Oceania's

**FIGURE 14.28 Tonga Village** Although the economic and technological effects of globalization have arrived in Tonga, village life remains important in many Polynesian settings. Most village housing in these tropical environments reflects the use of locally available construction materials.

cultural geography and its place in the larger world. The result is a modern setting where Pidgin English has mostly replaced native languages, Hinduism is practiced on remote Pacific Islands, and traditional fishing peoples now work at resort hotels and golf course complexes.

European colonialism transformed the cultural geography of the Pacific world by introducing new political and economic systems. In addition, the region's cultural makeup was changed by new people migrating into the Pacific islands. Hawaii illustrates the pattern. By the mid-19th century, Hawaii's King Kamehameha was already entertaining a varied assortment of whalers, Christian missionaries, traders, and navy officers from Europe and the United States. A small elite group of **haoles**, or light-skinned European and American foreigners, were successfully profiting from commercial sugarcane plantations and Pacific shipping contracts.

Labor shortages on the islands, however, led to the importation of Chinese, Portuguese, and Japanese workers who further complicated the region's cultural geography. By 1900, the Japanese had become a dominant part of the island workforce. The United States formally annexed the islands in 1898. The cultural mix revealed in the Hawaiian census of 1910 suggests the magnitude of change: More than 55 percent of the population was Asian (mostly Japanese and Chinese), native peoples made up another 20 percent, and about 15 percent (mostly imported European workers) were white. By the end of the 20th century, the Asian population was less dominant but more ethnically varied, about 40 percent of Hawaii's residents were white, and the small number of remaining native Hawaiians had been joined by an increasingly diverse group of other Pacific Islanders. In addition, ethnic mixing has produced a rich mosaic of Hawaiian cultures that offer a unique blend of North American, Asian, Pacific Island, and European influences (Figure 14.29).

Hawaii's story has also played out in many other Pacific Island locations. In the Mariana Islands, Guam was absorbed into America's Pacific empire as part of the Spanish-American War in 1898. Thereafter, not only did native peoples feel the effects of Americanization

**FIGURE 14.30 South Asians in Fiji**   British sugar plantation owners imported thousands of South Asian workers to Fiji during the colonial era. Today, almost half of Fiji's population is South Asian and is often in conflict with indigenous Fijians.

(the island remains a self-governing U.S. territory today), but thousands of Filipinos were moved there to supplement its modest labor force. To the southeast, the British-controlled Fiji Islands offered similar opportunities for redefining Oceania's cultural mix. The same sugar plantation economy that spurred changes in Hawaii prompted the British to import thousands of South Asian laborers to Fiji. The descendants of these Indians (most practicing Hinduism) now constitute almost half the island country's population and often come into sharp conflict with the native Fijians (Figure 14.30). In French-controlled portions of the realm, small groups of traders and plantation owners filtered into the Society Islands (Tahiti), but a larger group of French colonial settlers (many originally a part of a penal colony) had a major impact on the cultural makeup of New Caledonia. Still a French colony, New Caledonia's population is more than one-third French, and its capital city of Noumea reveals a cultural setting forged from French and Melanesian traditions.

Given the frequency of contact between different island cultures, it is no surprise that people have generated new forms of intercultural communication. For example, several forms of **Pidgin English** (also known simply as *Pijin*) are found in the Solomons, Vanuatu, and New Guinea, where it is the major language used between ethnic groups. In Pijin, a largely English vocabulary is reworked and blended with Melanesian grammar. Pijin's origin is commonly traced to 19th-century Chinese sandalwood traders ("pijin" is the Chinese pronunciation of the word for "business"). While of historical origin, Pijin is now becoming a globalized language of sorts in Oceania, as trade and political ties develop between different native island groups.

A tidal wave of outside influences since World War II has produced cultural changes as well as growing indigenous responses designed to preserve traditional values. Some groups, particularly in Melanesia, remain more isolated from the outside world, although even there growing demands for natural resources offer an avenue for increasing Western or Asian contacts. In many areas, however, the global growth of tourism has brought Oceania into the relatively easy reach of wealthier Europeans, North Americans, Asians,

**FIGURE 14.29 Multicultural Hawaiians**   Many residents of the Hawaiian Islands represent a blend of Pacific Island, Asian, and European influences. These young women express a mixture of Polynesian and Asian ancestors.

and Australians. The Hawaiian Islands, Fiji, French Polynesia, and American Samoa are being joined by an increasing number of other island tourist destinations. While offering tremendous economic benefits to certain places, the onrush of tourists and their consumer-driven values has often come into sharp conflict with native cultures.

## Geopolitical Framework: A Land of Changing Boundaries

Pacific geopolitics reflect a complex interplay of local, colonial-era, and global-scale forces (Figure 14.31). The complexities become apparent in the story of Micronesia's Marshall Islands. This sprinkling of islands and atolls (covering 70 square miles, or 180 square kilometers, of land) historically consisted of many ethnic groups that made up small political units. In 1914, the Japanese moved into the islands, and the area remained under their control until 1944, when

U.S. troops occupied the region. Following World War II, a United Nations trust territory (administered by the United States) was created across a wide swath of Micronesia, including the Marshall group. Demands for local self-government grew during the 1960s and 1970s, resulting in a new constitution and independence for the Marshall Islanders by the early 1990s. Today, still benefiting from U.S. aid, government officials in the modest capital city on Majuro Atoll struggle to unite island populations, protect large maritime sea claims, and resolve a generation of legal and medical problems that grew from U.S. nuclear bomb testing in the region. Similar stories are typical across the realm, suggesting a 21st-century political geography that is still very much in the making.

## Roads to Independence

The modern political states of the region have arrived at independence along many different pathways, while other political units remain colonial entities to this day. The newness and fluidity of the

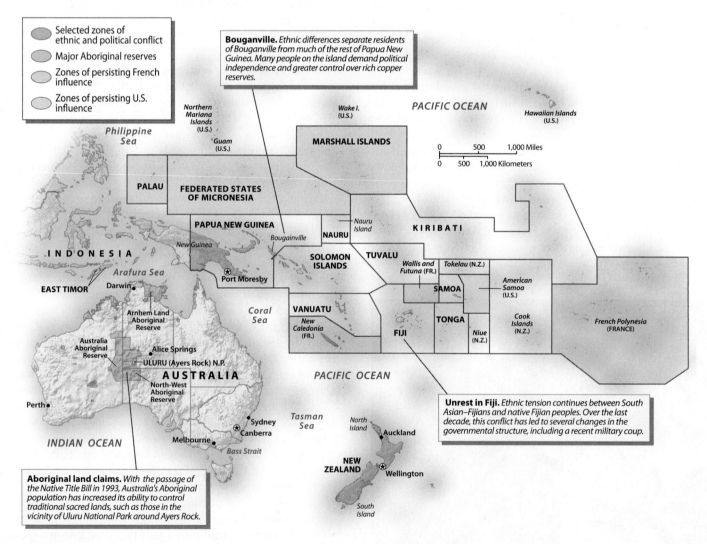

**FIGURE 14.31  Geopolitical Issues in Australia and Oceania**    Native land claim issues increasingly shape domestic politics in Australia and New Zealand. Elsewhere, ethnic conflicts have raised political tensions in settings such as Fiji and Papua New Guinea. Colonialism's impact endures as well: American and French interests remain particularly visible in the region, including a legacy of nuclear testing that continues to affect selected Pacific Island populations.

political boundaries are remarkable: The region's oldest independent states are Australia and New Zealand, and both were 20th-century creations that are only now considering whether they want to complete their formal political separation from the British Crown. Elsewhere, political ties between colony and mother country are even closer and more lasting. Even many of the newly independent Pacific **microstates**, with their tiny overall land areas, keep special political and economic ties to countries such as the United States.

Independent Australia (1901) and New Zealand (1907) gradually created their own political identities, yet both still struggle with the final shape of these identities. Although Australia became a commonwealth in 1901, it still acknowledges the British Crown as the symbolic head of its government. A national referendum in 1999 asked Australians to decide whether they would like their country to drop this remaining tie to Britain and instead become a genuine republic, with its own president replacing the British queen as head of state. However, a slight majority (55 percent) voted to retain Australia's ties to the Crown. Australia, like the United States, is a federal country, with each of its six states having significant powers. The Northern Territory, for unique historical reasons, however, remains directly under the authority of the central government. In New Zealand, formal legislative links with Great Britain were not broken until 1947. Today, New Zealand is discussing the same formal break with the British Crown being debated by the Australians.

Elsewhere in the Pacific, colonial ties were cut even more slowly, and the process has not yet been completed. In the 1970s, Britain and Australia began giving up their colonial empires in the Pacific. Fiji (Great Britain) gained independence in 1970, followed by Papua New Guinea (Australia) in 1975 and the Solomon Islands (Great Britain) in 1978. The small island nations of Kiribati and Tuvalu (Great Britain) also became independent in the late 1970s.

The United States has recently turned over most of its Micronesian territories to local governments, while still holding a large influence in the area. After gaining these islands from Japan in the 1940s, the U.S. government provided large monetary subsidies to islanders and also utilized a number of islands for military purposes. Bikini

Atoll was destroyed by nuclear tests, and the large lagoon of Kwajalein Atoll was used as a giant missile target. A major naval base, moreover, was established in Palau, the westernmost archipelago of Oceania. By the early 1990s, both the Marshall Islands and the Federated States of Micronesia (including the Caroline Islands) had gained independence. Their ties to the United States, however, remain close. A number of other Pacific islands remain under U.S. administration. Palau is a U.S. "trust territory," which gives Palauans some local autonomy. The people of the Northern Marianas chose to become a "self-governing commonwealth in association with the United States," a rather vague political position that allows them to become U.S. citizens. The residents of self-governing Guam and American Samoa are also U.S. citizens. Hawaii became a full-fledged U.S. state in 1959 and is now an integral part of the country.

Other colonial powers were less inclined to give up their oceanic possessions. New Zealand still controls substantial territories in Polynesia, including the Cook Islands, Tokelau, and the island of Niue. France has even more extensive holdings in the region. Its largest maritime possession is French Polynesia, which includes a large expanse of mid-Pacific territory. To the west, France still controls the much smaller territory of Wallis and Futuna in Polynesia and the larger island of New Caledonia in Melanesia.

## Persisting Geopolitical Tensions

Cultural diversity, colonial legacy, youthful states, and a rapidly changing political map contribute to ongoing geopolitical tensions in the Pacific world. Indeed, some of these conflicts have consequences that extend far beyond the boundaries of the region. Others are more locally based but are still reminders of the difficulties that occur as political space is redefined across varied natural and cultural settings.

**Native Rights in Australia and New Zealand**   Indigenous peoples in both Australia and New Zealand have used the political process to gain more control over land and resources in their two countries. Indeed, the strategies these native groups have used parallel efforts in the Americas and elsewhere. In Australia, Aboriginal groups are discovering newfound political power from both more effective lobbying efforts by native groups and a more sympathetic federal government. Because land treaties were generally not signed with Aborigines as whites conquered the continent, native peoples originally had no legal land rights whatsoever.

More recently, the Australian government established a number of Aboriginal reserves, particularly in the Northern Territory, and expanded Aboriginal control over sacred national parklands such as Uluru (Ayers Rock) (Figure 14.32). Further concessions to

**FIGURE 14.32 Aboriginals at Uluru National Park**   Australian Aborigines gathered recently at Uluru National Park to celebrate their increased political control over the region. Since then, the Native Title Bill has promoted numerous land cessions and further legal settlements.

indigenous groups were made in 1993, as the government passed the **Native Title Bill,** which compensated Aborigines for lands already given up, gave them the right to gain title to unclaimed lands they still occupied, and provided them with legal standing to deal with mining companies in native-settled areas.

However, efforts to expand Aboriginal land rights have met strong opposition. In 1996, an Australian court ruled that pastoral leases (the form of land tenure held by the cattle and sheep ranchers who control most of the Outback) do not necessarily negate or replace Aboriginal land rights. Grazing interests were infuriated, which led the government to respond that Aboriginal claims allow the visiting of sacred sites and some hunting and gathering but do not give native peoples complete economic control over the land (Figure 14.33).

In New Zealand, Maori land claims have generated similar controversies in recent years. The Maori constitute a far larger proportion of the overall population, and the lands they claim tend to be much more valuable, further complicating the issue. Recent protests include civil disobedience and demonstrations, growing Maori land claims over much of North and South islands, and a call to return the country's name to the indigenous **Aotearoa,** "Land of the Long White Cloud." The government response, complete with a 1995 visit from Queen Elizabeth, has been to acknowledge increased Maori land and fishing rights as well as to propose a series of financial and land settlements that have yet to be agreed to by the Maoris.

**Conflicts in Oceania** Geopolitical issues simmer elsewhere in the Pacific, periodically threatening to further redefine the region's territorial boundaries. As mentioned earlier, ethnic differences in Fiji have threatened to tear apart that small island nation.

Papua New Guinea also contends with ethnic tensions. The country is composed of different cultural groups, many of which have a long history of mutual hostility. Most of these peoples now get along with each other reasonably well, although tribal fighting occasionally breaks out in highland market towns. A much bigger problem for the national government has been the rebellion on Bougainville. This sizable island, which has large reserves of copper and other minerals, is located in the Solomon archipelago but belongs to Papua New Guinea because Germany colonized it in the late 1800s, and it thus became politically attached to the eastern portion of New Guinea. Many of Bougainville's native residents believe that their resources are being exploited by foreign interests and by an unsympathetic national government, and they demand local control. Papua New Guinea has reacted with military force; about 5 percent of the island's entire population has already been killed in the conflict, and recent efforts have failed to create a more stable regional government.

The continued French colonial presence in the Pacific region has also created political uncertainties, both in relations with native peoples and between the French and other independent states in the area. Continued French rule in New Caledonia has provoked much local opposition. This large island has large mineral reserves (especially of nickel) and sizable numbers of French colonists. French settlement and exploitation of mineral resources angered many native peoples. By the 1980s, a local independence movement was gaining strength, but in 1987 and 1998, the island's residents (indigenous and French immigrants alike) voted to remain under French rule, at least until 2018 (Figure 14.34).

Still, an underground independence movement continues to operate. Troubles have also arisen in French Polynesia. Although the region receives large subsidies from the French, and residents have voted to remain a colony, a large minority of the population opposes French control and demands independence.

## A Regional and Global Identity?

Australia and New Zealand have emerged to play key political roles in the South Pacific. Although these two countries sometimes disagree on strategic and military matters, their size, wealth, and collective political influence in the region make them important forces for political stability. Special colonial relationships still connect these nations with present and former Pacific holdings. Australia maintains close political ties to its former colony of Papua New Guinea, and New Zealand's continuing control over Niue, Tokelau, and the Cook Islands in Polynesia suggests that its political influence extends well beyond its borders. When political and ethnic conflicts arise elsewhere in Oceania, Australia and New Zealand are often involved in negotiating peace settlements. Recently, for example, both nations assisted in mediating ongoing disputes on Papua New Guinea's island of Bougainville. In general, the two countries enjoy close political and strategic relations and participate in joint military efforts in the region. Given the other global interests in the region, however, it remains unclear whether these nations can or wish to assert their political dominance across the entire South Pacific.

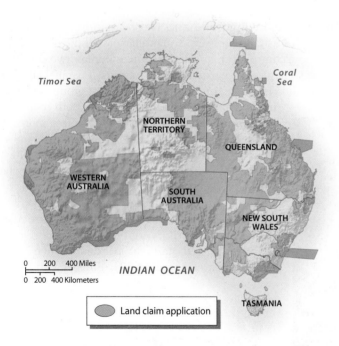

**FIGURE 14.33 Applications for Native Land Claims in Australia** This map shows the applications for native land claims in Australia filed by different Aboriginal groups as of 2004. Important to note is that these are applications only—not government-approved claims. Nevertheless, the widespread extent of the claims shows why the topic is so contentious and controversial.

**FIGURE 14.34 Unrest in New Caledonia** The indigenous Melanesians of New Caledonia, known as *kanaks*, recently voted overwhelmingly in favor of ending French colonial power and creating an independent government. France, however, is resisting this change. This photo is of protesters waving the flag of Kanaky independence.

## Economic and Social Development: A Difficult Path to Paradise

As with all other world regions, the Pacific realm contains a diversity of economic situations, resulting in both wealth and poverty. Even within affluent Australia and New Zealand, for example, there are pockets of pronounced poverty. Further, great economic disparities also exist between countries with numerous trade ties and small island nations lacking resources and external trade. While tourism offers some relief from abject poverty, the whims and fashions of foreign tourists can be fickle, resulting in a Pacific version of boom and bust economies. As a result, the economic future of the Pacific realm remains uncertain because of its own small domestic markets, its peripheral position in the global economy, and a diminishing resource base.

## The Australian and New Zealand Economies

Much of Australia's past economic wealth has been built on the cheap extraction and export of abundant raw materials. Export-oriented agriculture, for example, has long been one of the key supports of Australia's economy. Australian agriculture is highly productive in terms of labor input, and it produces a wide variety of both temperate and tropical crops, as well as huge quantities of beef and wool for world markets. While farm exports are still important to the economy, the mining sector has grown rapidly since 1970.

Today, Australia is one of the world's mining superpowers. The years since the 1850s gold rush in Victoria have seen a huge expansion in the nation's mineral output. This pattern has accelerated recently due to increased trade with China, an activity that has made Australia the world's largest exporter of iron and coal. Among Australia's many mineral resources are coal and iron ore, particularly in Western Australia, and an assortment of other metals, such as bauxite (for aluminum), copper, gold, nickel, lead, and zinc. Indeed, the New South Wales–based Broken Hill Proprietary Company (BHP) is one of the world's largest mining corporations.

Growing numbers of Asian immigrants and economic links with potential Asian markets also offer promise for the future. In addition, an expanding tourism industry is helping to diversify the economy. More than 7 percent of the nation's workforce is now devoted to serving the needs of more than 4 million visitors annually. Popular destinations include Melbourne and Sydney, as well as recreational settings such as Queensland's resort-filled Gold Coast, the Great Barrier Reef, and the vast, arid Outback. Along the Gold Coast, most luxury hotels are owned by Japanese firms and provide a bilingual resort experience for their Asian clientele (Figure 14.35).

New Zealand is also a wealthy country, but it is somewhat less well off than Australia. Before the 1970s, New Zealand relied heavily on exports to Great Britain, especially exports of agricultural products such as wool and butter. Problems with this strategy occurred, however, in 1973, when Britain joined the European Union, which had strict agricultural protection policies. Unlike Australia, New Zealand lacked a rich base of mineral resources to export to global markets. By the 1980s, the country had slipped into a serious recession. Eventually, the New Zealand government enacted drastic reforms. The country had previously been noted for its lofty taxes, high levels of social welfare, and state ownership of large companies. Suddenly, *privatization* (a change from state to private ownership) became the watchword, and most state industries were sold off to private parties. As a result, New Zealand has been transformed into one of the most market-oriented countries of the world. To diversify from its traditional export base, the nation has also encouraged more aggressive development of its timber resources, fisheries, and tourist industry.

## Oceania's Economic Diversity

Varied economic activities shape the Pacific Island nations. One way of life is oriented around subsistence-based economies, such as shifting cultivation or fishing. In other settings, the commercial extractive economy dominates, with large-scale plantations, mines, and timber activities often competing for land and labor with the traditional subsistence

**FIGURE 14.35 Queensland's Gold Coast**    Many of these luxury hotels in the Surfer's Paradise section of the Gold Coast are owned by Japanese firms specializing in accommodations for Asian tourists.

sector. Elsewhere, the huge growth in global tourism has transformed the economic geographies of many island settings, forever changing the way that people make a living. In addition, many island nations benefit from direct subsidies and economic assistance that come from present and former colonial powers, designed to promote development and stimulate employment.

Melanesia is the least-developed and poorest part of Oceania because these countries have benefited less from tourism and from subsidies from wealthy colonial and ex-colonial powers. Most Melanesians live in remote villages that remain somewhat isolated from the modern economy. The Solomon Islands, for example, with few industries other than fish canning and coconut processing, has a per capita gross national income (GNI) of only $1,710 per year. Similarly, Papua New Guinea's economy produces a per capita GNI of $1,870. Although traditional exports such as coconut products and coffee have increasingly been supplemented with the rapid development of tropical hardwoods, the economic returns remain low. Further, gold and copper mining have dramatically transformed the landscape, although political instability has often interfered with mineral production in settings such as Bougainville. In New Guinea's interior highlands, much of village life is focused on subsistence activities. In contrast, Samoa is the most prosperous Melanesian country, with a per capita GNI of $4,350, largely because of its tourist economy popular with North Americans and Japanese.

**The Economic Impact of Mining**    Among the smaller islands of Melanesia and Micronesia, mining economies dominate New Caledonia and Nauru. New Caledonia's nickel reserves, the world's second largest, are both a blessing and a curse While they currently sustain much of the island's export economy, income from nickel mining will lessen in the near future as the reserves dwindle. Dramatic price fluctuations for the industrial economy also hamper economic planning for the French colony. Other activities include coffee growing, cattle grazing, and tourism. To the north, the tiny, phosphate-rich island of Nauru also depended on mining; however, that day is now the past, as the deposits are exhausted and the future is uncertain (Figure 14.36; see also "Exploring Global Connections: Nauru and the Mixed Benefits of Globalization).

**Micronesian and Polynesian Economies**    Throughout Micronesia and Polynesia, economic conditions depend on both local subsistence economies or economic linkages to the wider world beyond. Many archipelagos export a few food products, but native populations survive mainly on fish, coconuts, bananas, and yams. Some island groups, though, enjoy large subsidies from either France or the United States, although such support often comes with a political price. Change is also occurring in some of these island settings: In 1999, Japan agreed to build a spaceport for its future shuttlecraft on Micronesia's Christmas Island (Kiribati), and the Marshall Islands are the site of a planned industrial park financed by mainland Chinese. Palau, however, is the Micronesian country

**FIGURE 14.36 The Globalization of Nauru**    A Nauruan local points out the scarred landscape left from decades of intensive phosphate mining by Australian mining companies. With no mineral riches left, the islanders have tried several different strategies to find a place in the globalized economy.

## TABLE 14.2    DEVELOPMENT INDICATORS

| Country | GNI per capita, PPP 2007 | GDP Average Annual %Growth 2000–07 | Human Development Index (2006)# | Percent Population Living Below $2 a Day | Life Expectancy* 2009 | Under Age 5 Mortality Rate 1990 | Under Age 5 Mortality Rate 2007 | Gender Equity 2007 |
|---|---|---|---|---|---|---|---|---|
| Australia | 33,400 | 3.2 | 0.965 | | 81 | 10 | 6 | 97 |
| Fed. States of Micronesia | 3,010 | | | | 68 | 58 | 40 | 102 |
| Fiji | 4,240 | | 0.743 | | 68 | 22 | 18 | 104 |
| French Polynesia | | | | | 74 | | | |
| Guam | | | | | 78 | | | |
| Kiribati | 2,040 | | | | 61 | 88 | 63 | 107 |
| Marshall Islands | | | | | 66 | 92 | 54 | 100 |
| Nauru | | | | | 56 | | | |
| New Caledonia | | | | | 76 | | | |
| New Zealand | 25,380 | 3.4 | 0.944 | | 80 | 11 | 6 | 103 |
| Palau | | | | | 69 | 21 | 10 | 102 |
| Papua New Guinea | 1,870 | 2.3 | 0.516 | 57.4 | 59 | 94 | 65 | |
| Samoa | 4,350 | | 0.760 | | 73 | 50 | 27 | 105 |
| Solomon Islands | 1,710 | | 0.591 | | 62 | 121 | 70 | 93 |
| Tonga | 3,880 | | 0.774 | | 71 | 32 | 23 | 99 |
| Tuvalu | | | | | 64 | | | |
| Vanuatu | 3,410 | | 0.686 | | 67 | 62 | 34 | |

*Source:* World Bank, *World Development Indicators 2009.*

United Nations, *Human Development Index, 2008.* #

Population Reference Bureau, *World Data Sheet, 2009**

Gender equity = Ratio of female to male enrollments in primary and secondary school. Numbers below 100 have more males in primary/secondary school, numbers above 100 have more females in primary/secondary schools.

with the highest GNI per capita ($8,000), a result of tourism, fishing, and—not unimportant—government assistance from the United States.

Other island groups have been completely transformed by tourism. In Hawaii, more than one-third of the state's economy flows directly from tourist dollars. With almost 7 million visitors annually (including more than 1.5 million from Japan), Hawaii represents all the classic benefits and risks of the tourist economy. While job creation and economic growth have reshaped the island realm, congested highways, high prices, and the unpredictable spending habits of tourists have put the region at risk for future problems. Elsewhere, French Polynesia has long been a favored destination of the international jet set. More than 20 percent of French Polynesia's GNI is derived from tourism, making it one of the wealthiest areas of the Pacific (Figure 14.37). More recently, Guam has emerged as a favorite destination of Japanese and Korean tourists, especially those on honeymoons. Indeed, on a smaller scale, tourism is on the rise across much of the island realm, and many economic planners see it as the avenue to future prosperity. Critics, however, warn that tourist jobs tend to be low paying, the local quality of life may in fact decline with the presence of tourists, and the natural environments of small islands can quickly be overwhelmed by a high number of demanding visitors.

**FIGURE 14.37    Tahitian Resort**    Luxury resort settings in Tahiti (near Papeete) exemplify a growth industry in the South Pacific. While bringing important investment capital, such ventures reorder the region's social and economic structure, as well as refashion its cultural landscape.

## The Global Economic Setting

Many international trade flows link the area to the far reaches of the Pacific and beyond. Australia and New Zealand dominate global trade patterns in the region. In the past 30 years, ties to Great Britain, the British Commonwealth, and Europe have weakened in comparison with growing trade links to Japan, East Asia, the Middle East, and the United States Australia, for example, now imports more manufactured goods from China, Japan, and the United States than it does from Britain and Europe (Figure 14.38). Other global economic ties have come in the form of capital investment in the region. U.S. and Japanese banks and other financial institutions now dot the South Pacific landscape from Sydney to Suva. Both Australia and New Zealand also participate in the **Asia-Pacific Economic Cooperation Group (APEC)**, an organization designed to encourage economic development in Southeast Asia and the Pacific Basin. The region's economic ties to Asia also carry risks, however; the Asian downturn in the late 1990s, for example, slowed the popular Korean tourist trade in New Zealand and lowered Asian demand for a variety of Australian raw material exports.

Governments have promoted economic integration within the region. In 1982, Australia and New Zealand signed the **Closer Economic Relations (CER) Agreement**, which successfully slashed trade barriers between the two countries. New Zealand benefited from the opening of larger Australian markets to New Zealand exports, and Australian corporate and financial interests gained new access to New Zealand business opportunities. Since the CER Agreement's signing, trade between the two countries has expanded almost 10 percent per year. Today, more than 20 percent of New Zealand's imports and exports come from Australia, and the pattern of regional free trade is likely to strengthen in the future. Smaller nations of Oceania, while often closely tied to countries such as Japan, the United States, and France, also benefit from their proximity to Australia and New Zealand. More than half of Fiji's imports come from those two nearby nations, and other countries, such as Papua New Guinea, Vanuatu, and the Solomon Islands, enjoy a similarly close trading relationship with their more developed Pacific neighbors.

## Continuing Social Challenges

Australians and New Zealanders enjoy high levels of social welfare but face some of the same challenges evident elsewhere in the developed world. Life spans average about 80 years in both countries, and rates of child mortality have fallen greatly since 1960. Paralleling patterns in North America and Europe, cancer and heart disease are leading causes of death, and alcoholism is a continuing social problem, particularly in Australia. Unfortunately, Australia's rate of skin cancer is among the world's highest, the result of having a largely fair-skinned, outdoors-oriented population from northwest Europe in a sunny, low-latitude setting. Overall, Australia's Medicare program (initiated in 1984) and New Zealand's system of social services provide high-quality health care to their populations. The position of

**FIGURE 14.38 Australia's Trade with China** Containers from Asia testify to the recent explosion of two-way trade between Australia and China. While raw materials, mainly iron ore, are exported to China, consumer goods flow from China into Australia, making that country a key beneficiary of China's recent economic growth.

women is also high in both countries, including participation in the workforce. Women have recently played key political roles in New Zealand, in particular.

Not surprisingly, the social conditions of the Aborigines and Maoris are much less favorable than those of the population overall. Schooling is irregular for many native peoples, and levels of postsecondary education for Aborigines (12 percent) and Maoris (14 percent) remain far below the national averages (32 to 34 percent). Many other social measures reflect the pattern, as well. For example, fewer than one-third of Aboriginal households own their own homes, while more than 70 percent of white Australian households homeowners. Furthermore, considerable discrimination against native peoples continues in both countries, a situation that has been aggravated and publicized with the recent assertion of indigenous political rights and land claims. As with North American African-American, Hispanic, and Native American populations, simple social policies do not yet exist as solutions to these lasting problems.

Levels of social welfare in Oceania are higher than one might expect, based on the region's economic situation. Many of its countries and colonies have invested heavily in health and education services and have achieved considerable success (Figure 14.39). For example, the average life expectancy in the Solomon Islands, one of the world's poorer countries as measured by per capita GNI figures, is a respectable 62 years. By other social measures as well, the Solomon Islands and a number of other Oceania states have reached higher levels of human well-being than exist in most Asian and African countries with similar levels of economic output. This is partly a result of successful policies, but it also reflects the relatively healthy natural environment of Oceania. Many of the tropical diseases that are so troublesome in Africa simply do not exist in the region.

Papua New Guinea is the major exception to the relatively high levels of social welfare in Oceania. Here the average life expectancy is only 59 years, and a recent study suggests that 34 percent of its young people suffer from malnutrition, particularly protein deficiencies. Papua New Guinea has found it difficult to provide even basic educational and health services to its people for two key reasons: The country possesses the largest expanse of land in Melanesia, and much of the population lives in relatively isolated villages in the rugged central highlands.

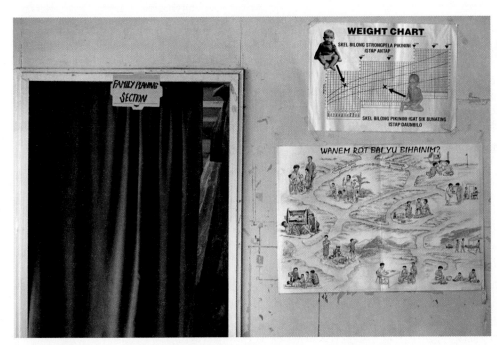

**FIGURE 14.39 Health Services in Oceania** A young boy peers from behind a curtain in the family planning section of a medical clinic in Port Moresby, the capital city of Papua New Guinea, a country with one of the highest infant mortality rates and the shortest life expectancies in Oceania.

# SUMMARY

- Globalization has brought fresh interconnections to Oceania, as Japanese-financed golf courses have popped up along tropical shores, Canadian and Chinese mining companies have invested in Australia iron ore and coal, and Korean newlyweds have honeymooned beneath coconut palms on tropical islands. How native cultures will respond and what new cultural hybrids will emerge as Pacific, European, North American, and Asian peoples mingle and interact are uncertain. Nonetheless, whatever the expression, these issues will give Oceania a new globalized face.

- The natural environment, which has been transformed over millennia by indigenous and colonial settlement, has witnessed accelerating change in the past 50 years, as urbanization, tourism, extractive economic activity, exotic species, and climate change from global warming have reconfigured the landscape and increased the vulnerability of island environments.

- The region's contemporary political geography reveals a fluid and changing character, as countries struggle to disentangle themselves from colonial ties by asserting their own political identities. Globalization complicates the process, with new economic linkages and alliances replacing historic colonial ties.

- If Australia sits at the edge of Asia, then New Zealand is on the edge of Polynesia, with its Maori population growing rapidly, thanks to immigrants from other parts of Polynesia. Further, New Zealand is taking an active role in the affairs of the entire Pacific Basin, thus becoming in many ways Oceania's leading state, economically and politically, as Australia becomes increasingly oriented to Asia.

- The global economic recession of 2008–09 has taken a heavy toll on the Pacific world, with Australia suffering particularly from China's reduced demand for coal and iron ore and tourist destinations such as Tahiti and Hawaii suffering from the dramatic downturn in global travel. Australian cities, as well, have seen a drop in Asian tourists.

# KEY TERMS

Aborigine (p. 416)
Aotearoa (p. 426)
archipelago (p. 407)
Asia-Pacific Economic
  Cooperation Group (APEC)
  (p. 429)
atoll (p.413)

Closer Economic Relations
  (CER) Agreement (p. 430)
haoles (p. 423)
high island (p. 413)
hot spot (p. 413)
kanakas (p. 420)
mallee (p. 412)

Maori (p. 407)
Melanesia (p. 407)
Micronesia (p. 407)
microstate (p. 425)
Native Title Bill (p. 426)
Oceania (p. 404)
Outback (p. 406)

Pidgin English (Pijin) (p. 423)
Polynesia (p. 407)
tsunami (p. 412)
uncontacted people (p. 421)
viticulture (p. 418)
White Australia Policy (p. 420)

# THINKING GEOGRAPHICALLY

1. Identify what you see as the three principal economic challenges facing Papua New Guinea. Discuss how the country's physical and human geography may affect its future prospects.

2. Should New Zealand welcome or be wary of increased economic and political ties with Australia? Include in your response perspectives from different segments of New Zealand's population, such as sheep ranchers, Maoris, dairy farmers, fishermen, and middle-class urbanites.

3. As an Australian Aborigine in touch with your North American counterparts, what are some of the similarities and differences in the plight of your people?

4. As a maker of educational films, you have been hired to film a three-part series titled *The Essential South Pacific*. Episodes are to be set in (a) Australia, (b) New Zealand, and (c) Polynesia. Because the budget is limited, you can film in only a single location in each region. Identify three specific locations (one in each region) and then describe what you would film to capture the essence of those countries.

5. You are part of a team responsible for a comprehensive economic and social development plan for a small South Pacific island. Choose a specific island, become conversant with its needs and resources, define your goals, and describe a plan for achieving them in three or four pages, complete with multimedia support.

Log in to **www.mygeoscienceplace.com** for videos, interactive maps, RSS feeds, case studies, and self-study quizzes to enhance your study of Australia and Oceania.

# GLOSSARY

**Aborigine**  An indigenous inhabitant of Australia.

**acid rain**  A harmful form of precipitation high in sulfur and nitrogen oxides. Caused by industrial and auto emissions, acid rain damages aquatic and forest ecosystems in regions such as eastern North America and Europe.

**African diaspora**  The forced removal of Africans from their native area.

**African Union (AU)**  A mostly political body that has tried to resolve regional conflicts. Founded in 1963, the organization grew to include all the states of the continent except South Africa, which finally was asked to join in 1994. In 2004, the body changed its name to the African Union.

**agrarian reform**  A popular but controversial strategy to redistribute land to peasant farmers. Throughout the 20th century, various states redistributed land from large estates or granted title from vast public lands in order to reallocate resources to the poor and stimulate development. Agrarian reform occurred in various forms, from awarding individual plots or communally held land to creating state-run collective farms.

**alluvial fan**  A fan-shaped deposit of sediments dropped by a river or stream flowing out of a mountain range.

**Altiplano**  The largest intermontane plateau in the Andes, which straddles Peru and Bolivia and ranges in elevation from 10,000 to 13,000 feet (3,000 to 4,000 meters).

**animism**  A wide variety of tribal religions based on the worship of nature's spirits and human ancestors.

**anthropogenic**  An adjective for human-caused change to a natural system, such as the atmospheric emissions from cars, industry, and agriculture that are causing global warming.

**anthropogenic landscape**  A landscape heavily transformed by human agency.

**Aotearoa**  Maori name for New Zealand, meaning "Land of the Long White Cloud."

**apartheid**  The policy of racial separateness that directed the separate residential and work spaces for white, blacks, coloureds, and Indians in South Africa for nearly 50 years. It was abolished when the African National Congress came to power in 1994.

**archipelago**  An island group. Many archipelagos are oriented in an elongated pattern.

**Asia-Pacific Economic Cooperation Group (APEC)**  An international group of Asian and Pacific Basin nations that fosters coordinated economic development within the region.

**Association of Southeast Asian Nations (ASEAN)**  A supranational geopolitical group linking together the 10 different states of Southeast Asia.

**asymmetrical warfare**  Military action between a superpower using strategies dependent on high-technology weapons and the low technology and guerilla tactics used by small insurgent groups.

**atoll**  A low, sandy island made from coral. Atolls are often oriented around a central lagoon.

**autonomous areas**  Minor political subunits created in the former Soviet Union and designed to recognize the special status of minority groups within existing republics.

**autonomous region**  In the context of China, provinces that have been granted a certain degree of political and cultural autonomy, or freedom from centralized authority, due to the fact that they contain large numbers of non-Han Chinese people. Critics contend that they have little true autonomy.

**Baikal-Amur Mainline (BAM) Railroad**  A key central Siberian railroad connection completed in the Soviet era (1984), which links the Yenisey and Amur rivers and parallels the Trans-Siberian Railroad.

**Berlin Conference**  A 1884 conference that divided Africa into European colonial territories. The boundaries created in Berlin satisfied European ambition but ignored indigenous cultural affiliations. Many of Africa's civil conflicts can be traced to ill-conceived territorial divisions crafted in 1884.

**biofuels**  Energy sources derived from plants or animals. Throughout the developing world, wood, charcoal, and dung are primary energy sources for cooking and heating.

**biome**  Ecologically interactive flora and fauna adapted to a specific environment. Examples are deserts and tropical rainforests.

**bioregion**  A spatial unit or region of local plants and animals adapted to a specific environment, such as a tropical savanna.

**Bolsheviks**  A faction within the Russian Communist movement led by Lenin that successfully took control of the country in 1917.

**boreal forest**  A coniferous forest found in a high-latitude or mountainous environment in the Northern Hemisphere.

**brain drain**  Migration of the best-educated people from developing countries to developed nations where economic opportunities are greater.

**brain gain**  The potential of return migrants to contribute to the social and economic development of a home country with the experiences they have gained abroad.

**British East India Company**  A private trade organization that acted as an arm of colonial Britain in ruling most of South Asia until 1857, when it was abolished and replaced by full governmental control.

**bubble economy**  A highly inflated economy that cannot be sustained. Bubble economies usually result from rapid influx of international capital into a developing country.

**buffer zone**  An array of nonaligned or friendly states that "buffer" a larger country from invasion. In Europe, keeping a buffer zone has been a long-term policy of Russia (and also of the former Soviet Union) to protect its western borders from European invasion.

**Bumiputra**  The name given to native Malay (literally, "sons of the soil"), who are given preference for jobs and schooling by the Malaysian government.

**bustees**  Settlements of temporary and often illegal housing in Indian cities, caused by rapid urban migration of poorer rural people and the inability of the cities to provide housing for this rapidly expanding population.

**capital leakage**   The gap between the gross receipts an industry (such as tourism) brings into a developing area and the amount of capital retained.

**Caribbean Community and Common Market (CARICOM)**   A regional trade organization established in 1972 that includes former English colonies as its members.

**Caribbean diaspora**   The economic flight of Caribbean peoples across the globe.

**caste system**   The complex division of South Asian society into different hierarchically ranked hereditary groups. The caste system is most explicit in Hindu society but is also found in other cultures to a lesser degree.

**Central American Free Trade Association (CAFTA)**   A trade agreement between the United States and Guatemala, El Salvador, Nicaragua, Honduras, Costa Rica, and the Dominican Republic to reduce tariffs and increase trade between member countries.

**centralized economic planning**   An economic system in which the state sets production targets and controls the means of production.

**centrifugal forces**   Cultural and political forces—such as linguistic minorities, separatists, and fringe groups—that pull away from and weaken an existing nation-state.

**centripetal forces**   Cultural and political forces—such as a shared sense of history, a centralized economic structure, and the need for military security—that promote political unity in a nation-state.

**chernozem soils**   A Russian term for dark, fertile soil, often associated with grassland settings in southern Russia and Ukraine.

**China proper**   The eastern half of the country of China, where the Han Chinese form the dominant ethnic group. The vast majority of China's population is located in China proper.

**clan**   A social unit that is typically smaller than a tribe or an ethnic group but larger than a family, based on supposed descent from a common ancestor.

**climate region**   A region of similar climatic conditions. An example is the marine west coast climate regions found on the west coasts of North America and Europe.

**climograph**   A graph of average annual temperature and precipitation data by month and season.

**Closer Economic Relationship (CER) Agreement**   An agreement signed in 1982 between Australia and New Zealand, designed to eliminate all economic and trade barriers between the two countries.

**Cold War**   An ideological struggle between the United States and the Soviet Union that was conducted between 1946 and 1991.

**Collective Security Treaty Organization (CSTO)**   A Russian-led military association that includes Belarus, Armenia, Kazakhstan, Kyrgyzstan, Tajikistan, and Uzbekistan. The CSTO and SCO work together to address military threats, crime, and drug smuggling.

**colonialism**   Formal, established (mainly historical) rule over local peoples by a larger imperialist government for the expansion of political and economic empire.

**coloured**   A racial category used throughout South Africa to define people of mixed European and African ancestry.

**command economy**   A centrally planned and controlled economy, generally associated with socialist or communist countries, in which all goods and services, along with agricultural and industrial products, are strictly regulated. This form of economy was used during the Soviet era in both the Soviet Union and its eastern European satellites.

**Commonwealth of Independent States (CIS)**   A loose political union of former Soviet republics (without the Baltic states) established in 1992 after the dissolution of the Soviet Union.

**concentric zone model**   A simplified description of urban land use in which a well-defined central business district (CBD) is surrounded by concentric zones of residential activity, with higher-income groups living on the urban periphery.

**Confucianism**   A philosophical system based on the ideas of Confucius, a Chinese philosopher who lived in the 6th century BCE. Confucianism stresses education and the importance of respecting authority figures, as well as the importance of authority figures acting in a responsible manner. Confucianism is historically significant throughout East Asia.

**connectivity**   The degree to which different locations are linked with one another through transportation and communication infrastructure.

**continental climate**   A climate region in a continental interior, removed from moderating oceanic influences, characterized by hot summers and cold winters. In such a climate, at least one month must average below freezing.

**core–periphery model**   A conceptualization of the world into two economic spheres. The developed countries of western Europe, North America, and Japan form the dominant core, with less-developed countries making up the periphery. Implicit in this model is that the core gained its wealth at the expense of peripheral countries.

**Cossacks**   Highly mobile Slavic-speaking Christians of the southern Russian steppe who were pivotal in expanding Russian influence in 16th- and 17th-century Siberia.

**counterurbanization**   The movement of people out of metropolitan areas toward smaller towns and rural areas.

**creolization**   The blending of African, European, and some Amerindian cultural elements into the unique sociocultural systems found in the Caribbean.

**crony capitalism**   A system in which close friends of a political leader are either legally or illegally given business advantages in return for their political support.

**cultural assimilation**   A process in which immigrants are culturally absorbed into the larger host society.

**cultural imperialism**   The active promotion of one cultural system over another, such as the implantation of a new language, school system, or bureaucracy. Historically, cultural imperialism has been primarily associated with European colonialism.

**cultural nationalism**   A process of protecting, either formally (with laws) or informally (with social values), the primacy of a certain cultural system against influences (real or imagined) from another culture.

**cultural syncretism or hybridization**   The blending of two or more cultures, which produces a synergistic third culture that exhibits traits from all cultural parents. Also called *cultural hybridization*.

**culture**   Learned and shared behavior by a group of people, empowering them with a distinct "way of life"; culture includes both material (technology, tools, etc.) and immaterial (speech, religion, values, etc.) components.

**culture hearth**   An area of historical cultural innovation.

**Cyrillic alphabet**   An alphabet based on the Greek alphabet and used by Slavic languages heavily influenced by the Eastern Orthodox Church. It is attributed to the missionary work of St. Cyril in the 9th century.

**Dalit**   The currently preferred term used to denote the members of India's most discriminated against ("lowest") caste groups, those people previously referred to as "untouchables."

**decolonialization** The process of a former colony's gaining (or regaining) independence over its territory and establishing (or reestablishing) an independent government.

**demographic transition model** A four-stage model of population change derived from the historical decline of the natural rate of increase as a population becomes increasingly urbanized through industrialization and economic development.

**denuclearization** The process whereby nuclear weapons are removed from an area and dismantled or taken elsewhere.

**dependency theory** A popular theory to explain patterns of economic development in Latin America. Its central premise is that underdevelopment was created by the expansion of European capitalism into the region that served to develop "core" countries in Europe and to impoverish and make dependent peripheral areas such as Latin America.

**desertification** The spread of desert conditions into semiarid areas due to improper management of the land.

**diaspora** The scattering of a particular group of people over a vast geographic area. Originally, the term referred to the migration of Jews out of their homeland, but now it has been generalized to refer to any ethnic dispersion.

**dollarization** An economic strategy in which a country adopts the U.S. dollar as its official currency. A country can be partially dollarized, using U.S. dollars alongside its national currency, or fully dollarized, in which case the U.S. dollar becomes the only medium of exchange and the country gives up its own national currency. Panama fully dollarized in 1904; more recently, Ecuador fully dollarized in 2000.

**domestication** The purposeful selection and breeding of wild plants and animals for cultural purposes.

**domino theory** A U.S. geopolitical policy of the 1970s that stemmed from the assumption that if Vietnam fell to the Communists, the rest of Southeast Asia would soon follow.

**Dravidian language** A strictly South Asian language family that includes such important languages as Tamil and Telugu. Once spoken through most of the region, Dravidian languages are now largely limited to southern South Asia.

**Eastern Orthodox Christianity** A loose confederation of self-governing churches in eastern Europe and Russia that are historically linked to Byzantine traditions and to the primacy of the patriarch of Constantinople (Istanbul).

**economic convergence** The notion that globalization will result in the world's poorer countries gradually catching up with more advanced economies.

**El Niño** An abnormally large warm current that appears off the coast of Ecuador and Peru in December. During an El Niño year, torrential rains can bring devastating floods along the Pacific coast and drought conditions in the interior continents of the Americas.

**entrepôt** A city and port that specializes in transshipment of goods.

**ethnic religion** A religion closely identified with a specific ethnic or tribal group, often to the point of assuming the role of the major defining characteristic of that group. Normally, ethnic religions do not actively seek new converts.

**ethnicity** A shared cultural identity held by a group of people with a common background or history, often as a minority group within a larger society.

**Euroland** The 13 states that form the European Monetary Union, with its common currency, the euro. The euro completely replaced national currencies in July 2002.

**European Union (EU)** The current association of 27 European countries that are joined together in an agenda of economic, political, and cultural integration.

**exclave** A portion of a country's territory that lies outside its contiguous land area.

**exotic river** A river that issues from a humid area and flows into a dry area otherwise lacking streams.

**federal state** Nations that allocate considerable political power to units of government beneath the national level.

**Fertile Crescent** An ecologically diverse zone of lands in Southwest Asia that extends from Lebanon eastward to Iraq and that is often associated with early forms of agricultural domestication.

**fjords** Flooded, glacially carved valleys. In Europe, fjords are found primarily along Norway's western coast.

**fossil water** Water supplies that were stored underground during wetter climatic periods.

**free trade zone (FTZ)** A duty-free and tax-exempt industrial park created to attract foreign corporations and create industrial jobs.

**genocide** The deliberate and systematic killing of a racial, political, or cultural group by a state.

**gentrification** A process of urban revitalization in which higher-income residents displace lower-income residents in central city neighborhoods.

**geomancy** The traditional Chinese and Korean practice of designing buildings in accordance with the principles of cosmic harmony and discord that are thought to course through the local topography.

**glasnost** A policy of greater political openness initiated during the 1980s by then Soviet President Mikhail Gorbachev.

**global warming** An increase in the temperature of earth's atmosphere.

**globalization** The increasing interconnectedness of people and places throughout the world through converging processes of economic, political, and cultural change.

**grassification** The conversion of tropical forest into pasture for cattle ranching. Typically, this process involves introducing species of grasses and cattle, mostly from Africa.

**Great Escarpment** A landform that rims southern Africa from Angola to South Africa. It forms where the narrow coastal plains meet the elevated plateaus in an abrupt break in elevation.

**Greater Antilles** The four large Caribbean islands of Cuba, Jamaica, Hispaniola, and Puerto Rico.

**Green Revolution** Highly productive agricultural techniques developed since the 1960s that entail the use of new hybrid plant varieties combined with large applications of chemical fertilizers and pesticides. The term is generally applied to agricultural changes in developing countries, particularly India.

**greenhouse effect** The natural process of lower atmospheric heating that results from the trapping of incoming and reradiated solar energy by water moisture, clouds, and other atmospheric gases.

**gross domestic product (GDP)** The total value of goods and services produced within a given country (or other geographical unit) in a single year.

**gross national income (GNI)** The value of all final goods and services produced within a country's borders (gross domestic product) plus the net income from abroad (formerly referred to as gross national product).

**gross national income (GNI) per capita** The figure that results from dividing a country's GNI by the total population.

**Group of Eight (G8)**   A collection of powerful countries—United States, Canada, Japan, Great Britain, Germany, France, Italy, and Russia—that confers regularly on key global economic and political issues.

**guest workers**   Workers from Europe's agricultural periphery—primarily Greece, Turkey, southern Italy, and the former Yugoslavia—solicited to work in Germany, France, Sweden, and Switzerland during chronic labor shortages in Europe's boom years (1950s to 1970s).

**Gulag Archipelago**   A collection of Soviet-era labor camps for political prisoners, made famous by writer Aleksandr Solzhenitsyn.

**Hajj**   An Islamic religious pilgrimage to Makkah. One of the five essential pillars of the Muslim creed to be undertaken once in life, if an individual is physically and financially able to do it.

**haoles**   Light-skinned Europeans or U.S. citizens in the Hawaiian Islands.

**high islands**   Large, elevated islands, often focused around recent volcanic activity.

**Hindi**   An Indo-European language with more than 480 million speakers, making it the second-largest language group in the world. In India, it is the dominant language of the heavily populated north, specifically the core area of the Ganges Plain.

**Hindu nationalism**   A contemporary "fundamental" religious and political movement that promotes Hindu values as the essential—and exclusive—fabric of Indian society. As a political movement, Hindu nationalism appears to be less tolerant of India's large Muslim minority than do other political movements.

*hiragana*   The main Japanese syllabary, used for writing indigenous words. Each symbol stands for a particular vowel–consonant combination.

**Horn of Africa**   The northeastern corner of Sub-Saharan Africa that includes the states of Somalia, Ethiopia, Eritrea, and Djibouti. Drought, famine, and ethnic warfare in the 1980s and 1990s resulted in political turmoil in this area.

**hot spot**   A supply of magma that produces a chain of mid-ocean volcanoes atop a zone of moving oceanic crust.

**Human Development Index (HDI)**   For the past three decades, the United Nations has tracked social development in the world's countries through the Human Development Index (HDI), which combines data on life expectancy, literacy, educational attainment, gender equity, and income.

**hurricane**   A storm system with an abnormally low-pressure center, sustaining winds of 75 miles per hour (121 km/hour) or higher. Each year during hurricane season (July–October), a half dozen to a dozen hurricanes form in the warm waters of the Atlantic and Caribbean, bringing destructive winds and heavy rain.

**hydropolitics**   The interplay of water resource issues and politics.

**ideographic writing**   A writing system in which each symbol represents not a sound but a concept.

**indentured labor**   Foreign workers (generally South Asians) contracted to labor on Caribbean agricultural estates for a set period of time, often several years. Usually the contract stipulated paying off the travel debt incurred by the laborers. Similar indentured labor arrangements have existed in most world regions.

**Indian diaspora**   The historical and contemporary propensity of Indians to migrate to other countries in search of better opportunities. This has led to large Indian populations in South Africa, the Caribbean, and the Pacific islands, along with western Europe and North America.

**Indian subcontinent**   The name frequently given to South Asia in reference to its largest country. It forms a distinct landmass separated from the rest of the Eurasian continent by a series of sweeping mountain ranges, including the Himalayas—the highest mountains in the world.

**informal sector**   A much-debated concept that presupposes a dual economic system consisting of formal and informal sectors. The informal sector includes self-employed, low-wage jobs that are usually unregulated and untaxed. Street vending, shoe shining, artisan manufacturing, and self-built housing are considered part of the informal sector. Some scholars include illegal activities such as drug smuggling and prostitution in the informal economy.

**internally displaced persons (IDPs)**   Groups and individuals who flee an area due to conflict or famine but still remain in their country of origin. These populations often live in refugee-like conditions but are difficult to assist because they technically do not qualify as refugees.

**Iron Curtain**   A term coined by British leader Winston Churchill during the Cold War to define the western border of Soviet power in Europe. The notorious Berlin Wall was a concrete manifestation of the Iron Curtain.

**irredentism**   A state or national policy of reclaiming lost lands or those inhabited by people of the same ethnicity in another nation-state.

**Islamic fundamentalism**   A movement within both the Shiite and Sunni Muslim traditions to return to a more conservative, religious-based society and state. Often associated with a rejection of Western culture and with a political aim to merge civic and religious authority.

**Islamism**   A political movement within the religion of Islam that challenges the encroachment of global popular culture and blames colonial, imperial, and Western elements for many of the region's problems. Adherents of Islamism advocate merging civil and religious authority.

**isolated proximity**   A concept that explores the contradictory position of the Caribbean states, which are physically close to North America and economically dependent upon that region but also have strong loyalties to locality and limited economic opportunity.

**Jainism**   A religious group in South Asia that emerged as a protest against orthodox Hinduism around the 6th century BCE. Its ethical core is the doctrine of noninjury to all living creatures. Today, Jains are noted for their nonviolence, which prohibits them from taking the life of any animal.

**kanakas**   Melanesian workers imported to Australia, historically often concentrated along Queensland's "sugar coast."

*kanji*   The Chinese characters, or ideographs, used in Japanese writing.

**Khmer Rouge**   Literally, "Red (or Communist) Cambodians," the left-wing insurgent group that overthrew the royal Cambodian government in 1975 and subsequently created one of the most brutal political systems the world has ever seen.

**kibbutz**   A collective farm in Israel.

**kleptocracy**   A state where corruption is so institutionalized that politicians and bureaucrats siphon off a huge percentage of a country's wealth.

*laissez-faire*   An economic system in which the state has minimal involvement and in which market forces largely guide economic activity.

**latifundia**   A large estate or landholding.

**Lesser Antilles**   The arc of small Caribbean islands from St. Maarten to Trinidad.

**Levant**   The eastern Mediterranean region.

**lingua franca**   An agreed-upon common language to facilitate communication on specific topics such as international business, politics, sports, or entertainment.

**linguistic nationalism**  The promotion of one language over others that is, in turn, linked to shared notions of nationalism. In India, some Hindu nationalists promote Hindi as the national language, yet this is resisted by many other groups in which that language is either not spoken or does not have the same central cultural role as in the Ganges Valley. The lack of a national language in India remains problematic.

**location factor**  The various influences that explain why an economic activity takes place where it does.

**loess**  A fine, wind-deposited sediment that makes fertile soil but is very vulnerable to water erosion.

**Maghreb**  A region in northwestern Africa that includes portions of Morocco, Algeria, and Tunisia.

**maharaja**  Regional Hindu royalty, usually a king or prince, who ruled specific areas of South Asia before independence but who was usually subject to overrule by British colonial advisers.

**mallee**  A tough and scrubby eucalyptus woodland of limited economic value that is common across portions of interior Australia.

**Maori**  Indigenous Polynesian people of New Zealand.

**maquiladora**  Assembly plants on the Mexican border built by foreign capital. Most of their products are exported to the United States.

**marine west coast climate**  A moderate climate with cool summers and mild winters that is heavily influenced by maritime conditions. Such climates are usually found on the west coasts of continents between the latitudes 45 to 50 degrees.

**maritime climate**  A climate moderated by proximity to oceans or large seas. It is usually cool, cloudy, and wet and lacks the temperature extremes of continental climates.

**maroons**  Runaway slaves who established communities rich in African traditions throughout the Caribbean and Brazil.

**Marxism**  A philosophy developed by Karl Marx, the most important historical proponent of communism. Marxism, which has many variants, presumes the desirability and, indeed, the necessity of a socialist economic system run through a central planning agency.

**medieval landscape**  An urban landscapes from 900 to 1500 CE, characterized by narrow, winding streets, three- or four-story structures (usually in stone, but sometimes wooden), with little open space except for the market square. These landscapes are still found in the centers of many European cities.

**medina**  The original urban core of a traditional Islamic city.

**Mediterranean climate**  A unique climate, found in only five locations in the world, that is characterized by hot, dry summers with very little rainfall. These climates are located on the west side of continents, between 30 and 40 degrees latitude.

**megacity**  An urban conglomeration of more than 10 million people.

**Megalopolis**  A large urban region formed as multiple cities grow and merge with one another. The term is often applied to the string of cities in eastern North America that includes Washington, DC; Baltimore; Philadelphia; New York City; and Boston.

**Melanesia**  A Pacific Ocean region that includes the culturally complex, generally darker-skinned peoples of New Guinea, the Solomon Islands, Vanuatu, New Caledonia, and Fiji.

**Mercosur**  The Southern Common Market, established in 1991, which calls for free trade among member states and common external tariffs for nonmember states. Argentina, Paraguay, Brazil, and Uruguay are members; Chile is an associate member.

*mestizo*  A person of mixed European and Indian ancestry.

**Micronesia**  A Pacific Ocean region that includes the culturally diverse, generally small islands north of Melanesia. Micronesia includes the Mariana Islands, Marshall Islands, and Federated States of Micronesia.

**microstates**  Usually independent states that are small in both area and population.

**mikrorayons**  Large, state-constructed urban housing projects built during the Soviet period in the 1970s and 1980s.

**Millennium Development Goals**  A program of the United Nations, in collaboration with the World Bank, that aims to reduce extreme poverty by focusing resources on improving basic education, health care, and access to clean water in developing countries. The targeted goals are based on 1990 baselines and are supposed to be reached by 2015. Many countries in the developing world will reach their targets; it appears that many Sub-Saharan African countries will not.

**minifundia**  A small landholding farmed by peasants or tenants who produce food for subsistence and the market.

**mono-crop production**  Agriculture based on a single crop.

**monotheism**  A religious belief in a single God.

**Monroe Doctrine**  A proclamation issued by U.S. President James Monroe in 1823 that the United States would not tolerate European military action in the Western Hemisphere. Focused on the Caribbean as a strategic area, the doctrine was repeatedly invoked to justify U.S. political and military intervention in the region.

**monsoon**  The seasonal pattern of changes in winds, heat, and moisture in South Asia and other regions of the world that is a product of larger meteorological forces of land and water heating, the resultant pressure gradients, and jet-stream dynamics. The monsoon produces distinct wet and dry seasons.

**moraines**  Hilly topographic features that mark the path of Pleistocene glaciers. They are composed of material eroded and carried by glaciers and ice sheets.

**Mughal (or Mogul) Empire**  The powerful Muslim state that ruled most of northern South Asia in the 1500s and 1600s. The last vestiges of the Mughal dynasty were dissolved by the British following the rebellion of 1857.

**multinational**  Strictly speaking, "multinational" simply means "occurring in many different nations." A multinational company is one that establishes significant production operations outside of the country in which it is based.

**nation-state**  A relatively homogeneous cultural group (a nation) with its own political territory (the state).

**Native Title Bill**  A bill the Australian legislation signed in 1993 that provides Aborigines with enhanced legal rights over land and resources within the country.

**neocolonialism**  Economic and political strategies by which powerful states indirectly (and sometimes directly) extend their influence over other, weaker states.

**neoliberalism**  Economic policies widely adopted in the 1990s that stress privatization, export production, and few restrictions on imports.

**neotropics**  Tropical ecosystems of the Americas that evolved in relative isolation and support diverse and unique flora and fauna.

**net migration rate**  A statistic that depicts whether more people are entering or leaving a country.

**North American Free Trade Agreement (NAFTA)**  An agreement made in 1994 between Canada, the United States, and Mexico that established a 15-year plan for reducing all barriers to trade among the three countries.

**Oceania** A major world subregion that is usually considered to include New Zealand and the major island regions of Melanesia, Micronesia, and Polynesia.

**offshore banking** Financial services offered by islands or microstates that are typically confidential and tax exempt. As part of a global financial system, offshore banks have developed a unique niche, offering their services to individual and corporate clients for set fees. The Bahamas and the Cayman Islands are leaders in this sector.

**Organization of American States (OAS)** Founded in 1948 and headquartered in Washington, DC, an organization that advocates hemispheric cooperation and dialog. Most states in the Americas, except Cuba, belong to the OAS.

**Organization of the Petroleum Exporting Countries (OPEC)** An international organization (formed in 1960) of 12 oil-producing nations that attempts to influence global prices and supplies of oil. Algeria, Gabon, Indonesia, Iran, Iraq, Kuwait, Libya, Nigeria, Qatar, Saudi Arabia, the United Arab Emirates, and Venezuela are members.

**orographic rainfall** Enhanced precipitation over uplands that results from lifting and cooling of air masses as they are forced over mountains.

**Ottoman Empire** A large, Turkish-based empire (named for Osman, one of its founders) that dominated large portions of southeastern Europe, North Africa, and Southwest Asia between the 16th and 19th centuries.

**Outback** Australia's large, generally dry, and thinly settled interior.

**outsourcing** A business practice that transfers portions of a company's production and service activities to lower-cost settings, often located overseas.

**Palestinian Authority (PA)** A quasi-governmental body that represents Palestinian interests in the West Bank and Gaza.

**pastoral nomadism** A traditional subsistence agricultural system in which practitioners depend on the seasonal movements of livestock within marginal natural environments.

**pastoralists** Nomadic and sedentary peoples who rely on livestock (especially cattle, camels, sheep, and goats) for sustenance and livelihood.

**perestroika** A program of partially implemented, planned economic reforms (or restructuring) undertaken during the Gorbachev years in the Soviet Union and designed to make the Soviet economy more efficient and responsive to consumer needs.

**permafrost** A cold-climate condition in which the ground remains permanently frozen.

**physiological density** A population statistic that relates the number of people in a country to the amount of arable land.

**Pidgin English** A version of English that also incorporates elements of other local languages, often utilized to foster trade and basic communication between different culture groups.

**plantation America** A cultural region that extends from midway up the coast of Brazil, through the Guianas and the Caribbean, and into the southeastern United States. In this coastal zone, European-owned plantations, worked by African laborers, produced agricultural products for export.

**podzol soil** A Russian term for an acidic soil of limited fertility, typically found in northern forest environments.

**pollution exporting** The process of exporting industrial pollution and other waste material to other countries. Pollution exporting can be direct, as when waste is simply shipped abroad for disposal, or indirect, as when highly polluting factories are constructed abroad.

**Polynesia** A Pacific Ocean region, broadly unified by language and cultural traditions, that includes the Hawaiian Islands, Marquesas Islands, Society Islands, Tuamotu Archipelago, Cook Islands, American Samoa, Samoa, Tonga, and Kiribati.

**population pyramid** The structure of a population, which includes the percentage of young and old, is presented graphically as a population pyramid. This graph plots the percentage of all different age groups along a vertical. axis that divides the population into male and female.

**prairie** An extensive area of grassland in North America. In the more humid eastern portions, grasses are usually longer than in the drier western areas, which are in the rain shadow of the Rocky Mountain range.

**primate city** The largest urban settlement in a country that dominates all other urban places, economically and politically. Often, but not always, the primate city is also the country's capital.

**privatization** The process of moving formerly state-owned firms into the contemporary capitalist private sector.

**purchasing power parity (PPP)** An important qualification to these GNI per capita data is the concept of adjustment through PPP, an adjustment that takes into account the strength or weakness of local currencies.

**qanat system** A traditional system of gravity-fed irrigation that uses gently sloping tunnels to capture groundwater and direct it to low-lying fields.

**Quran (or Koran)** A book of divine revelations received by the prophet Muhammad that serves as a holy text in the religion of Islam.

**rain-shadow effect** A weather phenomenon in which mountains block moisture, producing an area of lower precipitation on the leeward side of the uplift.

**rate of natural increase (RNI)** The standard statistic used to express natural population growth per year for a country, a region, or the world, based on the difference between birthrates and death rates. RNI does not consider population change from migration. Though most often a positive figure (such as 1.7 percent), RNI can also be expressed as a negative (−.08 percent) figure for no-growth countries.

**refugee** A person who flees his or her country because of a well-founded fear of persecution based on race, ethnicity, religion, ideology, or political affiliation.

**remittances** Monies sent by immigrants working abroad to family members and communities in countries of origin. For many countries in the developing world, remittances often amount to billions of dollars each year. For small countries, remittances can equal 5 to 10 percent of a country's gross domestic product.

**Renaissance–Baroque landscape** An urban landscape generally constructed during the period from 1500 to 1800 that is characterized by wide, ceremonial boulevards; large monumental structures (palaces, public squares, churches, and so on); and ostentatious housing for the urban elite. This is a common landscape in European cities.

**rimland** The mainland coastal zone of the Caribbean, beginning with Belize and extending along the coast of Central America to northern South America.

**Russification** A policy of the Soviet Union designed to spread Russian settlers and influences to non-Russian areas of the country.

**rust belt** Regions of heavy industry that experience marked economic decline after their factories cease to be competitive.

**Sahel** The semidesert region at the southern fringe of the Sahara, and the countries that fall within this region, which extends from Senegal to Sudan. Droughts in the 1970s and early 1980s caused widespread famine and dislocation of population.

**salinization**  The accumulation of salts in the upper layers of soil, often causing a reduction in crop yields, resulting from irrigation using water with high natural salt content and/or irrigation of soils that contain a high level of mineral salts.

**Sanskrit**  The original Indo-European language of South Asia, introduced into northwestern India perhaps 4,000 years ago, from which modern Indo-Aryan languages evolved. Over the centuries, Sanskrit has become the classical literary language of the Hindus and is widely used as a scholarly second language, much like Latin in medieval Europe.

**Schengen Agreement**  The 1985 agreement between some—but not all—European Union member countries to reduce border formalities in order to facilitate free movement of citizens between member countries of this new "Schengenland." For example, today there are no border controls between France and Germany or between France and Italy.

**sectoral transformation**  The evolution of a labor force from being highly dependent on the primary sector to being oriented around more employment in the secondary, tertiary, and quaternary sectors.

**secularization**  The widespread movement in western Europe away from regular participation and engagement with traditional organized religions such as Protestantism or Catholicism.

**sediment load**  The amount of sand, silt, and clay carried by a river.

**Shanghai Cooperation Organization (SCO)**  Formed in 2001, a geopolitical group composed of China, Russia, Kazakhstan, Kyrgyzstan, Uzbekistan, and Tajikistan that focuses on common security threats and works to enhance economic cooperation and cultural exchange in Central Asia.

**shield**  A large upland area of very old exposed rocks. Shields range in elevation from 600 to 5,000 feet (200 to 1,500 meters). The three major shields in South America are the Guiana, Brazilian, and Patagonian.

**shield landscape**  Barren, mostly flat lands of southern Scandinavia that were heavily eroded by Pleistocene ice sheets. In many places, this landscape is characterized by large expanses of bedrock with little or no soil that resulted from glacial erosion.

**Shiites**  Muslims who practice one of the two main branches of Islam. Shiites are especially dominant in Iran and nearby southern Iraq.

**shogun, shogunate**  The true ruler of Japan before 1868. In contrast, the emperor's power was merely symbolic.

**Sikhism**  An Indian religion combining Islamic and Hindu elements, founded in the Punjab region in the late 15th century. Most of the people of the Indian state of Punjab currently follow this religion.

**Slavic peoples**  A group of peoples in eastern Europe and Russia who speak Slavic languages, a distinctive branch of the Indo-European language family.

**social and regional differentiation**  A process by which certain classes of people, or regions of a country, grow richer when others grow poorer.

**socialist realism**  An artistic style once popular in the Soviet Union that was associated with realistic depictions of workers in their patriotic struggles against capitalism.

**Spanglish**  A hybrid combination of English and Spanish spoken by Hispanic Americans.

**Special Economic Zones (SEZs)**  Relatively small districts in China that have been fully opened to global capitalism.

**spheres of influence**  In countries not formally colonized in the 19th and early 20th centuries (particularly China and Iran), limited areas gained by particular European countries for trade purposes and more generally for economic exploitation and political manipulation.

**special administrative region**  In China, a region of the country that temporarily maintains its own laws and own system of government. When Hong Kong was rejoined with China in 1997, it became a special administrative region, a position that it is scheduled to keep until 2047. In 1999 Macao passed from Portuguese rule to become China's second special administrative region.

**squatter settlement**  Makeshift housing on land not legally owned or rented by urban migrants, usually in unoccupied open spaces within or on the outskirts of a rapidly growing city.

**steppe**  Semiarid grasslands found in many parts of the world. Grasses are usually shorter and less dense in steppes than in prairies.

**structural adjustment programs**  Controversial yet widely implemented programs used to reduce government spending, encourage the private sector, and refinance foreign debt. Typically, these International Monetary Fund and World Bank policies trigger drastic cutbacks in government-supported services and food subsidies, which disproportionately affect the poor.

**subnational organizations**  Groups that form along ethnic, ideological, or territorial lines that can induce serious internal divisions within a state.

**Suez Canal**  A pivotal waterway connecting the Red Sea and the Mediterranean opened by the British in 1869.

**Sunnis**  Muslims who practice the dominant branch of Islam.

**superconurbation**  A massive urban agglomeration that results from the coalescing of two or more formerly separate metropolitan areas.

**supranational organizations**  Governing bodies that include several states, such as trade organizations, and often involve a loss of some state powers to achieve organizational goals.

**sweatshop**  Crude factories in developing countries in which workers perform labor-intensive tasks for extremely low wages.

**swidden**  Also called *slash-and-burn agriculture*, a form of cultivation in which forested or brushy plots are cleared of vegetation, burned, and then planted to crops, only to be abandoned a few years later as soil fertility declines. See also *shifting cultivation*.

**syncretic religions**  Religions that feature a blending of different belief systems. In Latin America, for example, many animist practices were folded into Christian worship.

**taiga**  The vast coniferous forest of Russia that stretches from the Urals to the Pacific Ocean. The main forest species are fir, spruce, and larch.

**theocracy**  A political state led by religious authorities. Also called a theocratic state.

**theocratic state**  A political state led by religious authorities. Also called a *theocracy*.

**tonal**  A language in which the same set of phonemes (or basic sounds) may have very different meanings, depending on the pitch in which they are uttered.

**total fertility rate (TFR)**  The average number of children who will be borne by women of a hypothetical, yet statistically valid, population, such as that of a specific cultural group or within a particular country. Demographers consider TFR a more reliable indicator of population change than the crude birthrate.

**transhumance**  A form of pastoralism in which animals are taken to high-altitude pastures during the summer months and returned to low-altitude pastures during the winter.

**transmigration**  The planned, government-sponsored relocation of people from one area to another within a state territory.

**transnational firm**    A firm or corporation that, although it may be chartered and have headquarters in one specific country, does international business through an array of global subsidiaries.

**Trans-Siberian Railroad**    A key southern Siberian railroad connection completed during the Russian empire (1904) that links European Russia with the Russian Far East terminus of Vladivostok.

**Treaty of Tordesillas**    A treaty signed in 1494 between Spain and Portugal that drew a north–south line some 300 leagues west of the Azores and Cape Verde islands. Spain received the land to the west of the line and Portugal the land to the east.

**tribal peoples**    Peoples who were traditionally organized at the village or clan level, without broader-scale political organization.

**tribalism**    Allegiance to a particular tribe or ethnic group rather than to the nation-state. Tribalism is often blamed for internal conflict in Sub-Saharan states.

**tribe**    A group of families or clans with a common kinship, language, and definable territory but not an organized state.

**tsar**    A Russian term (also spelled *czar*) for "Caesar," or ruler. Tsars were the authoritarian rulers of the Russian empire before its collapse in the 1917 revolution.

**tsetse fly**    A fly that is a vector for a parasite that causes sleeping sickness (typanosomiasis), a disease that especially affects humans and livestock. Livestock is rarely found in areas of Sub-Saharan Africa where the tsetse fly is common.

**tsunami**    A very large sea wave induced by earthquakes.

**tundra**    Arctic region with a short growing season in which vegetation is limited to low shrubs, grasses, and flowering herbs.

**typhoon**    A large tropical storm, similar to a hurricane, that forms in the western Pacific Ocean in tropical latitudes and can cause widespread damage to the Philippines and coastal Southeast and East Asia.

**uncontacted peoples**    Cultures that have yet to be contacted and influenced by the Western world.

**unitary state**    A political system in which power is centralized at the national level.

**universalizing religion**    A religion, usually with an active missionary program, that appeals to a large group of people, regardless of local culture and conditions. This contrasts with ethnic religions. Christianity and Islam both have strong universalizing components.

**urban decentralization**    A process in which cities spread out over a larger geographic area.

**urban primacy**    A state in which a disproportionately large city (for example, London, New York, Bangkok) dominates the urban system and is the center of economic, political, and cultural life.

**urban realms model**    A simplified description of urban land use, especially descriptive of the modern North American city. It features a number of dispersed, peripheral centers of dynamic commercial and industrial activity linked by sophisticated urban transportation networks.

**urbanized population**    The percentage of a country's population living in settlements characterized as cities. Usually, high rates of urbanization are associated with higher levels of industrialization and economic development, since these activities are usually found in and around cities. Conversely, lower urbanized populations (less than 50 percent) are characteristic of developing countries.

**Urdu**    One of Pakistan's official languages (along with English), Urdu is very similar to the Indian language of Hindi, although it includes more words derived from Persian and Arabic and is written in a modified form of the Persian Arabic alphabet. Although most Pakistanis do not speak Urdu at home, it is widely used as a second language and is extensively employed in education, the media, and government, thus giving Pakistan a kind of cultural unity.

**viticulture**    Grape cultivation.

**White Australia Policy**    Before 1973, a set of stringent Australian limitations on nonwhite immigration to the country. It has been largely replaced by a more flexible policy today.

**World Trade Organization (WTO)**    Formed as an outgrowth of the General Agreement on Tariffs and Trade (GATT) in 1995, a large collection of member states dedicated to reducing global barriers to trade.

# CREDITS

**Chapter 1 opening photo** Keren Su/Aurora Photos
**1.1** Rob Crandall **1.2** Sherwin Crasto/Corbis **1.3** Sean Sprague/The Image Works **1.4** Rob Crandall **1.5** Stephanie Kuykendal/Corbis **1.7** Rob Crandall **1.8** Tim Boyle/Getty Images **1.9** AP Photo **1.11** REUTERS/Bobby Yip/Landov **1.12** AP Photo/Thorvaldur Kristmundsson **1.13** Caroline Penn/Panos Pictures **1.14** Ramzi Stowers/Newscom **1.17** Chris Stowers/Newscom **1.18** Trygve Bolstad/Panos Pictures **1.21** Liz Gilbert/India Government Tourist Office **1.22a** Craig Aurness/CORBIS **1.22b** Tony Waltham/Robert Harding **1.24** Jay Directo/Getty Images **1.25** Ajit Solanki/AP Photo **1.26a** Rob Crandall **1.26b** Rob Crandall **1.26c** Rob Crandall **1.26d** Rob Crandall **1.27** Arko Datta Reuters/Corbis **1.29** Rob Crandall **1.31** Rob Crandall **1.32** Photodisc/Getty Images **1.34** Alberto Arzoz/Panos Pictures **1.35** Peter Macdiarmid/Getty Images **1.37** Catherine Karnow/Woodfin Camp **1.39** Ajay Verma/Reuters/Corbis **1.43** Ron Gilling/Panos Pictures

**Chapter 2 opening** photo Konrad Steffen/Landov Media
**2.1** Chris James/Peter Arnold **2.2** Erlend Kvalsvik/iStockphoto **2.6** ChinaFotoPress/Newscom **2.7** Claro Cortes IV/Reuters/Corbis **2.9** Gary Braasch/Woodfin Camp **2.10** James P. Blair/Getty Images **2.11** Wolfgang Kaehler/Alamy **2.12** Calvin Larsen/Photo Researchers **2.13** Andy Levin/Photo Researchers **2.14** Mark Edwards/Peter Arnold **2.15** Martin Wendler/Peter Arnold

**Chapter 3 opening photo** Chad Ehlers/Getty Images
**3.2** Canadian Tourism Commission **3.3a** NASA **3.3b** James M. Rubenstein, The Cultural Landscape, 8e, c 28. **3.3c** John McCusker/Landov **3.5** Alexander Lowry/Photo Researchers **3.6** Earth Satellite/SPL/Photo Researchers **3.7** Joe Sohm/The Image Works **3.13** Romare Bearden, "The Block, II". 1972. The Walter O. Evans Collection of African American Art. © Romare Bearden Foundation/Licensed by VAGA, New York, NY **3.14** Jeremy Woodhouse/Getty Images **3.16** Rob Crandall/The Image Works **3.18** Soffer Organization **3.19** Craig Aurness/Corbis **3.24** Jim Noelker/The Image Works **3.28** Alison Wright/Corbis **3.29** David McNew/Getty Images **3.31** Kim Steele/Getty Image **3.32** Scott Berner/Photolibrary **3.33** George Hall/Woodfin Camp **3.34** Dennis MacDonald/Photo Edit **3.35** Erica Lananer/Black Star **3.36** Cont/Frank Frames/Woodfin Camp **3.1.1** Phil Klein/CORBIS

**Chapter 4 opening photo** David R. Frazier/Alamy Images
**4.2** Rob Crandall **4.2.1** Larry Mangino/The Image Works **4.4** Rob Crandall **4.5a** Biological Resources Division, U.S. Geological Survey **4.5b** Biological Resources Division, U.S. Geological Survey **4.6** Rob Crandall **4.7** Jorge Uzon/AFP/Getty Images **4.8** Hubert Stadler/Corbis **4.9** Stephanie Maze/National Geographic Image Collection **4.10** Rob Crandall **4.13** Rob Crandall **4.15a** Rob Crandall **4.15b** Rob Crandall **4.15c** Rob Crandall **4.15d** Rob Crandall **4.16** Rob Crandall **4.18** AFP Photo/Toshifumi Kitamura/Newscom **4.19** Rob Crandall **4.20** Rob Crandall **4.21** Rob Crandall **4.23** Rob Crandall **4.24** Fernando Soutello/Reuters/CORBIS **4.26** Luis Acosta/Getty Images **4.29** AP Photo **4.30** Rob Crandall **4.31** John Maier Jr./The Image Works **4.32** Rob Crandall **4.35** Alejandro Pagni/AFP/Getty Images

**Chapter 5 opening photo** Michael Amme/Aurora Photos
**5.2** Rob Crandall **5.3** Biological Photo Service **5.5** Rob Crandall **5.6** Rob Crandall **5.7** European Space Agency **5.9** Zuma Press **5.11** Steve Bly/Alamy **5.13** Rob Crandall **5.14** Rob Crandall **5.15** Rob Crandall **5.16** Michael Holford **5.17** Steve Raymer/Corbis **5.19** Robert Caputo/Getty Images **5.20b** AFP/Getty Images **5.22** Rob Crandall **5.23** Rob Crandall **5.26** David Frazier/The Image Works **5.28** Frank Heuer/Redux Pictures **5.30** Steve Vidler/SuperStock **5.31** Rob Crandall **5.32** Edward Keating/Redux Pictures

**Chapter 6 opening photo** Rob Crandall
**6.2** Dung Vo Trung/Corbis **6.3** Rob Crandall **6.4** Altitude/Y. Arthus B./Peter Arnold **6.5** Robert Caputo/Aurora Photos **6.7** Nik Wheeler **6.8** Louis Gubb/The Image Works **6.10** Daniel Berehulak/Getty Images **6.11a** USGS Center for Earth Resources Observation and Science **6.11b** USGS Center for Earth Resources Observation and Science **6.12** Louis Gubb/Corbis **6.13** Roberto Schmidt/AFP/Getty Images **6.14** Issouf Sanogo/Getty Images **6.15** Nomi Baumgart/Bilderberg/Peter Arnold **6.17** Gideon Mendel/Corbis **6.19** Issouf Sanaogo/AFP/Getty Images **6.20** Kennan Ward/Corbis **6.21** Jeremy Horner/Corbis **6.22** Akintunde Akinleye/Reuters **6.24** Rob Crandall **6.26** Ed Kashi/Corbis **6.28** Greenshoots Communications/Alamy **6.30** Frans Lemmens/Alamy **6.31** Francois Xavier Marit/AFP/Getty Images **6.34** Victor Englebert **6.37** David Lewis/Reuters **6.38** Marc & Evelyne Bernheim/Woodfin Camp **6.39** Andre

# INDEX